AF683492

NONTRADITIONAL ACTIVATION METHODS IN GREEN AND SUSTAINABLE APPLICATIONS

Advances in Green and Sustainable Chemistry

NONTRADITIONAL ACTIVATION METHODS IN GREEN AND SUSTAINABLE APPLICATIONS

Microwaves; Ultrasounds; Photo-, Electro- and Mechanochemistry and High Hydrostatic Pressure

Series Editor

BÉLA TÖRÖK

Professor, University of Massachusetts Boston, Boston, MA, United States

CHRISTIAN SCHÄFER

Lecturer, University of Massachusetts Boston, Boston, MA, United States

ELSEVIER

Elsevier
Radarweg 29, PO Box 211, 1000 AE Amsterdam, Netherlands
The Boulevard, Langford Lane, Kidlington, Oxford OX5 1GB, United Kingdom
50 Hampshire Street, 5th Floor, Cambridge, MA 02139, United States

© 2021 Elsevier Inc. All rights reserved.

No part of this publication may be reproduced or transmitted in any form or by any means, electronic or mechanical, including photocopying, recording, or any information storage and retrieval system, without permission in writing from the publisher. Details on how to seek permission, further information about the Publisher's permissions policies and our arrangements with organizations such as the Copyright Clearance Center and the Copyright Licensing Agency, can be found at our website: www.elsevier.com/permissions.

This book and the individual contributions contained in it are protected under copyright by the Publisher (other than as may be noted herein).

Notices
Knowledge and best practice in this field are constantly changing. As new research and experience broaden our understanding, changes in research methods, professional practices, or medical treatment may become necessary.

Practitioners and researchers must always rely on their own experience and knowledge in evaluating and using any information, methods, compounds, or experiments described herein. In using such information or methods they should be mindful of their own safety and the safety of others, including parties for whom they have a professional responsibility.

To the fullest extent of the law, neither the Publisher nor the authors, contributors, or editors, assume any liability for any injury and/or damage to persons or property as a matter of products liability, negligence or otherwise, or from any use or operation of any methods, products, instructions, or ideas contained in the material herein.

Library of Congress Cataloging-in-Publication Data
A catalog record for this book is available from the Library of Congress

British Library Cataloguing-in-Publication Data
A catalogue record for this book is available from the British Library

ISBN: 978-0-12-819009-8

For information on all Elsevier publications
visit our website at https://www.elsevier.com/books-and-journals

Publisher: Susan Dennis
Acquisitions Editor: Anneka Hess
Editorial Project Manager: Lena Sparks
Production Project Manager: Debasish Ghosh
Cover Designer: Alan Studholme

Working together to grow libraries in developing countries
www.elsevier.com • www.bookaid.org

Typeset by SPi Global, India
Transferred to Digital Printing 2021

Contents

Contributors

Ricardo Belloso
Department of Chemistry, Georgia Southern University, Statesboro, GA, United States

Clarence Charnay
ICGM, Univ Montpellier, CNRS, ENSCM, Montpellier, France

Pedro Cintas
Department of Organic and Inorganic Chemistry, Faculty of Sciences and IACYS-Green Chemistry and Sustainable Development Unit, University of Extremadura, Badajoz, Spain

Evelina Colacino
ICGM, Univ Montpellier, CNRS, ENSCM, Montpellier, France

Giancarlo Cravotto
Department of Drug Science and Technology, and NIS-Centre for Nanostructured Interfaces and Surfaces, University of Turin, Turin, Italy

Francesco Delogu
Dipartimento di Ingegneria Meccanica, Chimica, e dei Materiali, Università degli Studi di Cagliari, Cagliari, Italy

Franziska Emmerling
BAM Federal Institute for Materials Research and Testing, Berlin, Germany

Satoshi Fujii
Department of Information and Communication System Engineering, Okinawa College, Okinawa, Japan

Aharon Gedanken
Department of Chemistry, Bar Ilan University, Ramat Gan, Israel

Allen Gordon
Department of Chemistry and Biochemistry, Georgia Southern University, Statesboro, GA, United States

Norbert Hoffmann
CNRS, Université de Reims Champagne-Ardenne, ICMR, Equipe de Photochimie, UFR Sciences, Reims, France

István Jablonkai
MTA TTK Lendület Functional Organic Materials Research Group, Institute of Organic Chemistry, Research Centre for Natural Sciences, Budapest, Hungary

Anke Kabelitz
BAM Federal Institute for Materials Research and Testing, Berlin, Germany

Attila Kunfi
MTA TTK Lendület Functional Organic Materials Research Group, Institute of Organic Chemistry, Research Centre for Natural Sciences, Budapest, Hungary

Shainaz Landge
Department of Chemistry and Biochemistry, Georgia Southern University, Statesboro, GA, United States

Corentin Lefebvre
CNRS, Université de Reims Champagne-Ardenne, ICMR, Equipe de Photochimie, UFR Sciences, Reims, France

Gábor London
MTA TTK Lendület Functional Organic Materials Research Group, Institute of Organic Chemistry, Research Centre for Natural Sciences, Budapest, Hungary

Katia Martina
Department of Drug Science and Technology, and NIS-Centre for Nanostructured Interfaces and Surfaces, University of Turin, Turin, Italy

Adam A.L. Michalchuk
BAM Federal Institute for Materials Research and Testing, Berlin, Germany

Manisha Mishra
Department of Chemistry, University of Massachusetts Boston, Boston, MA, United States

Tara Mooney
Department of Chemistry, University of Massachusetts Boston, Boston, MA, United States

Maria Jesus Moran
Department of Drug Science and Technology, and NIS-Centre for Nanostructured Interfaces and Surfaces, University of Turin, Turin, Italy

Bilal Nişancı
Department of Chemistry, Faculty of Science, Atatürk University, Erzurum, Turkey

István Pálinkó
University of Szeged, Szeged, Hungary

Andrea Porcheddu
Dipartimento di Scienze Chimiche e Geologiche, Università degli Studi di Cagliari, Cittadella Universitaria, Monserrato, CA, Italy

Indra Neel Pulidindi
Department of Chemistry, Bar Ilan University, Ramat Gan, Israel

Da-Hui Qu
Key Laboratory for Advanced Materials, School of Chemistry and Molecular Engineering, East China University of Science and Technology, Shanghai, China

Christian Schäfer
Department of Chemistry, University of Massachusetts Boston, Boston, MA, United States

Abid Shaikh
Department of Chemistry, Georgia Southern University, Statesboro, GA, United States

Pál Sipos
University of Szeged, Szeged, Hungary

Márton Szabados
University of Szeged, Szeged, Hungary

Béla Török
Department of Chemistry, University of Massachusetts Boston, Boston, MA, United States

Shuntaro Tsubaki
School of Materials and Chemical Technology, Tokyo Institute of Technology, Yokohama, Japan

Gábor Varga
University of Szeged, Szeged, Hungary

Yuji Wada
Institute of Innovation Research, Tokyo Institute of Technology, Yokohama; Microwave Chemical Co. Ltd., Techno Alliance 3F, Suita, Osaka, Japan

Kazutaka Yamamoto
Head of Laboratory, Food Process Engineering, Division of Food Processing and Distribution, Food Research Institute, National Agriculture and Food Research Organization, Tsukuba, Ibaraki, Japan

CHAPTER ONE

Application of nontraditional activation methods in green and sustainable chemistry: Microwaves, ultrasounds, electro-, photo-, and mechanochemistry, and high hydrostatic pressure

Christian Schäfer and Béla Török
Department of Chemistry, University of Massachusetts Boston, Boston, MA, United States

1 Introduction

Many discoveries made in chemistry in the 20th century enabled strong developments in the preparation of a broad range of products. The developing chemical industry was mostly based on petroleum resources that provided a steady flow of raw materials. The continuous development of new fields such as medicinal chemistry, polymers and plastics, agrochemicals, and the likes resulted in booming industrial production. Although the new products had a largely positive impact on the society as a whole, several unintended issues began to appear. These included industrial disasters, drug side effects, and the damage caused to Earth's protective ozone layer among many others. These incidents made the society realize the potential harmful effects of large-scale production and the associated waste produced by it, some products themselves, and the inability of these products to decompose at the end of their lifecycle. In addition, the extensive use of fossil fuels for energy production and petroleum derivatives for industrial feedstock resulted in issues that can further threaten the planet's ecosystem. While there have been corrective measures implemented along the way in the forms of government regulations, a concentrated effort in the late 1990s resulted in the birth of the Green Chemistry movement[1] to address the fundamental nature of the problems. In fact, the current efforts often appear to be joint ventures by government, academia, and industry suggesting a much

Nontraditional Activation Methods in Green and Sustainable Applications
https://doi.org/10.1016/B978-0-12-819009-8.00016-5
© 2021 Elsevier Inc.
All rights reserved.

more effective tackling of the problems than when these entities were actively working against each other delaying the development of actual solutions. This ongoing partnership ensures the continuous expansion and advancement of environmentally friendly processes and products,[2–4] and green chemistry and environmental science made steady developments in the general science education as well.[5–7] There are many ways chemistry can contribute to these advances, replacing old outdated chemical processes with ones having less environmental impact, traditional petroleum based raw materials with sustainable ones, or notoriously stable conventional plastics and other chemicals with biodegradable alternatives.

This introductory chapter and the entire book focus on a special tool set that is a great contributor to the development of green synthetic methods; the so-called nontraditional activation methods that use alternative energy forms to conventional convective heating to initiate chemical reactions. A large majority of chemical reactions require activation energy in order to occur. Traditionally this energy was introduced to a reaction as heat mostly via convective heating using an external heat source. Although effective, convective heating is slow and essentially dependent on heat diffusion (mixing) in the system. Another problem is that the temperature of the wall of the reaction vessel is commonly 10–20°C higher than that in the center of the liquid, often causing side reactions or decomposition on or around the surface of the wall. To alleviate these issues, several nontraditional activation methods have been developed. Although different in nature, the common feature of these methods is that they are based on the direct energy transfer in the reaction mixture. In this chapter, we will only attempt to describe the basic features of these methods, such as microwave- and ultrasonic irradiation, photo-, electro- and mechanochemistry and high hydrostatic pressure with a few representative examples, which will be followed by state-of-the-art specific applications in the upcoming chapters.

2 Microwave-assisted organic synthesis

The first observations of the microwave heating effect occurred in the 1940s when the magnetron, now the common microwave source, was applied during the development of the RADAR. Following up on the original discovery the application of microwave ovens in the food and airline industry began in the 1950s.[8] Although the first synthetic application in 1969[9] largely went unnoticed, the independent publication of microwave-activated synthetic applications by two teams, Gedye et al.[10]

and Giguere et al.[11] initiated an exponential development of microwave-assisted synthesis. Since then microwave-assisted organic synthesis (MAOS) has become a household name in laboratories, with thousands of papers published in this area. The contemporary microwave instruments and reactors provide the essential elements that are required in synthetic chemistry, such as temperature control, reliability, reproducibility, and in special cases, scalability. The theories of how microwaves interact with materials have been a subject of a sometimes intense debate[12–14] and numerous reports, and reviews.[15–17] Here, we only focus on the basic information related to this issue.

The frequency range for microwave radiation is between 1 and 300 GHz appearing in the relatively low energy region of the electromagnetic spectrum. Microwaves possess lower energy than infrared irradiation, but higher than radio waves. Given their limited energy, microwave irradiation is not able to excite electron transitions, break chemical bonds, or even activate changes within individual molecules such as vibration. The rotational motion of molecules is the physical phenomenon that microwave energy can excite to generate internal heat formation. This entails that the microwave effect is limited to liquids and solids. Although, gaseous materials can absorb microwaves, due to the limited interaction between the molecules it does not generate heat, rather results in sharp peaks in a fingerprint fashion, which is the foundation of microwave spectroscopy. In more dense materials, the rotating molecules will collide with each other, and due to the restricted movement, dielectric polarization will occur, which is the major source of microwave heating. Particularly two features, the dielectric constant (ε') and dielectric loss (ε''), and their ratio, the so-called loss tangent delta ($\text{tg}\,\delta = \varepsilon''/\varepsilon'$) are important to determine how a certain compound or material is able to convert the microwave irradiation to heat. Materials of significant dipole moments (polar solvents, salts, ionic liquids, certain metal oxides, etc.) are the best media for microwave-assisted chemistry. Nonpolar materials (hydrocarbons and most plastics) are weak microwave absorbers (or completely transparent) and thus do not generate internal heat formation.

Although the application of microwave irradiation as an activation method carries multiple benefits, the major advantages are rapid reactions, and therefore significantly reduced reaction times. This could be quite significant; in our experience for example, the condensation of trifluoroacetophenones with a variety of benzylamines require 168 h (1 week!) heating at 130°C, while the same reaction heated by microwaves is complete within 45 min producing even somewhat better yields.[18] In many other cases, the hours long reaction time can be reduced to few minutes.[19, 20] The reduced reaction times are also beneficial due to the diminished side reactions that result in higher selectivity and overall higher product purity. It certainly carries significant green advantages and often the product purification is not necessary, which can save a large amount of solvents or adsorbents for chromatography. A significant disadvantage of microwave reactors is the limited penetration depth that, depending on the actual system could be no more than about 2.5 cm. It, commonly, is not a major issue for laboratory-scale instruments; however, even a relatively moderate size (up to 20 kg) batch reactor has much greater dimensions (about a 50–60 cm diameter). This problem can be addressed in two different ways; (i) a batch system with highly effective agitation system[21] or (ii) the application of microwave flow reactors.[22, 23] Depending on the scale of the production, microwave flow systems appear more beneficial for large-scale operation than the batch reactors, as they allow continuous flow operations and thus uninterrupted preparation of the product (depending on catalyst stability). These flow reactors gained increasing popularity as with an even limited size microwave cavity the scale-up of reactions can easily be carried out. In the majority of applications MAOS also appears to be more energy efficient than the traditional convective heating, however, several papers pointed out that lower energy consumption is not an automatic feature for all microwave-assisted reactions and every reaction should be separately evaluated.[24–26]

The application possibilities for MAOS are truly unlimited. Over the years several thousands of reactions have been described from solvent-free acid catalysis or multicomponent reactions to metal-catalyzed transformations both in homogeneous and heterogeneous catalytic systems. The following few examples are intended to show a glimpse of applications of this broad field.

Several examples of MAOS involve heterogeneous catalysis, as the use of solid catalysts often enables the use of solvent-free conditions. Kokel and Török showed that microwave activation can be a good way to achieve transformations that are typically associated with harmful and dangerous chemicals. They used a solvent-free approach to synthesize benzotriazoles from *o*-phenylenediamines using microwave heating.[20] The reactions were performed using montmorillonite K10 as a catalyst and $NaNO_2$ as nitrogen source. Substrates with electron-donating and electron-withdrawing substituents gave the corresponding benzotriazoles in excellent yields (Scheme 1).

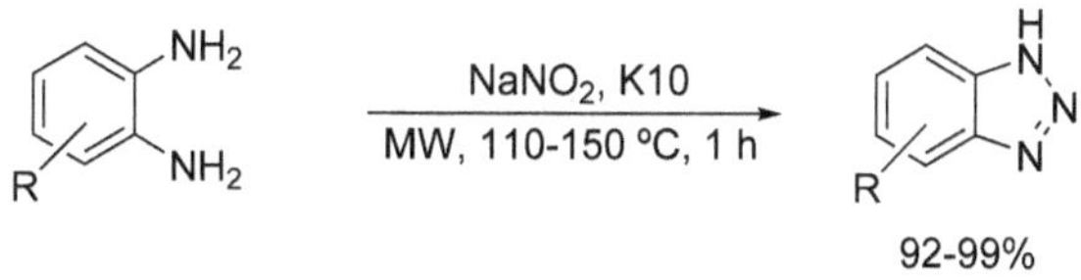

Scheme 1 Microwave-assisted solvent-free synthesis of benzotriazoles via a solid phase diazotization.

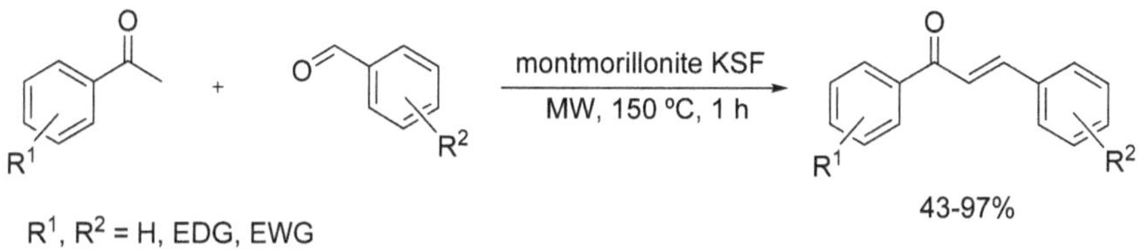

Scheme 2 Synthesis of chalcones by a microwave-assisted condensation.

The application of microwaves in condensation reactions was reported by the group of Menéndez. The authors used montmorillonite KSF as a catalyst for the formation of chalcones from acetophenones and benzaldehydes.[27] The products were obtained in medium-to-good yields, with electron-withdrawing substituents generally giving lower product amounts than electron-donating substituents (Scheme 2).

The chemoselective reduction of chalcones under microwave conditions was reported by Sinha et al.[28] Using $PdCl_2$ anchored to silica gel they showed the successful transformation into the saturated carbonyl compound using a MeOH, HCOOH, H_2O system as a hydrogen source. A variety of electron-donating and -withdrawing substituents were tolerated in the reaction (Scheme 3).

The application of microwaves in multicomponent reactions is illustrated by an example published by Shinde et al.[29] The synthesis of multisubstituted pyrroles was achieved by combining a β-ketoester, an amine, an aldehyde, and nitromethane (Scheme 4). A heterogenous acid catalyst (*p*-toluenesulfonic acid doped polystyrene, PTSA-PS) was used and the authors could show that microwave heating gave higher yields in shorter reaction times compared to conventional heating.

The use of microwave chemistry is not limited to the use in synthetic organic chemistry but can also be used in a prior stage. Ros-Lis and coworkers recently summarized the use of microwaves in the synthesis covalent organic frameworks, a class of compounds that are often used as catalysts in microwave chemistry.[30]

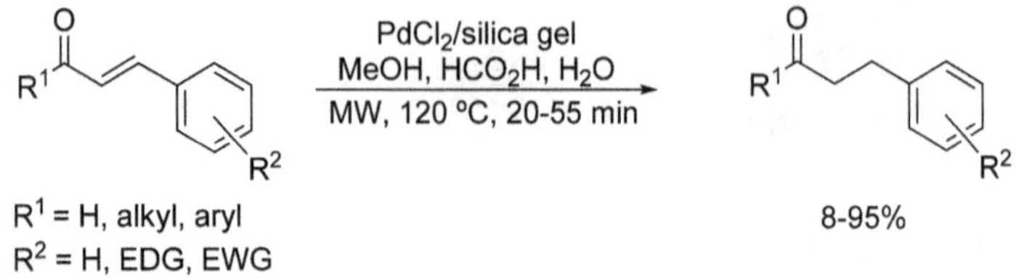

Scheme 3 Transfer hydrogenation of α,β-unsaturated carbonyl compounds under microwave conditions.

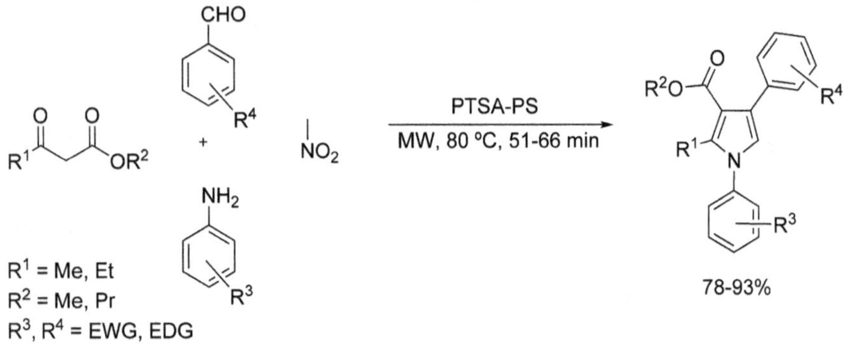

Scheme 4 A microwave-assisted multicomponent synthesis of pyrroles.

3 Ultrasonic activation

The application of ultrasounds in chemical synthesis has a nearly 100 years old history.[31] The first application dates back to 1927,[32, 33] with the early mechanistic theories appearing in 1945.[34] The widespread use of the technique began in around 1980s when the dedicated reliable instruments appeared on the market. The medical applications such as noninvasive diagnostics further initiated the instrument design and by the 1990s ultrasonic activation became a mainstream technique in biology and chemistry laboratories, with several industrial applications.[35, 36] Sonochemistry is a significant contributor to green synthesis design as well.[37–42]

The frequency range of ultrasounds is beyond that of the human hearing, which ranges from 16 Hz to 18 kHz. It is a broad range with different uses. The most common frequency span for the synthetic use of ultrasounds is between 20 and 100 kHz, also known as power ultrasounds.[43] The 100 kHz–2 MHz region is commonly referred to as high frequency ultrasounds.[44] Finally, the 5–10 MHz range gained widespread application possibilities in noninvasive medical diagnostics and other treatments.[45] Overall, the overarching ultrasounds term covers the 20 kHz–10 MHz frequency range.

Unlike MAOS, the applications of ultrasounds are limited to the liquid medium. When ultrasound waves are passed through a liquid medium they initiate the oscillation of the molecules of the medium. Due to their periodic nature, the oscillations will result in a series of compressions and rarefactions in the liquid, effectively breaking the secondary forces that keep the molecules together. By this, ultrasounds will generate cavities (or small bubbles) within the liquid, which is commonly referred to as acoustic cavitation. Due to the nature of the cavitation, there is a considerable vacuum inside the bubbles. Large bubbles appear to be quite stable and remain in the liquid for extended time (stable cavitation). In contrast, the smaller bubbles tend to collapse, which releases a significant amount of energy locally (transient cavitation). The energy released at the spot of the bubble collapse will result in high temperature and pressure in the range of 1000–3000 K and 10^5 kPa according to theoretical estimates.[46] The experimental evidence, described by Suslick et al. confirmed this assumption, reporting even higher temperature values up to 4000–5000 K.[47] Due to the localized nature of the phenomenon it has been named as *hot spot* model.[48] Although, there are certainly other theories,[49] such as the electric model; the hot spot theory provides the broadest interpretation of the experimental observations in solvents of different polarity.

Similarly to microwave chemistry, experimental variables affect the efficacy of ultrasonic activation. The major parameters are; frequency, power, temperature, external pressure, the gas that fills the experimental vessel, e.g., a regular lab air or protective gas. Last, but not least, the solvent will also have a significant effect on modifying cavitation. Many of the parameters with numerical descriptors (e.g., frequency, power, temperature, and pressure) usually possess an optimum value, essentially showing a maximum curve as a function of product yields vs increasing value of the variable.[50] While the frequency and power affect the system via the energy input, the temperature, and pressure modify the pressure inside the cavities and thus have a significant effect on the intensity of the bubble collapse. The protecting gas can also modify the efficacy of the system in two ways; monoatomic gases, often improve the rates, while reactive gases (O_2, H_2, N_2, etc.) can participate in the reactions. The solvent must be stable under the reaction conditions, which is a rare case, as most solvents undergo some form of radical degradation under sonochemical conditions. As shown by this brief description, sonochemical reactions can be very complex and while generally useful, often they require a thorough optimization process. Similarly to MAOS, here also the internally and locally generated heat is the main driving force for reactions, thus resulting in shorter reaction times and higher yields. Unlike microwaves, however, cavitation also contributes to reactions by the extremely effective mixing and, in heterogeneous catalytic systems, by the powerful surface cleaning effect. Representative applications of sonochemistry include a wide variety of reactions as described below.

An example for a condensation reaction performed under sonochemical conditions was reported by Vidal et al.[51] Using simple K_2CO_3 as a catalyst and water as the solvent, the synthesis of pyrimidinols from β-ketoesters and amidines was achieved in moderate yields (Scheme 5).

The application of ultrasounds in C—C coupling chemistry will be demonstrated by two examples. The Claisen-Schmidt reaction between acetophenone and different benzaldehydes can be achieved using KF/alumina as a catalyst.[52] The products were obtained in good yields using ethanol as the solvent (Scheme 6). The application of ultrasounds allowed for a reduction in reaction time.

The use of ultrasounds in palladium-catalyzed cross-coupling chemistry was illustrated by Das and coworkers.[53] The Suzuki-coupling of several aryl iodides with phenylboronic acid gave the desired products in high yields using a water-methanol solvent mixture (Scheme 7).

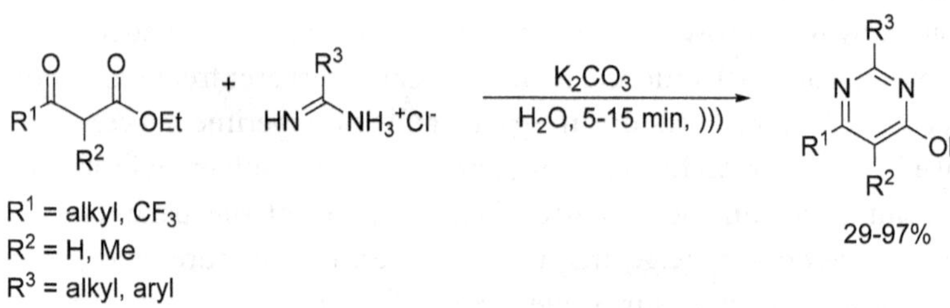

Scheme 5 Sonochemical condensation of β-ketoesters and amidines to pyrimidinols.

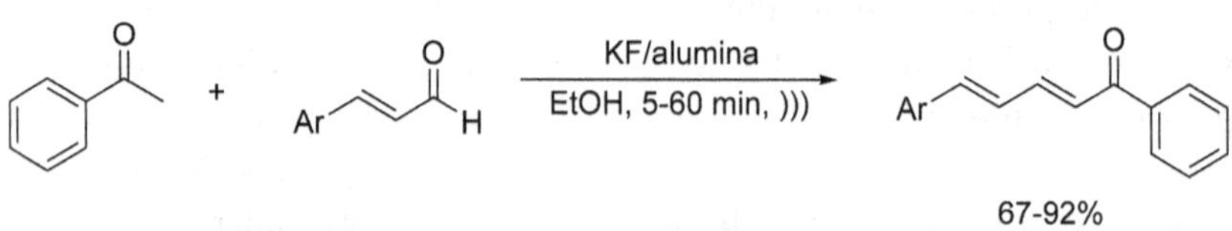

Scheme 6 A sonochemical Claisen-Schmidt reaction.

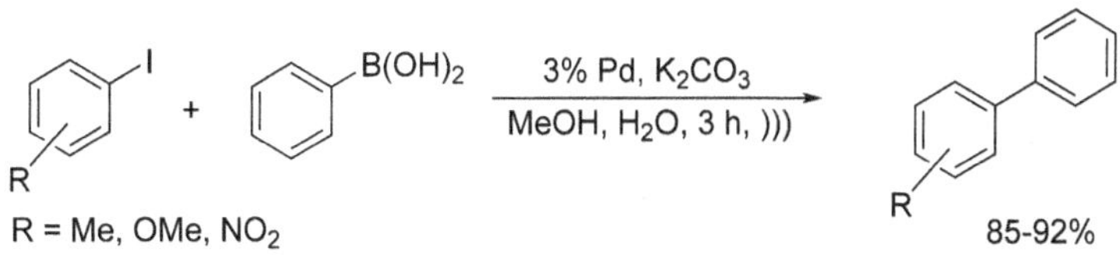

Scheme 7 A sonochemical Suzuki-coupling.

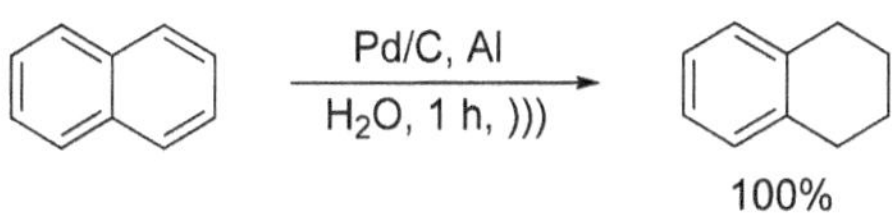

Scheme 8 Ultrasound-assisted reduction of naphthalene to tetrahydronaphthalene.

Ultrasounds have also been used for the reduction of various compounds using the aluminum-water system as the hydrogen source.[54, 55] As an example, the selective reduction of one of the aromatic rings in naphthalene is shown in Scheme 8.[56] The reduction reactions are performed for a variety of C—C and C-heteroatom bonds under environmentally friendly conditions and give the desired products in high yields and selectivities using a simple experimental setup.

4 Photochemical activation

The initiation of chemical reactions by light, commonly known as photochemical activation is one of the oldest, most established nontraditional activation methods.[57] The first photochemical reaction was reported in the early 1900s when Ciamician found that sunlight could initiate chemical reactions.[58] As the early light source in photochemistry, the solar energy provided essentially unlimited and rather broad-spectrum radiation that was used to initiate a number of early photochemical reactions.[59] The solar spectrum covers a wide range of wavelengths, from the ultraviolet, visible to the infrared region.[60] Later, the basic rules of photochemical activation that drive the development of new processes were established.[61] The theoretical basis of the photochemical activation is the same as it was with the previous methods; in order to activate a reaction, at least one component of the system must absorb the light also known as the first law of photochemistry (Grotthuss-Draper law).[62] The law of photoequivalence, which is based on the dual wave/particle nature of light, establishes that each photon that is absorbed in a chemical system can activate one molecule only

(Stark-Einstein law).[63] Extensive efforts have been made to describe the efficiency of photochemical processes. The primary efficiency is described by quantum yield (Φ) that is the ratio of the absorbed photons that initiate change vs all absorbed photons. Depending on the process, Φ can be calculated by many observed variables, such as emitted light (fluorescence) or decomposed molecules (in a decomposition reaction), etc. In theory, when every absorbed photon generates a change $\Phi = 1$, any case when $\Phi < 1$ indicates that at least part of the absorbed photons was wasted. In real systems, however, the quantum yield could also be higher than one, sometimes even up to a million. This suggests that the absorption of a single photon initiates a chain reaction that can produce a larger number of product molecules. The simplest example for this phenomenon is the chlorination of methane. In practical photochemical syntheses, the light source is limited to ultraviolet (UV) and/or visible (Vis) lamps. This determines that at least one component of a system (substrate, solvent, catalyst, etc.) must be able to absorb UV-Vis light.

The major goal of green and sustainable chemistry is to reduce the environmental impact of a process. From this point of view, photochemical activation possesses several advantages. The energy is introduced to the reaction mixture in the form of photons that are the cleanest reagents (in addition to electrons, see electrochemical activation, vide infra). Although side reactions can occur, photons themselves will not generate any waste.[64] In addition, the appropriate selection of photons with specific energy will only activate the corresponding chemical bonds, thus carefully designed photochemical reactions mostly occur with high chemoselectivity.[65] Furthermore, these reactions do not require conductive heating and commonly occur at room temperature (or lower). Therefore, thermal processes that potentially can generate isomeric or by-products, especially in stereoselective reactions, are less of a factor in photochemical processes, which often result in exclusive stereoselectivity.[66]

Although based on the above, photochemical activation has several advantages from the green chemistry point of view, it does have disadvantages as well. One such issue is the scale-up of photochemical processes that is difficult to do at the industrial level. One of the major issues is the significant loss of energy that is due to multiple reasons;[67] (i) ineffective photon absorption (e.g., luminescence), (ii) the wall of the reactor if transparent, (iii) most light sources have a broad spectrum while the absorption occurs at a specific wavelength, resulting in often over 50% loss. Another serious issue is the limited penetration depth of the light[68] as already discussed with microwaves. As a potential solution, large reaction vessels are set up with multiple light sources, which make the reactor design complicated and expensive. In addition, none of the light sources has 100% efficiency, thus they generate considerable heat as well. Working with often low boiling, flammable solvents, this represents a safety issue and the system may require continuous cooling especially when immersion lamps are used.[69] Although these shortcomings somewhat limit the application of photochemical activation, it is still widely used in a plethora of organic synthesis protocols. In fact, it is the best way to proceed with many radical and pericyclic reactions and for substrates with electron-rich double bonds that can be excited by UV or visible light (alkenes, dienes, and carbonyl compounds).[70]

One of the most recent exciting new applications in organic synthesis is in photochemistry, namely photocatalysis that attracted great attention.[71] In these processes, an originally inactive material (procatalyst) is activated by light and becomes a viable catalyst for different processes. This concept may contribute to solving several of the problems outlined above and holds significant promise for the development of potential green technologies.

Several applications are listed below.

Beeler et al. showed that the [2 + 2] cycloaddition, a typical photochemical reaction, can be performed under stereo-controlled conditions using an organocatalyst in a flow reactor (Scheme 9).[72]

Pandey et al. achieved the formation of six-membered nitrogen-containing rings by a photocatalyzed reaction.[73, 74] Dicyanonaphthalene was used as the photocatalyst for the transformation which can be performed in isopropanol as the solvent (Scheme 10).

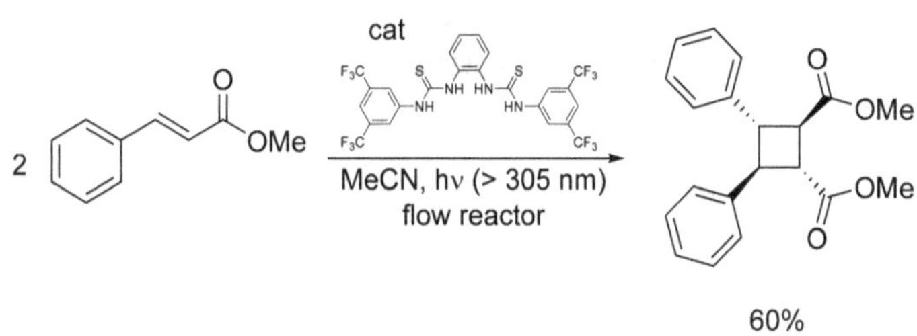

Scheme 9 An organocatalytic photochemical [2 + 2]cycloaddition.

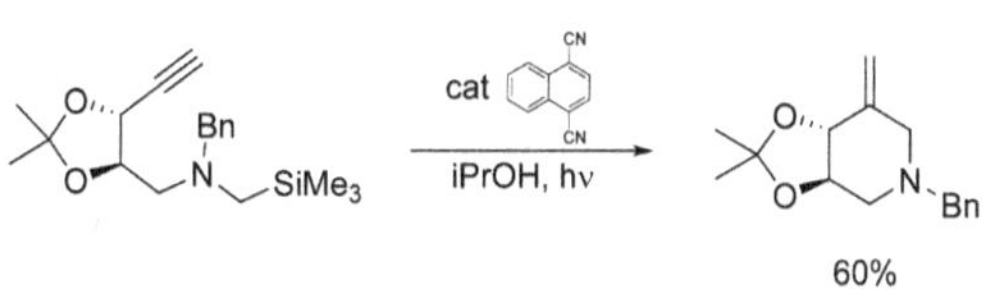

Scheme 10 A photocatalytic cyclization.

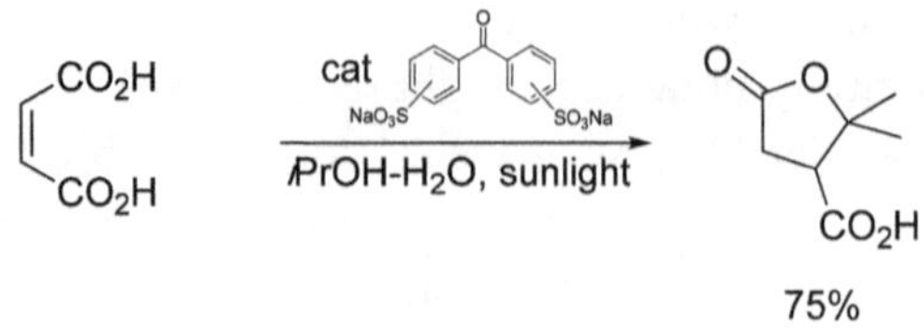

Scheme 11 Cyclialkylation of maleic acid by a benzophenone-based photocatalyst.

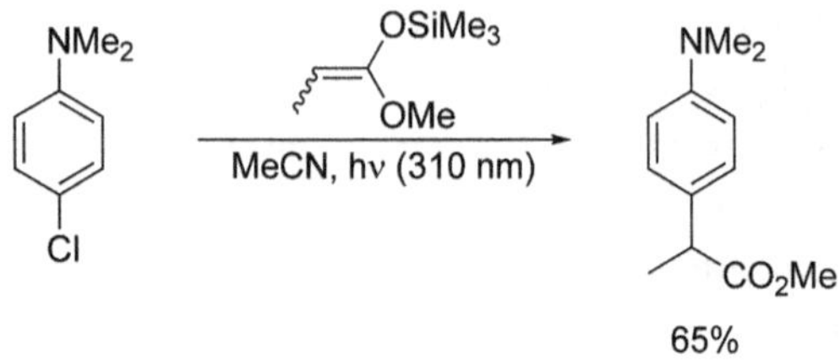

Scheme 12 Photochemical alkylation of chloroarenes.

Using a substituted benzophenone photocatalyst the alkylation of maleic acid was proven to be possible in an aqueous reaction mixture using sunlight as the energy source (Scheme 11).[75]

Albini and coworkers showed that the alkylation of chloroarenes is possible under photochemical conditions with ketene silyl acetals as an alkylating agent.[76] The authors could show that the reaction can be applied to the gram-scale synthesis of the products (Scheme 12).[77]

5 Electrochemical activation

The electrochemical activation is the oldest and most established way of initiating chemical reactions without using external heat, starting with Volta's groundbreaking experiments in 1799.[78] In 1830, Faraday introduced electrolysis and the terms cathode and anode.[79] The first practical application was described by Kolbe who observed the dimerization of radicals via the decarboxylation of carboxylic acids.[80] Since then, electrochemistry became a mainstream activation method and was used in synthetic chemistry extensively.[81, 82] It is now applied in major industrial processes, including chlorine production or aluminum manufacturing by the reduction of aluminum oxide and its theoretical foundations are well established.[83] In addition to the production of materials and chemicals, electrochemistry is also heavily involved in energy production and storage (batteries and fuel cells), toxic material destruction, soil remediation among other areas. Since 2000 we are witnessing an exciting resurgence in synthetic applications of this method, which became one of the most active research areas in organic synthesis.[84, 85] The nature of this activation method is based on the direct addition/removal of electrons to/from a susceptible substrate.[86] The process is carried out using metal or graphite electrodes.[87]

The application of electrochemical activation in synthetic processes carries major benefits when it comes to the design of green synthetic protocols.[88] Similarly to photons discussed under photochemical activation, electrons are also clean reagents, they do not contribute to waste formation.[89, 90] Commonly the transformations are carried out by a direct electron transfer at one of the electrodes and the reactions do not involve any other activation, e.g., heat. This entails that most electrochemical reactions occur at or near room temperature ensuring mild conditions that indeed bring about high selectivities. Due to the use of electrons as reagents, the atom economy of the electrochemical reaction is high when compared to traditional processes that employ high molecular weight added reagents.[91] Although building reactors (such as electrochemical cells) for these transformations is somewhat complicated and requires technical expertise, due to the long history of electrochemistry such reactors are also commercially available. In addition, when compared to the typical cost of microwave or ultrasonic reactors, the available electrochemical cells are relatively inexpensive. Besides the common green benefits of using electrochemical activation certain disadvantages also apply. Many older systems include mercury-containing amalgam-based electrodes, however, as the use of mercury is phased out gradually, this represents relatively low and quickly diminishing problem. Another potential drawback is the use of solvents. The redox potential of the solvent plays a crucial role, when it is too close to that of the substrate it will result in significant wasted energy, and the solvent itself may undergo electrochemical transformations resulting in by-product (waste) formation.[92] As mentioned above, several mainstream industrial processes focus on the production of inorganic chemicals and electrochemical organic synthesis is one of the most dynamically growing fields in synthesis design. Common examples include reduction, oxidation, or cleavage reactions among others.[93] Representative examples are listed below.

Aromatic systems exhibit high reactivity toward electrochemical synthesis methods and are therefore a highly researched area in electrochemistry.[94] A very interesting example was reported by Yoshida and coworkers who

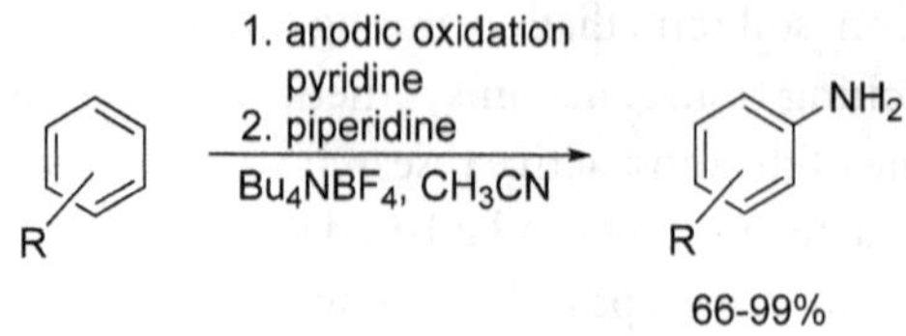

Scheme 13 Direct electrochemical amination of aromatics.

demonstrated the direct amination of aromatic compounds via anodic oxidation and a pyridinium ion intermediate.[95] The desired products were obtained in mostly high yields for a variety of compounds bearing different substituents (Scheme 13).

Anodic oxidation was also used to perform a critical cyclization step in the synthesis of several natural products. As an example, the formation of the central ring in alliacol A is shown in Scheme 14.[96] Not only the ring formation is interesting but also the reaction consists of an Umpolung reaction as both reactive sites are usually considered as nucleophiles.

Waldvogel and his group were able to show the use of cathodic reduction for the debromination of dibromocyclopropanes.[97] The dehalogenation occurs in good yield using DMF as the solvent (Scheme 15). The authors could also show that the reaction can be used on the industrially useful multigram scale for the production of an NSSA inhibitor key intermediate.[98]

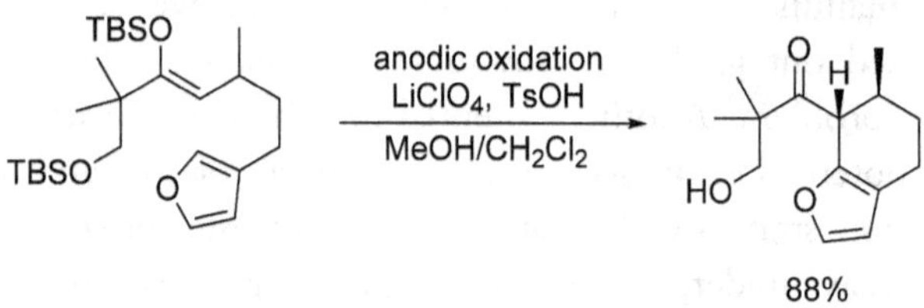

Scheme 14 A cyclization reaction via anodic oxidation.

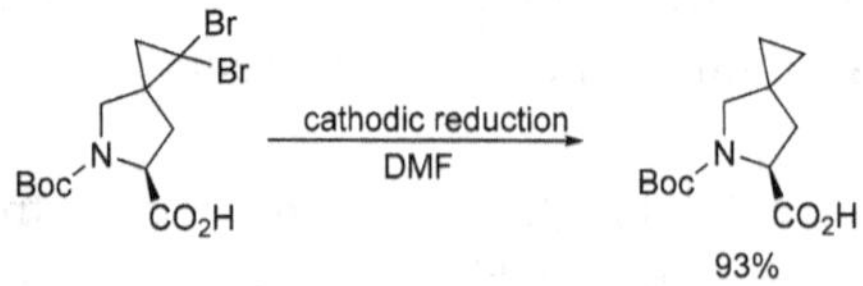

Scheme 15 Debromination of a complex dibromocyclopropane by cathodic reduction.

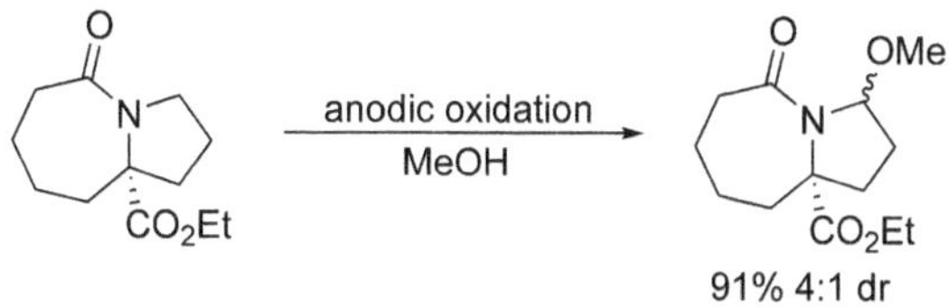

Scheme 16 Anodic oxidation of a cyclic amide.

While special electrochemical equipment is often used to perform electrosynthesis, Aubé et al. showed that the anodic oxidation of cyclic amides is possible using a repurposed phone charger as a power source and pencil lead as an electrode.[99] The desired product was obtained in high yield with methanol acting both as a solvent and as a reagent (Scheme 16).

6 Activation by mechanochemistry

The typical synthetic reactions occur in solvents, often putting a strain on process design teams to select an appropriate solvent that conforms to the recent standards of environmental safety and sustainability. While in many cases even that is not possible, the problem of solvent handling is key to the greenness of every process. If recycling is not possible, the spent solvents contribute to the increase of chemical waste a process produces. Even if the solvent can be recycled, it takes effort, time, and energy. In summary, although the use of solvents is traditional, it can cause serious environmental issues. As a consequence, scientists have been exploring the possibility of producing chemicals under solvent-free conditions.[100, 101] Although the product isolation and potential purification will consume some solvent, not having a solvent during a process carries multiple advantages. Efforts were made in several areas, for example, in microwave-assisted synthesis.[8, 17] However, no other field produced more advances in solvent-free synthesis than mechanochemistry. Mechanochemistry is an activation method to carry out chemical processes where the energy is provided by mechanical means, such as compression, friction, grinding, ball milling, or shear.[102, 103] One of the earliest examples goes back to 1820, when Faraday described the reduction of Ag(I) ions by other metals using grinding, however, the number of applications continuously grew over the 19th and 20th centuries.[104] Since then mechanochemistry grew to a flourishing field and it became a mainstream activation method in chemical synthesis. Several recent reviews summarize the development up to 2020.[105–108] Despite the extensive efforts the theoretical aspects of this technique is much less understood than the other methods discussed in this chapter and the book itself. The most common execution of mechanochemical synthesis is by the application of ball mills, where the kinetic energy of the milling balls will be transferred to the materials to initiate transformations. There are several specific application fields, such as ion- and liquid-assisted grinding, polymer-assisted grinding, accelerated and reactivated aging, catalysis, enzymatic reactions or gas-phase chemistry.[108, 109] Often the method is combined with other activation forms, such as photochemistry,[110] or ultrasounds.[111] The examples below clearly indicate the versatility and utility of the mechanochemical activation.

The group of Bolm reported that mechanochemistry can outcompete traditional methods for the proline-catalyzed enantioselective aldol addition.[112, 113] Compared to conventional solvent-based chemistry, performing the reaction under solvent-free conditions in a ball mill allowed to obtain the products in significantly shorter reaction times and with equal or better yields and selectivities (Scheme 17).

The application of ball milling in cross-coupling chemistry has been demonstrated, for example, by Jiang et al.[114] The authors reported that using a liquid-assisted grinding, the transformation of aryl chlorides was possible in high yields with a variety of substituents (Scheme 18).

Ball milling was also successfully applied in a variety of metathesis reactions. The example given below was reported by Friščić et al. showing the

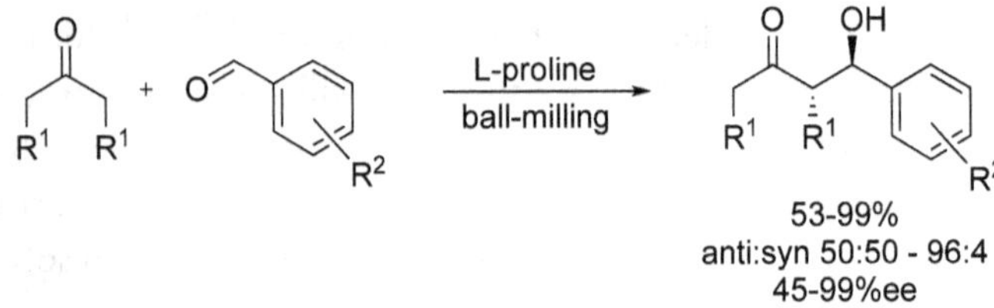

Scheme 17 Aldol addition under ball milling conditions.

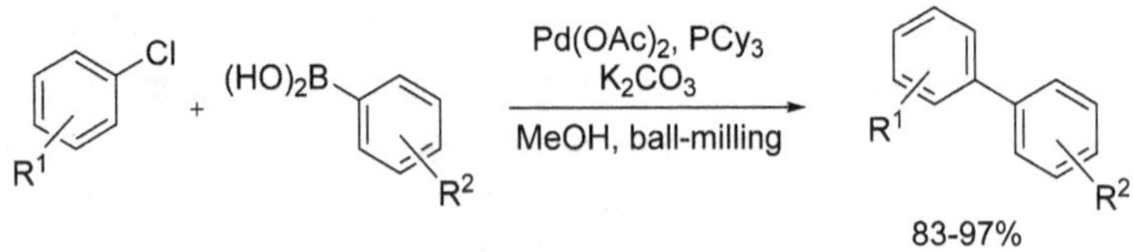

Scheme 18 Suzuki coupling under mechanochemical conditions.

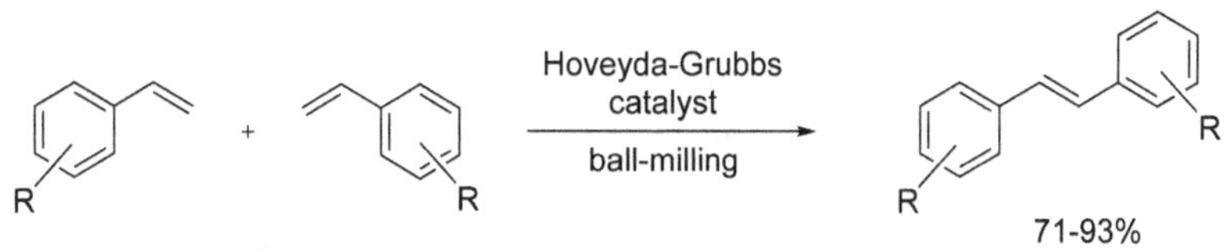

Scheme 19 Alkene metathesis by ball milling activation.

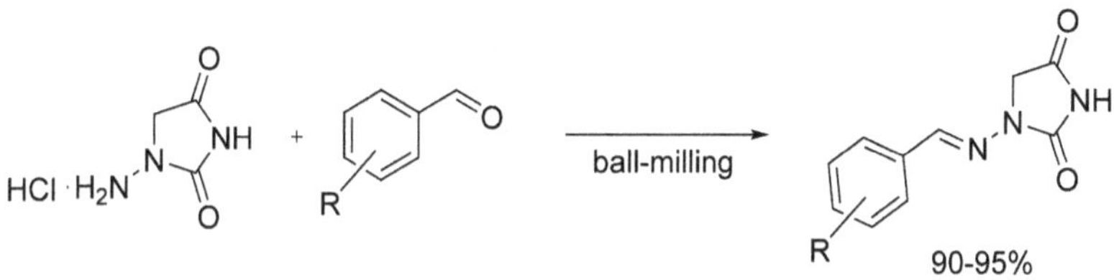

Scheme 20 Mechanochemical condensation of 1-amino hydantoins and aldehydes.

coupling of two styrene derivatives.[115] The products were obtained in high yields and short reaction times using no or only a minimal amount of solvent (Scheme 19).

The high activating potential of mechanochemistry was demonstrated by the formation of the condensation product between 1-amino hydantoins and aldehydes.[116] The remarkable feature of the reaction is that the nonnucleophilic hydrochloride salt of the amine was used and no base addition was needed (Scheme 20). The obtained molecules showed interesting biological activity.

7 Activation by high hydrostatic pressure

Similarly to the above discussed mechanochemical activation, high hydrostatic pressure (hhp) also applies a mechanical force to initiate reactions. Generally, it has been overlooked as a method to activate or modify chemical reactions although its beneficial effect in deactivating microbes in the food industry is known since the 1980s.[117] High-pressure gases are also commonly used in chemical synthesis, hydrogenation being the most common example.[118] However, the term high-pressure gases covers a completely different pressure range than hhp. Hydrogenation that occurs at 100 bar hydrogen pressure is considered a high-pressure reaction. In contrast, the hhp conditions most commonly involve 2–20 kbar range namely, the pressure used is higher by about two orders of magnitude. As discussed above, many nontraditional activation methods are based on the internal heat generation in the reaction mixture. In contrast, hhp generally does not initiate or accelerate chemical reactions, however, it changes the physical features of materials. It can break secondary, noncovalent bonds, in order to make the molecules tightly ordered by filling empty spaces. In addition, gas/liquid phase reactions appear to run favorably under these conditions, as the hhp dissolves the gases in the liquid component. The application of hhp does not result in active heating, although depending on the system, adiabatic heating or cooling can occur.[119] Although the application of hhp in chemical processes is somewhat limited, its applications in biology and biochemistry are widespread. These studies date back to the early 1900s.[120] One of the early examples is related to the denaturation of egg white albumin by hhp.[121] In the field of chemical reactions, the favorable effects of hhp were observed in the late 19th century,[122] despite, the first synthetic applications did not occur until the 1970s.[123, 124] The physicochemical analysis of hhp promoted reactions concluded that hhp is able to accelerate a chemical reaction when the sum of the volume of the starting components is higher than that of the transition state and the final product. Although the underlying phenomena are still poorly understood, essentially the reactions are driven by the decreasing volume throughout the reaction coordinate diagram. One of the most important parameters that describe the changes is the activation volume ($\Delta V^{\ddagger}$). As long as $\Delta V^{\ddagger}$ is negative the pressure will have a positive effect on a reaction. Extended efforts were made to determine the $\Delta V^{\ddagger}$ values for several transformations in order to be able to predict the effect of pressure on them.[125–127] Based on these observations the following reactions appear to be sensitive to pressure and thus hhp is a potential method to activate them: (i) combination reactions when two or more units form a product, for example, gas additions (e.g., hydrogenation), cycloaddition reactions, or multicomponent reactions; (ii) reactions with transition states that are cyclic such as rearrangements; (iii) reactions with dipolar transition state, such as Ar_{SE}; and (iv) reactions where the steric factors may inhibit the product formation, for example, enzyme reactions where hhp can help to achieve a more open, sterically available active center of the enzyme. In the following examples, some of these possibilities are illustrated by practical examples.

Török et al. used high hydrostatic pressure to achieve the selective reduction of ketones to secondary alcohols.[128] A Ni-Al alloy was used to generate both the hydrogen needed for the reaction as well as the hydrogenation catalyst. The yields and selectivities obtained are excellent in general (Scheme 21).

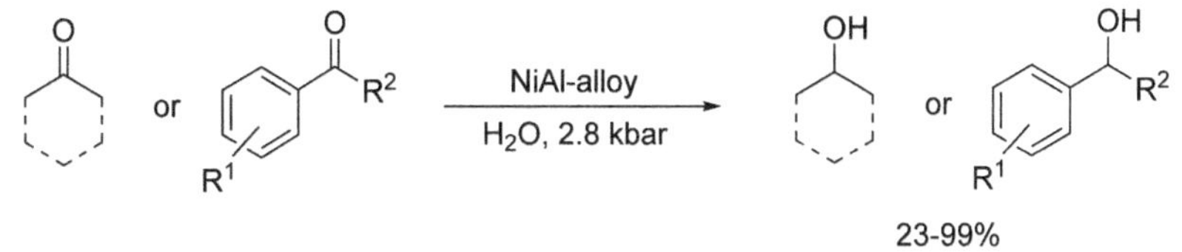

Scheme 21 NiAl initiated reduction of ketones in water under high hydrostatic pressure.

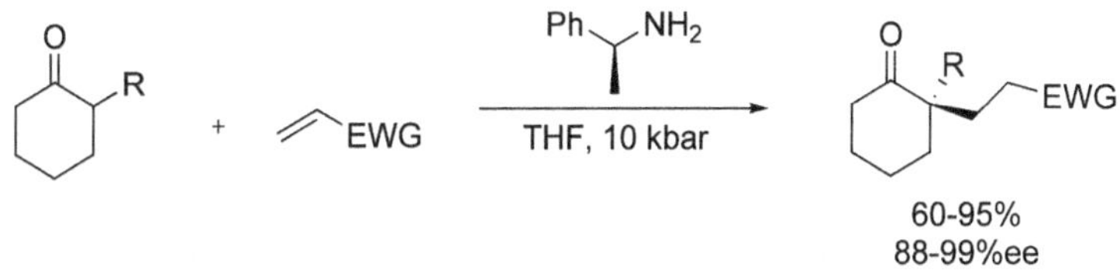

Scheme 22 An enantioselective Michael addition under high hydrostatic pressure.

Horinouchi et al. reported that the addition of enamines to Michael acceptors is strongly accelerated under high hydrostatic pressure and applied this observation in the enantioselective synthesis of cyclohexanones with a quartenary α-carbon (Scheme 22).[129]

Hayashi and coworkers revealed that the application of high hydrostatic pressure is beneficial for the enantioselective Mannich reaction (one example shown in Scheme 23).[130, 131] Using hhp conditions, the researchers obtained the desired products in higher yields and selectivities as well as with a broader substrate scope.

Eisenmenger and Reyes-De-Corcuera showed that hhp can enhance the lipase-catalyzed formation of isoamyl acetate, an important compound in the food industry. When the reaction is performed with *Candida antartica* lipase B (CALB) in a room temperature ionic liquid (RTIL) the authors reported

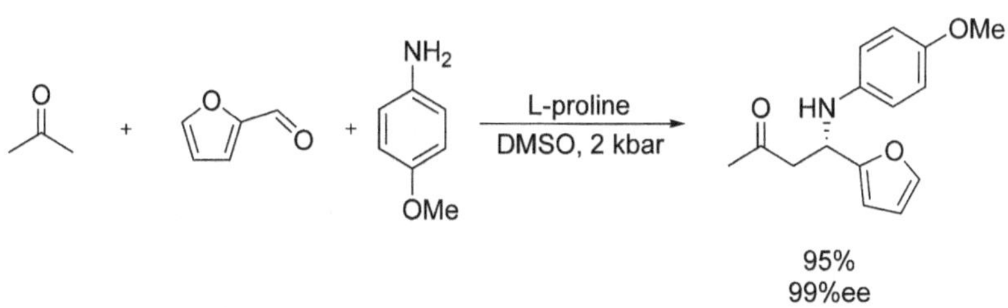

Scheme 23 A high hydrostatic pressure-activated enantioselective Mannich reaction.

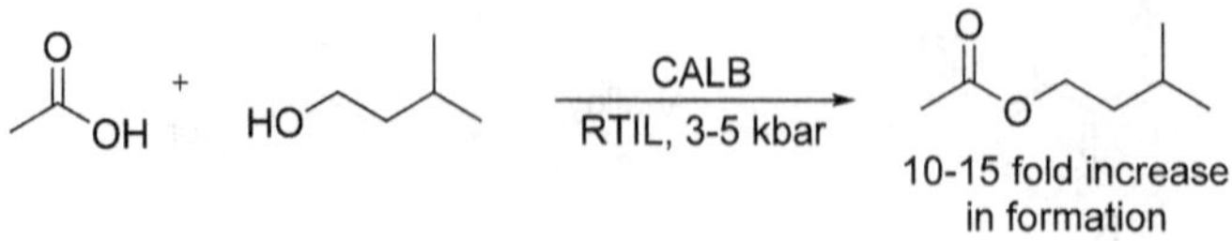

Scheme 24 Enzymatic esterification under high hydrostatic pressure.

that the product formation can be increased 10–15-fold (Scheme 24).[132] Unfortunately, no absolute yield was given by the authors.

8 Conclusions

This chapter summarizes six established and emerging methods that can be used to activate chemical reactions. Some of them involve electromagnetic radiation (microwaves and photochemistry), some sound waves (ultrasounds) or electricity (electrochemistry), or mechanical effects such as mechanochemistry or high hydrostatic pressure. We have provided an overview of each method with a set of representative recent applications in green or sustainable synthesis. The benefits and disadvantages of each method were also discussed. While in this chapter we have described the fundamentals of these activation methods, we encourage the reader to explore them in more detail by reading the state-of-the-art reviews in the book.

References

1. Anastas, P. T.; Warner, J. C. *Green Chemistry: Theory and Practice.* Oxford University Press: New York, 1998.
2. Leitner, W.; Jessop, P. G.; Li, C. J.; Wasserscheid, P.; Stark, A., Eds.; *Handbook of Green Chemistry–Green Solvents*; Wiley-VCH: Weinhheim, 2010.
3. Meyers, R. A., Ed.; *Encyclopedia of Sustainable Science and Technology*; Springer-Nature, 2018https://doi.org/10.1007/978-1-4939-2493-6_1008-1.
4. Török, B.; Dransfield, T., Eds.; *Green Chemistry: An Inclusive Approach*; Elsevier: Oxford-Cambridge, MA, 2018.
5. Matlack, A. S. *Introduction to Green Chemistry*; 2nd ed.; CRC Press, Taylor & Francis: New York, 2010.
6. Lancaster, M. *Green Chemistry—An Introductory Text*; 3rd ed.; RSC: Cambridge, 2016.
7. Baird, C.; Cann, M. *Enviromental Chemistry*; 5th ed.; W. H. Freeman and Co.: New York, 2012
8. Kokel, A.; Schäfer, C.; Török, B. Microwave-Assisted Reactions in Green Chemistry. In: *Encyclopedia of Sustainable Science and Technology;* Meyers, R. A., Ed.; Springer-Nature, 2018https://doi.org/10.1007/978-1-4939-2493-6_1008-1.
9. Vanderhoff, J. W. *Method for Carrying Out Chemical Reactions Using Microwave Energy.* US Patent 3432413A, 1969.

10. Gedye, R.; Smith, F.; Westaway, K.; Ali, H.; Baldisera, L.; Laberge, L.; Rousell, J. The Use of Microwave Ovens for Rapid Organic Synthesis. *Tetrahedron Lett.* **1986**, *27*, 279–282.
11. Giguere, R. J.; Bray, T. L.; Duncan, S. M.; Majetich, G. Application of Commercial Microwave Ovens to Organic Synthesis. *Tetrahedron Lett.* **1986**, *27*, 4945–4948.
12. Kappe, C. O.; Pieber, B.; Dallinger, D. Microwave Effects in Organic Synthesis: Myth or Reality? *Angew. Chem. Int. Ed.* **2013**, *52*, 1088–1094.
13. Dudley, G. B.; Richert, R.; Stiegman, A. E. On the Existence of and Mechanism for Microwave-Specific Reaction Rate Enhancement. *Chem. Sci.* **2015**, *6*, 2144–2152.
14. Ma, J. Master Equation Analysis of Thermal and Nonthermal Microwave Effects. *J. Phys. Chem. A* **2016**, *120*, 7989–7997.
15. de la Hoz, A.; Diaz-Ortiz, A.; Moreno, A. Microwaves in Organic Synthesis. Thermal and Non-thermal Microwave Effects. *Chem. Soc. Rev.* **2005**, *34*, 164–178.
16. Díaz-Ortiz, Á.; Prieto, P.; de la Hoz, A. A Critical Overview on the Effect of Microwave Irradiation in Organic Synthesis. *Chem. Rec.* **2019**, *19*, 85–97.
17. Kokel, A.; Schäfer, C.; Török, B. Application of Microwave-Assisted Heterogeneous Catalysis in Sustainable Synthesis Design. *Green Chem.* **2017**, *19*, 3729–3751.
18. Abid, M.; Savolainen, M.; Landge, S.; Hu, J.; Prakash, G. K. S.; Olah, G. A.; Török, B. Synthesis of Trifluoromethyl-Imines by Solid Acid/Superacid Catalyzed Microwave Assisted Approach. *J. Fluor. Chem.* **2007**, *128*, 587–594.
19. Kulkarni, A.; Török, B. Microwave-Assisted Multicomponent Domino Cyclization-Aromatization: an Efficient Approach for the Synthesis of Substituted Quinolines. *Green Chem.* **2010**, *12*, 875–878.
20. Kokel, A.; Török, B. Microwave-Assisted Solid Phase Diazotation: a Method for the Environmentally Benign Synthesis of Benzotriazoles. *Green Chem.* **2017**, *19*, 2515–2519.
21. Schmink, J. R.; Kormos, C. M.; Devine, V. G.; Leadbeater, N. E. Exploring the Scope for Scale-Up of Organic Chemistry Using a Large Batch Microwave Reactor. *Org. Process Res. Dev.* **2010**, *14*, 205–214.
22. Matsuzawa, M.; Togashi, S. Pilot Plant for Continuous Flow Microwave-Assisted Chemical Reactions. In: *Microwaves in Catalysis: Methodology and Applications;* Horikoshi, S.; Serpone, N., Eds.; Wiley-VCH: Weinheim, 2016; pp 141–154.
23. Mooney, T.; Török, B. Microwave-Assisted Flow Systems in the Green Production of Fine Chemicals. In: *Non-traditional Activation Methods in Green and Sustainable Applications: Microwaves, Ultrasounds, Photo-, Electro-, and Mechanochemistry, and High Hydrostatic Pressure;* Török, B.; Schäfer, C., Eds.; Elsevier: Cambridge, Oxford, 2020; (in press).
24. Razzaq, T.; Kappe, C. O. On the Energy Efficiency of Microwave-Assisted Organic Reactions. *ChemSusChem* **2008**, *1*, 123–132.
25. Moseley, J. D.; Kappe, C. O. A Critical Assessment of the Greenness and Energy Efficiency of Microwave-Assisted Organic Synthesis. *Green Chem.* **2011**, *13*, 794–806.
26. Cho, H.; Török, F.; Török, B. Energy Efficiency of Heterogeneous Catalytic Microwave-Assisted Organic Reactions. *Green Chem.* **2014**, *16*, 623–3634.
27. Rocchi, D.; González, J. F.; Menéndez, J. C. Montmorillonite Clay-Promoted, Solvent-Free Cross-Aldol Condensations Under Focused Microwave Irradiation. *Molecules* **2014**, *19*, 7317–7326.
28. Sharma, A.; Kurma, V.; Sinha, A. K. A Chemoselective Hydrogenation of the Olefinic Bond of α,β-Unsaturated Carbonyl Compounds in Aqueous Medium under Microwave Irradiation. *Adv. Synth. Catal.* **2006**, *348*, 354–360.
29. Shinde, V. V.; Lee, S. D.; Jeong, Y. S.; Jeong, Y. T. *p*-Toluenesulfonic Acid Doped Polystyrene (PS-PTSA): Solvent-Free Microwave Assisted Cross-Coupling-Cyclization–Oxidation to Build One-Pot Diversely Functionalized Pyrrole from Aldehyde, Amine, Active Methylene, and Nitroalkane. *Tetrahedron Lett.* **2015**, *56*, 859–865.

30. Diaz de Grenu, B.; Torres, J.; Garcia-Gonzalez, J.; Munoz-Pina, S.; de los Reyes, R.; Costero, A. M.; Amoros, P.; Ros-Lis, J. V. Microwave-Assisted Synthesis of Covalent Organic Frameworks: A Review. *ChemSusChem* **2020**https://doi.org/10.1002/cssc.202001865 (ahead of print).
31. Mason, T. J.; Lorimer, J. P. *Applied Sonochemistry—The Uses of Power Ultrasound in Chemistry and Processing*. Wiley-VCH: Weinheim, 2002.
32. Wood, R. W.; Loomis, A. L. Physical and Biologic Effects of High Frequency Sound Waves of Great Intensity. *Phil. Mag.* **1927**, *4*, 414–436.
33. Richards, W. T.; Loomis, A. L. The Chemical Effects of High Frequency Sound Waves I Preliminary survey. *J. Am. Chem. Soc.* **1927**, *49*, 3086–3100.
34. Grieser, F.; Choi, P. K.; Enomoto, N.; Harada, H.; Okitsu, K.; Yasui, K., Eds.; *Sonochemistry and the Acoustic Bubble*; Elsevier: Amsterdam, 2015.
35. Boudjouk, P. (1988). In: *Ultrasounds, Its Chemical, Physical, and Biological Effects;* Suslick, K. S., Ed.; VCH: Weinheim, 1988.
36. Suslick, K. S. Sonocatalysis. In: *Handbook of Heterogeneous Catalysis;* Ertl, G.; Knozinger, H.; Weitkamp, J., Eds.; Vol. 3; Wiley-VCH: New York, 1997.
37. Draye, M.; Kardos, N. Advances in Green Organic Sonochemistry. *Top. Curr. Chem.* **2016**, *374*, 1–29.
38. Chatel, G. How Sonochemistry Contributes to Green Chemistry? *Ultrason. Sonochem.* **2018**, *40*, 117–122.
39. Cintas, P. Ultrasound and Green Chemistry - Further Comments. *Ultrason. Sonochem.* **2016**, *28*, 257–258.
40. Mason, T. J. Sonochemistry and the Environment - Providing a "Green" Link between Chemistry, Physics and Engineering. *Ultrason. Sonochem.* **2007**, *14*, 476–483.
41. Török, B. Sustainable Synthesis. In: *Green Chemistry: An Inclusive Approach;* Török, B.; Dransfield, T., Eds.; Elsevier: Oxford-Cambridge, MA, 2018; pp 48–90 (chapter 2.2).
42. Ellstrom, C. J.; Török, B. Application of Sonochemical Activation in Green Synthesis. In: *Green Chemistry: An Inclusive Approach;* Török, B.; Dransfield, T., Eds.; Elsevier: Oxford-Cambridge, MA, 2018; pp 673–693 (chapter 3.19).
43. Luche, J.-L., Ed.; *Synthetic Organic Sonochemistry*; Springer: New York, 1998.
44. Cousin, T.; Chatel, G.; Kardos, N.; Andrioletti, B.; Draye, M. High Frequency Ultrasound as a Tool for Elucidating Mechanistic Elements of Cis-cyclooctene Epoxidation with Aqueous Hydrogen Peroxide. *Ultrason. Sonochem.* **2019**, *53*, 120–125.
45. Lockwood, G. R.; Turnball, D. H.; Christopher, D. A.; Foster, F. S. Beyond 30 MHz [Applications of High-Frequency Ultrasound Imaging]. *IEEE Eng. Med. Biol. Mag.* **1996**, *15*, 60–71.
46. Tudela, I.; Sáez, V.; Esclapez, M. D.; Díez-García, M. I.; Bonete, P.; González-García, J. Simulation of the Spatial Distribution of the Acoustic Pressure in Sonochemical Reactors with Numerical Methods: a Review. *Ultrason. Sonochem.* **2014**, *21*, 909–919.
47. Didenko, Y. T.; Suslick, K. S. The Energy Efficiency of Formation of Photons, Radicals and Ions during Single-Bubble Cavitation. *Nature* **2002**, *418*, 394–397.
48. Suslick, K. S.; Hammerton, D. A.; Cline, R. E. Sonochemical Hot Spot. *J. Am. Chem. Soc.* **1986**, *108*, 5641–5642.
49. Margulis, M. A. Sonoluminescence and Sonochemical Reactions in Cavitation Fields. A review. *Ultrasonics* **1985**, *23*, 157–169.
50. Török, B.; Balázsik, K.; Török, M.; Felföldi, K.; Bartók, M. Asymmetric Reactions in Sonochemistry. *Ultrason. Sonochem.* **2001**, *8*, 191–200.
51. Vidal, M.; Garcia-Arriagada, M.; Rezende, M. C.; Dominguez, M. Ultrasound-Promoted Synthesis of 4-Pyrimidinols and Their Tosyl Derivatives. *Synthesis* **2016**, *48*, 4246–4252.

52. Jin, H.; Xiang, L.; Wen, F.; Tao, K.; Liu, Q.; Hou, T. Improved Synthesis of Chalconoid-Like Compounds Under Ultrasound Irradiation. *Ultrason. Sonochem.* **2008**, *15*, 681–683.
53. Shil, A. K.; Guha, N. R.; Sharma, D.; Das, P. A Solid Supported Palladium(0) Nano/ Microparticle Catalyzed Ultrasound Induced Continuous Flow Technique for Large Scale Suzuki Reactions. *RSC Adv.* **2013**, *3*, 13671–13676.
54. Cho, H.; Török, F.; Török, B. Selective Reduction of Condensed N-Heterocycles Using Water as Solvent and Hydrogen Source. *Org. Biomol. Chem.* **2013**, *11*, 1209–1215.
55. Schäfer, C.; Nisanci, B.; Bere, M. P.; Dastan, A.; Török, B. Heterogeneous Catalytic Reductive Amination of Carbonyl Compounds with Ni-Al Alloy in Water as Solvent and Hydrogen Source. *Synthesis* **2016**, *48*, 3127–3133.
56. Schäfer, C.; Ellstrom, C. J.; Cho, H.; Török, B. Pd/C-Al-Water Facilitated Selective Reduction of a Broad Variety of Functional Groups. *Green Chem.* **2017**, *19*, 1230–1234.
57. Ngo, K. T.; Rochford, J. Principles of Photochemical Activation Toward Artificial Photosynthesis and Organic Transformations. In: *Green Chemistry: An Inclusive Approach, Chp. 3.21.;* Török, B.; Dransfield, T., Eds.; Elsevier: Oxford-Cambridge, MA, 2018; pp 729–752.
58. Ciamician, G. The Photochemistry of the Future. *Science* **1912**, *36*, 385–394.
59. Paterno, E. I nuovi orizzonti della sintesi in chimica organica. *Gazz. Chim. Ital.* **1909**, *39*, 213–219.
60. (a) *Standard Tables for Reference Solar Spectral Irradiances: Direct Normal and Hemispherical on 37 Degree Tilted Surface*; ASTM International, 2012; (b) *Standard Solar Constant and Zero Air Mass Solar Spectral Irradiance Tables*; ASTM International, 2014.
61. Evans, R. C.; Douglas, P.; Burrows, H. D., Eds.; *Applied Photochemistry*; Springer: Dodrecht, 2013.
62. Albini, A. Some Remarks on the First Law of Photochemistry. *Photochem. Photobiol. Sci.* **2016**, *15*, 319–324.
63. Balzani, V.; Ceroni, P.; Juris, A. *Photochemistry and Photophysics: Concepts, Research, Applications*. Wiley-VCH Verlag, 2014.
64. Bonfield, H. E.; Knauber, T.; Levesque, F.; et al. Photons as a 21st Century Reagent. *Nat. Commun.* **2020**, *11https://doi.org/10.1038/s41467-019-13988-4*.
65. Kisch, H. Semiconductor Photocatalysis for Chemoselective Radical Coupling Reactions. *Acc. Chem. Res.* **2017**, *50*, 1002–1010.
66. Woźniak, L.; Magagnano, G.; Melchiorre, P. Enantioselective Photochemical Organocascade Catalysis. *Angew. Chem. Int. Ed.* **2018**, *57*, 1068–1072.
67. Cambie, D.; Noël, T. Solar Photochemistry in Flow. *Top.Curr. Chem* **2018**, *37645*.
68. Weiss, G. H.; Nossal, R.; Bonner, R. F. Statistics of Penetration Depth of Photons Re-emitted From Irradiated Tissue. *J. Mod. Opt.* **1989**, *36*, 349–359.
69. Knowles, J. P.; Elliott, L. D.; Booker-Milburn, K. I. Flow Photochemistry: Old Light through New Windows. *Beilstein J. Org. Chem.* **2012**, *8*, 2025–2052.
70. Turro, N. J.; Schuster, G. Photochemical Reactions as a Tool in Organic Synthesis. *Science* **1975**, *187*, 303–312.
71. Corrigan, N.; Shanmugam, S.; Xu, J.; Boyer, C. Photocatalysis in Organic and Polymer Synthesis. *Chem. Soc. Rev.* **2016**, *4*, 6165–6212.
72. Telmesani, R.; Park, S. H.; Lynch-Colameta, T.; Beeler, A. B. [2 + 2] Photocycloaddition of Cinnamates in Flow and Development of a Thiourea Catalyst. *Angew. Chem. Int. Ed.* **2015**, *54*, 11521–11525.
73. Pandey, G.; Dumbre, S. G.; Khan, I. M.; Shabab, M. Convergent Approach Toward the Synthesis of the Stereoisomers of C-6 Homologues of 1-Deoxynojirimycin and

Their Analogues: Evaluation as Specific Glycosidase Inhibitors. *J. Org. Chem.* **2006**, *71*, 8481–8488.
74. Pandey, G.; Dumbre, S. G.; Khan, I. M.; Shabab, M.; Puranik, V. G. A β-lactam-azasugar hybrid as a competitive potent galactosidase inhibitor. *Tetrahedron Lett.* **2006**, *47*, 7923–7926.
75. Dondi, D.; Protti, S.; Albini, A.; Carpio, S. M.; Fagnoni, M. Synthesis of γ-Lactols, γ-Lactones and 1,4-Monoprotected Succinaldehydes under Moderately Concentrated Sunlight. *Green Chem.* **2009**, *11*, 1653–1659.
76. Fraboni, A.; Fagnoni, M.; Albini, A. A Novel α-Arylation of Ketones, Aldehydes, and Esters Via a Photoinduced S_N1 Reaction through 4-Aminophenyl Cations. *J. Org. Chem.* **2003**, *68*, 4886–4893.
77. Dichiarante, V.; Fagnoni, M.; Albini, A. *Solar*ylations Via 4-Aminophenyl Cations. *J. Org. Chem.* **2010**, *75*, 1271–1276.
78. Fabrizzi, L. Strange Case of Signor Volta and Mister Nicholson: How Electrochemistry Developed as a Consequence of an Editorial Misconduct. *Angew. Chem. Int. Ed.* **2019**, *58*, 5810–5822.
79. Ross, S. Faraday Consults the Scholars: The Origins of the Terms of Electrochemistry. *Notes Rec. R. Soc. Lond.* **1961**, *16*, 187–220.
80. Kolbe, H. Beobachtungen über die oxydirende Wirkung des Sauerstoffs, wenn derselbe mit Hülfe einer elektrischen Säule entwickelt wird. *J. Prakt. Chem.* **1847**, *41*, 137–139.
81. Lund, H. A Century of Organic Electrochemistry. *J. Electrochem. Soc.* **2002**, *149*, S21–S33.
82. Horn, E. J.; Rosen, B. R.; Baran, P. S. Synthetic Organic Electrochemistry: An Enabling and Innately Sustainable Method. *ACS Cent. Sci.* **2016**, *2*, 302–308.
83. Bard, A. J.; Faulkner, L. R. Electrochemical Methods: Fundamentals and Applications. In: *Electrochemistry. I;* Harris, D.; Swain, E.; Robey, C.; Aiello, E., Eds.; Wiley: New York, 2000.
84. Fuchigami, T.; Atobe, M.; Inagi, S. *Fundamentals and Applications of Organic Electrochemistry. Synthesis, Materials, Devices.* Wiley: Hoboken, NJ, 2014.
85. Yan, M.; Kawamata, Y.; Baran, P. Synthetic Organic Electrochemical Methods since 2000: On the Verge of a Renaissance. *Chem. Rev.* **2017**, *117*, 13230–13319.
86. Grimshaw, J. *Electrochemical Reactions and Mechanisms in Organic Chemistry.* Elsevier: Amsterdam, 2000.
87. Bard, A. J.; Faulkner, L. R.; Leddy, J.; Zoski, C. G. *Electrochemical Methods: Fundamentals and Applications.* Wiley: New York, 1980; Vol. 2.
88. McKinnon, M.; Rochford, J. Principles of Electrocatalysis. In: *Green Chemistry: An Inclusive Approach;* Török, B.; Dransfield, T., Eds.; Elsevier: Oxford-Cambridge, MA, 2018; pp 695–727 (chapter 3.20).
89. Walsh, F. C. Electrochemical Technology for Environmental Treatment and Clean Energy Conversion. *Pure Appl. Chem.* **2001**, *73*, 1819–1837.
90. Frontana-Uribe, B. A.; Little, R. D.; Ibanez, J. G.; Palma, A.; Vasquez-Medrano, R. Organic Electrosynthesis: A Promising Green Methodology in Organic Chemistry. *Green Chem.* **2010**, *12*, 2099–2119.
91. Schäfer, H. J. Contributions of Organic Electrosynthesis to Green Chemistry. *C. R. Chim.* **2011**, *14*, 745–765.
92. Gritzner, G.; Kuta, J. Recommendations on Reporting Electrode Potentials in Non-aqueous Solvents. *Pure Appl. Chem.* **1984**, *56*, 461–466.
93. Pollok, D.; Waldvogel, S. R. Electro-Organic Synthesis—A 21st Century Technique. *Chem. Sci.* **2020**https://doi.org/10.1039/d0sc01848a.
94. Waldvogel, S. R.; Lips, S.; Selt, M.; Riehl, B.; Kampf, C. J. Electrochemical Arylation Reaction. *Chem. Rev.* **2018**, *118*, 6705–6765.

95. Morofuji, T.; Shimizu, A.; Yoshida, J. Electrochemical C – H Amination: Synthesis of Aromatic Primary Amines Via N-Arylpyridinium Ions. *J. Am. Chem. Soc.* **2013**, *135*, 5000–5003.
96. Mihelcic, J.; Moeller, K. D. Oxidative Cyclizations: The Asymmetric Synthesis of (−)-Alliacol A. *J. Am. Chem. Soc.* **2004**, *126*, 9106–9111.
97. Gutz, C.; Selt, M.; Bänziger, M.; Bucher, C.; Romelt, C.; Hecken, N.; Gallou, F.; Galvao, T. R.; Waldvogel, S. R. A novel cathode material for cathodic dehalogenation of 1,1-dibromo cyclo-propane derivatives. *Chem. A Eur. J.* **2015**, *21*, 13878–13882.
98. Gutz, C.; Banziger, M.; Bucher, C.; Galvao, T. R.; Waldvogel, S. R. Development and scale-up of the electrochemical dehalogenation for the synthesis of a key intermediate for NS5A inhibitors. *Org. Process Res. Dev.* **2015**, *19*, 1428–1433.
99. Frankowski, K. J.; Liu, R.; Milligan, G. L.; Moeller, K. D.; Aubé, J. Practical Electrochemical Anodic Oxidation of Polycyclic Lactams for Late Stage Functionalization. *Angew. Chem. Int. Ed.* **2015**, *54*, 10555–10558.
100. Toda, F. Solid State Organic Chemistry: Efficient Reactions, Remarkable Yields, and Stereoselectivity. *Acc. Chem. Res.* **1995**, *28*, 480–486.
101. Tanaka, K.; Toda, F. Solvent-Free Organic Synthesis. *Chem. Rev.* **2000**, *10*, 1025–1074.
102. Nič, M.; Jirát, J.; Košata, B.; Jenkins, A.; McNaught, A., Eds.; *IUPAC Compendium of Chemical Terminology*; IUPAC, 2009.
103. Margetić, D.; Štrukil, V. *Mechanochemical Organic Synthesis*. Elsevier, 2016.
104. Takacs, L. The Historical Development of Mechanochemistry. *Chem. Soc. Rev.* **2013**, *42*, 7649–7659.
105. James, S. L.; Friščić, T. Mechanochemistry. *Chem. Soc. Rev.* **2013**, *42*, 7494–7496.
106. Tan, D.; Friščić, T. Mechanochemistry for Organic Chemists: An Update. *Eur. J. Org. Chem.* **2018**, ;18–33.
107. Howard, J. L.; Cao, Q.; Browne, D. L. Mechanochemistry as an Emerging Tool for Molecular Synthesis: What Can it Offer? *Chem. Sci.* **2018**, *9*, 3080–3094.
108. Friščić, T.; Mottillo, C.; Titi, H. M. Mechanochemistry for Synthesis. *Angew. Chem. Int. Ed.* **2020**, *59*, 1018–1029.
109. Al-Terkawi, A.-A.; Lamaty, F.; Métro, T.-X. Efficient CO_2 Sequestration by a Solid–Gas Reaction Enabled by Mechanochemistry: The Case of L-Lysine. *ACS Sustain. Chem. Eng.* **2020**, *8*, 13159–13166.
110. Sokolov, A. N.; Bučar, D. K.; Baltrusaitis, J.; Gu, S. X.; MacGillivray, L. R. Supramolecular Catalysis in the Organic Solid State through Dry Grinding. *Angew. Chem. Int. Ed.* **2010**, *49*, 4273–4277.
111. Cravotto, G.; Gaudino, E. C.; Cintas, P. On the Mechanochemical Activation by Ultrasound. *Chem. Soc. Rev.* **2013**, *42*, 7521–7534.
112. Rodríguez, B.; Rantanen, T.; Bolm, C. Solvent-Free Asymmetric Organocatalysis in a Ball Mill. *Angew. Chem. Int. Ed.* **2006**, *45*, 6924–6926.
113. Rodríguez, B.; Bruckmann, A.; Bolm, C. A Highly Efficient Asymmetric Organocatalytic Aldol Reaction in a Ball Mill. *Chem. A Eur. J.* **2007**, *13*, 4710–4722.
114. Jiang, Z. J.; Li, Z. H.; Yu, J. B.; Su, W. K. Liquid-Assisted Grinding Accelerating: Suzuki-Miyaura Reaction of Aryl Chlorides under High-Speed Ball-Milling Conditions. *J. Org. Chem.* **2016**, *81*, 10049–10055.
115. Do, J. L.; Mottillo, C.; Tan, D.; Strukil, V.; Friscic, T. Mechanochemical ruthenium-catalyzed olefin metathesis. *J. Am. Chem. Soc.* **2015**, *137*, 2476–2479.
116. Colacino, E.; Porcheddu, A.; Halasz, I.; Charnay, C.; Delogu, F.; Guerra, R.; Fullenwarth, J. Mechanochemistry for "No Solvent, No Base" Preparation of Hydantoin-Based Active Pharmaceutical Ingredients: Nitrofurantoin and Dantrolene. *Green Chem.* **2018**, *20*, 2973–2977.

117. Hayashi, R. Possibility to Apply High Pressure to Cooking, Processing, Pasteurization, and Preservation. *Shokuhin to Kaihatsu* **1987**, *Food Process. Ingred. 22*, 55–62 (in Japanese).
118. Augustine, R. L. Catalytic Hydrogenation. In: *Techniques and Applications in Organic Synthesis;* Dekker: New York, 1965.
119. Grauwet, T.; Van der Plancken, I.; Vervoort, L.; Hendrickx, M.; Van Loey, A.; Balasubramaniam, V. M.; Barbosa-Cánovas, G. V.; Lelieved, H. L. M. High-Pressure Processing Uniformity. In: *High Pressure Processing of Foods;* Springer: New York, 2016; pp 253–268.
120. Hite, B. H. *The Effect of Pressure in the Preservation of Milk*. Bull. West Virginia Univ. Agricult. Exp. Station, 1899; Vol. 5815–35.
121. Bridgman, P. W. The Coagulation of Albumen by Pressure. *J. Biol. Chem.* **1914**, *19*, 511–512.
122. Röntgen, W. C. Kurze Mittheilung von Versuchen über den Einfluss des Druckes auf einige physikalische Erscheinungen. *Annu. Rev. Phys. Chem.* **1892**, *281*, 98.
123. Holzapfel, W.; Isaacs, N. *High Pressure Techniques in Chemistry and Physics. A Practical Approach.* Oxford University Press, 1997;;400.
124. Chataigner, I.; Maddaluno, J. (2019). In: *High-Pressure Synthesis: An Eco-friendly Chemistry in Activation Methods: Sonochemistry and High Pressure;* Goddard, J.-P.; Malacria, M.; Ollivier, C., Eds.; ISTE Ltd and John Wiley & Sons, Inc., 2019
125. Asano, T.; Le Noble, W. J. Activation and Reaction Volumes in Solution. *Chem. Rev.* **1978**, *78*, 407–489.
126. Van Eldik, R.; Asano, T.; Le Noble, W. J. Activation and Reaction Volumes in Solution. 2. *Chem. Rev.* **1989**, *89*, 549–688.
127. Drljaca, A.; Hubbard, C. D.; Van Eldik, R.; Asano, T.; Basilevsky, M. V.; Le Noble, W. J. Activation and reaction volumes in solution. 3. *Chem. Rev.* **1998**, *98*, 2167–2290.
128. Tomin, A.; Lazarev, A.; Bere, M. P.; Redjeb, H.; Török, B. Selective Reduction of Ketones Using Water as a Hydrogen Source under High Hydrostatic Pressure. *Org. Biomol. Chem.* **2012**, *10*, 7321–7326.
129. Horinouchi, R.; Kamei, K.; Watanabe, R.; Hieda, N.; Tatsumi, N.; Nakano, K.; Ichikawa, Y.; Kotsuki, H. Enantioselective Synthesis of Quaternary Carbon Stereogenic Centers through the Primary Amine-Catalyzed Michael Addition Reaction of α- Substituted Cyclic Ketones at High Pressure. *Eur. J. Org. Chem.* **2015**, *20*, 4457–4463.
130. Hayashi, Y.; Tsuboi, W.; Shoji, M.; Suzuki, N. Application of High Pressure Induced by Water-Freezing to the Direct Catalytic Asymmetric Three-Component List–Barbas – Mannich Reaction. *J. Am. Chem. Soc.* **2003**, *125*, 11208–11209.
131. Hayashi, Y. The Asymmetric Catalytic Mannich Reaction Catalyzed by Organocatalyst—A Personal Account. *J. Synt. Org. Chem.* **2014**, *72*, 1228–1238.
132. Eisenmenger, M. J.; Reyes-De-Corcuera, J. I. Enhanced Synthesis of Isoamyl Acetate Using an Ionic Liquid–Alcohol Biphasic System at High Hydrostatic Pressure. *J. Mol. Catal. B: Enzym.* **2010**, *67*, 36–40.

CHAPTER TWO

Activation of chemical reactions on solid catalysts under microwave irradiation

Yuji Wada[a,b], **Satoshi Fujii**[c], **and Shuntaro Tsubaki**[d]
[a]Institute of Innovation Research, Tokyo Institute of Technology, Yokohama, Japan
[b]Microwave Chemical Co. Ltd., Techno Alliance 3F, Suita, Osaka, Japan
[c]Department of Information and Communication System Engineering, Okinawa College, Okinawa, Japan
[d]School of Materials and Chemical Technology, Tokyo Institute of Technology, Yokohama, Japan

1 Introduction: What are the issues in applying microwave irradiation to promoting and controlling heterogeneous catalysis?

"Microwave special effects" are observed as acceleration of chemical reactions, shortening of the completion time of chemical synthesis, and increase in the chemoselectivity in both homogeneous and/or heterogeneous chemical reaction systems. These unusual phenomena have attracted strong interests of not only chemists but also engineers working in the fields of chemical and materials production, because these special effects have a potential to change the conventional production processes into energy-efficient and space-saving ones, realizing the ideas of "green and sustainable chemistry" and "sustainable development goals (SDGs)" in chemical and materials production. A large number of review articles can be found on these topics, which were described not only on the positive basis for the existence of microwave special effects but also even on the negative and sceptic attitude against the existence of especially nonthermal effects of microwaves.[2–12] Especially it should be emphasized that Loupy et al. described "microwave nonthermal effects" in organic synthesis and explained those special effects by postulating "stabilization of polar transition states" through interacting with the vibrating electric field of microwaves as demonstrated in Fig. 1.[1] Loupy's idea was welcomed by organic synthetic chemists, who were eager to find new techniques to promote and control chemical reactions, even though Loupy's hypothesis was not supported by the physical

https://orcid.org/0000-0003-3717-0559

Nontraditional Activation Methods in Green and Sustainable Applications
https://doi.org/10.1016/B978-0-12-819009-8.00011-6
© 2021 Elsevier Inc.
All rights reserved.

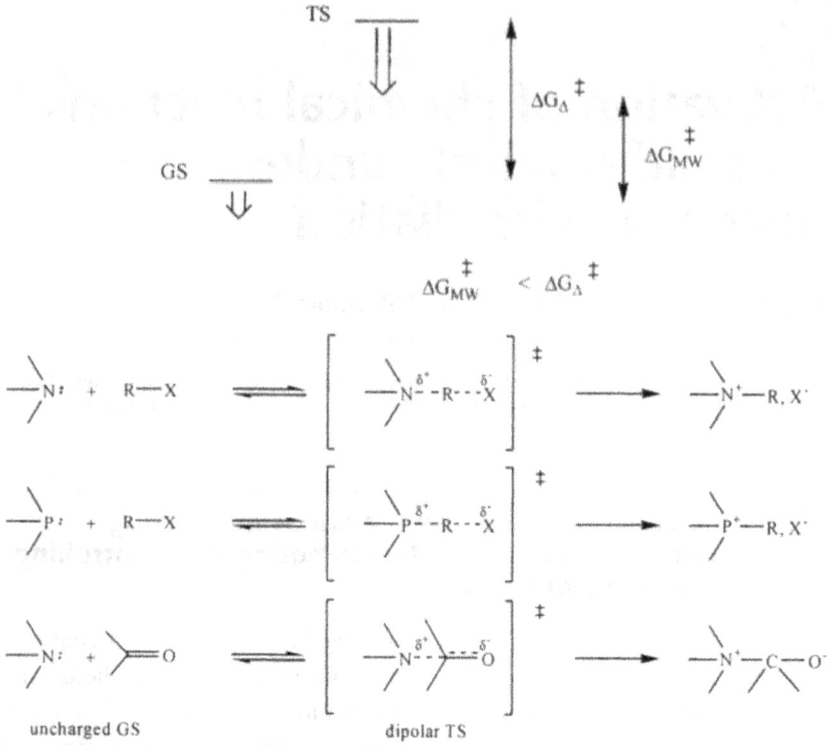

Fig. 1 Loupy's hypothesis for explaining "the microwave nonthermal effects" observed as the enhancement of the reaction rates of organic reactions. He postulated that the $\Delta G^{\neq}$ is lowered for the reaction with the more polar transition states than the ground states by the interaction with the electric field of microwaves.[1]

background and experimental data. Similar idea-based but more theoretical works were also reported by Jhung[13] and Ma[14] later. Yamada et al. reported extraordinarily novel works, in which they carried out some organic reactions with temperature-sensitive chemoselectivities and were successful in obtaining the data proving the enhancement of the reaction rates under microwave irradiation without losing the chemoselectivities, indicating that the reacting molecules did experience no change in the reaction temperatures.[15–17] Yamada's excellent results are very persuasive for the existence of microwave nonthermal effects.

However, microwave special effects and their mechanisms have not been confirmed yet, and the questions how these special effects occur are not well understood on the basis of physical science. Kappe et al. claimed that so-called microwave special effects are originating from the misinterpretation of incorrectly carried out the experiments, mostly temperature measurements. The experimentally obtained data were essentially the same for the experiments performed under microwave irradiation and conventional heating when the temperature measurements and control were performed in the proper ways.[18] Kappe started a critical and fruitful discussion on the existence of "microwave special effects" from the view point of the accuracy of temperature measurements of reaction solutions under microwave irradiation. To be involved in such discussion is not a purpose of this chapter, but it is important to refer to this issue that is still under discussion in the scientific community. We will review a number of research articles here, exhibiting their affirmative standing points for "microwave special effects," but our standing point in this account is not to take a side in this debate, rather to be neutral for both and try to stay scientifically rational. The "positive" and "negative" standing points of both sides for "microwave special effects" possess their own reasonable arguments for supporting their claims. We will describe these research works and try to organize them according to their values in science and technology.

We should recognize that there are a lot of negative reports regarding the "existence of microwave special effects" found in various areas before starting our review. In the research fields containing more complex subjects such as biorelated systems, microwave-specific effects are also a target for the scientific discussion as shown in the succeeding text. Shimoni et al. tested the Maillard reaction, protein denaturation, polymer solubility, mutagenesis of bacteria, mutarotation equilibrium of α/β-D-glucose, and saturation solubility of NaCl for checking the possibility of an nonthermal effect in a very well-controlled, high radiation intensity system (2.45 GHz, up to 1000 W kg, with continuous radiation up to 48 h).[19] They describe that all data failed to show any significant nonthermal effects when compared with conventional heating. Wang et al. mentioned that no nonthermal effect of microwave irradiation was revealed for D,L-lactide synthesis from D,L-lactic acid by the addition of $SnCl_2$ and Cat-A as a catalyst under continuous microwave irradiation in less than 1 h ["Cat A" used therein could not be identified according to the information in reference 20]. Xu et al. investigated the change of the conformation and activity of wheat germ lipase (WGL) to the lipolysis of triglyceride giving fatty acid treated by microwaves and conventional heating and concluded that the activity decrease was based on the thermal effects of microwaves.[21] They described that the inactivation efficiency of microwave radiation was about 10% higher than that of conventional heating based on the thermal effects. The authors concluded that microwave-inactivated WGL can be explained based on thermal effects rather than nonthermal effect and that microwave irradiation was more effective in deactivating the enzyme than conventional heating.

In addition to the earlier discussed works, we can also find research interests from the negative standing points against the "special effects" of microwave irradiation in chemistry-related areas as described in the succeeding text. Koifman et al. performed a comparative study of radical polymerization of styrene under thermal and microwave-heating conditions and concluded that nonthermal microwave effects are not manifested during styrene polymerization, because styrene is a weakly polar substance.[22] Dai's group reported on pyrolysis of wet sewage sludge using conventional pyrolysis by electric heating and microwave pyrolysis by microwave irradiation separately to obtain the emission characteristics of polychlorinated dibenzo-*p*-dioxins and dibenzofurans during pyrolysis.[23] Microwave pyrolysis produced equivalent quantity of polychlorinated dibenzo-*p*-dioxins and dibenzofurans as compared with conventional pyrolysis, which indicated the absence of a special effect (nonthermal effect). Santos et al. performed carefully designed and executed experiments on microwave porcelain firing by measuring the temperature with metal-shielded thermocouples, optical fibers, and infrared pyrometers and concluded that the data did not support the microwave nonthermal effects in microwave-induced porcelain firing.[24]

Beside the earlier described works with the negative standing points against "microwave special effects," especially in heterogeneous systems, we can find publications in which the existence of "microwave special effects" is positively demonstrated and discussed. Examples include the hydrodesulfurization of thiophene (exothermic) and decomposition of hydrogen sulfide (endothermic) on MoS_2/Al_2O_3,[25] the enhanced carbon-carbon dioxide reaction (Boudouard reactions)[26] and the enhanced reduction of copper oxide under the vibrating magnetic field irradiation selectively separated from the electromagnetic field of microwaves, which can be understood as coupling of H-field with the Fermi level electrons of CuO, resulting in cleavage of Cu—O bond.[27, 28]

Here, we need to define the terminologies of "microwave special effects," "microwave thermal effects," and "microwave nonthermal effects" for the clarity in our chapter. We can define "microwave thermal effects" as the effects, which can be derived from the special characteristics of microwave heating, that is, (1) quick rise of substance temperature called as "rapid heating," "fast heating," or "flash heating"; (2) rise of temperature in the inside of a bulk substance called as "internal heating"; and (3) rise of substance temperature depending on the physical properties of substance such as dielectric loss and electric conductivity called as "substance-selective heating." These should be separated as "microwave thermal effects" from normal heating through heat conduction and convection. Microwaves interact with substances, and the energy is converted into heat through the main mechanisms, such as joule heating, dielectric heating, and magnetic heating as expressed in Eq. (1):

$$P = \frac{1}{2}\sigma|E|^2 + pf\varepsilon_0\varepsilon_r''|E|^2 + p'f\mu_0\mu_r''|H|^2 \tag{1}$$

where P is the heat generated in the microwave energy loss, σ is the resistivity, $|E|$ is the electric field strength, f is the frequency of microwaves, ε_0 is the permittivity of vacuum, ε_r'' is the loss factor in the permittivity of the substance, μ_0 is the permeability of vacuum, and μ_r'' is the loss factor in the permeability of the substance. Each term of Eq. (1) depends on substances, resulting in the specific mechanisms of microwave heating. These specific mechanisms lead to specific and unusual heating modes only achieved by microwave irradiation as described above in (1), (2), and (3).

Compared with the observable "microwave thermal effects" as the temperature rises, it is very difficult to extract "microwave nonthermal effects" from the investigated phenomena since we cannot avoid the temperature rise for the substances irradiated by microwaves. "Microwave nonthermal effects" should be investigated and discussed on the basis of careful experimentation in reports. Now we propose that we let "microwave special effects" cover both "microwave thermal effects" and "microwave nonthermal effects." We would use a terminology "microwave chemistry" in this text to express a broad area containing not only organic synthesis, inorganic synthesis, but also other materials chemistry using "microwaves" therein in order to promote or control the chemical reactions included in those processes.

We can appreciate the excellent characteristics of microwave heating observed as "microwave thermal effects," which cannot be achieved by conventional heating using an electric heater, fuel burning, a gas turbine, and others. It should be noted that these "special heating modes" of microwaves have been used not only in a domestic microwave oven but also in traditional production processes, such as rubber vulcanization and drying plants, fruits, and vegetables.

We describe in this chapter how we can understand chemistry driven under microwave irradiation and what we can do using microwaves in solid surface reactions in solid catalysis, gas-solid reactions, liquid-solid reactions, and solid-solid reactions. Although heterogeneous systems containing solid particles are more complicated than homogeneous solution systems when we try to in situ observe the reaction fields under microwave irradiation and discuss "microwave special effects," there are some tools for investigating it. On the other hand, when we think about the in situ observation of homogeneous systems under microwave irradiation, we are lost in a way to find proper tools. The present tools described in the succeeding text of this chapter applied to heterogeneous systems are not appropriate for applying to homogeneous systems.

We start our review from the research articles dealing with "temperature measurements" in the next section.

2 Temperature measurements

We would introduce two important review articles on temperature measurements of chemical systems under microwave irradiation. Kappe's tutorial review with a title "How to measure reaction temperature in microwave-heated transformations" describes how difficult and tricky the accurate measurement of reaction temperature in single-mode microwave reactors is. He proposes some favorable ways to monitor the accuracy of relatively common available methods that measure the temperatures in microwave reaction, to discuss the existence of "microwave nonthermal effects."[29] He also describes the structures and working mechanisms of the reactors, the temperature sensors, and how to employ the sensors to correctly measure the temperatures. On the other hand an article by Stefanidis et al. "On the accuracy and reproducibility of fiber optic and infrared temperature measurements of solid materials in microwave applications" discussed the temperature measurements of solid materials.[30] They compared three techniques of temperature determination, that is, a radiation pyrometer, a fiber-optic sensor, and a thermocouple, and discussed the advantages and disadvantages in using these methods. They especially focused on the former two as the most commonly used and reliable techniques and concluded the importance of using a glass probe guide to lengthen the lifetime of a fiber-optic sensor and ensure the reproducibility of the measurements. The authors also concluded that one cannot forget that we are measuring the temperature of the surface of the glass container wall when we use an IR probe. Significant temperature nonuniformities were found by using fiber-optic sensors in both the axial and radial directions despite the use of a monomode cavity and the small amounts of samples tested. Their final conclusion is that temperature measurements at multiple coordinates are required for obtaining the temperature map in the reactor by multiple sensors.

Let us start discussing an interesting work by Dudley et al. dealing with an organic synthetic reaction carried out in a homogeneous solution under microwave irradiation.[31] They examined thermally promoted Friedel-Crafts benzylation of arene solvents under both conventional convective heating in an oil bath and heating using microwave energy. The authors describe an observation that is central to MW-specific thermal phenomena, namely, the difference between heat and temperature. Temperature is a measure of the ensemble average kinetic molecular energy of all solution components, but temperature does not provide information about solute-specific energy differences that may arise as a consequence of selective MW heating on the basis of their observation of the rate enhancements at a given temperature. They proposed the presence of the "transient solvent cage domain" model for selective MW heating of strongly MW-absorbing solutes in an otherwise MW-transparent system as illustrated in Fig. 2. Macquarrie et al. performed demethylation of (3-methoxyphenyl)methylammonium bromide producing (3-hydroxyphenyl)methylammonium bromide(3-methoxyphenyl)methylammonium bromide, which need to be removed from the reaction system during the reaction.[32] They proposed a model where microwave heating occurs inhomogeneously around gas/liquid interfaces of the bubbles of the product as illustrated in Fig. 3. They observed very pronounced zero-order kinetics, displaying a much higher reaction rate than observed for the conventionally heated thermal reaction, leading to an idea where they related this observation to a different gas product removal mechanism, changing from classic nucleation into gaseous bubbles to a facilitated removal through escaping gases/vapors. These investigations of organic reactions carried out in homogeneous solutions initiate us to observe the temperature gradient or distribution in nanoscale in the reaction systems, but unfortunately, appropriate techniques are not available at present especially for homogeneous solutions.

The papers described earlier dealt with organic chemical reactions in solutions. To measure and control the temperatures of these solution reaction systems, a fiber-optic sensor and/or an infrared sensor is employed. While Macquarrie et al. used two commercially available MW apparatuses equipped with a fiber-optic sensor, Dudley et al. employed two types of thermosensors in parallel. They describe that they took care to suspend and secure the tip of the fiber-optic probe close to the middle of the stirred solution but outside of the stirring vortex. Furthermore, they employed

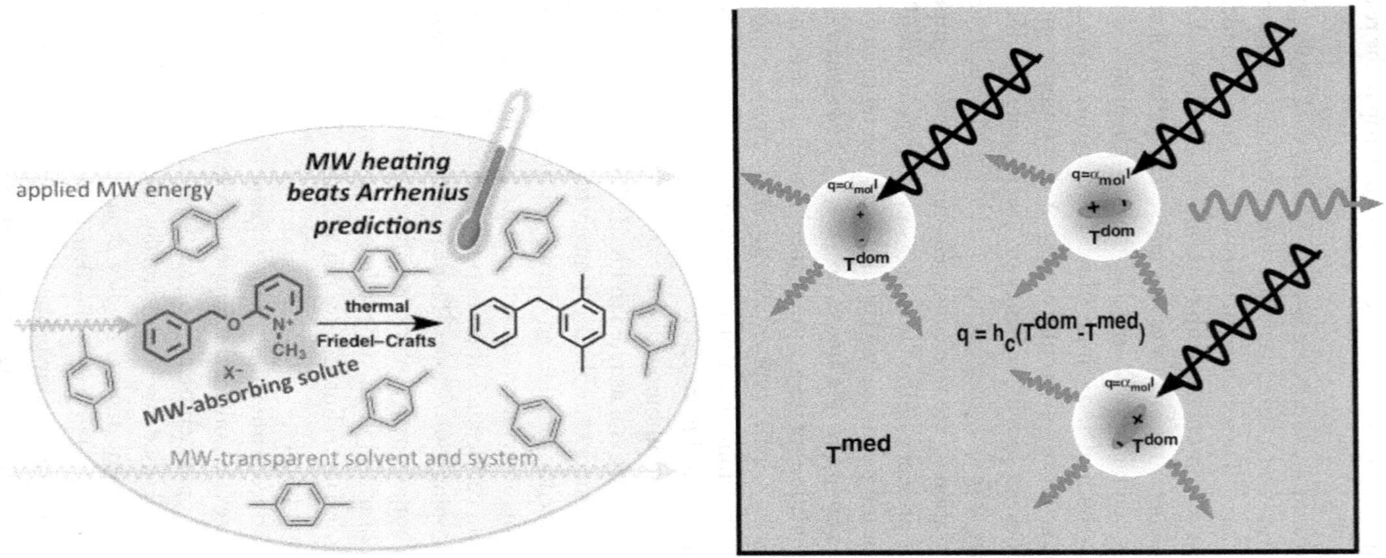

Fig. 2 Dudley postulated an idea that MW-absorbing solute generating heats forms the "transient solvent cage domain" around it in the MW-transparent solvent with high temperature, causing the "microwave nonthermal effects."[31] *Reprinted (adapted) with permission from Rosana, M. R.; Hunt, J.; Ferrari, A.; Southworth, T. A.; Tao, Y.; Stiegman, A. E.; Dudley, G. B.,* J. Org. Chem., ***2014****, 79, 7437–7450. Copyright 2014 American Chemical Society.*

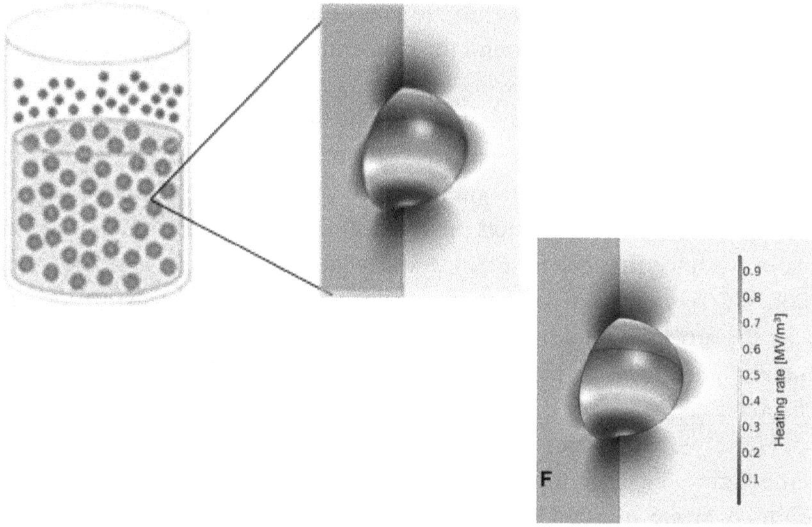

Fig. 3 Macquarrie calculated the concentrated distribution of heat generation rates at the surface and its vicinity of the bubbles, causing localized over heating.[32]

both external and internal sensors to collect their thermometric data, as there is value in validating observations using both types of temperature measurements. Both have a place in microwave chemistry, although one can always argue that the limitations of one and/or the other may make it difficult to interpret the observed results.

Conner et al. attempted to analyze the aqueous solution system containing zeolite precursor under microwave irradiation.[33] They constructed a special vessel to allow the measurement of the temperature at different heights and radial distances from the wall in a zeolite precursor solution under microwave heating. They drilled several holes into the Teflon-made reactor and inserted the fiber-optic sensors through a Teflon-coated glass tubes acting as "thermowells" to hold the probes. It was found that the measured temperature variations were significant, depending on the dielectric permittivity of the reaction medium and their changes with temperature. The authors described that NaY synthesis solution exhibited the smallest penetration depth; as a result the solution temperature near the wall increased by 338 K over the target temperature when the temperature at the center of the solution was targeted to be 333 K. The system investigated in this work, conventionally used for hydrothermal synthesis of zeolite, is an extreme case, where no stirring and high dielectric loss of the solution due to the ionic components contained in the precursor are employed. However, the clear and drastic temperature gradients observed even in a homogeneous solution should be kept in mind for performing accurate temperature measurements in experimental works.

An important review article published by Stankiewicz et al. in 2009 summarized the research papers published related to solid catalysis carried out under microwave irradiation and described the discoveries and new aspects reported therein.[34] The authors provided an extensive discussion for describing importance in measuring temperatures under microwave conditions and modeling approaches in microwave-assisted heterogeneous catalysis. We consider this article as a milestone in the research works of microwave chemistry and also put a focus on these two topics in our own analysis described in this chapter.

Here, we review several typical works that attempt to determine the temperature distribution in solid reaction systems under microwave irradiation. A group of Stankiewicz and Stefanidis published a paper reporting noncontact high-temperature measurements in microwave-assisted catalytic reactors.[35] They carried out methane dry reforming under microwave heating on Pt-supported active carbon in a tailor-made reactor for microwave irradiation and determined the temperature of the catalyst bed by both a thermal camera and a thermocouple. They concluded that the thermal camera-thermocouple dual temperature measurement method developed in their work allows for real-time high-temperature measurements in MW-heated catalytic reactors as demonstrated in Fig. 4. The authors insisted that it was necessary to have at least one contact sensor to ensure the correct temperature recording by the thermal camera to avoid an erroneous measurement due to the multiparameter emissivity dependence in such a complex reaction system. They succeeded in detecting hot spot generation and providing an explanation for the nonhomogeneous heating profiles during MW processing.

Stiegman et al. developed a catalyst of nanoscale magnetic spinel oxides, MCr_2O_4, which were used in an aqueous methanol solution for converting methanol into formalin by oxidation under microwave conditions.[36] This is an investigation of the aqueous solution system containing nanoparticles as a catalyst and methanol as a reactant. Both nanoparticles and methanol/water absorb microwaves strongly and convert their energy into heat. The authors investigated the temperature distribution of the dispersion solution under microwave irradiation and showed relatively uniform distribution of the temperature across the surface of the solution. Sajiki et al. also investigated a solution reaction system catalyzed by solid catalysts under microwave

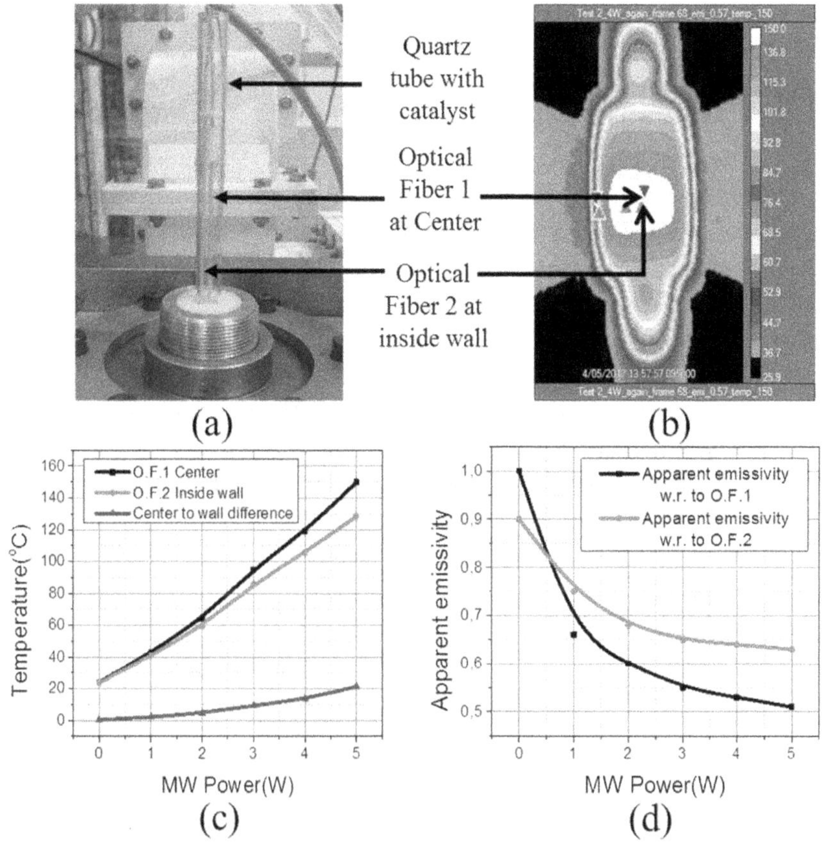

Fig. 4 Temperature distribution observed for Pt/carbon catalyst by means of 2-D camera in the combination with FO and thermocouples. Two-dimensional temperature distribution is visible as in the image. The local temperatures go up, and their difference gets wider at the high temperature. The apparent emissivity changes differently at the two locations depending on the temperatures.[35] *https://pubs.acs.org/doi/10.1021/acs.iecr.7b02091, further permissions related to the material excerpted should be directed to the ACS.*

irradiation. A reaction investigated therein is the dehydrogenative aromatization of saturated cyclic hydrocarbons (e.g., methyl cyclohexane) on Pt supported on carbon beads.[37] Carbon is a strong microwave-absorbing material and heated up vigorously under microwave conditions. Then a steep temperature gradient at the outer surface of the catalyst bed was observed under microwave irradiation due to effective cooling by the flowing reactant. They studied the change of this remarkable temperature distribution by a thermal camera as a function of the irradiation time.

Kappe's tutorial review article claimed a warning on how difficult and tricky the accurate temperature measurements are under microwave conditions as described at the beginning of this section.[29] Stefanidis discussed the difficulty of accurate temperature measurements of even more complicated systems containing solids.[30] Their idea to measure accurate temperatures is to employ multiple techniques to detect the temperatures of the system, that is, a combination of fiber-optic sensor and thermal camera. Furthermore, Santamaria introduced fiber-optic probes at the multipositions for temperature measurements during their investigations of a hydrothermal synthetic system of NaY zeolite.[38] After the review article of Stankiewicz et al. in 2009,[34] researchers have recognized the importance of accurate temperature measurements of chemical systems under microwave irradiation. Multiposition temperature measurements using fiber-optic probes and 2-D temperature measurements giving the temperature distribution on surfaces can complementarily provide a more accurate information on the actual temperature distribution of a system.

Now we will analyze the temperature distribution obtained by these methods and understand how the distribution is formed. Theoretical simulations for the series analysis of electromagnetic fields and heat flow would be helpful to understand the temperature distribution and change during microwave irradiation to understand the distribution and the time-dependent change. This would be one of the best methods we can employ in our research works as described in the next section.

3 New trends: Temperature measurements combined with theoretical simulation

As pointed out at the end of the last section, temperature measurements at multiple positions by using a combination of fiber-optic sensors and 2-D IR cameras can provide the most accurate 2-D distribution of surfaces of solids under microwave irradiation currently. If the earlier data can be understood by assistance with theoretical simulations, we will be able to understand the mechanism ensuring the temperature distribution and even be able to control it.

Mallada and Santamaria et al. reported a theoretical analysis of microwave heating cavity containing NaY zeolite in a fixed-bed reactor. The investigation was carried out solving the energy/heat momentum equation together with Mawell's electromagnetic field equations using COMSOL MULTIPHYSICS simulation software.[38] They obtained a high-level prediction of the temperature rise of the NaY fixed bed that was experimentally observed, through the theoretical simulation of electric field distribution, axial and radial temperature profiles, and temperature evolution with the irradiation time as seen in Fig. 5. In another example, Chen et al. investigated the microwave drying kinetics and pumping phenomenon of lignite spheres by numerical simulation based on the Maxwell equation and compared with the temperature distribution observed by an IR camera.[39] This work shows how useful the combined technique of investigation by a thermocamera and theoretical analysis is for the analysis even for soft materials. They successfully analyzed the drying and pumping phenomenon of this soft biomaterial. The readers would be impressed by the remarkable combination of the temperature measurements with the theoretical simulation in the above two works, which enabled the high-level discussion on the "microwave thermal effects" occurring in these systems. Now we focus on this new method by reviewing our two works published recently as a typical work for solid catalysis but giving the profound insight into "microwave thermal effects" in heterogeneous catalysis.

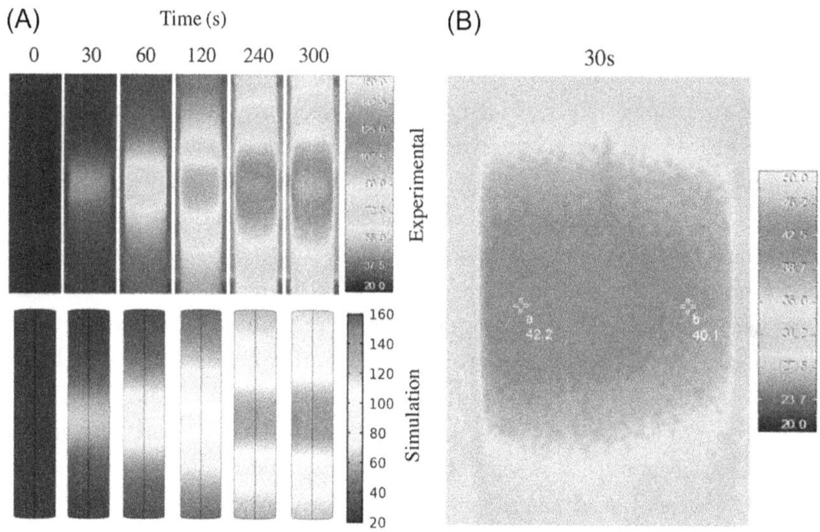

Fig. 5 Images obtained by numerical analysis of microwave heating cavity for fixed bed NaY zeolite.[38]

Haneishi et al. published a work on high-temperature solid catalysis under microwave irradiation, in which the theoretical simulation showed a remarkable ability to analyze the temperature distribution on the catalyst surface, leading to a hypothesis of the occurrence of a new type of "microwave thermal effects".[40] The authors investigated the dehydrogenation of ethylbenzene to styrene on magnetite working as a catalyst under microwave irradiation. The temperatures of the catalyst bed were monitored by using ceramic optic sensors at the several fixed points inside the catalyst bed and by an IR radiation thermometer at the side surface of the catalyst bed. The yield of styrene was higher than that obtained at 873 K under conventional heating when the temperature of the side surface was controlled at that temperature but that was lower when the temperature of the center was controlled at that temperature at the center of the catalyst bed due to the temperature gradient in the catalyst bed.

A theoretical simulation using COMSOL Physics was performed to analyze numerically the temperature distribution in the catalyst bed and the heat flow therein. The electromagnetic waves, frequency-domain module, and heat-transfer module were used to calculate the electromagnetic field distributions, electromagnetic power-loss density, and temperature in the cavity modeled for the system. They encountered several problems in performing the simulation. The complex permittivity and complex permeability of the magnetite catalysts were obtained using perturbative methods only at room temperature. These parameters at the reaction temperatures were regarded as the same as those at room temperature, which might be justified by a fact that the resonance state was maintained without impedance matching during the heating process to the reaction temperature. They used the electric conductivity set to 0, since the permittivity should include a conductivity contribution. The catalyst bed was modeled as a uniform and continuous cylindrical material. The value of the heat conductivity could not be accurately obtained due to the presence of the voids in the catalyst bed. Then they used the value of the heat conductivity determined by fitting the actual measured value of the temperature at two points in the catalyst bed (side surface and core). Fig. 6 shows the temperature distribution in the catalyst bed estimated by the simulations. The temperature at the core was highest and decreased gradually toward the side surface and the upper and bottom surface. The reader can see the good coincidence of the temperature distribution obtained by the simulation with the experimentally measured

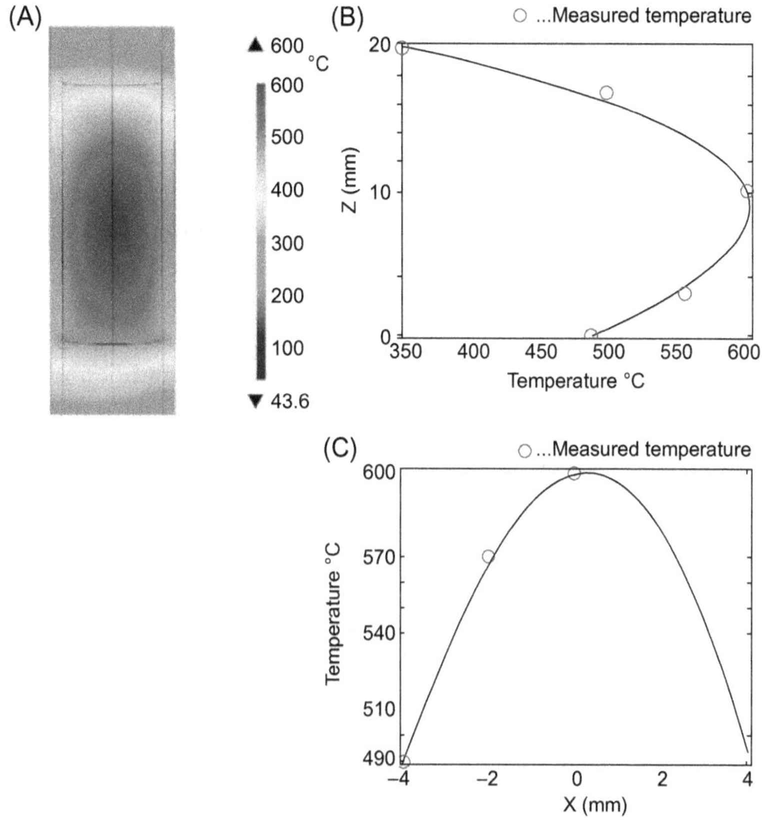

Fig. 6 Temperature distribution in the magnetite catalyst bed obtained by the numerical simulation. The local temperatures measured by the fiber-optic sensors are indicated by the red circles, showing the good coincidence with the simulated ones.[40]

temperatures, justifying the high reproducibility of the temperature distribution by the simulation method employed in this work.[40]

Haneishi estimated the rate constant of dehydrogenation of ethylbenzene for the catalyst bed under microwave irradiation by the summation of the rate constants for all the meshes with different temperatures by using the Arrhenius parameters obtained in the experiments carried out under conventional heating.[40] The resulting rate constant (4.20×10^{-2} L kg^{-1} s^{-1}) obtained earlier for microwave irradiation was higher than the measured reaction rate (3.12×10^{-2} L kg^{-1} s^{-1}), suggesting the enhancement in the reaction rate under microwave irradiation. This work still contains a lot of problems, in which unavailable physical parameters were used on the postulated basis, that is, the complex permittivity and permeability at room temperatures were used for the calculation of the system at 873 K, but this approach would lead to a proper way to understand the microwave chemistry of heterogeneous catalysis.

These results encouraged Haneishi to proceed to the next step for examining the presence of nonequilibrium local heating in the catalyst bed. Now let us describe the work that revealed the presence of high-temperature area at the contact point between the catalyst particles, which cannot be identified in the earlier work employing the continuous model of the catalyst bed.[41]

In a follow-up work, Haneishi et al. decided to obtain more insight into the detailed analysis of the temperature distribution at the μ size scale in the catalyst bed.[41] The authors carried out the theoretical simulation using the catalyst bed model reflecting the particles packing.[41] Fig. 7 shows the model employed in the simulation work consisting of 160 spheres with 1.6 mm in their diameters. The spheres were arranged in a simple cubic lattice with 19 mm in height and 8 mm in diameter. To investigate the effect of the vibration direction of the electric field on its distribution, the spheres were arranged so that their contact direction was completely perpendicular or parallel to the vibration direction of electromagnetic field. During the actual experiments the antenna and the model catalyst bed in the quartz tube were placed at the focal points of the elliptical applicator.

Fig. 8 shows the electric field distribution in the applicator and inside the catalyst bed. One can see the bright areas at the surfaces of the spheres and the vicinity at the contact points between the catalyst spheres, indicating the accumulation of the electric field at those areas and points. Especially high loss density of electromagnetic power is observed at the contact points, indicating the local heat generation there. For example, the electric field was less than 500 V m^{-1} at the surface of the spheres, except for the contact points. The concentrated distribution of electric field and its energy loss can be seen in the enlarged images in Fig. 9. The electric field at the contact point between C15 and C16 was 45,500 V m^{-1} when 5-W microwaves were input. This concentration of the electric field leads to the extensive heat generation (1010 W m^{-3}) at the contact point compared with the average value at the surface (106 W m^{-3}). Fig. 10 shows the temperature distribution along the X-axis and Z-axis in the cross section of the catalyst bed. The readers can see that the temperature spikes localized at the contact points of the spheres both for the X-axis and Z-axis. The temperatures at the contact points are 80–100 K higher than other areas. Now looking at the temperature distribution at the contact points of the spheres in the smaller scale, one can observe the increase in temperature by 90 K at the contact point compared with the average surface temperature.

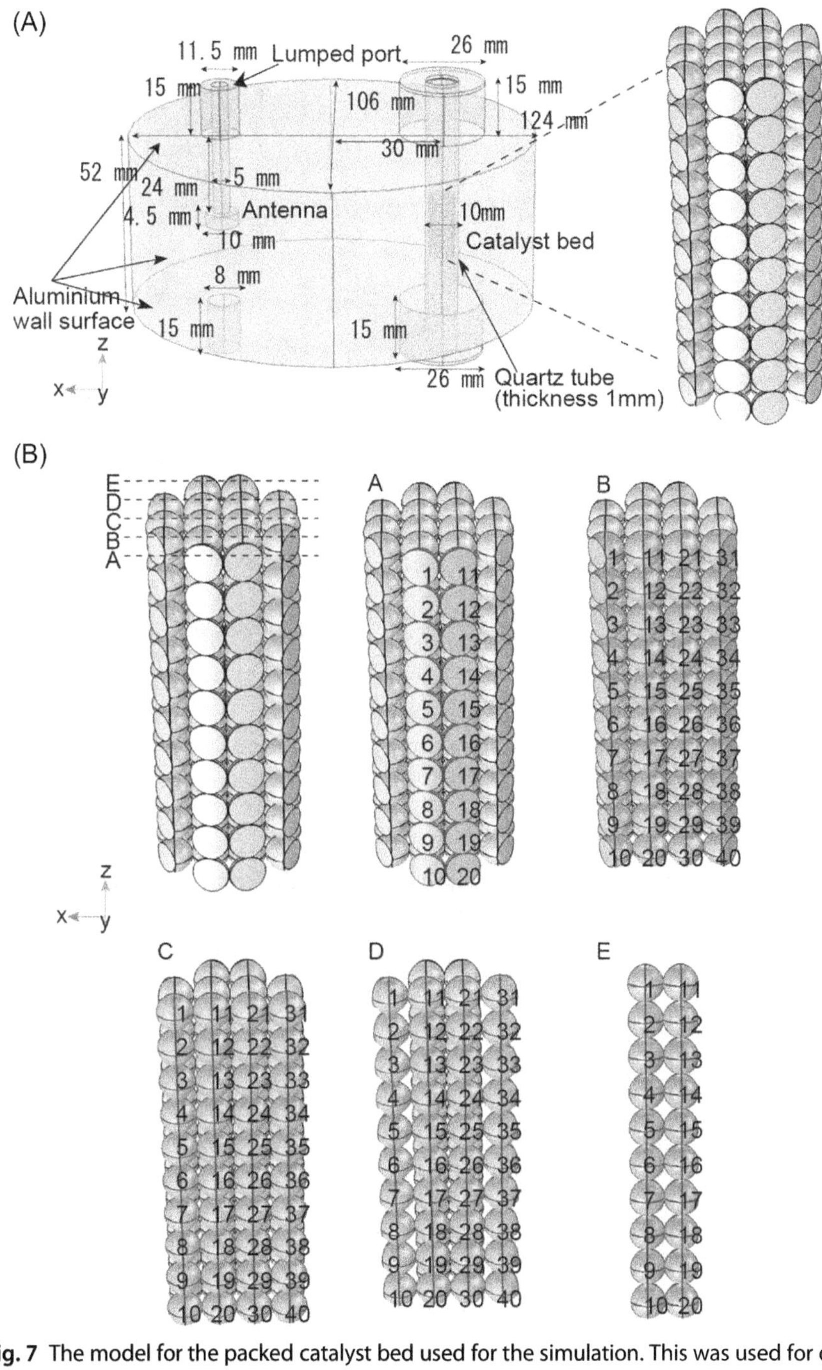

Fig. 7 The model for the packed catalyst bed used for the simulation. This was used for discussing the accumulation of electric field at the contact point of the catalyst particles and the local heating occurring to the points.[41]

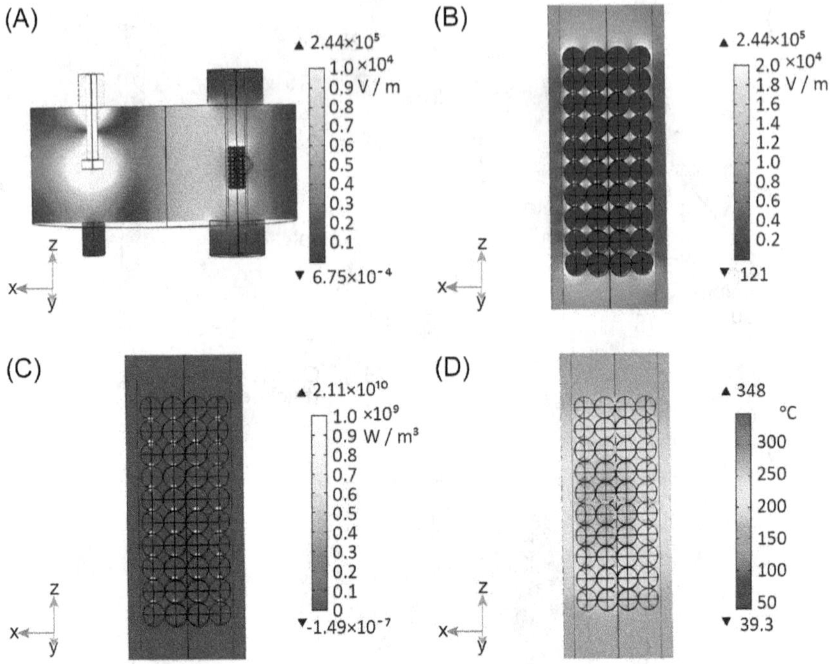

Fig. 8 Distributions of electromagnetic field in the cavity (A), in the catalyst bed (B), and energy loss in the catalyst bed (C) and temperature in the catalyst bed. The bright spots indicate the contact points of the catalysts.[41]

The occurrence of the high temperature at the contact points of the particles was confirmed by a model experiment in which SiC spheres were employed instead of magnetite, since both have similar dielectric constants. Haneishi successfully observed the high-temperature area at the contact point of two SiC spheres under microwave irradiation by measuring the emission spectra from the surface of the SiC spheres.[41] The area with the temperature 100–200 K higher than the average temperature on the surface was observed at the contact points under microwave irradiation.[41]

Haneishi attributed the enhancement in the reaction rate of dehydrogenation of 2-propol on magnetite observed in this work[41] under microwave irradiation to the occurrence of the local heating at the contact points induced by the concentration of the electric field, which is observed only under microwave irradiation and can never occur under conventional heating. This is one good example of "microwave thermal effects."

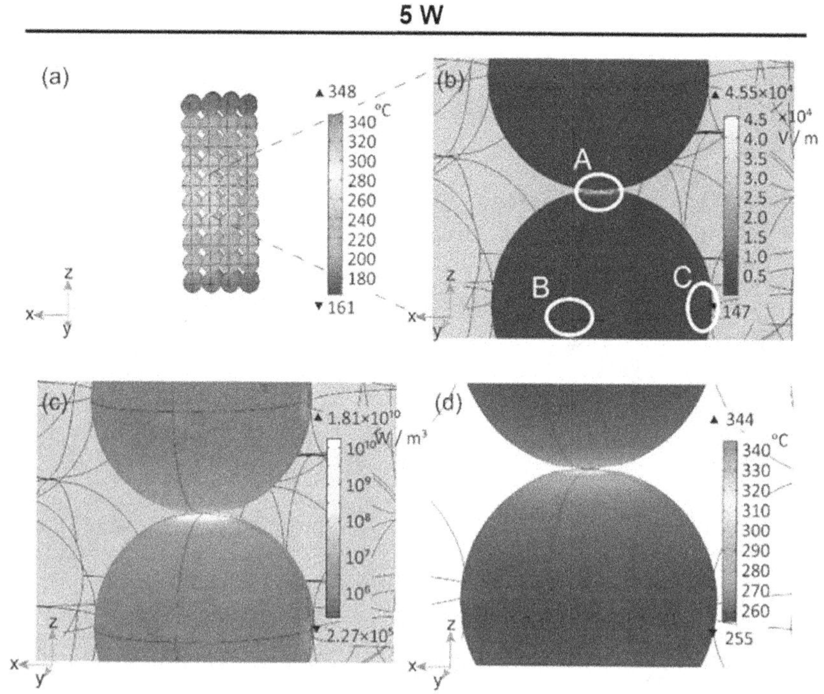

Fig. 9 Enlarged images of distributions of electromagnetic field in the cavity (A), in the catalyst bed (B), and energy loss in the catalyst bed (C) and temperature in the catalyst bed. The bright spots indicate the contact points of the catalysts.[41]

4 New trends: Operando observation

One can gain in situ information of a system under microwave irradiation in addition to temperature distribution, if we can observe the system in situ under working conditions. As commonly called, "operando in situ observation" has been attempted for chemical systems under microwave irradiation for that purpose. We start from a work on in situ observation of "nonequilibrium local heating" reported by Tsukahara et al.[42] The authors succeeded in in situ observing the higher temperature of the DMSO molecules that closely surrounded Co metal particles under microwave irradiation. A key tool to succeed in measuring the local temperature of only the DMSO molecules close to Co particles is a Raman spectroscopy probe specially designed for in situ collecting signals in solution under microwave irradiation. Raman spectroscopy can be a tool not only for determining the

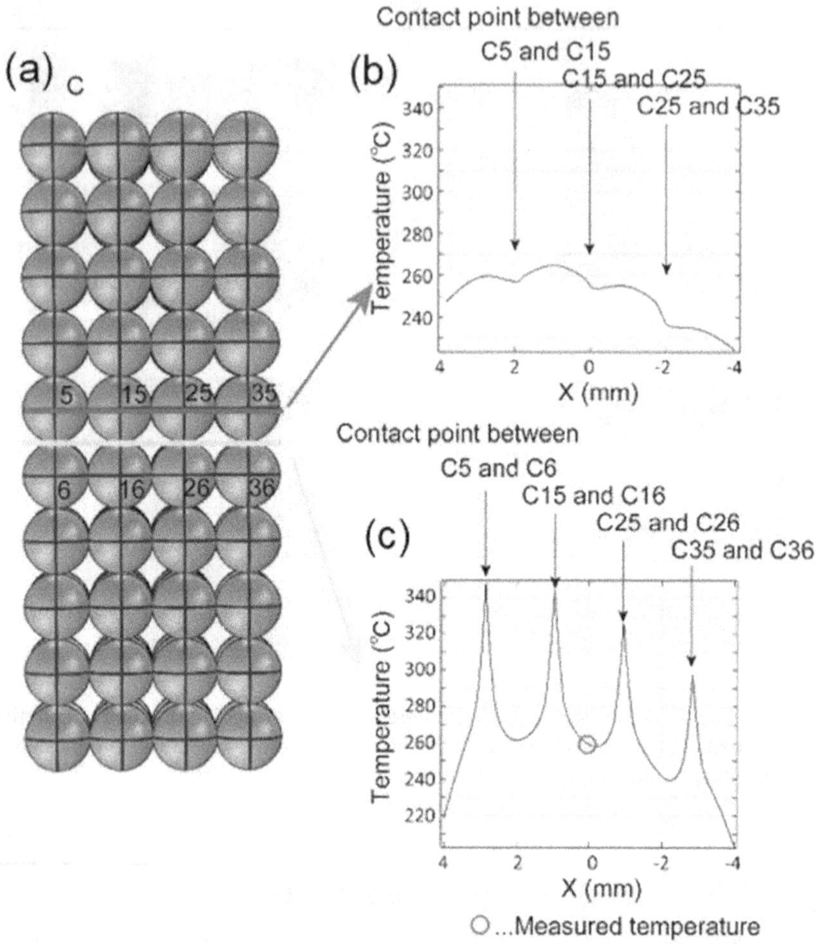

Fig. 10 Temperature distribution along the *X*-axis and *Z*-axis in the cross section of the catalyst bed. The temperature spikes can be observed at the contact points of the catalyst particles only in the direction perpendicular to the vibrating electric field.[41]

temperature of the targeted molecules estimated from the ratio of the Stokes line and anti-Stokes line depending on the form of the Boltzmann equation but also for obtaining the chemical bond structures. The Raman spectroscopy probe was made of Teflon to be inserted into organic solvents under microwaves, since Teflon scarcely has interaction with microwaves. This Raman probe was inserted into a DMSO solution containing dispersed Co particles under stirring and collected the Raman spectra continuously under microwave treatment. The average temperature of the dispersion solution was monitored with a fiber-optic thermometer in parallel with the Raman measurements. It is important to recognize that the temperature obtained in this method is assigned to the identified chemical species because of the identified spectrum used for the temperature determination. Spikes, observed only for the temperature of the dispersion solution under microwave irradiation estimated from Raman spectroscopy during the microwave irradiation, demonstrate the occurrence of the high temperatures in the small area observed by the Raman probe. Considering that the detection size of the Raman probe was 10–20 μm, this high-temperature area should be around this size or smaller, attributed to DMSO molecules heated to higher temperature than the average temperature of the solution.

This epoch-making work shows how important a practical and smart technique is to in situ revealing the local temperature in a system under microwave irradiation. Now we describe how the researchers solved this important but challenging task. Our list of the related works contains Raman spectroscopy, infrared absorption spectroscopy, x-ray and neutron spectroscopy, analysis using microwave absorption, and dynamic laser scattering measurements. Let us overview what were done in "operando in situ observation" in the last 20 years in microwave chemistry research.

Raman spectroscopy has been employed for *operando* in situ observation of chemical synthetic systems by several groups because this technique uses UV or Vis light transmitting into the solution, giving the information of chemical bonds.[43] Tompsett et al. employed Raman spectroscopy in the combination with X-ray scattering for in situ observation of the synthesis of a zeolite in solution, enhanced under microwave irradiation. They succeeded in recording the change of the Raman spectra of the aqueous solution during the preparation of a silicalite zeolite under microwave irradiation. They observed many distinctive bands attributed to amorphous silicate species, and the tetrapropyl ammonium ion template used for the synthesis was followed to observe the changes of these bands in relation with the progress of the synthesis. Synthesis of zeolites has been one of interesting chemical syntheses especially enhanced under microwave irradiation; therefore this group's observations would be very attractive for researchers working in the corresponding field.

On the other hand, Leadbeater et al. employed Raman spectroscopy to examine the occurrence of the local heating at the molecular level. For example, they observed no selective heating of benzaldehyde molecules in various solvents and obtained similar observations with chromium arene complexes (which act as a metal catalyst and absorb microwaves strongly) in solvents. They concluded that the local temperature at the molecular level is no higher than the bulk temperature of the reaction mixture and the polar molecules employed in their work are not at a temperature greater than that of the bulk, that is, there is not any localized superheating. Leadbeater also showed a way to use Raman spectroscopy to monitor the reaction progress for esterification of acetic acid and 1-butanol and succeeded in scaling up the reaction system under Raman spectroscopy monitoring and even further scaled it up using an automated stop-flow apparatus.[44] Leadbeater et al. insisted that use of Raman spectroscopy is excellent as a tool for in situ obtaining precise kinetic data for organic transformations in this work.[45] As Raman spectroscopy does not interfere with microwaves, therefore it must be also excellent for in situ observing chemical reaction systems under microwave irradiation.

In situ surface-enhanced Raman scattering (SERS) was employed for the in situ observation of the Ag particle formation in a precursor solution under microwaves by Tsukahara et al.[46] The authors used SERS to observe Rhodamine 6G in its interaction with the Ag nanoparticles to monitor the nanoparticle formation process and succeeded in monitoring the progress of the particle formation reaction. This is a very unique information in in situ observation of chemical synthesis under microwave irradiation.

Another traditional technique of infrared spectroscopy was applied to study heterogeneous catalysts under microwave irradiation by McDougall et al.[47] They employed a transmission infrared microreactor cell that held a pressed disc in a controlled atmosphere and allowed the data collection under microwave irradiation and conventional heating, respectively. The authors carried out CO oxidation on the standard catalyst EUROPT-1 (Pt/SiO_2), in which the behavior of the adsorbed CO on the catalyst surface was monitored in situ under microwave irradiation. Silverwood et al. found similar activity under conventional and microwave heating.[48]

There are several reports describing the in situ observation of material synthesis systems by means of neutron diffraction and X-ray diffraction. Whittaker et al. reported a new apparatus for the study of the microwave-driven growth of particles in solution by in situ small angle neutron scattering (SANS) and tried to observe the iron oxide particles in the sample of iron chloride in heavy water.[49] Harrison et al. developed an apparatus that enables X-ray or neutron scattering measurements to be performed on powders or single crystals under microwave irradiation with controlled powder level.[50] The authors in situ observed a single crystal of the molecular solid aspirin and a powder of the microwave-susceptible ionic material

$BaTiO_3$, during microwave irradiation. They determined the structure of aspirin at various temperatures under microwave irradiation in comparison with that observed under conventional heating. The authors tried their apparatus to analyze $BaTiO_3$ at various temperatures under microwave irradiation in comparison with conventional heating and found clear difference in the unit cell parameters and the major phase fraction. Synchrotron radiation was very effectively utilized for in situ investigation of titanium powder sintering[51] and phase transitions of Al-Cu-Fe alloys during microwave heating.[52] Xu et al. showed a clear image of the time-dependent formation of the necking in the power by computed tomography in their elegant work. Reeja-Jayan discussed the instrumentation available to study field-matter interaction mechanisms and identify future needs in synchrotron characterization to better understand how electromagnetic fields can contribute to the engineering of advanced materials in their review articles related to synchrotron X-ray characterization of materials prepared under microwave conditions.[53]

Let us introduce some unique works in which fluid or solution behavior were in situ observed under microwave irradiation from the view point of chemical engineering. Asakuma's group investigated the surface tension, bubble size, and convection for a mixture of water and ethylene glycol under microwave irradiation.[54] The authors were dependent on the content of ethylene glycol and concluded that microwave influence in aqueous solution can be manipulated by varying the fluid properties via addition of a cosolvent. Asakuma's group applied dynamic laser scattering technique to monitor the nanoparticle formation in solutions under microwave irradiation.[55] The major finding in this work is that adding ethylene glycol (called as antisolvent in the articles) is able to suppress superheating and swift bubble growth and consequently stable nanoparticle formation could be achieved. In a further study dealing with "two-stage irradiation" and "antisolvent addition," Asakuma et al. proposed the beneficial combined effect of these two ways on the observation and confirmation of finer nanoparticle formation.[56]

5 Structural design of the solid catalyst to control the temperature distribution

We reviewed a number of papers on measurements of temperature distribution and *operando* in situ observation of catalysts under microwave irradiation. These *operando* in situ observations have been challenged not only for understanding the chemistry but also for controlling the chemical reactions occurring in the targeted systems. Now we are ready to discuss how we can design the structure of a catalyst to control the temperature distribution based on these accomplishments so that we will be able to control the performance of the catalysts.

Horikoshi's group published a series of articles on "control of microwave-generated hot spots".[57, 58] The authors monitored the reaction solution containing the starting materials of Suzuki-Miyaura coupling together with Pd supported on active carbon in toluene and identified the occurrence of hot spot as the visible blinking areas at the surface of the catalyst particles located through the aggregates of the catalyst particles by a high-speed camera. The theoretical simulation also showed that high electric field was accumulated within the spatial gap between two catalyst particles, resulting in the hot spot formation. Furthermore, they suggested a role of the reagents to firmly connect the catalyst particles to get closer. Another article of the same group proposed the factors that determine the hot spot generation, that is, (i) mass transfer of reactants to the Pd/C catalyst, (ii) continuous versus pulsed microwave irradiation, (iii) presence of a standing wave versus a nonstanding wave, and (iv) microwave input power level.[58] These factors affect the generation of hot spots between the catalyst particles; in other words, they can be used to control the generation of the hot spots. The authors detected the blinking arrange-colored light or white light, which they attributed to the black body radiation. This abnormal light blinking might be attributed to not the black body radiation but discharge. It might be interesting to identify the phenomenon by observing the spectra of the emitted light from the system.

An interesting article by Wang et al.[59] was found, which deals theoretically and experimentally with the generation of hot spots for a system of SiC particles dispersed in paraffin oil. This system might be closely related to Horikoshi's work.[57, 58] According to their article, hot spots are definitely shown to exist and may trigger temperature gaps between surrounding substances at the magnitude of several hundred degrees or even higher. The authors describe that the extent of hot spot effects substantially depends on how much and how quickly the heat generated by the strong microwave absorbing media can get transferred to the weak ones. They refer to the discharges occurring under microwave irradiation but do not differentiate it from the occurrence of the high-temperature areas. This article is recommended to be read together with Horikoshi's articles.[57, 58]

Solid-gas biphasic systems were also under investigation by scientists and engineers having interests in microwave techniques. Santanaria et al. measured the temperatures of the solid and gas in the catalyst system containing a silver-copper oxide catalyst for the ethylene epoxidation reaction as exhibited in Fig. 11.[60] Infrared thermographic techniques were used to determine real-time gas and solid temperatures, as well as gas-solid temperature gradients in the microwave-heated structured reactors. The authors investigated the temperature distribution in the monolith-shaped catalyst and succeeded in determining the temperature difference, which was obtained as higher by 50 K between the catalyst monolith wall and the reaction gas mixture for an average catalyst temperature of 200 K. This was the first in situ observation of the temperature gradient between the solid catalyst surface and the reacting gas under microwave irradiation. The systematic work of this type certainly leads to a new way to control chemical reactions of gases proceeding on the surface of solid catalysts under microwave irradiation. The next article deals with efforts made on avoiding undesired side reactions in reactions of gases in heterogeneous catalytic reactors.[61] This can be executed by the in situ observation of the temperature gradient between the catalyst surface and the gaseous reactants on the basis of the previous work by Santamaria.[60] The authors claimed that they took extra precautions to ensure that (i) the solid phase is free from significant hot spots and (ii) an accurate estimation of both solid and gas temperatures was obtained. Direct microwave heating allows operating with a catalyst that is significantly hotter than the surrounding gas, achieving a high conversion on the catalyst while reducing undesired homogeneous gas phase reactions. The aforementioned concept was demonstrated using the CO_2-mediated oxidative dehydrogenation of isobutane, resulting in the selective dehydrogenation predominantly proceeding on the catalyst surface to isobutylene together with the diminished formation of the decomposition products such as propylene, ethylene, and methane formed in the gas phase.

Another way to control the chemical reaction on solid catalysts under microwave irradiation is described by Chaouki et al.[62] Microwave irradiation was utilized for accomplishing a significantly higher temperature for the solid (as microwave susceptors) as compared with the gas-phase temperature according to the principles of microwave irradiation approach.[62] The authors mention in their article that to highlight the influence of microwave heating on the overall performance of a gas-solid fluidized bed reactor, a simulation study was attempted for a model reaction system. They selected catalytic oxidation of n-butane over the fluidized vanadium phosphorous

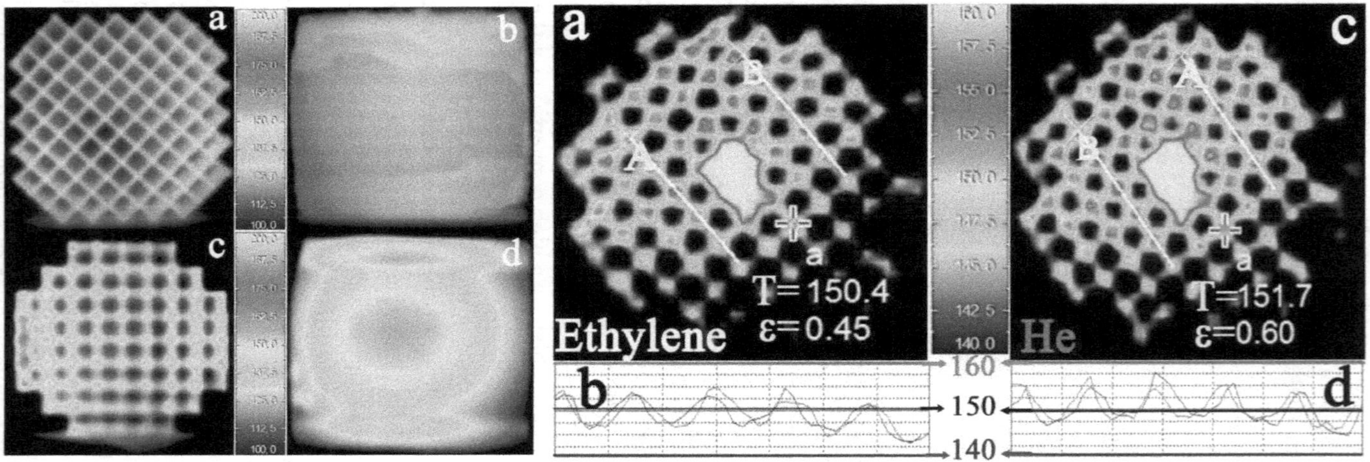

Fig. 11 Temperature distribution in the monolith structure catalyst covered with silver-copper oxide catalyst for ethylene epoxidation. Side reaction occurring in the gas phase is hindered by the temperature gradient between the catalyst and the gas phase.[60]

oxide (VPO) catalyst to produce maleic anhydride and employed the bed hydrodynamics described by a dynamic two-phase flow model. The simulation study indicated that when competitive gas- and solid-phase reactions are occurring in a gas-solid fluidized bed reactor, the application of microwave-selective heating approach can significantly enhance the overall performance of the reactor in comparison with the conventional heating, where solids, bulk, and bed temperatures are identical. This excellent work could open new approaches in chemical process design using fluidized catalyst bed reactor and broaden the range of applications of microwave techniques.

An interesting new idea from the structural view point in utilizing microwave techniques for chemical processes was reported in continuous cleaning of VOC-containing air streams using solid catalyst. Santamaria described the construction of a double fixed-bed reactor configuration consisting of adsorptive dealuminated Y-type (DAY) zeolite + catalytic PtY zeolite as shown in Fig. 12.[63] The first bed (DAY) acting as an adsorbing reagent and the second bed (PtY) acting as a catalyst for oxidation of VOC were heated up selectively under microwave pulse irradiation, and both beds played their respective role efficiently. This configuration

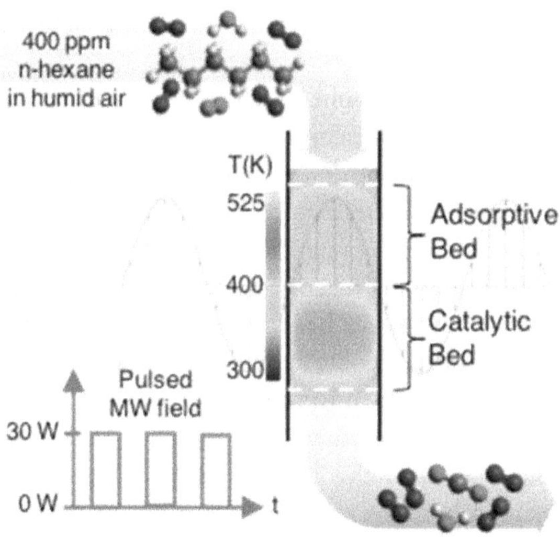

Fig. 12 A smart double-bed structure of zeolite catalyst bed designed for microwave chemistry. The first bed is for adsorbing reagent, and the second bed (PtY) acts a catalyst for oxidation of VOC. Both play the independent roles separately under microwave irradiation.[63]

provided the same performance with only half of the catalyst load compared with a single bed of PtY.

Now we exhibit an example in our research works in designing a catalyst structure in a nanosize scale for utilizing microwaves to control catalytic reactions on solid catalysis that fully exploits the effects of microwave non-equilibrium local heating.[64] A zeolite can be formed as one particle with a core part and a shell part by multistage synthesis, in which different function can be designed for each part. Mochizuki et al. designed and synthesized a zeolite possessing the two separate parts in a particle, that is, a carbon-filled core for providing heat through the microwave energy loss by the strong absorption of microwaves and a shell acting as a reaction field, which is heated by the core carbon selectively.[64] The framework of Y zeolite used was relatively transparent to microwaves compared with carbon, as carbon is well known as one of the substances that is strongly and selectively heated by microwaves. In the actual experimental procedures, first, core zeolite crystals were prepared, washed, dried, and separated as NaY zeolite. After filling the core zeolite with carbon by carbonization of furfuryl alcohol preadsorbed in the core zeolite at high temperature under nitrogen atmosphere, crystal overgrowth of shell zeolite was carried out by treating the carbon-filled core under hydrothermal conditions, resulting in the core-shell zeolite. The shell zeolite was made to function as a reaction field for acid catalysis by ion exchange of Na^+ in the shell zeolite with protons to form the acidic HY form.

Dehydration reaction of 1-phenylethanol under microwave irradiation in *n*-heptane as a solvent was examined as a test reaction under microwave irradiation in comparison with an identical reaction under conventional heating. This reaction generates two products, that is, olefin by intramolecular dehydration and ether by intermolecular dehydration. The core-shell zeolite gives higher amounts of both reaction products under the boiling point of heptane in microwave heating than in conventional heating. On the other hand the normal zeolite shows the similar yields both by microwave heating and conventional heating.

Mochizuki et al. estimated the actual temperature in the shell acting as the reaction field by using Arrhenius plots of the reactions using the kinetic parameters obtained by the experiments carried out under conventional heating.[64] The enhancements of the reactions under microwave heating were estimated to correspond to the temperature rise of +6 K in the case of the olefin and +11 K in the case of the ether compared with conventional heating. It was concluded that the shell part acting as the reaction field was warmed up 6–11 K higher by the supplied heat from the core part under microwave heating than the same system under conventional heating.

The second example is related to an inconvenience of substance-selective heating under microwave irradiation. Microwaves can be absorbed by substances with high dielectric loss and therefore do not heat the materials transparent to microwaves. This peculiar feature of microwave heating critically limits its area of application. Maitani et al. investigated this issue to find a way in which they can heat up substances with low dielectric and magnetic loss factors under microwave irradiation.[65] The authors found that substances with low loss factors under MW irradiation occasionally enhance their ability of MW heating in case those substances are contacted in the layered configuration.[65] They found a heating phenomenon of TiO_2 layer with low dielectric loss (low microwave-absorbing substance) by making a contact with fluorine-doped SnO_2 conductive glass plate. As an application of this interfacial MW-heating phenomenon, they firstly applied this MW heating to the annealing process of TiO_2 nanoporous films on transparent conductive glass (F/SnO_2) for a photovoltaic device, dye-sensitized solar cells (DSSCs).[65, 66] The results consistently exhibited better device characteristics in addition to one order of magnitude shorter annealing process time. The same technique was demonstrated to be useful for fabricating a scaffold layer of a perovskite solar cell.[67] It is important to emphasize that TiO_2 powder has a low dielectric loss factor and is not heated under microwaves well. However, a layer consisting of TiO_2 particles is put into contact with a fluorine-doped SnO_2 layer; the layer is efficiently heated up to 723 K necessary for calcining the TiO_2 powder to convert it to the sintered structure. Furthermore, Maitani insists that the atomic level connection between TiO_2 and SnO_2 layers gets more intimate at lower temperature under microwave irradiation than under conventional heating. The authors proposed the mechanism of the heating phenomenon based on the Maxwell-Wagner-type heating attributed to the perturbation of interfacial electric dipole by incident oscillating electric field. This result provides significant potential to design and engineer the heated objectives for implementation of nonequilibrium local heating at the interface of materials for novel applications in nanomaterial processing, such as solid-state catalysts, nanocomposites, and layered film device fabrications.

The application of the interfacial MW heating to nanosized constructions opened a new direction of research and provided a novel strategy for designing a protocol for solid-solid chemical processes.

6 Precise control of electromagnetic field achieved by solid-state generator of microwaves

Dielectric constants of substances change the wavelength of microwaves, and loss factors determine the heat generation through the microwave energy loss that is dependent on temperatures. Both are essentially important in controlling the temperature of solid catalysts used under microwave irradiation. Complex dielectric constants increase with the rise of temperature, causing a thermal runaway of metal oxides under microwave irradiation.[68] To control the temperature rise of solid catalysts under microwave irradiation, the accurate measurements of temperature-dependent complex dielectric constants are important, but it is not easy to measure these constants at high temperatures. Various methods were attempted for the measurements using a cylindrical cavity perturbation method or a coaxial transmission line method under conventional heating of samples.[69–72]

According to the rapid development of telecommunication technology, availability of a voltage controlled high-frequency oscillator (VCO) with microwave band (>1 GHz) and a high power amplifier (HPA) with the output power above 100 W at ultrahigh-frequency (UHF) band (0.3–3 GHz) enabled semiconductor amplifiers with low phase noise, spurious free, and temperature stability to be used readily for chemical processing and materials science and engineering, substituting magnetrons in the future to share their roles in the practical applications. Recent development of microwave sources with excellent frequency and phase stability in addition to VCO and HPA has realized a new stage in which pulse modulation and frequency autotracking are practically possible and have been applied in microwave heating technology.[73–75] We can control the microwave-irradiating conditions according to the desired temperature distribution in a catalyst by frequency autotracking and vary the energy input according to the reaction rate in situ monitored during the reaction. We can select and control the heating area by using multimicrowave generators with controlled phase-modulation modes. These new hardware and software units are leading the researchers to a new stage of microwave chemistry using a single-mode cavity as a main instrumentation. Civera et al. employed a combination of a dual mode cylindrical cavity with two modes and a high power amplifier for measuring dielectric properties using a cylindrical cavity perturbation method with TM010 mode under microwave heating with TE111mode and established a method for in situ measurements under microwave irradiation.[76, 77]

These techniques were employed for monitoring the volume change of Pt/C catalyst by dehydration at 573–623 K[35] and the changes of the dielectric properties of the zeolite catalyst beds piled into the double structure with the different coking situations under microwave irradiation.[78] These works demonstrate how useful the new techniques of detecting slight and delicate changes in the catalysts and their properties are even at the high working temperatures under microwave irradiation. The distribution of electromagnetic fields in microwaves changes when they interact with catalysts whose properties change depending on the temperature.

Nishioka et al. designed a flow reactor for the synthesis of Ag nanoparticles that are used in polyol processing under microwave irradiation.[79] They employed a cylindrical cavity with TM010 mode under frequency autotracking mode for this process. Tsubaki et al. also succeeded in temperature control in ultrafast pyrolysis of lignocellulose by a combined use of a cylindrical cavity resonator and a frequency autotracking solid-state generator.[80]

Santamaria et al. proposed two strategies of a pulsed microwave irradiation and a monolith-supported structure for regeneration of a catalyst.[63, 81] These are smart strategies for maximizing the microwave-heating characteristics. The pulsed heating mode is available only for microwave heating, but not for conventional heating. They carried out experiments on coke formation in the catalytic valorization of methane on Mo-ZSM-5 zeolite coated on silicon carbide monolith and succeeded to ensure low amount of coke formation in the pulsed microwave irradiation due to the large temperature gradient between the catalyst surface and the gas.

Multimicrowave sources controlled by the phase array beam-forming technique are effective for the large area irradiation by microwaves.[82] Iyomasa et al. made an array of GaN pallet amplifiers with 15 semiconductor microwave sources to apply phase-controlled microwave irradiation to a large-scale reactor.[83, 84] These new hardware and software units for controlling the irradiation modes of microwaves rapidly developed in the telecommunication technology will be promising in applying to microwave chemistry as well.

7 Toward applications in industries

A local heating at the contact points of the particles in a catalyst packed bed was described as an origin earlier in this chapter.[41] This common phenomenon attracts extensive number of scientists and engineers working in various fields. We need to refer to an interesting work reported by Qiao et al., showing this behavior, namely, the accumulation of electric field at the contact points around two adjacent dielectric spheres.[85] They also predicted the accumulation of the strongest electric field at the gap between the dielectric spheres by the theoretical simulation. Qiao's original report demonstrated that their discovery could be extended to applications to sintering process of dielectric powders and metal powders.

On the other hand, we applied the extensive local heating observed for solid-solid contact points to metal smelting, in which metal oxide is reduced by solid reducing reagent. During the smelting of magnesium metal in which dolomite (a mixed oxide of MgO and CaO) is reduced by ferrosilicon (intermetalic compound of Fe and Si) at high temperature above 1473 K under vacuum.[86]

Currently the smelting of magnesium metal is mainly performed by using the Pidgeon method (thermal reduction method) where the material temperature is raised by burning a large amount of coal as the heat source.[87] A large amount of coal is consumed for smelting, resulting in the generation of the air pollutant PM 2.5 (fine particulate matter) and the release of carbon dioxide to the atmosphere, which are major problems. The Pidgeon method is a technique driven by heating a mixture of dolomite ore and ferrosilicon to high temperatures (1473–1673 K) maintained by coal burning. Metal magnesium is deposited at the cooled wall of the reactor according to the following chemical reaction equation:

$$2MgO\,(s) + 2CaO\,(s) + (Fe)Si\,(s) \rightarrow 2Mg\,(g) + Ca_2SiO_4\,(s) + Fe\,(s)$$

s: solid, g: gas; dolomite mineral: MgO, CaO; ferrosilicon: FeSi; heat source: coal.

Fujii et al. developed a novel Pidgeon process driven by microwaves, called the microwave Pidgeon process, to produce Mg metal with less energy consumption and no direct CO_2 emission.[86] Normally, dolomite is a poor absorber of microwave energy and does not generate heat under microwave irradiation. However, when a mixture of powdered dolomite and powdered ferrosilicon was pressed into a briquette, the briquette was found to be heated up very efficiently under microwave irradiation. We would remind the readers of "occurrence of high-temperature areas at the contact points between particles" for understanding this heating behaviors of the briquette, in which the dolomite particles and ferrosilicon particles are adjacent to each other. It is reasonable to presume that the particles of dolomite and ferrosilicon with mutual contact in the briquette experience occurrence of high-temperature areas at the contact points between particles. Actually, we confirmed lowering the temperature necessary for the reaction by 100–200 K under microwave irradiation compared with conventional heating.

Furthermore, Fujii et al. constructed an antenna-like structure with conductive ferrosilicon powder at the center of each briquette as shown in the photo (cross-sectional and diagonal views) in Fig. 13 and made a polar array by piling the five briquettes into a pillar.[86] The readers see a schematic image of a pillar of the five briquettes, which corresponds to 1/2 wavelength of 66 mm of microwave, acting as an antenna for accumulating microwaves into the briquette in Fig. 13. Thanks to the microwave thermal effects, that is, internal heating and contact point heating, the average reaction temperature for this smelting was lowered from the temperature necessary for the conventional process, 1473–1273 K.

1.7 g of magnesium metal was obtained in a laboratory reactor using a multimode type microwave cavity. A scale-up system was constructed to accurately estimate the energy consumption in the smelting process as a demonstration furnace about five times larger than the laboratory furnace, resulting in the successful smelting of about 7.62 g of magnesium metal. The data provided by this scale-up system showed that the energy consumption was reduced by 68.6% in the microwave smelting compared with the conventional method.

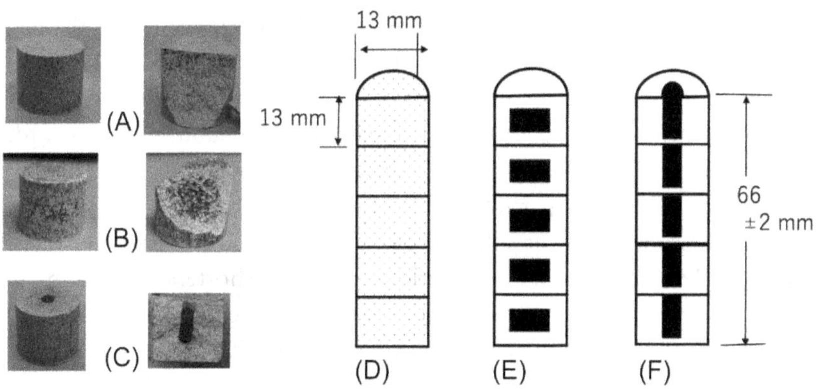

Fig. 13 A pillar made of five briquettes accumulating microwaves as an antenna. Black-colored parts are ferrosilicon-rich parts. (A) Homogeneous, (B) briquettes with ferrosilicon rich at the core, and (C) those with continuously ferrosilicon rich at the center. (C) was found to show the best performance.[86]

8 New horizon in special effects: Acceleration of electron transfer on solid surfaces

We want to differentiate "microwave thermal effects" and "microwave nonthermal effects" observed as the "microwave special effects," but it is extremely difficult to extract only "microwave nonthermal effect" from the overall phenomenon observed for a substance under microwave irradiation, because substances cannot avoid heating phenomenon under microwave irradiation. The scale of heated area can be as large as meter size, but it also can be at even nanoscale size depending on the physical properties and structures and the microwave irradiation conditions. There had been no reports describing nonthermal microwave effects in solid-surface reactions except for our works reporting on the acceleration of electron transfer on solid surfaces. In contrast, several articles describing "nonthermal effects" in homogeneous organic reactions are available and are still continuing the discussion on the nature of these phenomena as described in Dudley's[8, 10] and Yamada's[15–17] works. We believe that it is worth describing, even if briefly, our discoveries on acceleration of electron transfer at the surface of solids under microwave irradiation. This work can be the starting point for a future discussion on microwave nonthermal effects on solid surfaces.

Kishimoto et al. observed microwave-enhanced photocatalytic reduction of *N,N′*-bis(3-sulfonatopropyl)-4,4′-bipyridinium (PVS) on CdS nanoparticles anchored on SiO_2 particles as a photocatalyst under visible light.[88] The generation rate of the anion radicals under microwaves was two times faster than that under conventional heating, exhibiting the clear enhancement of the photocatalytic activity of CdS nanoparticles at the same average temperature of the solutions.

Photoemission decays of CdS nanoparticles in a subnanosecond order have been measured under conventional heating and microwave irradiation.[88] The CdS nanoparticles adsorbed on a porous SiO_2 layer were immersed into an aqueous solution of PVS. The emission lifetime of CdS nanoparticles immersed in PVS solution was clearly shortened under microwave irradiation compared with that measured for the system without PVS, demonstrating that the microwaves can accelerate the photoinduced electron transfer reaction from the CdS nanoparticles to the PVS in the liquid phase. The effect of local heating of CdS nanoparticles by microwave absorption was excluded by the experiments carried out at 315, 325, and 335 K under conventional heating, showing no change in the emission decay. These results led to the conclusion that the shortening of the lifetime is not caused by the changes of the local temperatures close to the CdS nanoparticles, that is, the microwave thermal effects, but attributed to the microwave nonthermal effects shown by the acceleration of excited electron transfer from CdS nanoparticles to PVS. It can be expressed in a general statement as "the transfer of the excited electrons in the conduction band in a semiconductor to an electron acceptor is accelerated in the interaction with microwaves".[87]

The rate constants of the electron transfer rates were estimated from the decay curves as $2.6 \times 10^{11}\ M^{-1}\ s^{-1}$ and $1.0 \times 10^{11}\ M^{-1}\ s^{-1}$ under microwaves and under conventional heating, respectively.[88] Kishimoto employed the Marcus theory to analyze these parameters, leading to an idea that the electronic coupling between CdS nanoparticles and PVS is modified under microwave irradiation, leading to the collective vibration of electrons in CdS induced by microwaves.[88]

Another example for acceleration of electron transfer as a microwave nonthermal effects was demonstrated by Kishimoto et al.[89] The authors described that the anodic current for water oxidation on a hematite (α-Fe_2O_3) electrode was increased under microwave conditions.[88] The anodic current for the oxidation reaction of water observed above 1.5 V versus reference hydrogen electrode (RHE) showed the increase synchronized with the microwave irradiation as demonstrated in Fig. 14. This is attributed to the enhancement of the water oxidation reaction on the surface of the hematite electrode by the oscillating electric field of microwaves. In contrast, Kishimoto et al. recently found that the water reduction current at the surface of a Pt disc electrode was accelerated due to the thermal effects of microwave irradiation.[90] Microwaves can penetrate into the solid α-Fe_2O_3, while microwaves are reflected on the surface of the Pt electrode, probably suggesting the importance of the interaction of microwaves with α-Fe_2O_3 for the acceleration of electron transfer. Tsubaki et al. reported the enhanced anodic current for the aq. solution of $[Ru_4(\mu\text{-}O)_4(\mu\text{-}OH)_2(H_2O)_4(\gamma\text{-}SW_{10}O_{36})_2]^{10-}$ under irradiation with radio frequency (RF) on a fluorine-doped SnO_2 working electrode.[91] The effects of radio frequency irradiation are related with the promoted diffusion of ions in a solution. The enhancement in the anodic current observed for this system not only should be explained in a different way from the aforementioned two examples but also can be interpreted as an example for microwave nonthermal effect.

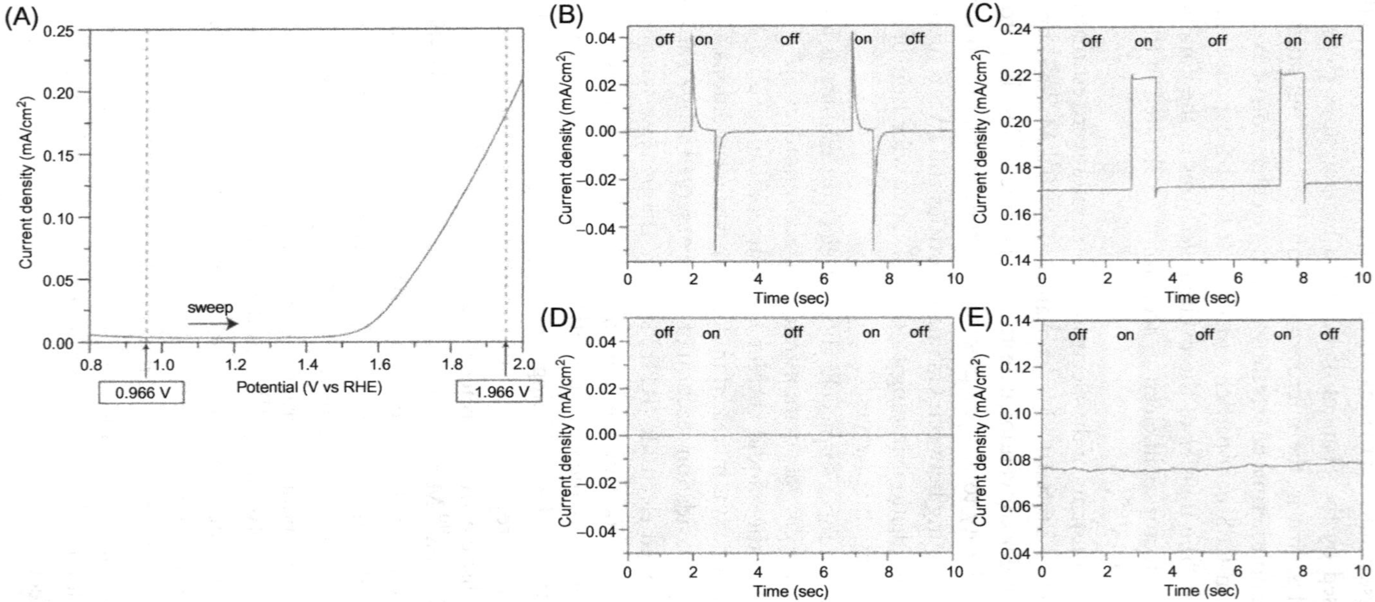

Fig. 14 Linear sweep voltammetry anode and potentiostatic amperometry on α-Fe_2O_3 in water. The readers can observe the increase in the anodic current by the pulsed irradiation of microwaves, showing the microwave-induced acceleration of electron transfer.[88]

9 Summary

This chapter demonstrates promising novel directions showing on how we can apply microwave chemistry to new chemical production. It should be emphasized that nonequilibrium local heating observed in solid-solid, solid-gas, and solid-liquid systems is an important phenomenon derived by microwave irradiation only and never obtained under conventional heating or any other traditional technologies. We should understand the mechanism of this phenomenon depends on the reaction systems and systematize the knowledge so that we can control chemical reactions, leading to industrial applications. Furthermore, it is our hope that we will be able to achieve acceleration of electron transfer on solid surfaces, which can be very interesting in a new idea that we will be able to enhance redox reactions at the solid surfaces. We hope that this account will help the readers to get an idea about microwave chemistry as one of the candidates to renew industrial production for achieving ideal production systems for the future society according to sustainable development goals.

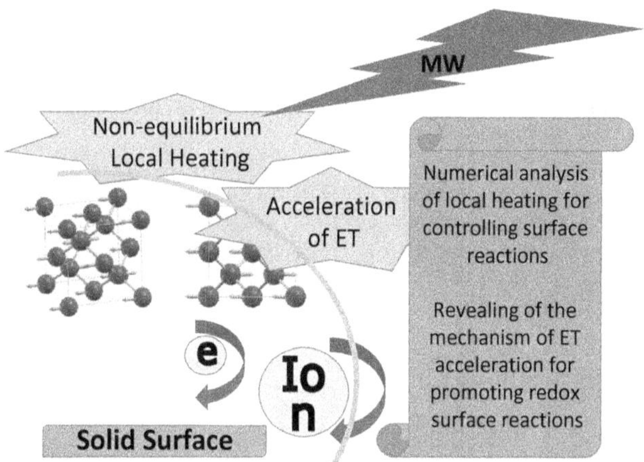

Acknowledgments

The authors thank very much for the financial supports by JSPS Grant-in-Aid for Scientific Research (S) 17H06156, JSPS Grant-in-Aid for Young Scientists (A) 17H05049, and JST PRESTO Grant Number JPMJPR19T6, which enabled us to accomplish the research works on microwave chemistry among the science world described in this review article. Y. W. would express our sincere gratitude to my students and my research lab scientists for focusing their interests on this topic and to Ms. Tomoko Ikegaya for collecting and categorizing the papers for this article.

References

1. Perreux, L.; Loupy, A.Â. A Tentative Rationalization of Microwave Effects in Organic Synthesis According to the Reaction Medium, and Mechanistic Considerations. *Tetrahedron* **2001**, *57*, 9199–9223.
2. Nuechter, M.; Ondruschka, B.; Bonrath, W.; Gun, A. Microwave Assisted Synthesis—A Critical Technology Overview. *Green Chem.* **2004**, *6*, 128–141.
3. de la Hoz, A.; Diaz-Ortiz, A.; Moreno, A. Microwaves in Organic Synthesis. Thermal and Non-thermal Microwave Effects. *Chem. Soc. Rev.* **2005**, *34*, 164–178.
4. Bykov, Y. V.; Rybakov, K. I.; Semenov, V. E. Hight-Temperature Microwave Processing of Materials. *J. Phys. D: Appl. Phys.* **2001**, *34*, R55–R75.
5. Suegra, D. Microwave-Material Interaction and Dielectric Properties, Key Ingredients for Mastery of Chemical Microwave Processes. In *Microwaves in Organic Synthesis*; Loupy, A., Ed.; 2nd ed.; Wiley-VCH Verlag GmbH & Co. KGaA: Weinheim, 2006; pp. 1–61. ISBN: 3-527-31452-0, Copyright © 2006.
6. Kappe, C. O.; Pieber, B.; Dallinger, D. Microwave Effects in Organic Synthesis: Myth or Reality? *Angew. Chem. Int. Ed.* **2013**, *52*, 1088–1094.
7. Horikoshi, S.; Serpone, N. Role of Microwaves in Heterogeneous Catalytic Systems. *Cat. Sci. Technol.* **2014**, *4*, 1197–1210.
8. Dudley, G. B.; Richert, R.; Stiegman, A. E. On the Existence of and Mechanism for Microwave-Specific Reaction Rate Enhancement. *Chem. Sci.* **2015**, *6*, 2144–2152.
9. Nomanbhay, S.; Ong, M. Y. A Review of Microwave-Assisted Reactions for Biodiesel Production. *Bioengineering* **2017**, *4*, 57.
10. Dudley, G. B.; Stiegman, A. E. Changing Perspectives on the Strategic Use of Microwave Heating in Organic Synthesis. *Chem. Rec.* **2018**, *18*, 381–389.
11. Stankiewicz, A.; Sarabi, F. E.; Baubaid, A.; Yan, P.; Nigar, H. Perspectives of Microwaves-Enhanced Heterogeneous Catalytic Gas-Phase Processes in Flow Systems. *Chem. Rec.* **2019**, *19*, 40–50.
12. Priecel, P.; Lopez-Sanchez, J. A. Advantages and Limitations of Microwave Rectors: from Chemical Synthsis to the Catalytic Valorization of Biobased Chemicals. *ACS Sustainable Chem. Eng.* **2019**, 7, 3–21.
13. Haque, E.; Khan, N. A.; Kim, C. M.; Jhung, S. H. Syntheses of MetalOrganic Frameworks and Aluminophosphatesunder Microwave Heating: Quantitative Analysis of Accelerations. *Cryst. Growth Des.* **2011**, *11*, 4413–4421.
14. Ma, J. Master Equation Analysis of Thermal and Nonthermal Microwave Effects. *J. Phys. Chem. A* **2016**, *120*, 7989–7997.
15. Nushiro, K.; Kikuchi, S.; Yamada, T. Microwave Effect on Catalytic Enantioselective Claisen Rearrangement. *Chem. Commun.* **2013**, *49*, 8371–8373.
16. Nushiro, K.; Kikuchi, S.; Yamada, T. Extraordinary Microwave Effect on Atropo-Enantioselective Catalytic Reduction of Biaryl Lactones. *Chem. Lett.* **2013**, *42*, 165–167.
17. Tashima, S.; Sawada, T.; Saito, K.; Yamada, T. Microwave Specific Effect on Catalytic Enantioselective Conia-Ene Reaction. *Chem. Lett.* **2016**, *45*, 649–651.
18. Herrero, M. A.; Kremsner, J. M.; Kappe, C. O. Nonthermal Microwave Effects Revisited: On the Importance of Internal Temperature Monitoring and Agitation in Microwave Chemistry. *J. Org. Chem.* **2008**, *73*, 36–47.
19. Shazman, A.; Mizrahi, S.; Cogan, U.; Shimoni, E. Examining for Possible Non-thermal Effects during Heating in a Microwave Oven. *Food Chem.* **2007**, *103*, 444–453.
20. Zhang, Y. M.; Wang, P.; Han, N.; Lei, H. F. Microwave Irradiation: A Novel Method for Rapid Synthesis of D,L-Lactide. *Macromol. Rapid Commun.* **2007**, *28*, 417–421.
21. Chen, Z.; Li, Y.; Wang, L.; Liu, S.; Wang, K.; Sun, J.; Xu, B. Evaluation of the Possible Non-thermal Effect of Microwave Radiation on the Inactivation of Wheat Germ Lipase. *J Food Process Eng.* **2017**, *40*, e12506.

22. Mudrova, A. N.; Tishchenkoa, A. S.; Ageevaa, T. A.; Koifman, O. I. Kinetic Features of Radical Polymerization of Styrene under Microwave Irradiation Conditions. *Russ. J. Gen. Chem.* **2016**, *86*, 1510–1514.
23. Dai, Q.; Wen, J.; Jing, X.; Dai, L.; Jin, Y.; Wang, F.; Yan, Y. J. Distribution of PCDD/Fs Over Three Product Phases in Wet Sewage Sludge Pyrolysis. *J. Anal. Appl. Pyrolysis* **2018**, *133*, 169–175.
24. Santos, T.; Hennetier, L.; Costa, V. A. F.; Costa, L. C. Microwave Versus Conventional Porcelain Firing: Temperature Measurement. *J. Manuf. Proc.* **2019**, *41*, 92–100.
25. Zhang, X.; Hayward, D. O.; Mingos, D. M. P. Effects of Microwave Dielectric Heating on Heterogeneous Catalysis. *Catal. Lett.* **2003**, *88*, 33–38.
26. Hunt, J.; Ferrari, A.; Lita, A.; Crosswhite, M.; Ashley, B.; Stiegman, A. E. Microwave-Specific Enhancement of the Carbon–Carbon Dioxide (Boudouard) Reaction. *J. Phys. Chem. C* **2013**, *117*, 26871–26880.
27. Fukushima, J.; Kashimura, K.; Takayama, S.; Sato, M.; Sano, S.; Hayashi, Y.; Takizawa, H. In-Situ Kinetic Study on Non-thermal Reduction Reaction of CuO Duringmicrowave Heating. *Mater. Lett.* **2013**, *91*, 252–254.
28. Fukushima, J.; Takizawa, H. Enhanced Reduction of Copper Oxides Via Internal Heating, Selectiveheating, and Cleavage of Cu-O Bond by Microwave Magnetic-Fieldirradiation. *Mater. Chem. Phys.* **2016**, *172*, 47–53.
29. Kappe, C. O. How to Measure Reaction Temperature in Microwave-Heated Transformations. *Chem. Soc. Rev.* **2013**, *42*, 4977–4990.
30. Durka, T.; Stefanidis, G. D.; Van Gerven, T.; Stankiewicz, A. On the Accuracy and Reproducibility of Fiber Optic (FO) and Infrared (IR) Temperature Measurements of Solid Materials in Microwave Applications. *Meas. Sci. Technol.* **2010**, *21*, 045108 (8pp).
31. Rosana, M. R.; Hunt, J.; Ferrari, A.; Southworth, T. A.; Tao, Y.; Stiegman, A. E.; Dudley, G. B. Microwave-Specific Acceleration of a Friedel–Crafts Reaction: Evidence for Selective Heating in Homogeneous Solution. *J. Org. Chem.* **2014**, *79*, 7437–7450.
32. De Bruyn, M.; Budarin, V. L.; Sturm, G. S. J.; Stefanidis, G. D.; Radoiu, M.; Stankiewicz, A.; Macquarrie, D. J. Subtle Microwave-Induced Overheating Effects in an Industrial Demethylation Reaction and Their Direct Use in the Development of an Innovative Microwave Reactor. *J. Am. Chem. Soc.* **2017**, *139*, 5431–5436.
33. Gharibeh, M.; Tompsett, G.; Lu, F.; Auerbach, S. M.; Yngvesson, K. S.; Conner, W. C. Temperature Distributions Within Zeolite Precursor Solutions in the Presence of Microwaves. *J. Phys. Chem. B* **2009**, *113*, 12506–12520.
34. Durka, T.; Van Gerven, T.; Stankiewicz, A. Microwaves in Heterogeneous Gas-Phase Catalysis: Experimental and Numerical Approaches. *Chem. Eng. Technol.* **2009**, *32* (9), 1301–1312.
35. Gangurde, L. S.; Sturm, G. S. J.; Devadiga, T. J.; Stankiewicz, A. I.; Stefanidis, G. D. Complexity and Challenges in Noncontact High Temperature Measurements in Microwave-Assisted Catalytic Reactors. *Ind. Eng. Chem. Res.* **2017**, *56* (56), 13379–13391.
36. Crosswhite, M.; Hunt, J.; Southworth, T.; Serniak, K.; Ferrari, A.; Stiegman, A. E. Development of Magnetic Nanoparticles as Microwave-Specific Catalysts for the Rapid, Low-Temperature Synthesis of Formalin Solutions. *ACS Catal.* **2013**, *3*, 1318–1323.
37. Ichikawa, T.; Matsuo, T.; Tachikawa, T.; Yamada, T.; Yoshimura, T.; Yoshimura, M.; Takagi, Y.; Sawama, Y.; Sugiyama, J.; Monguchi, Y.; Sajiki, H. Microwave-Mediated Site-Selective Heating of Spherical-Carbon-Bead-Supported Platinum for the Continuous, Efficient Catalytic Dehydrogenative Aromatization of Saturated Cyclic Hydrocarbons. *ACS Sustainable Chem. Eng.* **2019**, 7, 3052–3061.
38. Nigar, H.; Sturm, G. S. J.; Garcia-Baños, B.; Peñaranda-Foix, F. L.; Catalá-Civera, J. M.; Mallada, R.; Stankiewicz, A.; Santamaría, J. Numerical Analysis of Microwave

Heating Cavity: Combining Electromagnetic Energy, Heat Transfer and Fluid Dynamics for a NaY Zeolite Fixed-Bed. *Appl. Therm. Eng.* **2019**, *155*, 226–238.
39. Fu, B. A.; Chen, M. Q.; Song, J. J. Investigation on the Microwave Drying Kinetics and Pumping Phenomenon of Lignite Spheres. *Appl. Therm. Eng.* **2017**, *124*, 371–380.
40. Haneishi, N.; Tsubaki, S.; Maitani, M. M.; Suzuki, E.; Fujii, S.; Wada, Y. Electromagnet and Heat-Transfer Simulation of the Catalytic Dehydrogenation of Ethylbenene under Microwave Irradiation. *Ind. Eng. Chem. Res.* **2017**, *56*, 7685–7692.
41. Haneishi, N.; Tsubaki, S.; Abe, E.; Maitani, M. M.; Suzuki, E.; Fujii, S.; Fukushima, J.; Takizawa, H.; Wada, Y. Enhancement of Fixed-Bed Flow Reactions Under Microwave Irradiation by Local Heating at the Vicinal Contact Points of Catalyst Particles. *Sci. Rep.* **2019**, *9* (9), 222.
42. Tsukahara, Y.; Higashi, A.; Yamauchi, T.; Nakamura, T.; Yasuda, M.; Baba, A.; Wada, Y. In Situ Observation of Nonequilibrium Local Heating as an Origin of Special Effect of Microwave on Chemistry. *J. Phys. Chem. C* **2010**, *114*, 8965–8970.
43. Tompsett, G. A.; Panzarella, B.; Conner, W. C.; Yngvesson, K. S.; Lu, F.; Suib, S. L.; Jones, K. W.; Bennett, S. *In situ* Small Angle X-Ray Scattering, Wide Angle X-Ray Scattering, and Raman Spectroscopy of Microwave Synthesis. *Rev. Sci. Instrum.* **2006**, 77, 124101.
44. Schmink, J. R.; Leadbeater, N. E. Probing "Microwave Effects" Using Raman Spectroscopy. *Org. Biomol. Chem.* **2009**, 7, 3842–3846.
45. Schmink, J. R.; Holcomb, J. L.; Leadbeater, N. E. Use of Raman Spectroscopy as an In Situ Tool to Obtain Kinetic Data for Organic Transformations. *Chem. A Eur. J.* **2008**, *14*, 9943–9950.
46. Tsukahara, Y.; Nakamura, T.; Kobayashi, T.; Wada, Y. Homogeneous Ag Particle Formation Confirmed by Real-Time In Situ Surface-Enhanced Raman Scattering Measurements Under Microwave Irradiation. *Chem. Lett.* **2006**, *35* (12), 1396–1397.
47. Silverwood, I. P.; McDougall, G. S.; Whittaker, A. G. A Microwave-Heated Infrared Reaction Cell for the In Situ Study of Heterogeneous Catalysts. *Phys. Chem. Chem. Phys.* **2006**, *8*, 5412–5416.
48. Silverwood, I.; McDougall, G.; Whittaker, G. Comparison of Conventional Versus Microwave Heating of the Platinum Catalysed Oxidation of Carbon Monoxide Over EUROPT-1 in a Novel Infrared Microreactor Cell. *J. Mol. Catal. A Chem.* **2007**, *269*, 1–4.
49. Whittaker, A. G.; Harrison, A.; Oakley, G. S.; Youngson, I. D.; Heenan, R. K.; King, S. M. Preliminary Experiments on Apparatus for *In Situ* Studies of Microwave-Driven Reactions by Small Angle Neutron Scattering. *Rev. Sci. Instrum.* **2001**, *72*, 173–176.
50. Harrison, A.; Ibberson, R.; Robb, G.; Whittaker, G.; Wilsonb, C.; Youngson, D. In Situ Neutron Diffraction Studies of Single Crystals and Powders During Microwave Irradiation. *Faraday Discuss.* **2002**, *122*, 363–379.
51. Xiao, Y.; Xu, F.; Hu, X.; Li, Y.; Liu, W.; Dong, B. In Situ Investigation of Titanium Powder Microwave Sintering by Synchrotron Radiation Computed Tomography. *Metals* **2016**, *6*, 9.
52. Vauchera, S.; Nicula, R.; Català-Civera, J.-M.; Schmitt, B.; Patterson, B. In Situ Synchrotron Radiation Monitoring of Phase Transitions During Microwave Heating of Al–Cu–Fe Alloys. *J. Mater. Res.* **2008**, *23* (1), 170–175.
53. Nakamura, N.; Reeja-Jayana, B. Synchrotron X-Ray Characterization of Materials Synthesized under Microwave Irradiation. *J. Mater. Res.* **2019**, *34* (1), 194–205.
54. Asakuma, Y.; Matsumura, S.; Asada, M.; Phan, C. In Situ Investigation of Microwave Impacts on Ethylene Glycol Aqueous Solutions. *Int. J. Thermophys.* **2018**, *39*, 21.
55. Asakuma, Y.; Matsumura, S.; Saptoro, S. A. Method for Suppressing Superheating Behavior During Microwave Assisted Nanoparticle Formation by Ethylene Glycol Addition. *Chem. Eng. Process. Process Intensif.* **2018**, *132*, 11–15.

56. Shibatani, A.; Matsumura, S.; Asakuma, Y.; Saptoro, A. Promoting Nucleation in Microwave-Assisted Nanoparticle Synthesis Using Combined Two-Stage Irradiation and Anti-Solvent Addition. *Cryst. Res. Technol.* **2019**, *54*, 1800205.
57. Horikoshi, S.; Kamata, M.; Mitani, T.; Serpone, N. Control of Microwave-Generated Hot Spots. 6. Generation of HotSpots in Dispersed Catalyst Particulates and Factors That AffectCatalyzed Organic Syntheses in Heterogeneous Media. *Ind. Eng. Chem. Res.* **2014**, *53*, 14941–14947.
58. Horikoshi, S.; Kamata, M.; Mitani, T.; Serpone, N. Control of Microwave-Generated Hot Spots. 6. Generation of Hot Spots in Dispersed Catalyst Particulates and Factors That Affect Catalyzed Organic Syntheses in Heterogeneous Media. *Ind. Eng. Chem. Res.* **2014**, *53*, 14941–14947.
59. Wang, W.; Wang, B.; Sun, J.; Mao, Y.; Zhao, X.; Song, Z. Numerical Simulation of Hot-Spot Effects in Microwave Heating Due to the Existence of Strong Microwave-Absorbing Media. *RSC Adv.* **2016**, *6*, 52974–52981.
60. Ramirez, A.; Hueso, J.; Mallada, R.; Santamaria, J. In Situ Temperature Measurements in Microwave-Heated Gas-Solid Catalytic Systems. Detection of Hot Spots and Solid-Fluid Temperature Gradients in the Ethylene Epoxidation Reaction. *Cem. Eng. J.* **2017**, *316*, 50–60.
61. Ramirez, A.; Hueso, J. L.; Abian, M.; Alzueta, M. U.; Mallada, R.; Santamaria, J. Escaping Undesired Gas-Phase Chemistry: Microwave-Driven Selectivity Enhancement in Heterogeneous Catalytic Reactors. *Sci. Adv.* **2019**, *5*, eaau9000.
62. Hamzehlouia, S.; Shabanian, J.; Latifi, M.; Chaouki, J. Effect of Microwave Heating on the Performance of Catalytic Oxidation of n-Butane in a Gas-Solid Fluidized Bed Reactor. *Chem. Eng. Sci.* **2018**, *192*, 1177–1188.
63. Nigar, H.; Julián, I.; Mallada, R.; Santamaría, J. Microwave-Assisted Catalytic Combustion for the Efficient Continuous Cleaning of VOC-Containing Air Streams. *Environ. Sci. Technol.* **2018**, *52*, 5892–5901.
64. Mochizuki, D.; Sasaki, R.; Maitani, M. M.; Okamoto, M.; Suzuki, E.; Wada, Y. Catalytic Reactions Enhanced Under Microwave-Induced Local Thermalnon-Equilibrium in a Core–Shell, Carbon-Filled Zeolite@Zeolite. *J. Catal.* **2015**, *323*, 1–9.
65. Maitani, M. M.; Inoue, T.; Tsukushi, Y.; Hansen, N. D. J.; Mochizuki, D.; Suzuki, E.; Wada, Y. Collaborational Effect of Heterolytic Layered Configuration for Enhancement of Microwave Heating. *Chem. Commun.* **2013**, *49*, 10841–10843.
66. Maitani, M. M.; Tsukushi, Y.; Hansen, N. D. J.; Sato, Y.; Mochizuki, D.; Suzuki, E.; Wada, Y. Low-Temperature Annealing of Mesoscopic TiO_2 Films by Interfacial Microwave Heating Applied to Efficiency Improvement of Dye-Sensitized Solar Cells. *Sol. Energy Mater. Sol. Cells* **2016**, *147*, 198–202.
67. Maitani, M. M.; Iso, D.; Kim, J.; Tsubaki, S.; Wada, Y. Microwave Application to Efficient Annealing Process of $CH_3NH_3PbI_3$ Perovskite Crystalline Films. *Electrochemistry* **2017**, *85* (5), 236–240.
68. Kenkre, V. M.; Skala, L.; Skala, M. W. W.; Weiser, M. W. Theory of Microwave Interactions in Ceramic Materials: the Phenomenon of Thermal Runaway. *J. Mater. Sci.* **1991**, *26*, 2483–2489.
69. Sugawara, H.; Kashimura, K.; Hayashi, M.; Matsumuro, T.; Watanabe, T.; Mitani, T.; Shinohara, N. Temperature Dependence and Shape Effect in High-Temperature Microwave Heating of Nickel Oxide Powders. *Physica B* **2015**, *458*, 35–39.
70. Peng, Z.; Hwang, J.-Y.; Park, C.-L.; Kim, B.-G.; Andriese, M.; Wang, X. Microwave Permittivity, Permeability, and Absorption Capability of Ferric Oxide. *ISIJ Int.* **2012**, *52* (9), 1535–1538.
71. Hotta, M.; Hayashi, M.; Nagata, K. Complex Permittivity and Permeability of a-Fe_2O_3 and $Fe_{1-x}O$ Powders in the Microwave Frequency Range Between 0.2 and 13.5 GHz. *ISIJ Int.* **2010**, *50* (10), 1514–1516.

72. Wypych, A.; Bobowska, I.; Tracz, M.; Opasinska, A.; Kadlubowski, S.; Krzywania-Kaliszewska, A.; Grobelny, J.; Wojciechowski1, P. Dielectric Properties and Characterisation of Titanium Dioxide Obtained by Different Chemistry Methods. *J. Nanomater.* **2014**, 124814. 9 p.
73. Fujii, S.; Kujirai, H.; Mochizuki, D.; Maitani, M. M.; Suzuki, E.; Wada, Y.; Mayama, N. Chemical Reaction Under Highly Precise Microwave Irradiation. *J. Microw. Power Electromagn. Energy* **2014**, *48*, 89–103.
74. Atuonwu, J. C.; Tassou, S. A. Quality Assurance in Microwave Food Processing and the Enabling Potentials of Solid-State Power Generators: A review. *J. Food Eng.* **2018**, *234*, 1e15.
75. Horikoshi, S.; Watanabe, T.; Narita, A.; Suzuki, Y.; Serpone, N. The Electromagnetic Wave Energy Effect(S) in Microwave–Assisted Organic Syntheses (MAOS). *Sci. Rep.* **2018**, *8*, 5151.
76. Catalá-Civera, J. M.; Canós, A. J.; Plaza-González, P.; Gutiérrez, J. D.; García-Baños, B.; Peñaranda-Foix, F. L. Dynamic Measurement of Dielectric Properties of Materials at High Temperature During Microwave Heating in a Dual Mode Cylindrical Cavity. *IEEE Trans. Microw. Theory Tech.* **2015**, *63* (9), 2905–2914.
77. Peñaranda-Foix, F. L.; Janezic, M. D.; Catala-Civera, J. M. Full-Wave Analysis of Dielectric-Loaded Cylindrical Waveguides and Cavities Using a New Four-Port Ring Network. *IEEE Trans. Microw. Theory Tech.* **2012**, *60*, 2730–2740.
78. Glasnov, T. N.; Kappe, C. O. The Microwave-to-Flow Paradigm: Translating High-Temperature Batch Microwave Chemistry to Scalable Continuous-Flow Processes. *Chem. A Eur. J.* **2011**, *17*, 11956–11968.
79. Nishioka, M.; Miyakawa, M.; Kataoka, H.; Koda, H.; Sato, K.; Suzuki, T. M. Continuous Synthesis of Monodispersed Silver Nanoparticles Using a Homogeneous Heating Microwave Reactor System. *Nanoscale* **2011**, *3*, 2621–2626.
80. Tsubaki, S.; Nakasako, Y.; Ohara, N.; et al. Ultra-Fast Pyrolysis of Lignocellulose Using Highly Tuned Microwaves: Synergistic Effect of a Cylindrical Cavity Resonator and a Frequencyauto- Tracking Solid-State Microwave Generator. *Green Chem.* **2019**,. https://doi.org/10.1039/c9gc02745a.
81. Juliana, I.; Ramireza, H.; Huesoa, J. L.; Malladaa, R.; Santamaria, J. Non-oxidative Methane Conversion in Microwave-Assisted Structured Reactors. *Chem. Eng. J.* **2019**, *377*, 119764.
82. Horikoshi, S.; Yamazaki, S.; Narita, A.; Mitani, T.; Shinohara, N.; Serponec, N. A Novel Phased Array Antenna System for Microwave-Assisted Organic Syntheses Under Waveguideless and Applicatorless Setup Conditions. *RSC Adv.* **2016**, *6*, 113899–113902.
83. Iyomasa, K.; Shode, T.; Yamanaka, K.; Mori, K.; Fukumoto, H.; Tsukahara, Y.; Wada, Y. High Efficiency Chemical Reactions Induced by Concentrated Microwave Heating Using GaN Amplifier Modules. *J. Japan Petrol. Inst.* **2018**, *61*, 163–170.
84. Sugitani, T.; Iyomasa, K.; Hangai, M.; Kawamura, Y.; Nishihara, J. Shinjo, S., 2018. 2.45 GHz ISM-Band 450W High Efficiency GaN Pallet Amplifier for Microwave Heating, Proceedings of 2018 Asia-Pacific Microwave Conference, FR3-IF-48.
85. Qiao, X.; Xie, X. The effect of electric field intensification at interparticle contacts in microwave sintering. *Sci. Rep.* **2016**, *6*, 32163. https://doi.org/10.1038/srep32163.
86. Wada, Y.; Fujii, S.; Suzuki, E.; Maitani, M. M.; Tsubaki, S.; Chonan, S.; Fukui, M.; Inazu, N. Smelting Magnesium Metal Using a Microwave Pidgeon Method. *Sci. Rep.* **2017**, 7, 46512. https://doi.org/10.1038/srep46512.
87. Ramakrishnan, S.; Koltun, P. Global Impact of the Magnesium Produced in China Using the Pidgeon Process. *Resour. Conserv. Recycl.* **2004**, *42*, 49–64.

88. Kishimoto, F.; Imai, T.; Fujii, S.; Mochizuki, D.; Maitani, M. M.; Suzuki, E.; Wada, Y. Microwave-Enhanced Photocatalysis on CdS Quantum Dots—Evidence of Acceleration of Photoinduced Electron Transfer. *Sci. Rep.* **2015**, *5*, 11308. https://doi.org/10.1038/srep11308.
89. Kishimoto, F.; Matsuhisa, M.; Kawamura, S.; Fujii, S.; Tsubaki, S.; Maitani, M. M.; Suzuki, E.; Wada, Y. Enhancement of Anodic Current Attributed to Oxygen Evolution on α-Fe_2O_3 Electrode by Microwave Oscillating Electric Field. *Sci. Rep.* **2016**, *6*, 35554. https://doi.org/10.1038/srep35554.
90. Kishimoto, F.; Leong, K. H.; Kawamura, S.; Haneishi, N.; Tsubaki, S.; Wada, Y. Acceleration of Water Electrolysis by Accumulation of Microwave Energyat a Pt Disk Electrode. *Chem. Lett.* **2017**, *46*, 1593–1596.
91. Tsubaki, S.; Hayakawa, S.; Ueda, T.; Fujii, S.; Suzuki, E.; Zhang, J.; Bond, A.; Wada, Y. Radio Frequency Alternating Electromagnetic Field Enhanced Tetraruthenium Polyoxometalate Electrocatalytic Water Oxidation. *Chem. Commun.* **2019**, *55*, 1032–1035.

CHAPTER THREE

Recent advances in multicomponent microwave-assisted Ugi reactions

Allen Gordon and Shainaz Landge
Department of Chemistry and Biochemistry, Georgia Southern University, Statesboro, GA, United States

1 Introduction

The Ugi reaction was first reported by Ivar Ugi in 1959 by conducting the reaction on isonitriles.[1, 2] The traditional Ugi four-component reaction generally has a carbonyl component (aldehyde or ketone moiety), an amine, an isonitrile, and a carboxylic acid in a polar protic solvent to result in amido amides (Scheme 1A). The mechanistic pathway involves the generation of imine (condensation of carbonyl and amine) followed by an attack by the isonitrile, resulting in the nitrilium intermediate. This intermediate is then attacked by the carboxylic acid, followed by an intramolecular rearrangement to yield the bis-amide Ugi product (Scheme 1B).[1, 2] This bis-amide structure has five different positions of attachment for alkyl (—R) groups, giving opportunity for a highly tailored functionalized molecule with broad applications.[3, 4]

The alkyl or aromatic groups attached to the starting materials for the Ugi reaction can lead to the post-Ugi reaction such as Ugi-Heck,[3, 5, 6] Ugi-Diels-Alder,[7–9] Ugi-click,[10–12] and Ugi-Buchwald-Hartwig[13, 14] reactions. Numerous studies have used Ugi reactions for further intermolecular coupling reactions such as the double Ugi[15] or sequential Ugi reaction[16–19] and the Ugi-Mumm/Ugi-Smile reactions.[20, 21] The Ugi products are also referred to as peptoids and are used in various linear and cyclic peptide-based[22, 23] syntheses. The diversity in the final products can be achieved by varying substituents present in the starting material.[3, 4]

Multicomponent reactions (MCR) have achieved prominence in modern organic chemistry.[8] They have a tremendous edge over linear synthesis approaches by saving time, energy, having a great atom economy, providing one-pot synthetic approach, and needing less labor. These attributes truly fit

Nontraditional Activation Methods in Green and Sustainable Applications
https://doi.org/10.1016/B978-0-12-819009-8.00014-1
© 2021 Elsevier Inc.
All rights reserved.

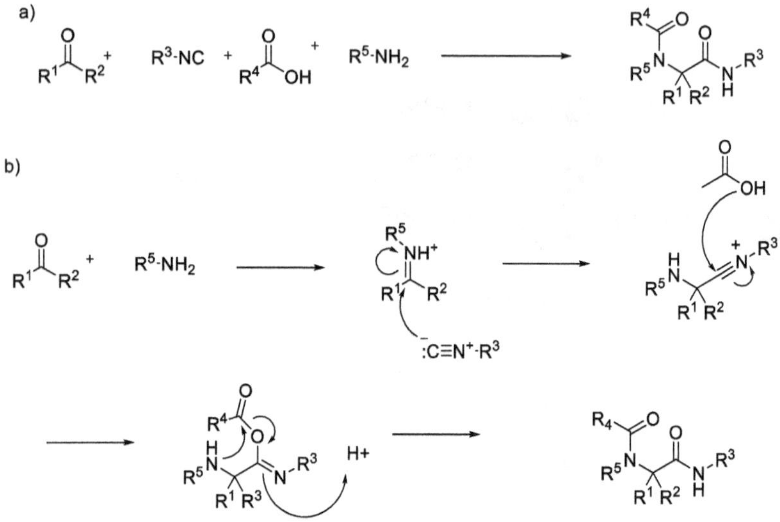

Scheme 1 (A) A standard Ugi four-component reaction and (B) its mechanistic pathway.[1]

the bill for a perfectly green reaction.[24] Microwave (MW) energy synergistically adds more to the multicomponent reactions by decreasing reaction times and increasing efficiency, thus providing high product yields and becoming eco-friendly. Due to the bulk solution reaching its desired temperature at a faster rate, the yield is increased by limiting the formation of side products.[25, 26] Together, multicomponent reactions and microwave-assisted organic synthesis provide a library of compounds with structural diversity and flexibility. It has shown an exponential growth in its importance and has been regularly utilized in the last few decades. These two important green tools bring the best out of organic chemistry synthesis.[25–27]

This chapter will be divided into topics such as Ugi component coupling reaction (Ugi-4CC, Ugi-3CC, and Ugi-5CC), sequential three- and four-component coupling reactions, and post-Ugi reactions specifically published in the last 5 years.

2 The Ugi-four component reaction (U-4CR)

The Ugi-4CR is the most popular and efficient in the Ugi reaction sequence, and it can provide five new centers for functional diversity. The following are a few of the selected examples from recent years.

A one-pot reaction through Ugi four-component (Ugi-4C) condensation of lipoic acid, cyclohexyl isocyanide, substituted aniline, and 2-methoxyquinoline-3-carbaldehyde derivatives under microwave irradiation resulted in the quinoline-based peptides in short reaction time and excellent yields. The scope of the reaction is optimized through the effects of microwave irradiation and solvent. The synthesized peptides are then screened for antimicrobial and antioxidant properties and evaluated for toxicity and are discussed in detail by Thangaraja et al.[22] Through this green multicomponent microwave-assisted organic synthesis (MCMAOS), complex heterocycles with great medicinal value[22] were synthesized (Scheme 2).

A Ugi-4CR using microwave irradiation was utilized to result in the synthesis of steroid-amino acid conjugates. This new class of peptidomimetic hybrid compounds consisting of alanine and valine methyl esters with secosteroids (A, B, and D ring cleavage) has diverse biological properties. These compounds were synthesized in good yields and short reaction times. For example, the steroid (**1**) when reacted with L-alanine methyl ester hydrochloride paraformaldehyde, benzyl isocyanide, and triethylamine at 400 W resulted in the corresponding steroid-amino acid conjugate (**2**) with good yield (Scheme 3).[28]

Through double multicomponent reactions (MCR), Rivera et al.[15] discussed a bidirectional access to novel thiadiazine hybrid molecules (Fig. 1). The synthesis of novel 1,3,5-thiadiazine peptides relies on the Ugi-4CR through reaction of bis-carboxylic acid functionalized 1,3,5-thiadiazine along with aldehydes, amines, and isocyanides, resulting in derivatives of thiadiazine peptide-like macrocycles. In this work the thiazidine

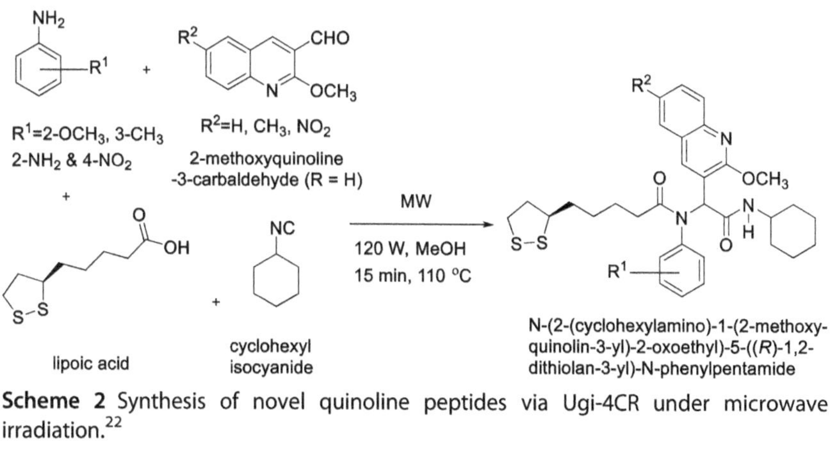

Scheme 2 Synthesis of novel quinoline peptides via Ugi-4CR under microwave irradiation.[22]

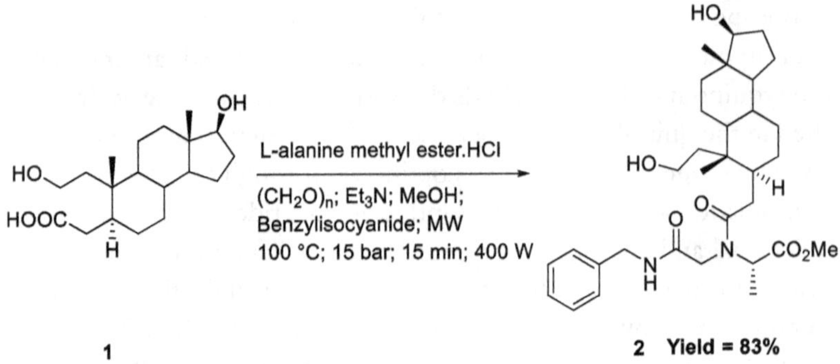

Scheme 3 Microwave (MW)-irradiated Ugi four-component reaction (Ugi-4CR) in synthesis of steroid-amino acid conjugates.[28]

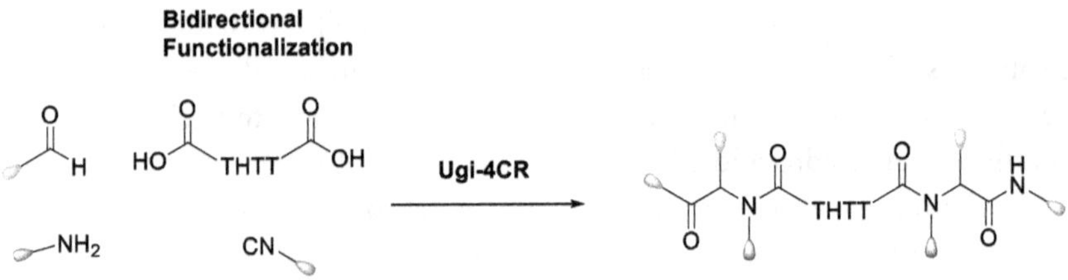

Fig. 1 Thiadiazine-MCR strategies to hybrid molecule synthesis.[15]

dicarboxylic acid (**3**) reaction with paraformaldehyde, benzylamine, and cyclohexyl isocyanide was chosen as a model reaction to result in the 1,3,5-thiadiazinane-2-thione (THTT) hybrids (**4**) through a double Ugi-4CR[15] (Fig. 1 and Scheme 4).

A synthesis of dimeric peptide nucleic acid (PNA)-ribonucleic acid (RNA) using microwave-assisted Ugi four-component reaction (Ugi-4CR) is described for the first time. The four components used in this reaction were a functionalized amine (**5**), a carboxymethyl nucleobase (**6**), paraformaldehyde (**7**), and *N*4,2′-*O*,3′-*O*-tribenzoyl-5′-deoxy-5′-isocyanocytidine (**8**), which,

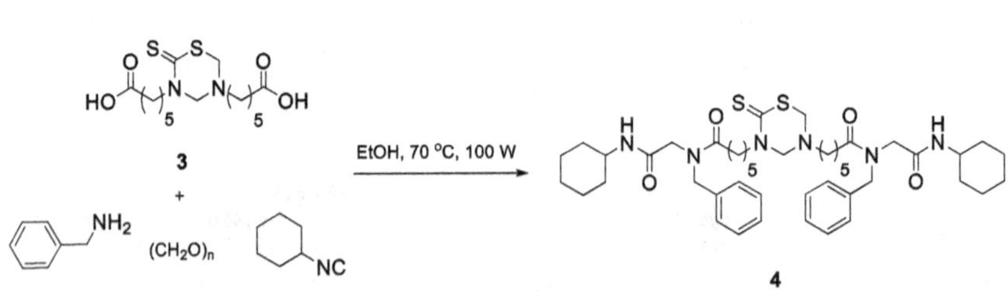

Scheme 4 Reaction scope of THTT hybrids by a double Ugi-4CR.[15]

when combined, generate the dimeric structures in one step (**9**). This simple methodology resulted in a quick reaction to generate derivatives of five PNA-RNA chimeras. The scope of the reaction was studied in detail using various carboxymethyl nucleobases with a "RNA unit" isocyanide. In future the researchers aim to develop PNA-DNA dimers, which can help in peptide chemistry research[29] (Scheme 5).

To anchor the peptides to a solid support, the novel one-pot Ugi-4CR method was researched and developed using an isocyanide-based reaction. Microwave energy was utilized to increase the rate of the reaction. The first step of the reaction included the condensation of the N-protected amino acid and an amino ester, which was then bonded with the supported aldehyde (**10**). This synthesized backbone was then used in the solid-phase peptide synthesis to generate head-to-tail macrocyclic peptides (**11**) through a reaction with *tert*-butyl isocyanide, Fmoc-Gly-OH (Fmoc-Xaa2-OH), and H-Gly-OtBu (H-Xaa1-OPG) hydrochloride salt in MeOH/MeCN (1:1) in combination with microwave (MW) energy. The anchoring of the peptide was released from the resin in the presence of trifluoroacetic acid. The overall Ugi-4CR process was efficient because of the use of microwave energy in less time and reagent equivalents. Various parameters such as temperature, reagent equivalents, and irradiation time were assessed to find optimal reaction conditions[30] (Scheme 6).

In an effort to synthesize steroid-peptide conjugates, microwave energy was used in the Ugi-4CR, resulting in an important hybrid system based on amino steroids. This high-level synthesis has gained attention in drug development using expedite microwave-assisted synthesis. The reaction of 3-α-amino cholesterol (**12**) with *N*-(*tert*-butoxycarbonyl)-L-phenylalanine (**13**)

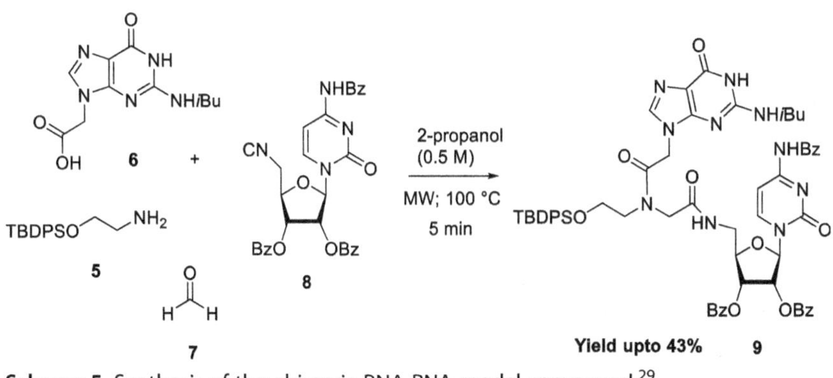

Scheme 5 Synthesis of the chimeric PNA-RNA model compound.[29]

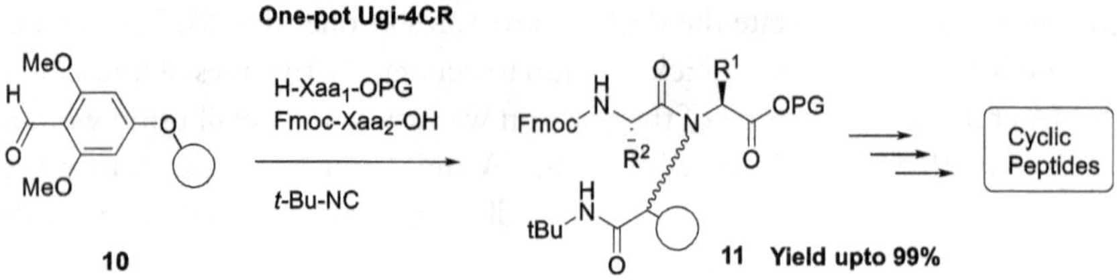

Scheme 6 General solid-phase traceless Ugi-4CR ligation and backbone anchoring.[30]

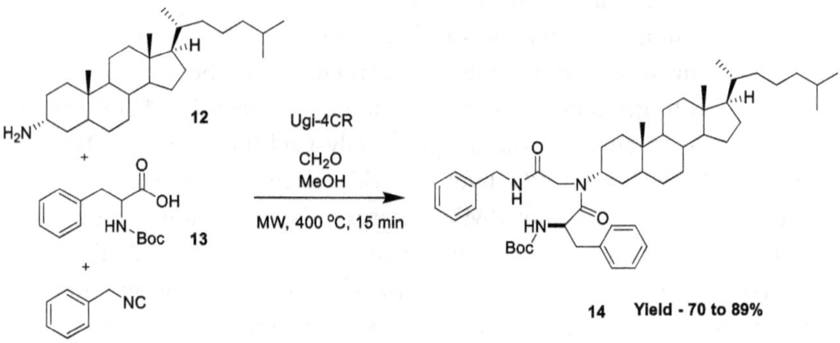

Scheme 7 Synthesis of steroid-peptide conjugates based on various aminosteroids.[31]

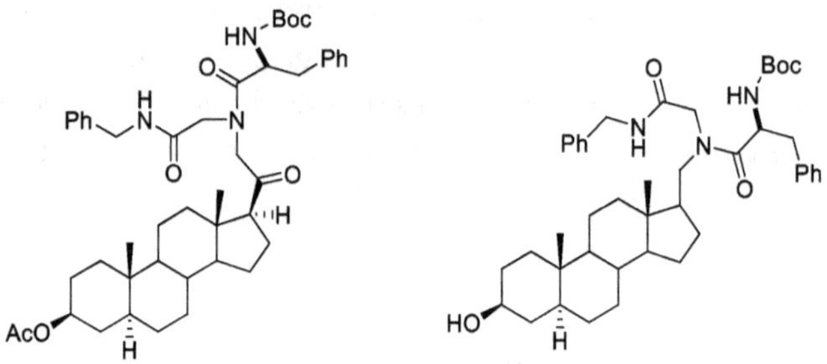

Scheme 8 Synthesis of steroid-amino acid conjugates based on various steroidal amines.[32]

in the presence of benzyl isocyanide, paraformaldehyde, and triethylamine under Ugi-4CR furnished steroid-peptide conjugate (**14**) within 15 min. Both the classical and MW approach were used and compared to synthesize this conjugate, and MW energy helped to accelerate the conjugation process in excellent yields[31] (Scheme 7). Few aminosteroids through this synthesis are shown in Scheme 8.

The same research group reported a high yield synthesis of another important hybrid system based on aminosteroids for the first time. The use of microwave energy with Ugi-4CR resulted in an expedited and clean reaction process.[32]

Another example used a microwave-assisted multicomponent synthesis of dihydroquinoxalinones on a polymer support. This one-pot reaction used aldehydes, Fmoc-protected α-amino acid, isocyanide, and soluble polymer-supported 4-fluoro-3-aminobenzoate ester (**15**) as starting material for the Ugi-4CR to synthesize the key intermediate diamide (**16**) (Scheme 9). The diamide (**16**) was further Fmoc deprotected, followed by an intramolecular cyclization through an ipso displacement to result in a fused ring of quinoxalin-2-ones. The substrate scope of the reaction was studied for the dihydroquinoxalinone product with good yields (86%). Through a simple purification process and straightforward synthesis, this study provided 21 compounds with potential biological active properties and high functional group tolerance.[26]

A new methodology was developed to introduce picolinamide (PA) directing group in aliphatic ketones by using the ammonia-Ugi-4-CR. The Ugi reaction was performed on 2-pyridinecarboxylic acid (**17**), isocyanide, and ketone (cyclic or aliphatic) with an excess of ammonia solution (7 N in MeOH) under microwave heating conditions. Utilizing this Ugi adduct in the Pd-mediated γ-C(sp^3)-H bond activation resulted in a series of γ-arylated α-aminoamides (**18**), with excellent yields. The practicability and flexibility of the approach were demonstrated on other directing groups as well. This method is the first of a kind example of using the

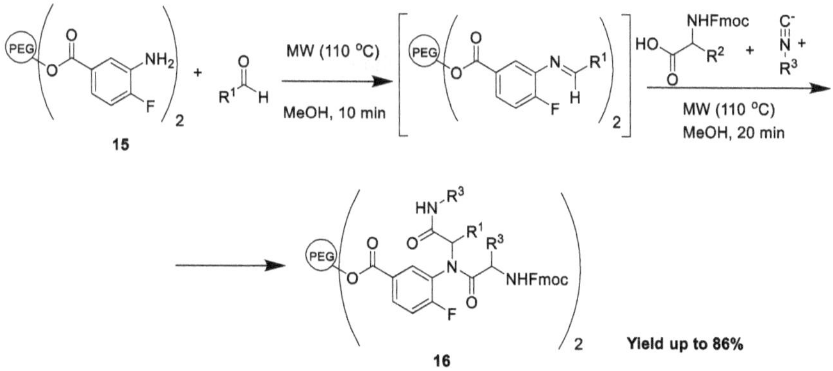

Scheme 9 Two-step, one-pot synthesis of PEG-supported diamide via Ugi-4CR.[26]

isocyanide-based multicomponent reaction (IMCR) and picolinamide as a directing group[33] (Scheme 10).

A series of functionalized peptoids (**19**) were synthesized via commonly used starting materials for Ugi-4CR and MAOS to test against antileishmanial activity. Using inexpensive reagents, this efficient reaction provided decent to good yield (55%–80%) with less than 10 min reaction time. A detailed study about the biological antileishmanial activities are reported in this paper[34] (Scheme 11).

A study conducted by Jobin et al.[23] reported a traceless isocyanide-based Ugi-4-CR to couple peptide fragments on a solid support. This multicomponent reaction uses microwave-assisted synthesis on-resin to attach a carboxyl-free peptide (**21**) to a supported peptide bearing a free N-terminal amine via the formation of an N-protected amide bond (**20**). Through the synthesized backbone amide, the protecting group can be removed by an acidolysis reaction with trifluoroacetic acid. Various lengths of oligopeptides (**22**) were synthesized using this approach with moderate yield and purity. Microwave irradiation not only improved the yield of 4-CR but also

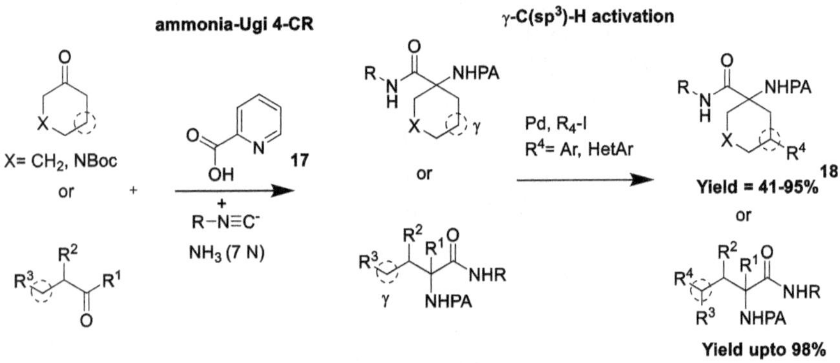

Scheme 10 γ-C(sp^3)-H bond activation in aliphatic systems.[33]

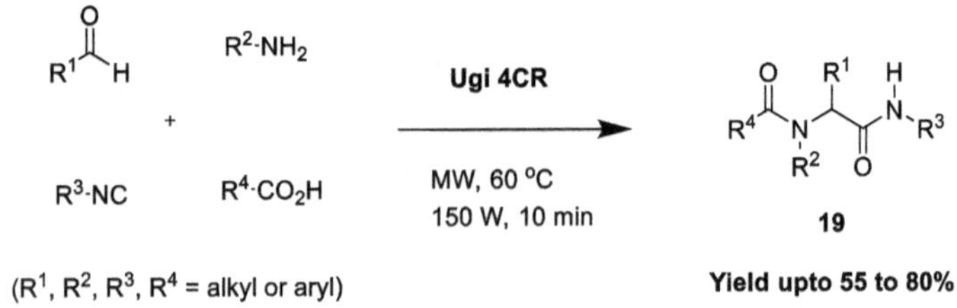

Scheme 11 Preparation of linear peptoids by the Ugi-4CR-MAOS.[34]

provided a backbone amide deprotection step in less time. This protocol does not require any particular modifications on the C- or N-terminal before the ligation process. The optimization of the reaction conditions was performed in detail to know the scope of the overall process.[23] (Scheme 12).

Using starting materials such as derivatives of an aldehyde, an amine, isocyanide, and an ethanoic acid in protic ionic liquid (ethylammonium nitrate [EAN] and propylammonium nitrate [PAN]) through the Ugi-4CR afforded derivatives of α-phenylacetamido amides. The use of methanol as a solvent showed moderate yields, but changing it to ionic liquid such as EAN and PAN resulted in excellent yields. Microwave energy was used to synthesize multiple libraries based on the parent isocyanide, cyclohexyl isocyanide, benzyl isocyanide, and ethyl isocyanoacetate with EAN as the ionic liquid, resulting in good to excellent product yields. The general reaction of aniline (**23**), benzaldehyde (**24**), cyanoacetic acid (**24**), and cyclohexyl isocyanide (**26**) afforded the linear Ugi- (**26**) product (Scheme 13). Various reaction conditions such as the effects of microwave

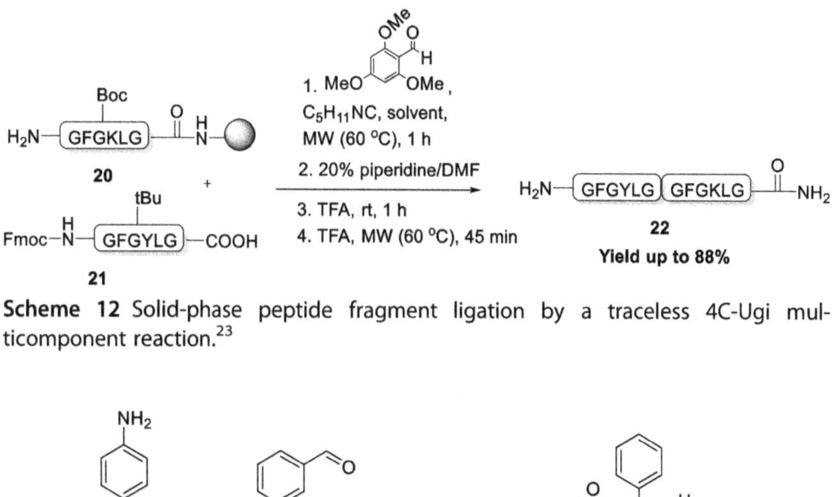

Scheme 12 Solid-phase peptide fragment ligation by a traceless 4C-Ugi multicomponent reaction.[23]

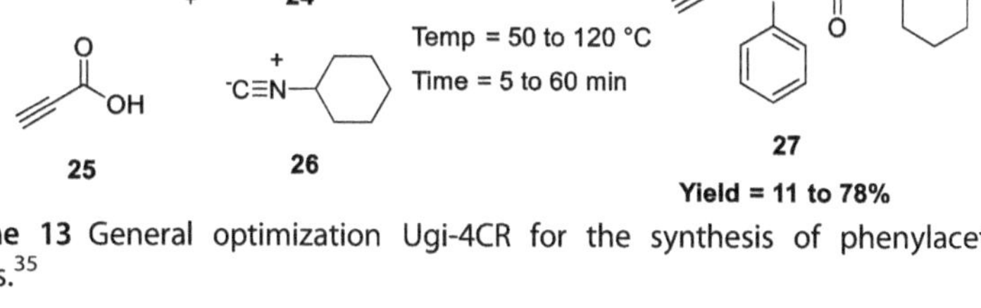

Scheme 13 General optimization Ugi-4CR for the synthesis of phenylacetamido amides.[35]

irradiation, time, and temperature were studied to get the best yield. The Ugi-Knoevenagel product was observed in the presence of an acid with an acidic methylene group. The synthesized products were used in cancer cell lines in the development of potential anticancer leads.[35]

3 Synthesis of the Boronate derivatives

Through Ugi-4CR, various new boron-containing dipeptide analogues (**29**) are synthesized. These compounds are specifically designed for the boron neutron capture therapy (BNCT) to deliver boron to tumor cells. For the optimization reaction, derivatives of aminophenylboronic acid pinacol ester (**28**), formylphenylboronic acid pinacol ester, 4-carboxyphenylboronic acid pinacol ester, and *tert*-butyl isocyanide are used as starting materials. Through facile MAOS, various peptide derivatives resembling basic cell metabolites are synthesized and are used as boron carriers in BNCT. At low temperature (50°C), general Ugi reaction (Scheme 14) takes place, whereas higher temperatures (60°C) are required for the fluoro-substituted derivatives to yield better results.[36]

A similar protocol using microwave irradiation was used by Hsiao et al.[37] to synthesize a broad range of boronate analogues via Ugi-4CR under mild conditions. Their method involved the use of benzylamine, 4-(4,4,5,5-tetramethyl-1,3,2-dioxaborolan-2-yl)benzaldehyde (**30**), propanoic acid, and cyclohexyl isocyanide as starting materials at 45°C in methanol as solvent and generate a broad range of boronate-containing derivatives (**31**). For the optimization step, varieties of solvent ranging from methanol to water were used. The temperature and time studies were carried out to receive a general trend in this type of reaction to maximum yield of ~78%. A variety of Bpin-phenyl derivatives were tested to study the scope of the reaction; 4-Bpin-phenyl derivatives increase the product yield by 93%. The post-Ugi modification of Pd-mediated Suzuki-Miyaura cross-coupling of boron

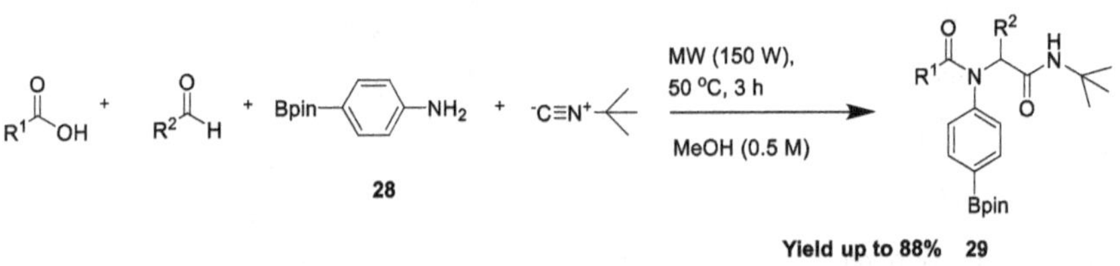

Scheme 14 Synthesis of multiple boron-containing compounds.[36]

was also successfully shown to generate compounds of great importance[37] (Scheme 15).

However, one-pot facile MAOS of two fused piperazine-benzimidazoles is described by Gang-Xu et al.[38] in good yields. Two series of derivatives of fused piperazine-benzimidazole scaffolds (**32**) have been targeted through Ugi/deprotection/cyclization strategy (UDC) followed by an intermolecular nucleophilic substitution reaction. The nucleophilic reaction (pathway A) resulted in a maximum of 89% yield, and the one-pot procedure (pathway B) resulted in a maximum of 70% yield of product **32**. This UDC process helped to synthesize diverse molecules targeting medicinal values[38] (Scheme 16).

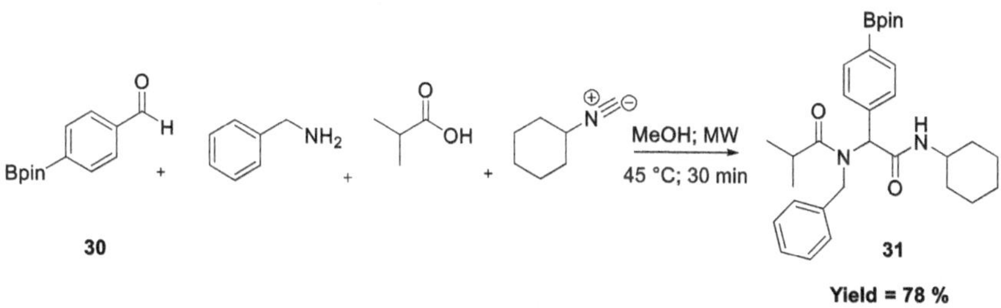

Scheme 15 Optimization of Ugi-4CR reaction for the synthesis of boronate analogues.[37]

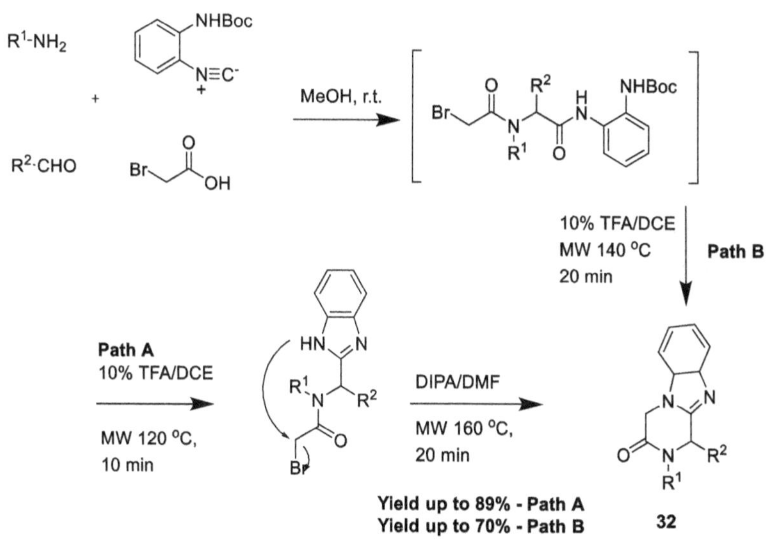

Scheme 16 Synthetic route for the synthesis of fused piperazine-benzimidazole.[38]

4 Sequential Ugi reactions: Four- or three-component coupling reaction

In one of the unique approaches, a sequential Ugi multicomponent/copper-catalyzed Click reaction is discussed by Kappe and coworkers.[39] This continuous flow multistep strategy is used for the synthesis of linear peptoids, leading to macrocyclization followed by click chemistry. This approach avoids exposure of toxic limiting isocyanide derivatives by the dehydration of the corresponding amide using Burgess reagent. The linear peptide (**33**) is synthesized by the combination of an isocyanide containing alkyne, carboxylic acid with azide group, an amine, and an oxo component. The resulting peptides have alkyne and azide at the terminal groups, which undergoes intramolecular copper-mediated azide-alkyne cycloaddition chemistry leading to cyclic peptide (**34**). The reaction resulted in clean compounds in a single process with good to excellent yields. A detailed study revealed that the macrocyclization reaction is quantitatively produced within 25 min at 140°C. Depending upon the ring strain, the dimeric or the monomeric form of the final cyclic product is formed[39] (Schemes 17 and 18).

Synthesis of cyclic heptapeptoids are well known through a series of Ugi reactions and microwave irradiation involving isocyanide derivatives. Isocyanide-based multicomponent reactions (IMCRs) benefited the preparation of cyclic peptides in fewer steps. This efficient reaction uses four successive Ugi reactions, three being four-centered four-component reactions and one set for four-centered three-component reactions. In the reaction, methyl isocyanoacetate (**35**), paraformaldehyde, pentylamine, and N-Cbz-glycine (**36**) in methanol were subjected to microwave irradiation at 80°C to obtain peptoid (**37**) in 88% yield. It was then subjected to hydrolysis reaction with LiOH at 60°C in the presence of THF/H_2O followed by treatment with $NaHSO_4$ providing the corresponding acid (**38**) (Scheme 19). This strategy involves eight steps in which seven were microwave-assisted with less than

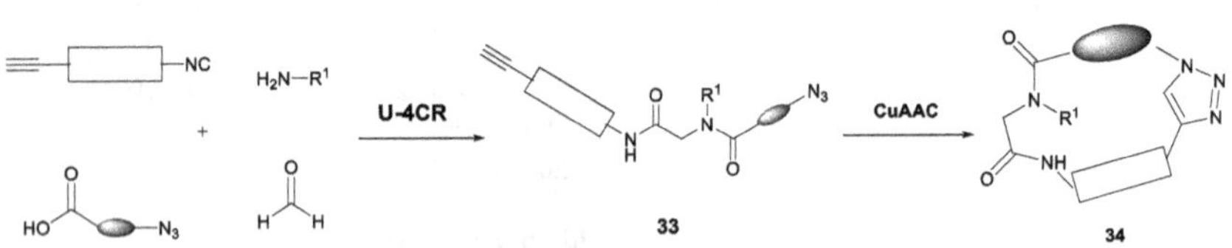

Scheme 17 General synthetic strategy for the sequential synthesis of linear and cyclic peptidomimetic scaffolds.[39]

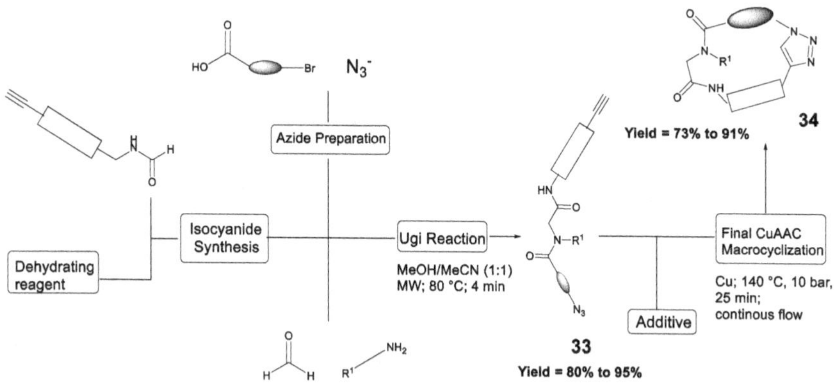

Scheme 18 Continuous strategy for the modular synthesis of linear and cyclic peptoids consisting of isocyanide synthesis, azide preparation, Ugi reaction, and final CuAAC macrocyclization.[39]

5 min reaction time, resulting in good to excellent yields (98%). This study provides an easy methodology to create macroheterocycles with high atom economy coupled with microwave radiation[40] (Scheme 19).

The synthesis of monomeric and dimeric peptide nucleic acid (PNA) by multicomponent reaction microwave-assisted organic reaction (MCMAOR) was reported in solution phase. Functionalized amine, a carboxymethyl nucleobase (**39**), paraformaldehyde, and an isocyanide (**40**) are used as starting building materials to generate PNA monomers (**41**). The monomers are then partially deprotected and sequentially used in second Ugi-4CR, leading to dimers of PNA (**42**). The optimization reactions were carried out on PNA monomer model through various solvents, time, incubation treatment, and microwave power to get better yields. The simplicity, great atom economy, and molecular diversity are some key features of this particular project, which results in various derivatives of PNA in moderate yields[41] (Scheme 20).

Biron et al.[17] utilized the sequential Ugi and intramolecular substitution nucleophilic aromatic (SNAr) reactions to synthesize 1,4-diazepin-3-ones and 1,5-diazocin-4-ones in two steps (**45**). The group used Ugi-4CR to condense readily available *N*-α-Fmoc-amino acids (**43**), amines, and isocyanides with a 2-fluorobenzaldehyde derivative (**44**) followed by a one-pot Fmoc-group removal, intramolecular S_NAr for ring closure, and side-chain deprotection. This strategy resulted in benzo-fused 7- and 8-membered rings bearing a wide variety of functionalized substituents

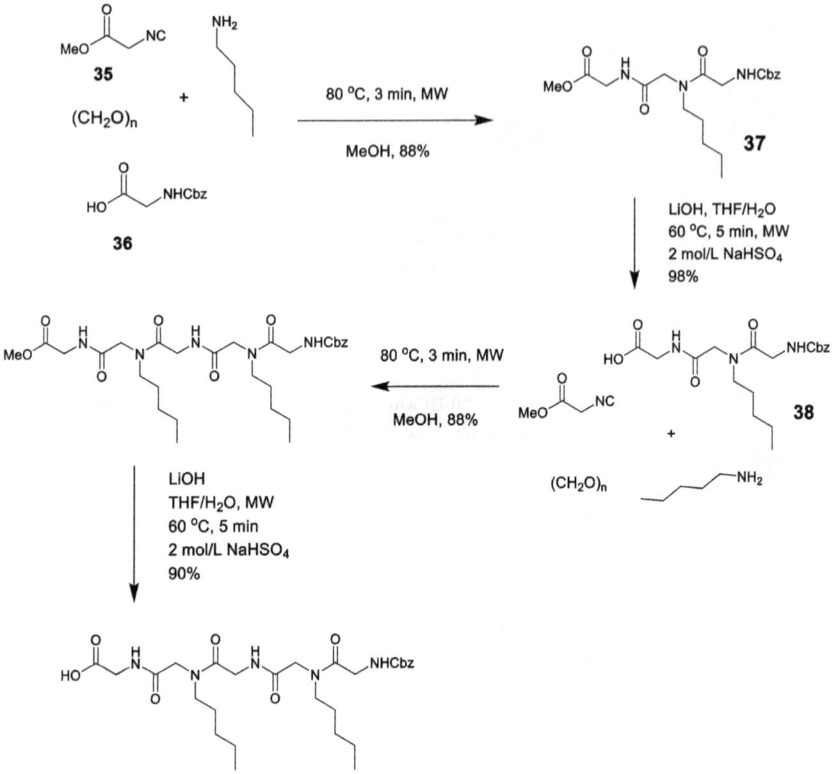

Scheme 19 Synthesis of the intermediate with sequential Ugi reaction with only two steps of Ugi-4CR shown.[40]

leading to the targeted compounds in high yields. The nitro or cyano substituent at the 4-position was necessary to efficiently displace the fluorine and allow cyclization by S_NAr (Scheme 21).

4.1 The Ugi three-component coupling reaction (Ugi-3CC)

In this Ugi coupling reaction, three components are used as the starting materials instead of the standard four reactants.

Using Ugi-3CC reactions combined with microwave energy, potentially active benzimidazole scaffolds, which are of interest in medical research, were synthesized in good yields. Two new derivatives of the fused benzimidazole scaffolds (**47**) have been developed via a one-pot Ugi/deprotection/cyclization (UDC) strategy. This reaction did not use any kind of metal catalyst, and the product was synthesized using inexpensive and commercially available starting materials (2-aminopyridine, Boc-protected

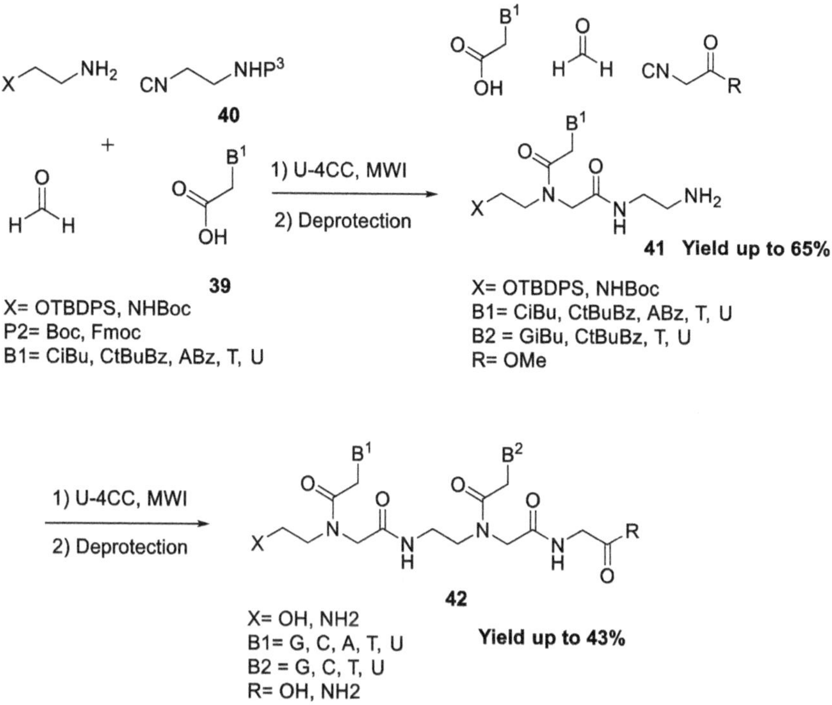

Scheme 20 Synthetic scheme of PNA dimers via two consecutive Ugi-4CC reactions.[41]

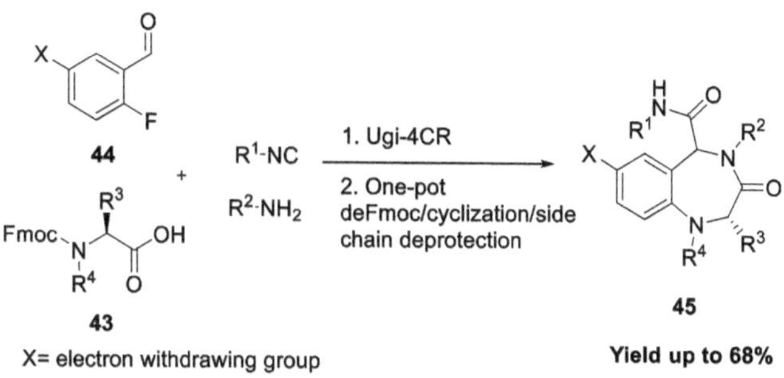

Scheme 21 Synthesis of polysubstituted 1,4-benzodiazepin-3-ones from readily available building blocks by a Ugi/deprotection/cyclization (UDC) approach.[17]

isonitrile, and an aromatic ester aldehyde in the presence of a catalytic amount of $HClO_4$). Optimization of the reaction conditions was carried out for imidazo[1,2-*a*]pyridine (**46**) to obtain the target compounds with a variety of solvents, different temperatures, reaction times, etc. The mechanistic details of the synthesis are also explained in this study[42] (Scheme 22).

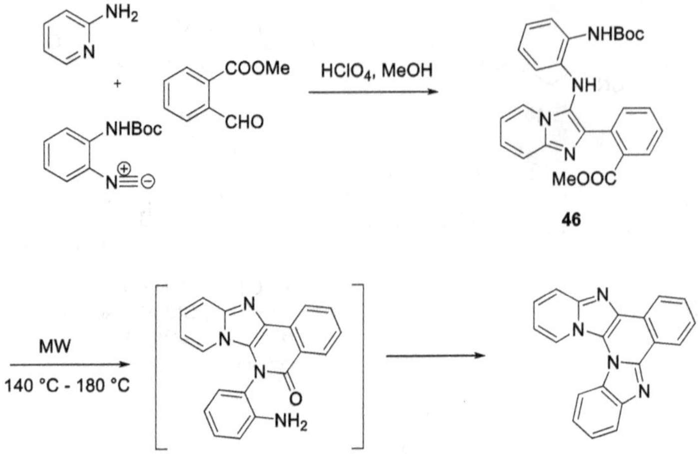

Yield up to 87% **47**

Scheme 22 Optimization of the reaction conditions to obtain target compound.[42]

Through an improved one-pot microwave-assisted cascade reaction, a series of 5-aryl-benzo[*f*][1,7]naphthyridines (**51**) were synthesized in low to moderate (17%–64%) yields. These Ugi three-component reactions/ intramolecular aza-Diels-Alder cycloadditions were coupled through the aromatization process under mild conditions. The starting materials for this reaction include the trifunctional dienophile-containing ester-anilines (**48**), substituted benzaldehydes (**49**), and the chain-ring tautomerizable to 2-isocyano-1-morpholino-3-phenylpropan-1-one (**50**). Further, these Ugi adducts were utilized to synthesize polysubstituted 5-aryl-polyheterocycles (**51**), which are known bioactive molecules, through an in situ aza-Diels-Alder cycloaddition/aromatization (dehydration/oxidation) process[9] (Scheme 23).

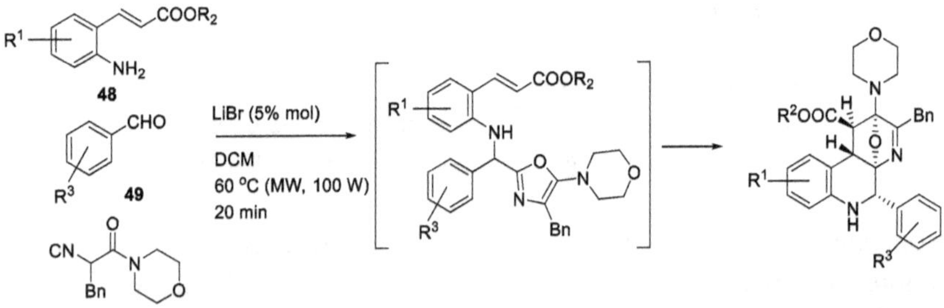

50 Yield up to 17-64 % **51**

Scheme 23 Synthesis of the new oxa-bridged 5-aryl-containing benzo[*f*][1,7] naphthyridines.[9]

4.2 Ugi five-component coupling (Ugi-5CC)

This Ugi coupling component reaction uses five reactants as the starting materials instead of the commonly used four reactants. Huang et al.[43] proposed the one-pot synthesis of 5-hydroxy hydantoins through a five-component (amine, isocyanide, alcohol, and carbon dioxide) Ugi condensation reaction along with an aryl carboxylate at the α-position (**52**). This reaction was carried out via microwave heating under basic conditions and resulted in moderate to good yields (38%–70%). This research provided a direct one-step approach to create synthetically challenging pharmacophores without the isolation of the 5-hydroxy hydantoin amides (**53**) as the products (Scheme 24).

4.3 Post-Ugi modification

Ugi adducts/products are used as the substrate in many substantial reactions such as the Ugi-Heck, the Ugi-Buchwald, or the Ugi-Diels-Alder processes. A selected few (after 2015) post-modifications reactions are discussed in the following text.

The Ugi/reductive Heck sequence was utilized to synthesize derivatives of the 3-benzazepine scaffold via propargylamine substrates. Peshkov et al.[3] used a two-step sequence involving the Ugi reaction followed by reductive Heck cyclization to synthesize a 3-benzazepine framework. The scope of the reaction was studied using different substrate and functional groups to construct a library of benzazepine compounds (**54a–d**). Four different types of three benzodiazepine skeletons were assembled, showing that the overall efficiency of the process significantly depends upon the initial distribution of the required functional groups in the Ugi reaction (Scheme 25).

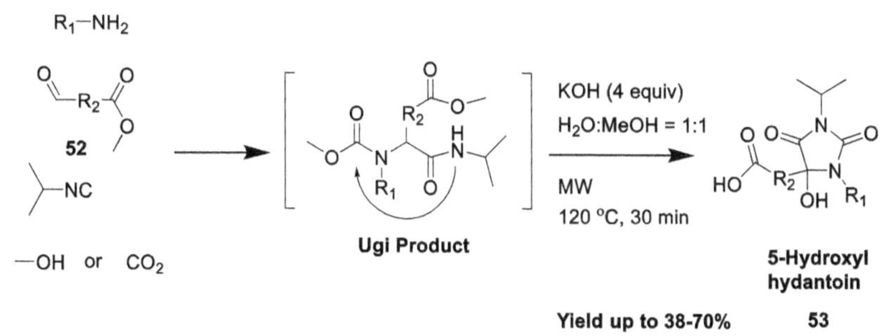

Scheme 24 Proposed synthetic route to 5-hydroxyl hydantoins via Ugi/cyclization approach.[43]

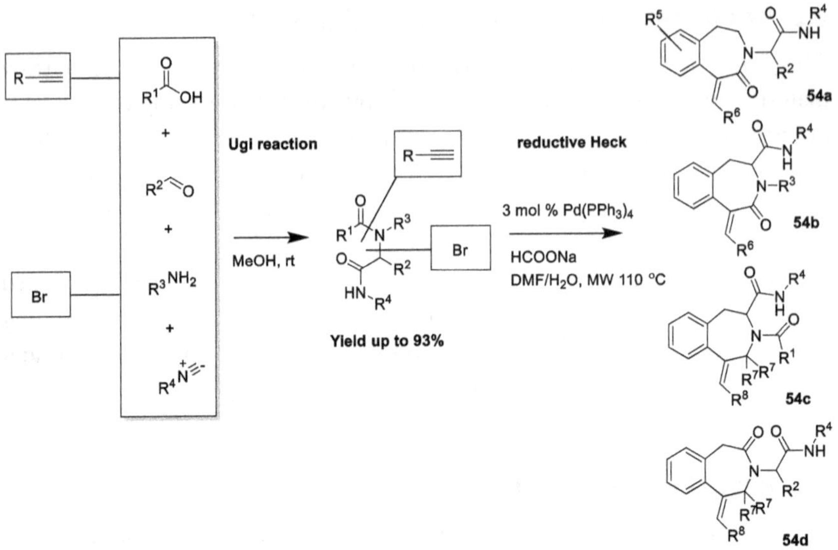

Scheme 25 Synthesis of 3-benzazepines through reductive Heck cyclization.[3]

A one-pot synthetic procedure involving the Ugi reaction followed by tautomerization and an intramolecular nucleophilic aromatic substitution reaction resulted in the benzoxazepine-isoquinoline (**55**) derivatives in excellent yields. The tautomerization reaction was between the keto and enol forms needed to create the fused-ring structure. Xu and coworkers have carried out detailed optimization reaction condition studies to synthesize the final product. This multicomponent reaction has the potential to bring diverse new compounds to drug discovery[44] (Scheme 26).

The diversity-oriented efficient synthesis of 4*H*-benzo[*f*]imidazo[1,4] diazepin-6-ones (**60**) via a post-Ugi copper-catalyzed intramolecular Ullmann coupling reaction is described in this study. The scope of this

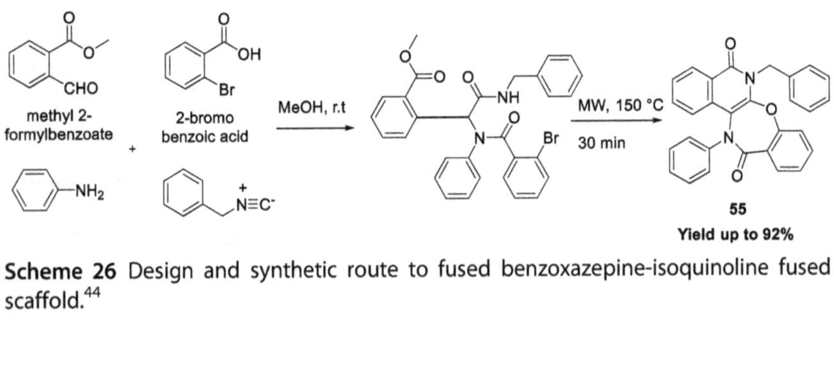

Scheme 26 Design and synthetic route to fused benzoxazepine-isoquinoline fused scaffold.[44]

reaction was tested on imidazole-4-carbaldehyde (**56**), *p*-methoxybenzylamine (**57**), 2-iodobenzoic acid (**58**), and *tert*-butyl isonitrile (**59**). The optimization study was carried out at different times, temperatures, bases, catalyst loadings, and solvents for this reaction under microwave conditions. It was found that the reaction gives the highest yields with 10 mol% of CuI at 100°C[45] (Scheme 27).

A post-Ugi cascade reaction was employed to synthesize fused quinoxaline-quinolinones as reported by Li et al.[46] This strategy includes an Ugi/deprotection/cyclization (UDC) reaction followed by a nucleophilic aromatic substitution reaction. This MAOS under acid-free conditions provides good yields in a short reaction time to generate quinolino [3,4-*b*]quinoxalin-6(5*H*)-ones (**63**), which are of great biological importance[46] (Scheme 28). Derivatives of phenylacetaldehyde, substituted isonitrile, *N*-Boc-protected-phenylenediamine (**61**), and *N*-Boc amino acetic acid (**62**) were used to generate the Ugi adduct.

Isoindolinone derivatives were synthesized via an Ugi reaction for the purpose of being used as biologically active Kv1.5 ion channel blockers. In some cases, the sequential Ugi reaction was followed by Suzuki and Diels-Alder reactions to give a diverse set of compounds. The starting materials for the Ugi reaction include 2-formylbenzoic acids, an amine, and an isocyanide with methanol as the solvent. This reaction is carried out at room temperature and results in a racemic molecule in one step. These racemic mixtures (E1 and E2) were separated using chiral chromatography. The Ugi reaction allowed for a high-throughput synthesis of various analogues, and the compounds of interest were evaluated for biological their activity[8] (Scheme 29).

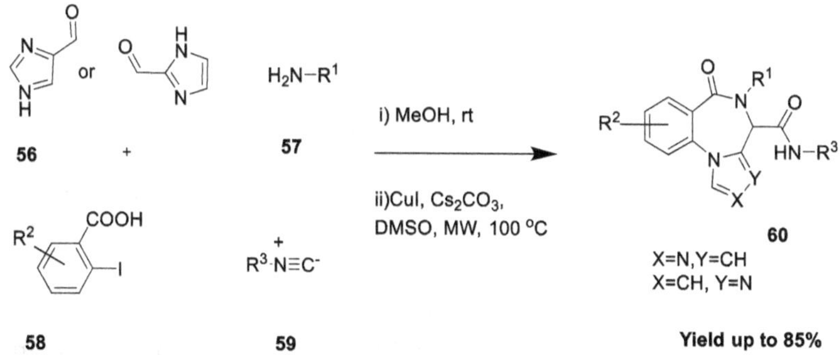

Scheme 27 Synthesis of 4*H*-benzo[*f*]imidazo[1,4]diazepin-6-ones by post-Ugi copper-catalyzed intramolecular Ullmann coupling.[45]

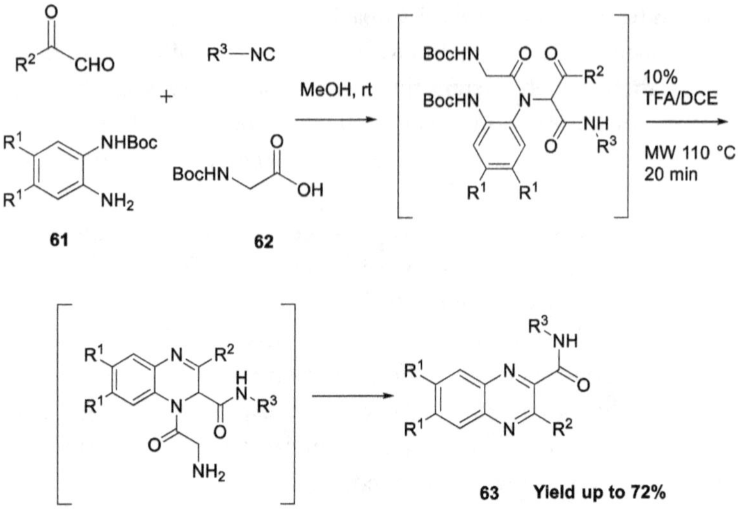

Scheme 28 Scope of the Ugi/deprotection/cyclization strategy for the synthesis of quinolino[3,4-*b*]quinoxalin-6(5*H*)-ones.[46]

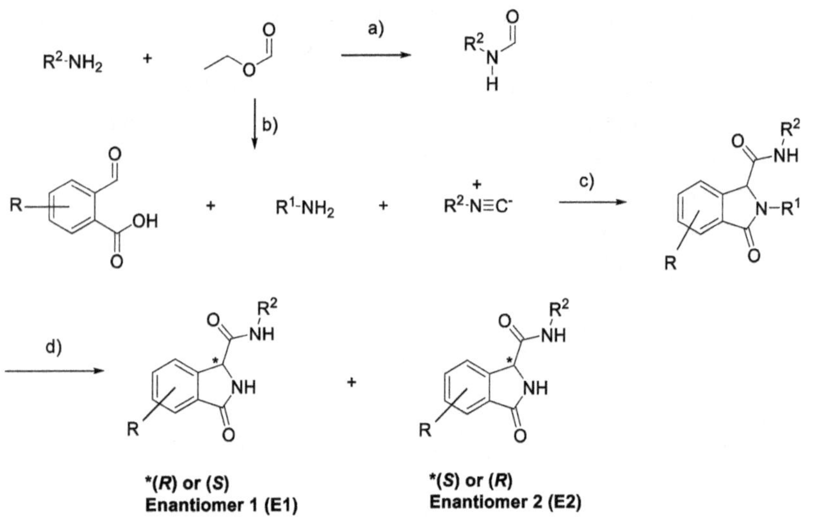

Scheme 29 Reagents and conditions: (A) neat, 100–120°C microwave heating, 20–30 min, 57%–99%; (B) Burgess reagent, CH_3CN, 50°C, 4 h, quantitative; (C) MeOH, rt., overnight, 4%–64%; and (D) chiral separation.[8]

A series of novel imidazotetrazolodiazepinones (**64**) are synthesized via postcondensation modifications of the Ugi-azide reaction assisted by multicomponent microwave irradiation. The reaction proceeds in three steps through rigidification of the $TMSN_3$-Ugi condensation product by treatment with isocyanates to afford hydantoins. The reaction is tolerant to a wide variety of functional groups and generates a diverse set of biologically active compounds. The product consists of fused-ring structures by choice of different isonitrile derivatives and primary amines, resulting in a varied group of novel compounds with good yields, which can be subjected to high throughput synthesis for future drug discovery.[47] (Scheme 30).

To prepare derivatives of antimalarial agents, a diverse group of novel benzimidazole compounds were synthesized via a post-Ugi multicomponent reaction. The methodology utilized an isocyanide-based Ugi reaction followed by an acid-catalyzed condensation cyclization promoted by microwave irradiation. To synthesize the benzimidazole derivatives, pharmacophoric moieties of Boc-protected *o*-phenylenediamine (**65**) is reacted with benzoic acid, benzaldehyde, and *tert*-butyl isocyanide. The results showed that the majority of the synthesized compounds exhibited moderate to good antimalarial activities against chloroquine as a reference compound[4] (Scheme 31).

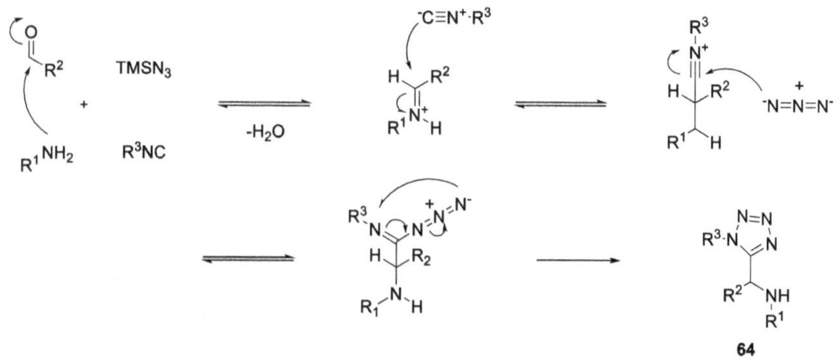

Scheme 30 Mechanism of the Ugi-azide reaction with primary amines.[47]

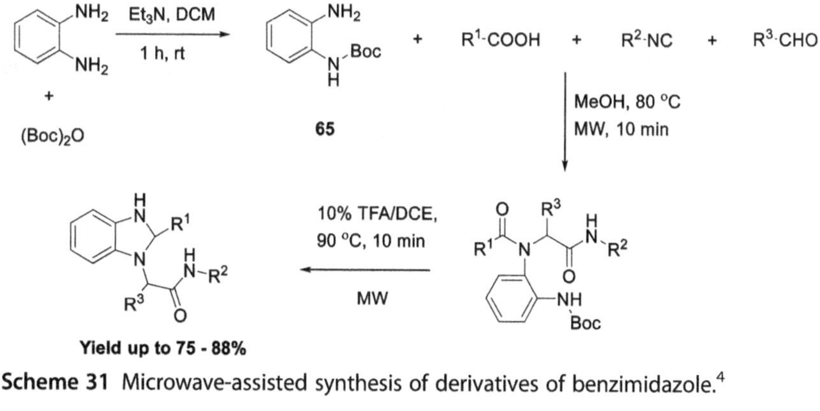

Scheme 31 Microwave-assisted synthesis of derivatives of benzimidazole.[4]

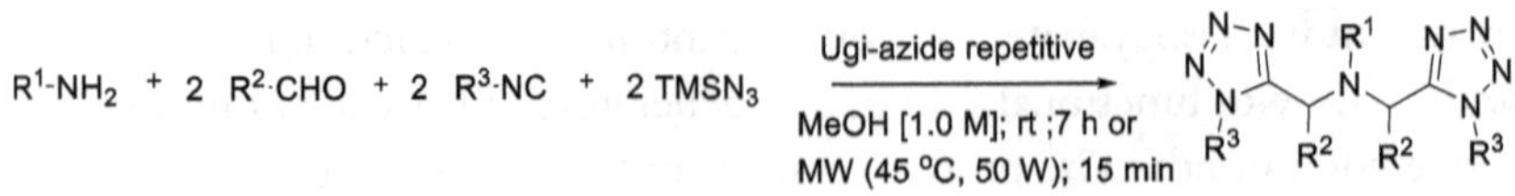

Yield up to 80-90% **66**

Scheme 32 Synthesis of novel *bis*-1,5-disubstituted-1*H*-tetrazoles.[48]

Galindo et al. focused on the synthesis of a series of *bis*-1,5-disubstituted-1*H*-tetrazoles (**66**) through a catalyst-free repetitive Ugi-azide reaction. The MAOS resulted in 80%–90% yields. The reaction was also carried out at room temperature, resulting in good yields[48] (Scheme 32). These tetrazoles are used as excellent chelating agents and organocatalysts. This Ugi-azide repetitive reaction uses various derivatives of amine, aldehydes (2 eq.), isocyanides (2 eq.), and azidotrimethylsilane (2 eq.) in a polar protic solvent to help generate final products.

The one-pot Ugi-indole microwave-assisted cyclization reaction was used to synthesize pyrrolopyridinone ring scaffolds. These derivatives were synthesized using *N*-Boc-pyrrole-2-carbaldehyde (**67**) along with aniline, 2-oxo-2-phenylacetic acid (**68**), and benzyl isocyanide (**69**) as starting materials through an acid-mediated indole cyclization. The targeted pyrrolopyridinone (**70**) was expected, but instead, the Ugi/indole cyclization product was obtained (**71**). The scope of the indole cyclization reaction was optimized using various solvents, acids, bases, temperatures, and reaction times to get maximum yields. The methodology included a nucleophilic substitution reaction and an Ugi/deprotection/cyclization (UDC) strategy, which led to two new heterocyclic fused-ring structures[49] (Scheme 33).

An intramolecular Ullmann diaryl etherification, as the post-Ugi annulation reaction, was utilized for the synthesis of dibenz[*b*,*f*][1,4]oxazepines (**75a** and **b**). This reaction was scaffolded through microwave irradiation in two steps. The Ugi-4CR used 2-aminophenols (**72**) and 2-bromobenzoic acids (**73**), or 2-bromobenzaldehydes (**74**), as the starting material. The MAOS was used by intramolecular Ullmann etherification to result in 6/7/6 fused tricyclic heterocycles. The optimization of this reaction was carried out using a variety of bases, MW times, temperatures, solvents, and ligands. The Ullmann diaryl ether formation was catalyzed by CuI and *N*,*N*-dimethylglycine hydrochloride (DMG·HCl) in the presence of Cs_2CO_3 to yield final products in good to excellent yields (64%–100%)[50] (Scheme 34).

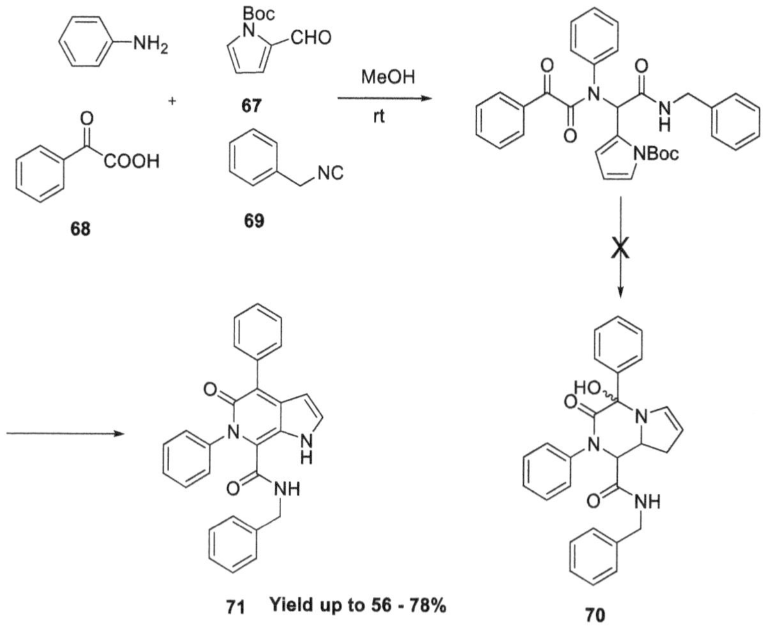

Scheme 33 Synthesis of pyrrolopyridinone ring scaffolds using an Ugi/indole cyclization reaction.[49]

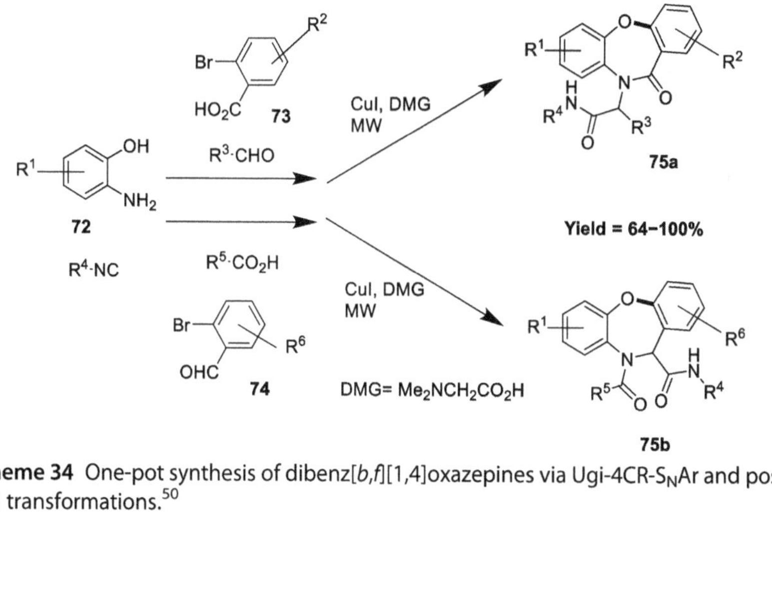

Scheme 34 One-pot synthesis of dibenz[*b*,*f*][1,4]oxazepines via Ugi-4CR-S_NAr and post-Ugi transformations.[50]

Libraries of 1,5-disubstituted tetrazoles (**77**) containing a fragment of the anticancer drug imatinib were synthesized via the Ugi-azide reaction for the time first time using microwave energy. These novel compounds were obtained with a variety of substituents using derivatives of benzaldehyde, isocyanide, trimethylsilyl azide, and imatinib as starting reagents. These compounds were synthesized in moderate to good yields and were assayed for their cytotoxicity. Their anticancer potency was compared with the imatinib (**76**). Computational studies were also carried out for this project to understand the stereoelectronic factors involved for a molecule with multiple basic sites[51] (Scheme 35).

The two paths for the MAOS post-Ugi cascade reaction result in two important heterocyclic compounds, namely, benzimidazopyrazinone (**81**) and quinoxalinone (**82**). The Ugi product was synthesized using bromoacetic acid (**78**), benzaldehyde, Boc-protected phenylenediamines (**79**), and isocyanides. By way of the single Ugi adduct (intermediate, **80**), the order of the reaction can be altered by (1) a nucleophilic substitution cyclization step and (2) a deprotection and cyclization step, resulting in two different products. The nucleophilic substitution reaction was optimized by varying solvents, choice of bases, time, temperatures, and different equivalent amounts of reactants to get a maximum yield of 92% with K_2CO_3 as the base and 10 min as optimized reaction time. These compounds are easily synthesized with clean purification steps in good yields. The bromoacetic acid group in the Ugi adduct provides various cyclization routes through the intramolecular nucleophilic substitution's reaction giving easy access to a library of compounds. The mild conditions allow the opportunity for functionalization at multiple positions in an effort to yield derivatives demonstrating the great benefits of combinatorial synthesis[52] (Scheme 36).

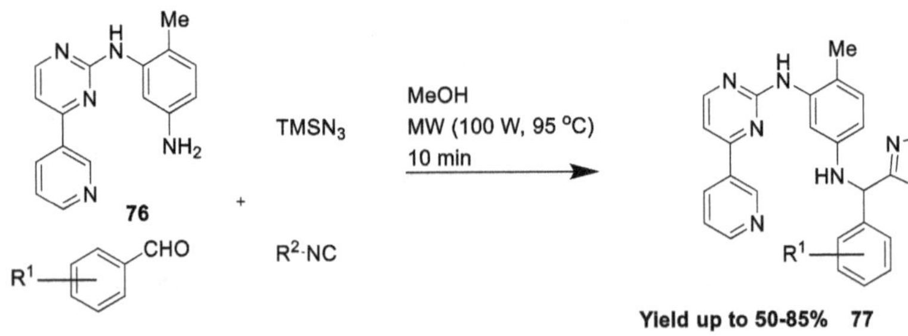

Scheme 35 Synthesis of 1,5-disubstituted tetrazoles containing a fragment of the anticancer drug imatinib.[51]

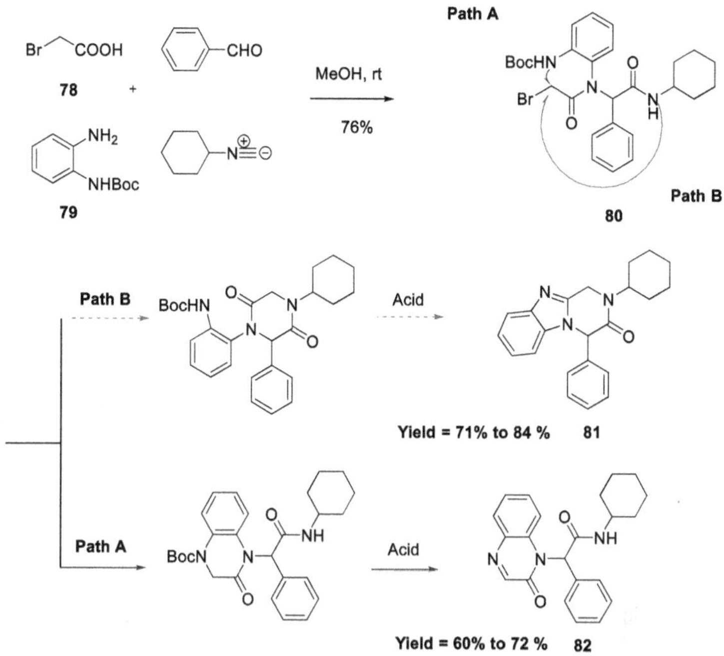

Scheme 36 Design of synthetic routes for quinoxalin-2(1*H*)-ones (**Path A**: (I) K_2CO_3, MeCN, MW, 80°C, 10 min; (II) 10% TFA/DCE; MW; 60°C; 10 min) and benzimidazopyrazinones (**Path B**: (I) 10% TFA/DCE; MW; 60°C; 10 min; (II) K_2CO_3, MeCN, MW, 80°C, 10 min).[52]

The use of the sequential two-step Groebke-Blackburn-Bienaymé (GBB)/Ugi reaction resulted in the feasible and efficient synthesis of novel bis-heterocyclic (imidazo[2,1-*b*][1,3]thiazinyl/benzothiazoyl-phenyl)benzamide scaffolds (**86**). This reaction takes advantage of multiple substitution sites leading to diversity in creation of a wide array of medicinally important compounds. The one-pot microwave-assisted procedure for the synthesis of dipeptide scaffolds used bis-aldehyde (**83**), 2-amino-4*H*-1,3-thiazin-4-one (84), and *N*-*tert*-butyl isonitrile (**85**) for the first GBB reaction step. The addition of aniline and aromatic acids resulted in the sequential Ugi reaction step to give products in good to excellent yields. The room temperature Ugi synthesis did not proceed as expected, while the use of microwave irradiation resulted in the final compounds in excellent yields. The bis-aldehyde used in the reaction was tolerant and did not require the additional protection/deprotection steps[16] (Scheme 37).

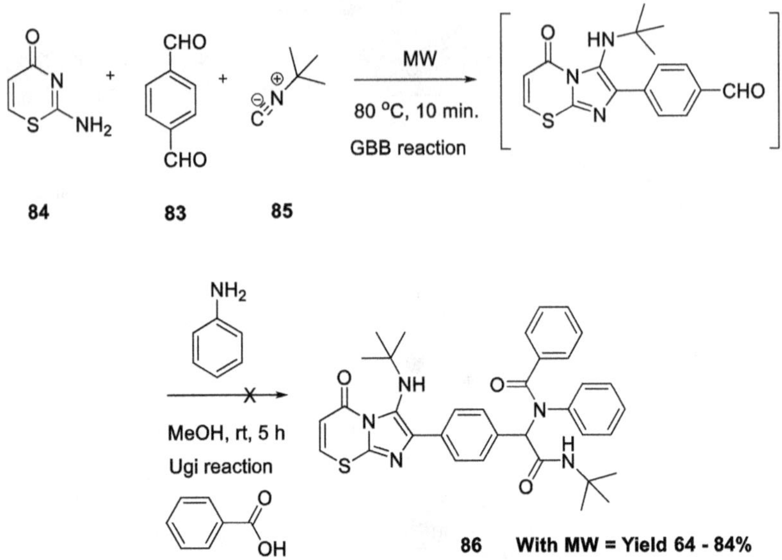

84 83 85

86 With MW = Yield 64 - 84%

Scheme 37 One-pot microwave-assisted procedure for the synthesis of imidazo[2,1-*b*][1,3]thiazinyl-dipeptide scaffolds.[16]

5 Conclusions

The Ugi reaction has found a prominent position in organic synthesis, and its combination with microwave energy has advanced its potential even further. Carrying out quick and facile reactions, avoiding multiple steps, following safety protocols, and, most importantly, providing structural diversity has helped the Ugi reaction to gain its importance, and broadly applied. Both these multicomponent and microwave-assisted reactions have a bright future in organic synthesis research because reactions such as Ugi provides a pathway to generate entire *libraries* of biologically important small-molecule compounds (combinatorial synthesis approach) needed for structure-activity relationship studies.

References

1. Váradi, A.; Palmer, C. T.; Notis Dardashti, R.; Majumdar, S. Isocyanide-Based Multicomponent Reactions for the Synthesis of Heterocycles. *Molecules* **2016**, *21*, 19.
2. Ugi, I. The α-Addition of Immonium Ions and Anions to Isonitriles Accompanied by Secondary Reactions. *Angew. Chem. Int. Ed. Engl.* **1962**, *1*, 8–21.
3. Peshkov, A. A.; Peshkov, V. A.; Pereshivko, O. P.; Van der Eycken, E. V. Diversification of the 3-Benzazepine Scaffold Applying Ugi/Reductive Heck Sequence. *Tetrahedron* **2015**, *71*, 3863–3871.

4. Dandia, A.; Potlia, P.; Khan, I.; Kumar, V.; Verma, V. P.; Karnatak, M.; Dhasmana, A.; Pandey, S.; Shrimali, T.; Parewa, V.; Khan, S.; Singh, J. Diversity-Oriented Synthesis of Novel Benzimidazoles as Antimalarial Agents Via Post Ugi MCR. *Chem. Biol. Interact.* **2019**, *9*, 216–224.
5. Kalinski, C.; Umkehrer, M.; Schmidt, J.; Ross, G.; Kolb, J.; Burdack, C.; Hiller, W.; Hoffmann, S. D. A Novel One-Pot Synthesis of Highly Diverse Indole Scaffolds by the Ugi/Heck Reaction. *Tetrahedron Lett.* **2006**, *47*, 4683–4686.
6. Gracias, V.; Moore, J. D.; Djuric, S. W. Sequential Ugi/Heck Cyclization Strategies for the Facile Construction of Highly Functionalized N-Heterocyclic Scaffolds. *Tetrahedron Lett.* **2004**, *45*, 417–420.
7. Islas-Jacome, A.; Gonzalez-Zamora, E.; Gamez-Montano, R. A Short Microwave-Assisted Synthesis of Tetrahydroisoquinolin-Pyrrolopyridinones by a Triple Process: Ugi-3CR-Aza Diels-Alder/S-Oxidation/Pummerer. *Tetrahedron Lett.* **2011**, *52*, 5245–5248.
8. Kajanus, J.; Jacobson, I.; Åstrand, A.; Olsson, R. I.; Gran, U.; Björe, A.; Fjellström, O.; Davidsson, Ö.; Emtenäs, H.; Dahlén, A.; Löfberg, B.; Yuan, Z.-Q.; Sundell, J.; Cassel, J.; Gyll, J.; Iliefski, T.; Högberg, Å.; Lindhardt, E.; Malmberg, J. Isoindolinone compounds active as Kv1.5 blockers identified using a multicomponent reaction approach. *Bioorg. Med. Chem. Lett.* **2016**, *26*, 2023–2029.
9. Vázquez-Vera, Ó.; Segura-Olvera, D.; Rincón-Guevara, A. M.; Gutiérrez-Carrillo, A.; García-Sánchez, A. M.; Ibarra, A. I.; Lomas-Romero, L.; Islas-Jácome, A.; González-Zamora, E. Synthesis of New 5-Aryl-Benzo[*f*][1,7]Naphthyridines Via a Cascade Process (Ugi-3CR/Intramolecular Aza-Diels-Alder Cycloaddition)/Aromatization. *Molecules* **2018**, *23*, 2039.
10. Yang, B.; Zhao, Y.; Fu, C.; Zhu, C.; Zhang, Y.; Wang, S.; Wei, Y.; Tao, L. Introducing the Ugi Reaction into Polymer Chemistry as a Green Click Reaction to Prepare Middle-Functional Block Copolymers. *Poly. Chem.* **2014**, *5*, 2704–2708.
11. Akrami, S.; Firoozpour, L.; Goli-Garmroodi, F.; Moghimi, S.; Mahdavi, M.; Zonouzi, A.; Foroumadi, A. One-Pot Synthesis of Oxoisoindoline-1,2,3-Triazole Hybrid by a Ugi–Click Reaction. *Synth. Commun.* **2016**, *46*, 1708–1712.
12. Zakharova, E. A.; Shmatova, O. I.; Kutovaya, I. V.; Khrustalev, V. N.; Nenajdenko, V. G. Synthesis of Macrocyclic Peptidomimetics Via the Ugi-Click-Strategy. *Org. Biomol. Chem.* **2019**, *17*, 3433–3445.
13. Sharma, N.; Li, Z.; Sharma, U. K.; Van der Eycken, E. V. Facile Access to Functionalized Spiro[Indoline-3,2′-Pyrrole]-2,5′-Diones Via Post-Ugi Domino Buchwald–Hartwig/Michael Reaction. *Org. Lett.* **2014**, *16*, 3884–3887.
14. Bonnaterre, F.; Bois-Choussy, M.; Zhu, J. Rapid Access to Oxindoles by the Combined Use of an Ugi Four-Component Reaction and a Microwave-Assisted Intramolecular Buchwald-Hartwig Amidation Reaction. *Org. Lett.* **2006**, *8*, 4351–4354.
15. Echemendía, R.; Rabêlo, W. F.; López, E. R.; Coro, J.; Suárez, M.; Paixão, M. W.; Rivera, D. G. A Bidirectional Access to Novel Thiadiazine Hybrid Molecules by Double Multicomponent Reactions. *Tetrahedron Lett.* **2018**, *59*, 4050–4053.
16. Kaur, T.; Gautam, R. N.; Sharma, A. Assembly of New Heterocycles Through an Effective Use of Bisaldehydes by Using a Sequential GBB/Ugi Reaction. *Chem. Asian J.* **2016**, *11*, 2938–2945.
17. Vézina-Dawod, S.; Gerber, N.; Liang, X.; Biron, E. Convenient Two-Step Synthesis of Highly Functionalized Benzo-Fused 1,4-Diazepin-3-Ones and 1,5-Diazocin-4-Ones by Sequential Ugi and Intramolecular SNAr Reactions. *Tetrahedron* **2017**, *73*, 6347–6355.
18. Masquelin, T.; Bui, H.; Brickley, B.; Stephenson, G.; Schwerkoske, J.; Hulme, C. Sequential Ugi/Strecker Reactions Via Microwave Assisted Organic Synthesis: Novel 3-Center-4-Component and 3-Center-5-Component Multi-Component Reactions. *Tetrahedron Lett.* **2006**, *47*, 2989–2991.

19. Chavez-Acevedo, L.; Miranda, L. Synthesis of Novel Tryptamine-Based Macrocycles Using an Ugi 4-CR/Microwave Assisted Click-Cycloaddition Reaction Protocol. *Org. Biomol. Chem.* **2015**, *13*, 4408–4412.
20. Iacobucci, C.; Reale, S.; Gal, J.-F.; De Angelis, F. Insight into the Mechanisms of the Multicomponent Ugi and Ugi–Smiles Reactions by ESI-MS(/MS). *Eur. J. Org. Chem.* **2014**, *2014*, 7087–7090.
21. Jaffett, V. A.; Nerurkar, A.; Cao, X.; Guzei, I. A.; Golden, J. E. Telescoped Synthesis of C3-Functionalized (E)-Arylamidines Using Ugi–Mumm and Regiospecific Quinazolinone Rearrangements. *Org. Biomol. Chem.* **2019**, *17*, 3118–3128.
22. Thangaraj, M.; Gengan, R. M.; Ranjan, B.; Muthusamy, R. Synthesis, Molecular Docking, Antimicrobial, Antioxidant and Toxicity Assessment of Quinoline Peptides. *J. Photochem. Photobiol. B Biol.* **2018**, *178*, 287–295.
23. Jobin, S.; Méjean, A.; Galindo, S.-M.; Liang, X.; Vézina-Dawod, S.; Biron, E. Toward Solid-Phase Peptide Fragment Ligation by a Traceless-Ugi Multicomponent Reaction Approach. *Org. Biomol. Chem.* **2016**, *14*, 11230–11237.
24. Hügel, M. H. Microwave Multicomponent Synthesis. *Molecules* **2009**, *14*, 4936–4972.
25. Jiang, B.; Shi, F.; Tu, S.-J. Microwave-Assisted Multicomponent Reactions in the Heterocyclic Chemistry. *Curr. Org. Chem.* **2010**, *14*, 357–378.
26. Dalvi, P. B.; Lin, S.-F.; Paike, V.; Sun, C.-M. Microwave-Assisted Multicomponent Synthesis of Dihydroquinoxalinones on Soluble Polymer Support. *ACS Comb. Sci.* **2015**, *17*, 421–425.
27. Mohanram, I.; Meshram, J.; Shaikh, A.; Kandpal, B. Microwave-Assisted One-Pot Synthesis of Bioactive UGI-4CR Using Fluorite as Benign and Heterogeneous Catalyst. *Synth. Commun.* **2013**, *43*, 3322–3328.
28. Borah, P.; Borah, J.; Chowdhury, P. Microwave (MW) Irradiated Ugi Four-Component Reaction (Ugi-4CR): Expedited Synthesis of Steroid-Amino Acid Conjugates—A Novel Class of Hybrid Compounds. *Steroids* **2015**, *98*, 49–57.
29. Ovadia, R.; Mondielli, C.; Vasseur, J.; et al. Contribution to PNA-RNA Chimera Synthesis: One-Pot Microwave-Assisted Ugi Reaction to Obtain Dimeric Building Blocks. *Eur. J. Org. Chem.* **2017**, 469–475.
30. Jobin, S.; Beaumont, C.; Biron, E. Development of a Solid-Phase Traceless-Ugi Multicomponent Reaction for Backbone Anchoring and Cyclic Peptide Synthesis. *Pept. Sci.* **2018**, *111*, e24044.
31. Borah, P.; Chowdhury, P.; Begum, A. Microwave Energy in Ugi Four-Component Reaction (Ugi-4CR): Expedient Synthesis of Steroid-Peptide Conjugates Based on Aminosteroids—An Important Class of Hybrid Systems. *Curr. Org. Synth.* **2017**, *14*, 120–126.
32. Borah, D. P.; Chowdhury, D. P. Steroid Hybrid Systems: New Molecular Entities With Potential Therapeutical Spectrum. *Curr. Drug Ther.* **2017**, *12*, 3–22.
33. de Alemán-Ponce León, D.; Sánchez-Chávez, A. C.; Polindara-García, L. A. Pd-Mediated γ-C(sp3)–H Bond Activation in Ammonia–Ugi 4-CR Adducts by Using Picolinamide as Directing Group. *J. Org. Chem.* **2019**, *84*, 12809–12834.
34. Previdi, D.; Rodrigues, S.; Coelho, M.; Candido, A.; Magalhães, L.; Donate, P. Synthesis and Antileishmanial Activity of Some Functionalized Peptoids. *J. Braz. Chem. Soc.* **2019**, *30*.
35. Al Otaibi, A.; Deane, F. M.; Russell, C. C.; Hizartzidis, L.; McCluskey, S. N.; Sakoff, J. A.; McCluskey, A. A Methanol and Protic Ionic Liquid Ugi Multicomponent Reaction Path to Cytotoxic α-Phenylacetamido Amides. *RSC Adv.* **2019**, *9*, 7652–7663.
36. Chen, Y.-W.; Liao, P.-C.; Zhang, Y.-X.; Yeh, S.-Y.; Wu, Y.-H.; Qiu, S.-B.; Tsai, P.-N.; Xin, Z.; Ting, Y.-Y.; Chen, H.-C.; Cheung, S.-F.; Hsu, C.-Y.; Lien, W.-H.; Pan, P.-S. Synthesis of Multiple Boron-Containing Analogs Via Ugi-4CR. *Res. Chem. Intermed.* **2019**, *45*, 103–118.

37. Hsiao, C.; Huang, S.; Lien, W.; Hsu, C.; Hsieh, K.; Lin, M.; Wu, M.; Fu, C.; Chen, H.; Fang, H.; Li, C.; Pan, P. Synthesis of Ugi-4CR Boronate Analogues Via Microwave Irradiation. *Res. Chem. Intermed.* **2017**, *43*, 3585–3597.
38. Song, G.-T.; Li, S.-Q.; Yang, Z.-W.; Yuan, J.-H.; Wang, M.-S.; Zhu, J.; Chen, Z.-Z.; Xu, Z.-G. Microwave-Assisted Synthesis of Fused Piperazine-Benzimidazoles Via a Facile, One-Pot Procedure. *Tetrahedron Lett.* **2015**, *56*, 4616–4618.
39. Salvador, C. E. M.; Pieber, B.; Neu, P. M.; Torvisco, A.; Andrade, C. K. Z.; Kappe, C. O. A Sequential Ugi Multicomponent/Cu-Catalyzed Azide–Alkyne Cycloaddition Approach for the Continuous Flow Generation of Cyclic Peptoids. *J. Org. Chem.* **2015**, *80*, 4590–4602.
40. Barreto, A.; Andrade, C. Microwave-Mediated Synthesis of a Cyclic Heptapeptoid Through Consecutive Ugi Reactions. *Tetrahedron* **2018**, *74*, 6861–6865.
41. Ovadia, R.; Lebrun, A.; Barvik, I.; Vasseur, J.-J.; Baraguey, C.; Alvarez, K. Synthesis and Structural Characterization of Monomeric and Dimeric Peptide Nucleic Acids Prepared by Using Microwave-Promoted Multicomponent Reactions. *Org. Biomol. Chem.* **2015**, *13*, 11052–11071.
42. Song, G.; Xu, Z.; Tang, D.; Li, S.; Xie, Z.; Zhong, H.; Yang, Z.; Zhu, J.; Zhang, J.; Chen, Z. Facile Microwave-Assisted Synthesis of Benzimidazole Scaffolds Via Ugi-Type Three-Component Condensation (3CC) Reactions. *Mol. Divers.* **2016**, *20*, 575–580.
43. Huang, A.; Wo, K.; Pierson, A.; Lee, S.; Yang, E.; Johnson, C. 5-Hydroxyl Hydantoins Via One-Pot Microwave-Assisted Air Oxidation of Ugi Products. *J. Heterocyclic Chem.* **2016**, *53*, 1499–1504.
44. Xu, J.; Li, Y.; Meng, J.; Lei, J.; Chen, Z.; Tang, D.; Zhu, J.; Xu, Z. Efficient Microwave-Assisted Synthesis of Fused Benzoxazepine-Isoquinoline Derivatives Via an Ugi Reaction/Tautomerization/Intramolecular SNAr Reaction Sequence. *Tetrahedron Lett.* **2017**, *58*, 1640–1643.
45. Li, Z.; Legras, L.; Kumar, A.; Vachhani, D.; Sharma, S.; Parmar, V.; Van der Eycken, E. Microwave-Assisted Synthesis of 4H-Benzo[*f*]Imidazo[1,4]Diazepin-6-Ones Via a Post-Ugi Copper-Catalyzed Intramolecular Ullmann Coupling. *Tetrahedron Lett.* **2014**, *55*, 2070–2074.
46. Li, Y.; Lei, J.; Xu, J.; Tang, D.-Y.; Chen, Z.-Z.; Zhu, J.; Xu, C. A Facile Method for Building Fused Quinoxaline-Quinolinones Via an Acidless Post-Ugi Cascade Reaction. *Chin. Chem. Lett.* **2017**, *28*, 541–545.
47. Medda, F.; Martinez Ariza, G.; Hulme, C. A Facile and Concise Route toward the Synthesis of Novel Imidazo-Tetrazolodiazepinones Via Post-Condensation Modifications of the Ugi-Azide Adduct. *Tetrahedron Lett.* **2015**, *56*, 5295–5298.
48. Cárdenas-Galindo, E. L.; Islas-Jácome, A.; Colmenero-Martínez, M. K.; Martínez-Richa, A.; Gámez-Montaño, R. Synthesis of Novel Bis-1,5-Disubstituted-1H-Tetrazoles by an Efficient Catalyst-Free Ugi-Azide Repetitive Process. *Molecules* **2015**, *20*, 1519–1526.
49. Li, Y.; Lei, J.; Chen, Z.; Tang, D.; Yuan, H.; Wang, M.; Zhu, J.; Xu, Z. Microwave-Assisted Construction of Pyrrolopyridinone Ring Systems by Using an Ugi/Indole Cyclization Reaction. *Eur. J. Org. Chem.* **2016**, 5770–5776.
50. Shi, J.; Wu, J.; Cui, C.; Dai, W. Microwave-Assisted Intramolecular Ullmann Diaryl Etherification as the Post-Ugi Annulation for Generation of Dibenz[*b,f*][1,4]oxazepine Scaffold. *J. Org. Chem.* **2016**, *81*, 10392–10403.
51. Cortes-Garcia, C.; Islas-Jacome, A.; Renteria-Gomez, A.; Gamez-Montano, R. Synthesis of 1,5-Disubstituted Tetrazoles Containing a Fragment of the Anticancer Drug Imatinib Via a Microwave-Assisted Ugi-Azide Reaction. *Monatsh. Chem.* **2016**, *147*, 1277–1290.
52. Song, G.; Li, Y.; Xu, J.; Xu, Z.; Ding, Y.; Lei, J.; McConnell, N.; Zhu, J.; Chen, Z. - Microwave-Assisted Facile Construction of Quinoxalinone and Benzimidazopyrazinone Derivatives Via Two Paths of Post-Ugi Cascade Reaction. *Mol. Divers.* **2019**, *23*, 137–145.

CHAPTER FOUR

Microwave-assisted flow systems in the green production of fine chemicals

Tara Mooney and Béla Török
Department of Chemistry, University of Massachusetts Boston, Boston, MA, United States

1 Introduction

Due to the numerous benefits that microwave-assisted organic synthesis (MAOS)[1–4] offers, it has become a mainstream activation method in organic synthesis and is often considered in green or environmentally benign applications.[5–7] The first use of microwave heating in synthesis dates back to 1969 when it was applied for the aqueous emulsion polymerization of acrylic acid, methacrylic acid, and butyl acrylate.[8] However, the widespread extension on the use of this technique and the de facto establishment of microwave-assisted chemical synthesis started with the publications by Gedye et al.[9] and Giguere et al.[10] in 1986. It was not long after these first reports that microwave activation was coupled to flow synthesis. The early efforts on the development of a continuous microwave reactor (CMR) focused on laboratory-scale applications.[11–13] The first microwave-assisted flow systems were a significant step toward scaled-up synthesis; many microwave-assisted reactions use heterogeneous catalysts, which are more suitable for continuous flow processes that are commonly used in large-scale industrial practice.[14–17] While microwave synthesis is also suitable for small batch sizes, its use for scale-up batch synthesis is limited due to the short penetration depth of the microwaves.[18] Scale-up efforts were made in both batch and flow reactor systems.[19–21] While the improvements in the batch systems are considerable,[22, 23] the reactor size and the produced amount are still far below most industrial requirements. As the majority of industrial processes apply continuous flow systems scale-up efforts were mostly focused in this direction. In addition to industrial applicability, practical considerations also play a significant role; the design of tubular reactors of low diameter can take into account the low penetration depth of the microwave

Nontraditional Activation Methods in Green and Sustainable Applications
https://doi.org/10.1016/B978-0-12-819009-8.00015-3
© 2021 Elsevier Inc.
All rights reserved.

irradiation.[24–27] This advantage was highlighted by Damm et al. in comparative studies of three typical synthetic reactions using both large size batch reactors and continuous flow system.[28] In addition to the obvious advantages in reactor design, most solid catalysts applied in these processes are excellent microwave absorbers ensuring uniform internal heating upon microwave irradiations.[29–36] Over the past two decades, microwave-based flow chemistry became a separate research field on its own. Several reviews have been published regarding instrumentation,[37] heterogeneous gas-phase reactions,[38] micro and meso-fluidic systems,[39] enzymatic reactions, and nanoparticle synthesis.[40]

In this chapter, recent advances made in microwave-assisted flow synthesis will be described including the major types of reactors that have been developed for various protocols in organic and inorganic synthesis, environmental chemistry, and other industrial-scale applications.

2 Microwave reactors

Since 1947, the International Telecommunication Union has regulated the use of the electromagnetic spectrum used for telecommunications in order to avoid interference among different devices.[41] As part of this early regulation, microwaves used for heating were designated the frequencies between 2.4 and 2.5 GHz, which results in an average wavelength of roughly 12 cm. This wavelength puts limitations on the size of microwave reactors. In the decades since their use for microwave-assisted chemical synthesis, microwave reactors have become more sophisticated to overcome the limitations of the designated wavelength and frequency. There are three major components of the reactor that should be considered in order to optimize efficiency: the oscillator, the cavity, and the reaction vessel itself.

The *oscillator* is responsible for producing the microwaves in the 2.4–2.5 GHz region of the electromagnetic spectrum. The two most common oscillators used for the generation of microwaves for heating are the magnetron or solid-state device. The general descriptions and major differences between the two oscillators are described in this section. (i) The magnetron[42] will produce the full array of frequencies between 2.4 and 2.5 GHz emitted at various orientations, or modes, that travel through the waveguide that transfers them to the reactor cavity. The multifrequencies with random directionalities in the reactor are referred to as multimode. The output power of the magnetron will be the sum of the output power of each frequency released by the magnetron. It can be tuned to a single-frequency traveling in a specific locale using a waveguide that has been made to select for a specific frequency. The single frequency applied in the reactor is referred to as single-mode, or mono-mode. While the output power of a single frequency is less, it is concentrated in a specific area so that the total absorbed energy is comparable to the multimode device. (ii) Solid-state devices[43] are semiconductor transistors that emit a single, specified frequency. All solid-state devices are therefore single-mode. The advantage of the solid-state device is that as the temperature of the reaction medium increases resulting in changing its frequency due to the dielectric loss of the medium, the solid-state device can be automatically retuned to resonate with the reaction medium frequency as it changes. While magnetrons can be automatically tuned to assigned frequencies using inductive tunable elements, the solid-state device has superior power and frequency control.[42]

The *cavity*[42] of the reactor is the space where the reaction vessel is located. In this space, the microwaves will pass through the vessel and be reflected off the sides of the cavity. There are two major types of microwave cavities, multimode and single mode, which refers to the frequency and spatial orientation of the waves. There are two subtypes of single-mode cavities differentiated by how the waves flow through the vessel, traveling vs standing. (i) All multimode cavities[44] are coupled with a magnetron. The resulting broadband of microwaves is randomly dispersed throughout the cavity. This results in uneven heating, "hot spots" of constructive interference and "cold spots" of destructive interference. This is usually mitigated by mixing of the reaction medium. Cavity size is not limited in multimode and these devices are generally larger. Domestic microwave ovens are the simplest multimode devices. Many commonly used "homemade" microwave-assisted flow-reactors are modified domestic ovens. (ii) Single-mode cavities are based on either traveling or standing waves. In a single-mode traveling wave reactor, also called nonresonant,[45] the direction of the traveling wave is perpendicular to the reaction vessel. The traveling wave flows through the reaction medium, is reflected off the cavity wall, passes through the reaction medium once more, and is then reabsorbed by the oscillator. Due to the oscillatory motion of the traveling wave, the cavity or vessel size is not dependent on the wavelength. Vessels of various sizes and shapes can be used. In a single-mode standing wave reactor (resonant),[42, 45] the microwaves are focused by an iris so that the maxima of the waves resonate at a specific location. This maxima locale specificity limits the size and shape of the cavity and vessel that one can use with a single-mode standing wave reactor due to the path length of the wave. Single-mode reactors, particularly standing wave, are more suitable for smaller reaction vessels due to the short path length. Single-mode standing waves result in the most rapid heating due to the concentration and resonance at a fixed locale of the reaction vessel. This can lead to runaway heating if the power is not controlled. As the temperature of the medium changes, the resonance in the cavity changes due to the dielectric properties of the substance. Only the solid-state devices allow for auto-tuning during the flow synthesis process, while the magnetron coupled with a waveguide will lose the efficiency provided by resonance. This also means that the machine's frequency must be initially tuned specifically with the medium's dielectric properties in mind.

The *reaction vessel* is the only easily modifiable part of a microwave reactor. One can exchange vessels of different sizes, shapes, and materials to optimize any reaction. In this subsection, the effects of size, shape, and material will be discussed.

(i) The penetration depth of microwaves limits the possible diameter of the vessel; however, the size and shape can be varied in other ways.[46] During the optimization process, the residence time (time that the reaction mixture spends under microwaves as it flows through the vessel) and flow rate are often determined by reaction kinetics. In order to maximize residence time under the microwave, one can choose tubes of longer lengths, U-shaped tubes, or most commonly, helical tubes. The shape of the tube can also influence mixing. If the reaction occurs more efficiently with mixing, a tube with a bend could be considered. Straight tubes result in a laminar flow, with no mixing, unless the flow rate is increased to the point that the flow becomes turbulent. This would require a balance between residence time and flow rate, and one may be sacrificing time under microwave to increase mixing. Helical and U-shaped tubes, with a bend, afford mixing naturally. Fig. 1 depicts the natural mixing that occurs in the laminar flow of a U-shaped tube.[46] In addition to shape, tubes can be created to have striations, ridges, etching, etc. to promote turbulence.

(ii) The material of the vessel[42] is another way to develop a broad variety of applications in microwave-assisted flow chemistry. Most microwave reactor vessels are made of microwave transparent materials such as borosilicate, quartz, or PTFE. In these cases, the reaction mixture itself; solvent, reactants or catalyst, will be the target for absorbing the microwaves. Alternatively, the vessel can be made of microwave absorbing material such as silicon carbide or include a metal coating and thus would be responsible for

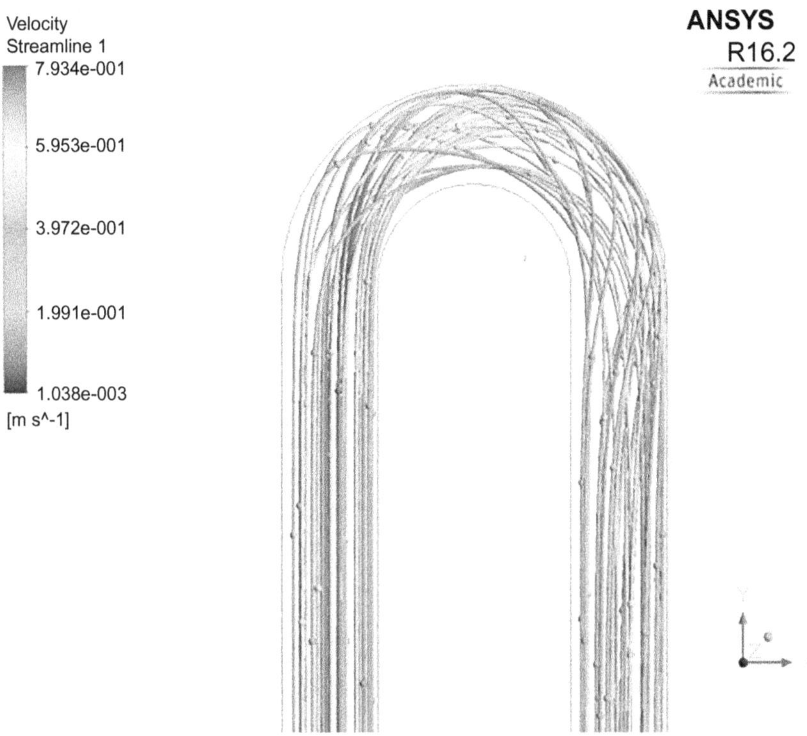

Fig. 1 Natural mixing of laminar flow in a U-bend tube. *(Reproduced from Estel, L.; Poux, M.; Benamara, N.; Polaert, I. Continuous Flow-Microwave Reactor: Where Are We?* Chem. Eng. Proc. ***2017****,* 113*, 56–64, with permission of the corresponding author, Dr. Lionel Estel.)*

transforming the microwave energy to heat. In this case, the reaction medium will be heated by convection rather than the medium itself, which can cause uneven heating unless accompanied by effective mixing (e.g., turbulence) since the fluid closer to the vessel walls will be heated more than the middle of the flow.

3 Applications of microwave-assisted flow systems

3.1 Organic synthesis

Similar to batch reactor-based static MAOS, the majority of applications for microwave-assisted flow chemistry have also been developed in the field of organic synthesis. Several solvents, organic reagents, and catalysts that promote these reactions are considered excellent microwave absorbers, which certainly contribute to the popularity of these applications. The most common synthetic uses are summarized below based on the traditional reaction categories.

3.1.1 Esterification

In 2018, Balint et al.[47] optimized the catalyst-free alcoholysis of dialkyl-*H*-phosphonates using a reactor with a magnetron oscillator that produces a single-mode standing wave. The reactor was fitted with a straight, 7-mL borosilicate glass flow vessel. Their aim was to optimize the power, temperature, flow rate, and residence time of the reaction in the microwave-assisted flow system. In this reaction mixture, the alcohol was the microwave target as the vessel was transparent to microwaves. Scheme 1 shows the optimal reaction conditions for both the single (A) and the double transesterification (B). While the yield of the single transesterification seems low, the conversion under the optimized conditions favored the single transesterification of dimethyl *H*-phosphonate with *n*-pentanol at 55%–13% of the double esterification, and the authors were able to isolate most of the single transesterification product (51%) after column chromatography.

A follow-up work by the same group to the above alcoholysis of the phosphonates describes a continuous flow method optimization for the catalyst-free synthesis of α-aryl-α-aminophosphonates using the same reactor system and vessel.[48] To avoid transesterification of the phosphonate, the authors performed the reaction in alcohol that had the same aliphatic group as the phosphonate ester. They were successful in the optimization for both the Kabachnik-Fields condensation shown in Scheme 2 as well as the Aza-Pudovik reaction shown in Scheme 3. Both reaction types afforded 100% conversion for certain substituents, and isolated yields over 90%. It is worth mentioning that both reactions occurred without a catalyst which adds to the green advantages of the protocol.

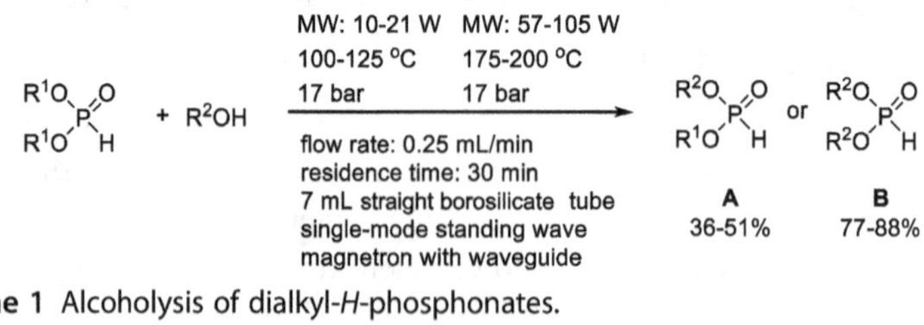

Scheme 1 Alcoholysis of dialkyl-*H*-phosphonates.

Another example of esterification using microwave-assisted flow synthesis has been reported by Daviot et al.[49] in which the authors used sulfuric acid to catalyze the esterification of succinic acid to dialkyl succinate. They used a reactor with a solid-state device oscillator that produced a single-mode standing wave. The reactor was fitted with a helical, 10 mL borosilicate glass reaction tube. The absorber of the microwave irradiation is the alcohol used as both a reactant for esterification and a solvent. The reaction was optimized first for the catalyst, and then the microwave conditions seen in Scheme 4. These conditions provided excellent conversion, 95%–100%, for all of the alkyl substituents, with modest-to-excellent yields of the double esterification.

The same group went further and reported the transesterification of dialkyl succinate using the enzyme Cal B. The optimized conditions can be seen in Scheme 5.[49] The transesterification reaction was carried out in

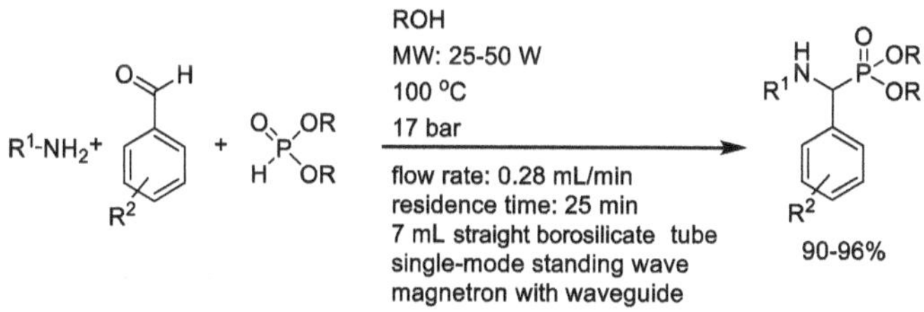

Scheme 2 The Kabachnik-Fields condensation in a microwave-assisted flow system.

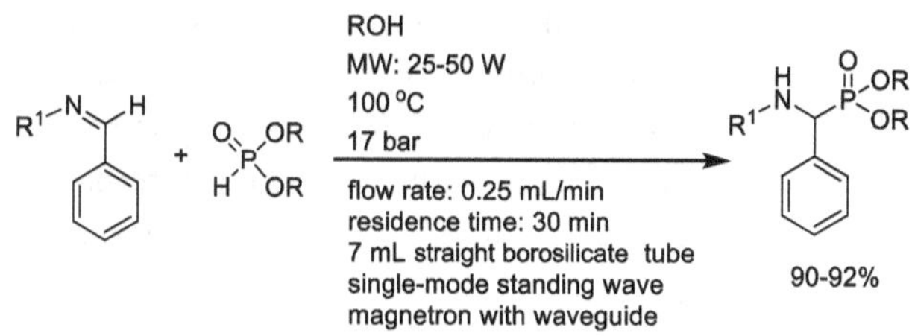

Scheme 3 The Aza-Pudovik reaction in a microwave-assisted flow system.

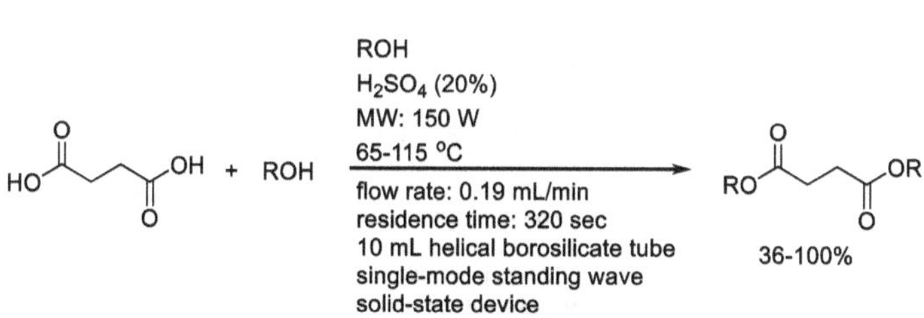

Scheme 4 A sulfuric acid-catalyzed diesterification of succinic acid.

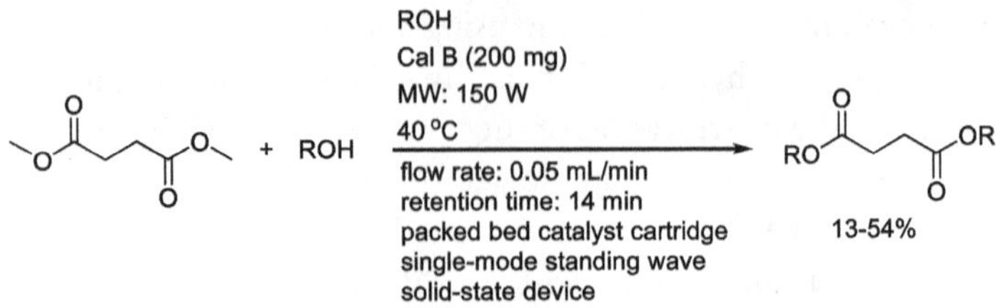

Scheme 5 Transesterification of dialkyl succinate using enzyme Cal B.

the same reactor. However, the reaction was instead passed through a microwave transparent cartridge that contained a packed bed containing the enzyme Cal B. From green synthesis point of view, this reaction is preferable to the use of corrosive mineral acids, such as H_2SO_4.

A recent approach for the esterification of L-leucine with *n*-butanol combined the effects of flow and batch processes by a stop-flow system. Nagahata et al.[50] first determined that the rate of temperature increase in the microwave activated system (166 °C/min) was far superior to that of the oil bath (27 °C/min). The authors then determined the penetration depth of their reaction mixture under microwave irradiation to be between 10 and 14 mm. Next, they determined that the flow rate of 0.25 mL/min resulted in the maximum yield, though not reaching 100%. Finally, they decided to pump the reaction mixture into the rotating borosilicate tube at the optimal rate until the reaction filled the tube to about 13 mm in height as determined by the penetration depth. After more than 60 min of the reaction under microwave conditions, 100% yield was observed for butanol and 36% for ethanol. Under batch conditions, the afforded yields were 64% and 4% respectively. Nagahata et al. used a multimode magnetron reactor fitted with a straight borosilicate tube measuring 40 mm in internal diameter and 300 mm in length. A summary of the reaction conditions is shown in Scheme 6.

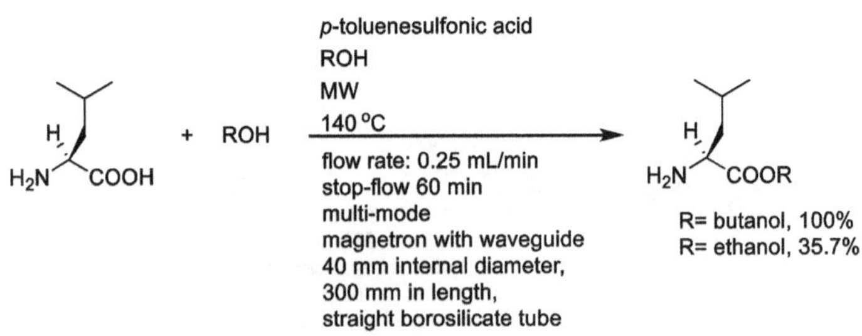

Scheme 6 Microwave-assisted esterification of L-leucine.

In continuation of the earlier work on transesterification of phosphonates by Bálint et al.[47], Kiss et al.[51] have reported the direct esterification of phenyl-*H*-phosphinic acid and the transesterification of ethyl-*H*-phenylphosphonate and dibenzyl phosphite with a wide range of aliphatic alcohols from ethanol, propanols, butanols, and pentanols to octanol isomers. For the direct esterification of phenyl-*H*-phosphinic acid, the optimum temperature was dependent on the alcohol being used, varying between 160 °C and 200 °C. The microwave reactor being used could autotune the power output to maintain the required temperature. The lower temperature reactions required a slower flow rate and higher residence time, the higher temperature reactions needed less time in the microwave irradiation, so the flow rate was higher. The optimized conditions can be seen in Scheme 7. The authors also determined that half as much 1-butyl-3-methylimidazolium hexafluorophosphate ([bmim][PF_6]) catalyst is required for the flow synthesis compared to the batch process. In addition, the flow process could afford the esterification products 4.5–6.9-fold higher in quantity than the batch protocol.

The same paper reports the transesterification of ethyl-*H*-phenylphosphonate, which was also successful with 100% conversion and 82%–89% isolated yields. The butanols required a slightly higher temperature at 225 °C than the pentanols at 220 °C, but the same flow rate of 0.15 mL/min. The optimized conditions can be seen in Scheme 8.

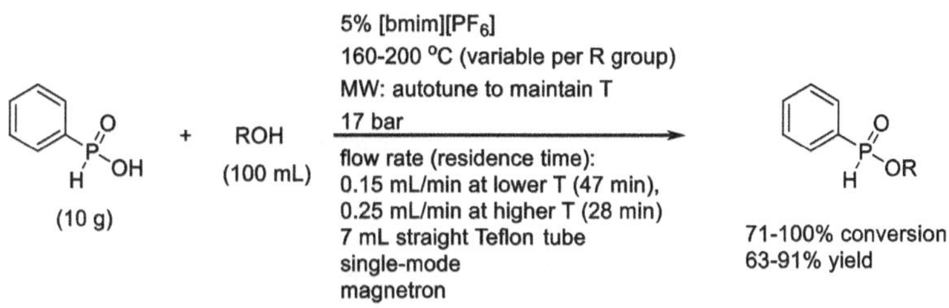

Scheme 7 Direct esterification of phenyl-*H*-phosphinic acid.

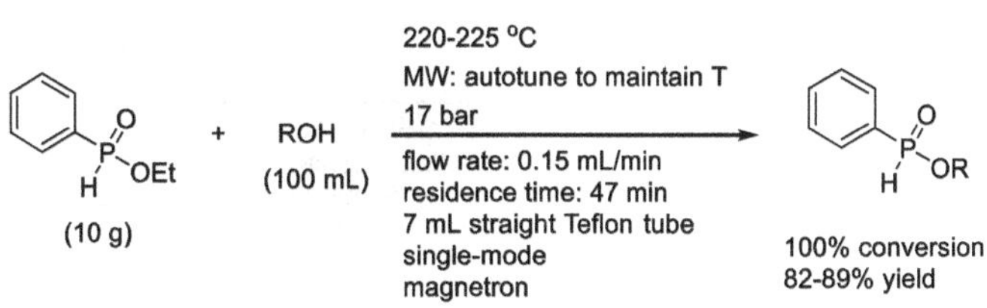

Scheme 8 Transesterification of ethyl-*H*-phenylphosphonate in a microwave-assisted flow reactor.

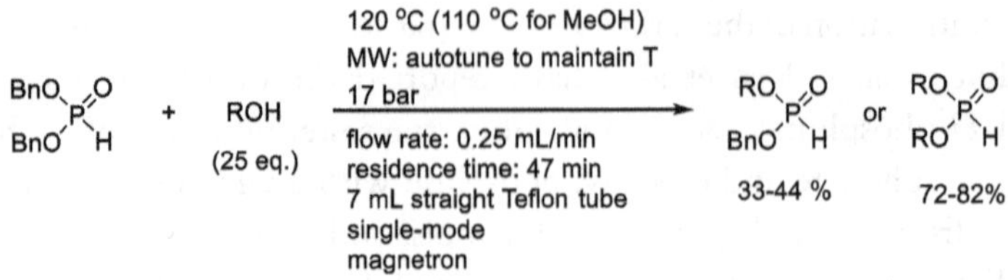

Scheme 9 Transesterification of dibenzyl phosphite.

Lastly, the transesterification of dibenzyl phosphite was optimized for the single transesterification leading to modest yields. The optimized conditions can be seen in Scheme 9.

3.1.2 Pericyclic reactions

In 2015, Yokozawa et al.[52] tested the application of their new 200 W solid-state microwave reactor that produced single-mode standing waves on the temperature-sensitive Diels-Alder reaction between diethyl acetylenedicarboxylate and furan. The vessel used was a helical, 6 mL borosilicate glass tube that had a 3.6 mm internal diameter and 20 mm in length, which provided microwave transparency and adequate penetration depth of the microwaves into the solvent. The optimized reaction conditions are shown in Scheme 10.

Vámosi et al.[53] used the same solid-state reactor to further optimize the above described Diels-Alder reaction during the development of their 9 + 4 + 1 rapid optimization technique. Though they reported using the same reactor, they fitted it with an unspecified microwave transparent straight tubular vessel rather than the helical vessel. The 9 + 4 + 1 technique moderates one variable at a time for nine reaction conditions followed by four reactions curtailed around the conditions that resulted in the best yield of the first nine reactions. The final optimized reaction uses the values determined by the maxima of the final surface approximation of the first 13

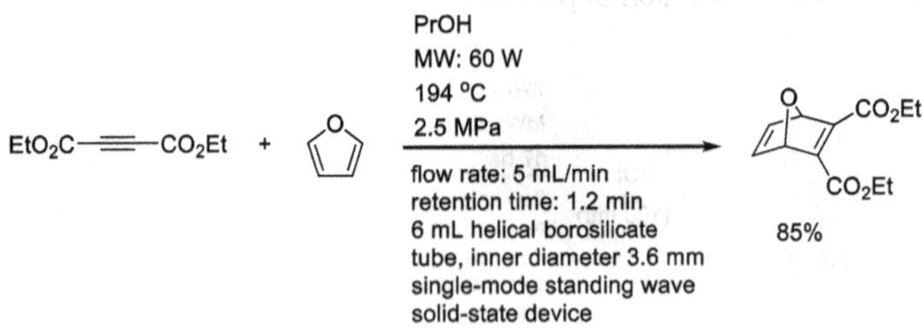

Scheme 10 Microwave-assisted Diels-Alder reaction between diethyl acetylenedicarboxylate and furan.

reactions. The optimized variables were the microwave power and flow rate, starting with three values for each parameter for the first nine reactions. For the Diels-Alder reaction optimization, the conditions that led to a temperature of 210 °C or greater resulted in the retro-Diels-Alder process and a tubular surface approximation. As a result, values below the threshold were eliminated. The final optimized conditions can be seen in Scheme 11.

Additionally, in those same two studies using the same solid-state reactor, Yokozawa[52] and Vámosi[53] optimized the Fischer indole synthesis using cyclohexanone and phenylhydrazine as substrates as shown in Scheme 12.

Due to the high importance of indole derivatives in biomedical applications, the development of such syntheses attracted extensive attention. A new synthetic method was developed by Xu et al.[54] for an alternative Fischer indole synthesis of 7-ethyltryptophol in 2017. The authors also used a microwave-assisted flow system, though the present system used a multimode magnetron reactor. The optimized conditions can be seen in Scheme 13.

Egami et al.[55] applied the same solid-state reactor as Yokozawa[52] to perform their Johnson-Claisen rearrangement and optimized the process using

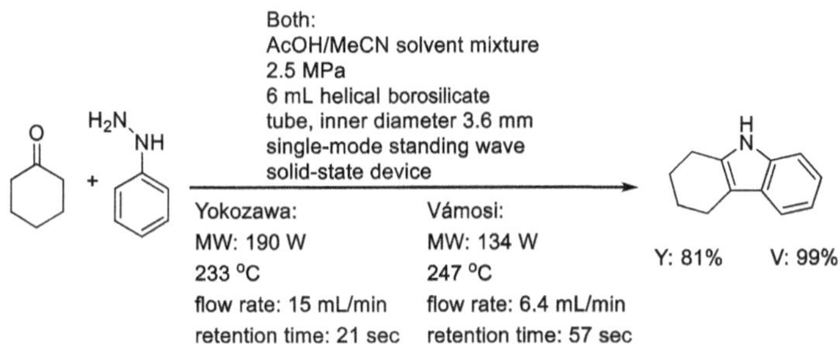

Scheme 12 Microwave-assisted Fischer indole synthesis from cyclohexanone and phenylhydrazine.

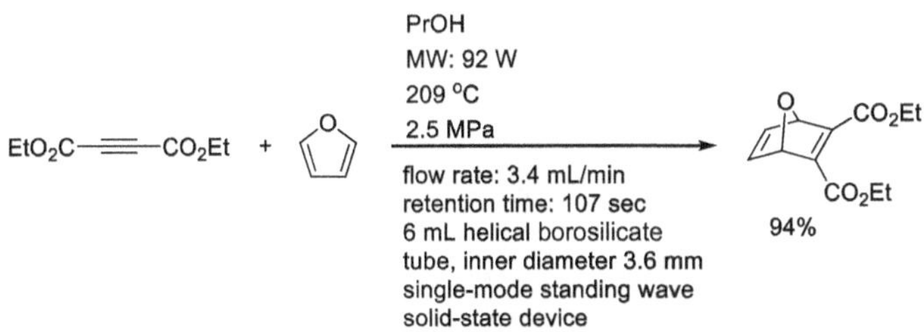

Scheme 11 Diels-Alder reaction between diethyl acetylenedicarboxylate and furan.

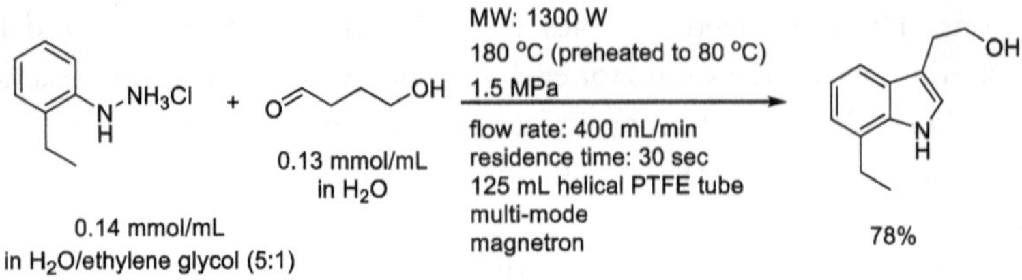

Scheme 13 The Fischer indole synthesis of 7-ethyltryptophol in a microwave-assisted flow reactor.

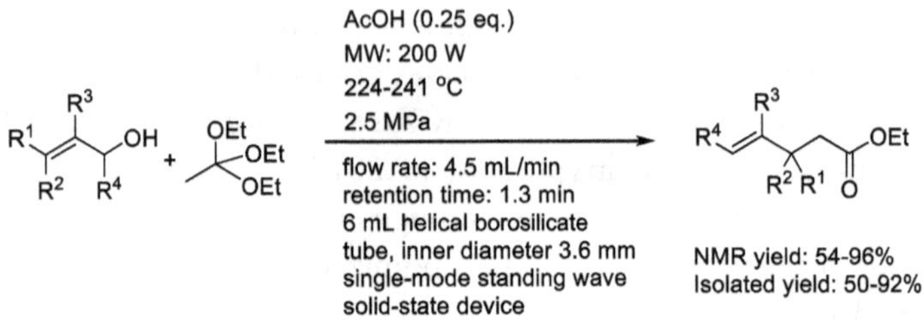

Scheme 14 Johnson-Claisen rearrangement of allyl alcohol and triethyl orthoacetate.

allyl alcohol and triethyl orthoacetate obtaining nearly quantitative yield. The best conditions are summarized in Scheme 14 for the Johnson-Claisen rearrangement with various substituted allyl alcohol derivatives and trialkyl orthoacetates.

3.1.3 Acetylation

In addition to the Diels-Alder and Fischer indole syntheses, Vámosi et al.[53] performed the *O*-acetylation of benzyl alcohol by acetic anhydride and obtained nearly quantitative yield. As part of the same study mentioned above, the apparatus used was the same. The optimized conditions are shown in Scheme 15.

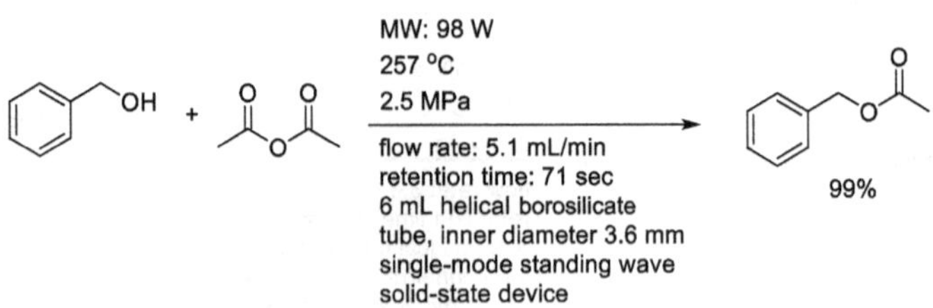

Scheme 15 Acetylation of benzyl alcohol by acetic anhydride in a microwave-assisted flow system.

3.1.4 Williamson ether synthesis

Egami et al.[56] used the Yokozawa[52] setup again for the Williamson ether synthesis using 4-hydroxy-cinnamic acid and 5-bromo-1-pentene as starting materials. The authors were able to reduce the reaction time down from 24 h to roughly 10 min using the optimized conditions in Scheme 16.

3.1.5 Hydrazinolysis

Noshita et al.[57] wanted to prove that they could deprotect unactivated amines by hydrogenative cleavage of amide bonds. With process chemistry in mind, they applied microwave-assisted flow synthesis. While this group also used the same solid-state reactor as Yokozawa,[52] they used a 5.5 mL borosilicate reaction vessel but did not specify its shape. They optimized the synthesis for biologically relevant paracetamol and *N*-benzylacetamide as seen in Scheme 17.

3.1.6 α-Alkylation

Barham et al.[58] have developed a protocol for the α-C alkylation of amides using styrene as an alkylating agent. After optimizing the catalyst and solvent

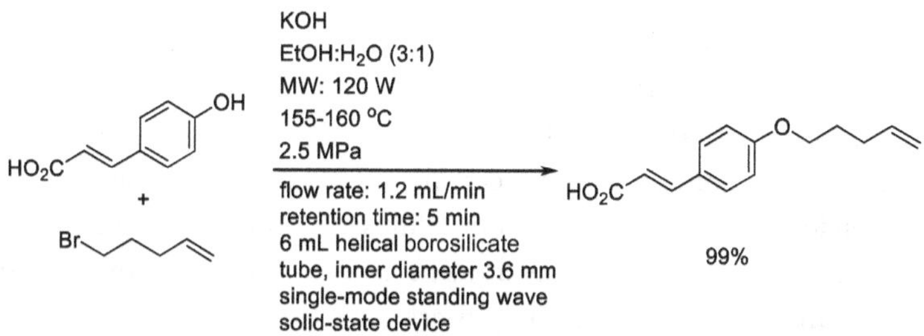

Scheme 16 Microwave-assisted Williamson ether synthesis using 4-hydroxy-cinnamic acid with 5-bromo-1-pentene.

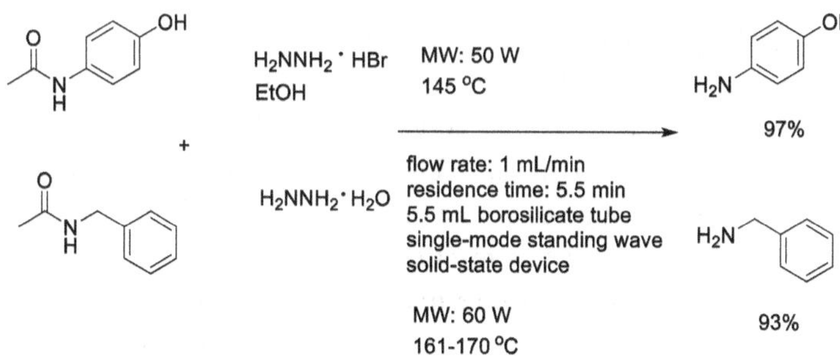

Scheme 17 Hydrazinolysis of paracetamol and *N*-benzylacetamide under microwave-assisted flow conditions.

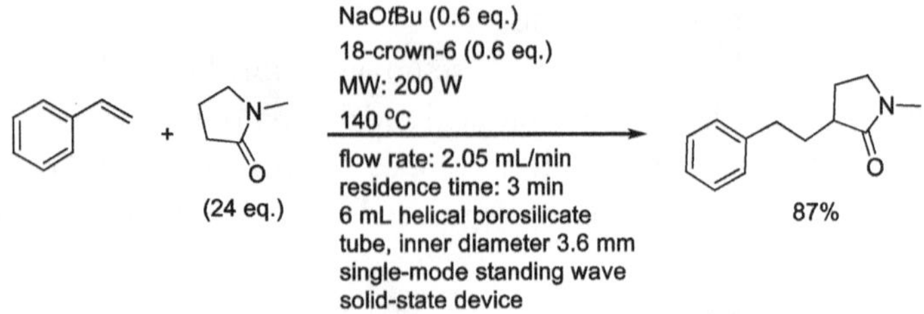

Scheme 18 C-Alkylation of *N*-methyl-2-pyrrolidone with styrene under continuous microwave-assisted flow conditions.

conditions, the authors adjusted their flow parameters to the best performing setup as shown in Scheme 18. The single alkylation was primarily observed; however, the double alkylation did occur to a small degree, 8% even under the optimized conditions. The authors also used the solid-state reactor setup as described by Yokazawa.[52]

3.1.7 Metal-catalyzed coupling reactions

Metal-catalyzed coupling reactions are popular applications and are also well known for their use of metal catalysts homogeneous and heterogeneous alike. The interaction of metals with microwaves is well-known. The common phenomenon called arcing could represent a fire hazard in organic solvents. Due to its importance, the issue was widely studied, and it has been concluded that soluble metal complexes and small metal particles (nanoparticles or nanosize metal crystallites of supported metal catalysts) do not generate arcing and thus are safe to use under microwave conditions.[59, 60] In addition, metal catalysts are excellent at transforming microwaves into heat under the right conditions, primarily in the absence of oxygen to avoid combustion. Metal-catalyzed coupling reactions have been a popular target for microwave-assisted flow synthesis, a recent review summarized the past developments in 2019.[61]

Ichikawa et al.[62] have described the Mizoroki-Heck coupling between 4-iodoacetophenone and *n*-butyl acrylate using the solid-state microwave reactor described by Yokozawa.[52] The heterogeneous Pd/WA30 catalyst was prepared in a glass tube cartridge on a packed bed. Dimethylacetamide, a polar solvent was used for this reaction due to two main reasons; first, it is an excellent microwave absorbing material and it also has a high boiling point. From a safety perspective, it is important to choose a solvent with a high boiling point when performing a reaction involving a metal catalyst to avoid rapid boiling so that control of the reaction can be maintained. This

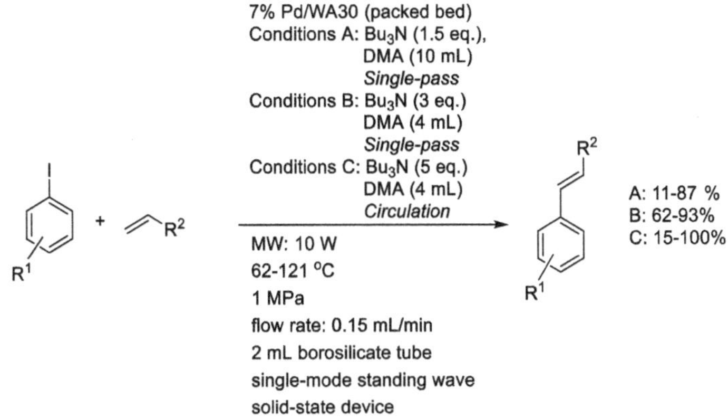

Scheme 19 Microwave-assisted flow Mizoroki-Heck coupling of 4-iodoacetophenones and *n*-butyl acrylate.

reaction was also performed in the absence of oxygen. As seen in Scheme 19, the authors explored the scope of the reaction and had varying success with several iodoacetophenone and alkene substrates.

Engen et al.[63] have performed the Suzuki-Miyaura cross-coupling reaction between a brominated spirooxindole dihydroquinazoline and *p*-toluene boronic acid using a solid-state microwave reactor with a nonresonant cavity. The authors performed the optimization of catalyst, base, and solvent under batch conditions carried out with the same microwave reactor. They then switched to flow using a straight borosilicate tube vessel to optimize the temperature and flow conditions. The flow rate of 0.5 mL/min was optimal, as higher flow rates did not allow long enough residence times. The full optimized conditions can be seen in Scheme 20. As in many other cases, the microwave target was the solvent and the free-floating metal catalyst. The

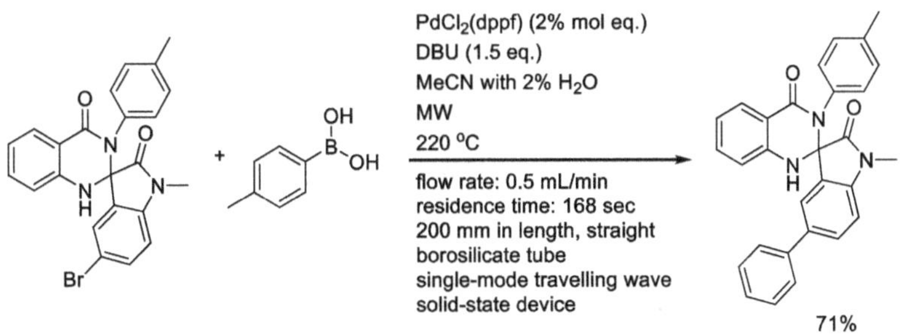

Scheme 20 Suzuki-Miyaura cross-coupling of a brominated spirooxindole dihydroquinazoline with *p*-toluene boronic acid under microwave-assisted flow conditions.

authors reported that they did not observe any clogging or combustion due to arcing.

3.1.8 Multicomponent reactions

Multicomponent reactions have been given special attention in green synthesis due to their usually high atom economy and coupling of several steps into a sequence that eliminates the need for the isolation and purification of intermediate products. Many multicomponent reactions respond well to microwave activation and these reactions have been a frequent target of earlier studies, mainly in batch reactors, however, more recent applications involve flow systems.[18]

Engen et al.[63] also used the same microwave-assisted flow system for the multicomponent reaction to produce brominated spirooxindole dihydroquinazolines that was used for their Suzuki-Miyaura coupling (vide supra, Scheme 20). However, the authors found that a microwave absorbent silicon carbide vessel performed better than the microwave transparent borosilicate vessel. The reaction vessel in this case is the microwave target. The reaction was optimized for reactant equivalents and temperature. It was found that there was a temperature optimum, temperatures too low did not afford conversion and temperatures too high caused product degradation. Scheme 21 shows the optimized conditions for this multicomponent reaction.

Christiaens et al.[64] performed the Hantzsch reaction to form 1,4-dihydropyridines under microwave-assisted flow conditions using a single-mode solid-state reactor fitted with a 40 mL helical borosilicate reaction vessel. The authors were able to observe the disappearance of the reactants in situ using Raman spectroscopy. It was found that using ammonia in methanol rather than aqueous ammonia helped avoid a biphasic reaction medium as well as the precipitation of product in the reaction

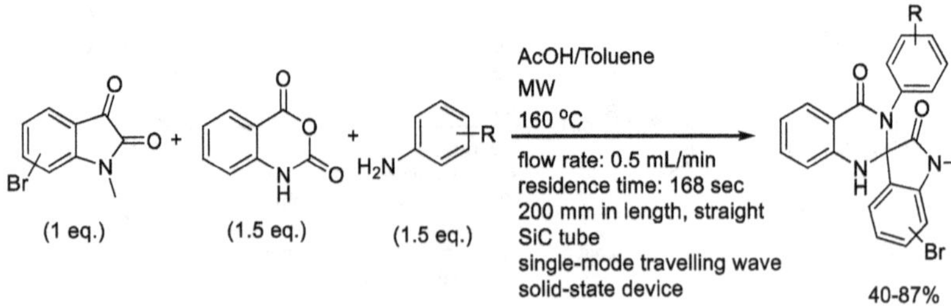

Scheme 21 Multicomponent reaction for the synthesis of brominated spirooxindole dihydroquinazoline by a microwave-assisted flow system.

vessel. The importance of this was that the authors were able to take accurate Raman spectra and avoid clogging the flow system. The conditions they used afforded 50 g, or 80% yield as shown in Scheme 22.

He et al.[65] performed a similar Hantzsch reaction to form 1,4-dihydropyridines, though they utilized a fixed bed catalyst comprised of 56.6 nm γ-Fe_2O_3 nanoparticles under solvent-free conditions. The authors had also found that certain aqueous solutions caused precipitation leading to clogging of the flow system. They employed a custom-built reactor using a magnetron and a series of waveguides to produce a single-mode cavity. Their optimization of reactants was performed under batch conditions as well as conventionally heated flow. The catalyst loading, temperature, and residence time were optimized under the microwave-assisted flow conditions with the optimal conditions being shown in Scheme 23.

3.2 Microwave-assisted flow systems in the syntheses of inorganics and materials

In addition to organic synthesis, microwave activation was heavily involved in the synthesis of inorganic materials, such as organometallic compounds, as

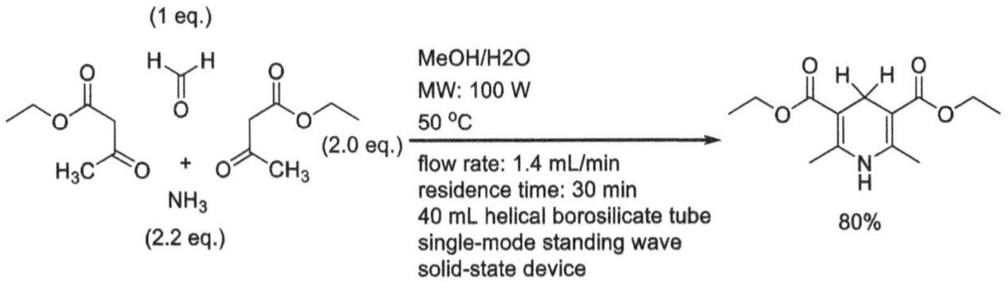

Scheme 22 Christiaens's microwave-assisted flow synthesis of a Hantzsch 1,4-dihydropyridine.

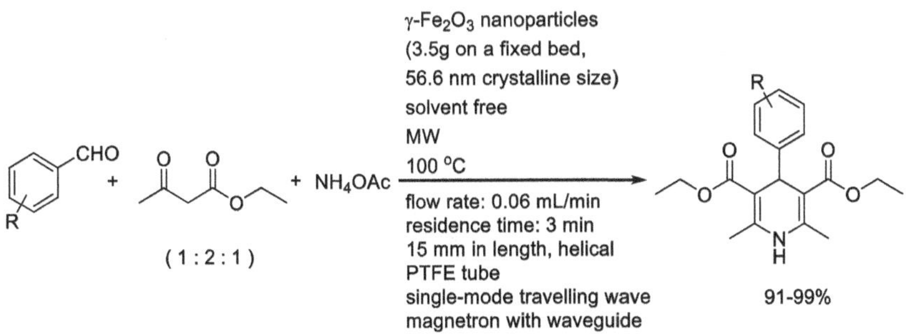

Scheme 23 He's synthesis of Hantzsch 1,4-dihydropyridines under microwave-assisted flow conditions.

well as in the synthesis of a broad array of materials, focused on the preparation of nanomaterials.

3.2.1 Nanomaterials

In 2014, Kim et al.[66] optimized the synthesis of colloidal nanocrystals to obtain the optimal stoichiometry of 1:1:2 of Cu:In:Se. The conditions applied during the microwave synthesis are summarized in Scheme 24. The reaction was carried out in a single-phase flow, but deposition of the nanocrystals on the wall of the PTFE tubing resulted in sparking. The group decided to introduce argon gas into the flow to segment the reaction medium. This was successful to avoid any further arcing. The segmented flow also improved reaction uniformity by introducing recirculation of the reaction medium. The reaction mixture was transferred from the microwave reactor into a 210 °C oil bath for 10 min before crystal isolation and characterization. The resulting nanoparticles had an average diameter of 16.6 ± 3.2 nm and moderate monodispersity with a coefficient of variation value of 19.3%.

In 2012, Dahal et al.[67] determined that under batch conditions, microwave irradiation was preferable for the nucleation of rhodium nanoparticles (RhNP) that were monodisperse and selective for cuboctahedral and icosahedral morphologies. The authors attributed the beneficial crystallization effects to the localized "hot spots" created by microwaves. The same group[68] followed up on this observation by investigating RhNP synthesis under flow conditions. They started with a one-phase flow system which afforded monodisperse RhNP cuboctahedra with an average diameter of 3.62 ± 0.71 nm. The reaction conditions can be found in Scheme 25. The flow conditions afforded results similar to the batch conditions, but at a lower temperature and shorter reaction time.

The authors were also interested in a segmented flow approach. In this protocol, droplets of the reaction mixture with an average diameter of

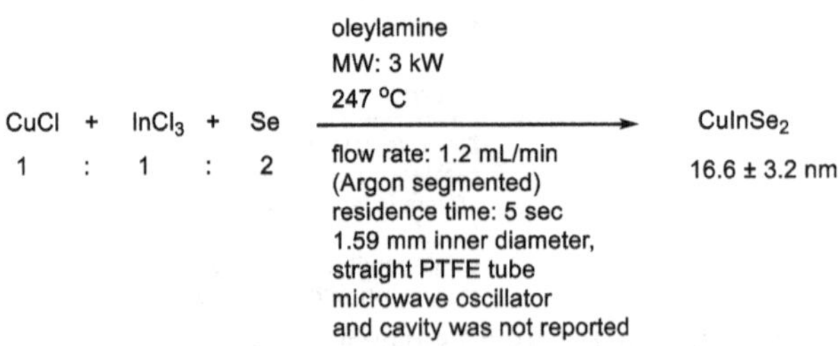

Scheme 24 Synthesis of colloidal nanocrystals by a microwave-assisted flow method.

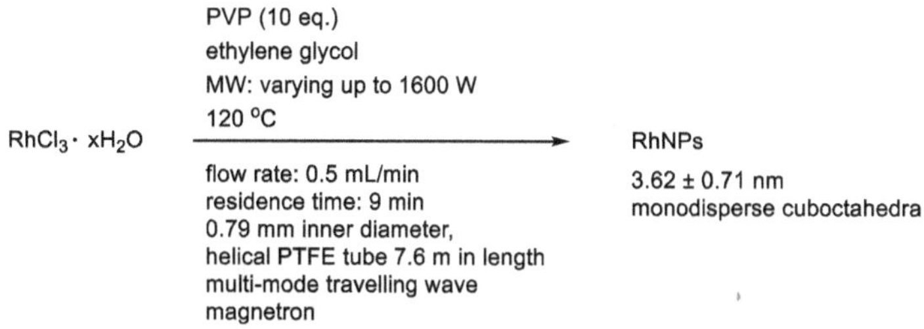

Scheme 25 Microwave-assisted one-phase flow synthesis of cuboctahedral RhNPs.

0.79 nm were introduced into the system separated by 10 equivalents of inert carrier oil by volume. The advantage of segmented flow comes from the circulatory intra-droplet mixing unavailable to continuous flow, which relies on dispersion for mixing. The two-phase approach resulted in the formation of tetrapods. Fig. 2 depicts the addition of $RhCl_3 \cdot xH_2O$ to four of the eight triangular faces of the cuboctahedra to form tetrapods. Scheme 26 details the reaction conditions for tetrapod formation.

Finally, the authors successfully synthesized alloy RhAgNPs using the two-phase microwave-assisted flow synthesis. The resulting $Rh_{70}Ag_{30}$NPs had a mixture of isotropic morphologies. This was attributed to the different

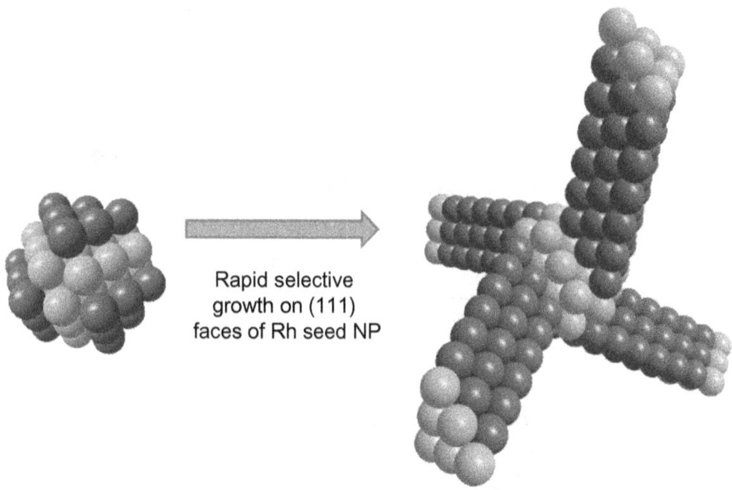

Fig. 2 Representation of Rh tetrapod formation by selective 111 facet growth. *(Reproduced from Kunal, P.; Roberts, E. J.; Riche, C. T.; Jarvis, K.; Malmstadt, N.; Brutchey, R. L.; Humphrey, S. M. Continuous Flow Synthesis of Rh and RhAg Alloy Nanoparticle Catalysts Enables Scalable Production and Improved Morphological Control.* Chem. Mater. ***2017****, 26, 4341–4350, with permission of the corresponding author, Dr. Simon M. Humphrey.)*

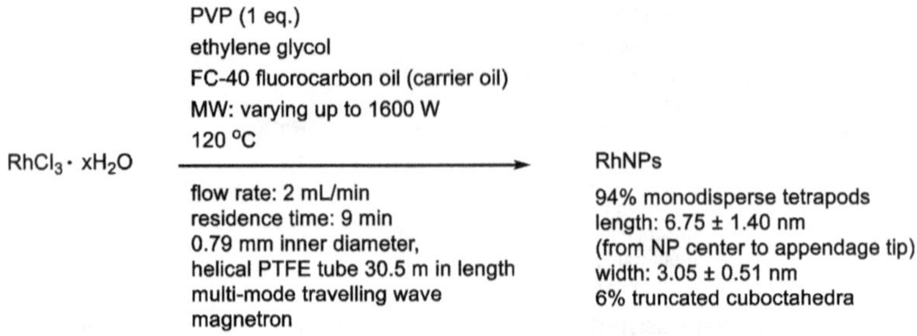

Scheme 26 Microwave-assisted two-phase flow synthesis of RhNP tetrapods.

electronic properties of Rh and Ag, specifically the weaker binding of Ag to the capping PVP monomers. The reaction conditions for the synthesis of the RhAgNP alloy can be seen in Scheme 27.

Akram et al.[69] were successful at demonstrating the use of microwave-assisted flow for the synthesis of ZnO nanoparticles. Using a modified domestic microwave oven, the authors investigated the effect of residence time (5–20 min) on the formation of ZnO nanoparticles by varying the flow rate of the system (125.6–31.4 mL/min respectively). It was observed that decreasing the flow rate to increase residence time resulted in increased particle size and crystallinity and decreased surface oxygen vacancies and defects. In addition, at 20 min residence time, the ZnO particles had a red-shift in the energy band gap, from 3.19 to 2.91 eV, which can be beneficial for use in optoelectronics or solar cells. The reaction conditions are shown in Scheme 28.

Anwar and Akbar[70] have reported a fast and facile synthesis of the biologically important Mn-substituted hydroxyapatite using microwave-assisted flow synthesis. Hydroxyapatites make up 70 wt% of human bone and are thus used in bone grafting materials and dental repair. Manganese not only increases the density of the hydroxyapatite without phase

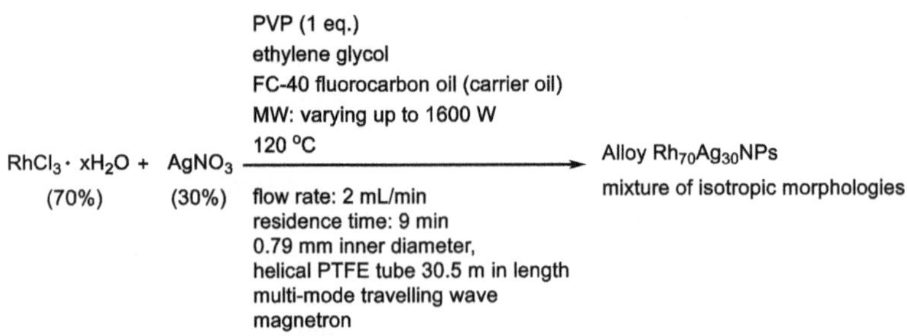

Scheme 27 Two-phase microwave-assisted flow synthesis of alloy RhAgNPs.

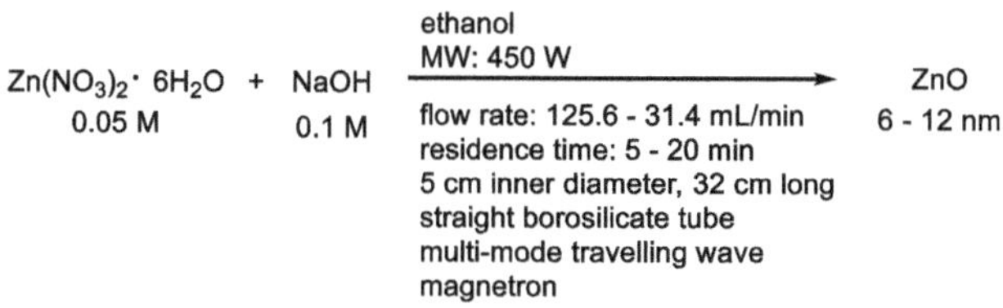

Scheme 28 Synthesis of ZnO nanoparticles by a microwave-assisted flow protocol.

transformation but also increases the coordination ability of integrin and enhances cell adhesion.[71] By using microwave heating, the reaction time is greatly reduced from 26 h (conventional heating) to minutes, and the nanoparticles produced are well dispersed and of fine particle size, with morphologies ranging from rod-like at 2 wt% Mn concentration (85 ± 15 nm) to semispherical at 4 wt% Mn concentration (75 ± 15 nm). The optimized conditions afforded a white, free-flowing, fine powder with an 85% yield and are shown in Scheme 29.

Anwar and Akbar[72] also reported a rapid synthesis of biologically important Ca-deficient hydroxyapatite (CDHA). CDHA has a greater dissolution rate in vivo and specific surface area that could be advantageous for bone regeneration compared to pure hydroxyapatite. The ratio of Ca/P was 1.5, which translates to 9 Ca for every 6 P rather than 10 Ca for every 6 P in the pure hydroxyapatite. The Ca-deficient hydroxyapatite product was well dispersed in a rod-like morphology and fine crystal size (90 ± 10 nm in length, 15 ± 4 nm in width). The optimized conditions afforded a very fine white nano-powder with 85% yield and can be seen in Scheme 30.

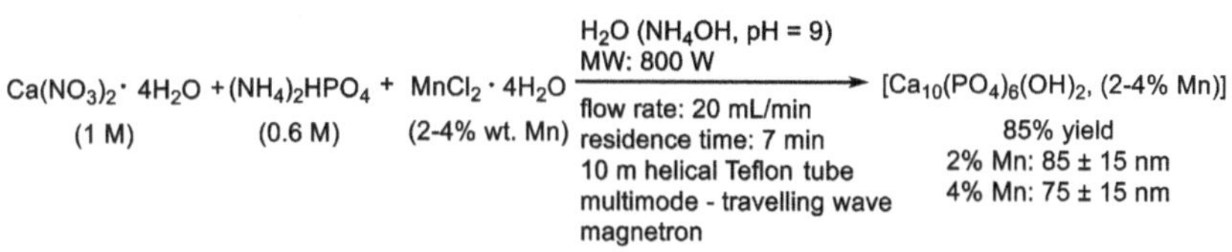

Scheme 29 Microwave-assisted flow synthesis of Mn-substituted hydroxyapatite.

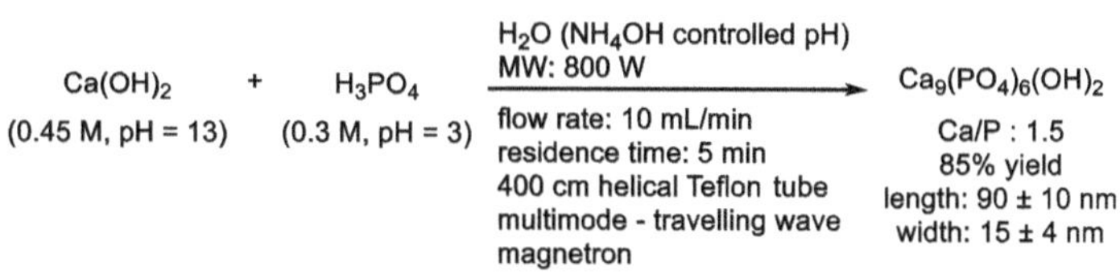

Scheme 30 Synthesis of Ca-deficient hydroxyapatite by a microwave-assisted flow protocol.

Cong et al.[73] have employed a microwave-assisted flow reactor to synthesize Pt/Pd and Pt/Rh nanoparticles with a 1:1 molar ratio in both cases. The resulting nanoparticles were well-dispersed colloidal bimetallic particles and had a cluster-in-cluster structure. Scheme 31 shows the optimized conditions with average particle size of the products.

Albuquerque et al.[74] have developed a method to synthesize Ag nanoparticles selectively aiming for cubic structures. The continuous flow approach included segmenting the liquid flow with Ar gas. The method and conditions are summarized in Scheme 32. The reactor was made of Teflon, thus the components of the reaction mixture, including the ionic units as well as the solvent were responsible for the absorption of the microwaves.

Kunze et al.[75] were interested in investigating the effect of inlet velocity and residence time on a pilot plant scale-up of the synthesis for silicon nanoparticles. The authors concluded that particle size increased with decreasing flow rates up to 69 m/s. Flow rates slower than 69 m/s did not further increase particle size. It was also found that the flow rate did not significantly affect the aggregation of particles. However, the authors did find that increasing the silane concentrations more than 16% did cause more aggregation due to increased nucleation events. They were able to achieve a production rate of 67.3 g/h of 26 nm, spherical, soft-agglomerated

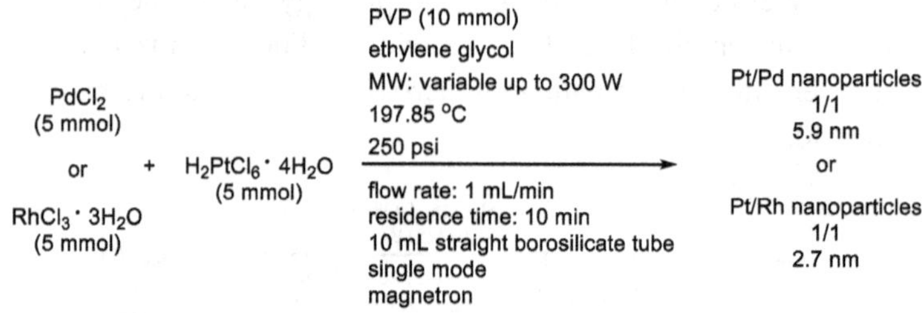

Scheme 31 Microwave-assisted flow synthesis of Pt/Pd or Pt/Rh bimetallic nanoparticles.

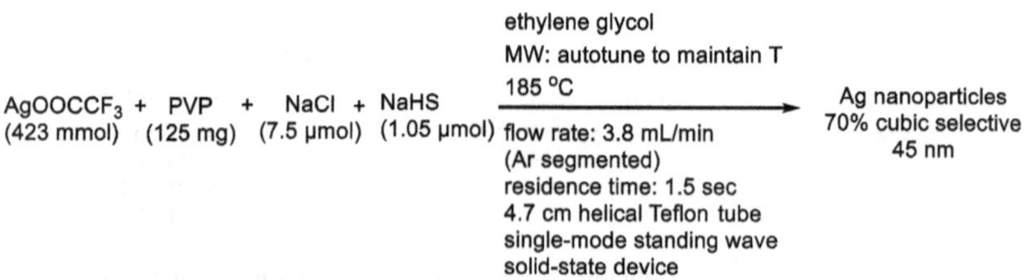

Scheme 32 Microwave-assisted flow synthesis of cubic Ag nanoparticles.

silicon nanoparticles at 16% silane content and a flow rate of 69 m/s. The reaction conditions can be found in Scheme 33.

Miyakawa et al.[76] have developed a microwave-assisted flow synthesis for Pd nanoparticles supported on carbon as the first step in the synthesis of Pd@Pt core-shell nanoparticles. The resulting Pd nanoparticles were well dispersed along the carbon surface with an average diameter of 3 nm. Scheme 34 depicts the optimized conditions for this process.

3.2.2 Organometallics

Taddei et al.[77] were interested in the microwave-assisted flow synthesis of three different metal-organic-frameworks (MOF). The first was UiO-66 or a framework of clusters of $[Zn_6O_4(OH)_4]$ with 1,4-benzenedicarboxylic acid ligands. The authors were able to obtain 94% yield with calculated space-time yields of up to 7163 kg/m^3/day. The reaction conditions can be found in Scheme 35.

The second MOF synthesized by the same group was MIL-53(Al), or a framework of clusters of $[AlO_4(OH)_2]$ with 1,4-benzenecarboxylic acid ligands. The reaction conditions shown in Scheme 36 afforded 90% yield and calculated space-time yields of up to 3618 kg/m^3/day.

The final MOF synthesized by the same group was HKUST-1, or a framework of copper nodes with coordinating waters and 1,3,5-benzenetricarboxylic acid ligands. The reaction conditions shown in

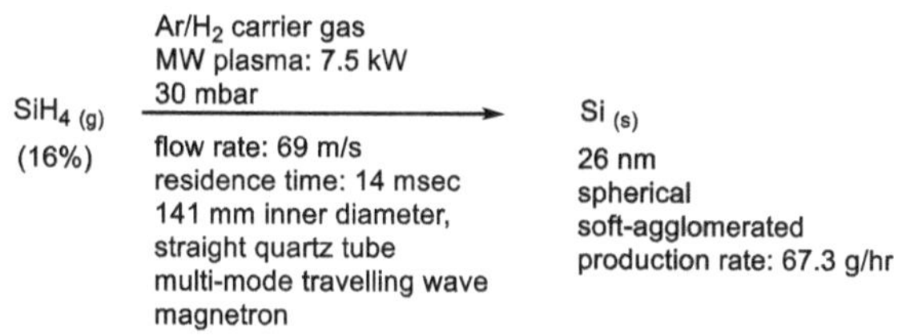

Scheme 33 Microwave-assisted plasma flow scale-up synthesis of silicon nanoparticles.

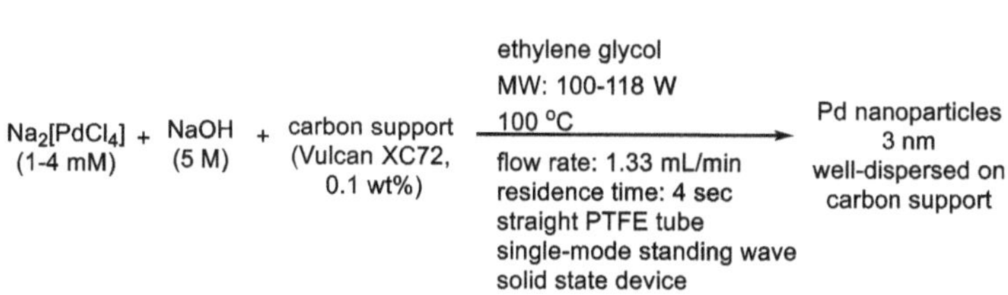

Scheme 34 Synthesis of carbon-supported Pd nanoparticles by a microwave-assisted flow protocol.

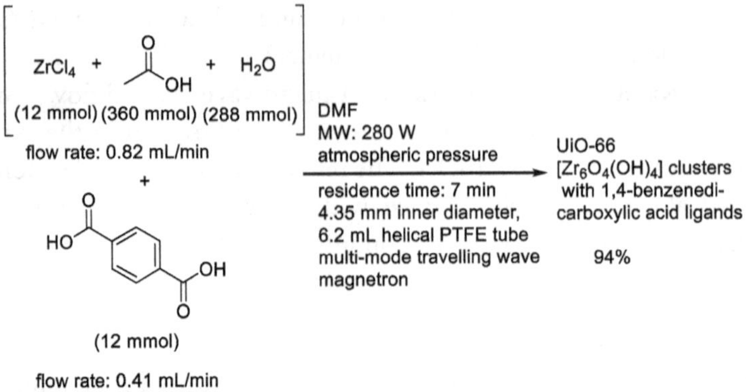

Scheme 35 Synthesis of UiO-66 metal-organic framework by a microwave-assisted flow protocol.

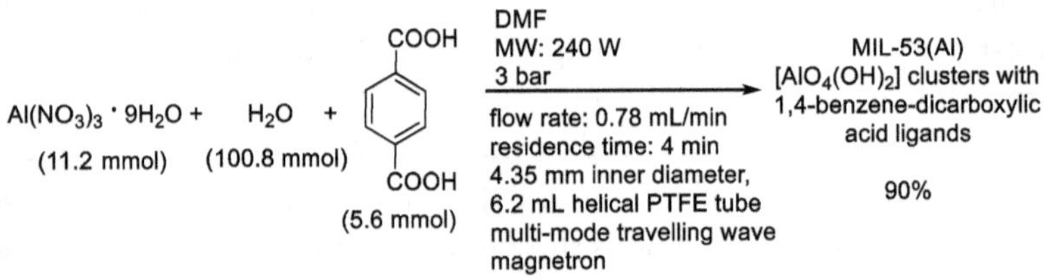

Scheme 36 Microwave-assisted flow synthesis of MIL-53(Al).

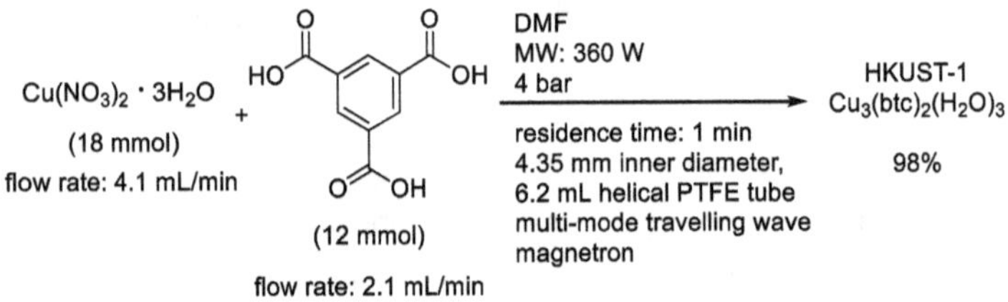

Scheme 37 Taddei's synthesis of HKUST-1.

Scheme 37 afforded 98% yield and calculated space-time yields of up to 66,650 kg/m^3/day.

Vo et al.[78] were interested in the volatile organic compound capture capabilities of UiO-66(Zr) with mixtures of amine-functionalized ligands. They developed a microwave-assisted flow synthesis for the production of UiO-66(Zr) with 0%, 25%, 50%, 75%, and 100% NH_2 functionality, respectively, by varying the amounts of 2-amino-1,4-benzenedicarboxylic acid and 1,4-benzenedicarboxylic acid used in the synthesis. It was determined that the structure of the 25%–75% ligand mixtures of UiO-66(Zr) contained randomly distributed mixed linkers rather than a mixture of the pure UiO-66(Zr) and UiO-66(Zr)-NH_2 products. The authors ultimately concluded that the pure UiO-66(Zr)-NH_2 performed the best to absorb 95% of the VOC (toluene), having a BET surface area of 935 m^2/g and a pore volume of 1.39 cm^3/g. Scheme 38 depicts the reaction conditions for the pure UiO-66(Zr)-NH_2 product. The same conditions were used for pure UiO-66(Zr), except 4.0 g of 1,4-benzenedicarboxylic acid were used. Similarly, the mixtures were prepared using the relative wt% of 2-amino-1,4-benzenedicarboxylic acid and 1,4-benzenedicarboxylic acid.

Le et al.[79] were interested in optimizing a microwave-assisted flow synthesis for MIL-100(Fe) as a precursor to CuO impregnation for selective CO capture. The optimized conditions in Scheme 39 afforded not only a high yield but a product with high crystallinity and porosity. Temperatures higher than 109 °C and residence times longer than 50 min resulted in metal deposition onto the tubing which resulted in clogging, while lower temperatures and shorter run times resulted in lower crystallinity and porosity.

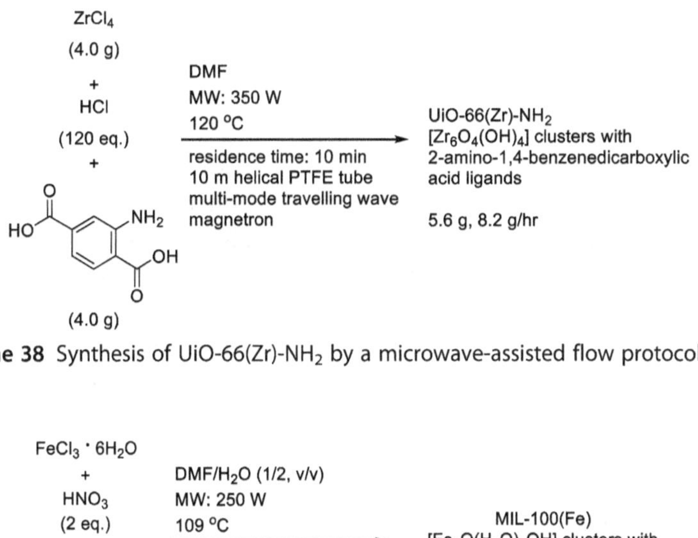

Scheme 38 Synthesis of UiO-66(Zr)-NH_2 by a microwave-assisted flow protocol.

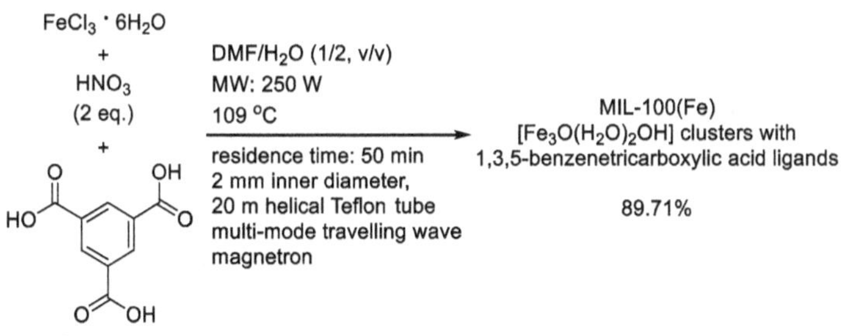

Scheme 39 Microwave-assisted flow synthesis of MIL-100(Fe).

3.3 Applications using bio-based feedstock and in environmental chemistry

Although an overwhelming majority of the above applications represent a green alternative to conventional synthetic pathways, one must not overlook that those reactions are still based on raw materials that were derived from mineral oil and its natural gas components. This is a step forward, however, in an ideal case, the starting materials for reactions are produced from nontoxic sustainable resources, such as biomass, or food waste.[80, 81] In addition to producing feedstock for the preparation of fine chemicals and other industrial products, focusing on environmental issues such as removal of volatile organic compounds (VOCs) is also in the forefront of research activities in this field.

3.3.1 Biodiesel production

One of the leading uses of microwave-assisted flow synthesis for an environmental application is the transesterification of waste cooking oils using methanol to produce fatty acid methyl esters. Although free fatty acids could be used for combustion, their mild acidity causes corrosion of metals over time, in addition to having a lower cetane number, a measure of quality in diesel fuels. Fatty acid methyl esters, or FAMEs, are not corrosive and have a higher cetane number, meaning better quality for fuel.[82, 83] Waste cooking oil has a mixture of triglycerides and free fatty acids. Upon transesterification with methanol, triglycerides give FAMEs and glycerol while free-floating fatty acids also give FAMEs and water, as shown in Scheme 40. Given that this process is carried out on the million-ton scale worldwide, making it more efficient and greener is of high importance.

In 2015, Choedkiatsakul et al.[84] measured the energy consumption of their optimized microwave-assisted flow transesterification of palm oil

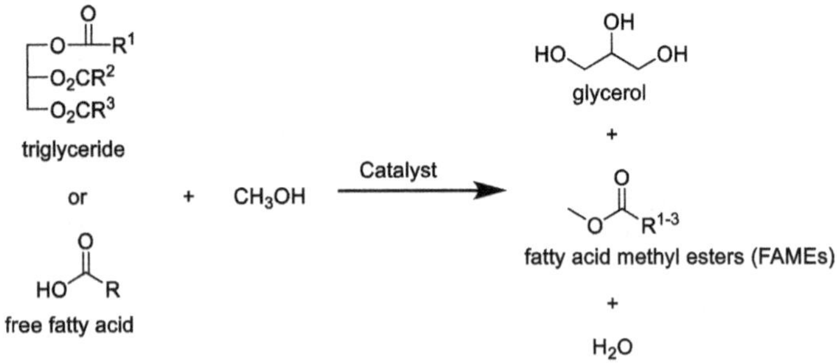

Scheme 40 Production of fatty acid methyl esters from triglycerides and free fatty acids.

and found that the new protocol required half the energy of its conventionally heated counterpart, 0.1167 kWh/L vs 0.222 kWh/L, respectively. Similarly, in 2017 Tangy et al.[85] performed energy consumption measurements and found that their microwave process required 0.1132 kWh/L vs the conventional heating which required 0.222 kWh/L. Several other groups[86–88] also optimized the transesterification, but did not report energy consumption. Table 1 consolidates the optimized reaction conditions of all five reports, whereas Table 2 summarizes the microwave conditions used in each research group's optimizations.

Table 1 Optimized reaction conditions for microwave-assisted transesterifications.

Starting material	Methanol equivalent	Catalyst (eq.)	Solvent	Temperature (°C)	Pressure (bar)	Yield (%)	Reference
Palm oil	12	NaOH (1 wt%)	Heptane	70	–	97	84
Waste cooking oil	12	SrO on silicon beads (1.25 wt %)	None	65	–	98	85
Soybean oil	9	$NaOCH_3$ (1 wt%)	None	–	–	100	86
Waste cooking oil	6	SrO (3 wt %)	None	66	–	93	87
Palm oil	6	KOH (1 wt%)	Heptane	200	7	97	88

3.3.2 Glycerol conversion

This subsection is very closely attached to the above; the primary by-product of biodiesel production from the transesterification of common waste cooking oils is glycerol (Scheme 40). Given that glycerol is produced in the biodiesel process in enormous amounts, finding new applications for its use, as well as developing green technologies for its transformations attracts extensive attention. While glycerol has its own wide range of uses in industry, it can also be carbonylated to another commonly used starting material with a wide range of uses, glycerol carbonate. According to Ochoa-Gómez et al.[89] glycerol carbonate is extremely versatile due to its unique physical properties. It is not flammable and nontoxic, and is a water-soluble, readily biodegradable, and viscous liquid with a very low evaporation rate.

Table 2 Optimized microwave conditions for several microwave-assisted flow transesterification protocols.

Power (W)	Residence time	Flow rate (mL/min)	Vessel material	Vessel shape	MW oscillator	MW mode	Reference
400	1.75 min	100	Teflon	Straight	Magnetron	Multi	84
207	~1 min	200	Teflon	Straight	Magnetron	Multi	85
Up to 900	3 cycles of 3 min each	540	Borosilicate	Straight	Magnetron	Multi	86
300	0.63 min	160	Unspecified MW transparent material	Straight	Magnetron	Multi	87
780	4 s	100	Teflon	Straight	Magnetron	Multi	88

Glycerol carbonate can be used as a reagent or a solvent. It is currently used in several processes and applications in the cosmetic, pharmaceutical, polymer, and battery industry.[89]

Nogueria et al.[90] have optimized the carbonylation of glycerol to afford 84% yield of glycerol carbonate using microwave-assisted flow synthesis. The optimized conditions can be seen in Scheme 41.

3.3.3 Biochar activation for water desalination

For a chemical product to be truly green, the process by which it is made should also be green. Tang et al.[91] have posited that most routes to synthesize activated biochars for the desalination of water apply multistep, long

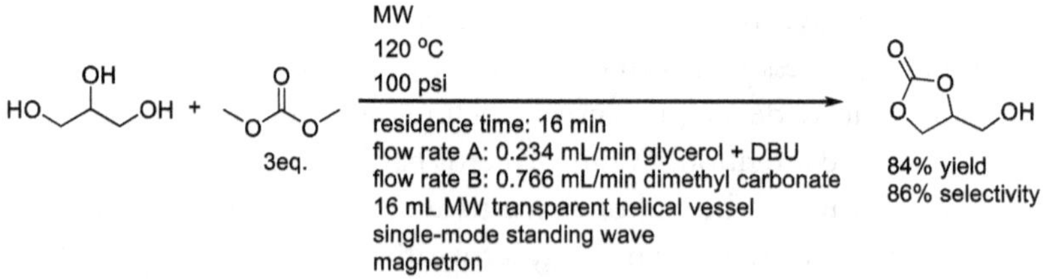

Scheme 41 Carbonylation of glycerol to glycerol carbonate by a microwave-assisted flow protocol.

synthesis, and generate a large amount of potentially toxic and hazardous waste. To address the problem, they have designed a process using microwave-assisted flow synthesis to activate biochar made from sugarcane bagasse (food waste) and carbon dioxide. After the bagasse is mixed with 3 eq. of KOH solution and dried, it is packed into a straight quartz vessel. The quartz vessel with the bagasse and KOH is placed into the multimode microwave cavity and the magnetron supplies 700 W of power over 5 min while CO_2 flows through the system at 300 mL/min. The resulting activated biochar had a surface area of 1019 m^2/g and a ratio of mesopores to total pore volume (V_{meso}/V_{total}) of 56.7%. Tang et al. tested the electrochemical properties of their biochar electrode and found that they had a superlative specific capacitance of 208 F/g and an excellent electrosorption capacity of 28.9 mg/g.

3.3.4 Green ethylene production

Jermolovicius et al.[92] were interested in designing a process to produce ethylene gas by ethanol dehydration without producing carbon dioxide, an inevitable consequence of using fossil fuels as a source of heat. The authors were successful in producing a similar yield of ethylene using microwave-assisted flow synthesis (60.8%) as compared to the sample that was produced using conventional heating; petrochemical cracking (58%) and ethanol dehydration (minimum 41%, maximum 99.8%). The research team created a catalyst bed with 8.0 g of zeolite ZSM-5 catalyst doped with 7.6 g of zinc chloride and 0.2 g of manganese chloride and silicon carbide pellets. The catalyst bed with a height of 150 mm, an equivalent volume of 47.1 mL, was packed inside a quartz reaction vessel, which was placed inside a multimode microwave cavity. Ethanol gas, 87.96% in a carrier gas of N_2, was passed through the system at a flow rate of 1.7 g ethanol/min under microwave irradiation of 119 W and a resulting temperature of 509 °C.

3.3.5 Volatile organic compound removal

Volatile organic compounds, or VOCs, are common anthropogenic pollutants. One such pollutant is toluene. Bo and Sun[93] have proposed a method of removal for toluene from air by catalytic oxidation using a novel Cu-Mn-Ce/cordierite honeycomb catalyst in a microwave-assisted flow system. The mass ratio of components of the catalyst are as follows: 3:3:1:120 to Cu:Mn:Ce:cordierite. The fixed bed catalyst (6.7 active wt%) was loaded into a quartz reaction tube. The packed quartz vessel was then placed in the modified domestic microwave. Treated air carrying toluene in a concentration of 1 mg/L was passed through the system with a flow rate of 2 L/min (residence time of 4 s) under 117 W microwave power and a temperature of 200 °C. This process resulted in 98% removal of toluene from the air although the authors did not discuss the products of the oxidation.

3.3.6 Solar cell material synthesis

Indene-C_{60} monoadduct ($IC_{60}MA$) and Indene-C_{60} bisadduct ($IC_{60}BA$) are key components of polymer-based solar cells. Barham et al.[94] remark that their application had been limited due to high reaction temperatures, extended reaction times, and the use of polychlorinated aromatic solvents due to low solubility in most organic solvents. To overcome these barriers, the authors have developed a method that increases the production of $IC_{60}MA$ and $IC_{60}BA$ while also employing a greener synthesis, as shown in Scheme 42.

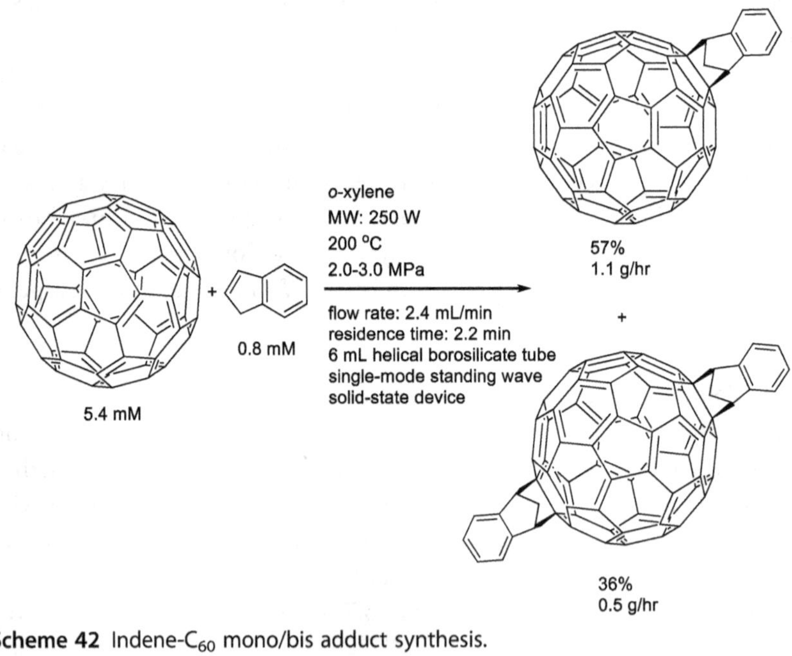

Scheme 42 Indene-C_{60} mono/bis adduct synthesis.

4 Conclusions

The explosion-like development of flow reactors and protocols in organic synthesis has been combined with microwave activation in the past 20 years. The early examples published in the late 1990s were based on domestic microwave ovens that served as the heating component of the microwave-assisted flow system. Since then the field underwent a significant development and sophisticated microwave instruments have been applied along with special reactor types. The application possibilities appear as broad as they are in batch reactor-based MAOS. This survey was intended to provide a broad overview of applications from typical organic synthesis to the production of nanoparticles or environmental applications. Based on the pace of developments in the past decade, there is every reason to assume that the number of applications for microwave-assisted flow synthesis will continuously grow and likely we will witness industrial applications in the near future.

References

1. Baig, R. N. B.; Varma, R. S. Alternative Energy Input: Mechanochemical, Microwave and Ultrasound-Assisted Organic Synthesis. *Chem. Soc. Rev.* **2012**, *41*, 1559–1584.
2. Kokel, A.; Schäfer, C.; Török, B. Microwave-Assisted Reactions in Green Chemistry. In *Encyclopedia of Sustainable Science and Technology*; Meyers, R. A., Ed.; Springer-Nature, 2018. https://doi.org/10.1007/978-1-4939-2493-6_1008-1.
3. Coquerel, Y.; Colacino, E.; Rodriguez, J.; Martinez, J.; Lamaty, F. In *Microwaves-Assisted Stereoselective Synthesis in Stereoselective Synthesis of Drugs and Natural Products*; Andrushko, V., Andrushko, N., Eds.; Wiley: Hoboken, 2013; pp. 145–166 (chapter 5).
4. Díaz-Ortiz, Á.; Prieto, P.; de la Hoz, A. A Critical Overview on the Effect of Microwave Irradiation in Organic Synthesis. *Chem. Rec.* **2019**, *19*, 85–97.
5. Strauss, C. R.; Varma, R. S. Microwaves in Green and Sustainable Chemistry. In *Microwave Methods in Organic Synthesis*; Larhed, M., Olofssonq, K., Eds.; Topics in Current Chemistry, Vol. 266; Springer: Berlin, Heidelberg, 2006; pp. 199–231.
6. Török, B. Sustainable Synthesis. In *Green Chemistry: An Inclusive Approach*; Török, B., Dransfield, T., Eds.; Elsevier: Oxford, 2018; pp. 49–89.
7. Horikoshi, S.; Serpone, N., Eds. *Microwaves in Catalysis: Methodology and Applications*; Wiley-VCH: Weinheim, 2016.
8. Vanderhoff, J. W. *Method for Carrying Out Chemical Reactions Using Microwave Energy*. US Patent 3432413A, 1969.
9. Gedye, R.; Smith, F.; Westaway, K.; Ali, H.; Baldisera, L.; Laberge, L.; Rousell, J. The Use of Microwave Ovens for Rapid Organic Synthesis. *Tetrahedron Lett.* **1986**, *27*, 279–282.
10. Giguere, R. J.; Bray, T. L.; Duncan, S. M.; Majetich, G. Application of Commercial Microwave Ovens to Organic Synthesis. *Tetrahedron Lett.* **1986**, *27*, 4945–4948.
11. Strauss, C. R.; Trainor, R. W. Developments in Microwave-Assisted Organic Chemistry. *Aust. J. Chem.* **1995**, *48*, 1665–1692.

12. Cablewski, T.; Faux, A. F.; Strauss, C. R. Development and Application of a Continuous Microwave Reactor for Organic Synthesis. *J. Org. Chem.* **1994**, *59*, 3408–3412.
13. Chen, S.-T.; Chiou, S.-H.; Wang, K.-T. Preparative Scale Organic Synthesis Using a Kitchen Microwave Oven. *J. Chem. Soc., Chem. Commun.* **1990**, 807–809.
14. Thomas, J. M.; Thomas, W. J. *Principles and Practice of Heterogeneous Catalysis*; Wiley-VCH: New York, Weinheim, 1996.
15. Horvath, I., Ed. *Encyclopedia of Catalysis*; Wiley: New York, 2003.
16. Ertl, G., Ed. *Handbook of Heterogeneous Catalysis*; 2nd ed.; Wiley-VCH: Weinheim, New York, 2008.
17. Somorjai, G. A.; Li, Y. *Introduction to Surface Chemistry and Catalysis*, 2nd ed.; Wiley: New York, 2010.
18. Kokel, A.; Schäfer, C.; Török, B. Application of Microwave-Assisted Heterogeneous Catalysis in Sustainable Synthesis Design. *Green Chem.* **2017**, *19*, 3729–3751.
19. Strauss, C. R. On Scale Up of Organic Reactions in Closed Vessel Microwave Systems. *Org. Process Res. Dev.* **2009**, *13*, 915–923.
20. Dallinger, D.; Lehmann, H.; Moseley, J. D.; Stadler, A.; Kappe, C. O. Scale-Up of Microwave-Assisted Reactions in a Multimode Bench-Top Reactor. *Org. Process Res. Dev.* **2011**, *15*, 841–854.
21. Lee, D. S.; Amara, Z.; Poliakoff, M.; Harman, T.; Reid, G.; Rhodes, B.; Brough, S.; McInally, T.; Woodward, S. Investigating Scale-Up and Further Applications of DABAL-Me3 Promoted Amide Synthesis. *Org. Process Res. Dev.* **2015**, *19*, 831–840.
22. Kim, D.; Seol, S. K.; Chang, W. S. Energy Efficiency of a Scaled-up Microwave-Assisted Transesterification for Biodiesel Production. *Korean J. Chem. Eng.* **2016**, *33*, 527–531.
23. Schmink, J. R.; Kormos, C. M.; Devine, V. G.; Leadbeater, N. E. Exploring the Scope for Scale-Up of Organic Chemistry Using a Large Batch Microwave Reactor. *Org. Process Res. Dev.* **2010**, *14*, 205–214.
24. Protasova, L. N.; Bulut, M.; Ormerod, D.; Buekenhoudt, A.; Berton, J.; Stevens, C. V. Latest Highlights in Liquid-Phase Reactions for Organic Synthesis in Microreactors. *Org. Process Res. Dev.* **2013**, *17*, 760–791.
25. Porta, R.; Benaglia, M.; Puglisi, A. Flow Chemistry: Recent Developments in the Synthesis of Pharmaceutical Products. *Org. Process Res. Dev.* **2016**, *20*, 2–25.
26. Matsuzawa, M.; Togashi, S. Pilot Plant for Continuous Flow Microwave-Assisted Chemical Reactions. In *Microwaves in Catalysis: Methodology and Applications*; Horikoshi, S., Serpone, N., Eds.; Wiley-VCH: Weinheim, 2016; pp. 141–154.
27. Nagahata, R.; Takeuchi, K. Encouragements for the Use of Microwaves in Industrial Chemistry. *Chem. Rec.* **2019**, *19*, 15–39.
28. Damm, M.; Glasnov, T. M.; Kappe, C. O. Translating High-Temperature Microwave Chemistry to Scalable Continuous Flow Processes. *Org. Proc. Res. Dev.* **2010**, *14*, 215–224.
29. Pálinkó, I. Heterogeneous Catalysis: A Fundamental Pillar of Sustainable Synthesis. In *Green Chemistry: An Inclusive Approach*; Török, B., Dransfield, T., Eds.; Elsevier: Oxford, 2018; pp. 415–447.
30. Bag, S.; Dasgupta, S.; Török, B. Microwave-Assisted Heterogeneous Catalysis: An Environmentally Benign Tool for Contemporary Organic Synthesis. *Curr. Org. Synth.* **2011**, *8*, 237–261.
31. Varma, R. S. Clay and Clay-Supported Reagents in Organic Synthesis. *Tetrahedron* **2002**, *58*, 1235–1255.
32. Polshettiwar, V.; Varma, R. S. Microwave-Assisted Organic Synthesis and Transformations Using Benign Reaction Media. *Acc. Chem. Res.* **2008**, *41*, 629–639.
33. Abid, M.; Török, B.; Huang, X. Microwave-Assisted Tandem Processes for the Synthesis of N-Heterocycles. *Aus. J. Chem.* **2009**, *62*, 208–222.

34. Daştan, A.; Kulkarni, A.; Török, B. Environmentally Benign Synthesis of Heterocyclic Compounds by Combined Microwave-Assisted Heterogeneous Catalytic Approaches. *Green Chem.* **2012**, *14*, 17–37.
35. Cho, H.; Schäfer, C.; Török, B. Microwave-assisted Solid Acid Catalysis. In *Microwaves in Catalysis—Fundamental Research and Scale-up Technology*; Horikoshi, S., Serpone, Eds.; Wiley-VCH: Weinheim, 2016; pp. 193–212 (chapter 10).
36. Gawande, M. B.; Shelke, S. N.; Zboril, R.; Varma, R. S. Microwave-Assisted Chemistry: Synthetic Applications for Rapid Assembly of Nanomaterials and Organics. *Acc. Chem. Res.* **2014**, *47*, 1338–1348.
37. Kappe, C. O. My Twenty Years in Microwave Chemistry: From Kitchen Ovens to Microwaves That Aren't Microwaves. *Chem. Rec.* **2019**, *19*, 15–39.
38. Stankiewicz, A.; Sarabi, F. E.; Baubaid, A.; Yan, P.; Nigar, H. Perspectives of Microwaves-Enhanced Heterogeneous Catalytic Gas-Phase Processes in Flow Systems. *Chem. Rec.* **2019**, *19*, 40–50.
39. Tagliapietra, S.; Gaudino, E. C.; Martina, K.; Barge, A. Giancarlo Cravotto, Microwave Irradiation in Micro- Meso-Fluidic Systems; Hybrid Technology Has Issued the Challenge. *Chem. Rec.* **2019**, *19*, 98–117.
40. Horikoshi, S.; Serpone, N. Microwave Flow Chemistry as a Methodology in Organic Syntheses, Enzymatic Reactions, and Nanoparticle Syntheses. *Chem. Rec.* **2019**, *19*, 118–139.
41. http://search.itu.int/history/HistoryDigitalCollectionDocLibrary/1.41.48.en.101.pdf (accessed 17 August 2020).
42. Barham, J. P.; Koyama, E.; Norikane, Y.; Ohneda, N.; Yoshimura, T. Microwave Flow: A Perspective on Reactor and Microwave Configurations and the Emergence of a Tunable Single-Mode Heading Toward Large-Scale Applications. *Chem. Rec.* **2019**, *19*, 188–203.
43. Jerby, E.; Schwartz, E.; Gerling, J. F.; Werner, K.; Durnan, G.; Yakovlev, V. V.; Achkasov, K.; Meir, Y.; Metaxas, A. C. *Ampere Newsletter*, July 7, 2016.
44. Chandrasekaran, E.; Ramanathan, S.; Basak, T. Microwave Material Processing—A Review. *J. AIChE* **2011**, *25*, 330–363.
45. Metaxas, A. C.; Meredith, R. J. *Industrial Microwave Heating*; Peter Peregrinus Ltd: London, 1983.
46. Estel, L.; Poux, M.; Benamara, N.; Polaert, I. Continuous Flow-Microwave Reactor: Where Are We? *Chem. Eng. Proc.* **2017**, *113*, 56–64.
47. Bálint, E.; Tajti, Á.; Tóth, N.; Keglevich, G. Continuous Flow Alcoholysis of Dialkyl *H*-Phosphonates with Aliphatic Alcohols. *Molecules* **2018**, *23*, 1618–1633.
48. Bálint, E.; Tajti, Á.; Ladányi-Pára, K.; Tóth, N.; Mátravölgyi, B.; Keglevich, G. Continuous Flow Synthesis of α-Aryl-α-Aminophosphonates. *Pure Appl. Chem.* **2019**, *91*, 67–76.
49. Daviot, L.; Len, T.; Lin, C.; Len, C. Microwave-Assisted Homogeneous Acid Catalysis and Chemoenzymatic Synthesis of Dialkyl Succinate in a Flow Reactor. *Catalysts* **2019**, *9*, 272–282.
50. Nagahata, R.; Nakamura, T.; Mori, Y.; Takeuchi, K. Microwave-Assisted Facile and Rapid Esterification of Amino Acids I: Esterification of L-Leucine from Batch to Flow Processes and Scale-Up. *Nat. Sci.* **2017**, *9*, 110–122.
51. Kiss, N. Z.; Henyecz, R.; Keglevich, G. Continuous Flow Esterification of a H-Phosphinic Acid, and Transesterification of H-Phosphinates and H-Phosphonates Under Microwave Conditions. *Molecules* **2020**, *25*, 719.
52. Yokozawa, S.; Ohneda, N.; Muramatsu, K.; Okamoto, T.; Odajima, H.; Ikawa, T.; Sugiyama, J.; Fujita, M.; Sawairi, T.; Egami, H.; Hamashima, Y.; Egi, M.; Akai, S. Development of a Highly Efficient Single-Mode Microwave Applicator with a Resonant Cavity and its Application to Continuous Flow Syntheses. *RSC Adv.* **2015**, *5*, 10204–10210.

53. Vámosi, P.; Matsuo, K.; Masuda, T.; Sato, K.; Narumi, T.; Takeda, K.; Mase, N. Rapid Optimization of Reaction Conditions Based on Comprehensive Reaction Analysis Using a Continuous Flow Microwave Reactor. *Chem. Rec.* **2019**, *19*, 77–84.
54. Xu, J.; Yu, J.; Jin, Y.; Li, J.; Yu, Z.; Lv, Y. A Continuous Flow Microwave-Assisted Fischer Indole Synthesis of 7-Ethyltryptophol. *Chem. Eng. Proc.* **2017**, *121*, 144–148.
55. Egami, H.; Tamaoki, S.; Abe, M.; Ohneda, N.; Yoshimura, T.; Okamoto, T.; Odajima, H.; Mase, N.; Takeda, K.; Hamashima, Y. Scalable Microwave-Assisted Johnson–Claisen Rearrangement with a Continuous Flow Microwave System. *Org. Process Res. Dev.* **2018**, *22*, 1029–1033.
56. Egami, H.; Sawairi, T.; Tamaoki, S.; Ohneda, N.; Okamoto, T.; Odajima, H.; Hamashima, Y. (E)-3-[4-(Pent-4-en-1-yloxy)phenyl]Acrylicc Acid. *Molbank* **2018**, *2018*, M996.
57. Noshita, M.; Shimizu, Y.; Morimoto, H.; Akai, S.; Hamashima, Y.; Ohneda, N.; Odajima, H.; Ohshima, T. Ammonium Salt-Accelerated Hydrazinolysis of Unactivated Amides: Mechanistic Investigation and Application to a Microwave Flow Process. *Org. Process Res. Dev.* **2019**, *23*, 588–594.
58. Barham, J.; Tamaoki, S.; Egami, H.; Ohneda, N.; Okamoto, T.; Odajima, H.; Hamashima, Y. *C*-Alkylation of *N*-Alkylamides with Styrenes in Air and Scale-up Using a Microwave Flow Reactor. *Org. Biomol. Chem.* **2018**, *16*, 7568–7573.
59. Whittaker, A. G.; Mingos, D. M. P. Arcing and Other Microwave Characteristics of Metal Powders in Liquid Systems. *J. Chem. Soc. Dalton Trans.* **2000**, 1521–1526.
60. Dressen, M. H. C. L.; van de Kruijs, B. H. P.; Meuldijk, J.; Vekemans, J. A. J. M.; Hulshof, L. A. Vanishing Microwave Effects: Influence of Heterogeneity. *Org. Process Res. Dev.* **2007**, *11*, 865–869.
61. Monguchi, Y.; Ichikawa, T.; Yamada, T.; Sawama, Y.; Sajiki, H. Continuous-Flow Suzuki-Miyaura and Mizoroki-Heck Reactions Under Microwave Heating Conditions. *Chem. Rec.* **2019**, *19*, 3–14.
62. Ichikawa, T.; Mizuno, M.; Ueda, S.; Ohneda, N.; Odajima, H.; Sawama, Y.; Monguchi, Y.; Sajiki, H. A Practical Method for Heterogeneously-Catalyzed Mizoroki-Heck Reaction: Flow System With Adjustment of Microwave Resonance as an Energy Source. *Tetrahedron* **2018**, *74*, 1810–1816.
63. Engen, K.; Sävmarker, J.; Rosenström, U.; Wannberg, J.; Lundbäck, T.; Jenmalm-Jensen, A.; Larhed, M. Microwave Heated Flow Synthesis of Spiro-oxindole Dihydroquinazolinone Based IRAP Inhibitors. *Org. Process Res. Dev.* **2014**, *18*, 1582–1588.
64. Christiaens, S.; Vantyghem, X.; Radoiu, M.; Vanden Eynde, J. Batch and Continuous Flow Preparation of Hantzsch 1,4-Dihydropyridines under Microwave Heating and Simultaneous Real-Time Monitoring by Raman Spectroscopy. An Exploratory Study. *Molecules* **2014**, *19*, 9986–9998.
65. He, W.; Fang, Z.; Zhang, K.; Tu, T.; Lv, N.; Qiu, C.; Guo, K. A Novel Micro-Flow System Under Microwave Irradiation for Continuous Synthesis of 1,4-Dihydropyridines in the Absence of Solvents Via Hantzsch Reaction. *Chem. Eng. J.* **2018**, *331*, 161–168.
66. Kim, K. J.; Oleksak, R. P.; Hostetler, E. B.; Peterson, D. A.; Chandran, P.; Schut, D. M.; Paul, B. K.; Herman, G. S.; Chang, C. H. Continuous Microwave Gas-Liquid Segmented Flow Reactor for Controlled Nucleation and Growth of Nanocrystals. *Cryst. Growth Des.* **2014**, *14*, 5349–5355.
67. Dahal, N.; García, S.; Zhou, J.; Humphrey, S. M. Beneficial Effects of Microwave-Assisted Heating *versus* Conventional Heating in Noble Metal Nanoparticle Synthesis. *ACS Nano* **2012**, *6*, 9433–9446.
68. Kunal, P.; Roberts, E. J.; Riche, C. T.; Jarvis, K.; Malmstadt, N.; Brutchey, R. L.; Humphrey, S. M. Continuous Flow Synthesis of Rh and RhAg Alloy Nanoparticle Catalysts

Enables Scalable Production and Improved Morphological Control. *Chem. Mater.* **2017**, *26*, 4341–4350.
69. Akram, M.; Razali, I. R.; Ghafoor, F.; Ur-Rahman, S.; Butt, F. K.; Hussain, R. Continuous Facile Synthesis of Nano-Sized Zinc Oxide and its Optical Properties. *Mater. Res. Express.* **2018**, *5*, 075901.
70. Anwar, A.; Akbar, S. Novel Continuous Microwave Assisted Flow Synthesis of Nanosized Manganese Substituted Hydroxyapatite. *Ceram. Int.* **2018**, *44*, 10878–10882.
71. György, E.; Toricelli, P.; Socol, G.; Iliescu, M.; Mayer, I.; Mihailescu, I. N.; Bigi, A.; Werckman, J. Biocompatible Mn^{2+}-Doped Carbonated Hydroxyapatite Thin Films Grown by Pulsed Laser Deposition. *J. Biomed. Mater. Res. A.* **2004**, *71*, 353–358.
72. Anwar, A.; Akbar, S. Continuous Microwave Assisted Flow Synthesis and Characterization of Calcium Deficient Hydroxyapatite Nanorods. *Adv. Powder Tech.* **2018**, *29*, 1493–1498.
73. Cong, C.; Nakayama, S.; Maenosono, S.; Harada, M. Microwave-Assisted Polyol Synthesis of Pt/Pd and Pt/Rh Bimetallic Nanoparticles in Polymer Solutions Prepared by Batch and Continuous-Flow Processing. *Ind. Eng. Chem. Res.* **2018**, *57*, 179–190.
74. Albuquerque, G. H.; Squire, K.; Wang, A. X.; Herman, G. S. Continuous Synthesis of Monodisperse Ag Nanocubes. *Cryst. Growth Des.* **2018**, *18*, 119–125.
75. Kunze, F.; Kuns, S.; Spree, M.; Hülser, T.; Schulz, C.; Wiggers, H.; Schnurre, S. Synthesis of Silicon Nanoparticles in a Pilot-Plant-Scale Microwave Plasma Reactor: Impact of Flow Rates and Precursor Concentration on the Nanoparticle Size and Aggregation. *Powder Technol.* **2019**, *342*, 880–886.
76. Miyakawa, M.; Hiyoshi, N.; Koda, H.; Watanabe, K.; Kunigami, H.; Miyazawa, A.; Nishioka, M. Continuous Syntheses of Carbon-Supported Pd and Pd@Pt Core–Shell Nanoparticles Using a Flow-Type Single-Mode Microwave Reactor. *RSC Adv.* **2020**, *10*, 6571–6575.
77. Taddei, M.; Steitz, D. A.; van Bokhoven, J. A.; Ranocchiari, M. Continuous-Flow Microwave Synthesis of Metal-Organic Frameworks: A Highly Efficient Method for Large-Scale Production. *Chem. Eur. J.* **2016**, *22*, 3245–3249.
78. Vo, T. K.; Le, V. N.; Nguyen, V. C.; Song, M.; Kim, D.; Yoo, K. S.; Park, B. J.; Kim, J. Microwave-Assisted Continuous-Flow Synthesis of Mixed-Ligand UiO-66(Zr) Frameworks and Their Application to Toluene Adsorption. *J. Ind. Eng. Chem.* **2020**, *86*, 178–185.
79. Le, V. N.; Kwon, H. T.; Vo, T. K.; Kim, J. H.; Kim, W. S.; Kim, J. Microwave-Assisted Continuous Flow Synthesis of Mesoporous Metal-Organic Framework MIL-100 (Fe) and Its Application to Cu(I)-Loaded Adsorbent for CO/CO_2 Separation. *Mat. Chem. Phys.* **2020**, *253*, 123278–123284.
80. Kokel, A.; Török, B. Sustainable Production of Fine Chemicals and Materials Using Nontoxic Renewable Sources. *Toxicol. Sci.* **2018**, *161*, 214–224.
81. Brigham, C. Biopolymers: Biodegradable Alternatives to Traditional Plastics. In *Green Chemistry: An Inclusive Approach*; Török, B., Dransfield, T., Eds.; Elsevier: Oxford, 2018; pp. 753–770.
82. Schobert, H. Chemistry of Fossil Fuels and Biofuels. *Cambridge Series in Chemical Engineering*; Cambridge University Press: Cambridge, 2013; pp. 53–68.
83. Brigham, C.; Bader, J. From Ethanol to Biodiesel: A Survey of Green Fuels. In *Green Chemistry: An Inclusive Approach*; Török, B., Dransfield, T., Eds.; Elsevier: Oxford, 2018; pp. 861–880.
84. Choedkiatsakul, I.; Ngaosuwan, K.; Assabumrungrat, S.; Mantegna, S.; Cravotto, G. Biodiesel Production in a Novel Continuous Flow Microwave Reactor. *Renewable Energy.* **2015**, *83*, 25–29.

85. Tangy, A.; Pulidindi, I. N.; Perkas, N.; Gedanken, A. Continuous Flow Through a Microwave Oven for the Large-Scale Production of Biodiesel from Waste Cooking Oil. *Bioresour. Technol.* **2017**, *224*, 333–341.
86. Hung, C. H.; Chew, Y. M.; Fu, Y. C.; Chen, C. S. Multi-Stage Microwave-Assisted Transesterification of Soybean Oil—An Economic Design for High Yield Biodiesel Production. *Int. J. Chem. Eng. App.* **2018**, *9*, 32–37.
87. Lee, H.; Liao, J. D.; Yang, J. W.; Hsu, W. D.; Liu, B. H.; Chen, T. C.; Sivashanmugan, K.; Gedanken, A. Continuous Waste Cooking Oil Transesterification with Microwave Heating and Strontium Oxide Catalyst. *Chem Eng. Technol.* **2017**, *41*, 192–198.
88. Ardebili, S. M. S.; Ge, X.; Cravotto, G. Flow-Mode Biodiesel Production From Palm Oil Using a Pressurized Microwave Reactor. *Green Process Synth.* **2019**, *8*, 8–14.
89. Ochoa-Gómez, J. R.; Gómez-Jiménez-Aberasturi, O.; Ramírez-López, C.; Belsué, M. A Brief Review on Industrial Alternatives for the Manufacturing of Glycerol Carbonate, a Green Chemical. *Org. Process Res. Dev.* **2012**, *16*, 389–399.
90. Nogueria, D. O.; de Souza, S. P.; Leão, R. A. C.; Miranda, L. S. M.; de Souza, S. P.; Leão, R. A. C.; Miranda, L. S. M.; de Souza, R. O. M. A. Process Intensification for Tertiary Amine Catalyzed Glycerol Carbonate Production: Translating Microwave Irradiation to a Continuous-Flow Process. *RSC Adv.* **2015**, *5*, 20945–20950.
91. Tang, Y. H.; Liu, S. H.; Tsang, D. C. W. Microwave-Assisted Production of CO_2-Activated Biochar From Sugarcane Bagasse for Electrochemical Desalination. *J. Haz. Mat.* **2020**, *383*, 121192–121200.
92. Jermolovicius, L. A.; Pouzada, E. V. S.; do Nascimento, R. B.; de Castro, E. R.; Senise, J. T.; Mente, B. B.; Martins, M. C.; Yamaguchi, S. M.; Sanchez, V. C. Greening the Green Ethylene With Microwaves. *Chem. Eng. Proc.* **2018**, *127*, 238–248.
93. Bo, L.; Sun, S. Microwave-Assisted Catalytic Oxidation of Gaseous Toluene With a Cu-Mn-Ce/Cordierite Honeycomb Catalyst. *Front. Chem. Sci. Eng.* **2019**, *13*, 385–392.
94. Barham, J. P.; Tanaka, S.; Koyama, E.; Ohneda, N.; Okamoto, T.; Odajima, H.; Sugiyama, J.; Norikane, Y. Selective, Scalable Synthesis of C_{60}-Fullerene/Indene Monoadducts Using a Microwave Flow Applicator. *J. Org. Chem.* **2018**, *83*, 4348–4354.

CHAPTER FIVE

Application of sonochemical activation in synthetic organic chemistry

Márton Szabados, Pál Sipos, and István Pálinkó
University of Szeged, Szeged, Hungary

1 Introductory remarks

In contrast with several mammals like dolphins, whales, bats or a closer relative, the Philippine tarsier, the human race did not need to acquire the capability of generating and perceiving ultrasound waves to survive. Since the development of the first ultrasound transducer by Chilowsky and Langevin in 1917[1] and ten years later the first description of physical and biological effects of ultrasound by Wood and Loomis,[2] the utilization of high-frequency sound waves exerted serious impact on various fields of research and appeared in military and medical applications and very soon in the everyday life. A nonexhaustive list of applications includes sound navigation and ranging (abbreviated as sonar) for submarine operations; diagnostic imaging and therapeutic technique in healthcare; nondestructive scanning method for defects in welds, metals, composites, and woods; cleaning devices for laboratory utensils; dental and jewelery applications; the first models of television remote controls; and one of the latest, the ultrasonic toothbrushes for daily domestic use. Many branches of chemical science like organic, inorganic, material, biological, polymer, food and environmental chemistry employ ultrasonic devices. The fundamental properties, generation, detection, and some chemical applications of ultrasound have been described in several series of books[3–7]; therefore our aim here was to give an overview of the most recent (last 5 years) endeavors on the uses of ultrasonic treatment in the syntheses and transformations of organic chemicals. For a start, let us familiarize the reader with the physical and chemical aspects of ultrasound propagation in liquid media.

Nontraditional Activation Methods in Green and Sustainable Applications
https://doi.org/10.1016/B978-0-12-819009-8.00007-4
© 2021 Elsevier Inc.
All rights reserved.

2 Sonochemistry

2.1 Scientific background of the ultrasonic radiation

Sound waves with frequency higher than 20 kHz are called ultrasound, while the range between a few hundred megahertz and several gigahertz is frequently defined as ultrahigh frequency ultrasound. As longitudinal waves, the propagation of the ultrasound results in the displacement of the atoms/ions/molecules of the medium (gas, fluid, and even solid) in the direction of the movement of the waves. The propagated waves induce compressed and evacuated spaces of molecular size with increased and decreased pressure, respectively. In liquids the rapid changes of the pressure results in the phenomenon called cavitation, the source of the physicochemical effects of ultrasound irradiation. When the pressure reaches a sufficiently low level, the distance between the molecules becomes high enough to break the consistency of the liquid medium. The thus continuously forming and collapsing cavities (they are often called as bubbles or voids) can create locally several hundreds of atmosphere pressure and few thousands of degrees Kelvin. There are two types of cavitation bubbles depending on the intensity of the ultrasound employed. One is the stable cavitation, when the cavities of equilibrium size oscillate steadily during many acoustic cycles. They can be frequently observed with the naked eye, because the bubbles with a few hundred micrometers size are very robust due to their moderate pulsation. The other is the transient cavitation, where the size of the voids gradually grows for a few acoustic cycles until they violently collapse into smaller daughter cavities. They get dissolved in the medium or coalesce with other cavities generating a larger one. The stable cavitation occurs at low ($<3\ W/cm^2$) ultrasonic intensities, while the formation of transition bubbles requires higher ($>10\ W/cm^2$) transmitted energy.[3, 7]

For the chemist, transient cavitation is more relevant than stable cavitation, since it serves as microreactor due to the chemical and mechanical aspects generated by the collapses. The high temperature results in the partial thermal decomposition of the medium producing reactive radicals. In aqueous medium, the formation of gaseous hydrogen and oxygen, hydrogen peroxide, and hydroxyl radicals is the most frequent. In addition, under ambient conditions (presence of air at atmospheric pressure), the main gas components of the bubbles are argon (from air), NO_x, and HNO_x due to the chemical reaction between N_2, O_2, and water molecules. In the final stage of the collapse, only argon is present, since the chemically active nitrogen oxide species are gradually dissolved in the medium.[8] Besides the thermal and chemical aspects, the potential energy of the cavities transforms into kinetic energy as well, when shock waves are emitted from the collapsed bubbles (the velocity of the moving liquid wall can exceed the speed of the sound in the gas inside the void). Moreover, rapid (100–200 m/s) liquid microjets penetrate into the voids, and impact on the solid substrates resulting in their disintegration to even nanometer sizes or circular erosion spots on the surface, if the voids are collapsed near the solid surface.[9]

Sonochemistry is the branch of chemistry where these effects of the continuously formed and destroyed cavitation bubbles are attempted to be applied in as many ways as possible. Interestingly, sono- and mechanochemical processes have several similar features like the presence of local hot spots with extreme conditions, fast heating and cooling rates ($>10^9$ K/s) and their inhomogeneous distribution in the reaction medium due to the inhomogeneous distribution of the cavitation regions.[10–12] Therefore, by now, it became clear that sonochemistry, similarly to mechanochemistry, has great potential in inducing chemical transformations, which do not take place under conventional conditions.

2.2 Sonoluminescence

Beyond the various chemical, kinetic, and thermal forms of energy transfer, the collapse of the bubbles frequently results in light emission as well. Even though this phenomenon has been well known for a long time, and is directly connected to the extreme conditions inside the cavities, it is worthwhile to introduce this field separately, because the quantized energy character of the emitted photons reveal in situ information about the inner environment of the bubbles. There are two types of sonoluminescence. One is the multibubble sonoluminescence (MBSL), when the plethora of the formed and collapsed bubbles emit only a few photons. The other is the single-bubble sonoluminescence (SBSL), when only one bubble is illuminated with high intensity. It is observable as bluish spots in a dark space lasting for several minutes. While MBSL frequently follows common ultrasonic treatments, the occurrence of SBSL is limited due to the strict terms of the stable emission regarding the size, shape, location, and composition of the bubbles.[13]

Both SBSL and MBSL spectra can extend from the ultraviolet to the infrared range. The photons stem from the relaxation of the excited ions and radicals and from the decomposition and the reaction of the molecules inside the bubbles. Since only the noble gases are capable of remaining intact at this high temperature and pressure, the quality (ionization energy, thermal conductivity, and solubility) of the dissolved noble gases has strong influence on the intensity of the emitted light.[14] The key factor is the temperature in the center of the cavities; therefore every component of the liquid medium contributes to sonoluminescence. Larger concentration of the volatile compounds results in higher density inside the bubbles; therefore lower temperature is reached, since the energy is used for their decomposition. The presence of alcohols or hydrocarbons or enhanced water evaporation decreases the intensity of light emission, while ultrasonic irradiation of highly concentrated aqueous sulfuric acid solution (low vapor pressure) results in higher temperature.[15, 16]

Finally, the ultrasonically enhanced chemiluminescence processes are often called as sonochemiluminescence, where the generated radicals can strengthen the emission intensity of the light during the chemical reactions; this was observed during the oxidation of luminol[17] or lucigenin.[18]

2.3 The most important parameters of ultrasonic treatments

Ultrasonic irradiation is commonly used to accelerate the mass and energy transfer in the reaction space. Therefore several processes including extraction, dispersion, aggregation, emulsification, crystallization, polymerization and cell disruption can be facilitated in sonochemical way. The energy induced by cavitation and thus the efficiency of sonication mainly depend on the pressure and temperature generated inside the bubbles and the frequency of bubble formation. In most cases, the sonochemical reactions take place under atmospheric pressure and in aqueous environment, commonly between 273 and 373 K. However, as it has already been shown for sonoluminescence, the chemical composition and temperature of the medium, the frequency and intensity of the ultrasonic irradiation and even the external pressure have serious influence on the formation of the cavities.[4, 19]

The choice of the **medium** and **temperature** applied is essential because of their already mentioned effect on the gas composition and thus the temperature reached inside the cavitation bubbles. In addition, the number of the generated cavities can be enhanced by selecting the media with low cohesive forces. The presence of van der Waals forces and hydrogen bonding, low surface tension, viscosity and vapor pressure result in enhanced cavity generation. Under carefully chosen conditions, the addition of surface active agent or increase in the temperature also favors the evolution of stable bubbles.[20] As far as temperature is concerned, enhanced temperature favors the formation of the bubbles due to the weaker intermolecular forces of the media, but more evaporated molecules can enter into the cavities. Furthermore, ultrasonic treatments always increase the temperature of the irradiated medium, which accelerates the reactions in general; therefore the separation of the effect of ultrasonic treatment and the applied temperature often requires special attention.

The **frequency** and the **intensity** of sonication are limited by the instrument used; however, nowadays, these parameters can be finely adjusted to the goals of the actual research project. The frequency of the ultrasound waves generated is precisely determined by the type of the ultrasonic transducer, and thus its variation can be problematic. Fortunately the duration of the acoustic cycles is in direct connection with the probability of cavity formation. At higher frequency, the waves are repeated more frequently, and the molecules cannot move far enough from each other to discontinue the consistency of the liquid phase; this facilitates bubble formation. The intensity of the ultrasonic treatment can be increased by increasing the amplitude of the vibration of the transducer. The intensified sonication can enhance the efficiency of oxidation reactions[21] or light emission[22]; however, at the cost of increased amortization of the transducer. At higher acoustic amplitude, the maximum size of the bubbles (R_{max}) increases, while the minimum sizes (R_{min}) decreases during the rarefaction and compression states of the acoustic cycles, respectively. Finally the increased R_{max}/R_{min} ratio results in lower pressure and higher temperature inside the cavitation bubbles and thus more effective ultrasonic treatment.

By raising the **external pressure**, the cohesive forces in the liquid phase are strengthened, and the R_{max} and R_{min} values, the time required for the collapse of the bubble, and the composition of the gas phase inside the bubbles change. This results in a maximum curve when one plots the power for one collapse (change in internal energy divided by the duration of bubble collapse) versus the external pressure.[19] Moreover the dissolved gas molecules can serve as nucleation spots for the cavities; however, the gas bubbles and solid particles in large amounts can influence the propagation of the ultrasound waves, and can induce the formation of microjets; therefore their impacts on a sonication process are scarcely predictable.

Obviously the efficiency of sonochemical treatments primarily depends on the goals set; therefore finally, the **time interval** and the **pulse characteristics** of the emission of the ultrasound radiation (i.e., continuous radiation and silent periods follow each other) need to be considered as well.

2.4 Technical aspects of implementing sonochemical reactions

Recently, the number of commercially available ultrasonic devices is growing gradually for industrial, laboratory, and even domestic uses, but the principles of their design, operation, and installation are quite similar.[4, 5] They consist of an electric oscillator to generate periodic, oscillating signals driving the transducer, which transforms the electric energy into mechanical energy (sound vibrations). There are two types of transducers: the magnetostriction types are made of ferromagnetic materials like ferrites, while the piezoelectric ones, more commonly used these days, are made of piezoelectric ceramics, polymers, films, or singe crystals from ZnO, $BaTiO_3$, etc. The diversity of the ultrasonic devices is connected to the different requirements in applications like cleaning, homogenizing, atomizing or disintegrating and, therefore, from the connection of the transducer and the vessel, containing the medium in which the ultrasound propagates. Generally, in the bath-type devices (ultrasonic bath and several sonoreactors), the transducer is firmly glued to the bottom or the outer surface of the tank. The uniform or focused propagation of the ultrasound waves is often complicated in these systems. For the horn-type devices, the transducer is directly connected to a vibration plate facing toward a metal rod called horn or probe, and its tip is submerged into the sonicated medium. This arrangement is capable of generating ultrasound with higher intensities at lower frequencies as well. The shape and the composition of the probes are also relevant regarding the amplitude of the transmitted ultrasound waves and the contamination of the medium. While the most widely used probes are made of titanium alloy and thus contaminating the products of the chemical reaction by Ti, Al, Cr, or V is hardly avoidable, the silica probes possess reasonably high chemical resistance. By using two or more probes at the same time in the reaction vessel, the output energy of the ultrasonic treatment can be efficiently increased.

Finally, it is worthwhile to classify the sonochemical processes by the way how the ultrasound is driven into the reaction mixture. When the wall of the sample tank transports the ultrasound radiation to the sample, we talk about indirect treatment. When there is no transmitter phase, direct ultrasonic treatment takes place. Using direct sonication, the risk of the contamination is enhanced; however, applying the indirect way, the intensity of the ultrasound is always lower than expected, because part of the energy is absorbed by the wall and/or the transmitting medium. For both cases, beyond the fixed operation parameters, temperature control, the geometry of the sonicated vessel, and the arrangement of the samples are crucial for the reproducibility and comparability of the experimental results and the determination of the efficiency of the sonochemical treatment. For the latter, the KI method is a generally accepted chemical technique.[23] The ultrasound irradiation of the aqueous solution of the KI salt produces I_2 molecules in a reaction between the in situ generated oxidizing agents (radicals, O_2, H_2O_2, NO_x, and HNO_x) and the iodide ion. Then the I_2 molecules and the iodide anions in excess form triiodide ions. The amount of the generated I_3^- ions calculated from data obtained by UV–Vis spectroscopy measurements provides with numerical data to map and compare the efficiencies of the ultrasonic treatments.

3 Using ultrasonic irradiation in synthetic organic chemistry

Ultrasound irradiation has numerous effects on the chemical reactions. Mechanical, thermal, translational, compressional and even photoinduced events occur in parallel, and the interplay among them determines the transformation pathway(s) and the outcome(s) of the reactions. Therefore it is extremely complicated to classify the ultrasound modified syntheses in light of the hardships in separating the various contributing factors of the ultrasonic treatment. Instead of doing this, it was decided to concentrate on the synthetic organic chemistry applications of sonochemistry. In addition, the recent trend of combining ultrasonic and microwave irradiations will also be discussed.

3.1 Syntheses with the assistance and control of ultrasound waves

The huge potential in sonochemistry is well indicated by the number of the published reviews only dealing with the organic chemistry area in the last five years. These works generally emphasize the "green and sustainable" advantages of the ultrasonic treatments, such as the chance to apply green solvents like water, ethylene glycol, or glycerol.[24, 25] The sonochemical activation also enhances the yields and selectivity during the shortened reaction times[26] avoiding the formation of toxic byproducts and complex workup procedures in multicomponent noncatalytic reactions.[27, 28] The advantages of sonochemistry can also be exploited in the preparation and/or the use of catalysts.[29] The low energy consumption aspect of the ultrasound-assisted methods is also an often referred popular and green benefit. However, it should be taken into consideration that the determination of the ratio of the energy consumption of conventional and sonochemically activated processes is extremely complicated due to the various effects of the propagating ultrasound waves; it is especially so at scaled-up industrial applications.[30, 31]

To demonstrate the progress in organic sonochemistry, we separated the original papers into two main groups. In the first one, attention is focused on the reaction under ultrasound irradiation in noncatalytic and in the second, in catalytic systems. The papers are further grouped in terms describing expected (higher reactivity, conversion and yield, milder conditions, simple setup, etc.) and specific (changes in chemo- and/or regioselectivity, solventless conditions, etc.) features.

3.2 Noncatalytic transformations

3.2.1 Ultrasound-accelerated reactions

Most of the papers referenced in this section belong to the field of multicomponent reactions and/or the preparation of azoles under ultrasound irradiation. Their popularity stems from the intense industrial interests due to the atom and step economic aspects of the multicomponent reactions and the biological importance of the azole derivatives. Shabalala et al. developed a one-pot four-component reaction to produce functionalized 1,4-dihydropyridine derivatives (biologically active heterocyclic systems) testing the polarity of the solvent with different conditions (Scheme 1).[32] However, the best results were obtained without any additive and in 1:1 ethanol/water medium at room temperature. Under these mild conditions, the reaction was completed after twelve hours (<90% yield); however, with ultrasound assistance (generated by a commercially available ultrasonic cleaner of 40 kHz frequency and 250 W output power), in addition to significantly shortening the reaction time, the yield exceeded 90%.

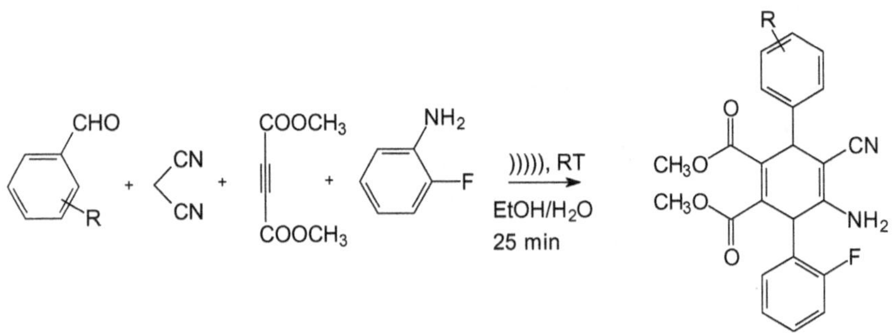

Scheme 1 Synthesis of functionalized 1,4-dihydropyridine derivatives (scope: 11 examples; yield: 89%–96%; 15 min at 40 kHz).

Similar enhancements were observed for the more complicated six-component[33] (Scheme 2) and three-component reactions (Scheme 3) in ethanol or aqueous ethanol.[34, 35]

To demonstrate the advantageous features of using ultrasound, it is very instructive to compare the reaction time, temperature, and yield under reflux and ultrasound-assisted conditions (Scheme 4).[35] The reaction proceeds faster under mild conditions resulting in higher yield in the ultrasonically-assisted reaction than under the "classical" reflux conditions.

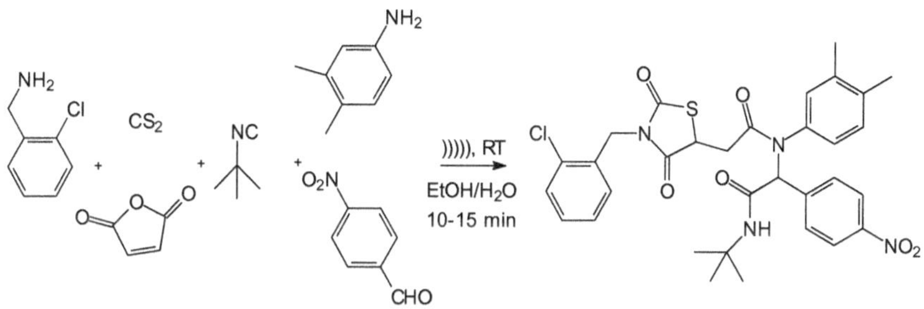

Scheme 2 One-pot six-component synthesis of a pseudopeptide containing rhodanine scaffold (scope: 12 aliphatic amines, 6 aromatic amines, 2 anhydrides, 13 aldehydes, 6 isocyanides; yield: ~70% in water; irradiation time: 10–15 min at 45 kHz).

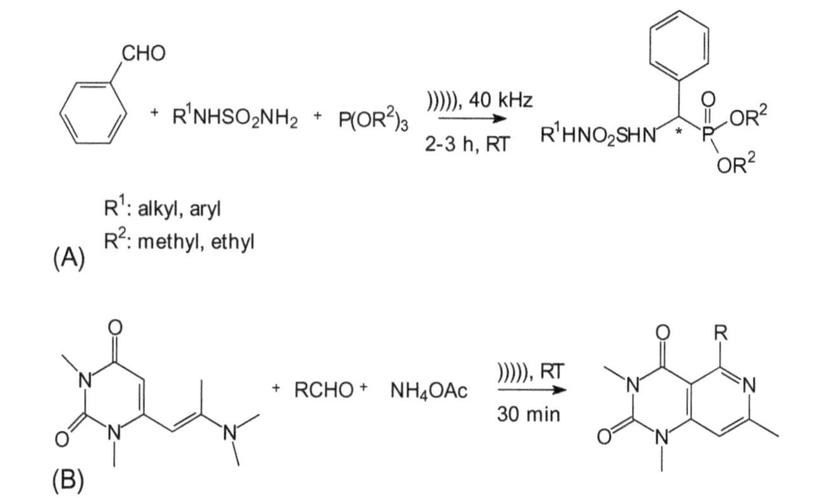

Scheme 3 One-pot syntheses of (A) α-sulfamidophosphonates (scope: 17 examples; yield: 65%–95%) and (B) 7-substituted pyrido[4,3-*d*]pyrimidines (scope: 13 examples; yield: 78%–85%) under ultrasound irradiation.

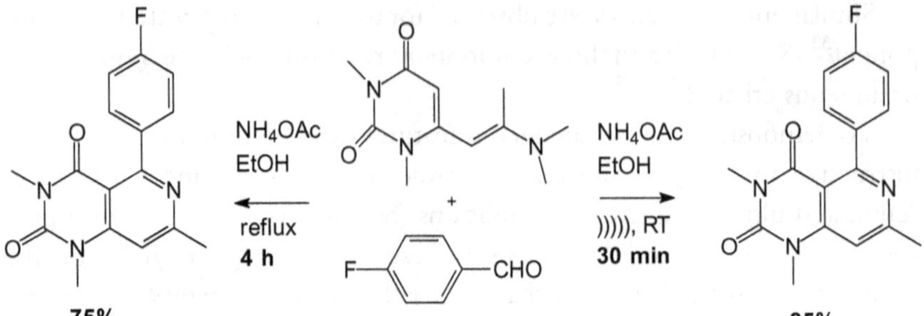

Scheme 4 The synthesis of the pyrido[4,3-*d*]pyrimidine derivative under ultrasound irradiation and under "conventional" reflux conditions.

In the ultrasound-assisted one-pot, three-component synthesis of indole derivatives containing 1,3,4-oxadiazole and 1,2,4-triazole moieties (Scheme 5), both the reaction times and the yields were higher than those observed in the conventional thermal reactions, and the reaction times were comparable in the two methods, but the yields were even higher in the microwave-assisted reaction (Scheme 6).[36]

In another three-component ultrasonically assisted route, new C—Se bond-forming coupling reaction was performed in one pot using polyethylene glycol (PEG) as nonhazardous, water-soluble, easily reusable, and excellent ultrasound mediating solvent.[37] Unfortunately, this work contained no information about the technical data of the sonication: the type of the device and the ultrasound transmission as well as the applied frequency and intensity, which makes extremely difficult to reproduce the data by others.

In another much better documented paper, a simple method was described for the preparation of C—Se bonds (Scheme 7). Here a horn-type ultrasound device was successfully applied to aid the oxidation of the

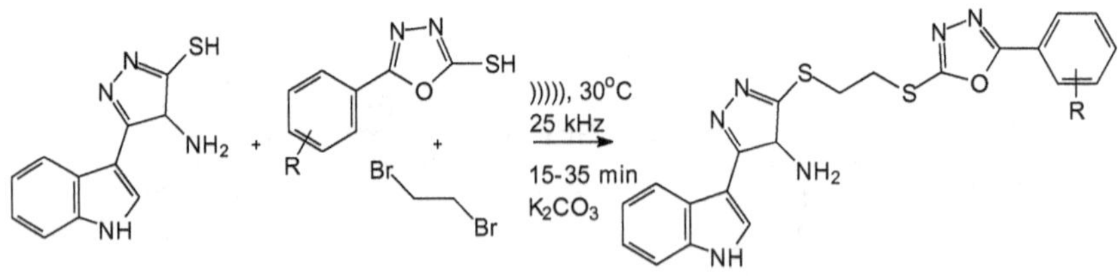

Scheme 5 The ultrasound-assisted, three-component one-pot reaction of indole derivatives containing 1,3,4-oxadiazole and 1,2,4-triazole components (scope: 13 examples; yield: 82%–92%).

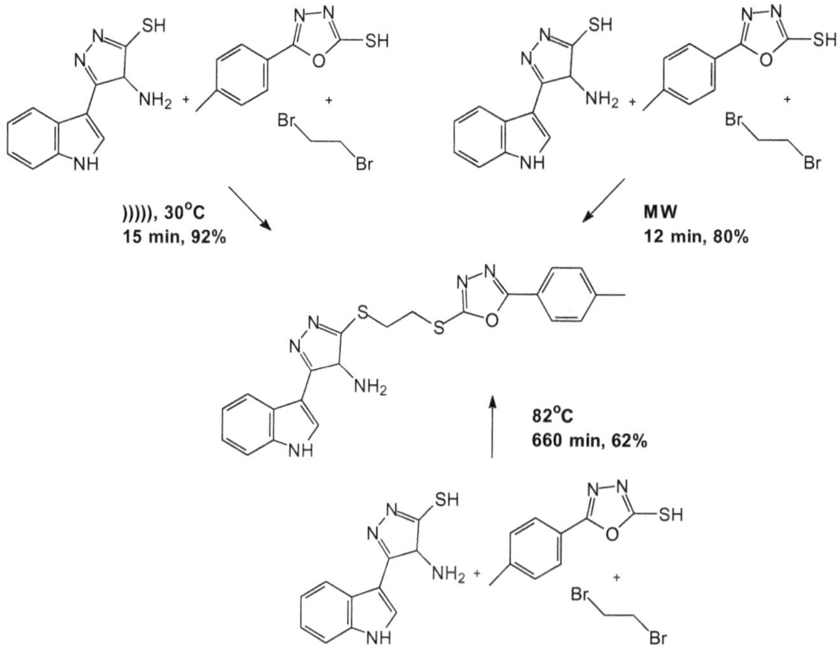

Scheme 6 The comparison of some reaction parameters (reaction temperature, duration, and yield) in the synthesis of an indol derivative under ultrasound-assisted, microwave (MW), and "conventional" conditions.

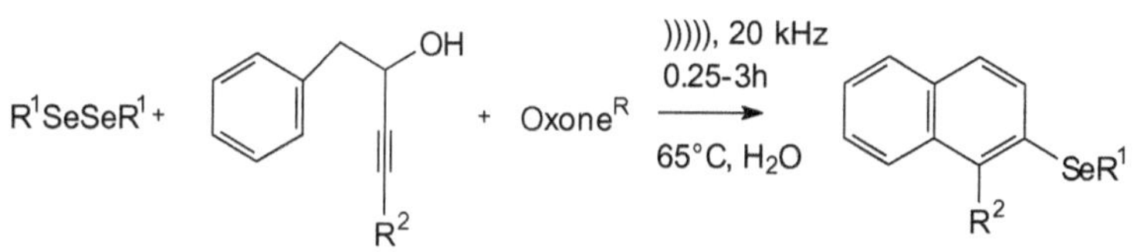

Scheme 7 The ultrasound-assisted synthesis of 2-organoselanyl-naphthalenes (scope: 9 examples; yield: 56%–94%).

diselenide starting material by the commercially available potassium peroxymonosulfate (Oxone^R) oxidizing agent.[38]

In the rapid synthesis (10–90 min) of pyrazole derives, a mixture of PEG and water proved to be an efficient, environmentally friendly solvent and suitable medium for transmitting ultrasonic energy.[39]

Furthermore, the Passerini tetrazole reaction (for preparing bioactive scaffolds) was successfully enhanced by aiding the reaction in an ultrasonic cleaner in 1:1 methanol/water mixture avoiding the use of excess azide source (Scheme 8).[40]

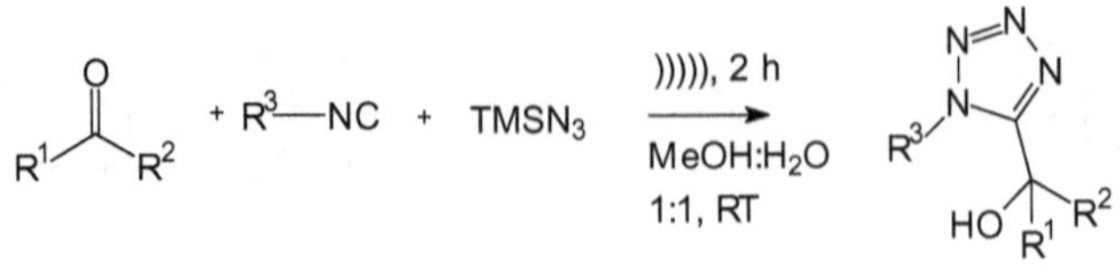

Scheme 8 The Passerini tetrazole reaction under ultrasonic conditions (scope: 9 examples; yield: 39%–98%).

Numerous studies dealing with the formation of new C—N bond were published in the last few years in which the application of ultrasound irradiation resulted in significant decrease in reaction times,[42–48] even better than those carried out under the microwave irradiation (Scheme 9).[41] In several cases the reaction time was less than 10–15 min providing excellent yields.[49–55]

In addition, the effects of the intensity of the generated ultrasound in isocyanide-based reactions have been assessed. It was found that the ultrasound irradiation had optimum output power operation parameters, the highest yields were obtained at 100 W, and at weaker and stronger sonication the yields were the same or lower.[56, 57] This observation has been explained by the efficiency of the mass transfer. However, the higher reaction rate was also claimed to be due to the faster formation of the conjugate

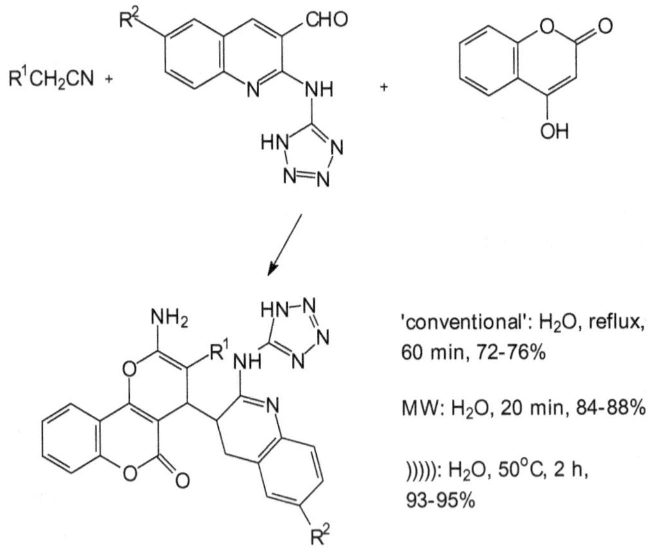

Scheme 9 The syntheses of benzopyrano[4,3-*b*]chromene derivatives (4 examples) under "conventional," microwave (MW), and ultrasonic conditions.[41]

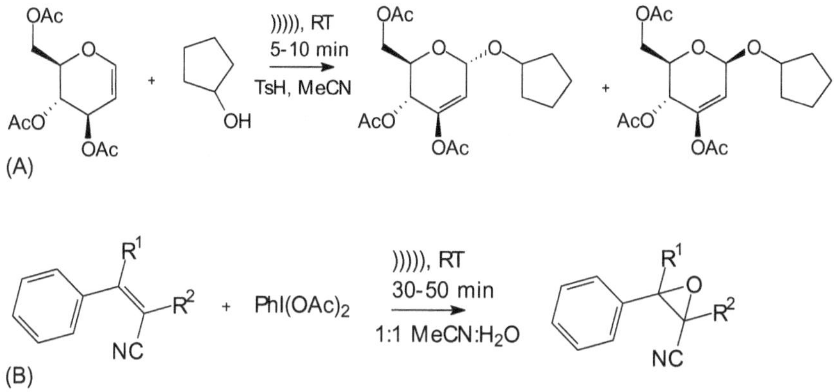

Scheme 10 The ultrasound-promoted syntheses of (A) 2,3-unsaturated pseudoglycosides (α:β = 81–89:19–11; scope: 6 examples; yield: 72%–92%)[59] and (B) β-cyanoepoxides (scope: out of 13, 11 working examples; yield: 79%–94%).[60]

bases of the various carboxylic acid starting materials. The latter is most probably due the molecular fragmentation caused by the hot spots with high transient temperature and pressure and the occurrence of reactive species.[57] Similar mechanism was proposed for the ultrasonic treatment of the biomass to transform the lignocellulosic materials into useful petrochemical products by ultrasonic assistance.[58] It was observed that the cup horn ultrasonic transducer resulted in yields higher than those obtained when mechanical stirring was used.

Finally, several recent reports describe the ultrasound-promoted synthesis of new C—O bonding for glycosidation[59] and epoxidation[60] reactions as well (Scheme 10).

3.2.2 Alteration of reaction pathways by ultrasonic irradiation

Among the ultrasonically assisted noncatalytic reactions, the most frequently experienced and unexpected changes observed are related to the chemoselectivity of the reactions. Due to the influence of cavitation, the induced hot spots with extreme physical conditions, to the enhanced mass transfer and to the reactive species thus generated, the reaction path can differ from that of the silent reaction. In the field of multicomponent reactions, the synthesis of unique products can be selectively exploited.

The outcome of the three-component reactions of alkyl pyruvates, pyridazines, and aldehydes proved to be highly diverse and easily modifiable in the sonochemical applications; the ultrasound-assisted reaction led to a product different from that obtained under "conventional" or microwave conditions (Scheme 11).[61]

In another detailed report the reaction of arylpyruvic acid with imine derivates resulted in the formation of tetrahydropyrimidines under ultrasonic irradiation. In contrast, pyrrolone derivatives were formed under microwave irradiation or conventional heating.[62]

In the formation of new C—S bond (2-sulfonylquinoline derivatives), in the sonication-assisted first step, the reduction of the sulfonyl chloride by zinc dust takes place to form a reactive intermediate. Through the enhanced formation of this species, the product was obtained with higher yield because of minimized side reactions.[63]

Finally, there are two recent examples demonstrating the immense influence of ultrasound treatment on the rate of reactions, where even the sterically hindered pyrazole[64] and chromene[65] derivatives were synthesized in good yields (Scheme 12).

In the field of polymer chemistry, the application of ultrasound was proven to be advantageous in various steps of the polymerization and even in the posttreatments to modify the physicochemical parameters of the polymer molecules. The frequency of the ultrasonic radiation is a decisive parameter during the polymerization due to the sonolysis (that is, transforming the water molecules into reactive radicals) mainly occurring at higher frequencies (above 100–200 kHz). The low frequencies (around 20 kHz) increase the decomposition of the initiator molecules releasing free radicals and accelerating the rate of the polymerization. Thus higher monomer conversion

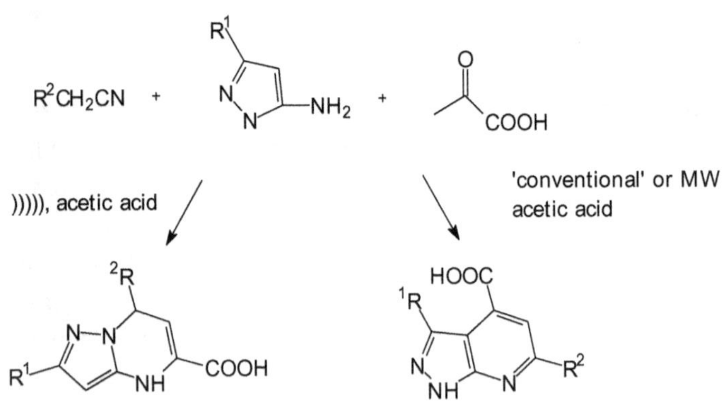

Scheme 11 The products in the three-component reactions of alkyl pyruvates, pyridazines, and aldehydes depending on the reaction conditions applied (10 examples).

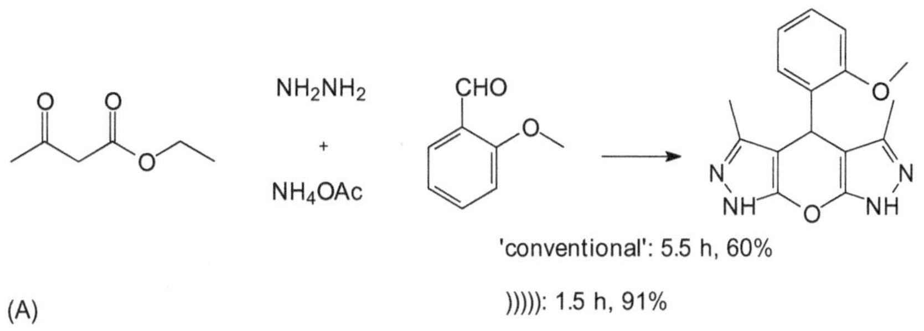

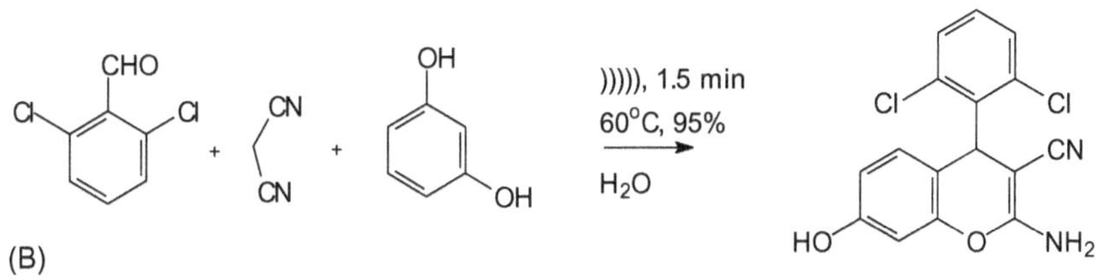

Scheme 12 Yields for the sterically hindered products in the ultrasonically assisted syntheses of (A) pyrazole[64] and (B) chromene[65] derivatives.

was observed compared with the silent polymerization for the emulsion polymerization of polyacrylates.[66] As a further improvement, with the help of sonication, the necessary initiator/monomer ratio and temperature of the polymerization could be reduced.

In another interesting publication a novel approach was applied to sonochemical control of the polymerization mechanism.[67] Piezoelectric $BaTiO_3$ nanoparticles absorbed the ultrasound; therefore the mechanical energy of the waves could be transformed into electrochemical energy. Cu(I) species, in situ generated from ligand-stabilized Cu(II) precursors by the induced piezocatalytic reduction, were used as activator. In this setup, the length of the polymer chain formed was in linear correlation with the applied time of the activation besides the excellent monomer conversion and narrow molecular weight distribution.

Using high-frequency ultrasound irradiation means that in the absence of added initiator, the polymerization could be initiated and conducted with species formed from the sonolysis of the water molecules due to the chain-breaking effect of the shear forces from the continuously forming and collapsing cavities. Moreover, the intensity of the ultrasound irradiation has remarkable control over the rate of the polymerization, and with water applied as initiator and solvent at the same time, this technique can be a green alternative for conventional polymerization techniques.[68, 69]

The application of ultrasound irradiation for the posttreatments of macromolecules to break bonds and to form shorter/smaller chains due to the fragmentation effect of the cavitation phenomenon is a well-investigated area. In a recent paper, the determining factor of the ultrasound-induced chain scission was investigated in detail.[70] The work with polystyrene and polynorbornene macromolecules attested that the influence of the contour length was crucial; while the effect of the molecular weight proved to be minimal, the ultrasonic treatment generated similar weight reduction for the polymers with similar contour length.

There are some original papers describing observations that cannot be fitted to any of the previous groups; therefore these works are presented at the end of this chapter.

The ultrasonic energy dissipation capability of various solvents has been recently investigated. It was found that the ionic liquid 1-butyl-3-methylimidazolium tetrafluoroborate ([BMIM][BF_4]) displayed higher energy absorption capability than even water. The key parameter was the viscosity of the solvents, while the effect of their vapor pressure and surface tension was negligible in the process of transforming the ultrasonic energy into thermal energy. Thus, the ionic liquid proved to be ideal in increasing the heating efficiency of the conventionally used organic solvents. The authors used this effect in the ultrasonically aided N-alkylation reaction of pyrazoles (Scheme 13).[71]

It is worth mentioning the sonochemical synthesis of polyhydroxylated carbon nanomaterials. A detailed summary of the knowledge gathered about the modification of graphene and buckminsterfullerene (C_{60}) with OH groups was released recently.[72] The review describes the effect of the ultrasonic treatments, from the fundamental parameters (temperature, frequency, intensity, and duration) to the specific ones (flow rate of inert gas, dimension of the reactor, position of the sonotrode tip and quality of the dissolved gas). The hydroxylation of these materials results in the enhancement of water solubility increasing their biocompatibility—the sonochemical process

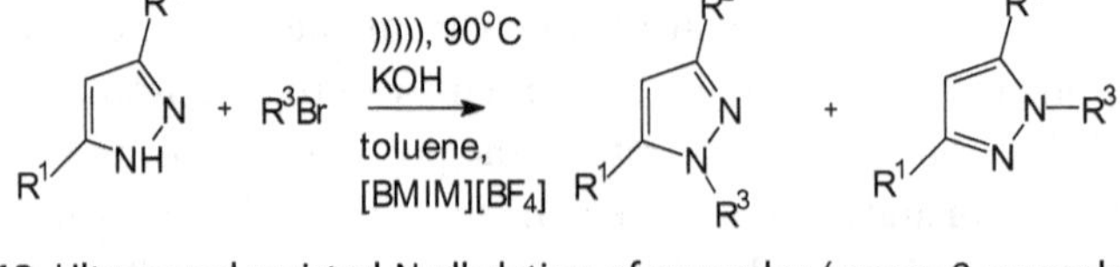

Scheme 13 Ultrasound-assisted N-alkylation of pyrazoles (scope: 9 examples).

appears to be the fastest and most convenient method even at the industrial scale.

At the end of this section, let us mention an experimental study representing the plausible role of the cavities in the genesis of the life on Earth.[73] Australian researchers probed the feasibility of amino acid formation in water mimicking the primordial environment of Earth under ultrasonic irradiation. They observed the formation of glycine, ethylglycine, alanine, serine, proline, valine, threonine and leucine molecules from the reaction of radicals generated during the sonolysis of the water solution with acetic acid, methane, carbon dioxide, and nitrogen gas. This work is a good addition to confirm the hypothesis that the first complex organic molecules are connected to cavitation chemistry. As initiators the hydrodynamic cavitation in the ocean waves confronting onto solid surface of the coasts could offer a plausible system for the formation of the organic molecules. Later, these species adsorbing on minerals with catalytic properties could transform further into more complex molecules becoming eventually the building blocks of life.

3.3 Catalyzed organic transformations

It is reasonable to say that the catalytic reactions make up the largest group of the sonochemical organic syntheses. Therefore, here, our aim was to list and present these reactions regarding the chemical nature of the applied catalyst and not only in terms of the final products. In the field of the multicomponent reactions, the magnetic catalysts have received substantial attention recently due to their easy and controlled separation from the reaction mixture. Iron oxide-based materials are excellent candidates in these applications; for instance, chitosan-functionalized Fe_3O_4 magnetic nanoparticles (MNP) were prepared and tested in the synthesis of chromene derivatives (Scheme 14).[74]

The ultrasonic irradiation helped the formation of the product molecules with higher yield and shorter synthesis time than under reflux conditions. The effect of the solvents was also investigated. The best ones proved to be water and ethanol; moreover, the intensity of the ultrasonic treatment

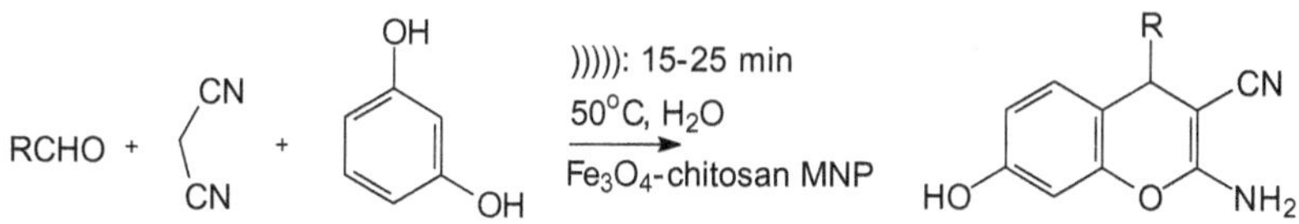

Scheme 14 Ultrasonically-assisted, Fe_3O_4-chitosan magnetic MNP catalyzed syntheses of chromene derivatives (scope: 18 examples; yield: 94%–99%).

had significant influence on the obtained conversions. The reactions were conducted by using 40, 80, 120, 160, and 200 W power. The necessary time and the yields were found to steadily increase with the increasing power; however, above 160 W, a small decline was observed in the yield. The authors obtained similar results in the syntheses of the pyrimidines by $H_3PW_{12}O_{40}$-functionalized chitosan@Co_3O_4 nanocomposites as well.[75]

Similar intensity (or power) dependence regarding the yields obtained was observed in the preparation of dipyrimidines on $CuFe_2O_4$ magnetic nanoparticles (Scheme 15); the best performance was connected to the use of ultrasound irradiation with low power (40 W).[76]

In several further studies, ultrasound treatment proved to be an excellent activation method to improve the performance of the iron oxide-based magnetic catalysts. All these reports successfully demonstrated that the sonochemical organic reactions are feasible in environmentally benign aqueous/alcoholic[77–81] media or under solvent-free conditions.[82]

Another popular catalyst family is the Cu(I)-based one, frequently applied in cycloadditions, due to their high chemical stability and their efficiency even under mild conditions. An innovative method was designed and probed for the synthesis of L-ascorbic acid derivatives, in which sonochemistry and flow chemistry were combined (Scheme 16).[83] The microreactor part was

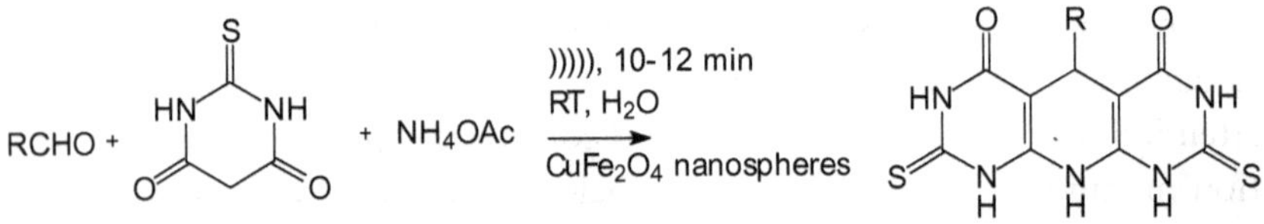

Scheme 15 Ultrasonically-assisted, $CuFe_2O_4$ nanosphere catalyzed syntheses of dipyrimidine derivatives (scope: 10 out 12 successful examples; yield: 9%–99%).

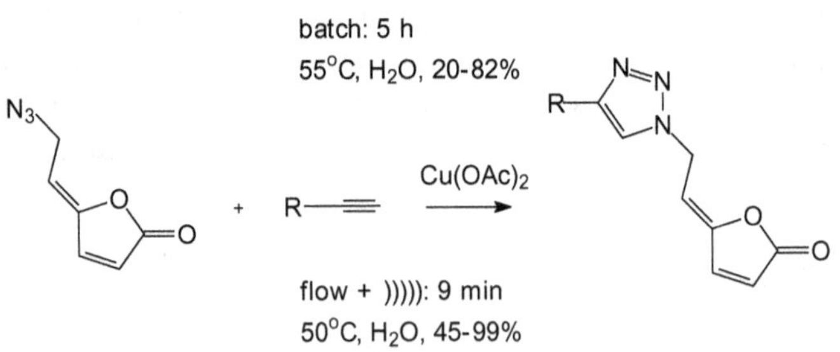

Scheme 16 Demonstrating the synergetic effect of the combination of ultrasound irradiation and flow chemistry on the syntheses of L-ascorbic acid derivatives in a click reaction (14 examples).

immersed into the tank of a commercially available ultrasonic bath; with this setup, the reaction times were reduced significantly, from hours to minutes, well under the boiling point of the water (at 50–55 °C) serving as solvent as well as the ultrasound propagating medium.

In graphene oxide and/or Cu(I)-catalyzed syntheses of substituted triazoles, the individual effect of the intensity of the ultrasound radiation was tested again: the increase of the energy input aided the formation of the triazole molecules, but only up to 70 W power.[84, 85]

Sonochemical treatments could also verify their raison d'être in the field of Cu(I)-catalyzed preparations of organic molecules containing new C—S[86] and C—Se bonds[87, 88] (Scheme 17). In both cases the applied methodology became greener due to the significantly reduced reaction times even compared with the microwave-assisted protocols.

As alternative energy transmission source, ultrasound radiation was regularly used in several organocatalytic reactions such as enamine-azide cycloaddition[89] or one-pot multicomponent reactions.[90, 91] In the latter studies, sonication was employed in the most efficient way, that is, by using a

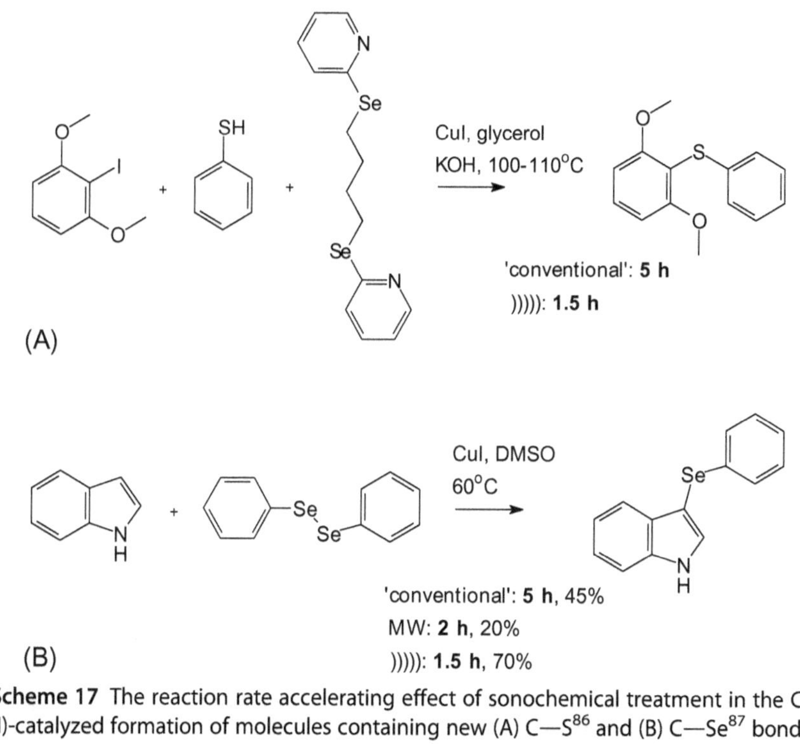

Scheme 17 The reaction rate accelerating effect of sonochemical treatment in the Cu (I)-catalyzed formation of molecules containing new (A) C—S[86] and (B) C—Se[87] bonds.

mixture of water and various alcohols as solvents. Beyond these examples, there are several additional reports in which water, ethanol and glycerol were applied as solvents in the ultrasonic treatments independently from the type of the heterogeneous catalysts.[92–98] The continuously forming and collapsing cavities result in intense mass transfer at the liquid-liquid and solid-liquid interfaces by disrupting the interfacial regions and increasing the contacting surfaces; therefore the nontypical organic solvents like the aforementioned examples can be successfully utilized in sonochemical reactions. Similar effect was observed when phase-transfer catalysts were used under ultrasound irradiation, and thus higher reaction rates could be achieved.[99, 100] The authors tested the influence of the enhancement of the intensity and the frequency on the conversion values. Generally, increasing conversion was found; however, at nearly continuous irradiation, the conversion started to decrease.

In the previous chapter, ionic liquids were presented as doping agents for the frequently used organic solvents to enhance the transmission of the ultrasonic energy; nevertheless, their catalytic applications are also noteworthy. By combining their use with ultrasonic treatment, one can develop a synthetic method for the rapid preparation of organic compounds where the ionic liquids serve as easily reusable solvents and catalysts at the same time.[101–104] As a remarkable example, an ultrasonically assisted four-component Hantzsch condensation of the polyhydroquinolines (Scheme 18) can be mentioned.[102]

Sulfonic acid functionalized ionic liquids were applied as catalyst as well as solvent under ultrasound irradiation giving the products in extremely fast reactions (in few minutes) and with excellent yields (88%–97%). Moreover, the recycling test of the ionic liquid showed that the obtained yields only decreased slightly even after 10 consecutive runs.

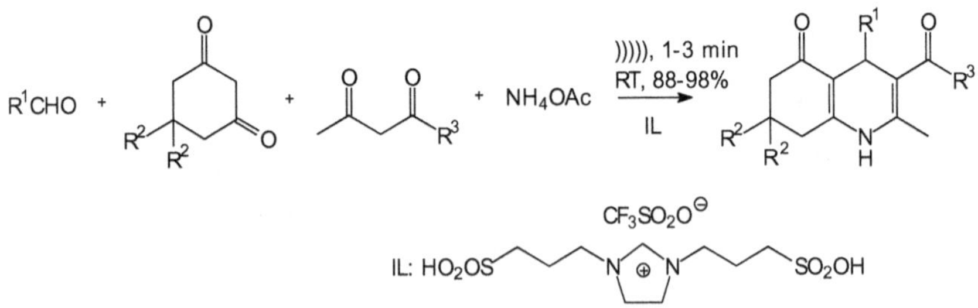

Scheme 18 Ultrasonically-assisted Hantzsch condensation forming polyhydroquinolines using an ionic liquid as catalyst and solvent (scope: 21 examples).

The performance of enzymes is also frequently improved by ultrasound; this was mainly demonstrated in esterification reactions.[105–108] These reports revealed again the specific effects of the intensity and pulse motif of the radiation: The increased ultrasonic power aided the synthesis through the larger interfacial area, but above a certain point, decrease in the reaction rates was registered because of the denaturation and/or the defragmentation of the enzymes due to the high ultrasound power. The adverse effect of applying too intense ultrasound radiation was well demonstrated in an original paper dealing with the whole-cell bioconversion of 1-(3-hydroxyphenyl)-2-(methyamino)ethanone (HPMAE) to (*R*)-phenylephrine under ultrasound assistance (Scheme 19).[109]

The effect of ultrasonic treatment on living cells was examined at 90, 120, and 150 W power. By increasing the intensity the number of destroyed cells also increased; at 90 W, the concentration of the living cells was 7.3×10^7 cells/mL, while at 150 W, it was reduced to 5.2×10^7 cells/mL. This observation directly highlights the reason of the recorded lower bioconversion rate at high-power ultrasound irradiation and the importance of mapping the effects related to the applied ultrasonic powers in each reaction.

It is of relevance to mention that the sonochemical transesterification for biodiesel production is also a promising field of green fuel production on mineral catalyst.[110, 111] Interestingly, after reaching an optimum, higher ultrasonic power did not enhance the conversions; lower yields were calculated for CaO and for dolomite $CaMg(CO_3)_2$ at above 40 and 45 W, respectively.[110]

In metal-mediated reactions, the applied elemental additives serve as reducing agents, their oxidation state gradually and permanently changes during the synthesis, and they are frequently used in more than the catalytic amount. Nevertheless, they resemble catalytic behavior; therefore some pieces of research are presented concentrating at the ultrasonically promoted Mg-,[113] In-,[112] and Pd-mediated[114] procedures at the end of this chapter.

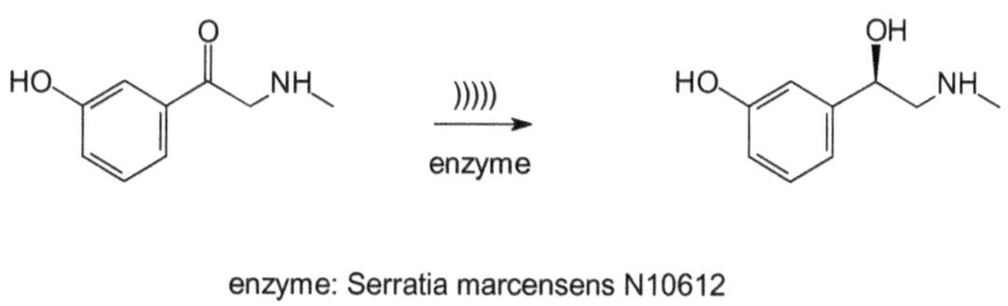

Scheme 19 Ultrasonically-assisted, enzyme-catalyzed enantioselective synthesis of (*R*)-phenylephrine.

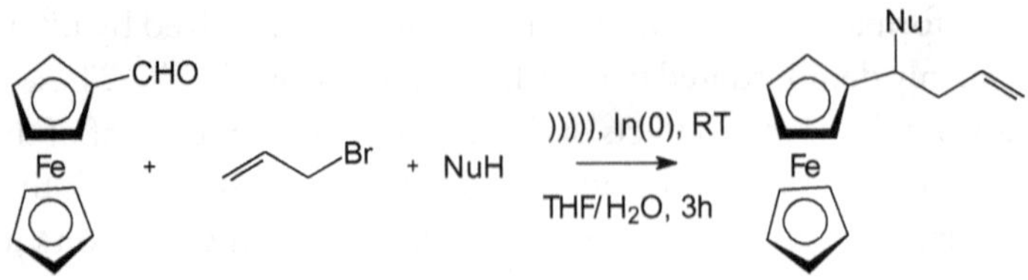

Scheme 20 Three-component indium-mediated domino allylation of ferrocene carbaldehyde in the presence of nucleophiles (scope: 18 examples; yield: 27%–88%).[112]

It is worthwhile to treat the indium-mediated reaction in detail, since in this process, a special cup horn-type ultrasonic devices was used. These devices have similar arrangement to ultrasonic bathes, but with smaller tank dimension and intensified ultrasound output power; therefore the energy transfer of the sonication could be higher, more concentrated, and more efficient. The authors developed a one-pot three-component reaction for the synthesis of ferrocenyl derivatives, and in this Barbier-type indium-mediated allylation (Scheme 20), the ultrasound treatment proved to be an excellent way to remove the oxide layer from the metal particles and thus to help the formation of the organoindium intermediate.

Three different sonication setups were investigated: In the generally used ultrasonic bath (60 W, 20.3 kHz), the isolated yield was 48%, but in the cup horn test (75 W, 19.9 kHz), the yield almost doubled: 86% was achieved. Interestingly, the cup horn system with high frequency (70 W, 300 kHz) resulted in lower yield (70%); presumably the regression was connected to the decrease in the mass transfer due to the less intense cavitation.

3.3.1 Alteration of reaction pathways by ultrasonic irradiation

In spite of the huge potential in this area, the notably smaller number of publications well indicates the challenge in the development of an ultrasonic treatment in which synergistic effects are generated and utilized among the catalyst, the ultrasound waves, and the cavities. The most conceivable way of how the ultrasonic irradiation and the catalysts can work together is based on the mechanical effect of the cavities modifying the surface properties and the internal structure of the solid catalyst. These modifications do not always result in the increase in the isolated yields, but they can open new reaction paths toward unusual chemo- and/or stereoselectivities.

The synthesis of pyrano-oxazine derivatives is an example to a not just ultrasonically-assisted (irradiation enhances the reaction rate), but to an ultrasonically-induced (irradiation initiates the reaction) organocatalytic reaction (Scheme 21).[115]

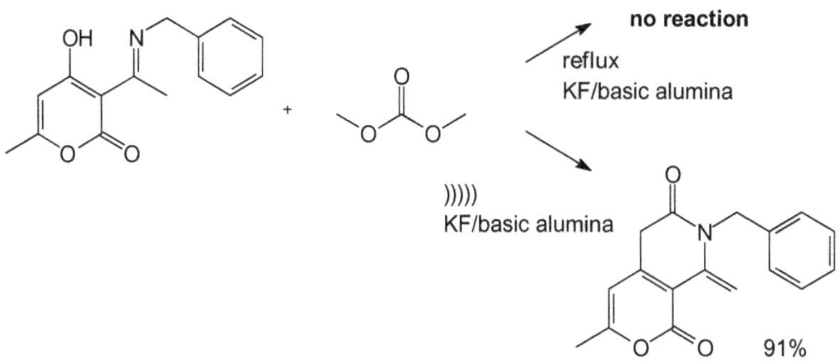

Scheme 21 Synthesis of pyrano[3,4-e][1,3]oxazine-2,5-dione.

The paper reveals that there were no signs of product formation in the absence of the catalyst, whether ultrasonic treatment was applied or not. The preparations were only successful when ultrasound and the catalyst were applied together. In the light of these observations, the authors presumed that the ultrasonic irradiation provided extra energy to overcome the activation barrier of the reaction through shockwave damages. Due to the erosions and deformations caused by the microjets, the number of the dislocations increased, and the total energy of the active sites was also raised.

In a study, the influence of ultrasonic treatment on the physicochemical attributes of the Lindlar catalyst particles during hydrogenation was investigated in detail to explain the slower reaction rates observed under conventional stirring.[116] The authors revealed that the cavitation was capable to remove the loosely bound Pd/Pb particles from the fragmented $CaCO_3$ carrier, especially at low frequency. This report points out the relevance of the characterization of the solid catalyst not just before their applications, but following each recycling test.

The combination of enzymatic synthesis and ultrasonic treatment has also potential in the diversion from the original reaction path. At a lipase-catalyzed acylation of *rac* alcohols, the microwave and the ultrasound irradiations resulted in significantly enhanced reactivity and enantioselectivity.[117] The selectivity was increased from 8 (conventional heating and stirring) to more than 200 in the course of the two unconventional activation techniques in the reaction with 4-chromanol. The possible synergic effects of the microwave and ultrasound irradiations were connected mainly to their extreme and localized heating capabilities, which could modify the enzyme tertiary structure and thus its active sites.

The sonochemical route proved to be an efficient alternative for enzymatic esterification of fatty acids and short-chain alcohols by phospholipase Lecitase-Ultra.[118] Beyond the sonochemically improved reaction rates, the researchers observed that higher substrate concentration could be applied compared with the case of the mechanical stirring, in spite of the harmful effects of the alcohols and acids on the performance of the enzyme. As a conceivable explanation, the intense mixing effect of the cavitation was emphasized; thus the concentration of the substrates or products could be lower in the pores of the enzyme support, and/or the formation of water cages around the enzyme molecules was restrained avoiding the accumulation of the acid and the decrease of the pH to deleterious level. Similar water cluster disturbing effect was presumed to explain the changes recorded in the reaction rates and selectivities of the olefin metathesis on ruthenium-based catalysts under ultrasound or even microwave radiation.[119]

However, under suitable conditions and with the outstanding chemical/thermal stability of the enzyme molecules, one can find an example of the growing enantioselectivity due to higher ultrasonic intensities paralleled with excellent reusability.[120]

Last, but not least, it is of relevance to present an original paper dealing with the *tert*-butyl hydroperoxide-mediated trifluoromethylation of caffeine under ultrasonic irradiation (Scheme 22).[121]

The initial aim of the work was to study the influence of ultrasound irradiation on the kinetic properties of the reaction, but the investigation revealed a new component that supplemented well the synthetic process. The authors recognized that the extreme decrease in the reaction time, from 40 h by stirring to 2 h when ultrasonic irradiation was used, cannot be solely explained in terms of the alternative energy transfer aspect of the ultrasound waves. They made test reactions with fluoropolymer-coated tip of the ultrasound transducer of the applied horn-type device, and found that the rate of the reaction became significantly lower. This result led them to investigate the effect of the different metallic elements, which potentially leached from the bulk of the titanium alloy tip during the direct ultrasonic treatment.

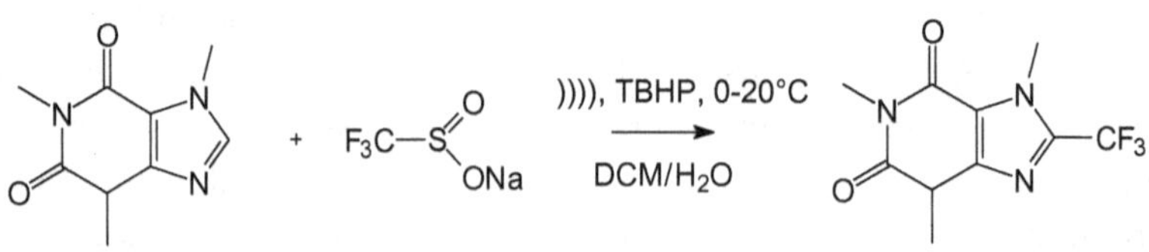

Scheme 22 The trifluoromethylation of caffeine under ultrasonic conditions.

After extensive research, vanadium was identified as the reaction promoter providing a suitable environment to the controlled *tert*-butoxy radical formation. This work demonstrates well the necessity of increased attention when direct sonication is directly applied in the reaction mixture without a transmitting medium.

3.4 The advantages of using the ultrasound-microwave irradiation hybrid technique

In the light of the aforementioned studies, the use of microwave irradiation is frequently the competitor of the ultrasound treatment in synthetic organic chemistry. However, there is a great potential in their simultaneous application with several synergetic and complementary effects. Both irradiations result in localized hot spots, and thus are capable of generating reactive species. Sonication enhances mass transfer; microwave irradiation has more efficient heating due to the direct connection between the molecules and electromagnetic field.[122, 123] In spite of the technical complexities and limitations of the hybrid reactors, in numerous cases, the combination of the micro and ultrasound treatments can be beneficial for organic reactions of various kinds like transesterification, forming new C—C and C—N bonds as well as reductions.[123] This combination is advantageous in food chemistry[124, 125] and separation processes.[126]

In a recent report, biologically active phenoxyacetic acid derivatives were prepared, and the combined use of ultrasound and microwave irradiation aided the formation of the product in most instances.[127] As the highlighted example the synthesis of phenoxyacetic acid (Scheme 23) was presented in detail; the effect of the applied solvents, the ultrasound, and microwave treatments applied separately as well as together were all investigated.

Sonication was capable to reduce the reaction time (30 min) without significantly enhancing the yield (27%), while microwave radiation resulted in increased yield (68%) as well as reduced reaction time (10 min). However, for the highest yield (86%, 5–10 min), the simultaneous utilization of the ultrasound and microwave application proved to be necessary.

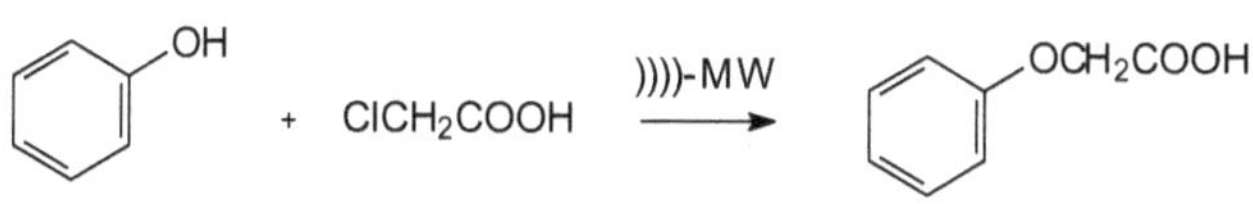

Scheme 23 The synthesis of phenoxyacetic acid applying the combination of microwave heating and ultrasonic irradiation.

4 Conclusions and outlook

The previously detailed example of the irradiation with ultrasound is a very useful tool not only in enhancing reaction rates, influencing the selectivities of the reactions, but also in opening entirely new reaction pathways. However, finding novel ways for utilizing the potential synergic effects between the solid particles and the propagated ultrasound in liquid media is naturally a cumbersome and time- and energy-consuming challenge. But it is unavoidable for the success and fruitful future of sonochemistry; therefore we describe three concepts as promising starting points for future innovations. The first is based on the radical formation in the cavities, which changes the conductivity of the medium; therefore it is conceivable that electrochemical methods can profit from the ultrasound waves.[128] Another field is connected to the ultrasonic capillary effect. The penetration of the fluid into the capillary channels can be aided remarkably by ultrasound, and thus the pores, the micro- or nanoducts, or even the interlamellar galleries of the materials become more approachable. The solid-liquid interfacial area is increased, and chemical reaction can be induced in these special, controlled, and limited spaces.[128, 129] Finally, the use of the energy stored in the multi- or single-bubble sonoluminescence photons to conduct photochemical reactions is also an exciting field. Due to the wide energy range of the emitted photons, the number of their potential applications can be large in spite of the low intensity of the emission, which can be increased by ultrasonic frequency or chemical additives[130] or supported in sonochemiluminescence way.

References

1. Chilowsky, C.; Langevin, P. *Production of Submarine Signals and the Location of Submarine Objects*; United States Patent Office, 1917. US1471547A.
2. Wood, R. W.; Loomis, A. L. The Physical and Biological Effects of High-Frequency Sound-Waves of Great Intensity. *Philos. Mag.* **1927**, *4*, 417–436.
3. Mason, T.; Peters, D. *Practical Sonochemistry*; Woodhead Publishing, 2003.
4. Martínez, J.-L. C., Ed. *Ultrasound in Chemistry: Analytical Applications*; Wiley-VCH, 2009.
5. Grieser, F.; Harada, H.; Okitsu, K.; Yasui, K.; Enomoto, N.; Choi, P.-K., Eds. *Sonochemistry and the Acoustic Bubble*; Elsevier, 2015.
6. Ashokkumar, M., Ed. *Handbook of Ultrasonics and Sonochemistry*; Springer, 2016.
7. Pollet, B. G.; Ashokkumar, M., Eds. *Springer Briefs in Molecular Science: Ultrasound and Sonochemistry*; Springer, 2018.
8. Brenner, M. P.; Hilgenfeldt, S.; Lohse, D. Single-Bubble Sonoluminescence. *Rev. Mod. Phys.* **2002**, *74*, 425–484.

9. Lamminen, M. O.; Walker, H. W.; Weavers, L. K. Mechanisms and Factors Influencing the Ultrasonic Cleaning of Particle-Fouled Ceramic Membranes. *J. Membr. Sci.* **2004**, *237*, 213–223.
10. Suslick, K. S. Sonochemistry. *Science* **1990**, *247*, 1439–1445.
11. Boldyrev, V. V. Mechanochemistry and Sonochemistry. *Ultrason. Sonochem.* **1995**, *2*, S143–S145.
12. Cintas, P.; Cravotto, G.; Barge, A.; Martina, K. Interplay Between Mechanochemistry and Sonochemistry. *Top. Curr. Chem.* **2014**, *369*, 239–284.
13. Gaitan, D. F.; Crum, L. A.; Church, C. C.; Roy, R. A. Sonoluminescence and Bubble Dynamics for a Single, Stable, Cavitation Bubble. *J. Acoust. Soc. Am.* **1992**, *91*, 3166–3183.
14. Okitsu, K.; Suzuki, T.; Takenaka, N.; Bandow, H.; Nishimura, R.; Maeda, Y. Acoustic Multibubble Cavitation in Water: a New Aspect of the Effect of a Rare Gas Atmosphere on Bubble Temperature and its Relevance to Sonochemistry. *J. Phys. Chem. B* **2006**, *110*, 20081–20084.
15. Jarman, P. Measurements of Sonoluminescence From Pure Liquids and Some Aqueous Solutions. *Proc. Phys. Soc.* **1959**, *73*, 628–640.
16. Chen, W.; Huang, W.; Liang, Y.; Gao, X.; Cui, W. Time-Resolved Spectra of Single-Bubble Sonoluminescence in Sulfuric Acid with a Streak Camera. *Phys. Rev. E* **2008**, *78*. 035301(R).
17. Hatanaka, S.; Mitome, H.; Yasui, K.; Hayashi, S. Single-Bubble Sonochemiluminescence in Aqueous Luminol Solutions. *J. Am. Chem. Soc.* **2002**, *124*, 10250–10251.
18. Matsuoka, M.; Takahashi, F.; Asakura, Y.; Jin, J. Sonochemiluminescence of Lucigenin: Evidence of Superoxide Radical Anion Formation by Ultrasonic Irradiation. *Jpn. J. Appl. Phys.* **2016**, *55*, 07KB01.
19. Merouani, S.; Hamdaoui, O.; Rezgui, Y.; Guemini, M. Energy Analysis during Acoustic Bubble Oscillations: Relationship between Bubble Energy and Sonochemical Parameters. *Ultrasonics* **2014**, *54*, 227–232.
20. Lee, J.; Vakarelski, I. U.; Yasui, K.; Tuziuti, T.; Kozuka, T.; Towata, A.; Lida, Y. Variations in the Spatial Distribution of Sonoluminescence Bubbles in the Presence of an Ionic Surfactant and Electrolyte. *J. Phys. Chem. B* **2010**, *114*, 2572–2577.
21. Entezari, M.; Kruus, P. Effect of Frequency on Sonochemical Reactions. II: Temperature and Intensity Effects. *Ultrason. Sonochem.* **1966**, *3*, 19–24.
22. Torres, R. A.; Pétrier, C.; Combet, E.; Carrier, M.; Pulgarin, C. Ultrasonic Cavitation Applied to the Treatment of Bisphenol A. Effect of Sonochemical Parameters and Analysis of BPA by-Products. *Ultrason. Sonochem.* **2008**, *15*, 605–611.
23. Koda, S.; Kimura, T.; Kondo, T.; Mitome, H. A Standard Method to Calibrate Sonochemical Efficiency of an Individual Reaction System. *Ultrason. Sonochem.* **2003**, *10*, 149–156.
24. Cravotto, G.; Borretto, E.; Oliverio, M.; Procopio, A.; Penoni, A. Organic Reactions in Water or Biphasic Aqueous Systems under Sonochemical Conditions. A Review on Catalytic Effects. *Cat. Com.* **2015**, *63*, 2–9.
25. Lupacchini, M.; Mascitti, A.; Giachi, G.; Tonucci, L.; d'Alessandro, N.; Martinez, J.; Colacino, E. Sonochemistry in Non-conventional, Green Solvents or Solvent-Free Reactions. *Tetrahedron* **2017**, *73*, 609–653.
26. Draye, M.; Kardos, N. Advances in Green Organic Sonochemistry. *Top. Curr. Chem. (Z)* **2016**, *374*, 74.
27. Banerjee, B. Recent Developments on Ultrasound-Assisted One-Pot Multicomponent Synthesis of Biologically Relevant Heterocycles. *Ultrason. Sonochem.* **2017**, *35*, 15–35.
28. Banerjee, B. Recent Developments on Ultrasound Assisted Catalyst-Free Organic Synthesis. *Ultrason. Sonochem.* **2017**, *35*, 1–14.

29. Chatel, G. Sonochemistry in Nanocatalysis: The Use of Ultrasound from the Catalyst Synthesis to the Catalytic Reaction. *Curr. Opin. Green Sustain. Chem.* **2019**, *15*, 1–6.
30. Cintas, P. Ultrasound and Green Chemistry—Further Comments. *Ultrason. Sonochem.* **2016**, *28*, 257–258.
31. Chatel, G. How Sonochemistry Contributes to Green Chemistry? *Ultrason. Sonochem.* **2018**, *40*, 117–122.
32. Shabalala, N.; Maddila, S.; Jonnalagadda, S. B. Catalyst-Free, One-Pot, Four-Component Green Synthesis of Functionalized 1-(2-Fluorophenyl)-1,4-Dihydropyridines Under Ultrasound Irradiation. *New J. Chem.* **2016**, *40*, 5107–5112.
33. Shaabani, A.; Hooshmand, S. E. Diversity-Oriented Catalyst-Free Synthesis of Pseudopeptides Containing Rhodanine Scaffolds Via a One-Pot Sequential Isocyanide-Based Six-Component Reactions in Water Using Ultrasound Irradiation. *Ultrason. Sonochem.* **2018**, *40*, 84–90.
34. Belhani, B.; Berredjem, M.; Le Borgne, M.; Bouaziz, Z.; Lebreton, J.; Aouf, N.-E. A One-Pot Three-Component Synthesis of Novel α-Sulfamidophosphonates Under Ultrasound Irradiation and Catalyst-Free Conditions. *RSC Adv.* **2015**, *5*, 39324–39329.
35. Sarmah, M. M.; Prajapati, D. Ultrasound-Mediated One-Pot Synthesis of 7-Methyl-Substituted Pyrido[4,3-d]Pyrimidine Scaffolds by a Catalyst-Free Protocol. *Synlett* **2015**, *26*, 91–94.
36. Shi, Z.; Zhao, Z.; Huang, M.; Fu, X. Ultrasound-Assisted, One-Pot, Three-Component Synthesis and Antibacterial Activities of Novel Indole Derivatives Containing 1,3,4-Oxadiazole and 1,2,4-Triazole Moieties. *C. R. Chim.* **2015**, *18*, 1320–1327.
37. Khan, M. N.; Karamthulla, S.; Choudhury, L. H.; Faizi, M. S. H. Ultrasound Assisted Multicomponent Reactions: a Green Method for the Synthesis of Highly Functionalized Selenopyridines Using Reusable Polyethylene Glycol as Reaction Medium. *RSC Adv.* **2015**, *5*, 22168–22172.
38. Perin, G.; Araujo, D. R.; Nobre, P. C.; Lenardao, E. J.; Jacob, R. G.; Silva, M. S.; Roehrs, J. A. Ultrasound-Promoted Synthesis of 2-Organoselanyl-Naphthalenes Using Oxone® in Aqueous Medium as an Oxidizing Agent. *PeerJ* **2018**, *6*, e4706.
39. Nemati, F.; Nikkhah, S. H.; Elhampour, A. An Environmental Friendly Approach for the Catalyst-Free Synthesis of Highly Substituted Pyrazoles Promoted by Ultrasonic Radiation. *Chin. Chem. Lett.* **2015**, *26*, 1397–1399.
40. Chandgude, A. L.; Dömling, A. An Efficient Passerini Tetrazole Reaction (PT-3CR). *Green Chem.* **2016**, *18*, 3718–3721.
41. Gohil, J. D.; Patel, H. B.; Patel, M. P. Comparative Study on the Use of Conventional, Microwave and Ultrasound-Irradiation for the Synthesis of Pyrano[3,2-c]Chromene and Benzopyrano[4,3-b]Chromene Derivatives in Water. *Heterocycl. Lett.* **2016**, *6*, 123–132.
42. Ramazani, A.; Rouhani, M.; Nasrabadi, F. Z.; Gouranlou, F. Ultrasound-Promoted Three-Component Reaction of N-Isocyaniminotriphenyl-Phosphorane, (E)-Cinnamic Acids, and Biacetyl. *Phosphorus Sulfur Silicon Relat. Elem.* **2015**, *190*, 20–28.
43. Gohil, J. D.; Patel, H. B.; Patel, M. P. Ultrasound Assisted Synthesis of Triazole/Tetrazole Hybrids Based New Biquinoline Derivatives as a New Class of Antimicrobial and Antitubercular Agents. *Indian J. Adv. Chem. Sci.* **2016**, *4*, 102–113.
44. Tan, S. H.; Chuah, T. S.; Chia, P. W. An Improved Protocol on the Synthesis of Thiazolo[3,2-a]Pyrimidine Using Ultrasonic Probe Irradiation. *J. Korean Chem. Soc.* **2016**, *60*, 245–250.
45. Crawford, D. E. Solvent-Free Sonochemistry: Sonochemical Organic Synthesis in the Absence of a Liquid Medium. *Beilstein J. Org. Chem.* **2017**, *13*, 1850–1856.

46. Dofe, V. S.; Sarkate, A. P.; Shaikh, Z. M.; Jadhav, C. K.; Nipte, A. S.; Gill, C. H. Ultrasound-Assisted Synthesis of Novel Pyrazole and Pyrimidine Derivatives as Antimicrobial Agents. *J. Heterocyclic Chem.* **2018**, *55*, 756–762.
47. Gudi, Y.; Mangali, M. S.; Gundala, S.; Venkatapuram, P.; Adivireddy, P. Synthesis, Characterization, and Bioassay of a New Class of Pyrazolyl/Isoxazolyl Oxadiazoles. *Monatsh. Chem.* **2018**, *149*, 2311–2326.
48. Mane, S. N.; Gadalkar, S. M.; Rathod, V. K. Intensification of Paracetamol (Acetaminophen) Synthesis from Hydroquinone Using Ultrasound. *Ultrason. Sonochem.* **2018**, *49*, 106–110.
49. Pattarawarapan, M.; Jaita, S.; Wangngae, S.; Phakhodee, W. Ultrasound-Assisted Synthesis of Substituted Guanidines From Thioureas. *Tetrahedron Lett.* **2016**, *57*, 1354–1358.
50. Duangkamol, C.; Pattarawarapan, M.; Phakhodee, W. Ultrasonic-Assisted Synthesis of Carbodiimides From N,N′-Disubstituted Thioureas and Ureas. *Monatsh. Chem.* **2016**, *147*, 1945–1949.
51. Phakhodee, W.; Duangkamol, C.; Wiriya, N.; Pattarawarapan, M. Ultrasound-Assisted Synthesis of Substituted 2-Aminobenzimidazoles, 2-Aminobenzoxazoles, and Related Heterocycles. *Tetrahedron Lett.* **2016**, *57*, 5290–5293.
52. Costa, G. P.; Seus, N.; Roehrs, J. A.; Jacob, R. G.; Schumacher, R. F.; Barcellos, T.; Luque, R.; Alves, D. Ultrasound-Promoted Organocatalytic Enamine–Azide [3+2] Cycloaddition Reactions for the Synthesis of ((Arylselanyl)Phenyl-1H-1,2,3-Triazol-4-Yl)Ketones. *Beilstein J. Org. Chem.* **2017**, *13*, 694–702.
53. Chettibi, N.; Bentoumi, H.; Liacha, M. Conventional and Ultrasound Assisted Synthesis and Antibacterial Evaluation of some N′-(Nitro-Substituted Benzylidene)-2(3-Oxo-2H-Benzo[b][1,4]Oxazin-4(3H)-YL)Acetohydrazide. *UPB Sci. Bull. B* **2018**, *80*, 147–158.
54. Fujita, M.; Furusho, Y. Ultrasound-Assisted Synthesis of Substituted Guanidines Using 1H-Pyrazole-1-Carboxamidine and S-Methylisothiouronium Sulphate Under Solvent-Free Conditions. *Tetrahedron* **2018**, *74*, 4339–4342.
55. Mishra, A.; Singh, S.; Quraishi, M. A.; Srivastava, V. A Catalyst-Free Expeditious Green Synthesis of Quinoxaline, Oxazine, Thiazine, and Dioxin Derivatives in Water under Ultrasound Irradiation. *Org. Prep. Proced. Int.* **2019**, *51*, 345–356.
56. Rouhani, M.; Ramazani, A.; Joo, S. W. Ultrasonics in Isocyanide-Based Multicomponent Reactions: A New, Efficient and Fast Method for the Synthesis of Fully Substituted 1,3,4-Oxadiazole Derivatives under Ultrasound Irradiation. *Ultrason. Sonochem.* **2015**, *22*, 391–396.
57. Ramazani, A.; Rouhani, M.; Joo, S. W. Catalyst-Free Sonosynthesis of Highly Substituted Propanamide Derivatives in Water. *Ultrason. Sonochem.* **2016**, *28*, 393–399.
58. Bizzi, C. A.; Santos, D.; Sieben, T. C.; Motta, G. V.; Mello, P. A.; Flores, E. M. M. Furfural Production From Lignocellulosic Biomass by Ultrasound-Assisted Acid Hydrolysis. *Ultrason. Sonochem.* **2019**, *51*, 332–339.
59. Regueira, J. L. L. F.; Dantas, C. R.; de Freitas, J. J. R.; Santos da Silva, A. J. F.; Freitas Filho, J. R.; Menezes, P. H.; Freitas, J. C. R. Stereoselective Synthesis of 2,3-Unsaturated Pseudoglycosides Promoted by Ultrasound. *Synthesis* **2016**, *48*, 1069–1078.
60. Singh, F. V.; Mangaonkar, S. R.; Kole, P. B. Ultrasound-Assisted Rapid Synthesis of b-Cyanoepoxides Using Hypervalent Iodine Reagents. *Synth. Commun.* **2018**, *48*, 2169–2176.
61. Sakhno, Y. I.; Kozyryev, A. V.; Desenko, S. M.; Shishkina, S. V.; Musatov, V. I.; Sysoiev, D. O.; Chebanov, V. A. Features of Two- and Multicomponent Heterocyclization Reactions Involving 3,4-Disubstituted 5-Aminopyrazoles and Alkyl Pyruvates. *Tetrahedron* **2018**, *74*, 564–571.

62. Murlykina, M. V.; Sakhno, Y. I.; Desenko, S. M.; Shishkina, S. V.; Shishkin, O. V.; Sysoiev, D. O.; Kornet, M. N.; Schols, D.; Goeman, J. L.; Van der Eycken, J.; Van der Eycken, E. V.; Chebanov, V. A. Study of the Chemoselectivity of Multicomponent Heterocyclizations Involving 3-Amino-1,2,4-Triazole and Pyruvic Acids as Key Reagents, and Biological Activity of the Reaction Products. *Eur. J. Org. Chem.* **2015**, 4481–4492.
63. Xie, L.-Y.; Li, Y.-J.; Qu, J.; Duan, Y.; Hu, J.; Liu, K.-J.; Cao, Z.; He, W.-M. A Base-Free, Ultrasound Accelerated One-Pot Synthesis of 2-Sulfonylquinolines in Water. *Green Chem.* **2017**, *19*, 5642–5646.
64. Shabalala, N. G.; Pagadala, R.; Jonnalagadda, S. B. Ultrasonic-Accelerated Rapid Protocol for the Improved Synthesis of Pyrazoles. *Ultrason. Sonochem.* **2015**, *27*, 423–429.
65. Safari, J.; Heydarian, M.; Zarnegar, Z. Synthesis of 2-Amino-7-Hydroxy-4H-Chromene Derivatives under Ultrasound Irradiation: A Rapid Procedure without Catalyst. *Arab. J. Chem.* **2017**, *10*, S2994–S3000.
66. Hasanzadeh, I.; Barikani, M.; Mahdavian, A. R. Ultrasound-Assisted Emulsion Polymerization of Poly(Methyl Methacrylate-Co-Butyl Acrylate): Effect of Initiator Content and Temperature. *Polym. Eng. Sci.* **2016**, *56*, 214–221.
67. Mohapatra, H.; Kleiman, M.; Esser-Kahn, A. P. Mechanically Controlled Radical Polymerization Initiated by Ultrasound. *Nat. Chem.* **2017**, *9*, 135–139.
68. McKenzie, T. G.; Colombo, E.; Fu, Q.; Ashokkumar, M.; Qiao, G. G. Sono-RAFT Polymerization in Aqueous Medium. *Angew. Chem. Int. Ed.* **2017**, *56*, 12302–12306.
69. Piogé, S.; Tran, T. N.; McKenzie, T. G.; Pascual, S.; Ashokkumar, M.; Fontaine, L.; Qiao, G. Sono-RAFT Polymerization-Induced Self-Assembly in Aqueous Dispersion: Synthesis of LCST-Type Thermosensitive Nanogels. *Macromolecules* **2018**, *51*, 8862–8869.
70. Schaefer, M.; Icli, B.; Weder, C.; Lattuada, M.; Kilbinger, A. F. M.; Simon, Y. C. The Role of Mass and Length in the Sonochemistry of Polymers. *Macromolecules* **2016**, *49*, 1630–1636.
71. Frizzo, C. P.; Bacim, C.; Moreira, D. N.; Rodrigues, L. V.; Zimmer, G. C.; Bonacorso, H. G.; Zanatta, N.; Martins, M. A. P. Sonochemical Heating Profile for Solvents and Ionic Liquid Doped Solvents, and their Application in the N-Alkylation of Pyrazoles. *Ultrason. Sonochem.* **2016**, *32*, 432–439.
72. Afreen, S.; Muthoosamy, K.; Manickam, S. Sono-Nano Chemistry: A New Era of Synthesising Polyhydroxylated Carbon Nanomaterials with Hydroxyl Groups and their Industrial Aspects. *Ultrason. Sonochem.* **2019**, *51*, 451–461.
73. Dharmarathne, L.; Grieser, F. Formation of Amino Acids on the Sonolysis of Aqueous Solutions Containing Acetic Acid, Methane, or Carbon Dioxide, in the Presence of Nitrogen Gas. *J. Phys. Chem. A* **2016**, *120*, 191–199.
74. Safari, J.; Javadian, L. Ultrasound Assisted the Green Synthesis of 2-Amino-4H-Chromene Derivatives Catalyzed by Fe_3O_4-Functionalized Nanoparticles with Chitosan as a Novel and Reusable Magnetic Catalyst. *Ultrason. Sonochem.* **2015**, *22*, 341–348.
75. Safari, J.; Tavakoli, M.; Ghasemzadeh, M. A. Ultrasound-Promoted an Efficient Method for the One-Pot Synthesis of Indeno Fused Pyrido[2,3-d]Pyrimidines Catalyzed by $H_3PW_{12}O_{40}$ Functionalized Chitosan@Co_3O_4 as a Novel and Green Catalyst. *J. Organomet. Chem.* **2019**, *880*, 75–82.
76. Naeimi, H.; Didar, A. Efficient Sonochemical Green Reaction of Aldehyde, Thiobarbituric Acid and Ammonium Acetate Using Magnetically Recyclable Nanocatalyst in Water. *Ultrason. Sonochem.* **2017**, *34*, 889–895.
77. Eidi, E.; Kassaee, M. Z.; Nasresfahani, Z. Synthesis of 2,4,5-Trisubstituted Imidazoles over Reusable $CoFe_2O_4$ Nanoparticles: an Efficient and Green Sonochemical Process. *Appl. Organomet. Chem.* **2016**, *30*, 561–565.

78. Safaei-Ghomi, J.; Eshteghal, F. Nano-Fe_3O_4/PEG/Succinic Anhydride: A Novel and Efficient Catalyst for the Synthesis of Benzoxanthenes under Ultrasonic Irradiation. *Ultrason. Sonochem.* **2017**, *38*, 488–495.
79. Maleki, A.; Aghaei, M. Ultrasonic Assisted Synergetic Green Synthesis of Polycyclic Imidazo (Thiazolo)Pyrimidines by Using Fe_3O_4@Clay Core-Shell. *Ultrason. Sonochem.* **2017**, *38*, 585–589.
80. Elhamifar, D.; Ramazani, Z.; Norouzi, M.; Mirbagheri, R. Magnetic Iron Oxide/Phenylsulfonic Acid: A Novel, Efficient and Recoverable Nanocatalyst for Green Synthesis of Tetrahydrobenzo[b]Pyrans under Ultrasonic Conditions. *J. Colloid Interface Sci.* **2018**, *511*, 392–401.
81. Maleki, A.; Rahimi, J.; Demchuk, O. M.; Wilczewska, A. Z.; Jasiński, R. Green in Water Sonochemical Synthesis of Tetrazolopyrimidine Derivatives by a Novel Core-Shell Magnetic Nanostructure Catalyst. *Ultrason. Sonochem.* **2018**, *43*, 262–271.
82. Ghomi, J. S.; Akbarzadeh, Z. Ultrasonic Accelerated Knoevenagel Condensation by Magnetically Recoverable $MgFe_2O_4$ Nanocatalyst: A Rapid and Green Synthesis of Coumarins under Solvent-Free Conditions. *Ultrason. Sonochem.* **2018**, *40*, 78–83.
83. Mescic, A.; Salic, A.; Gregoric, T.; Zelic, B.; Raic-Malic, S. Continuous Flow-Ultrasonic Synergy in Click Reactions for the Synthesis of Novel 1,2,3-Triazolyl Appended 4,5-Unsaturated L-Ascorbic Acid Derivatives. *RSC Adv.* **2017**, 7, 791–800.
84. Naeimi, H.; Dadashzadeh, S.; Moradian, M. Facile and Efficient Sonochemical Synthesis of 1,4-Disubstituted 1,2,3-Triazole Derivatives Catalyzed by CuI under Mild Conditions. *Res. Chem. Intermed.* **2015**, *41*, 2687–2695.
85. Naeimi, H.; Shaabani, R. Ultrasound Promoted Facile One Pot Synthesis of Triazole Derivatives Catalyzed by Functionalized Graphene Oxide Cu(I) Complex under Mild Conditions. *Ultrason. Sonochem.* **2017**, *34*, 246–254.
86. Cargnelutti, R.; Lang, E. S.; Schumacher, R. F. Bis(2-Pyridyl)Diselenoethers as Versatile Ligands for Copper-Catalyzed C–S Bond Formation in Glycerol. *Tetrahedron Lett.* **2015**, *56*, 5218–5222.
87. Vieira, B. M.; Thurow, S.; Brito, J. S.; Perin, G.; Alves, D.; Jacob, R. G.; Santi, C.; Lenardão, E. J. Sonochemistry: An Efficient Alternative to the Synthesis of 3-Selanylindoles Using CuI as Catalyst. *Ultrason. Sonochem.* **2015**, *27*, 192–199.
88. Fonseca, S. F.; Padilha, N. B.; Thurow, S.; Roehrs, J. A.; Savegnago, L.; de Souza, M. N.; Fronza, M. G.; Collares, T.; Buss, J.; Seixas, F. K.; Alves, D.; Lenardão, E. J. Ultrasound-Promoted Copper-Catalyzed Synthesis of Bis-Arylselanyl Chrysin Derivatives with Boosted Antioxidant and Anticancer Activities. *Ultrason. Sonochem.* **2017**, *39*, 827–836.
89. Xavier, D. M.; Goldani, B. S.; Seus, N.; Jacob, R. G.; Barcellos, T.; Paixão, M. W.; Luque, R.; Alves, D. Sonochemistry in Organocatalytic Enamine-Azide [3+2] Cycloadditions: A Rapid Alternative for the Synthesis of 1,2,3-Triazoyl Carboxamides. *Ultrason. Sonochem.* **2017**, *34*, 107–114.
90. Chen, L.-H.; Chung, T.-W.; Narhe, B. D.; Sun, C.-M. A Novel Mechanistic Study on Ultrasound-Assisted, One-Pot Synthesis of Functionalized Benzimidazo[2,1-b] Quinazolin-1(1H)-Ones. *ACS Comb. Sci.* **2016**, *18*, 162–169.
91. Tabassum, S.; Govindaraju, S.; Pasha, M. A. Sonochemistry—An Innovative Opportunity towards a One-Pot Three-Component Synthesis of Novel Pyridylpiperazine Derivatives Catalysed by Meglumine in Water. *New J. Chem.* **2017**, *41*, 3515–3523.
92. Doustkhah, E.; Rostamnia, S.; Hassankhani, A. The Raise of SBA-SO_3H Catalytic Activity by Inducing Ultrasound Irradiation in the Multicomponent Syntheses. *J. Porous. Mater.* **2016**, *23*, 549–556.
93. Verma, D.; Sharma, V.; Okram, G. S.; Jain, S. Ultrasound-Assisted High-Yield Multicomponent Synthesis of Triazolo[1,2-a]Indazole-Triones Using Silica-Coated ZnO Nanoparticles as a Heterogeneous Catalyst. *Green Chem.* **2017**, *19*, 5885–5899.

94. Nile, S. H.; Kumar, B.; Park, S. W. Chemo Selective One-Pot Synthesis of 2-Aryl-1-Arylmethyl-1H-Benzimidazoles Using Amberlite IR-120. *Arab. J. Chem.* **2015**, *8*, 685–691.
95. Mendes, S. R.; Thurow, S.; Penteado, F.; da Silva, M. S.; Gariani, R. A.; Perin, G.; Lenardão, E. J. Synthesis of Bis(Indolyl)Methanes Using Ammonium Niobium Oxalate (ANO) as an Efficient and Recyclable Catalyst. *Green Chem.* **2015**, *17*, 4334–4339.
96. Sancheti, S. V.; Gogate, P. R. Intensification of Heterogeneously Catalyzed Suzuki-Miyaura Cross-Coupling Reaction Using Ultrasound: Understanding Effect of Operating Parameters. *Ultrason. Sonochem.* **2018**, *40*, 30–39.
97. Brahmachari, G.; Mandal, M.; Karmakar, I.; Nurjamal, K.; Mandal, B. Ultrasound-Promoted Expedient and Green Synthesis of Diversely Functionalized 6-Amino-5-((4-Hydroxy-2-Oxo-2H-Chromen-3-Yl)(Aryl)Methyl)Pyrimidine-2,4(1H,3H)-Diones Via One-Pot Multicomponent Reaction under Sulfamic Acid Catalysis at Ambient Conditions. *ACS Sustain. Chem. Eng.* **2019**, 7, 6369–6380.
98. Verma, D.; Sharma, V.; Jain, S.; et al. Ultrasound-Assisted Synthesis of 1, 8-Dioxodecahydroacridine Derivatives in Presence of Ag Doped CdS Nanocatalyst. *J. Dispers. Sci. Technol.* **2020**, *41*, 1145–1158.
99. Abimannan, P.; Selvaraj, V.; Rajendran, V. Sonication Effect on the Reaction of 4-Bromo-1-Methylbenzene with Sodium Sulfide in Liquid–Liquid Multi-Site Phase-Transfer Catalysis Condition—Kinetic Study. *Ultrason. Sonochem.* **2015**, *23*, 156–164.
100. Diwathe, D. C.; Gogate, P. R. Ultrasound Assisted Intensified Synthesis of 1-Benzyloxy-4-Nitrobenzene in the Presence of Phase Transfer Catalyst. *Chem. Eng. J.* **2018**, *346*, 438–446.
101. Reddy, B. P.; Rajesh, K.; Vijayakumar, V. Ionic Liquid [Tbmim]Cl_2/$AlCl_3$ Under Ultrasonic Irradiation towards Synthesis of 1,4-DHP's. *Arab. J. Chem.* **2015**, *8*, 138–141.
102. Li, B. L.; Zhong, A. G.; Ying, A. G. Novel SO_3H-Functionalized Ionic Liquids—Catalyzed Facile and Efficient Synthesis of Polyhydroquinoline Derivatives Via Hantzsch Condensation under Ultrasound Irradiation. *J. Heterocyclic Chem.* **2015**, *52*, 445–449.
103. Reddy, M. V.; Reddy, A. V. S.; Jeong, Y. T. Di-N-Butyl Ammonium Chlorosulfonate Ionic Liquids as an Efficient and Recyclable Catalyst for the Synthesis of 1,4-Dihydrobenzo[4,5]Imidazo[1,2-a]Pyrimidine-3- Carboxylates under Solvent-Free Ultrasound Irradiation. *Res. Chem. Intermed.* **2016**, *42*, 4893–4906.
104. Dadhania, H. N.; Raval, D. K.; Dadhania, A. N. Sonochemical Synthesis of 2,3-Dihydro-4(1H)-Quinazolinones and 1-Amidoalkyl-2-Naphthols Using Magnetic Nanoparticle-Supported Ionic Liquid as a Heterogeneous Catalyst. *Res. Chem. Intermed.* **2018**, *44*, 117–134.
105. Tomke, P. D.; Rathod, V. K. Ultrasound Assisted Lipase Catalyzed Synthesis of Cinnamyl Acetate Via Transesterification Reaction in a Solvent Free Medium. *Ultrason. Sonochem.* **2015**, *27*, 241–246.
106. Waghmare, G. V.; Vetal, M. D.; Rathod, V. K. Ultrasound Assisted Enzyme Catalyzed Synthesis of Glycerol Carbonate from Glycerol and Dimethyl Carbonate. *Ultrason. Sonochem.* **2015**, *22*, 311–316.
107. Zhao, X.; Bansode, S. R.; Ribeiro, A.; Abreu, A. S.; Oliveira, C.; Parpot, P.; Cogate, P. R.; Rathod, V. K.; Cavaco-Paulo, A. Ultrasound Enhances Lipase-Catalyzed Synthesis of Poly (Ethylene Glutarate). *Ultrason. Sonochem.* **2016**, *31*, 506–511.
108. Zenevicz, M. C. P.; Jacques, A.; Silva, M. J. A.; Furigo, A., Jr.; Oliveira, V.; de Oliveira, D. Study of a Reactor Model for Enzymatic Reactions in Continuous Mode Coupled to an Ultrasound Bath for Esters Production. *Bioprocess Biosyst. Eng.* **2018**, *41*, 1589–1597.

109. Zang, C.-Z.; Kan, S.-C.; Yeh, C.-W.; Lin, C.-C.; Shieh, C.-J.; Liu, Y.-C. Ultrasound-Assisted (R)-Phenylephrine Whole-Cell Bioconversion by S. Marcescens N10612. *Ultrason. Sonochem.* **2015**, *26*, 415–421.
110. Korkut, I.; Bayramoglu, M. Ultrasound Assisted Biodiesel Production in Presence of Dolomite Catalyst. *Fuel* **2016**, *180*, 624–629.
111. Varghese, R.; Henry, J. P.; Irudayaraj, J. Ultrasonication-Assisted Transesterification for Biodiesel Production by Using Heterogeneous ZnO Nanocatalyst. *Environ. Prog. Sustain. Energy* **2018**, *37*, 1176–1182.
112. Cappelletti, L.; Vaghi, L.; Rinaldi, L.; Rotolo, L.; Palmisano, G.; Cravotto, G.; Penoni, A. One-Pot Sonochemical Synthesis of Ferrocenyl Derivatives Via a Three-Component Reaction in Aqueous Media. *Ultrason. Sonochem.* **2015**, *27*, 30–36.
113. Rabiei, K.; Naeimi, H. Ultrasonic Assisted Synthesis of Gem-Dichloroaziridine Derivatives Using Mg/CCl_4 under Neutral Conditions. *Ultrason. Sonochem.* **2015**, *24*, 150–154.
114. Rezaei, S. J. T. PEDOT Nanofiber/Pd(0) Composite-Mediated Aqueous Mizoroki–Heck Reactions under Ultrasonic Irradiation: an Efficient and Green Method for the C–C Cross-Coupling Reactions. *J. Iran. Chem. Soc.* **2017**, *14*, 585–594.
115. Saleh, T. S.; Al-Bogami, A. S.; Mekky, A. E. M.; Alkhathlan, H. Z. Sonochemical Synthesis of Novel Pyrano[3,4-e][1,3]Oxazines: A Green Protocol. *Ultrason. Sonochem.* **2017**, *36*, 474–480.
116. Tripathi, B.; Paniwnyk, L.; Cherkasov, N.; Ibhadon, A. O.; Lana-Villarreal, T.; Gómez, R. Ultrasound-Assisted Selective Hydrogenation of C-5 Acetylene Alcohols with Lindlar Catalysts. *Ultrason. Sonochem.* **2015**, *26*, 445–451.
117. Melais, N.; Boukachabia, M.; Aribi-Zouioueche, L.; Riant, O. Easy Preparation of Enantiomerically Enriched Heteroaromatic Alcohols through Lipase-Catalyzed Acylation with Succinic Anhydride under Unconventional Activation. *Bioprocess Biosyst. Eng.* **2015**, *38*, 1579–1588.
118. Alves, J. S.; Garcia-Galan, C.; Danelli, D.; Paludo, N.; Barbosa, O.; Rodrigues, R. C.; Fernandez-Lafuente, R. Use of Lecitase-Ultra Immobilized on Styrene-Divinylbenzene Beads as Catalyst of Esterification Reactions: Effects of Ultrasounds. *Catal. Today* **2015**, *255*, 27–32.
119. Gułajski, Ł.; Tracz, A.; Urbaniak, K.; Czarnocki, S. J.; Bieniek, M.; Olszewski, T. K. Ammonium-Tagged Ruthenium-Based Catalysts for Olefin Metathesis in Aqueous Media under Ultrasound and Microwave Irradiation. *Beilstein J. Org. Chem.* **2019**, *15*, 160–165.
120. An, B.; Fan, H.; Wu, Z.; Zheng, L.; Wang, L.; Wang, Z.; Chen, G. Ultrasound-Assisted Enantioselective Esterification of Ibuprofen Catalyzed by a Flower-like Nanobioreactor. *Molecules* **2016**, *21*, 565.
121. Demaerel, J.; Govaerts, S.; Paul, R. R.; van Gerven, T.; de Borggraeve, W. M. Non-innocent Probes in Direct Sonication: Metal Assistance in Oxidative Radical C−H Functionalization. *Ultrason. Sonochem.* **2018**, *41*, 134–142.
122. Wu, Z.; Borretto, E.; Medlock, J.; Bonrath, W.; Cravotto, G. Effects of Ultrasound and Microwaves on Selective Reduction: Catalyst Preparation and Reactions. *ChemCatChem* **2014**, *6*, 2762–2783.
123. Martina, K.; Tagliapietra, S.; Barge, A.; Cravotto, G. Combined Microwaves/Ultrasound, a Hybrid Technology. *Top. Curr. Chem.* **2016**, *374*, 79.
124. Huang, M.; Zhang, M.; Bhandari, B. Synergistic Effects of Ultrasound and Microwave on the Pumpkin Slices Qualities during Ultrasound-Assisted Microwave Vacuum Frying. *J. Food Process Eng.* **2018**, *41*, e12835.
125. Szadzińska, J.; Łechtańska, J.; Pashminehazar, R.; Kharaghani, A.; Tsotsas, E. Microwave- and Ultrasound-Assisted Convective Drying of Raspberries: Drying Kinetics and Microstructural Changes. *Drying Technol.* **2019**, *37*, 1–12.

126. Liew, S. Q.; Ngoh, G. C.; Yusoff, R.; Teoh, H. Microwave Assisted Extraction (MAE) and Microwave-Ultrasound Assisted Extraction (MUAE) of Pectin from Pomelo Peels. *Int. J. Food Eng.* **2017**, *3*, 117–121.
127. Pawełczyk, A.; Sowa-Kasprzak, K.; Olender, D.; Zaprutko, L. Microwave (MW), Ultrasound (US) and Combined Synergic MW-US Strategies for Rapid Functionalization of Pharmaceutical Use Phenols. *Molecules* **2018**, *23*, 2360.
128. Mason, T. J. Some Neglected or Rejected Paths in Sonochemistry—A Very Personal View. *Ultrason. Sonochem.* **2015**, *25*, 89–93.
129. Aljbour, S.; Yamada, H.; Tagawa, T. Ultrasound-Assisted Phase Transfer Catalysis in a Capillary Microreactor. *Chem. Eng. Process.* **2009**, *48*, 1167–1172.
130. Nikitenko, S. I.; Pflieger, R. Toward a New Paradigm for Sonochemistry: Short Review on Nonequilibrium Plasma Observations by Means of MBSL Spectroscopy in Aqueous Solutions. *Ultrason. Sonochem.* **2017**, *35*, 623–630.

CHAPTER SIX

The catalytic production of biofuels (biodiesel and bioethanol) using sonochemical, microwave, and mechanical methods

Indra Neel Pulidindi and Aharon Gedanken
Department of Chemistry, Bar Ilan University, Ramat Gan, Israel

1 Introduction

A Web of Science search using key words "Biofuels, Biodiesel, and Bioethanol" yields search results of "10.806, 4796, and 1540," respectively, for the year 2018 alone. This indicates the intensity of research activity in the field of biofuels and more specifically biodiesel and bioethanol. Nontraditional methods of activation, for instance, ultrasound and microwave irradiation, are being employed for the acceleration of key chemical reactions like lipid or oil extraction from biomass, transesterification of triglycerides (lipids), waste cooking oil, and free fatty acids, leading to biodiesel production,[1–102] and vital reactions like the biomass pretreatment,[103–129] biomass hydrolysis,[130–139] and fermentation of sugars,[140–146] leading to the accelerated and high yield bioethanol production. However, the potential of nontraditional methods of activation, especially ultrasound and microwave irradiation, for biofuel production is not utilized to its fullest as evident from the relatively smaller number of publications dealing with these subjects (Fig. 1). The use of unconventional methods of activation is nearly two orders of magnitude lower than the traditional catalytic methods for bioethanol production and an order of magnitude lower for biodiesel production. This prompts a review on the role and potential of microwave and sonochemical irradiation on the conversion of biomass to biofuels, the challenges involved in the design of large-scale processes based on such nontraditional methods of activation, and the opportunities and rewards awaiting in this field of research and development.

Nontraditional Activation Methods in Green and Sustainable Applications
https://doi.org/10.1016/B978-0-12-819009-8.00013-X
© 2021 Elsevier Inc.
All rights reserved.

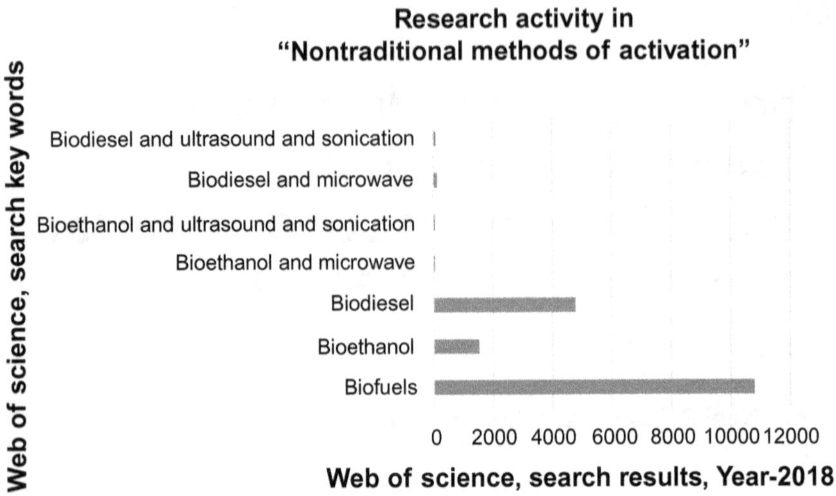

Fig. 1 Results of a Web of Science search on the use of nontraditional methods of activation for biofuel production in the year 2018.

2 Biodiesel production using nontraditional methods of activation

Biodiesel or fatty acid alkyl ester (FAAE) consists of methyl or ethyl esters produced by the transesterification reaction using feedstock like non-edible triglyceride sources like *Jatropha* and castor oil seeds, vegetable oil, waste cooking oil, microalgal lipids, or animal fat.[83–94] Use of biodiesel for transportation and industrial use not only reduces the dependence on fossil fuels but also reduces the rate of depletion of fossil resources and contributes significantly to the CO_2 emission reduction. CO_2 is one of the major greenhouse gases accelerating the climate change with serious implications on the sustenance of human societies.

Nontraditional methods of activation especially, microwave and sonochemical, play a key role in the future and upcoming industrial processes for the large-scale biodiesel production using feedstock like waste cooking oil and the lipids of algal biomass.[31] Microwave heating depends on the dielectric properties of the materials. When the polar molecular medium is placed in the microwave field, the reacting species rotate in coherence with the high-frequency electric field. As the field alternates, the molecules change the direction of rotation. The rotating molecules undergoing pushing and pulling collide with other molecules in the medium through electrical forces. As a result the energy is distributed to adjacent reacting species in the medium (atoms or molecules). This microscopic process of energy transfer from microwave to substrate molecules causes radiative heating. As more microwave interacts with the reacting medium, more of the increase in temperature of the reacting medium takes place.[12] In microwave-assisted heating, maximum absorption of microwave (MW) and a corresponding heat energy generation occur in all the material in the reacting system. MW forces dipoles to align with the oscillating electric field. Material heated by MW will have higher temperature at its center unlike conventional heating wherein the walls of the container are heated before the material is heated by the processes of conduction or convection loss.[147] Microwave thus activates all the catalytic sites and the reactant molecules in the system.[30] Microwave processing has evolved into a potential tool in organic synthesis and plays key role in the developing sustainable world. Microwave heating is better than conventional heating due to unique thermal and nonthermal effects contributing electrical and the magnetic field components of the electromagnetic radiation, namely, microwaves. Microwave heating is known to accelerate chemical reactions by a factor of 5 to 10^3 compared with conventional processes.[31] Another aspect that generated interest in microwave-driven processes among chemists, scientists, and industrialists is the fact that chemical reactions have been revolutionized with unexplainable results and unexpected results in some instances. Typical advantages of microwave-based processes include environmental friendliness, rigid heating, lower energy consumption (though cannot be generalized and dependent on the specific reaction and the method of heating being compared in the study), process controllability, shorter process time, and improved product and process quality.[31]

Microwaves are electromagnetic waves with energies and frequencies located in between the infrared and radio wave region of the electromagnetic spectrum. The frequency of microwaves, $0.3–0.3 \times 10^3$ GHz, translates to energy units, $1.24 \times 10^{-6}–1.24 \times 10^{-3}$ eV, following Plank's law, $E = h\upsilon$. This energy of microwaves is much lower than the bond energies in chemical compounds (covalent bond energy, 5 eV; hydrogen bond energy, 2 eV; van der Waals intermolecular interactions, <2 eV; and Brownian motion at 37°C, 2.7×10^{-3} eV) and biological systems (biological compounds, 13.6 eV). A comparison of the bond energies with the energies of microwave indicates that microwave energy is insufficient to break chemical bonds or induce chemical reaction.[31]

However, microwave irradiation, via thermal and nonthermal means, activates the materials in the reaction medium, like substrate and catalyst. Microwave heating utilizes the properties of certain materials, liquids or solids, to transform electromagnetic energy into heat energy. The electric field and magnetic field components interact differently with the material (either substrate or catalyst in different mechanisms). For instance, Liu and Jia investigated on the nonthermal effects of microwave irradiation upon the structure and dynamics of water and hydrogen bonding network through molecular dynamic simulations.[32] The nonthermal effects of microwave irradiation are of high significance for the activation of biomass, either by enhancing the release of lipids from algal biomass or by reducing the structural rigidity of biomass components like cellulose by altering the lifetime of hydrogen bonding network. The authors demonstrated in the specific case of aqueous NaCl solution how the tetrahedral structure of water is perturbed upon microwave irradiation (Fig. 2). There is a phenomenon called threshold effect, meaning beyond a specific value of microwave power, there will be drastic alteration in the structure of water, hydrogen bonding network leading to the reduction in the life time of hydrogen bonding (Fig. 3), and a decrease in the number of hydrogen bonds with increasing energy of microwaves. Biological systems, for instance, human body, are made up of more than 70% of water. Likewise, biomass, whether it is a terrestrial lignocellulosic biomass or algal biomass, has inherent water content whose dynamics will be significantly altered upon activation with microwave irradiation leading to accelerated conversion of biomass to desired products upon microwave irradiation.

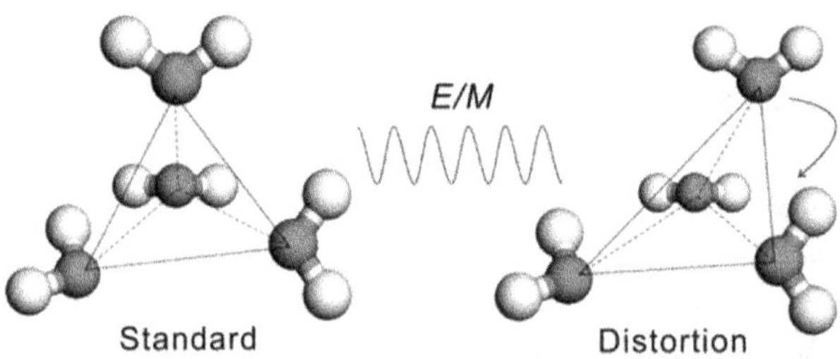

Fig. 2 Effect of microwaves (electromagnetic radiation) on the perturbation of the tetrahedral structure of water and the hydrogen bonding network. *Adapted with permission from Liu, J.; Jia, G. Non-thermal effects of microwave in sodium chloride aqueous solution: Insights from molecular dynamics simulations. J. Mol. Liquids* ***2017****, 227, 2017, 31–36. Elsevier.*

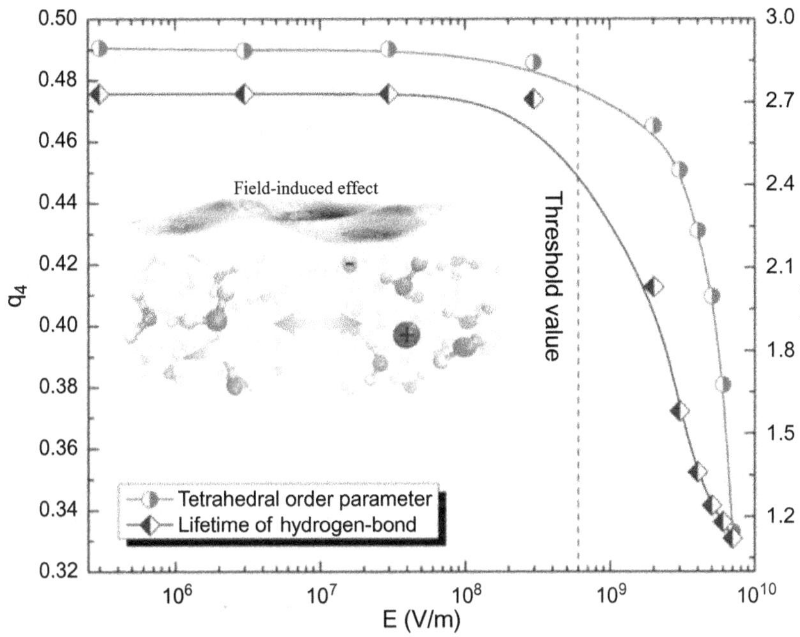

Fig. 3 Role of microwave field strength on the dynamics of water structure and hydrogen bonding lifetime in the specific example of aqueous NaCl solution. *Adapted with permission from Liu, J.; Jia, G. Non-thermal effects of microwave in sodium chloride aqueous solution: Insights from molecular dynamics simulations. Journal of Molecular Liquids,* ***2017****, 227, 31–36. Elsevier.*

In the specific example of the microwave irradiation of aqueous NaCl solution, microwave energy distribution is partly stored in the internal rearrangement of the hydrogen bonding network. From the perspective of the tetrahedral structure of water as shown in Fig. 2, the average value of the tetrahedral order parameter (q_4) was defined. This parameter, q_4, quantified the tetrahedral order of the system by measuring the derivative or distortion in the tetrahedral structure of water from the ideal tetrahedral structure upon exposure to microwave radiation. The dynamics of the aqueous systems did not vary monotonically with microwave frequency, even though the microwave field strength can affect the structure of water. The tetrahedral, hydrogen-bonded structure of water is perturbed by the static electric field. The lifetime of hydrogen bonding network is decreased as the field strength increases. A field-induced threshold effect was observed. Once the microwave field strength was greater than the threshold value, a significant distortion in the water structure was observed, which is named as threshold effect (Fig. 3). High-intensity field can cause a decline in lifetime of hydrogen bonding network (Fig. 4).[32]

Microwave-induced alteration of hydrogen bonding dynamics plays a crucial role in the biochemical transformations relevant to biofuel production via activation of biomass.

Patel et al. evaluated various pretreatment methods for the efficient extraction of lipids from *Oleaginous* cellular biomass.[26] The methods studied consist of oil expeller, bead milling, ultrasonication, microwave, high-speed and high-pressure homogenizer, laser, autoclaving, pulsed electric field, and nonmechanical methods such as enzymatic treatment. After the mechanical and nonmechanical pretreatments, solvent extraction is employed to extract the lipids from the pretreated biomass. Use of toxic and harmful solvents like chloroform and methanol could be replaced with supercritical fluid extraction. Drying of biomass is costly. Therefore drying steps before pretreatment

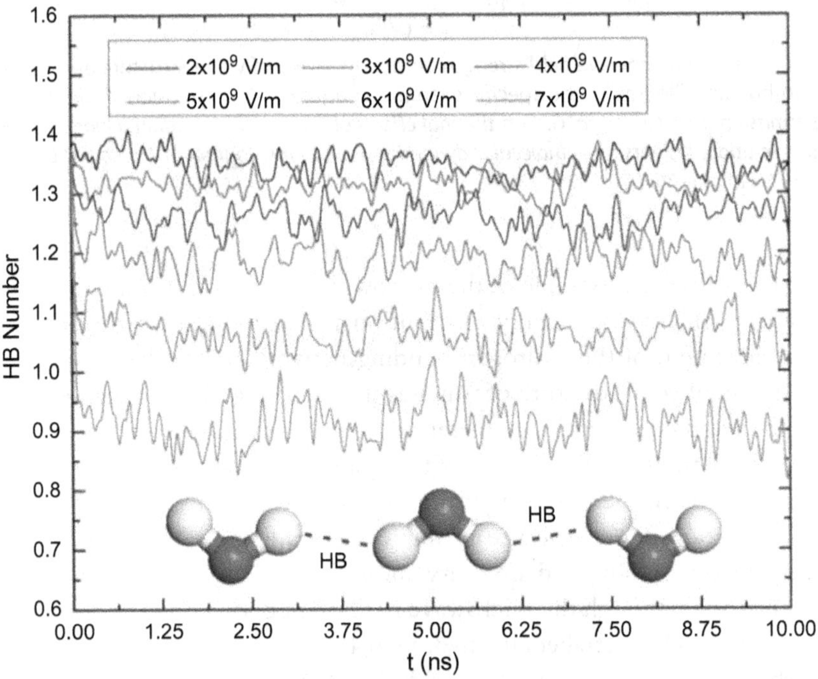

Fig. 4 Effect of the static electric field strength of microwaves on the hydrogen bond (HB) number (in aq. NaCl electrolyte). *Adapted with permission from Liu, J.; Jia, G., Non-thermal effects of microwave in sodium chloride aqueous solution: Insights from molecular dynamics simulations. Journal of Molecular Liquids,* ***2017****, 227, 31–36. Elsevier.*

should be avoided. Use of single pretreatment method may not be enough for the complete removal or extraction of lipids. So the use of multiple pretreatment methods like a combination of ultrasonication and microwave irradiation under wet conditions facilitates easy breakdown of cells and releases the lipids to external environment. After the initial ultrasonication step, the use of microwave treatment leads to the alignment of polar compounds in the cellular structure in the direction of the applied electric field and rotation of the lipid moieties in high speed as the microwave field alters. The duration of pretreatment and the efficiency of pretreatment (yield of lipids) are improved, and solvent consumption is decreased by the adoption of such combined strategy for pretreatment.[26] Bundhoo reviewed the application of microwave for biofuel production from biological and thermochemical conversion processes with emphasis on biodiesel production from transesterification. Challenges in microwave technology like energy efficiency of microwave-driven processes and large-scale implementation were discussed.[40] Mishra and Goswami reviewed the history, importance, properties, sources, and techniques to produce biodiesel.[44] Use of nonedible oil, genetically engineered plants, and microalgae as feedstock for biodiesel ensures sustainable biodiesel production. The authors viewed that the policy makers should support promotion of biodiesel fuel as alternative to existing transportation and industrial fuels by generously funding the research in this field that would lead to cost competitiveness with conventional energy sources like petrol and diesel. The byproduct of biodiesel production process, namely, glycerol, should be converted to value-added products leading to cost reduction of biodiesel as well as managing the waste from biodiesel industry.[44] Singh and Gaurav reviewed the advances in the catalysis (homogeneous, heterogeneous, and enzymatic catalysts) of transesterification reaction leading to biodiesel production.[45] The unique advantage of time and energy savings with improved product yield upon the use of nontraditional methods of activation, namely, microwave and sonochemical irradiations, in the specific example of conversion of waste cooking oil to biodiesel is shown pictorially in Fig. 5.[82] The process depicted in Fig. 5 uses NaOH (1%) as catalyst with waste cooking oil as feedstock (10 mL) and a methanol-to-oil ratio of 9:1. Conventional heating was performed on a hot plate at 500-W power; likewise, microwave and ultrasound irradiations were applied at 800 and 500 W, respectively. The figure is self-explanatory and highlights the necessity to shift to nontraditional methods of activation, especially, for biodiesel production at industrial scale. However, intense research is required toward the development of large-scale processes using

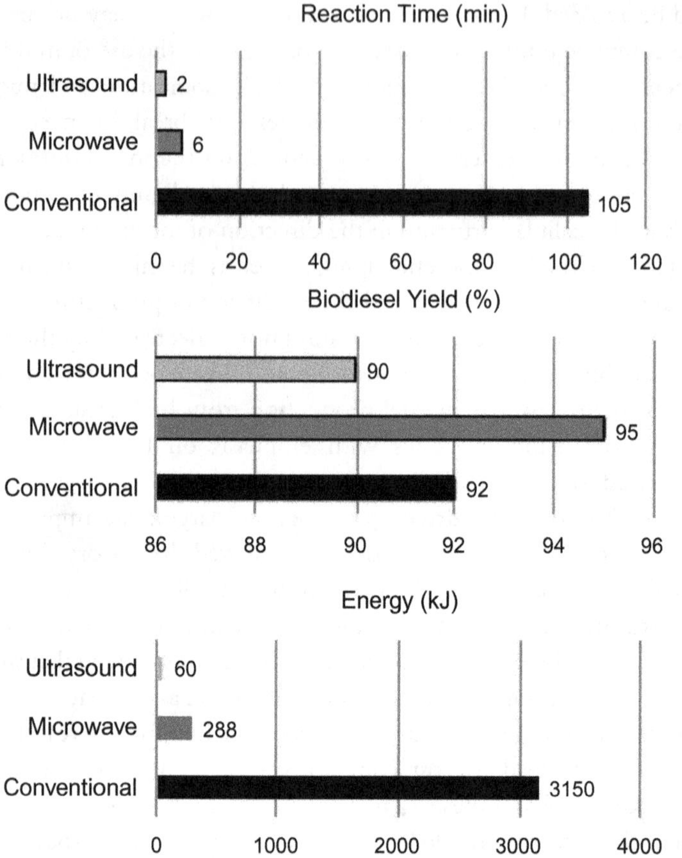

Fig. 5 Unique advantage of time and energy savings with enhanced product yield upon the use of microwave and sonochemical activation compared with conventional activation. *Adapted with permission from Gude, V. G.; Martinez-Guerra, E., Green chemistry with process intensification for sustainable biodiesel production, Environmental Chemistry Letters,* ***2018****, 16(2), 327–341. Springer Nature.*

diverse nonedible feedstock for biodiesel production utilizing the potential of such unconventional methods of activation to reach the goal of an industrial process. Advantages of using microwave (electromagnetic field) and sonochemical (acoustic cavitation field) activation of catalytic and substrate species are composed of process intensification, improvement in reaction efficiency, higher product recovery, reduced byproduct formation, and reduced energy consumption.[82] Use of such nontraditional methods of activation makes the biodiesel production process green by enhancing positively the green metrics like *E*-factor, atom economy, mass productivity, reaction mass efficiency, safe operation, and high efficiency. A hybrid technology that exploits the unique potentials of microwave and ultrasound activations, namely, extensive heating in short time and effective mixing, respectively, is proposed to have a synergistic effect on process intensification for biofuel production.[82]

Recent advances (the year 2018) in the use of nontraditional activation methods (microwave and sonochemical irradiation) for biodiesel production are highlighted in Table 1 (shown towards the end of this chapter).

Gedanken's group has contributed significantly for the expansion of the knowledge domain of the nontraditional methods of activation for biodiesel[83–94] and bioethanol production.[146, 148–157] Next few sections of this chapter would be devoted to the major and significant breakthroughs achieved in this field by Gedanken's research group.[158–161]

2.1 Novel solid base catalysts (modified SrO) for biodiesel production using microwave activation

SrO showed outstanding performance as solid acid catalyst for the conversion of a variety of nonedible lipid feedstocks, like microalgae (*Nannochloropsis* and *Chlorella vulgaris*),[85, 89, 95] castor and jatropha seeds,[86] and waste cooking oil,[84, 91, 93] into biodiesel in high yield in shortest reaction time upon microwave activation.[87] The acceleration of transesterification leading to fast production of biodiesel under microwave activation is due to the super heating and the synergistic activation of both the catalyst (SrO) and the substrate (oil and methanol) owing to their ionic and polar components that interact with the microwave at the molecular level leading to efficient superheating that was very different from the bulk temperature observed experimentally.[87]

For the effective utilization of SrO catalyst from the economics and process improvement perspective, a novel strategy was developed to deposit nanoparticles of SrO on millimetric silica beads (an inert support). Such a strategy reduced the amount of SrO catalyst utilization and facilitated the use of SrO catalyst in a continuous flow microwave reactor for large-scale processing of waste cooking oil to biodiesel in short reaction time and in high yield. Such a catalyst should be tested yet for pilot level biodiesel operation plant. Typical method of depositing nanoparticles of SrO on SiO_2 millimetric silica beads, SrO/SiO_2, in an aqueous phase reaction under microwave irradiation consists of reacting equimolar amounts of Sr $(NO_3)_2$ and Na_2CO_3 along with silica beads (3–6 mm) in water under vigorous stirring. Ethylenediamine was added to the reaction medium to

Table 1 Brief description of scientific advances on the use of nontraditional activation methods for biodiesel production (the year 2018).

S. No.	Application of microwave/ultrasound activation	Reference
1	Microalgal (*Chlorella* sp.) lipids converted to biodiesel under microwave heating with KOH as base catalyst. Maximum FAME yield of 32.18% was achieved using microwave activation of biomass as well as catalyst in 30 min	7
2	Neutral lipid extraction/FFA (free fatty acid) extraction using microwave method is 9.6 wt% compared with 13.2 wt% using Soxhlet extraction	10
3	96.89 wt% FAME yield obtained from kitchen food waste as feedstock using microwave heating (cell pressure, 2.2 MPa; T, 170°C; and t, 4 min.)	11
4	Heating of oleic acid (biodiesel feedstock) usually requires high temperature. A novel microwave-assisted heat transfer process is designed using porous SiC as heat transfer medium. Porous SiC absorbs the microwave energy, gets heated up, and then transfers the heat from the high loss material porous medium to oleic acid and facilitates accelerated production of biodiesel. Improved quality and instantaneous control are the advantages of microwave-based method compared with traditional heating method. Due to the unique heating principle microwave irradiation is used in various fields. Required guidelines for the design of larger microwave-based systems for the industrial-scale production of biodiesel are provided	12
5	Enhanced lipid extraction from microalgae (*Botryococcus braunii* LB 572) was achieved using microwave treatment (2450 MHz, 1250 W at 150°C for 20 min). C_{14}–C_{24} fatty acids extracted from lipid. Among them the major fatty acids were oleic (C18:1) acid, linolenic (C18:3), and palmitic acid (C16:0). Optimal conditions of transesterification were 70 mL of methanol, 6 mL of H_2SO_4, and 8 mL of chloroform, boiling at 100°C for 3 min	14
6	Microwave heating is not only used for the accelerated production of biodiesel from canola oil but also was used for the microwave-based combustion method for the synthesis of carbonated calcium aluminate catalyst with dextrose fuel, which showed promising activity for biodiesel production. KOH was impregnated on carbonated calcium aluminate to generate basic sites on the solid support and used for transesterification reaction	15

7	Incredibly high yield of biodiesel (42.22%) was achieved by the single step and direct transesterification of wet *Nannochloropsis* sp. by the synergistic effect of combining ionic liquid solvent and microwave heating under optimized reaction conditions	16
8	Microwave irradiation facilitated both lipid extraction from *Mucor circinelloides* URM 41F2 and the transesterification of the extracted lipids in the presence of Novozym 435 catalyst. Ethanol is used as solvent for the extraction of lipids. Both simultaneous or sequential approaches are tested. Two step method composed of lipid extraction followed by enzymatic transesterification reaction under microwave irradiation yield 98.5% ethyl ester in 10 h compared with conventional heating for 30 h. Simultaneous extraction and transesterification also resulted in promising results. However, the yield of FAME was slightly lower	19
9	Environmental wastes like palm empty fruit bunch (PEFB), coconut meal residue (CMR), and coconut coir husk (CH) were used for the preparation of sulfonated solid acid catalysts. Highest biodiesel yield of 95.5% was obtained using sulfonated PEFB solid acid catalyst under microwave heating for 60 min at 70°C with waste palm oil (WPO) as biodiesel feedstock	20
10	Calcium aluminate is proposed as a support material for KOH for the preparation of KOH/$Ca_{12}Al_{14}O_{33}$ solid base catalyst. The catalyst was prepared by microwave-driven combustion synthesis using different fuels, namely, glycine, sorbitol, urea, and citric acid. The catalyst support prepared using sorbital as fuel showed best performance upon loading KOH owing to high basicity, surface area, and porosity converting 93.4% of canola oil to biodiesel with a microwave irradiation (450 W) for 60 min. Calcium aluminate is a promising support material for the preparation of KOH-supported solid base catalyst for industrial-scale production of biodiesel	21
11	Nyamplung plants are most widely grown in Indonesia. The possibility of exploiting Nyamplung oil as feedstock for biodiesel production was exploited using CaO catalyst and microwave heating. Process parameters were optimized using Box-Behnken design. Employing parameters such as microwave power of 449.29 W, catalyst amount of 4.86%, and transesterification time of 10.07 min yields 92.73% biodiesel with 93.22% reliability	22

Continued

Table 1 Brief description of scientific advances on the use of nontraditional activation methods for biodiesel production (the year 2018)—cont'd

S. No.	Application of microwave/ultrasound activation	Reference
12	Esterification of fatty acids like oleic and stearic acids was carried out with homogeneous acid catalyst using microwave, conduction, and ultrasound as heating methods. Microwave heating was found to be the most efficient method with 52% yield of biodiesel with oleic acid as feedstock	23
13	Cotton oil is used as feedstock for biodiesel production via ultrasound activation with potassium methoxide as catalyst. Interesterification of cotton oil with methyl acetate using sonication for activating the substrate and catalytic species resulted in 14% higher conversion of triglycerides to biodiesel	24
14	Carbon-supported $KOH/Ca_{12}Al_{14}O_{33}$ nanocomposite was prepared by combustion synthesis using dextrose as carbon source and fuel. The solid base catalyst was used for the transesterification of canola oil with microwave heating. 98.8% conversion of canola oil was observed upon transesterification reaction for 30 min with methanol/oil molar ratio of 15 and 4 wt% nanocomposite as microwave absorbing material. A comparison was made between carbonated and noncarbonated KOH modified calcium aluminate catalysts for transesterification activity. Carbon-supported base catalyst showed twice the activity compared with noncarbon containing system even though both of the catalysts possessed same basicity. The enhancement of activity for transesterification reaction of carbon-supported KOH modified calcium aluminate was attributed to high specific surface area and porosity. Large pore size of the carbon-supported KOH/calcium aluminate facilitated the easy accessibility of catalytic active sites for the accelerated production of biodiesel from canola oil. The specific properties of the product biodiesel like density, kinematic viscosity (at 40°C), acid number, and flash point matched well with the standard values (EN 14214 and ASTM D-6751)	25, 34
15	Among KOH, NaOH, and LiOH, KOH and NaOH showed higher transesterification activity for the conversion of microalgae (*Chlorella vulgaris*) to biodiesel compared with LiOH irrespective of the method of activation, namely, microwave or ultrasound or conventional heating. However, the yield of FAME was nearly 5% higher, and the reaction time was reduced by 50% when nontraditional methods of activation like microwave and ultrasound were used as heating methods	27
16	Microwave heating is an easy and fast method for the extraction of lipids from the paste of microalgae. *Nannochloropsis oculata*, *Amphora coffeaeformis*, *Chaetoceros gracilis*, and *Isochrysis galbana*. The lipid contents obtained from the aforementioned algae are 33.6 ± 2.6, 16.27 ± 1.5, 22.31 ± 2.5, and	28

	19.87 ± 2.4%, respectively, upon microwave irradiation of the biomass for 1 min. The lipid yield is high, and the extraction time is low when microwave irradiation is used compared with ultrasound irradiation. Extensive cell damage resulting in cracking and fragmentation is achieved, as revealed from SEM analysis, upon microwave treatment leading to the release of lipids from the cells of algal biomass. Moreover the microwave irradiation does not degrade the fatty acids released from cell disruption. Thus microwave irradiation is a promising option for the extraction of lipids from biomass	
17	A maximum biodiesel yield of 94.86% was achieved from waste cooking oil using microwave irradiation with calcium diglyceroxide as catalyst. The yield of biodiesel was much lower with conventional heating (42.59%) with the same reaction time (15 min) as that of microwave irradiation ($T = 65°C$). However, the activity of the catalyst decreased upon reuse for consecutive cycles. The transesterification reaction was found to follow pseudo first-order reaction kinetics with an activation energy and preexponential factor values of 26.56 $kJmol^{-1}$ and 1.65×10^3 min^{-1}, respectively. The fuel properties of the biodiesel obtained matched well with ASTM-D6751 standards. Microwave irradiation promoted rapid mixing of heterogeneous phase and effective heating of reaction mixture, and as a result the rate of the transesterification reaction is enhanced with high yields of FAME compared with conventional heating	29
18	KF-modified hydrotalcite was used as solid base catalyst for the conversion of *Jatropha* oil to biodiesel using microwave heating and conventional heating using reflux method. Better yields of FAME were obtained with microwave heating, and the solid base catalyst was reusable without appreciable changes in the textural properties after the transesterification reaction	30
19	Microwave is used as an effective heating method for promoting transfer hydrogenation reaction for upgrading *Jatropha* oil biodiesel catalyzed by Raney-Ni with water as solvent. In 50 min the mass conversion ratio of methyl linoleate was found to be 91.98 wt%. High selectivity and high reaction rate for the partial hydrogenation reaction were observed using microwave irradiation. The reaction followed pseudo first-order kinetics. The activation energy values for the hydrogenation of C18:2 and C18:1 fractions were found to be 115.7 and 91.9 kJ/mol	33

Continued

Table 1 Brief description of scientific advances on the use of nontraditional activation methods for biodiesel production (the year 2018)—cont'd

S. No.	Application of microwave/ultrasound activation	Reference
20	High lipid containing larva, soldier fly, was subjected to microwave irradiation for extracting lipid that could serve as feedstock for biodiesel production. In the process of extraction, the diffusion of lipids into the solvent was found to be the rate limiting step with an activation energy value of 84.26 kJ/mol. The key thermodynamic parameters for the extraction of lipids from the larval fly were found to be 4.37–6.76 kJ/mol (ΔH), 18.73–28.01 J/mol (ΔS), and −2.57 to −1.31 kJ/mol (ΔG)	35
21	The potential of the synergistic activation of microalgae, *Nannochloropsis* sp., by simultaneously employing microwave and sonochemical irradiation for biodiesel production is exploited. The transesterification reaction followed first-order reaction kinetics with an activation energy value of 17.298 kJ/mol. The yield of FAME is 48.2% with a microwave energy dissipation of 140 W combined with 140 W ultrasound intensity	36
22	Methyl ester (yield, 33.84%) was produced by the transesterification of coconut oil using CaO as catalyst with microwave heating for 3 min. The transesterification reaction followed pseudo first-order kinetics with an activation energy value of 82.08 kJ/mol	37
23	Synergistic catalytic activity was observed between microwave heating and graphene oxide catalyst for the esterification of oleic acid to FAME in high yield (99.2%) in 3 min. with 1% catalyst. Mole ratio of methanol/oleic acid was crucial for the yield of FAME. Graphene oxide is a potential solid acid catalyst with excellent microwave absorption and thermal conversion efficiency	38
24	Biodiesel with improved was produced by the KOH-catalyzed microwave-driven process using a mixture of waste cooking oil and *Calophyllum inophyllum* oil with a biodiesel yield of 97.65%. The optimal conditions of the process are methanol/oil ratio, 58.60 (*v*/v); catalyst concentration, 0.774% (*w*/w); stirring speed, 600 rpm; and reaction time, 7.15 min. The biodiesel from a mixed feed (waste cooking oil and *Calophyllum inophyllum* oil) showed higher oxidation stability (18.03 h) compared with waste cooking oil-based biodiesel (4.61 h). Physicochemical properties of the biodiesel matched with specifications of ASTM D6751 and EN14214 standards	39

25	Eggshell waste was incinerated at 1000°C for 2 h to form CaO that was used as solid base catalyst for the transesterification of waste cooking oil to biodiesel with microwave heating. Under best reaction conditions (reaction time, 165 min; catalyst loading, 5 wt%, methanol-to-oil molar ratio, 9:1; and temperature, 65°C), the yield of FAME was 87.8%	41
26	Barium-deposited CaO was used as potential solid base catalyst for the transesterification of canola oil under microwave heating. Waste eggshells (major constituent, $CaCO_3$) were calcined at 900°C for 2 h to produce CaO that are used as support for Ba. Optimal conditions of biodiesel production with yield 97.68% composed of microwave irradiation (300 W) for 2 min with 15 wt% Ba/CaO (3 wt% catalyst loading), a reaction temperature of 60°C	42
27	Basic faujasite catalysts were used for the transesterification of rapeseed oil to biodiesel under microwave heating. Intracrystalline mesoporosity is generated in the faujasite zeolites X and Y by template-assisted modification of porosity. Basic sites were generated in the zeolite catalysts by ion exchange with Cs^+ and K^+ species. Increase in the concentration of basic sites (Cs and K species) and Al content, increased the conversion of triglycerides in the rapeseed oil to biodiesel. CsNaX zeolite exhibited highest activity for the transesterification with 90% conversion in 1 h	43
28	Synergistic catalytic activity (NaOH, 1 wt%) for the conversion of waste cooking oil to biodiesel was observed with the addition of microwave absorbing material, namely, activated carbon (1 wt%). Under optimal reaction conditions (methanol-to-oil ratio of 7:1; microwave power of 200 W; microwave irradiation time, 2 min; and reaction temperature, 65°C), a waste cooking oil conversion of 97.8 wt% was observed	46
29	In the pursuit of large-scale microwave-driven biodiesel production, the authors presented a model of reaction tube with cross structures that has improved biodiesel production rate and microwave heating uniformity. Scientific aspects of microwave heating, namely, Maxwell's equation, heat conduction equation, reaction kinetics equation, and Navier-Stokes equation, are coupled together to develop the continuous flow microwave-driven biodiesel production. The distance between each cross structure influenced greatly the microwave heating uniformity and the heating efficiency, while the angle was found to have little effect	47

Continued

Table 1 Brief description of scientific advances on the use of nontraditional activation methods for biodiesel production (the year 2018)—cont'd

S. No.	Application of microwave/ultrasound activation	Reference
30	Biodiesel from palm oil was produced using 25 wt% KOH/ZrO_2 bentonite under microwave heating. Eighty-one percent FAME yield was obtained under optimal reaction conditions (10-min microwave irradiation; microwave power, 700 W; temperature, 65°C; 3 h; and 3 wt% of catalyst loading)	48
31	Biomass (*Yarrowia lipolytica*) grown on waste cooking oil (WCO) was used as feedstock for the single-step (in situ) production of biodiesel (FAME yield, 0.22 g/g biomass). Biomass was pretreated by sonication (30°C, 50 kHz, 5 min) prior to transesterification reaction. Biodiesel production was carried out with conventional heating (thermostatic water bath). The optimal reaction conditions of transesterification process include 0.2 M, H_2SO_4 (1 mL/g), biomass (4 g), methanol (100 mL), chloroform (10 mL), reaction time (8 h), and reaction temperature (50°C). The amounts of saturated monounsaturated and polyunsaturated methyl esters in the FAME are 32.81%, 36.41%, and 30.59%, respectively. Apart from high yield production of biodiesel, the methodology provides an alternative for the productive utilization of environmental pollutant (*Yarrowia lipolytica*)	49
32	Unconventional cell disruption techniques, like microwave and ultrasound irradiation, were employed for the extraction of lipids from microalgae (*Chaetoceros calcitrans*). Amount of lipid extracted from ultrasound (5 min), microwave (40 s), and conventional chemical (2 M HCl, 1 h) treatments was 24.6 ± 1.3, 24.2 ± 0.9, and 24.2 ± 1.0%. Microwave and sonication methods yield lipid quantities similar to conventional chemical treatments however in less time using fewer chemicals. Moreover the saturated fatty acid and polyunsaturated fatty acid contents were higher using unconventional method of activation. Especially the energy consumption was lower with ultrasound treatment compared with microwave irradiation. Sonication technique is recommended for the extraction of lipids from algal biomass via cell disruption for large-scale applications	50
33	The promise of microalga as promising feedstock for biodiesel production arises owing to its high oil lipid content. *Chlorella* sp. with high oil content (28.32% based on dry weight) was employed as feedstock for the transesterification reaction. Microwave heating facilitated simultaneous extraction	51

	of lipids and their conversion to FAME (75.68%). Use of cosolvent like n-hexane improved yield of biodiesel. The optimal process parameters include microwave power, 450 W; reaction time, 60 min.; acid catalyst, 0.2 M H_2SO_4; and cosolvent, 10 mL. Palmitic and methyl ester and palmitoleic acid methyl ester were the major constituents of FAME	
34	A continuous flow fluid system with a focused microwave heating device was used for the conversion of waste cooking oil (WCO) and soybean oil into biodiesel. SrO was used as solid base catalyst for the transesterification reaction leading to a conversion value of 93% of WCO in a continuous flow fluid flow system. A pilot study for scaling up of the process is conducted and its potential is demonstrated. The rate-determining step of the transesterification process was found to be the breaking of ether bonds and the formation of a tetrahedral reaction intermediate. SrO catalyst lowers the activation energy of the transesterification process by promoting the decomposition of ester bonds leading to the formation of a tetrahedral intermediate. Gedanken's group was the first to introduce continuous flow microwave-driven process for the production of biodiesel using vegetable oil, namely, canola and sunflower oil as feedstock, with high yields of the product biodiesel, 92% and 89%, respectively	52, 53
35	A physical mixture of CaO and MgO (1:10) was used as solid base catalyst for the transesterification of *Jatropha curcas* using microwave heating. Under optimal reaction conditions (7.5% CaO-MgO and 10-min microwave irradiation), biodiesel yield of 92.39 ± 0.04% was produced from *Jatropha* oil. Typical FAME composition comprised of methyl palmitoleate 2.40%, methyl palmitate 21.20 methyl oleate 62.40% and methyl stearate 14.00%	54
36	Environmental and industrial waste, such as palm fatty acid distillate (PFAD), was used as a feedstock for the production of biodiesel. PFAD was esterified using sulfonated glucose acid catalyst (prepared using microwave heating method). Under best reaction conditions (methanol to PFAD ratio of 10:1, catalyst loading of 2.5%, and reaction temperature of 70°C), the yield of biodiesel was found to be 98.23%	55

Continued

Table 1 Brief description of scientific advances on the use of nontraditional activation methods for biodiesel production (the year 2018)—cont'd

S. No.	Application of microwave/ultrasound activation	Reference
37	Biodiesel was produced by the esterification of corn oil with methanol, driven by microwave irradiation. Under the best reaction conditions (*T*, 72°C; *t*, 17.5 min; catalyst dosage, 8.5 vol%; catalyst, conc. H_2SO_4; and microwave power, 200 W), the yield of biodiesel was 77.6%. The order of the esterification reaction is 1.28 with a very low activation energy value of 1.79 J/mol	56
38	Microalgae have the potential to be a sustainable feedstock for biodiesel production that could be a substitute to fossil fuels. However, the problems to be surmounted in this area of research are limited scale and high cost. The authors proposed possible technologies for enhancing the growth rate of algae and extraction of lipids from the microalgae, *Chlorella sorokiniana*. Suitable light conditions, consists of 24L:0D photoperiod with 5400 lx white light for the improved growth rate of *Chlorella sorokiniana*. Likewise, microwave heating, at 90°C with 1000-W power, is the effective method for lipid extraction from *Chlorella sorokiniana*. Major components of the fatty acid esters consist of C18:1, C18:2, and C18:3	57
39	Microwave activated, H_2SO_4 catalyzed in situ transesterification of *Chlorella vulgaris* resulted in a biodiesel yield of 31.56% with a reaction time of 70 min	58
40	KF-modified kaolinite was used as solid acid catalyst for the transesterification of *Jatropha* oil to biodiesel. Increasing the basicity with increasing the loading of KF on kaolinite increased the catalytic activity of transesterification reaction. Compared with conventional reflux method, microwave (MW)-assisted reaction shortened the reaction time, with higher turnover frequency (TOF) and better reusability of the catalyst, although the yield remained similar to that of a reflux method. 25 wt% loading of KF was optimum loading for the best catalyst performance	59
41	Cost and time of biodiesel production is reduced by using waste cooking oil (WCO) as feedstock and microwave irradiation as a method of activation of the substrate and the catalyst. CaO produced from the calcination of cockle shells is used as solid base catalyst for the transesterification of WCO. The optimal conditions of biodiesel production are composed of a reaction temperature of 65°C, microwave power of 445 W, stirring speed of 380 rpm, LHS of the feed of 71.5 h^{-1} resulting in waste cooking oil methyl ester (WCOME) yield of 72.5% with an energy consumption of 0.595 kWh for the production of 1 L of product	60

42	Waste date seed oil is used as feedstock using microwave activation. Under best reaction conditions (*T*, 60°C; methanol-to-oil ratio of 6:1; microwave irradiation time, 1.75 min; and the biodiesel yield, 98.72)	61
43	Sulfated iron oxide (SO_4^{2-}/Fe_2O_3) solid acid catalyst was prepared by conventional impregnation method followed by calcination. Fe_2O_3 in the sulfated iron catalyst transformed from amorphous to crystalline form when the calcination temperature is above 400°C. TPD of NH_3 revealed the presence of strong and super acidic sites on the solid catalyst. The solid acid (sulfated iron oxide) catalyst was used for the transesterification of castor oil with microwave activation. Under optimal reaction conditions (*T*, 65°C; alcohol/oil molar ratio, 30:1; catalyst loading, 20 wt%; reaction time, 3 h; and microwave power, 300 W), the yield of biodiesel was 65.3%. The catalytic activity was better under microwave irradiation conditions than under conventional heating condition	62
44	Ultrasound irradiation promoted the growth of algal biomass (*Ochromonas danica*). Waste-activated sludge (WAS) was sonicated leading to the release of bacteria—size volatile solids that facilitated/supported the growth of phagotrophic alga *Ochromonas danica*. Under optimal WAS treatment via sonication, the yield of algal biomass was 44.5%	63
45	Ultrasonication is used for the cell rupture and extraction of lipid from the marine microalgae *Nannochloropsis* sp. Slurry solids concentration has a profound effect on the yield of lipids from the algal biomass. Sonication under best process conditions (20% solids concentration), ultrasound frequency, 20 kHz at 3.8 W/mL for up to 5 min) resulted in 11%–70% enhancement in yields of lipids within 5 min of sonication compared with conventional lipid extraction method. Lower solids concentration (<20%) resulted in wastage of solid energy in heating the water. Likewise, higher solid concentrations (<20%) lead to attenuation of ultrasound due to increase in viscosity and decrease in the speed of sound	64
46	CaO derived from egg, oyster, and clam shells were activated for enhancing their textural properties. Eggshell-derived CaO powder upon ultrasound irradiation for 5 h showed 34% reduction in particle size and a corresponding 56% enhancement in transesterification activity for the conversion of soybean oil to biodiesel. Likewise, postcalcination treatment of CaO derived from oyster and clam shells by two cycles of hydration-dehydration-calcination enhanced their catalytic activity by a factor of 2.5 time (rate of biodiesel production is enhanced)	66

Continued

Table 1 Brief description of scientific advances on the use of nontraditional activation methods for biodiesel production (the year 2018)—cont'd

S. No.	Application of microwave/ultrasound activation	Reference
47	Single-step sonication strategy for the conversion of wet algal biomass (*Aurantiochytrium* sp.) to biodiesel in high yield (94.6%) catalyzed by K_2CO_3 is developed. This method is particularly advantageous because the wet algae can be directly used for the transesterification reaction without drying. However, the water content in the algae poses a problem of mass transfer constraint, which was surmounted by the use of ultrasonic irradiation. Moreover the problem of saponification was suppressed by the use of ethanol for the transesterification process	67
48	A two-stage process for the conversion of *Jatropha* oil to biodiesel by sonochemical activation is developed. The renewable and nonedible *Jatropha* oil contains high amount of free fatty acids (FFAs), which will cause soap formation in the presence of alkali. So as to avoid such saponification and loss of FFA as well as hindrance to transesterification of oil to biodiesel, an acid-catalyzed (1% H_2SO_4) esterification is carried out with methanol. In this step, 99% FFA are converted to esters. In the second step the *Jatropha* oil devoid of FFA is subjected to transesterification reaction with alkali catalyst (NaOH) under sonochemical activation (20 kHz) resulting in 92% conversion of *Jatropha* oil to biodiesel in 45 min. at 60°C. This two-stage sonochemical process for the conversion of oil to biodiesel is most promising and environmentally friendly	68
49	Best reaction conditions for the transesterification of microalgae *Monoraphidium* sp. (both dry and wet) activated by ultrasound (180 W, 5 min). The transesterification reaction is carried out using either acid catalyst (2% H_2SO_4) or base catalyst (0.5% NaOH) or both acid and base catalyzed in succession. The respective FAME yields in the aforementioned catalytic processes were ~150, 70, and 170 mg/g of wet biomass. The superiority of acid-catalyzed process over the base-catalyzed process is due to the ability of the acid catalyst to convert both the FFAs and triglycerides to biodiesel. Effect of reaction parameters, namely, volume of methanol, reaction temperature, reaction time, and ultrasound power on the yield of FAME, was examined, and methanol volume was found to be crucial for enhancing the FAME yield	70

50	Solid acid catalyst for biodiesel production was prepared from rubber deoiled seed cake upon treatment with sulfuric acid chlorosulfonic acids. Blended oil (a mixture of rubber seed oil, castor oil, *Jatropha* oil, palm oil, and waste cooking oil) was used as biodiesel feedstock. Using sonochemical activation, two methods of biodiesel production, namely, a single-stage process (3 h) and a two-stage process (1 h), were developed for biodiesel production. Under optimal process conditions (methanol-to-oil mole ratio of 12.8:1; catalyst loading, 8.18%; and T, 63°C), the yield from two-stage process is strongly higher (93 ± 1.3%) than the single-step process (91.2 ± 1.1%). In the two-step process, the activation energy in sonochemical activated system is much lower (27.47 kJ/mol) compared with mechanically activated process (49.07 kJ/mol). The transesterification process followed Eley-Rideal mechanism. The decrease in activation energy and the corresponding enhancement in reaction rate (reduction in reaction time) by the use of ultrasound irradiation are attributed to the intense mixing of the components in the reaction medium, namely, the oil, methanol, and the catalyst, through a unique cavitation phenomenon. The cavitation phenomenon is accompanied by the formation and transient collapse of bubbles leading to improved mixing of oil and methanol via formation of a fine emulsion. As a result the mass transfer barrier between methanol and oil is reduced, the interfacial area between the two phases is enhanced, and the active sites on the solid acid catalyst are regenerated. This facilitates the adsorption of methanol on the catalyst surface, while the triglyceride moiety reacts from the bulk with the adsorbed methoxide species on the catalyst surface forming biodiesel at increased rate	71
51	Ultrasound-activated enzyme-catalyzed continuous flow biodiesel production process was developed using Novazyme 435 as catalyst in a packed bed glass reactor immersed in bath sonicator. With best process parameters of esterification (soybean-to-ethanol molar ratio, 1:6; flow rate of the feed, 2.5 mL/min; and ultrasonic power output, 132 W, 6 min), biodiesel yield of 95% is achieved. The packed bed reactor (PBR) showed good performance for ethyl ester synthesis and could be used for the production of aromatic esters as well	72

Continued

Table 1 Brief description of scientific advances on the use of nontraditional activation methods for biodiesel production (the year 2018)—cont'd

S. No.	Application of microwave/ultrasound activation	Reference
52	Transesterification of waste cooking oil and fresh soybean oil was conducted using lipases as biocatalysts with sonochemical activation. Upon 18 h of biocatalytic reaction, the yields of biodiesel from the aforementioned oils were 70% and 90%, respectively. Lipases are hydrolyzed enzymes with potential to catalyze hydrolysis of triglycerides to fatty acids and glycerol. When the water content in the reaction medium is low, lipases promote esterification, transesterification, and interesterification reactions. Lipases are used in bioenergy industry for the production of biodiesel owing to the broad range of reactions they catalyze with high activity. Use of ultrasound made the process of enzymatic synthesis of biodiesel feasible. The biodiesel yields obtained under sonochemical activation are similar to those derived when the mechanical stirring is used. However, the reaction rate was 2.3 times higher in the first 12 h of the sonochemical activation process compared with mechanical stirring. The increased initial rate was attributed to better homogenization and reduction of enzyme agglomeration when ultrasound is used for activation. The bubbles formed by the effect of cavitation caused intense turbulence and fluid circulation at the microlevel leading to effective mixing of oil and alcohol phases disrupting the immiscible liquid layer. Moreover the mass transfer at the liquid-solid interface is promoted	73
53	Ultrasonication was used to alleviate the hindrance of growth and production of algal biomass, caused by the excretion of algal extracellular substances (AESs). AESs tend to inhibit the growth of algae and culture cycle. AESs are composed of protein-like substances and saccharides that restrict algal lipid biosynthesis. Ultrasound was effective for the degradation of AESs into small molecules that would no longer cause oxidative damage and suppress the cell activity. Ultrasound irradiation is a promising strategy for promoting the medium recycle for continuous microalgae production	74
54	The accelerating effect of ultrasound on the conversion of rapeseed oil to biodiesel is studied. The ultrasound-activated transesterification process is characterized by high yield of methyl ester and fast reaction rates. Even though an increase in methanol-to-oil ratio increased the biodiesel yield, the reaction rate was decreased. The reaction followed pseudo first-order kinetics with an activation energy value of 25.51 kJ/mol. The reaction rate was enhanced with increase in the power of ultrasound owing to an increase in reaction temperature. As the power of ultrasound irradiation is increased from 40%, 60%, 80%, to 100%, the reaction temperature increased as 60°C, 64°C, 68°C, and 70°C with a corresponding increase in the reaction rate as evident from the rate constants values of 0.1669, 0.1872, 0.2103, and 0.217 min^{-1}, respectively	75

55	Potassium (K) ion exchanged pumice powder was used as solid base catalyst for the conversion of microalgae (dried *Chlorella vulgaris*) to biodiesel with ultrasound activation. Under optimal process parameters (methanol-to-oil ratio of 12:1; catalyst loading of 10%; reaction time of 10 min and ultrasonic processor amplitude of 40%; output power, 400 W; and frequency, 24 kHz), the yield of biodiesel was 86%	76
56	Microalgae like *Scenedesmus obliquus* were considered as ideal candidates for biofuel (biodiesel) production due to their fast growth and high lipid content. Ultrasound-assisted solvent extraction (UASE) is developed as an effective methodology for the isolation of lipids from the microalgae. Parameters such as the resonance amplitude, n-hexane-and-isopropanol (HIP) ratio, and reaction time were varied for obtaining the high yield of lipids. Lipid yield as high as 26.63% was obtained under optimal process conditions (50-μm resonance amplitude, 4 (*v*/v) ratio of HIP). The lipids derived from the microalgae are particularly important owing to their high heating value (35.38 MJ kg^{-1}), high content of saturated fatty acids (79.38%), low content of nitrogen (1.03%), sulfur (0.16%), and ash (0.55%). The results demonstrated the potential of ultrasound-assisted solvent extraction method for the extraction of lipids from microalgae	77
57	Velocity/speed of ultrasound is used as a measure to evaluate the quality of biodiesel (amount of methyl ester) and to monitor the progress of the process of transesterification. In other words, change in velocity of ultrasound could be alternative to well-known analytical methods like gas chromatography and ^{1}H NMR for evaluating the quality of biodiesel. The transesterification process of soybean oil with methanol using KOH as catalyst is examined using ultrasound (change in velocity) as a tool to monitor the progress and quality of the biodiesel produced. A low-power, 1-MHz ultrasound signal with a pulse-echo method was used. Ultrasound was potential to identify changes occurring in the reaction medium, which were correlated to methyl ester production. Ultrasound velocity could distinguish different biodiesels produced. In real time the ultrasound-based analytical method could be used to monitor the progress of transesterification. Use of this technique is advantageous from energy and environmental perspectives as the waste of energy and reagents could be minimized by this method of analysis of biodiesel quality	78

Continued

Table 1 Brief description of scientific advances on the use of nontraditional activation methods for biodiesel production (the year 2018)—cont'd

S. No.	Application of microwave/ultrasound activation	Reference
58	The potential of biodiesel as substitute to petro diesel is underutilized. Ultrasound-based analytical method was developed for real-time monitoring of the transesterification process. For this purpose, low-power (<100 mW) ultrasound based on a pulse/echo scheme is employed. Time-grated ultrasound pulses (bursts) were applied to the reaction medium at 1 MHz. Time of flight and echo waveform amplitude were the two ultrasound-based analytical parameters for evaluating the quality (density) of biodiesel and rate of transesterification, respectively. The aforementioned parameters varied depending on the alcohol used for the transesterification reaction and also if or not a heating process is involved during the reaction. The thumb rule was that the stabilization of the ultrasonic parameters, namely, the time of flight and the echo waveform amplitude, imply the completion of the transesterification and that the reaction should be stopped. This analytical method based on the variation of ultrasound velocity has industrial utility for evaluating and monitoring the large-scale biodiesel production processes	79
59	Release of proteins held in the rigid cell walls of the microalgae is a challenge. Use of a combination of alkaline and ultrasonic treatment was found to have greatest cell disruption efficiency leading to the release of proteins from the algal biomass (*Chlorella sorokiniana*, *Chlorella vulgaris*) in high yields and their solubilization for large-scale utility	80
60	Ultrasound irradiation is used as an efficient mixing, emulsifying and heating method for producing biodiesel from *Jatropha* oil in two steps. The first stage is composed of reacting the free fatty acids with isopropanol in an acid (H_2SO_4) catalyzed esterification process driven by ultrasound. The use of methanol is avoided, and isopropyl alcohol (IPA) is proposed as a substitute to methanol. As a result of esterification, the free fatty acid (FFA) content in the *Jatropha* oil is reduced from 18.06 to 0.14 wt%. This will be advantageous in the subsequent stage of transesterification of triglycerides of oil to biodiesel (fatty acid isopropyl ester) by preventing saponification and also by reducing the amount of alkali catalyst for the process. Cavitation phenomena in the ultrasound activation process are vital for mixing and heating processes	81

61	A comparative study is made on the potential of acoustic activation and conventional mechanical stirring for the acid catalyst (SO_3H functionalized ionic liquids) for the esterification of free long-chain fatty acids (oleic acid) with methanol to produce fatty acid methyl esters (FAME). SO_3H-functionalized ionic liquids are proposed as a substitute to conventional mineral acids (H_2SO_4) for the esterification of FFA. The process of esterification was enhanced by a factor of 2.2 and 1.5 by the use of ultrasound, in the presence and absence of acid catalyst, compared with conventional mechanical stirring-driven process. Optimal process parameters are composed of a reaction temperature of 60°C for 1 h with a methanol-to-oil mole ratio of 9 and a catalyst amount of 0.4 mol %. After the reaction the product (methyl oleate) and catalyst (ionic liquids) were recovered using liquid-liquid extraction and centrifugation	95
62	Biocatalytic (fermase as biocatalyst) synthesis of biodiesel (methyl laurate) with stepwise addition of ethanol to lauric acid under solvent free conditions was accomplished at an accelerated rate by the simultaneous use of acoustic activation and mechanical stirring. Key thermodynamic parameters like ΔG (54.63 kJ/mol) and ΔH (−26.35 kJ/mol) and kinetic parameter like E_a (−23.92 kJ/mol) were found to decrease when a combination of ultrasound irradiation and mechanical stirring were used relative to the use of only ultrasound or only mechanical stirring. The best reaction conditions include mole ratio of lauric acid to ethanol, 1:2; 2% enzyme loading; T, 50°C; t, 40 min; power, 100 W; and yield of FAME, 96.87%). Reaction time was lowered ~6 times with the use of a combination of ultrasound and mechanical stirring compared with conventional heating technique	96
63	Ultrasound treatment reduced the mass transfer constraints; increased the miscibility between the immiscible phases; improved the interaction (contact) between the substrate and the catalytic species, leading to increased rate of transesterification; and shortened the time required for biodiesel production. Ultrasound is an efficient tool and novel technology for the conversion of microalgal biomass to biodiesel. Ultrasound-based lipid extraction from microalgae consumed the least energy (0.15 MJ/g dry microalgae). Ultrasound irradiation enhances the emulsification of immiscible liquid reactants by microturbulence caused by cavitation bubbles leading to reduction in processing time. Reduction in production cost and energy consumption enhances the possibilities of scale-up	97

Continued

Table 1 Brief description of scientific advances on the use of nontraditional activation methods for biodiesel production (the year 2018)—cont'd

S. No.	Application of microwave/ultrasound activation	Reference
64	Acid-catalyzed esterification of Karanja oil was carried out in a batch and continuous flow processes under ultrasound irradiation with simultaneous stirring. Batch process is carried out with horn-type ultrasonication (20 kHz, 120 W), and the continuous flow process was performed in an ultrasonic flow cell reactor (20 kHz, 1500 W) of 0.5-L volume. Highest conversion values of FFA to biodiesel in batch and flow processes were 77% and 74%, respectively. Carrying out the biodiesel production process in two stages, namely, esterification of FFA in stage 1 and transesterification of triglycerides of the oil in stage 2, resulted in biodiesel yields of 92% and 88% in batch and flow processes. On the contrary, when the process was carried out in a single-stage without prior esterification of FFA, the biodiesel yields were lower, namely, 83% and 80%, respectively, in a batch and flow process activated by ultrasound. The key role of ultrasound is to increase the contact area and mass transfer between two immiscible phases of oil and methanol leading to a decrease in reaction time. Process economics is satisfactory when the process is scaled up to 1-L volume in an ultrasound flow cell	98
65	Transesterification of canola oil with methanol was carried out using sonochemical irradiation (20 kHz, 200-W power, reaction volume of 250 mL, and pulsed ultrasound [0.9]) using solid base catalysts, namely, CaO, calcined dolomite, and calcium diglyceroxide yielding biodiesel of 98.7%, 95.9%, and 86.3%, respectively. Under optimal process conditions (catalyst loading, 5.35%; methanol/oil ratio of 7.48; ultrasonic power of 40 W; reaction time of 150 min; and a reaction time of 60°C), a maximum biodiesel yield of 99.4% was obtained using CaO catalyst	99
66	Using a statistical (Plackett-Burman) design, the best process parameters for the production of biodiesel from chicken fat were studied. Process parameters like molar ratio of fat to alcohol, ultrasound power, reaction time, and moisture level were found to have positive effects on biodiesel yield, whereas parameters like concentration of catalyst, mixing intensity, and temperature had negative effects on the biodiesel yield	100

67	Raw decorticated *Jatropha* seeds were used as feedstock for biodiesel production with sonochemical activation. Prior to transesterification reaction, the free fatty acid content of the feedstock was reduced (to below 3%) by esterification using H_2SO_4 catalyst. Use of sonication facilitated the extraction of triglyceride component from the *Jatropha* seeds. The extraction efficiency, esterification efficiency, and the FAME yield were 83.96%, 71.10%, and 38.58%, respectively, with an ultrasound amplitude of 60% in a pulse mode (5 s on/2 s off) for 150 min with a n-hexane-to-methanol volume ratio of 3:1. In the second stage of transesterification of esterified *Jatropha* oil, 92.29% FAME yield was obtained upon 15-min sonication under optimal process conditions (methanol-to-oil ratio of 12:1, KOH loading of 1.0 wt%, 60% ultrasonic amplitude, and pulse mode of ultrasound, 5 s on/2 s off). Biodiesel production process activated by ultrasound is economically competitive with reduction in energy consumption and production cost	101
68	Weak interactions and difficult mixing properties of the reactants (oil and alcohol) lead to low reaction rates and high reaction time in a typical transesterification process. Ultrasound activation is a most effective method for overcoming the aforementioned problem by the use of sound waves with unique chemical and physical effects. The chemical effects arise from the free radicals generated (H· and OH·) during the transient implosive collapse of bubbles, which would accelerate the chemical reactions in the bulk medium. The physical effects are a result of emulsification caused by microturbulence. The interfacial contact between oil and alcohol is increased upon sonication leading to enhancement in the reaction rate and shortening of reaction time. Rubber seed oil was effectively converted to biodiesel in a two stage (esterification followed by transesterification process). Using ultrasound, rubber seed oil feedstock with a wide range of free fatty acid content (14%–42%) could be converted to biodiesel in high yields (78.84%–92.5%) in a two-stage process with catalysts like H_2SO_4 and HCl for the initial esterification process and KOH and NaOH as catalysts for the second stage of transesterification process. Clinker-loaded methoxide is proposed as a solid base catalyst for the transesterification process under sonication	102

prevent aggregation of $Sr(CO_3)_2$ deposited on the SiO_2 surface. The reaction mixture was irradiated in a domestic microwave oven for 30 s equipped with stirring and water reflex facility. The resulting solid material separated from the supernatant by centrifugation was washed with ethanol thrice and subjected to vacuum drying overnight result in the formation of $Sr(CO_3)_2/SiO_2$. Subsequently the $Sr(CO_3)_2$ is decomposed to SrO by thermal treatment in air (calcination) at 900°C for 4 h in a tubular furnace yielding SrO/SiO_2. The amount of SrO deposited on SiO_2 beads was 41.3 wt% as revealed by ICP analysis. Pressure of SrO and its uniform distribution on SiO_2 surface is known from SEM and EDAX analysis. Spherical morphology of the SiO_2 beads (3–6 nm) is by and large retained and remained unperturbed after microwave irradiation (30 s) and high-temperature calcination (900°C for 4 h) (Fig. 6A). Presence of Sr and O throughout the surface of SiO_2 is disclosed by the elemental mapping (Fig. 6C and D). Relative amounts of Sr and O were seen in the EDAX spectrum (Fig. 6B, inset).[91]

The potential of SrO/SiO_2 catalyst for the transesterification reaction was tested in a batch and a continuous flow process for the conversion of waste cooking oil (WCO) to biodiesel by microwave activation. The free fatty acid (FFA) and water content of the waste cooking oil were removed by treating the WCO with KOH solution via soap formation and separating the oil content by centrifugation. By heating the KOH-treated WCO at 110°C, the water content is effectively removed. After this pretreatment the WCO was used as feedstock in a batch process by taking WCO (15 g), methanol (4 mL), and catalyst (SrO/SiO_2, 0.5 g) in a round bottom (RB) flask and irradiating the contents in a microwave oven for 10 s at 70% power under stirring at 60°C. After cooling the contents the catalyst could be separated easily by centrifugation. Oil and unreacted methanol (if any) and the reaction byproduct glycerol were separated via centrifugation. The top layer of biodiesel with unreacted methanol is collected by decantation. Solid catalyst recovered from glycerol by filtration and the reusability of the catalyst was tested. The conversion of WCO to biodiesel was analyzed by ^1H NMR with $CDCl_3$ as solvent. The ratio of the integral areas of the methoxy groups of FAME (3.65 ppm) and the α-carboxyl methylene protons of the triglyceride derivatives (2.26 ppm) is a measure of the conversion triglyceride in the WCO to biodiesel. A WCO conversion of 99.4% to biodiesel was observed with the fresh SrO/SiO_2 catalyst in the first reaction run. The recovered catalyst could be used for 10 consecutive reaction runs without appreciable loss in activity as evident from the change in conversion values from 99.4 to 95 wt% as the reaction runs varied from 1 to 10.[91]

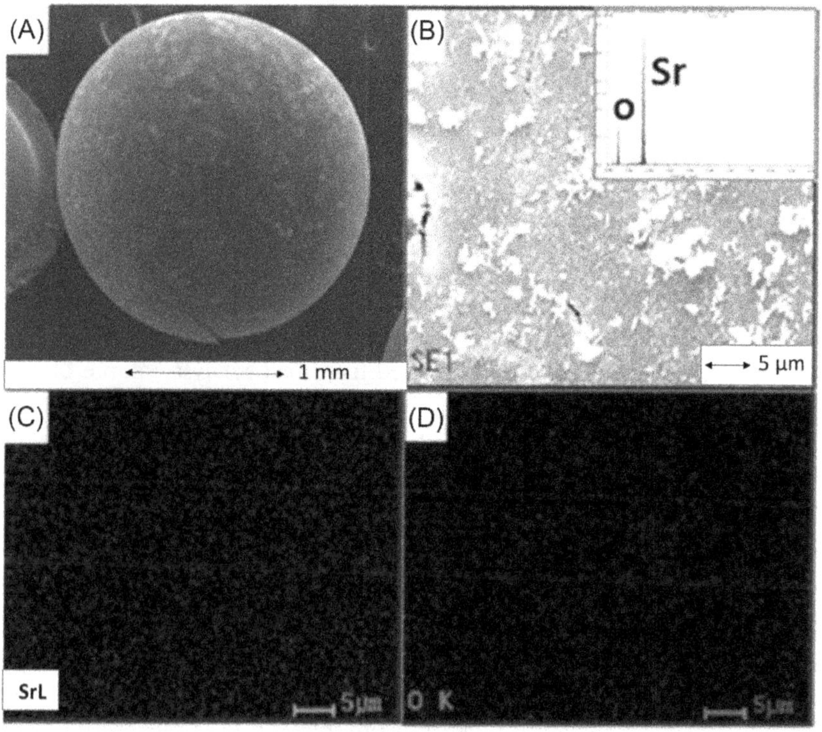

Fig. 6 SEM images (A and B), EDS spectrum (inset B), and elemental mapping of Sr and O (C and D) of SrO/SiO_2 solid base catalyst (SrO supported on millimetric silica beads). *Adapted with permission from Alex Tangy; Indra Neel Pulidindi; Aharon Gedanken. SiO_2 beads decorated with SrO nanoparticles for biodiesel production from waste cooking oil using microwave irradiation, Energy and Fuels* ***2016****, 30(4), 3151–3160. American Chemical Society.*

The batch process was further developed into a continuous flow process making use of the commercial microwave system (flow-SYNTH) depicted in Fig. 7, by clearly labeling various components of the system. The flow-SYNTH system is a continuous flow microwave system with 1000 W power, heating a vertically mounted flow through reactor of 190-mL working volume. The reaction mixture (WCO and methanol) was pumped by the high-pressure membrane pump from the top of the reactor packed with the solid base catalyst (SrO/SiO_2) as a fixed bed. Online sensors aid the continuous monitoring of the reaction temperature. The reactor is equipped with a magnetically driven paddle stirrer to facilitate the uniform distribution of temperature. The reaction mixture is mechanically stirred inside the microwave cavity by a stirring shaft equipped with three paddles. The pump connected to the reactor drives the flow of the reaction mixture through the reactor. The working volume of the reaction chamber in the microwave cavity is 190 mL, permitting maximum working conditions of the microwave flow system to be 200 mL, 30 bar, and 1000 W. The peristaltic pump connected to the microwave reactor regulates the flow rate of reaction mixtures (0.06–3400 mL/min). The best reaction conditions for the transesterification of WCO using SrO/SiO_2 catalyst in a continuous flow microwave system include a stirring shaft rotation velocity of 20%, mole ratio of WCO to methanol of 1:12, a reaction temperature of 65°C, a feed flow rate of 100 mL/min, and a packed fixed bed catalyst volume of 15 g (41.1 wt. 5 SrO/SiO_2). The active catalyst, SrO, in the 15 g supported catalyst (SrO/SiO_2) is only 6.195 g with the rest being millimetric silica beads. Under these modest operating conditions, WCO conversion as high as 99.2% is achieved during the first 8 min of feed flow, and the conversion value varied steadily with the reaction run time as the feedstock (WCO and methanol mixture) is passed through the catalyst bed in the microwave reactor. A plot of the effect of reaction run time on the conversion of WCO to biodiesel is depicted in Fig. 8.

The decrease in activity of the catalyst was further supported by the leaching of significant amount of SrO from the supported catalyst as evident from the presence of only 24.7 wt% of SrO in the spent catalyst (SrO/SiO_2) after 49.2 min of run time. The high performance of the SrO/SiO_2 catalyst was attributed to the unique structure-activity relationship as revealed from HR-SEM analysis. HR-SEM image of ~41 wt% SrO/SiO_2 catalyst is shown in Fig. 9A. Nanometric tubelike structures of SrO are formed on the millimetric silica beads under the microwave-driven synthesis ($Sr(CO_3)_2/SiO_2$) and the subsequent high-temperature calcination of ($Sr(CO_3)_2/SiO_2$) resulting in the formation of SrO/SiO_2. The average length and width of the nanotubules of SrO were 139 and 50 nm. These nanostructures of SrO were similar to the surface nanoarchitecture of lotus leaf (Fig. 9B) that imparts unique hydrophobic property to the lotus leaf (Fig. 9C). The nanoarchitecture of SrO prevailing on the SiO_2 surface was attributed to impart water repelling activity to the catalyst leading to high and sustained activity for 24.6 min under modest process conditions of microwave activation. The supported solid basecatalyst (SrO/SiO_2) has several advantages for industrial adaptability in processing large amounts of WCO as major challenges encountered in a

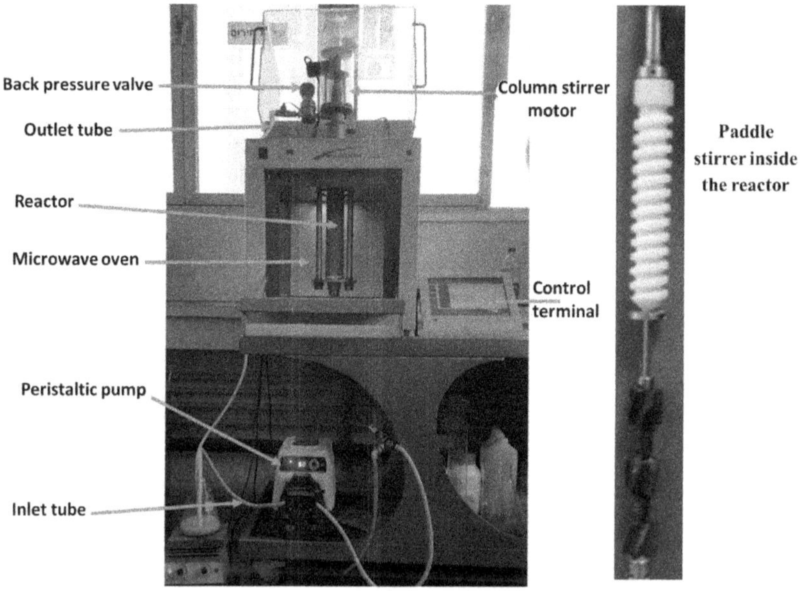

Fig. 7 Experimental setup for the continuous flow biodiesel production based on microwave irradiation. Stirring shaft with three Weflon15 blades for the tubular reactor of the FlowSYNTH (the outlet and inlet, as well as each of the components of the microwave reactor, are marked). *Adapted with permission from Alex Tangy; Indra Neel Pulidindi; Aharon Gedanken. Continuous flow biodiesel production from waste cooked oil using microwave irradiation and supported SrO catalyst, Bioresource Technol.* ***2017****, 224, 333–341. Elsevier Publishers.*

typical continuous flow process like mass flow constraints, pressure buildup, diffusion constraints, and temperature gradients could be effectively overcome by the use of the SrO/SiO_2 catalyst.

2.2 Eggshell-derived CaO catalyst as stable solid base catalyst for biodiesel production using sonochemical activation

Badnore et al. reported on the sonication-based reverse emulsion synthesis of nanocrystalline CaO.[65] Compared with commercial CaO the nanocrystalline CaO has a higher surface area (by a factor of 1.53) owing to smaller particle and crystallite sizes. The aforementioned catalysts were examined for transesterification activity for the conversion of soybean oil using sonochemical activation. Under best process conditions (catalyst loading

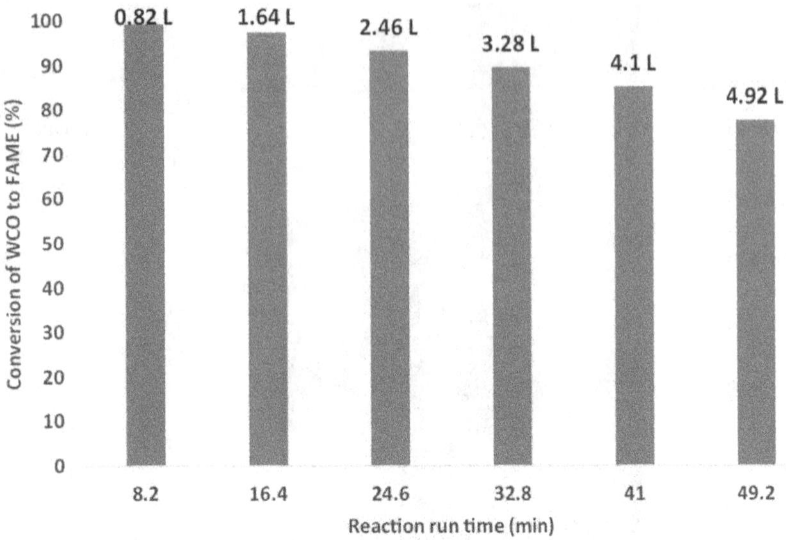

Fig. 8 Lifetime of the fixed bed catalyst (41 wt% SrO/SiO_2) for the conversion of waste cooking oil (WCO) to biodiesel. *Adapted with permission from Alex Tangy; Indra Neel Pulidindi; Aharon Gedanken. Continuous flow biodiesel production from waste cooked oil using microwave irradiation and supported SrO catalyst, Bioresource Technol.* ***2017****, 224, 333–341. Elsevier Publishers.*

of 4 wt% of oil; methanol-to-oil molar ratio of 6:1; $T = 60°C$, ultrasonic power, 212.89 kW/m^2; and reaction time—80 min), nanocrystalline CaO resulted in 18.18% higher conversion of soybean oil. The activation energy of the transesterification reaction of soybean oil activated by sonication and catalyzed by nanocrystalline CaO was 54.26 kJ/mol. The CaO catalysts derived from the calcination of $CaCO_3$ precursor were stored in nitrogen-filled glass containers to prevent them from reacting with atmospheric CO_2 and moisture. To evaluate the size, shape, and surface morphology of the nano-CaO and commercial CaO, field emission gun scanning electron micrographs were recorded. Samples were dispersed in hexane using bath sonicator before FEG-SEM analysis. The respective images were shown in Fig. 10. The particle size is larger, and their distribution is broader (300–900 nm) in commercial CaO compared with sonochemically prepared CaO (100–350 nm). Unlike the aggregation of CaO particles in the commercial CaO catalyst, reverse emulsion technique driven by sonication resulted in well-separated nanoparticles of CaO, which has led to better performance in the biodiesel production process.

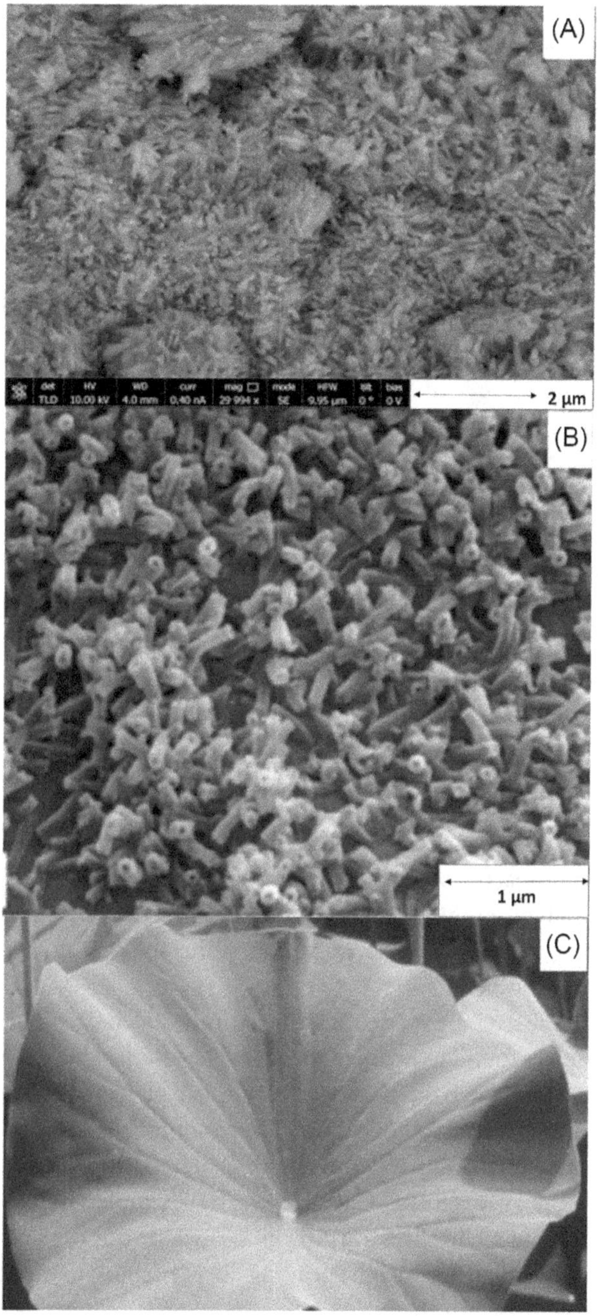

Fig. 9 HR-SEM image of SrO/SiO_2 (A), SEM image of the upper leaf side with nanotubules of wax (B), and lotus leaf with extraordinary hydrophobicity (C). *Adapted with permission from Alex Tangy; Indra Neel Pulidindi; Aharon Gedanken. SiO_2 beads decorated with SrO nanoparticles for biodiesel production from waste cooking oil using microwave irradiation, Energy and Fuels* ***2016****, 30(4), 3151–3160. American Chemical Society.*

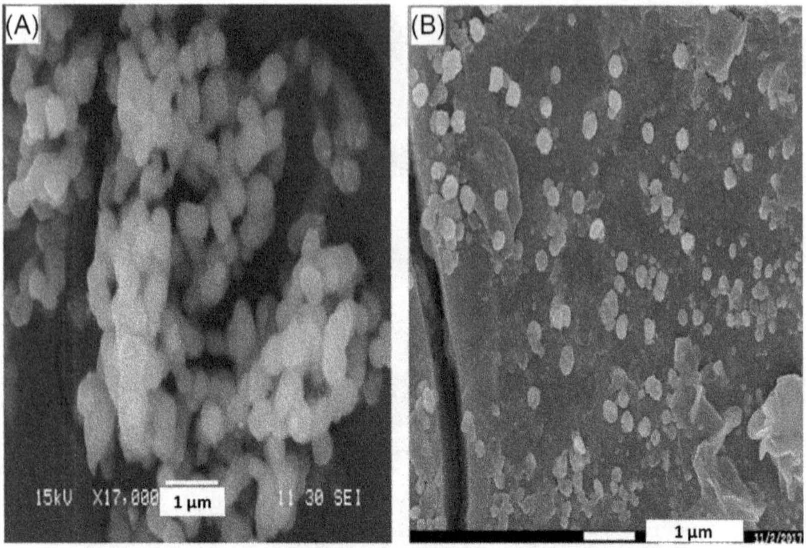

Fig. 10 Field emission gun scanning electron microscopy images of (A) commercial CaO and (B) nanocrystalline CaO prepared by sonication-driven reverse emulsion method. *Adapted with permission from Badnore, A. U.; Jadhav, N. L.; Pinjari, D. V.; Pandit, A. B. Efficacy of Newly Developed Nano-Crystalline Calcium Oxide Catalyst for Biodiesel Production. Chem. Eng. Process.,* ***2018****, 133, 312–319. Elsevier.*

HPLC trace of the biodiesel product obtained by the transesterification process via sonochemical activation using nanocrystalline CaO is shown in Fig. 11. The peaks of biodiesel (methyl ester, retention time from 3 to 12 min) and unreacted soybean oil (12.5–17 min) could be seen well separated from each other owing to the difference in their retention times. Without the catalyst, sonochemical activation alone showed no conversion of soybean oil to biodiesel.

From kinetic data the methanolysis reaction of soybean oil on the surface CaO catalyst was found to proceed through Eley-Rideal-type mechanism.[66] In this mechanism the basic sites on CaO surface play prominent role in the formation of methoxide anions upon adsorption of the substrate methanol molecules via formation of ion pair. Then a triglyceride molecule approaches the surface of CaO crystallite completely covered with methoxide anions. The three carboxylic groups of the triglyceride species interact with methoxide oxygen leading to methanolysis reaction almost completely as shown in the Fig. 12.[66]

The eggshell-based CaO-catalyzed biodiesel production at ambient conditions reported by Gedanken's group has an edge over the contemporary

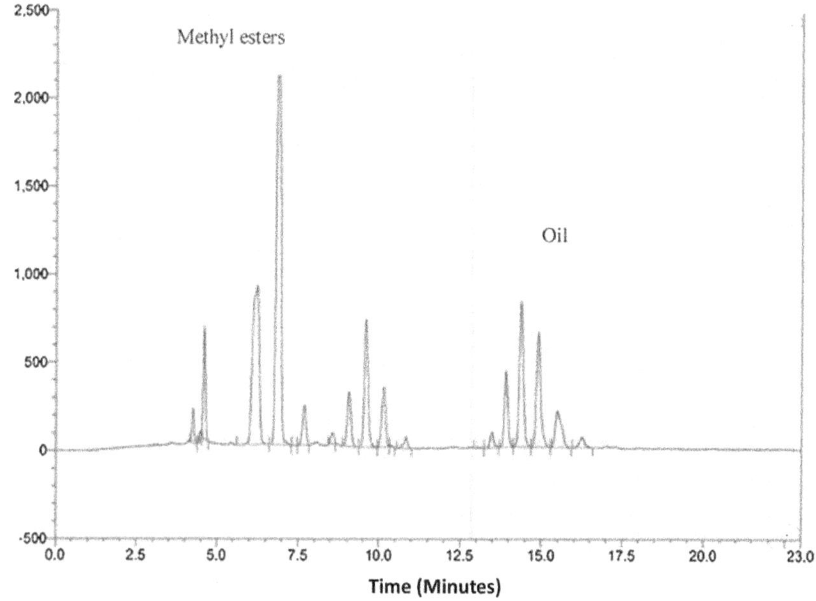

Fig. 11 HPLC trace of the biodiesel product from the transesterification of soybean oil using nanocrystallite CaO under sonication. *Adapted with permission from Badnore, A. U.; Jadhav, N. L.; Pinjari, D. V.; Pandit, A. B. Efficacy of Newly Developed Nano-Crystalline Calcium Oxide Catalyst for Biodiesel Production. Chem. Eng. Process.,* ***2018****, 133, 312–319. Elsevier.*

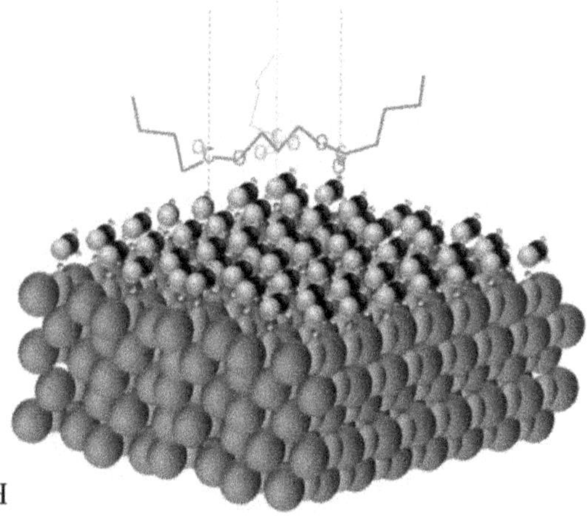

Fig. 12 Schematic depiction of the approaching of a triglyceride molecule toward the surface of a CaO crystallite covered with methoxide anion. *Adapted with permission from Risso, R.; Ferraz, P.; Meireles, S.; Fonseca, I.; Vital, J. Highly active CaO catalysts from waste shells of egg, oyster and clam for biodiesel production. Appl. Catal. A, General* ***2018****, 567, 56–64. Elsevier.*

reports in literature on similar subject from the perspective of scalability, simplicity, adaptability, and economics.[65] The eggshell-derived CaO catalyst has several unique features making their adaptability for industrial production of biodiesel from waste cooking oil and other lipid-containing feedstock like soybean oil advisable. Typical method of preparation of the solid base catalyst (CaO) is composed of washing the waste eggshells with water, drying, and then grinding in a blender for 5 min followed by calcination at ambient atmosphere at 900°C for 3 h. The obtained CaO catalyst was stored in a glass vial sealed with paraffin film and was found to exhibit sustainable activity even after 3 months of shelf life, and the yield of FAME only decreased 10% after 1 year of shelf life. In literature, CaO catalyst was often recommended to be stored under special anaerobic conditions for retaining its activity. The eggshell-derived CaO catalyst showed outstanding performance in the transesterification of soybean oil and WCO with FAME yields of 98 and 97 wt%, respectively, under best reaction conditions (11 h regular stirring, oil-to-methanol mole ratio of 1:6, catalyst amount of 5.8 wt %, ambient temperature, and pressure). Moreover the catalyst could be easily separated from the product mixture and reused without loss in activity for nine and four consecutive reaction runs with soybean and WCO feedstocks. The WCO was used as received without pretreatment. Loss in activity of CaO catalyst with WCO feedstock after the fourth reaction run (decrease in yield from 93 to 62 wt%) was attributed to the presence of FFA in the WCO poisoning the active basic sites in CaO. CaO derived from eggshell catalyst is recommended for testing for large-scale production of biodiesel in a pilot plant from WCO owing to the fact that such a process is eco-friendly minimizing the environmental wastes (egg shells and WCO), the process could be done under ambient conditions, the catalyst could be reused at least for four reaction runs without loss in activity and incurs no energy expenditure except for stirring, and the catalyst could be stored under ambient conditions for a long shelf life of 1 year with only 10% decrease in the FAME yield derivable compared with that of the fresh catalyst.

3 Can there be alternative to microwave and sonochemical activation for biodiesel production?

Solar energy is being used as a heating source in bioenergy conversion processes. It is anticipated that such solar energy-activated chemical and biochemical processes would play a central role in the upcoming renewable energy sector, on a par with photovoltaic cells involving the conversion of solar energy to electricity. Gedanken's group is the foremost to open a new avenue on the utilization of solar energy for biomass conversion for biofuels.[94, 153, 155, 157, 159]

3.1 SrO/activated carbon for biodiesel production using solar energy as heating source

A unique solar energy-activated biodiesel production process from feedstock like canola oil, soybean oil, and waste cooking oil is developed.[94] Success of the strategy is owed to the novel high surface area microporous-activated carbon-deposited SrO solid base catalyst, which showed nearly four- to fivefold enhancement in the activity depending on the feedstock used for the transesterification process. High FAME yield of 98.5 wt% was obtained from waste cooking oil under modest reaction conditions [t-60 min., oil to methanol mole ratio 6:1, and 7.1 wt% catalyst loading (24% SrO/activated carbon at 46°C). The catalyst could be used for four consecutive reaction runs without appreciable loss in activity. Owing to the unique advantages of the process, namely, effective utilization of the active component SrO, eco-friendly nature of the activated carbon support, renewable nature of the heating and activation source (solar energy), and relative fastness of the process compared with traditional reflux based processes, the current process is recommended for evaluating the feasibility for pilot-scale operation.[94]

The catalyst preparation process consists of a nontraditional method of sonochemical deposition of SrO (0.5 g) on the activated carbon surface (2.0 g), wherein the aforementioned components were dispersed separately in 60 and 20 mL of ethanol by stirring the solution for 2 h. The two solutions were then mixed and sonicated for 1 h using a high-intensity ultrasonic Ti-horn (20 kHz frequency, 100 W/cm^2 power, Sonics and Materials, Inc. Model VCX 750). Subsequently the solvent ethanol was evaporated by stirring on a hot plate under stirring. The catalyst thus obtained (SrO/C) was dried overnight in an air oven at 100°C. The presence of SrO in the supported catalyst was confirmed by XRD analysis. Typical x-ray diffraction patterns of activated carbon used as support, pristine SrO and SrO/C (SrO supported on activated carbon), were shown in Fig. 13. Amorphous nature of the activated carbon support is revealed from the broad peak at 2θ values in the range of 40–50 degrees.

Highly crystalline nature of pristine SrO was deduced from the sharp peaks with 2θ values of 30.09, 34.88, 50.16, 59.61, and 62.55 degrees, which were respectively indexed to the (111), (200), (220), (311), and (222) planes

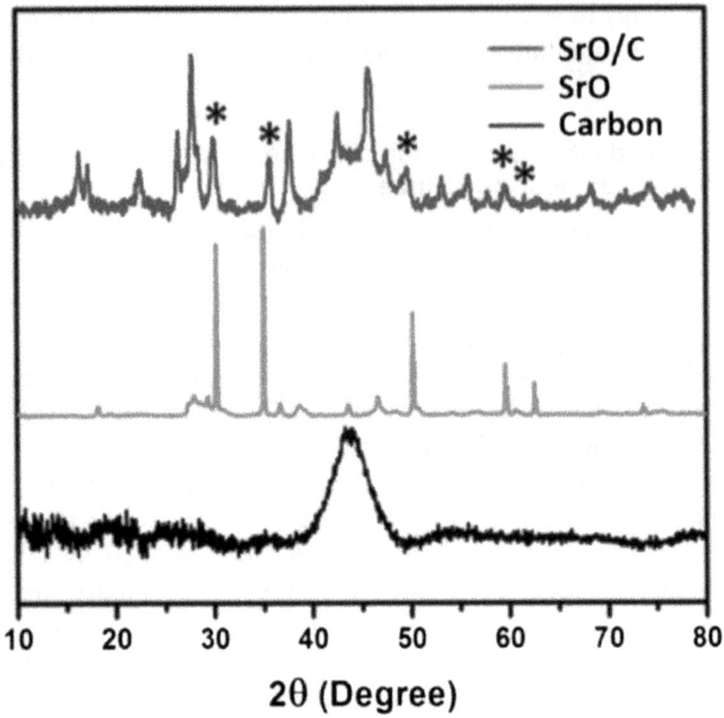

Fig. 13 XRD patterns of activated carbon support (C), pristine SrO and SrO/C (activated carbon supported SrO). *Adapted with permission from Tabah, B.; Nagvenkar, A. P.; Perkas, N.; Gedanken, A. Solar-heated sustainable biodiesel production from waste cooking oil using a sonochemically deposited SrO catalyst on microporous activated carbon. Energy and Fuels **2017**, 31, 6228–6239. American Chemical Society.*

of the cubic structure of the SrO (JCPDS powder diffraction file no. 75-0263). Additional weak signals in the XRD were attributed to the presence of hydroxides of strontium, $Sr(OH)_2$ (JCPDS power diffraction file no. 71-2365), present in the pristine SrO. Signal characteristics of SrO were retained upon the sonochemical deposition of SrO on activated carbon in the SrO/C catalyst revealing the crystalline nature of SrO on C. Broad peak around 45 degrees is attributed to the activated carbon support. Additional minor signals were attributed to strontium hydroxide species, $Sr(OH)_2$.

Presence and elemental state of Sr on activated carbon surface in the SrO/C catalyst is probed using XPS. The XPS traces of pristine SrO and the SrO/C are shown in Fig. 14. Pristine SrO showed two peaks at binding energy (BE) values of 131.5 and 133.4 eV originating from the $3d_{5/2}$ and $3d_{3/2}$ energy levels from Sr atom, which are indicative of the presence of Sr^{2+}.

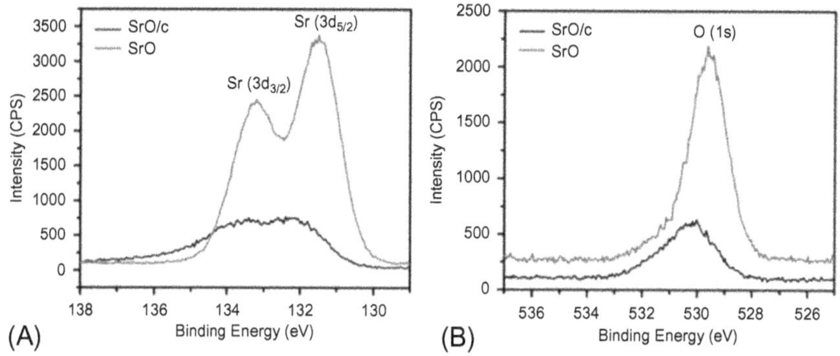

Fig. 14 XPS traces of (A) Sr 3d level and (B) O 1 s level of commercial SrO and SrO/C. *Adapted with permission from Tabah, B.; Nagvenkar, A. P.; Perkas, N.; Gedanken, A. - Solar-heated sustainable biodiesel production from waste cooking oil using a sonochemically deposited SrO catalyst on microporous activated carbon. Energy and Fuels* ***2017****, 31, 6228–6239. American Chemical Society.*

In the supported catalyst (SrO/C) too, similar peak pattern is observed, however, with much lower intensity and shift in the corresponding Sr^{2+} signals to higher BE values. While the lowering of intensity of the signals is consistent with the lower amount of SrO present on the activated carbon support (24 wt%) compared with the pristine SrO, the shift in the BE to higher values is characteristic of the strong interaction between SrO and the support. Such phenomena are well documented in literature as "strong metal support interaction (SMSI)."

The textural properties of the pristine and activated carbon-supported SrO were studied using BET sorptometry. The specific surface area values of activated carbon support, pristine SrO, and SrO/C were 1180, 0.42, and 432 m^2/g, respectively. The N_2 adsorption-desorption isotherms used to deduced the pore volume and the corresponding calculation of surface area values were shown in Fig. 15. As expected upon depositing the low specific surface area SrO over a much higher surface area-activated carbon material, the supported system exhibited fairly good textural property with a surface area value of 435 m^2/g indicative of the availability of more active sites (Sr^{2+}-O^{2-}) on the carbon surface.

The N_2 adsorption-desorption isotherm of activated carbon support exhibited typical type I isotherm feature characteristic of microporous materials. This is further supported by the pore size value of 2.16 nm of the activated carbon. According to the size of the pores, porous materials are characterized as microporous (pore size, <2), mesoporous (pore size, 2–50 nm), and

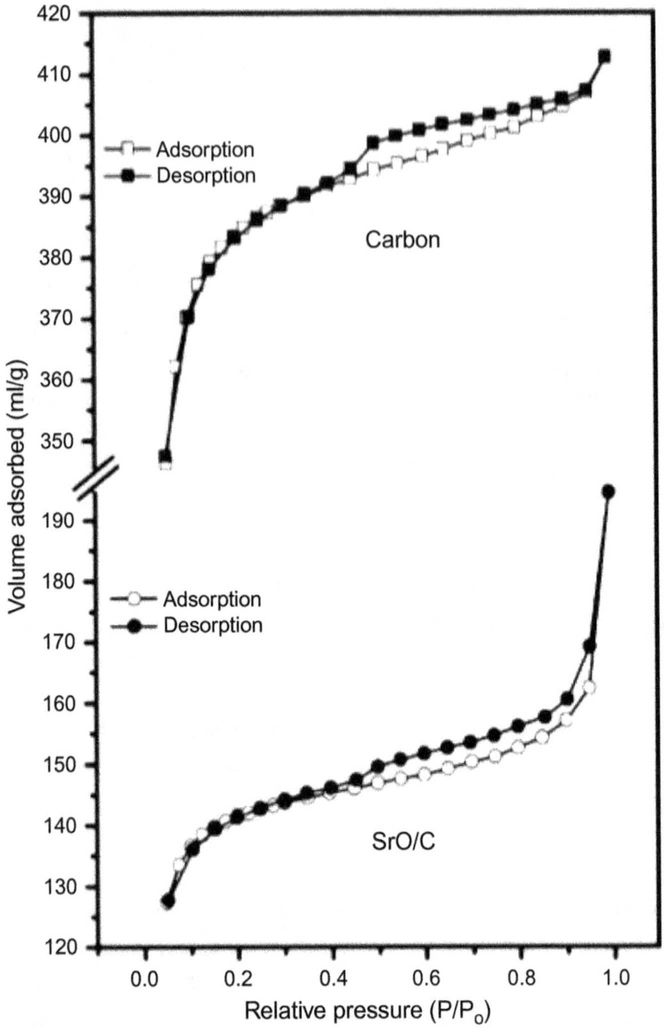

Fig. 15 Nitrogen adsorption-desorption isotherms of the SrO/C compared with activated carbon (C) support. *Adapted with permission from Tabah, B.; Nagvenkar, A. P.; Perkas, N.; Gedanken, A. Solar-heated sustainable biodiesel production from waste cooking oil using a sonochemically deposited SrO catalyst on microporous activated carbon. Energy and Fuels* ***2017****, 31, 6228–6239. American Chemical Society.*

macroporous (pore size, <50 nm) materials. A transition from microporous to mesoporous nature of the activated carbon support upon sonochemical deposition of SrO on activated carbon support is evident from the change in the shape of the isotherm from type I to type IV. Such a change in pore architecture by the activation of activated carbon by acoustic waves is further supported by the increase in the pore size from 2.16 to 2.76 nm. Enhancement in surface area of the supported system (SrO/C) plays a crucial role in the adsorption of methanol on the catalyst surface generating much higher number of methoxide species on the solid surface compared with pristine SrO, which will be approached by the triglyceride moieties from the bulk reaction medium following Eley-Rideal reaction mechanism, and thus the yield of FAME in the product is enhanced.

The potential of SrO/C catalyst is exploited for the conversion of canola, soybean, and waste cooking oil to biodiesel. The transesterification reaction conditions include 0.5 g, 24 wt% SrO/C (7.1 wt% loading w.r. to the amount of oil in the reaction medium), and solar heating for 60 min. FAME yields of 97.5, 96.5 and 98.5 wt% were obtained from canola, soybean, and waste cooking oil, respectively. The aforementioned products meet the EN14214 standards for biodiesel. The FAME yields derived from SrO/C from different feedstock (canola, soybean, and WCO) are comparable with those obtained using commercial SrO catalyst. However, there are typical advantages of using activated carbon-supported SrO catalyst, namely, the amount of active component SrO in the 0.5 g supported catalyst (24 wt% SrO/C) is only 0.12 g compared with the 0.5 g commercial SrO catalyst used for each of the transesterification reactions. This means that the SrO utilization is reduced ~4.2-fold upon supporting SrO on activated carbon. Moreover, with a given feedstock, the biodiesel produced using SrO/C catalyst contained higher saturated and monounsaturated FAMEs compared with the biodiesel produced with SrO catalyst. This shows that the biodiesel produced from SrO/C catalyst has higher oxidative stability and improved low-temperature performance. Mesoporous structure of SrO/C catalysts plays a vital role for the selective hydrogenation of the double bond resulting in the production of less polyunsaturated FAMEs.

A comparison of the reusability of SrO and SrO/C catalyst for the conversion of WCO to biodiesel is shown in Fig. 16.

In the first reaction run, high FAME yield of 98.3 wt% was obtained from WCO using SrO/C catalyst. The catalyst showed sustainable activity for four reaction runs with only 3% loss in activity. Upon subsequent reuses, 25.5% reduction in FAME yield was observed in the sixth run. SrO/C showed similar performance as that of unsupported SrO in reusability studies for six cycles, however, with lower amount of utilization of SrO and with improved quality of FAME. Catalyst (SrO/C) deactivation after the fourth reaction run could be attributed to the partial erosion (leaching) of SrO from the carbon support and to the poisoning of active sites (SrO) by the

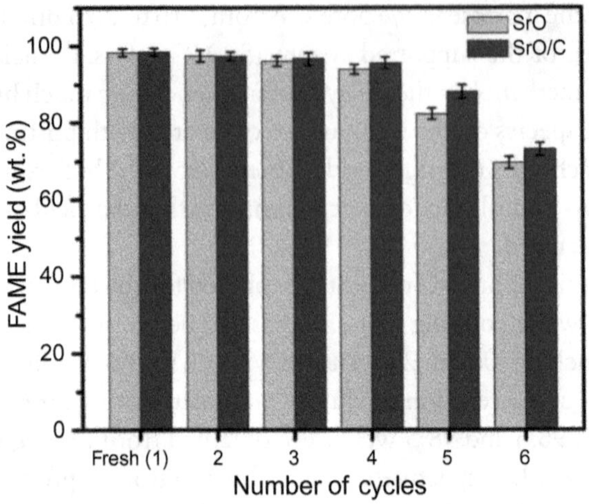

Fig. 16 Reusability of SrO and SrO/C catalysts for the transesterification of waste cooking oil. The error bars represent the standard deviation of three repeated measurement. *Adapted with permission from Tabah, B.; Nagvenkar, A. P.; Perkas, N.; Gedanken, A. - Solar-heated sustainable biodiesel production from waste cooking oil using a sonochemically deposited SrO catalyst on microporous activated carbon. Energy and Fuels* ***2017****, 31, 6228–6239. American Chemical Society.*

adsorption of reaction byproducts. Changes in the morphology of SrO/C upon reuses and partial blockage of mesopores by reactants/products/byproduct species could also contribute to the observed loss in catalytic activity beyond fourth reaction run.

To probe the surface functionalities of the activated carbon, SrO/C; to compare the bonding features of SrO and SrO/C; and to distinguish the product biodiesel from the byproduct glycerol, FT-IR spectra were recorded as shown in Fig. 17.

Sr—O bond stretching vibration is observed for SrO (commercial) at ~600 cm^{-1}. A broad peak at 3500 cm^{-1} was attributed to OH vibrations of $Sr(OH)_2$ in the commercial SrO catalyst (Fig. 17a). In the activated carbon-supported SrO (SrO/C) catalyst, similar features were observed confirming the deposition of SrO on activated carbon using sonochemical method. In addition, signals at ~1500 and 3000 cm^{-1} could be attributed to residual substrate triglyceride molecules (Fig. 17c). C—C and C=C stretching vibrations typical of aromatic ring structure were present in the spectrum of activated carbon support in the range of 1500–1650 cm^{-1} (Fig. 17b). FT-IR spectrum of transesterification byproduct glycerol matched with the spectral features of commercial glycerol indicting the effective

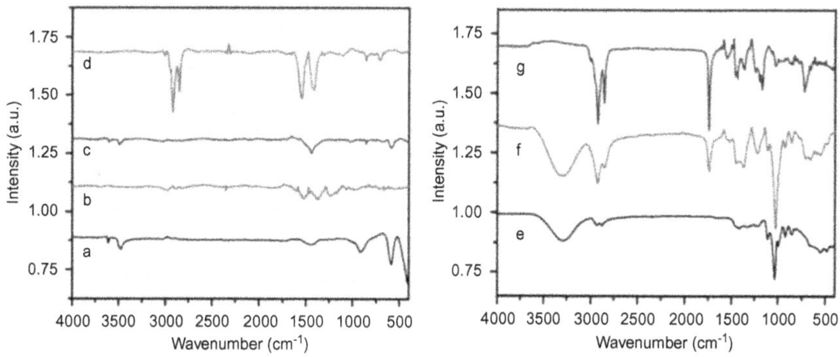

Fig. 17 FT-IR spectra of (a) commercial SrO, (b) activated carbon, (c) fresh SrO/C, (d) SrO/C catalyst used, (e) commercial glycerol, (f) glycerol byproduct in the transesterification of WCO, and (g) biodiesel from WCO (FAME). *Adapted with permission from Tabah, B.; Nagvenkar, A. P.; Perkas, N.; Gedanken, A. Solar-heated sustainable biodiesel production from waste cooking oil using a sonochemically deposited SrO catalyst on microporous activated carbon. Energy and Fuels* ***2017****, 31, 6228–6239. American Chemical Society.*

separation of the byproduct from the reaction medium (Fig. 17e and f). Peak characteristics of FAME were seen in the biodiesel produced from WCO (Fig. 17e). Evaluation of the solar energy-driven biodiesel production from WCO using SrO/C catalyst is recommended for industrial large-scale biodiesel production.[94]

Recently, Gupta et al. reported on the use of solar energy as a renewable source for the acceleration of biodiesel production from hempseed oil.[162] Solar-aided biodiesel production process uses Fresnel lens solar concentration (FSC) for the productive and efficient utilization of solar energy. In a short reaction time of 4 min., high yield of biodiesel (97.37%) was obtained in the solar energy-driven process under the optimal reaction conditions (molar ratio of methanol to oil of 4.5 to 1; catalyst loading, 0.9 wt%; reaction temperature, ~ 60°C; and stirring speed, 200 rpm). Solar-aided process outperformed the conventional heating method of biodiesel production, where the yield was only 21.3% under similar reaction conditions. For obtaining the yield as that in a solar-aided process (97%), conventional heating requires 70 min. KOH was used as catalyst in both processes. Solar-aided process (FSC method) was 73 times more energy efficient than the conventional heating method. The cost of biodiesel production is lowered by two orders of magnitude by the use of solar-aided process compared with conventional heating method.[162]

3.2 Supercritical CO_2 mediated microfluid reactor for biodiesel production

Quitain et al. designed a supercritical CO_2 (solvent)-mediated microfluid reactor for the noncatalytic production of biodiesel in high yield and in a short reaction time (<1 min) from the esterification of oleic acid and methanol yielding oleic acid methyl ester (OAME).[163] The developed method is superior to the advanced microwave-based biodiesel production. The main bottle neck of immiscibility of oil in methanol leading to high mass transfer resistance between the reactants resulting in longer reaction time and lower yield of FAME in conventional biodiesel production processes is surmounted by the effective mixing and molecular level interaction between the reactants (oleic acid and methanol) in the unique design of the microfluid reactor utilizing super critical CO_2 ($SCCO_2$) as solvent. As both oleic acid (OA) and methanol were soluble $SCCO_2$ even at low-pressure conditions, $SCCO_2$ was used as solvent for the reaction to overcome the problem of immiscibility. The solvent $SCCO_2$ could be separated by single simple depressurization step from the reaction medium after the esterification process making the use of $SCCO_2$ advantageous over conventional organic solvents like hexane. Unconventional methods of activation of reactant molecules and enhancing the reaction rates by improving molecular collisions and interactions include microwave irradiation, sonochemical irradiation, and microfluidic device (which is a relatively underexploited technique for biodiesel production). In a typical microwave irradiation method, reversed heating phenomena increase the probability of reacting molecules by collision interactions where heating is produced by the molecular motion of the microwave-absorbing polar molecules. Molecular collisions resulting in the acceleration of transesterification rate are considered to be the nonthermal component or effect of MW irradiation. In contrast to the microwave, thermal, and nonthermal activation method, the microfluid activation is based on the unique design of the microfluidic reactor with a few millimeters of internal diameter that facilitates continuous flow transesterification process. Such a reactor design facilitates processing in microscale, enhancing the heat released and molecular collisions due to short diffusion paths and large surface area values. Even under sudden temperature gradients, the process could be operated and controlled easily. Use of microfluidic reactors facilitate energy-efficient and cost-effective production of biodiesel. For large-scale production of biodiesel, numbering up of several microfluidic reactors, rather than scaling up, will result in the required throughput. Such microfluidic devices were exploited for processes such as direct synthesis of

H_2O_2 from H_2 and O_2, lignin oxidation, and for the direct conversion of syngas to dimethyl ether. Use of super critical fluids like $SCCO_2$ in the present context of oleic acid methyl ester further intensifies the process. Use of $SCCO_2$-mediated microfluidic reactor-based biodiesel production (esterification of oleic acid and methanol) was found to be superior to microwave activation process.[163]

Typical experimental setup used for the continuous flow production of OAME using $SCCO_2$-mediated microfluid rector is shown in Fig. 18. The apparatus consisted of a CO_2 pump, UFLC pump, microdevice, and an oven. Detailed description of the design of the microfluidic reactor is shown in Fig. 19. The internal diameter of the reactor is 0.9 mm, and the length is 500 mm with a semicircle shape cross section. The microfluidic device was housed inside a stainless steel casing of dimensions, 200 × 100 nm. The reactor could be operated up to a maximum pressure of 20 MPa.

Transesterification of OA and methanol (mole ratio of 1:4) was carried out by pumping the mixture into the reactor at a constant flow rate (0.02–0.1 mL/min). The flow rate of CO_2 was similar to that of the flow rate of the reaction mixture. The presence of $SCCO_2$ was varied from 8 to 12 MPa, and the temperature was varied in the range of 60–120°C. The products collected at the outlet were analyzed by GC equipped with FID detector and a capillary column (CP-SIL 8 CB-MS) with temperature programming mode, using He as carrier gas and 2,6-dimethyl naphthalene as internal standard.

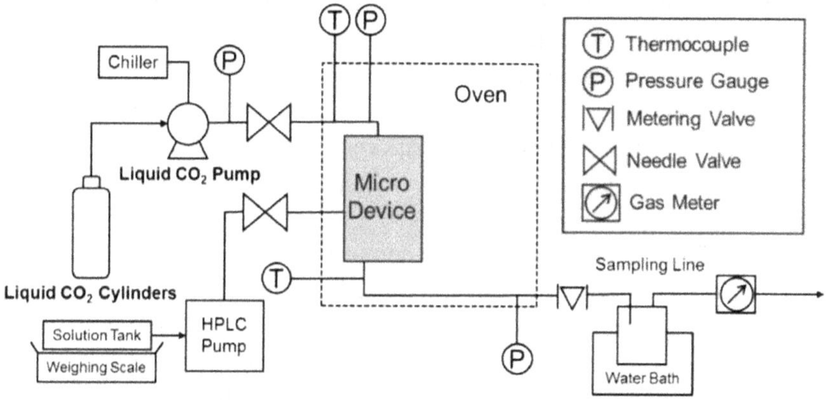

Fig. 18 Schematic diagram of the experimental apparatus for supercritical CO_2 ($SCCO_2$)-mediated microfluid reactor. *Adapted with permission from Quitain, A. T.; Mission, E. G.; Sumigawa, Y.; Sasaki, M. Supercritical carbon dioxide mediated esterification in a microfluidic reactor. Chem. Eng. Process. Process Intensificat.* ***2018****, 123, 168–173.*

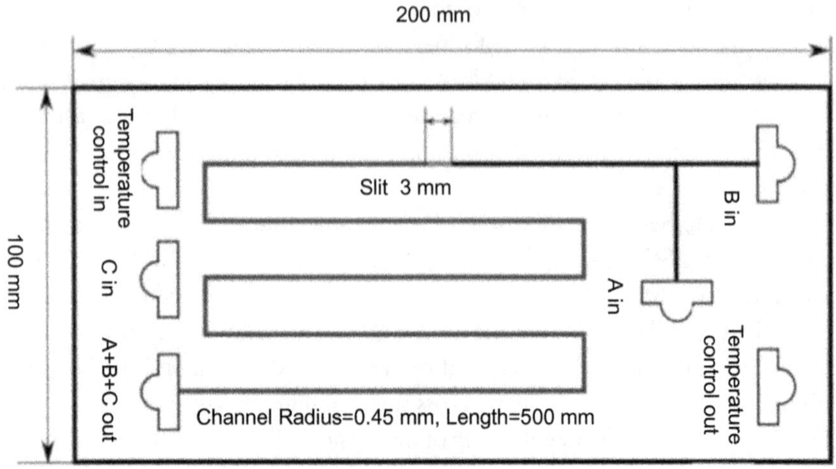

Fig. 19 Schematic diagram of the microfluidic reactor. *Adapted with permission from Quitain, A. T.; Mission, E. G.; Sumigawa, Y.; Sasaki, M. Supercritical carbon dioxide mediated esterification in a microfluidic reactor. Chem. Eng. Process. Process Intensificat.* ***2018****, 123, 168–173.*

Preliminary studies on the comparison of microwave and microfluidic noncatalytic activation methods for biodiesel production showed the superiority of microfluidic reactor-based continuous flow process for the production of OAME as depicted in Fig. 20. In a short duration of 60 s and at a $SCCO_2$ pressure of 10 MPa, 120°C, and a oleic acid-to-methanol molar ratio of 1:4, 19% FAME yield was obtained (Fig. 20-7) in a continuous flow microfluidic device. The data shown in Fig. 20(4–6) are the results adapted from the work of Melo-Junior et al. for comparison.[174]

As the esterification reaction is an endothermic reaction, increase in temperature drives the equilibrium toward product formation. So, Melo-Junior et al. carried out the uncatalyzed esterification process at 150 and 200°C for 5 min, which resulted in FAME yield of 2 and 25 wt%, respectively.[164] Further improvement in product yield to 60% was achieved with an increase in time from 5 to 30 min at 200°C. For comparison a batch esterification process under microwave activation was carried out. A noncatalytic process for 5 min at a temperature of 60–70°C resulted in low yield of OAME (0.62 wt %) (Fig. 20-1). Upon addition of a small amount of catalyst (0.1 wt%), a yield of 21.1% (30 s., 60°C) was achieved, which was further increased to 36% with an increase in time and temperature (60 s, 70°C). Even though the comparison of the results depicted in Fig. 20 (1–3), (4–6), and (7) is not logical, as none of the reactions were under identical experimental conditions,

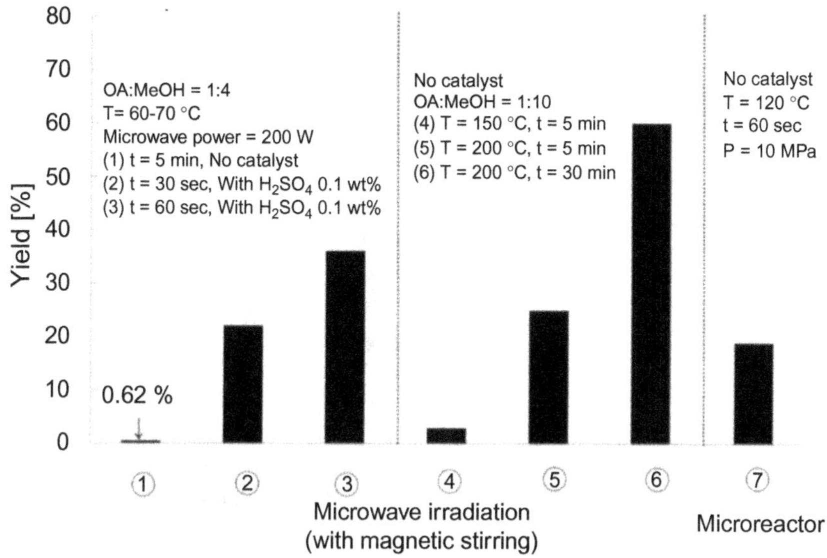

Fig. 20 Oleic acid methyl ester (OAME) yields using microwave and microfluidic device. *Adapted with permission from Quitain, A. T.; Mission, E. G.; Sumigawa, Y.; Sasaki, M. Supercritical carbon dioxide mediated esterification in a microfluidic reactor. Chem. Eng. Process. Process Intensificat.* ***2018****, 123, 168–173.*

an overview of Fig. 20(1) and Fig. 20(7) implies the promise of continuous flow through microfluidic method for high yield production of biodiesel, which was experimented further. Fourfold enhancement (~78%) in the biodiesel yield is observed, when the product derived in the first run (~19 wt% yield, in 60 s) is circulated for the second run, thereby increasing the residence time.

The effect of $SCCO_2$ pressure (8–12 MPa) on the yield of OAME was evaluated in a catalyzed (0.1 wt%, H_2SO_4) esterification process at 120°C, with a mole ratio of oleic acid to methanol of 4:1. The influence of $SCCO_2$ pressure was studied at a value slightly higher than the critical pressure of CO_2 (7.38 MPa). Yields of FAME were found to decrease as the pressure of SCCO2 was increased from 8 to 12 MPa as shown in Fig. 21.

A pressure of ~8 MPa resulted in high yield of the target product. However, depending on the operation temperature, flow rate, and OA-to-methanol molar ratio, the pressure is chosen to be 10 MPa for further optimization experiments. Moreover, working with high-pressure $SCCO_2$ was necessary to avoid the evaporation of reactants when the reaction temperatures were high. Reduction of product yield with increase in

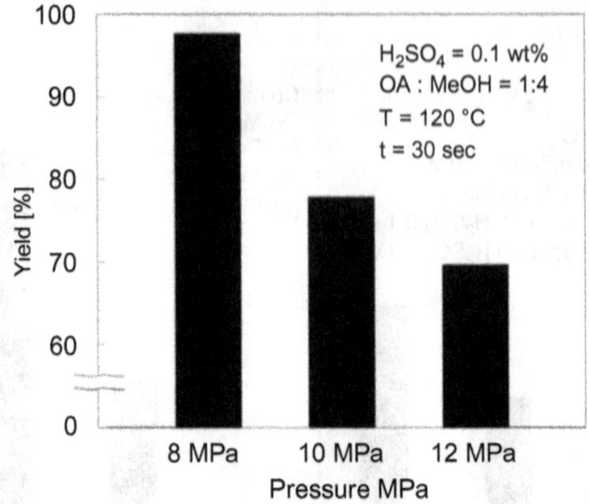

Fig. 21 Influence of supercritical carbon dioxide pressure on oleic acid methyl ester (OAME) yield. *Adapted with permission from Quitain, A. T.; Mission, E. G.; Sumigawa, Y.; Sasaki, M. Supercritical carbon dioxide mediated esterification in a microfluidic reactor. Chem. Eng. Process. Process Intensificat.* ***2018****, 123, 168–173.*

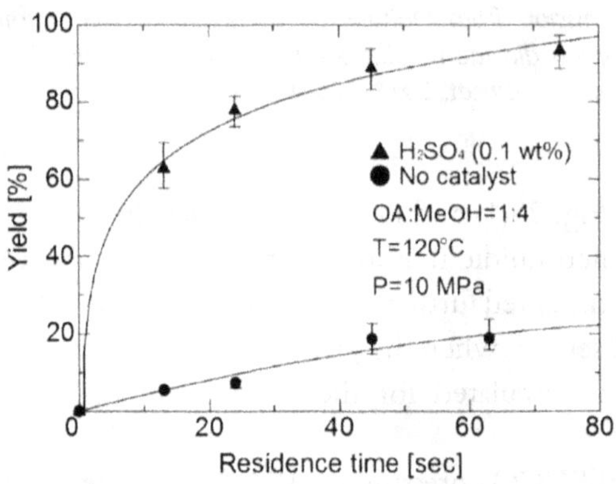

Fig. 22 Effect of 0.1 wt% sulfuric acid addition on oleic acid methyl ester yield. *Adapted with permission from Quitain, A. T.; Mission, E. G.; Sumigawa, Y.; Sasaki, M. Supercritical carbon dioxide mediated esterification in a microfluidic reactor. Chem. Eng. Process. Process Intensificat.* ***2018****, 123, 168–173.*

$SCCO_2$ pressure was attributed to the formation of two phases and difficulty in product recovery during depressurization step. Compared with an uncatalyzed process, addition of a small amount of catalyst (0.1 wt% H_2SO_4) increased the yield of OAME by four times (Fig. 22) at a residence time as short as 60 s.

The esterification reaction of oleic acid and methanol yielding OAME followed first-order reaction kinetics as shown in Fig. 23. An order of magnitude enhancement in the reaction rate is observed in a catalytic

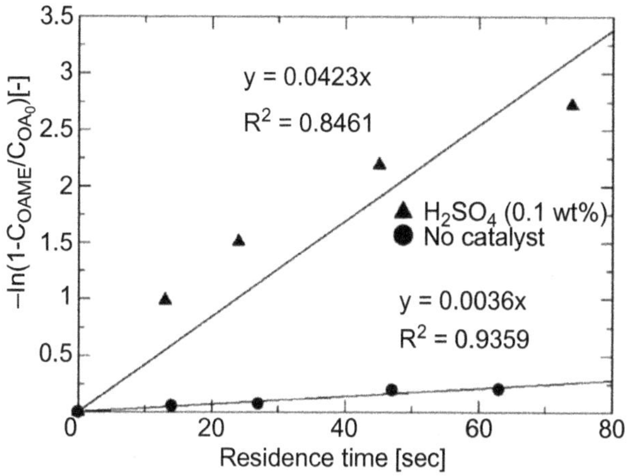

Fig. 23 Correlation between the rate of reaction and residence time in a microfluidic reactor with and without catalysts. *Adapted with permission from Quitain, A. T.; Mission, E. G.; Sumigawa, Y.; Sasaki, M. Supercritical carbon dioxide mediated esterification in a microfluidic reactor. Chem. Eng. Process. Process Intensificat.* ***2018****, 123, 168–173.*

microfluidized process ($k = 0.0634\ s^{-1}$) compared with an uncatalyzed process ($k = 0.0036\ s^{-1}$). Enhancement of yield of OAME (~78 wt%) in a noncatalyzed process was achieved by increasing the residence time by incorporating a recirculation loop in a continuous flow process. By recirculating the products obtained from the first pass (60 s), the yield of the biodiesel product was increased by fourfold. OAME formed in the first run acts as surfactant during the second pass, and the solubility of methanol and OA is improved in the second run leading to enhancement in the product yield.

3.3 Homogenization shows better yield efficiency than sonication with scaling up possibilities for biodiesel production[69]

Laosuttiwong et al. evaluated the performance efficiency of different cavitation and noncavitation reactors, namely, homogenizer, ultrasonic reactors (20 and 40 kHz), and mechanical stirrer for the production of biodiesel from palm oil with homogeneous (NaOH) and heterogeneous (CaO) catalysts.[69] The reactor efficiency was evaluated in terms of FAME yield efficiency, based on the yield of FAME, power consumption, and reaction time. Reaction volume (or scaling up of the process) significantly affected the reactor performance and efficiency of the respective processes. While the performance rate of FAME is not altered in the homogenizer as the batch volume is increased from 165 to 1000 mL, other reactor systems, namely, ultrasonic reactor (20 and 40 kHz) and the mechanical stirrer reactors, showed decrease in the rate of the transesterification reaction and the corresponding yield of FAME. Use of homogenizer outperformed ultrasonic reactors or mechanical stirrer for the large-scale (1000-mL reaction volume) biodiesel production. The yield efficiencies of the homogenized biodiesel production process from palm oil in homogeneous (NaOH) and heterogeneous (CaO) catalytic processes are 169.46×10^{-4} and 8.26×10^{-4} g/J for a reaction volume of 1000 mL.

Typical transesterification of palm oil was carried out by mixing desired amount of methanol with NaOH (1 wt%) or CaO (10 wt%) catalyst and transferring the same into a preheated palm oil (oil-to-methanol molar ratio of 1:6 for homogeneous and 1:9 for heterogeneous system) into a three-neck bottom flask (165, 500, and 1000 mL) immersed in an oil bath for controlling the reaction temperature at 60°C. The three-neck bottom flask was plugged with a condenser and a thermometer and had a provision for inserting the mechanical stirrer (MS) or the direct ultrasonic transducers (US, 20 and 40 kHz) or the homogenizer (Hz) for evaluating the effect of the reactor type on the FAME yield and the yield efficiency of the biodiesel production process. The process is monitored for 3 and 4 h, respectively, in homogeneously and heterogeneously catalyzed processes.

The effect of activation by various cavitation reactors, namely, homogenizer (HZ) and ultrasonicators (20 and 40 kHz), compared with conventional mechanical stirring (MS) on the yield of biodiesel in a 165-mL batch homogeneously catalyzed (NaOH) process is depicted in Fig. 24. Homogenizer showed outstanding performance, while all other reactors showed the requirement of initial induction time of a minimum of 5 min, and homogenizer showed high yield of 87.06% and 92.71% in 5 and 15 min, respectively. 90.98% FAME yield was observed with ultrasonicator (20 Hz for 30 min). 40-kHz US showed lower FAME yield compared with 20 kHz US (Fig. 24). As the ultrasonic frequency is increased from 20 to 40 kHz, the bubble size decreased. The input power of 80 W could be insufficient to generate cavitation collapse resulting in lower shear force and reduction in the cavitation intensity leading to decrease in activation and FAME yield. MS showed similar performance as that of 20 kHz US but a slower reaction rate compared with HZ.

Analogous to a homogeneous catalytic process (Fig. 24), in a heterogeneously (CaO) catalyzed process as well, US 40 kHz showed poor performance compared to either HZ or US 20 kHz or MS (Fig. 25). Moreover, HZ too required initial induction period on a par with US

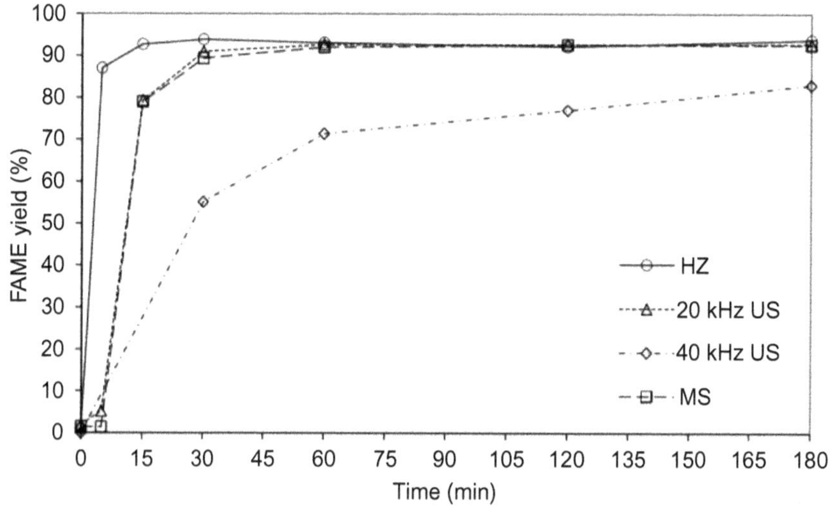

Fig. 24 Effect of type of mixing on FAME yield from palm oil feedstock using a reaction volume of 165 mL and NaOH as catalyst (*HZ*, homogenizer; *US*, direct ultrasonic; and *MS*, mechanical stirred reactors). *Adapted with permission from Laosuttiwong, T.; Ngaosuwan, K.; Kiatkittipong, W.; Wongsawaeng, D.; Kim-Lohsoontom, P.; Assabumrungrat, S. Performance comparison of different cavitation reactors for biodiesel production via transesterification of palm oil. J. Clean. Prod.* ***2018****, 205, 1094–1101. Elsevier.*

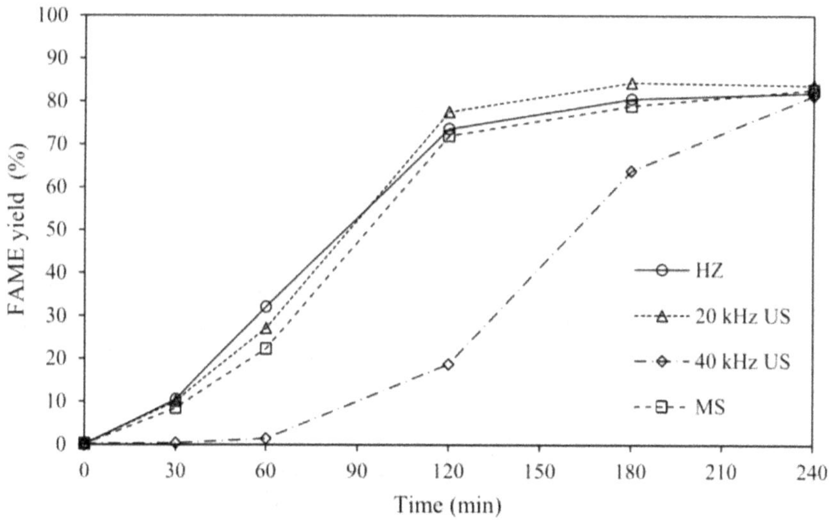

Fig. 25 Effect of type of mixing on FAME yield from palm oil feedstock using a reaction volume of 165 mL and CaO as solid base catalyst (*HZ*, homogenizer; *US*, direct ultrasonic; and *MS*, mechanical stirred reactors). *Adapted with permission from Laosuttiwong, T.; Ngaosuwan, K.; Kiatkittipong, W.; Wongsawaeng, D.; Kim-Lohsoontom, P.; Assabumrungrat, S. Performance comparison of different cavitation reactors for biodiesel production via transesterification of palm oil. J. Clean. Prod.* ***2018****, 205, 1094–1101. Elsevier.*

20 kHz or MS for a minimum of 30 min, unlike the homogeneous catalytic process wherein HZ showed superior performance. The FAME yield obtained as a function of time showed similar trend irrespective of the reactor type used for activating (via mixing the components) the reaction medium.

Highest yield of FAME (even slightly higher than HZ) was noticed in the case of US 20 kHz. Due to the unique cavitation phenomena, the level of mixing in the bulk fluid is enhanced with US 20 kHz by overcoming external and internal mass transfer constraints. On the contrary, use of solid catalyst CaO has reduced the performance of HZ compared with the homogeneous reaction (Fig. 24). According to the hydrodynamic cavitation principle on which the homogenizer works, small amounts of fluid are sucked through the gap between the rotor and the stator to provide the shear force to generate small bubbles. Use of CaO catalyst reduced the velocity of the reaction mixture in a HZ-activated process via decreasing the generation of cavitation leading to lower degree of mixing. At 240-min reaction time, FAME yield obtained in all reactors reached the similar equilibrium (Fig. 25).

Yield efficiency of biodiesel production process using various reactors for activation is calculated taking into consideration the three major process parameters, namely, the yield of FAME, power input, and the reaction time as shown in Eq. (1):

$$\text{Yield efficiency} = \frac{W_{FAME}}{P X t} \tag{1}$$

where W_{FAME} = amount of FAME produced (g), P = power supplied (J/s), and t = reaction time (s).

Yield of FAME is the major process parameter for evaluating the FAME production efficiency using various mixing types for the activation of the catalyst (NaOH or CaO) and the substrates. However, for the economically viable and profitable biodiesel production, in addition to FAME yield, other parameters, namely, reaction time and input energy, are to be considered. Therefore yield efficiency was used as a measure to evaluate the reactor performance based on the energy consumption, amount of FAME production, and the reaction time. The energy consumption in the process consisted of the electric power used for mixing and the heating facility. The effects of different reactor types (homogenizer, HZ; ultrasonicator, US; and mechanical stirrer, MS) on the yield efficiency using NaOH and CaO as catalysts at 90% and 80% FAME yield were depicted in Figs. 26 and 27, respectively.

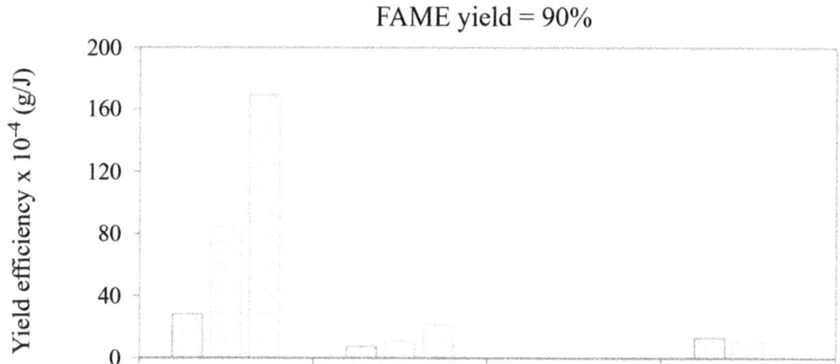

	HZ	20 kHz US	40 kHz US	MS
165 mL, NaOH	27.96	7.71		13.31
500 mL, NaOH	84.52	10.91		10.42
1000 mL, NaOH	169.46	21.52		

Type of reactor

Fig. 26 Effect of type of mixing on yield efficiency based on 90% FAME yield from palm oil feedstock using NaOH as catalyst (*HZ*, homogenizer; *US*, direct ultrasonic; and *MS*, mechanical stirred reactors). *Adapted with permission from Laosuttiwong, T.; Ngaosuwan, K.; Kiatkittipong, W.; Wongsawaeng, D.; Kim-Lohsoontom, P.; Assabumrungrat, S. Performance comparison of different cavitation reactors for biodiesel production via transesterification of palm oil. J. Clean. Prod.* ***2018****, 205, 1094–1101. Elsevier.*

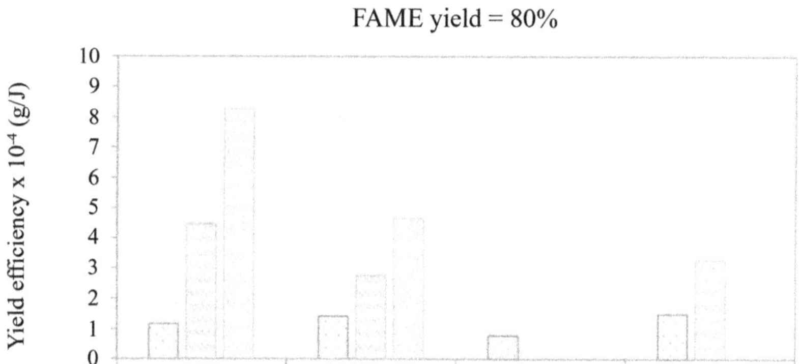

	HZ	20 kHz US	40 kHz US	MS
165 mL, CaO	1.16	1.41	0.77	1.49
500 mL, CaO	4.47	2.75		3.27
1000 mL, CaO	8.27	4.65		

Type of reactor

Fig. 27 Effect of type of mixing on the yield efficiency based on 80% FAME yield from palm oil feedstock using CaO as solid base catalyst (*HZ*, homogenizer; *US*, direct ultrasonic; and *MS*, mechanical stirred reactors). *Adapted with permission from Laosuttiwong, T.; Ngaosuwan, K.; Kiatkittipong, W.; Wongsawaeng, D.; Kim-Lohsoontom, P.; Assabumrungrat, S. Performance comparison of different cavitation reactors for biodiesel production via transesterification of palm oil. J. Clean. Prod.* ***2018****, 205, 1094–1101. Elsevier.*

Figs. 26 and 27 are self-explanatory. Irrespective of the catalyst used (whether the process is homogeneously or heterogeneously catalyzed), the kind of mixing employed (HZ, US, and MS), and the volume of the batch (165, 500, and 1000 mL), the use of homogenizer outperformed other kinds of reactors. For a 1000-mL batch of transesterification of palm oil, the values of yield efficiency for homogeneous and heterogeneous catalyzed process are 169.46×10^{-4} and 8.27×10^{-4} g/J, respectively. The enhancement of yield efficiency with increase in batch volume from 165 to 1000 mL with a homogenizer holds promise for further scaling up and building an industrial unit adapting this technology.

Though, beyond any doubt, microwave activation is the best available technology as of now for the industrial adaptability for the large scale production of biodiesel, additional technologies like solar energy, microfluidic reactor and homogenization reaction technologies should not be neglected for scaling up consideration as no single technology can cater to the growing demand for renewable fuels, especially, biodiesel.

4 Bioethanol production using nontraditional methods of activation

Mankind should always bear the image of ideal ecosystem in mind and strive to conserve the nature for the sustainable development, prosperity, and well-being of humanity. However, history has proved that any innovation or new technology has its own wells and ills. The current ecosystem reflects the damage caused to nature by indiscriminate use of natural resources leading to, for instance, climate change, global warming, water scarcity, and the menace of nondegradable plastic waste being dumped into water sources. The work of progress, technological innovation, should not be hindered, for fear of the accompanying menaces, as they are inevitable. However, appropriate pathways should be developed for managing the unwanted effects of any new technology. For instance, the advent of polymer materials has improved the standard of living and contributed much for catering to the needs of humanity for over 50 years. But now the problem of disposal of nondegradable plastic bottles and plastic waste has been gigantic, polluting the oceans and rivers and harming the ecosystem to great extent. Pollutants, like CO_2 and plastic wastes, should be turned into resources for producing biofuels, and such feedstock can yield fourth generation biofuels, with the first, second, and third being those from edible carbohydrates (corn and sugarcane), lignocellulosic, and algal biomass. Such a strategy will alleviate the energy and environmental crisis facing mankind.[161]

4.1 Microwave activation of biomass and catalytic species to produce bioethanol

Gedanken et al. developed microwave irradiation technology for the conversion of biomass (lignocellulosic, algal, and glycogen synthesized from CO_2) to glucose, the key economic driver for the upcoming biorefinery. The accelerating ability of microwave irradiation for the key reaction, namely, the hydrolysis of carbohydrate (cellulose, hemicellulose, and glycogen) to fermentable sugars like glucose and xylose, could be attributed to the weakening of extensive hydrogen bonding network prevailing in natural biopolymers, like cellulose available in biomass. The unique potential of microwave with its thermal and nonthermal modes of activating the substrate (biomass) and the catalyst (polar/ionic compounds) facilitating the single-step conversion of cellulose without any pretreatment or the use of special solvents like ionic liquids remains under utilized. The potential of microwave irradiation to weaken the hydrogen bonding network is pictorially shown in Fig. 28.

Even though the two key reactions, namely, biomass pretreatment and carbohydrate polymer hydrolysis, were activated by microwave irradiation, this technology has not found any utility in the third and major reaction, namely, fermentation of sugars (glucose and xylose) leading to bioethanol formation, and efforts in this direction are awaited.[148, 152, 154, 156, 158] Microwave-based glucose production from biomass in an acid catalyzed (homogeneous or heterogeneous) process is as promising as microwave-activated transesterification process for biodiesel production.[148,152,165,166]

4.2 Sonochemical activation of biomass and biocatalysts for the accelerated production of bioethanol

Bioethanol is a carbon neutral transportation fuel. Bioethanol is an energy efficient and clean alternative to fossil-based fuels in the transportation sector. Even though CO_2 evolution is inevitable, in the bioethanol production process and in the combustion of bioethanol in the engines of automobiles, the CO_2 emissions are compensated by the CO_2 absorbed during the growth and harvesting of biomass making the use of bioethanol a closed carbon cycle operation. Currently, use of biofuels is only restricted to 0.8% of the net energy consumption with the major portion of 78.4% contributed by

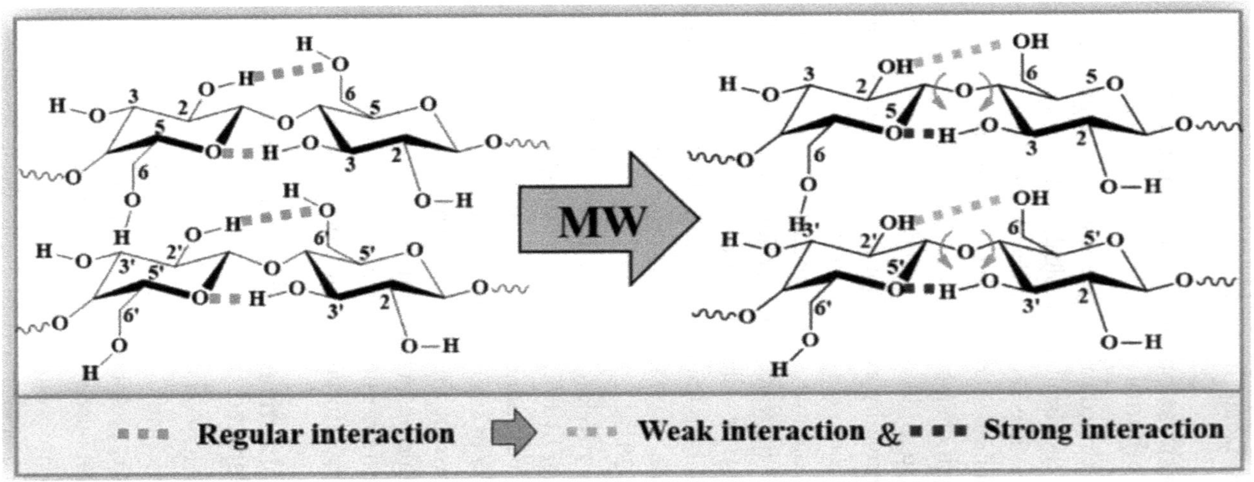

Fig. 28 Nonthermal effect of microwave irradiation on the relative strength of hydrogen bonding interactions in cellulose structure. *Adapted with permission from Zhai, C. K.; Teng, N.; Pan, B. H.; Chen, J.; Liu, F.; Zhu, J.; Na, H. N. Revealing the importance of cellulose by synchronous cooling assisted microwave driving. Carbohydrate Polymers* ***2018****, 197, 414–421. Elsevier.*

fossil fuels. In the next 50 years, the trend needs to be reverted to maintain sustainable ecosystem and to tackle the ongoing challenge of climate change. Biofuels are expected to contribute ~27% of the transportation fuels by 2050 globally.[154] This has renewed the interest in the development of bioethanol as an alternative fuel for transportation applications. With this motivation, Gedanken et al. developed the accelerated bioethanol production process using sonochemical activation of fermentation (glucose) and simultaneous saccharification and fermentation (marine macroalgae, *Ulva rigida*) processes.[146, 151, 158]

The key reaction in the conversion of carbohydrates to bioethanol is the fermentation reaction catalyzed by the yeast biocatalyst (fungal strain, *Saccharomyces cerevisiae*). There have been several shortcomings of this vital reaction, namely, slow kinetics and poisoning of the yeast strain by the ethanol product (beyond a concentration of 4 wt%). Gedanken et al. developed a mild sonication method (40-kHz bath sonicator) to improve the performance of yeast (*Saccharomyces cerevisiae*) for bioethanol production. When the fermentation of glucose (20 wt%) was carried out under mild sonication with *Saccharomyces cerevisiae* (Baker's yeast) at 30°C, the reaction rate was accelerated by a factor of 2.3 ± 0.2 compared with a conventional stirring process. The reaction rate constants for the glucose fermentation to bioethanol under stirring and sonication were 6.67×10^{-6} and $15.3 \times 10^{-6}\ s^{-1}$, respectively.[146] The sonochemical methodology was further exploited for the single-step conversion of marine macroalgae *Ulva rigida* to bioethanol.[151, 158] Much remains to be done in the arena of sonochemical activation of biomass conversion to bioethanol, and the sonochemical activation potential was underutilized for the application of biomass conversion to bioethanol.

4.3 Bioethanol production in a solar reactor

Gedanken et al. developed a unique solar reactor system for the conversion of biomass to bioethanol in a continuous flow process. No other energy except solar radiation is used for driving either the fermentation of glucose or the simultaneous saccharification and fermentation (SSF) of biopolymers like starch, cellulose, and algae (*Ulva rigida*).[157] Glucose concentrations as high as 40 wt% could be flown in to the solar reactor bed packed with yeast, and high bioethanol concentration (~19 wt%) close to the theoretical values was obtained. Moreover the formed bioethanol could be separated from the reaction chamber via evaporation condensation process prevailing in the system due to the unique geometry of the reactor.[155] The continuous flow process for the single-step conversion of *U. rigida* to bioethanol in the solar reactor is shown in Fig. 29. Aqueous suspension of *U. rigida* with the enzymes (amylases and cellulases) that were flown in to the reactor bed (packed with yeast) was prepared by ultrasound irradiation. The biomass

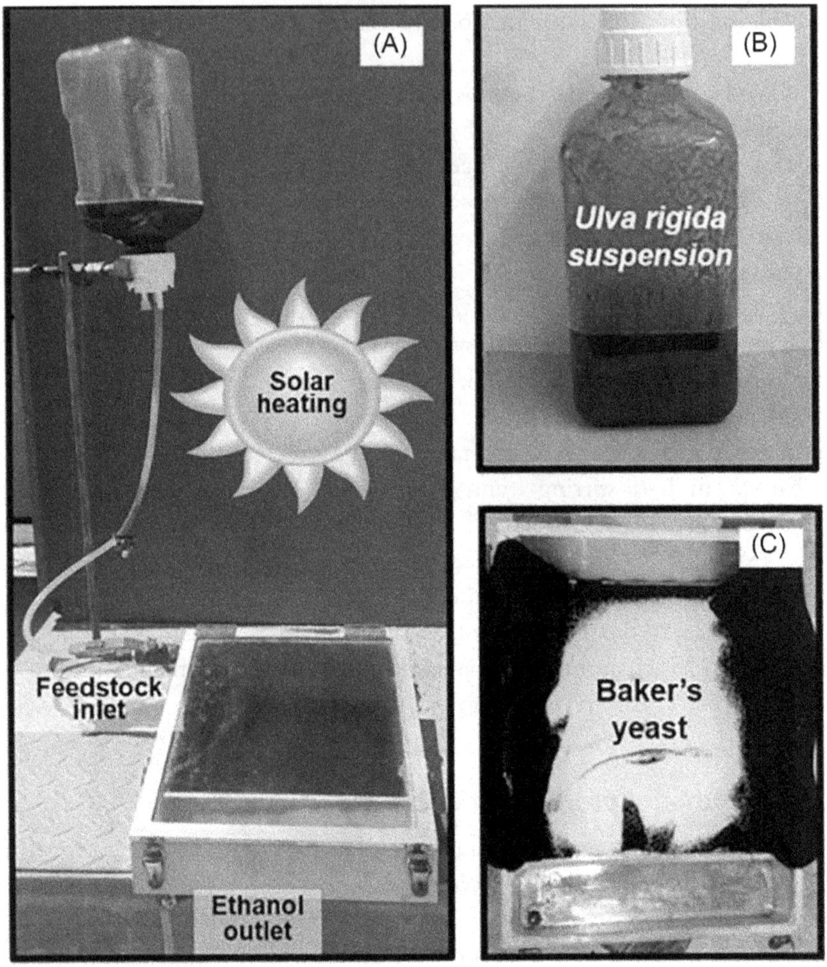

Fig. 29 Continuous flow solar reactor for the single-step conversion of marine macroalgae *Ulva rigida* to bioethanol; (A) complete experimental setup, (B) stable 5 wt% aqueous *Ulva rigida* suspension with enzymes, and (C) fermentation chamber loaded with baker's yeast on activated carbon cloth. *Adapted with permission from Gedanken, A.; Indra Neel, P.; Betina, T. Solar-aided conversion of marine algae and Biomass to bioethanol, US Provisional Patent Application No 62450107, 25th January 2017.*

suspensions were then flown into the solar reactor containing yeast and were converted to bioethanol. This process involves, three key reactions, namely, the biomass suspension in the reservoir will have sufficient residence time during which the biomass interacts with the enzymes and the partial saccharification reaction takes place within the reservoir from which the biomass with fermentable sugars is fed into the solar reactor. In the reactor, both the fermentation reaction of the hydrolyzed sugars by the action of yeast and the SSF of the algal biomass to bioethanol takes place. The formed ethanol is evaporated and condensed on the top inner glass surface of the reactor, which will be flown in to the separate bioethanol collection chamber and collected through the bioethanol outlet. The solar energy-driven bioethanol process is innovative, and industrial adaptation will be rewarding.

5 Conclusion

Biofuels, especially, biodiesel and bioethanol, should be produced on a war-footing using unconventional methods of activation, namely, microwave irradiation, sonochemical irradiation, solar irradiation, homogenization reactors, and microfluidic reactors. Use of solid base catalysts (SrO-based catalysts) for transesterification reaction and solid acid (polyoxometalates) catalysts for biomass hydrolysis reaction and biocatalysts like yeast (commercial Baker's yeast, *Saccharomyces cerevisiae*) for fermentation reaction should be promoted. Fossil-based fuels need to be systematically and completely replaced with renewable fuels. Major environmental pollutants like CO_2 and plastic are to be used as fourth generation feedstock for biofuel production. Such a path ahead will improve the ecosystem providing a clean, green, and healthy nature for future generations.

Acknowledgments

Grateful thanks to Dr. Eli Evan, the head of the research authority, Bar Ilan University, Israel. Dr. Eli Evan was as persistent as Professor Gedanken, in knocking at the doors of funding agencies at the face of rejections that has made possible winning several prestigious research grants including a huge 4-year-old Israel Science Foundation (ISF) grant (598/12) for which the authors thank the Israeli government and the funding agencies of Israel. Dr. Pulidindi wish to thank his present postdoctoral research advisor Professor Xinling Wang, Shanghai Jiao Tong University, China, for providing him with the employment as well as calm working environment that facilitated his creativity and writing skills to blossom and made possible the compilation of this book chapter. The authors are thankful to the editors of this book, Professors Torok and Schaefer, for inviting the authors for compiling this book chapter.

References

1. Lee, H.; Liao, J. D.; Lee, M. H.; Liu, B. H.; Fu, W. E.; Sivashanmugan, K.; Juang, Y. D. Strontium Oxide Deposited Onto a Load-Bearable and Porous Titanium Matrix as Dynamic and High Surface Contact Area Catalyst for Transesterification. *Nanomaterials* **2018**, *8* (973), 1–14.
2. Pinheiro, A. C.; Goncalves, A. B.; Baader, W. J.; Yamaguchi, F. L.; Malek, N. I.; Bastos, E. L.; Seoud, O. A. E. Biofuels From Coconut Fat and Soybean Oil: Microwave-Assisted Synthesis and Gas Chromatography/Mass Spectrometry Analysis. *Quim. Nova* **2018**, *41* (10), 1200–1204.
3. Anderson, E.; Zhou, J.; Fan, L.; Liu, S.; Zhou, N.; Peng, P.; Cheng, Y.; Chen, P.; Ruan, R. Mirowave-Assisted Pyrolysis as an Alternative to Vacuum Distillation for Methyl Ester Recovery From Biodiesel Vacuum Distillation Bottoms. *ACS Sustain. Chem. Eng.* **2018**, *6*, 14348–14355.
4. Harris, J.; Viner, K.; Champagne, P.; Jessop, P. G. Advances in Micro Algal Lipid Extraction for Biofuel Production: A Review. *Biofuels Bioprod. Biorefin.* **2018**, *12*, 1118–1135.
5. Yousefi, S.; Haghighi, M.; Vahid, B. R. Facile and Efficient Microwave Combustion Fabrication of Mg-Spinel Support for MgO Nanocatalyst Used in Biodiesel Production From Sunflower Oil: Fuel Type Approach. *Chem. Eng. Res. Des.* **2018**, *138*, 506–518.
6. Aghbashio, M.; Tabatabaei, M.; Amid, S.; Hosseinzadeh-Bandbafha, H.; Khoshnevisan, B.; Kianian, G. Life Cycle Assessment Analysis of an Ultrasound-Assisted System Converting Waste Cooking Oil Into Biodiesel. *Renew. Energy* **2020**, *151*, 1352–1564.
7. Kalsum, U.; Kusuma, H. S.; Roesyadi, A.; Mahfud, M. Production of Biodiesel Via In-Situ Transesterification From *Chlorella sp.* Using Microwave With Base Catalyst. *Korean Chem. Eng. Res.* **2018**, *56* (5), 773–778.
8. Rouchi, M. B.; Khanmohammadi, M.; Garmarudi, A. B.; Guardia, M. D. Synthesis and Optimization of Microwave-Assisted Exfoliated Functionalized Graphene as an Efficient Catalyst in Biodiesel Production. *Fullerenes Nanotubes Carbon Nanostruct.* **2018**, *26* (11), 729–738.
9. Howlader, M. S.; Rai, N.; French, W. T. Improving the Lipid Recovery From Wet Oleagionous Microorganism Using Different Pretreatment Techniques. *Bioresour. Technol.* **2018**, *267*, 743–755.
10. Viner, K. J.; Champagne, P.; Jessop, P. G. Comparison of Cell Disruption Techniques Prior to Lipid Extraction From *Scenedesmus sp.* Slurries for Biodiesel Production Using Liquid CO_2. *Green Chem.* **2018**, *20*, 4330–4338.
11. Priyadarshi, D.; Paul, K. K. Single Phase Blend: An Advanced Microwave Process for Improved Quality Low-Cost Biodiesel Production From Kitchen Food Waste. *Biochem. Eng. J.* **2018**, *137*, 273–283.
12. Ma, W.; Hong, T.; Xia, T.; Wang, F.; Luo, B.; Zhou, J.; Yang, Y.; Zhu, H.; Huang, K. Simulation and Analysis of Oleic Acid Pretreatment for Microwave-Assisted Biodiesel Production. *Processes* **2018**, *6* (142), 1–16.
13. Tabatabai, B.; Chen, H.; Lu, J.; Giwa-Otusajo, J.; Mckenna, A. M.; Shrivastava, A. K.; Sitther, V. *Fremyella diplosiphon* as a Biodiesel Agent: Identification of Fatty Acid Methyl Esters Via Microwave-Assisted Direct In Situ Transesterification. *Bioenergy Res.* **2018**, *11*, 528–537.
14. Gim, G. H.; Kim, S. W. Optimization of Cell Disruption and Transesterification of Lipids From *Botryococcus braunil* LB572. *Biotechnol. Bioprocess Eng.* **2018**, *23* (5), 550–556.
15. Nayebzadeh, H.; Haghighi, M. K.; Saghatoleslami, N.; Tabasizadeh, M.; Yousefi, S. Fabrication of Carbonated Alumina Doped by Calcium Oxide Via Microwave Combustion Method Used as Nanocatalyst in Biodiesel Production: Influence of Carbon Source Type. *Energ. Conver. Manage.* **2018**, *171*, 566–575.

16. Wahidin, S.; Idris, A.; Yusof, N. M.; Kamis, N. H. H.; Shalesh, S. R. M. Optimization of the Ionic Liquid-Microwave Assisted One-Step Biodiesel Production Process From Wet Microalgal Biomass. *Energ. Conver. Manage.* **2018**, *171*, 1397–1404.
17. Javalkar, P. D.; Krithika, S.; Hegde, N. S.; Memn, N. G.; Kamath, H. V. Use of Microwave Irradiation to Produce Biodiesel From *Simarouba glauca* Oil Seed: Rapid Extractive Transesterification Approach. *Res. J. Pharm., Biol. Chem. Sci.* **2018**, *9* (5), 1610–1616.
18. Efthymiopoulos, I.; Hellier, P.; Ladommatos, N.; Russo-Profili, A.; Eveleigh, A.; Alieve, A.; Kay, A.; Mills-Lamptey, B. Influence of Solvent Selection and Extraction Temperature on Yield and Composition of Lipids Extracted from Spent Coffee Grounds. *Ind. Crop Prod.* **2018**, *119*, 49–56.
19. Carvalho, A. K. F.; Bento, H. B. S.; Izario, H. J.; de Castro, H. F. Approaches to Convert *Mucor circinelloides* Lipid into Biodiesel by Enzymatic Synthesis Assisted by Microwave Irradiations. *Renew. Energy* **2018**, *125*, 747–754.
20. Thushari, I.; Babel, S. Preparation of Solid Acid Catalysts From Waste Biomass and Their Application for Microwave Assisted Biodiesel Production From Palm Oil. *Waste Manag. Res.* **2018**, *36* (8), 719–728.
21. Heydari, H.; Nayebzadesh, H.; Shahraki, M. Effect of Fuel Molecules on Properties and Activity of KOH/Calcium Alumunate Nanocatalyst for Biodiesel Production. *Int. J. Environ. Sci. Technol.* **2018**, *15*, 1719–1730.
22. Kusuma, H. S.; Ansori, A.; Wibowo, S.; Bhuana, D. S.; Mahfud, M. Optimization of Transesterification Process of Biodiesel From Nyamplung (*Calophyllum inophyllum Linn*) Using Microwave with Cao Catalyst. *Korean Chem Eng. Res.* **2018**, *56* (4), 435–440.
23. Souza, M. C. D.; Luis, D.; Caldeira, V. P. D.; Santos, A. G. D.; Beatrizz, A. Fatty Acid Esterification: Effects of Unsaturation and Synthesis With Thermal Heating Via Ultrasound and Microwaves. *Periodico Tche Quimica* **2018**, *15* (30), 447–462.
24. Medeiros, A. M.; Santos, E. R. M.; Azevedo, S. H. G.; Jesus, A. A.; Oliverira, H. N. M.; Sousa, E. M. B. D. Chemical Interesterification of Cotton Oil With Methyl-Acetate Assisted by Ultrasound for Biodiesel Production. *Braz. J. Chem. Eng.* **2018**, *35* (3), 1005–1017.
25. Nayebzadeh, H.; Saghatoleslamin, N.; Haghighi, M.; Tabasizadeh, M. Comparative Assessment of the Ability of a Microwave Absorber Nanocatalyst in the Microwave—Assisted Biodiesel Production Process. *C. R. Chim.* **2018**, *21* (7), 676–683.
26. Patel, A.; Mikes, F.; Matsakas, L. An Over View of Current Pretreatment Methods Used to Improve Lipid Extraction From Oleaginous Microorganisms. *Molecules* **2018**, *23*, 1–22.
27. Cercado, A. P. I.; Ballesteros, F. C.; Capared, S. C. Biodiesel From Three Microalgal Transesterification Processes Using Different Homogeneous Catalysts. *Int. J. Technol.* **2018**, *9* (4), 645–651.
28. de Moura, R. R.; Etges, B. J.; dos Santos, E. O.; Martins, T. G.; Roselet, F.; Abren, P. C.; Primel, E. G.; D'oca, M. G. M. Microwave-Assisted Extraction of Lipids From Wet Microalgae Paste: a Quick and Efficient Method. *Eur. J. Lipid Sci. Technol.* **2018**, *120*, 1–7.
29. Gupta, A. R.; Rathod, V. K. Calcium Diglyceroxide Catalyzed Biodiesel Production From Waste Cooking Oil in the Presence of Microwave: Optimization and Kinetic Studies. *Renew. Energy* **2018**, *121*, 757–767.
30. Fatimah, I.; Rubiyanto, D.; Nugraha, J. Preparation, Characterization and Modelling Activity of Potassium Fluoride Modified Hydrotalcite for Microwave Assisted Biodiesel Conversion. *Sustain. Chem. Pharm.* **2018**, *8*, 3–70.
31. Nomanbhay, S.; Ong, M. Y. A review of microwave-assisted reactions for biodiesel production. *Bioengineering* **2017**, *5*, 57. 1–21.

32. Liu, J.; Jia, G. Non-thermal Effects of Microwave in Sodium Chloride Aqueous Solution: Insights From Molecular Dynamics Simulations. *J. Mol. Liq.* **2017**, *227*, 31–36.
33. Wei, G. T.; Liu, Z. H.; Zhang, L. Y.; Li, Z. M. Catalytic Upgrading of Jatropha Oil Biodiesel by Partial Hydrogenation Using Raney-Ni as Catalyst under Microwave Heating. *Energ. Conver. Manage.* **2018**, *163*, 208–218.
34. Nayebzadeh, H.; Haghighi, M.; Saghatoleslami, N.; Tabasizadeh, M.; Binaeian, E. Influence of Carbon Source Content on the Structure and Performance of $KOH/Ca_{12}Al_{14}O_{33}$-C Nanocatalyst Used in the Transesterification Reaction Via Microwave Irradiation. *J. Alloys Compd.* **2018**, *743*, 672–681.
35. Feng, W. L.; Qian, L.; Wang, W. G.; Wang, T. L.; Deng, Z. K.; Yang, F.; Xiong, J.; Wang, C. W. Exploring the Potential of Lipids From Black Soldier Fly: New Paradigm for Biodiesel Production-(II)-Extraction Kinetics and Thermodynamics. *Renew. Energy* **2018**, *119*, 12–18.
36. Martinez-Guerra, E.; Howlader, M. S.; Shields-Menard, S.; French, W. T.; Gude, V. G. Optimization of Wet Microalgal FAME Production From *Nannochloropsis sp.* Under the Synergistic Microwave and Ultrasound Effect. *Int. J. Energy Res.* **2018**, *42* (5), 1934–1949.
37. Mahfud, M.; Suryanto, A.; Qadariyah, L.; Suprapto, S.; Kusuma, H. S. Production of Methyl Ester From Coconut Oil Using Microwave: Kinetics of Transesterification Reaction Using Heterogeneous Cao Catalyst. *Korean Chem. Eng. Res.* **2018**, *56* (2), 275–280.
38. Quitain, A. T.; Sumigawa, Y.; Mission, E. G.; Sasaki, M.; Assabumnangrat, S.; Kida, T. Graphene Oxide and Microwave Synergism for Efficient Esterification of Fatty Acids. *Energy Fuel* **2018**, *32*, 3599–3607.
39. Milano, J.; Ong, H. C.; Masjuki, H. H.; Silitonga, A. S.; Chen, W. H.; Kusumo, F.; Dharma, S.; Sebayang, A. H. Optimization of Biodiesel Production by Microwae Irradiation-Assisted Transesterification for Waste Cooking Oil-*Calophyllum inophyllum* Oil Via Response Surface Methodology. *Energ. Conver. Manage.* **2018**, *158*, 400–415.
40. Bundhoo, Z. M. A. Microwave-Assisted Conversion of Biomass and Waste Materials to Biofuels. *Renew. Sustain. Energy Rev.* **2018**, *82*, 1149–1177.
41. Peng, Y. P.; Amesho, K. T. T.; Chen, C. E.; Jhang, S. R.; Chou, F. C.; Lin, Y. C. Optimization of Biodiesel Production From Waste Cooking Oil Using Waste Egg Shell as a Base Catalyst Under a Microwave Heating System. *Catalysts* **2018**, *8*, 81. 1–16.
42. Jitjamnong, J. Effect of Barium Loading on CaO Derived from Waste Egg Shell Heterogeneous Catalyst for Canola Oil Biodiesel. In *MATEC Web of Conferences 2018, 192, 03006*; ICEAST, 2018.
43. Al-Ani, A.; Darton, R. J.; Scott, S.; Zholobenko, V. Nanostructured Zeolites: The Introduction of Intracrystalline Mesoporosity in Basic Faujasite-Type Catalysts. *ACS Appl. Nano Mater.* **2018**, *1*, 310–318.
44. Mishra, V. K.; Goswami, R. A Review of Production, Properties and Advantages of Biodiesel. *Biofuels* **2018**, *9* (2), 273–289.
45. Singh, A.; Gaurav, K. Advancement in Catalysts for Transesterification in the Production of Biodiesel: A Review. *J. Biochem. Technol.* **2018**, *9* (1), 17–27.
46. Lan, N. V.; Hsiao, M. C. Improvement of Biodiesel Transesterification Efficiency by Activated Carbon and Microwave Power. In *Proceedings of 2018, 4th International Conference on Green Technology and Sustainable Development (GTSD), Ho Chi Minh city, Vietnam*; 2018.
47. Zhu, H. C.; Hong, T.; Yang, F. Microwave-Assisted Continuous Flow Process of Biodiesel Production. In *Asia-Pacific Microwave Conference Proceedings (APMC), Kyoto, Japan, Nov. 06–09*; 2018.

48. Utubira, Y.; Wijaya, K.; Triyonol; Kunarti, E. S. Kalium Hydroxide/Zirconia Pillared Bentonite for Palm Oil Transesterification. *Orient. J. Chem.* **2018**, *34* (3), 1484–1491.
49. Katre, G.; Raskar, S.; Zinjarde, S.; Kumar, V. R.; Kulkarni, B. D. Optimization of the In Situ Transesterification Step for Biodiesel Production Using Biomass of *Yarrowia lipolytica* NCIM 3589 Grown on Waste Cooking Oil. *Energy* **2018**, *142*, 944–952.
50. Nogueira, D. A.; da Silverira, J. M.; Vidal, E. M.; Ribeiro, N. T.; Burkert, C. A. V. Cell Disruption of *Chaetoceros Calcitrans* by Microwave and Ultrasound in Lipid Extraction. *Int. J. Chem. Eng.* **2018**, 1–6.
51. Qadariyah, L.; Mahfud, P. R.; Mujaddid, F.; Kalsum, U. Biodiesel Production from Dry Microalga Biomass by Microwave-Assisted In-Situ Transesterification. In *MATEC Web of conferences 2018, 156, 06005, RSCE*; 2017.
52. Lee, H.; Liao, J. D.; Yang, J. W.; Hsu, W. D.; Liu, B. H.; Chen, T. C.; Sivashanmugan, K.; Gedanken, A. *Chem. Eng. Technol.* **2018**, *41* (1), 192–198.
53. Groisman, Y.; Gedanken, A. Continuous Flow, Circulating Microwave System and Its Application in Nano Particle Fabrication and Biodiesel Synthesis. *J. Phys. Chem. C* **2008**, *112*, 8802–8808.
54. Santoso, A.; Sumari, S.; Sukarianingsih, D.; Sari, R. M. Optimization of Synthesis of Biodiesel From Jatropha Curcas L. With Heterogeneous Catalyst of CaO and MgO by Transesterification Reaction Using Microwave. *J. Phys. Conf. Ser.* **2018**, *1093*, 1–7.
55. Saimon, N. N.; Eu, H. K.; Johari, A.; Ngadi, N.; Josoh, M.; Zakaria, Z. Y. Production of Biodiesel From Palm Fatty Acid Distillate by Microwave-Assisted Sulfonated Glucose Acid Catalyst. *Sains Malays.* **2018**, *47* (1), 109–115.
56. Wang, Z.; Yong, X.; Mengyu, L.; Kaibing, X.; Jingling, W.; Huanhuan, S.; Jiaheng, T. Preparation of Biodiesel From Corn Oil Assisted by Microwave and Reaction Kinetics of Esterification. *Chin. J. Process. Eng.* **2018**, *18* (4), 845–850.
57. Wang, Y.; Zhang, Q. T.; Sun, Y.; Yang, C. J. Study on Suitable Light Conditions and Efficient Lipid Extraction Technologies for Biodiesel Production Based on Microalgae. *Iop conf. Ser. Mater. Sci. Eng.* **2018**, *301*, 1–9.
58. Kalsum, U.; Mahmuddin; Mahfud, M.; Roesyadi, A. Biodiesel Production From *Chlorella Vulgaris* Via Homogenous Acid Catalyzed In Situ Transesterification With Microwave Irradiation. *Iop conf. Ser. Mater. Sci. Eng.* **2018**, *175*, 1–8.
59. Fatimah, I.; Rubiyanto, D.; Yudha, S. P. Effect of KF Modification to Kaolinite Catalytic Activity in Microwave-Assisted Biodiesel Conversion. *Egypt. J. Chem.* **2018**, *61* (1), 213–223.
60. Ani, F. N.; Said, N. H.; Said, M. F. M. Optimization of Biodiesel Production Using a Stirred Packed-Bed Reactor. *Int. J. Technol.* **2018**, *2*, 219–228.
61. Inayat, A.; Ghani, C.; Jamil, F.; Alobaidli, A. S.; Bawazir, H. M.; Ali, N. A. A. Biodiesel Production from Date Seeds via Microwave Assisted Technique. In *5th International conference on Renewable energy: Generation and applications (ICREGA)*; 2018; pp. 61–63.
62. Yuan, H.; Ma, X. Q.; He, J.; Dong, Z. Y. Surface Characterization of Sulfated Iron Oxide and its Synthesis of Biodiesel under Microwave Radiation. *Int. J. Chem. React. Eng.* **2018**, *16* (2), 1–13.
63. Xiao, S.; Ju, L. K. Phagotrophic Microalgae Production From Waste Activated Sludge under Non-Sterile Conditions. *Water Res.* **2018**, *145*, 190–197.
64. Yao, S.; Mettu, S.; Law, S. Q. K.; Ashok, K. M.; Martin, G. J. O. The Effect of High-Intensity Ultrasound on Cell Disruption and Lipid Extraction from High-Solids Viscous Slurries of Nannochloropsis sp. Biomass. *Algal Res.* **2018**, *35*, 341–348.
65. Badnore, A. U.; Jadhav, N. L.; Pinjari, D. V.; Pandit, A. B. Efficacy of Newly Developed Nano-Crystalline Calcium Oxide Catalyst for Biodiesel Production. *Chem. Eng. Process. Process Intensif.* **2018**, *133*, 312–319.
66. Risso, R.; Ferraz, P.; Meireles, S.; Fonseca, I.; Vital, J. Highly Active CaO Catalysts From Waste Shells of Egg, Oyster and Clam for Biodiesel Production. *Appl. Catal. A. Gen.* **2018**, *567*, 56–64.

67. Sung, M.; Han, J. I. Ultrasound-Assisted In-Situ Transesterification of Wet *Aurantiochytrium sp.* KRS 101 Using Potassium Carbonate. *Bioresour. Technol.* **2018**, *261*, 117–121.
68. Prasad, D. M. R.; Kong, V. V.; Sathiyamoorthy, D.; Sulaiman, A. Z.; Senthilkumar, R.; Yahya, N. W. Sonication Assisted Transesterification of Jatropha Oil for the Production of Biodiesel: A Kinetic Study. In *7th Brunei International Conference on Engineering and Technology 2018, BICET 2018, Bandar Seri Begawan, Bruner, 12-14, Nov*; 2018.
69. Laosuttiwong, T.; Ngaosuwan, K.; Kiatkittipong, W.; Wongsawaeng, D.; Kim-Lohsoontom, P.; Assabumrungrat, S. Performance Comparison of Different Cavitation Reactors for Biodiesel Production Via Transesterification of Palm Oil. *J. Clean. Prod.* **2018**, *205*, 1094–1101.
70. Chen, J. J.; Lee, Y. R. Optimization of the Transesterification Reaction of Microalga *Monoraphidium sp. Renew. Energy* **2018**, *129*, 717–723.
71. Malani, R. S.; Sardar, H.; Malviya, Y.; Goyal, A.; Moholkar, V. S. Ultrasound-Intensified Biodiesel Production From Mixed Nonedible Oil Feedstock Using Heterogeneous Acid Catalyst Supported on Rubber de-Oiled Cake. *Ind. Eng. Chem. Res.* **2018**, *58*, 14926–14938.
72. Zenevicz, M. C. P.; Jacques, A.; Silva, M. J. A.; Furigo, A., Jr.; Oliveira, V.; de Oliveira, D. Study of a Reactor Ultrasound Bath for Esters Production. *Bioprocess Biosyst. Eng.* **2018**, *41*, 1589–1597.
73. Poppe, J. K.; Matte, C. R.; Fernandez-Lafuente, R.; Rodrigues, R. C.; Ayub, M. A. Z. Transesterification of Waste Frying Oil and Soybean Oil by Combi-Lipases under Ultrasound-Assisted Reactions. *Appl. Biochem. Biotechnol.* **2018**, *186*, 576–589.
74. Yu, Z.; Pei, H. Y.; Hou, Q. J.; Nie, C. L.; Zhang, L. J.; Yang, Z. G.; Wang, X. D. The Effects of Algal Extracellular Substances on Algal Growth, Metabolism and Long-Term Medium Recycle, and Inhibition on Alleviation through Ultrasonication. *Bioresour. Technol.* **2018**, *267*, 192–200.
75. Encinar, J. M.; Pardal, A.; Sanchez, N.; Nogales, S. Biodiesel by Transesterification of Rapeseed Oil Using Ultrasound: A Kinetic Study of Base-Catalysed Reactions. *Energies* **2018**, *11* (9), 1–13.
76. Cercado, A. P.; Ballesteros, F., Jr.; Capareda, S. Ultrasound Assisted Transesterification of Microalgae Using Synthesized Novel Catalyst. *Sustain. Environ. Res.* **2018**, *28*, 234–239.
77. Ido, A. L.; de Luna, M. D. G.; Capareda, S. C.; Maglinao, A. L., Jr.; Nam, H. Application of Central Composite Design in the Optimization of Lipid Yield From *Scenedesmus obliquus* Microalgae by Ultrasound-Assisted Solvent Extraction. *Energy* **2018**, *157*, 949–956.
78. Baesso, R. M.; Oliveira, P. A.; Morais, G. C.; Alvarenga, A. V.; Costa-Felix, R. P. B. Using Ultrasonic Velocity for Monitoring and Analyzing Biodiesel Production. *Fuel* **2018**, *226*, 389–399.
79. Figueiredo, M. K. K.; Silva, C. E. R.; Alvarenga, A. V.; Costa-Felix, R. P. B. Relating Speed of Sound and Echo Amplitude With Biodiesel Manufacturing. *Chem. Eng. Res. Des.* **2018**, *136*, 825–833.
80. Phong, W. N.; Show, P. L.; Le, C. F.; Tao, Y.; Chang, J. S.; Ling, T. C. Improving Cell Disruption Efficiency to Facilitate Protein Release From Microalgae Using Chemical and Mechanical Integrated Method. *Biochem. Eng. J.* **2018**, *135*, 83–90.
81. Chang, C. C.; Teng, S.; Yuan, M. H.; Ji, D. R.; Chang, C. Y.; Chen, Y. H.; Shie, J. L.; Ho, C.; Tian, S. Y.; Andrade-Tacca, C. A.; Manh, D. V.; Tsai, M. Y.; Chang, M. C.; Chen, Y. H.; Huang, M.; Liu, B. L. Esterification of Jatropha Oil with Isopropanol Via Ultrasonic Irradiation. *Energies* **2018**, *11*, 1–15.
82. Gude, V. G.; Martinez-Guerra, E. Green Chemistry with Process Intensification for Sustainable Biodiesel Production. *Environ. Chem. Lett.* **2018**, *16* (2), 327–341.

83. Liao, J.; Gedanken, A.; Chen, T.; Cooley, R. C.; Liao, J. D.; Chen, T. C. *Generation of Biodiesel Used for Diesel Vehicle Engines Involves Adhering Nanoparticles Each Containing Alkali or Alkaline Earth Metal Compound on Support, Forming Composite, Contacting With Target and Performing Transesterification*; 2014. Patent Nos. US2015184096-A1; WO2015103341-A1; TW201525128-A; TW496882-B1.
84. Koberg, M.; Abu-Much, R.; Gedanken, A. Optimization of Biodiesel Production From Soybean and Wastes of Cooked Oil: Combining Dielectric Microwave Irradiation and a SrO Catalyst. *Bioresour. Technol.* **2011**, *102*, 1073–1078.
85. Koberg, M.; Cohen, M.; Ben-Amotz, A.; Gedanken, A. Biodiesel Production Directly From the Microalgae Biomass of *Nannochloropsis* by Microwave and Ultrasound Radiation. *Bioresour. Technol.* **2011**, *102*, 4265–4269.
86. Koberg, M.; Gedanken, A. Direct Transesterification of Castor and Jatropha Seeds for FAME Production by Microwave and Ultrasound Radiation Using a SrO Catalyst. *Bioenergy Res.* **2012**, *5*, 958–968.
87. Koberg, M.; Gedanken, A. Optimization of Biodiesel Production From Oils, Cooking Oil, Microalgae, and Castor and Jatropha Seeds: Probing Various Heating Sources and Catalysts. *Energy Environ. Sci.* **2012**, *5*, 7460–7469.
88. Priambodo, R.; Chen, T. C.; Lu, M. C.; Gedanken, A.; Liao, J. D.; Huang, Y. H. Novel Technology for Biodiesel Production from Cooking and Waste Cooking Oil by Microwave Irradiation. The 7th International Conference on Applied Energy—ICAE 2015, *Energy Procedia* **2015**, *75*, 84–91.
89. Naor, E. O.; Koberg, M.; Gedanken, A. Nonaqueous Synthesis of SrO Nanopowder and SrO/SiO_2 Composite and their Application for Biodiesel Production Via Microwave Irradiation. *Renew. Energy* **2017**, *101*, 493–499.
90. Piker, A.; Tabah, B.; Perkas, N.; Gedanken, A. A Green and Low-Cost Room Temperature Biodiesel Production Method From Waste Oil Using Egg Shells as Catalyst. *Fuel* **2016**, *182*, 34–41.
91. Tangy, A.; Pulidindi, I. N.; Gedanken, A. SiO_2 Beads Decorated With SrO Nanoparticles for Biodiesel Production From Waste Cooking Oil Using Microwave Irradiation. *Energy Fuel* **2016**, *30*, 3151–3160.
92. Tangy, A.; Kumar, V. B.; Pulidindi, I. N.; Kinel-Tahan, Y.; Yehoshua, Y.; Gedanken, A. In-Situ Transesterification of Chlorella vulgaris Using Carbon-Dot Functionalized Strontium Oxide as a Heterogeneous Catalyst under Microwave Irradiation. *Energy Fuel* **2016**, *30*, 10602–10610.
93. Tangy, A.; Pulidindi, I. N.; Perkas, N.; Gedanken, A. Continuous Flow Through a Microwave Oven the Large-Scale Production of Biodiesel From Waste Cooking Oil. *Bioresour. Technol.* **2017**, *224*, 333–341.
94. Tabah, B.; Nagvenkar, A. P.; Perkas, N.; Gedanken, A. Solar-Heated Sustainable Biodiesel Production From Waste Cooking Oil Using a Sonochemically Deposited SrO Catalyst on Microporous Activated Carbon. *Energy Fuel* **2017**, *31*, 6228–6239.
95. Masri, A. N.; Mutalib, M. I. A.; Aminuddin, N. F.; Leveque, J. M. Novel SO_3H-Functionalized Dicationic Ionic Liquids—A Comparative Study for Esterification Reaction by Ultrasound Cavitation and Mechanical Stirring for Biodiesel Production. *Sep. Purif. Technol.* **2018**, *196*, 106–114.
96. Gawas, D. S.; Rathod, V. K. Enhancement in Synthesis of Ethyl Laurate Catalyzed by Fermase by Combined Effect of Ultrasound and Stage Wise Addition of Ethanol. *Chem. Eng. Process. Process Intensif.* **2015**, *125*, 207–213.
97. Sivaramakrishnan, R.; Incharoensakdi, A. Microalgae as Feedstock for Biodiesel Production Under Ultrasound Treatment—A Review. *Bioresour. Technol.* **2018**, *250*, 877–887.
98. Joshi, S. M.; Gotage, P. R.; Kumar, S. S. Intensification of Esterification of Karanja Oil for Production of Biodiesel Using Ultrasound Assisted Approach with Optimization Using Response Surface Methodology. *Chem. Eng. Process. Process Intensif.* **2018**, *124*, 186–198.

99. Korkut, I.; Bayramoglu, M. Selection of Catalyst and Reaction Conditions for Ultrasound Assisted Biodiesel Production From Canola Oil. *Renew. Energy* **2018**, *116*, 543–551.
100. Thirugnanasambandham, K. Application of Plackett-Burman Design to Screen the Effective Process Parameters for Biodiesel Production. *Energy Sources, Part A* **2018**, *40*, 453–458.
101. Tan, S. X.; Ong, H. C.; Lim, S.; Pang, Y. L. In Situ Reactive Extraction of Jatropha Curcas L. Seeds Assisted by Ultrasound: Preliminary Studies. *Energy Sources, Part A* **2018**, *40* (14), 1772–1779.
102. Ulfah, M.; Mulyazmi, B.; Praputri, E.; Sundari, E.; Firdaus. Biodiesel production methods of rubber seed oil: a review. *Iop conf. Ser. Mater. Sci. Eng.* **2018**, *334*, 1–7.
103. Sewsynker-Sukai, Y.; Kana, E. B. G. Microwave-Assisted Alkali Salt Pretreatment of Corn Cob Wastes: Process Optimization for Improved Sugar Recovery. *Ind. Crop Prod.* **2018**, *125*, 284–292.
104. Amin, N.; Haritos, V. S.; Tanksale, A. Microwave Assisted Pretreatment of Eucalyptus Saw Dust Enhances Enzymatic Saccharification and Maximizes Fermentable Sugar Yield. *Renew. Energy* **2018**, *127*, 653–660.
105. Wu, M. J.; Liu, H. Y.; Guo, J. Y.; Yang, C. P. Enhanced Enzymatic Hydrolysis of Wheat Straw by Two-Step Pretreatment Combining Alkalization and Adsorption. *Appl. Microbiol. Biotechnol.* **2018**, *102* (22), 9831–9842.
106. Rigual, V.; Santos, T. M.; Dominguez, J. C.; Alonso, M. V.; Oliet, M.; Rodriguez, F. Evaluation of Hardwood and Softwood Fractionation Using Autohydrolysis and Ionic Liquid Microwave Pretreatment. *Biomass Bioenergy* **2018**, *117*, 190–197.
107. Boonsombuti, A.; Wanapiran, R.; Luengnaruemitchai, A.; Wongkasenijit, S. The Effect of the Addition of Acetic Acid to Aqueous Ionic Liquid Mixture Using Microwave-Assisted Pretreatment in the Saccharification of Napier Grass. *Waste Biomass Valoriz.* **2018**, *9* (10), 1795–1804.
108. Onumaegbu, C.; Mooney, J.; Alaswad, A.; Olabi, A. Pretreatment Methods for Production of Biofuel From Microalgae Biomass. *Renew. Sustain. Energy Rev.* **2018**, *93*, 16–26.
109. Hassan, S. S.; Williams, G. A.; Jaiswal, A. K. Emerging Technologies for the Pretreatment of Lignocellulosic Biomass. *Bioresour. Technol.* **2018**, *262*, 310–318.
110. Sindhu, R.; Binod, P.; Gnansounou, E.; Prabisha, T. P.; Thomas, L.; Mathew, A. K.; Abraham, A.; Pandey, A.; Faraco, V. Enzymatic Hydrolysis of Microwave Assisted Acid Pretreated Chili Postharvest Residue for the Production of Value Added Products. *Indian J. Exp. Biol.* **2018**, *56* (7), 479–483.
111. Laghari, S. M.; Tunio, M. M.; Laghari, A. Q.; Laghari, A. J.; Ali, Z. M. Delighnification of Rice-Husk by Microwave Assisted Chemical Pretreatment. *Eng. Technol. Appl. Sci. Res.* **2018**, *8* (3), 3084–3087.
112. See, T. Y.; Yusoff, R.; Chan, C. H.; Ngoh, G. C. A Solid-State Microwave Method to Disrupt Biomass Microstructure for Natural Product Extraction. *Food Bioprod. Process.* **2018**, *109*, 98–106.
113. Issac, A.; de Paula, J.; Viana, C. M.; Henriques, A. B.; Malachias, A.; Montoro, C. A. From Nano-to Micrometer Scale: The Role of Microwave—Assisted Acid and Alkali Pretreatments in the Sugar Cane Biomass Structure. *Biotechnology for Biofuels* **2018**, *11*, 73. 1–11.
114. Kamalini, A.; Muthusamy, S.; Ramapriya, R.; Muthusamy, B.; Pugazhendhi, A. Optimization of Sugar Recovery Efficiency Using Microwave Assisted Alkaline Pretreatment of Cassava Stem Using Response Surface Methodology and its Structural Characterization. *J. Mol. Liq.* **2018**, *254*, 55–63.
115. Rigual, V.; Santos, T. M.; Dominguez, J. C.; Alonso, M. V.; Oliet, M.; Rodriguez, F. Combining Autohydrolysis and Ionic Liquid Microwave Treatment to Enhance Enzymatic Hydrolysis of Eucalyptus Globulus Wood. *Bioresour. Technol.* **2018**, *251*, 197–203.

116. Mitani, T. Recent Progress on Microwave Processing of Biomass for Bioenergy Production. *J. Jpn. Pet. Inst.* **2018**, *61* (2), 113–120.
117. Chen, Z.; Wan, C. X. Ultrafast Fractionation of Lignocellulosic Biomass by Microwave-Assisted Deep Eutectic Solvent Pretreatment. *Bioresour. Technol.* **2018**, *250*, 532–537.
118. Pooja, N. S.; Sajeev, M. S.; Jeeva, M. L.; Padmaja, G. Bioethanol Production From Microwave Assisted Acid or Alkali Pretreated Agricultural Residues of Cassava Using Separate Hydrolysis and Fermentation (SHF). *3 Biotech* **2018**, *8* (69), 1–12.
119. Ravindran, R.; Jaiwal, S.; Abu-Ghannam, N.; Jaiswal, A. K. A Comparative Analysis of Pretreatment Strategies on the Properties and Hydrolysis of Brewer's Spent Grain. *Bioresour. Technol.* **2018**, *248*, 272–279.
120. Fatriasari, W.; Raniya, R.; Oktaviani, M.; Hermiati, E. The Improvement of Sugar and Bioethanol Production of Oil Palm Empty Fruit Bunches (*Elaeis guineensis Jacq*) Through Microwave-Assisted Maleic Acid Pretreatment. *BioResources* **2018**, *13* (2), 4378–4403.
121. Dai, L.; He, C.; Wang, Y.; Liu, Y.; Ruan, R.; Yu, Z.; Zhou, Y.; Duan, D.; Fan, L.; Zhao, Y. Hydrothermal Pretreatment of Bamboo Sawdust Using Microwave Irradiation. *Bioresour. Technol.* **2018**, *247*, 234–241.
122. Li, T.; Remon, J.; Jiang, Z.; Budarin, V. L.; Clark, J. H. Towards the Development of a Novel 'Bamboo-Refinery' Concept: Selective Bamboo Fractionation by Means of a Microwave-Assisted Acid-Catalyzed Organosolv Process. *Energ. Conver. Manage.* **2018**, *155*, 147–160.
123. Liu, C.; Zhao, Q.; Lin, Y.; Hu, Y.; Wang, H.; Zhang, G. Characterization of Aqueous Products Obtained from Hydrothermal Liquefaction of Rice Straw: Focus on Product Comparison Via Microwave-Assisted and Conventional Heating. *Energy Fuel* **2018**, *32*, 510–516.
124. Klinger, R.; Goroma, T. Growth Media Affects Microalgae Susceptibility to Disruption by Low-Frequency Power Ultrasound. *J. Energy Resour. Technol.* **2018**, *140* (12), 1–5.
125. Velazquez-Lucio, J.; Rodriguez-Jasso, R. M.; Colla, L. M.; Saenz-Galindo, A.; Cervantes-Cisneros, D. E.; Aguilar, C. N.; Fernandes, B. D.; Ruiz, H. A. Microalgal Biomass Pretreatment for Bioethanol Production: A Review. *Biofuel Res. J.* **2018**, *5* (1), 780–791.
126. Manasa, P.; Saroj, P.; Korrapati, N. Ultrasound-Assisted Alkaline Pretreatment to Intensify Enzymatic Saccharification of Crotalaria Using a Statistical Method. *Biomass Convers. Biorefin.* **2018**, *8*, 659–668.
127. Gaudino, E. C.; Tabasso, S.; Grillo, G.; Cravotto, G.; Dreyer, T.; Schories, G.; Altenberg, S.; Jashina, L.; Telysheva, G. Wheat Straw Lignin Extraction with Bio-Based Solvents Using Enabling Technologies. *C. R. Chim.* **2018**, *21* (6), 563–571.
128. Yu, X. J.; Bao, X. J.; Zhou, C. S.; Zhang, L.; Yagoub, A. E. A.; Yang, H. P.; Ma, H. L. Ultrasound-Ionic Liquid Enhanced Enzymatic and Acid Hydrolysis of Biomass Cellulose. *Ultrason. Sonochem.* **2018**, *41*, 410–418.
129. Subhedar, P. B.; Ray, P.; Gogate, P. R. Intensification of Delignification and Subsequent Hydrolysis for the Fermentable Sugar Production From Lignocellulosic Biomass Using Ultrasonic Irradiation. *Ultrason. Sonochem.* **2018**, *40*, 140–150.
130. Fan, J. J.; Santomauro, F.; Budarin, V. L.; Whiffin, F.; Abeln, F.; Chantasuban, T.; Gore-Lloyd, D.; Henk, D.; Scott, R. J.; Clark, J.; Chuck, C. J. The Additive Free Microwave Hydrolysis of Lignocellulosic Biomass for Fermentation to High Value Products. *J. Clean. Prod.* **2018**, *198*, 776–784.
131. Lin, Y. C.; Shangdiar, S.; Chen, S. C.; Chou, F. C.; Lin, Y. C.; Cho, C. A. Microwave Irradiation with Dilute Acid Hydrolysis Applied to Enhance the Saccharification Rate of Water Hyacinth (*Eichhornia Crassipes*). *Renew. Energy* **2018**, *125*, 511–517.

132. Shet, V. B.; Shetty, V. C.; Siddik, A.; Rakshith, K. G.; Shetty, N. J.; Goveas, L. C.; D'Mello, G.; Rao, C. V.; Ujwal, P.; Aparna, A. Optimization of Microwave Assisted H_2SO_4 Hydrolysis of Cocoa Pod Shells: Comparison Between Response Surface Methodology and Artificial Neural Network and Production of Bioethanol Thereof. *J. Microbiol. Biotechnol. Food Sci.* **2018**, 7 (5), 473–477.
133. Chen, J. Y.; Zhang, C.; Li, M. L.; Chen, J. M.; Wang, Y. D.; Zhou, F. F. Microwave-Enhanced Sub-Critical Hydrolysis of Rice Straw to Produce Reducing Sugar Catalyzed by Ionic Liquid. *J. Mater. Cycles Waste Manage.* **2018**, *20* (2), 1364–1370.
134. Ma, H. Y.; Zhao, Z. P.; Lu, P. Cellulose Hydrolysis by Acidic Ionic Liquids Enchanced With Microwave Heating. *Adv. Mat. Res.* **2018**, *1145*, 75–79.
135. Suzuki, S.; Takeoka, Y.; Rikukawa, M.; Yoshizawa-Fujita, M. Bronsted Acidic Ionic Liquids for Cellulose Hydrolysis in an Aqueous Medium: Structural Effects on Acidity and Glucose Yield. *RSC Adv.* **2018**, *8* (26), 14623–14632.
136. Zhai, C. K.; Teng, N.; Pan, B. H.; Chen, J.; Liu, F.; Zhu, J.; Na, H. N. Revealing the Importance of Non-thermal Effect to Strengthen Hydrolysis of Cellulose by Synchronous Cooling Assisted Microwave Driving. *Carbohydr. Polym.* **2018**, *197*, 414–421.
137. Hou, F.; Fan, L.; Ma, X.; Wang, D.; Wang, W.; Ding, T.; Ye, X.; Liu, D. Degradation of Carboxymethyl Cellulose Using Ultrasound and β-Glucanase: Pathways, Kinetics and Hydrolysates Properties. *Carbohydr. Polym.* **2018**, *201*, 514–521.
138. Grande, R.; Trovatti, E.; Pimenta, M. T. B.; Carvalho, A. J. F. Microfibrillated Cellulose From Sugarcane Bagasse as a Biorefinery Product for Ethanol Production. *J. Renew. Mater.* **2018**, *6* (2), 195–202.
139. Bundhoo, Z. M. A.; Mohee, R. Ultrasound-Assisted Biological Conversion of Biomass and Waste Materials to Biofuels: A Review. *Ultrason. Sonochem.* **2018**, *40*, 298–313.
140. Calinescu, L.; Vlaicu, A.; Chipurici, P.; Ighigeanu, D.; Lavric, V. Alcoholic Fermentation in the Presence of Microwaves. *Chem. Eng. Process. Process Intensif.* **2018**, *126*, 16–22.
141. Tizazu, B. Z.; Roy, K.; Moholkar, V. S. Ultrasonic Enhancement of Xylitol Production From Sugarcane Bagasse Using Immobilized *candida tropicalis* MTCC 184. *Bioresour. Technol.* **2018**, *268*, 247–258.
142. Tizazu, B. Z.; Roy, K.; Moholkar, V. S. Mechanistic Investigation in Ultrasound-Assisted Xylitol Fermentation. *Ultrason. Sonochem.* **2018**, *48*, 321–328.
143. Luo, D. L.; Wu, R. Y.; Zhang, J.; Zhang, K. Y.; Xu, B. C.; Li, P. Y.; Yuan, Y. X.; Li, X. Effects of Ultrasound Assisted Dough Fermentation on the Quality of Steamed Bread. *J. Cereal Sci.* **2018**, *83*, 147–152.
144. Leandro, A. D. G.; Ramiro, A. C. Fermentation Assisted by Pulsed Electric Field and Ultrasound: A Review. *Fermentation* **2018**, *4* (1), 1–12.
145. Celik, D. A.; Amer, M. A.; Novoa-Diaz, D. F.; Chavez, J. A.; Turo, A.; Garcia-Hernandez, M. J.; Salazar, J. Design and Implementation of an Ultrasonic Sensor for Rapid Monitoring of Industrial Malolactic Acid Fermentation Wines. *Instrum. Sci. Technol.* **2018**, *46* (4), 387–407.
146. Pulidindi, I. N.; Gedanken, A.; Schwarz, R.; Sendersky, E. Mild Sonication Accelerates Ethanol Production by Yeast Fermentation. *Energy Fuel* **2012**, *26*, 2352–2356.
147. Wada, Y.; Fujii, S.; Tsubaki, S. Activation of Chemical Reactions on Solid Catalysts Under Microwave Irradiation. In *Non-traditional Activation Methods in Green and Sustainable Applications: Microwaves, Ultrasound, Photo, Electro and Mechanochemistry and High Hydrostatic Pressure*; Torok, B., Schaefer, C., Eds.; Elsevier: Amsterdam, 2020.
148. Pulidindi, I. N.; Kimchi, B. B.; Gedanken, A. Can Cellulose Be a Sustainable Feedstock for Bioethanol Production? *Renew. Energy* **2014**, *71*, 77–80.
149. Korzen, L.; Pulidindi, I. N.; Israel, A.; Abelson, A.; Gedanken, A. Single Step Production of Bioethanol From the Seaweed *Ulva rigida* Using Sonication. *RSC Adv.* **2015**, *5*, 16223–16229.

150. Tabah, B.; Pulidindi, I. N.; Gedanken, A. A Study on Fermentation Kinetics for Accelerated Production of Bioethanol From Glucose, Sucrose, and Molasses. *J. Bioprocess. Biotech.* **2015**, *5* (232), 1–7.
151. Korzen, L.; Pulidindi, I. N.; Israel, A.; Abelson, A.; Gedanken, A. Marine Integrated Culture of Carbohydrate Rich *Ulva rigida* for Enhanced Production of Bioethanol. *RSC Adv.* **2015**, *5*, 59251–59256.
152. Victor, A.; Pulidindi, I. N.; Gedanken, A. Assessment of Holocellulose for the Production of Bioethanol by Conserving *Pinus radiata* Cones as Renewable Feedstock. *J. Environ. Manage.* **2015**, *162*, 215–220.
153. Tabah, B.; Pulidindi, I. N.; Chitturi, V. R.; Arava, L. M. R.; Gedanken, A. Solar Energy Driven Simultaneous Saccharification and Fermentation (SSF) of Starch to Bioethanol for Fuel Cell Applications. *ChemSusChem* **2015**, *8*, 3497–3503.
154. Kumar, V. B.; Pulidindi, I. N.; Kinel, Y.; Yehoshuva, Y.; Gedanken, A. Evaluation of the Potential of Chlorella Vulgaris for Bioethanol Production. *Energy Fuel* **2016**, *30*, 3161–3166.
155. Tabah, B.; Pulidindi, I. N.; Chitturi, V. R.; Arava, L. M. R.; Gedanken, A. Utilization of Solar Energy for Continuous Production of Bioethanol for Energy Applications. *RSC Adv.* **2016**, *6*, 24203–24209.
156. Klein, M.; Griess, O.; Pulidindi, I. N.; Perkas, N.; Gedanken, A. Bioethanol Production From *Ficus religiosa* Leaves. *J. Environ. Manage.* **2016**, *177*, 20–25.
157. Tabah, B.; Pulidindi, I. N.; Chitturi, V. R.; Arava, L. M. R.; Varvak, A.; Foran, E.; Banin, E.; Gedanken, A. Solar-Energy-Driven Conversion of Biomass to Bioethanol: A Sustainable Approach. *J. Mater. Chem. A* **2017**, *5*, 15486–15506.
158. Pulidindi, I. N.; Gedanken, A. Employing Novel Techniques (Microwave and Sonochemistry) in the Synthesis of Biodiesel and Bioethanol. In *Springer Book Series—Production of Biofuels and Chemicals: Ultrasound*; Fang, Z., Fan, L. S., Grace, J. R., Ni, Y. H., Scott, N. R., Smith, L. R., Jr., Eds.; Elsevier: Amsterdam, 2015; pp. 159–188.
159. Tabaha, B.; Pulidindi, I. N.; Chitturi, V. R.; Arava, L. M. R.; Gedanken, A. Solar-Energy Driven Bioethanol Production From Carbohydrates for Transportation Applications. In *Solar Energy: Systems, Performance and Recent Developments*; Carter, J. G., Ed.; Nova Science Publishers, Inc, 2017. ISBN:978-1-53610-408-0.
160. Pulidindi, I. N.; Korzen, L.; Tabah, B.; Victor, A.; Israel, A.; Abelson, A.; Gedanken, A. Potential Applications of *Ulva rigida* for Biofuels and Biochemicals Production. In *Seaweeds: Biodiversity, Environmental Chemistry and Ecological Impacts*; Newton, P., Ed.; Nova Science Publishers, Inc, 2017. ISBN:978-1-53611-895-7.
161. Pulidindi, I. N.; Gedanken, A. *Can Biofuels Alleviate the Energy and Environmental Crisis?*; Nova Science Publishers, Inc: New York, 2019. ISBN:978-1-53615-050-6.
162. Gupta, A. R.; Jalan, A. P.; Rathod, V. K. Solar Energy as a Process Intensification Tool for the Biodiesel Production From Hempseed Oil. *Energ. Conver. Manage.* **2018**, *171*, 126–132.
163. Quitain, A. T.; Mission, E. G.; Sumigawa, Y.; Sasaki, M. Supercritical Carbon Dioxide Mediated Esterification in a Microfluidic Reactor. *Chem. Eng. Process. Process Intensif.* **2018**, *123*, 168–173.
164. Melo-Junior, C. A.; Albuquerque, C. E.; Fortuny, M.; Dariva, C.; Egues, S.; Santos, A. F.; Ramos, A. L. Use of Microwave Irradiation in the Noncatalytic Esterification of C18 Fatty Acids. *Energy Fuel* **2009**, *23* (1), 580–585.
165. Gedanken, A.; Pulidindi, I. N.; Tabah, B. *Solar-Aided Conversion of Marine Algae and Biomass to Bioethanol*; US Provisional Patent Application No 62450107, 2017.
166. Victor, I. N.; Pulidindi, T. H. Design of Active Solid Acid Catalyst for the Optimization of Glucose Production From *Oryza sativa* Straw. *RSC Adv.* **2016**, *6*, 31–38.

CHAPTER SEVEN

A new color in green chemistry: Photochromic molecular switches as components of multifunctional catalytic systems

István Jablonkai[a], Attila Kunfi[a], Da-Hui Qu[b], and Gábor London[a]
[a]MTA TTK Lendület Functional Organic Materials Research Group, Institute of Organic Chemistry, Research Centre for Natural Sciences, Budapest, Hungary
[b]Key Laboratory for Advanced Materials, School of Chemistry and Molecular Engineering, East China University of Science and Technology, Shanghai, China

1 Introduction

Nature harnesses light energy to precisely control chemical events that drive processes such as photosynthesis or vision. Photochromic molecular switches[1, 2] are often regarded as basic components of man-made alternatives of light-responsive biological systems. Ultimately, the aim of the field is to mimic nature in its efficiency and controlling ability with small molecules designed and synthetized in the laboratory. Controlling chemical transformations reversibly with light using photoresponsive molecules dates back to the early 1980s; however, the field started to expand only in the last few years. Initially, the goal was to control catalytic activity (on/off, high/low) with light using catalysts bearing photochromic switching units. Later, although activity control remained a major motif of research, a range of systems were developed where switches contributed to selectivity control or more efficient separation and reusability of the catalysts.

One of the main principles of green chemistry is the use of catalysts instead of stoichiometric reagents.[3–5] The principle of catalysis, however, is highly integrated with other aims of green chemistry, such as the use of environmentally benign solvents (e.g., aqueous catalysis), reduction of waste (e.g., catalyst recycling), or avoidance of protecting groups (selective catalysis). Hence, when one designs a catalytic process in the context of green applications, one should not focus on a single factor but rather on a systems approach to fulfill as many of the requirements of green chemistry as possible. Using photochromic molecules as components of multifunctional

Nontraditional Activation Methods in Green and Sustainable Applications
https://doi.org/10.1016/B978-0-12-819009-8.00006-2
© 2021 Elsevier Inc.
All rights reserved.

catalytic systems has several advantages. The structure of these molecules can be reversibly controlled by light as an external stimulus, thereby not requiring any additional material that might turn to be a waste at the end of the process. Light-induced changes in molecular structure or shape can be associated with changes in function such as catalytic activity, solubility, aggregation, or selective binding.

Considering green chemistry principles that provide the fundamental basis of sustainability, molecular level control of function with light is undoubtedly a phenomenon that will be increasingly considered and used in green synthesis and catalysis.

2 Photochromic scaffolds used in catalytic applications

There are several types of molecular switches described to date[2] and new types are continually being developed.[6–8] Although the diversity of switching scaffolds is expanding, only a limited number of photoswitches are explored systematically in catalysis.[9–13] The basic features of those that are mostly used as components of catalytic systems are summarized below.

2.1 Azobenzenes

Azobenzenes[14] (Fig. 1) are among the most frequently used photochromic units. The origin of the popularity of azobenzenes in different applications is their relatively easy syntheses and fast and reversible *trans*-to-*cis* and *cis*-to-*trans* photoisomerization that is stable against photodegradation (fatigue resistance) accompanied by large changes in size and dipole moment. This large change in size that allows for controlling the distance between functional groups on the aryl rings is the key feature that is exploited in its catalytic applications.

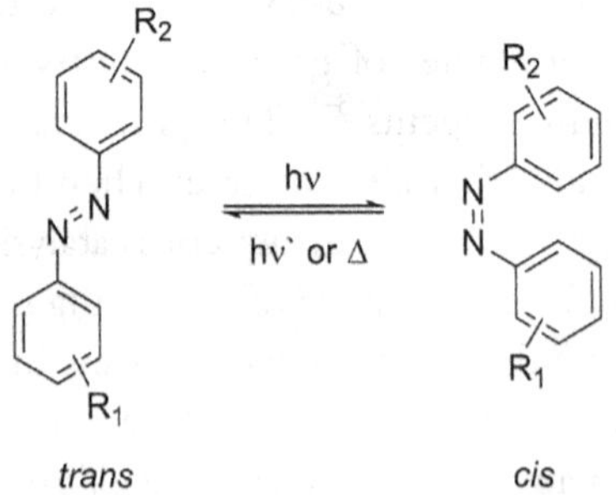

Fig. 1 Reversible photoisomerization of azobenzenes.

2.2 Dithienylethenes

1,2-Dithienylethenes[15, 16] (Fig. 2) undergo reversible photocyclization between an open form and a closed form by UV or visible light irradiation. Among the photochromic compounds, dithienylethenes have excellent thermal stability of both isomers and high fatigue resistance. The reversible change in their π-system upon photocyclization/ring opening is used to control the electronics of dithienylethene-based photochromic ligands.

2.3 Overcrowded alkene-based molecular motors

Light-powered molecular rotary motors are able to undergo controlled rotary motion which proceeds through four distinct steps in a unidirectional fashion[17, 18] (Fig. 3). Starting with the stable-*trans* state, a photochemical isomerization (step 1) generates unstable-*cis* in which the stereogenic methyl groups are forced into an unfavored pseudo-equatorial conformation. This conformational strain is released during the thermal helix inversion step upon which the molecule undergoes an isomerization to give the stable-*cis* form (step 2), in which the stereogenic methyl groups adopt the energetically favored axial orientation. A second photochemical double bond isomerization generates the unstable-*trans* form (step 3), which is followed by a second thermal helix inversion (step 4). This last step regenerates the stable-*trans* state and completes a full 360° rotation of the upper part relative to the lower part. The unidirectionality of the rotary cycle is dictated by the helicity of the overcrowded alkene and the absolute configuration of the stereogenic centers. In the majority of the catalytic applications, first-generation molecular motors are used due to the simplicity of their syntheses and functionalization.[19] Importantly, due to the complex stereochemistry that is encoded in the motors, stereoselectivity could be induced and controlled in several catalytic transformations.

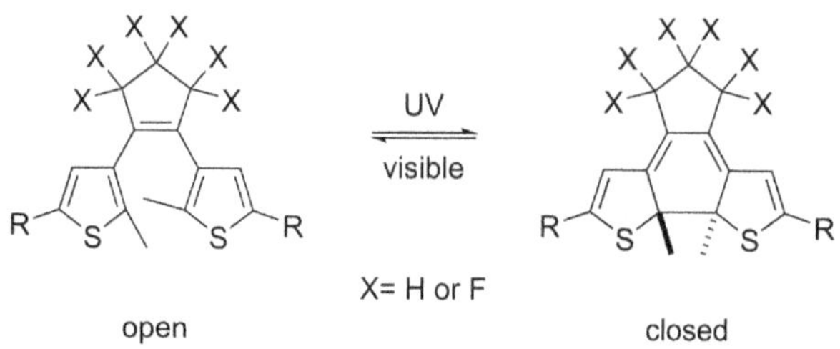

Fig. 2 Reversible photocyclization of 1,2-dithienylethenes.

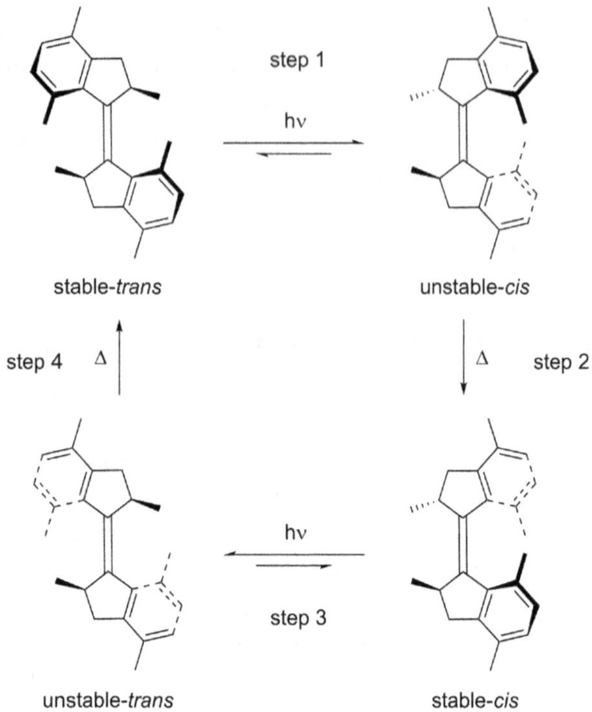

Fig. 3 The four-step rotary cycle of an overcrowded alkene-based molecular motor.

3 Homogeneous catalysis

3.1 Metal-free catalysis

3.1.1 *Early design*

The first report where the occurrence of a chemical reaction was linked to the reversible photoisomerization of an azobenzene dates back to 1980 by Ueno and coworkers.[20] The catalytic system consisted of a β-cyclodextrin (β-CD), which is active in ester hydrolysis under basic aqueous conditions (pH = 8.7, tris(hydroxymethyl)aminomethane (Tris) buffer), and an azobenzene carboxylate as a photoresponsive inhibitor (Fig. 4). Upon mixing the azobenzene and the macrocycle, the formation of an inclusion complex inhibited the hydrolysis of *p*-nitrophenyl acetate to *p*-nitrophenol which occurs preferentially inside the hydrophobic cavity of β-CD. Irradiating the mixture with UV light (320–390 nm) resulted in the formation of the *cis* form of the azobenzene (55%–65% at the photostationary state) that leaves the cavity due to its less preferential binding and provides the free

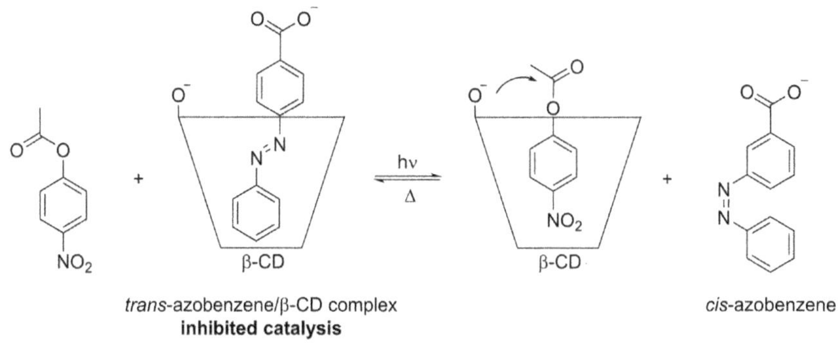

Fig. 4 Controlling ester hydrolysis by β-CD via reversible azobenzene binding.

catalyst for ester hydrolysis. The thermal isomerization of the *cis* form to the *trans* under dark conditions makes this process reversible.

Later, the same group approached the light-controlled complex formation through covalent strategies, where the inhibitor azobenzene was covalently bound to β-CD. Upon capping the β-CD macrocycle by the switch, the depth of the cavity could be controlled reversibly by light[21] (Fig. 5).

In this system, compared to the noncovalent approach, the catalysis could be better controlled as the half-life of the *cis*-to-*trans* reversion was 33 h under the experimental conditions (pH = 8.7, Tris buffer, 25°C). Although the shallow cavity (*trans*-azobenzene) provided a better geometry of the substrate for hydrolysis by the alkoxide group, due to its poorer

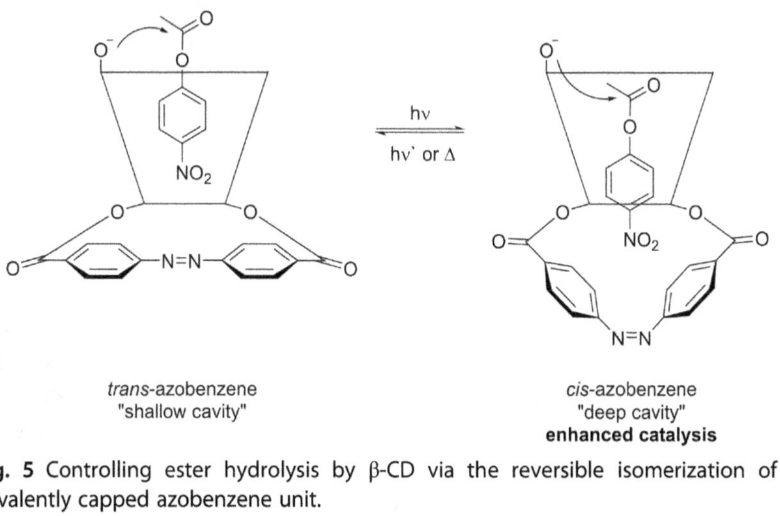

Fig. 5 Controlling ester hydrolysis by β-CD via the reversible isomerization of a covalently capped azobenzene unit.

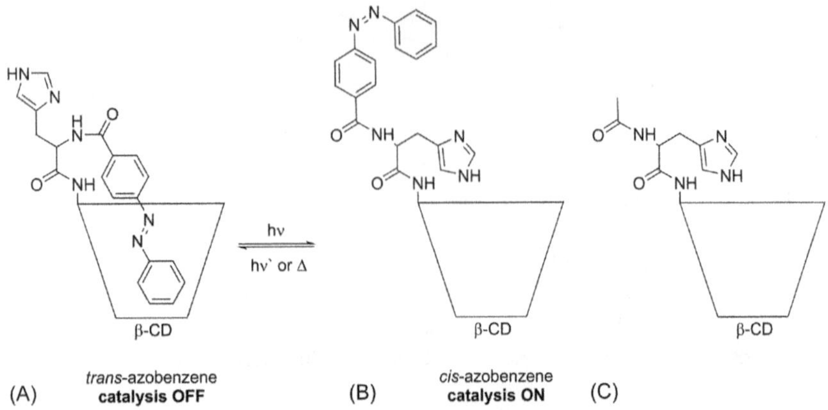

(A) (B) (C)

Fig. 6 (A) A covalent design for controlling ester hydrolysis by β-CD using light stimulus; (B) a control catalyst that is lacking the photochromic unit.

binding ability, an overall 5.5-fold rate enhancement was measured in the case of the deep cavity (*cis*-azobenzene). The catalyst was found to be selective to the hydrolysis of the *para*-nitro compound, as no reaction occurred with the *meta*-derivative. A more complex catalyst design for ester hydrolysis reactions was based on 6-amino-β-CD to which an azobenzene switch was connected via a histidine spacer[22] (Fig. 6A). Here, the catalytic function was provided by the imadozole moiety, while its activity was controlled by the photochromic unit. The authors also prepared a reference compound lacking the azobenzene moiety. The hydrolysis of *p*-nitrophenyl acetate, Boc-L-alanine-*p*-nitrophenyl ester, and Boc-D-alanine-*p*-nitrophenyl ester in phosphate buffer (pH = 7.2) at 25°C was tested.

The activity of the catalyst-containing *trans*-azobenzene was found even lower than simple imidazole in a phosphate buffer. The reason was that the azobenzene was fully included in the β-CD cavity blocking it from substrate inclusion. When the catalyst was irradiated with UV light (320–390 nm), a higher reaction rate was obtained in the ester hydrolysis reactions, indicating that substrate inclusion in the β-CD cavity was facilitated. This latter effect was due to the efficient removal of the photoinduced *cis* form of the azobenzene from β-CD. The *cis*-azobenzene exhibited a good thermal stability indicated by a half-life of 32 h at 40°C in the dark. When the *cis*-azobenzene-containing catalysts were compared to the reference catalyst that was lacking the photoisomerizable unit (Fig. 6B), similar activities were measured, supporting that the *cis*-azobenzene did not hinder the ester hydrolysis sterically. Interestingly, a stereochemical effect was also observed in the hydrolysis of Boc-L-alanine-*p*-nitrophenyl ester and Boc-D-alanine-*p*-nitrophenyl ester. The reference catalyst was found to be more active in the hydrolysis of Boc-L-alanine-*p*-nitrophenyl ester, while the *cis*-azobenzene-containing catalyst exhibited a 16-fold higher activity in the hydrolysis of Boc-D-alanine-*p*-nitrophenyl ester than the reference catalyst.

Another example for the "early design" of photoswitchable catalysis is the catalytic receptor developed by Würthner and Rebek.[23, 24] Their catalyst consisted of a photoisomerizable azobenzene spacer between two disubstituted carbazole ring systems (Fig. 7A). From their earlier studies on template effects on catalytic processes, they found that the coupling of aminoadenosine and a *p*-nitrophenyl ester derivative to give the corresponding amide (Fig. 7B) is catalyzed by similar but nonphotoactive ensembles of substituted carbazoles.[25] By incorporating the azobenzene, it was expected that the activity of the template catalyst can be controlled by light.

The azobenzene-containing receptor was found to undergo photochemical *trans*-to-*cis* isomerization upon exposure to UV light (366 nm). UV-irradiation yielded a 1:1 mixture of the two isomers, while upon visible light irradiation an equilibrium mixture of *trans*/*cis* 70:30 was obtained. These mixtures were found stable at room temperature in the dark and could be converted to the pure *trans* isomer at elevated temperature. The relatively poor switching efficiency of the azobenzene was attributed to the absorption of the coupled carbazole systems. Under the catalysis conditions, the effect of the *trans*-receptor on the amide formation was found moderate, while the mixture containing the *cis* isomer (1:1 ratio) showed a 10-fold increase in the coupling rate. Hence, a complete photostationary state was not a requirement for efficient reaction. Control experiments regarding the competitive binding ability of amide product to the active catalyst supported that the reaction occurs through a termolecular complex and the template effect is responsible for catalysis.

In the following decade, the field was dormant and became more active again with the advent of organocatalysis,[26, 27] a green approach to catalysis and synthesis. The novel catalyst structures and activation modes that were developed in the context of organocatalysis started to be integrated with photoswitchable molecules with the aim of controlling catalytic activity and, if possible, selectivity by light.

3.1.2 Photoswitchable H-bond catalysis

Imahori and coworkers developed a photoswitchable cooperative acid catalyst by merging the structures of azobenzene and an axially chiral

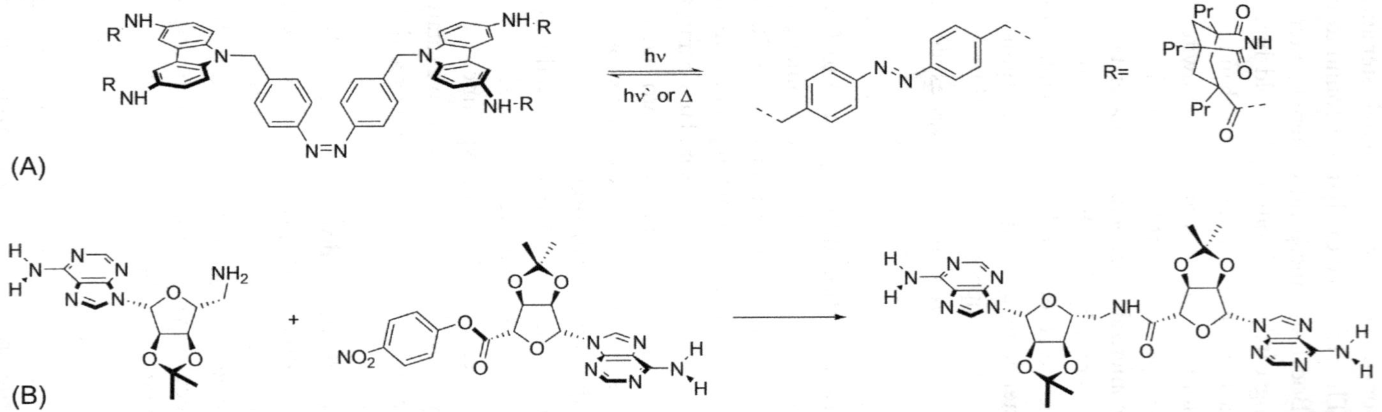

Fig. 7 (A) Structure of a photoswitchable synthetic receptor; (B) Amide formation reaction controlled by the photoswitchable synthetic receptor.

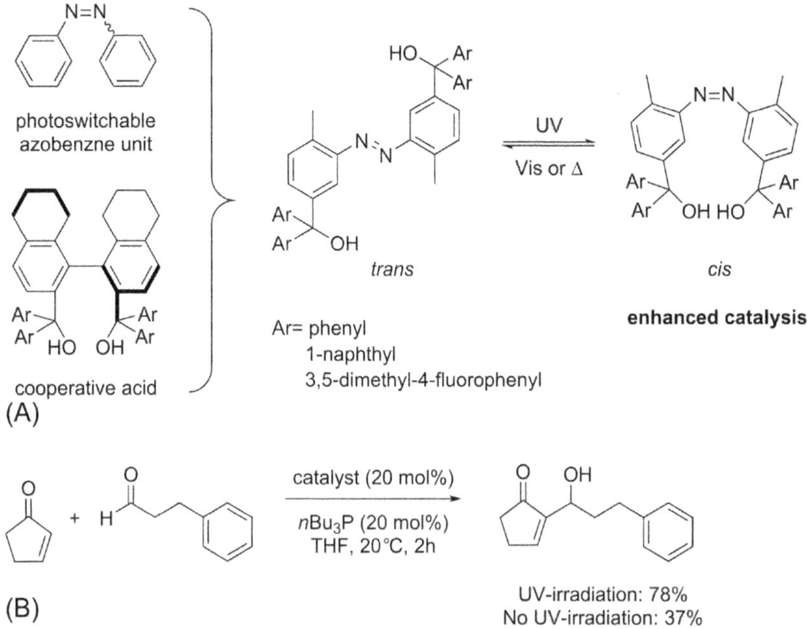

Fig. 8 (A) Design of a photoswitchable cooperative catalyst; (B) Morita-Baylis-Hillman reaction controlled by azobenzene photoisomerization.

bis(trityl alcohol)[28] (Fig. 8A). Axially chiral bis(trityl alcohol)s were described as H-bonding catalysts earlier.[29] According to the design, it was expected that in the *cis* state of the catalyst, when the two hydrogen-bonding OH-units are in close proximity (just as in the non-photoactive structure), cooperative acid catalysis will take place. On the other hand, when the OH-groups are positioned further apart from each other, as in the *trans* form of the catalyst, the activity could be turned off.

The synthesized catalyst was found to be photoisomerizable. Upon irradiation with UV light (365 nm), a photostationary composition of *trans*/*cis* = 5:95 could be reached (THF, 80 min). However, upon irradiation with visible light, the recovery of the *trans* isomer was not complete and only a *trans*/*cis* ratio of 48:52 could be obtained. Complete recovery could only be obtained under harsh thermal conditions (150°C, 24 h) indicating the stability of the *cis* form.

The effect of the different forms of the azobenzene catalyst was tested in the phosphine-catalyzed Morita-Baylis-Hillman reaction (Fig. 8B). This reaction proceeded with a 26.5% yield without the addition of H-bonding catalyst. However, when the catalyst was used, a clear difference was observed between the *trans* and the photochemically generated *cis* form. While the yield obtained with the *trans* (20 mol% catalyst) was only slightly higher (37%) than in the H-bonding catalyst-free reaction, the *cis* catalyst provided the product in 78% yield. Trityl alcohol as a control catalyst (40 mol%) gave similar yield (36%) as the *trans* catalyst, suggesting that the OH-groups in this molecule acted as independent H-bonding units. On the other hand, the increased activity in the case of the *cis* catalyst was explained to be due to a cooperative effect of the two proximal OH-groups. Changing the aryl groups from phenyl to 1-naphthyl resulted in worse switching ability, while the 3,5-dimethyl-4-fluorophenyl derivative showed high activity in both isomeric forms.

Among small molecule H-bond donors, thiourea derivatives are one of the most widely used and studied catalysts.[30, 31]

Pericàs and coworkers developed a diaryl thiourea-derived catalyst, where the H-bonding site, and thus the catalytic activity, could be reversibly blocked by the isomerization of a substituted azobenzene subunit within the molecule[32] (Fig. 9A). Among several H-bond acceptor groups on the azobenzene the NO_2-group was found be optimal showing the best photostationary state and fatigue resistance. It has to be noted though, that the composition of the photostationary state was still moderate (*cis/trans* = 56:44) and it was reached after a 2.5 h irradiation period. The ratio of the *cis* form

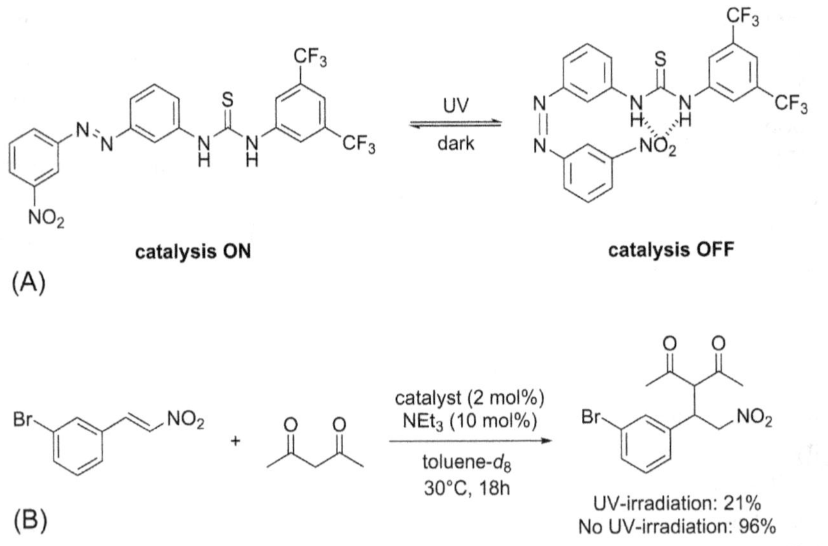

Fig. 9 (A) A photoswitchable diaryl thiourea-derived catalyst; (B) Light-controlled catalysis off a Michael-addition.

could be increased in more dilute solutions, however, those concentrations were too low for catalysis.

Nevertheless, the catalytic activity could be controlled via the reversible formation of intramolecular H-bonding interaction between the thiourea and the NO_2-group. In the *trans* form of the catalyst, where the H-bond donor and acceptor groups were not proximal, the Michael-addition of acetylacetone to *m*-bromonitrostyrene was proceeded with a nearly complete conversion with only 2 mol% catalyst loading (Fig. 9B). Upon irradiation of the catalyst with UV light, the *cis* form was generated, where the NO_2-group was able to interact with the thiourea so as to block this catalytic site and hinder the Michael-addition. Under this latter condition, only 21% conversion could be reached after 18 h.

Wang and Feringa brought thiourea-based H-bonding activation to a higher level by realizing a dynamically controlled cooperative Brønsted base/thiourea bifunctional catalyst based on a light-driven molecular motor.[33] The rotor and stator halves of the molecule were functionalized with a dimethylaminopyridine (DMAP) and a diaryl thiourea moiety, respectively, that provided the catalytic function (Fig. 10). The relative position of the catalytic units could be controlled through *cis*/*trans* isomerization by UV light irradiation (312 nm) and subsequent thermal isomerization. The geometrical isomerization brought about the cooperative catalytic

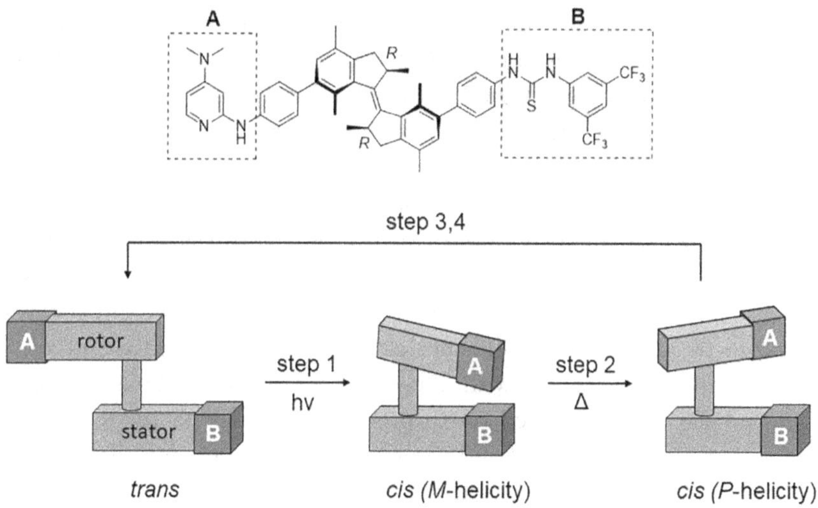

Fig. 10 Design of a cooperative Brønsted base/thiourea bifunctional catalyst based on a light-driven molecular motor.

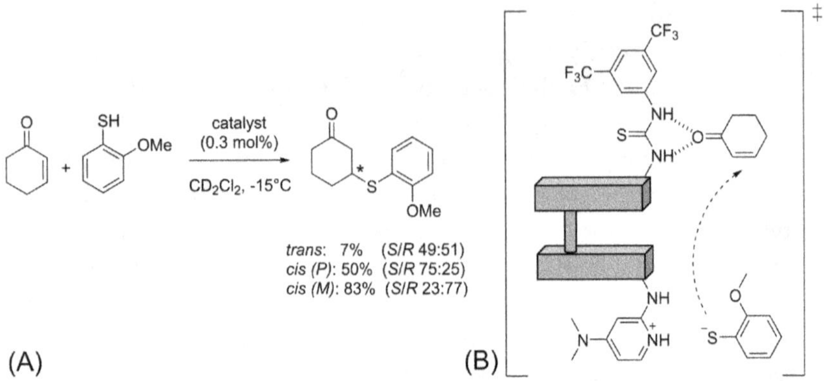

Fig. 11 (A) Light-controlled catalytic activity and stereoselectivity by a molecular motor-based catalyst in a conjugate addition reaction; (B) Proposed mechanism of the cooperative catalysis.

structure by positioning the DMAP and the thiourea units in proximity (Fig. 11B). In this way, the activated reactants would be close to each other which ensures more efficient transformation. On the other hand, the consecutive isomerization steps allowed for controlling the helicity of the catalyst dictated by the motor structure; hence, the chiral space around the reactants could be dynamically changed.

The catalyst design was tested in the conjugate addition of a thiophenol and cyclohexanone (Fig. 11A). When the catalyst in the *trans* orientation was applied, low catalytic activity was observed (7% yield after 15 h) and a racemic product mixture was isolated. This could be explained by the unfavorable relative orientation of the DMAP and the thiourea units preventing cooperative catalytic action. On the other hand, the photochemically generated *cis* form of the motor showed an increased catalytic efficiency giving the product in 50% yield under identical reaction conditions. Remarkably, in this case enantioenriched product was isolated with an enantiomer ratio of 75:25 (*S*/*R*) favoring the *S* enantiomer. The *cis* isomer having opposite helicity (*M*) that was generated via a thermal isomerization process was found to be an even more efficient catalyst with 83% yield (15 h) and a reversed enantioselectivity with a ratio of 23:77 (*S*/*R*). Thus, not only the catalyst could be turned on to promote the reaction in its *cis* form, but also the chiral space around the reactant could be controlled leading to helicity-dependent enantioselectivity.

Later, the catalyst was truncated by the removal of the aryl spacer between the motor core and the catalytic functional groups[34] (Fig. 12A).

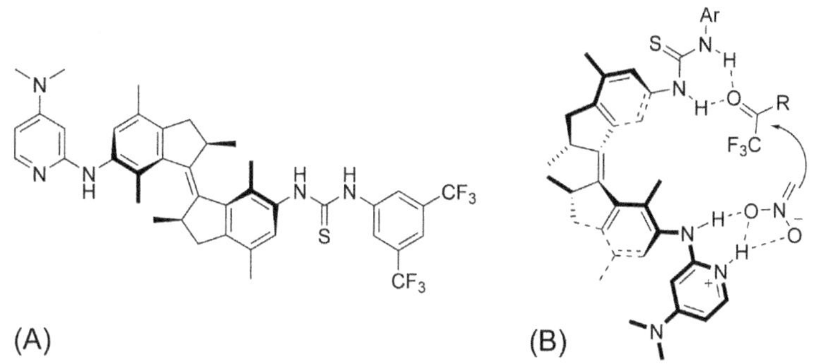

Fig. 12 (A) A more compact catalyst design based on molecular motors and (B) its effect in Henry reaction between trifluoromethyl ketones and nitromethane.

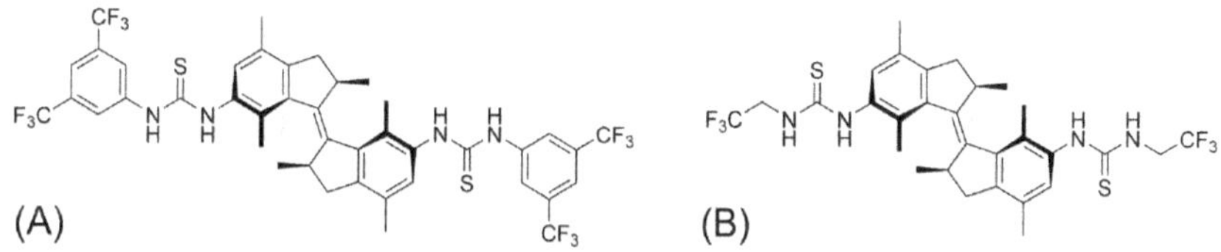

Fig. 13 (Bis)thiourea catalysts based on light-driven molecular motors.

This modification facilitated the synthesis of the molecule and provided a more constrained catalytic pocket, leading to a more versatile and efficient catalyst. The new design proved successful in a model Henry reaction between trifluoromethyl ketones and nitromethane (Fig. 12B).

During further research on the structure-property relationships of the catalyst, it was found that the base (DMAP) component is not a requirement for catalysis and it could be replaced by a H-bonding thiourea moiety[35] (Fig. 13). Importantly, the nature of the thiourea structure strongly influenced stereoselectivity. Although both the aromatic thiourea derivative (Fig. 13A) and the aliphatic derivative (Fig. 13B) showed activity in Henry reactions, only the aromatic thiourea derivative exhibited dual stereocontrol, while the aliphatic compound yielded racemic product mixtures.

Control of catalytic activity was also realized in Michael-additions by using second-generation molecular motors as catalysts.[36]

Hecht and coworkers reported a photoswitchable catalyst for controlling the polymerization of L-lactide via H-bonding activation.[37] The catalyst was composed of a photoswitchable diarylethene backbone having a phenolic OH group (Fig. 14). In the open form of the switchable catalyst, the phenolic OH was able to activate the carbonyl group of L-lactide by H-bonding towards polymerization of the molecule. A tertiary amine cocatalyst was

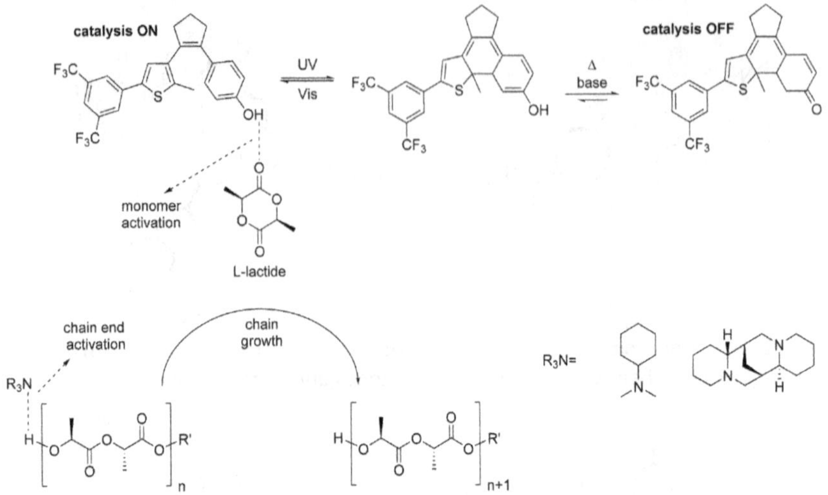

Fig. 14 Design of a diarylethene-based photoswitchable H-bonding catalyst and its application in the light-controlled polymerization of L-lactide.

required in combination with the phenol in order to activate the initiating alcohol or chain-end of the polymer.

When the catalyst system was irradiated with UV light, the ring closing of the switch yielded an enol that tautomerized to the corresponding keto-form. This latter structure did not promote the polymerization process due to the lack of the activating phenolic OH group. Irradiation with visible light restored the H-bonding activation and the initial catalytic activity. This light-controlled activation of polymerization could be extended to other monomers such as trimethylene carbonate and δ-valerolactone; however, in those cases DBU was used as the amine cocatalyst. Furthermore, the authors also demonstrated the photoregulated *co*-polymerization of the latter two monomers.

An alternative H-bonding catalyst was reported by the same group that was composed of a guanidine moiety and an azobenzene photoswitch; however, it did not show activity in the polymerization of lactide.[38]

3.1.3 Photoswitchable base and nucleophilic catalysis

Hecht and coworkers developed and studied photoswitchable organic base catalysts.[39, 40] The catalytically active unit was a conformationally restricted N-alkylated piperidine base, which was equipped with a photochromic azobenzene unit bearing bulky substituents (Fig. 15A). The steric bulk around the basic N atom can be controlled by the isomerization of the

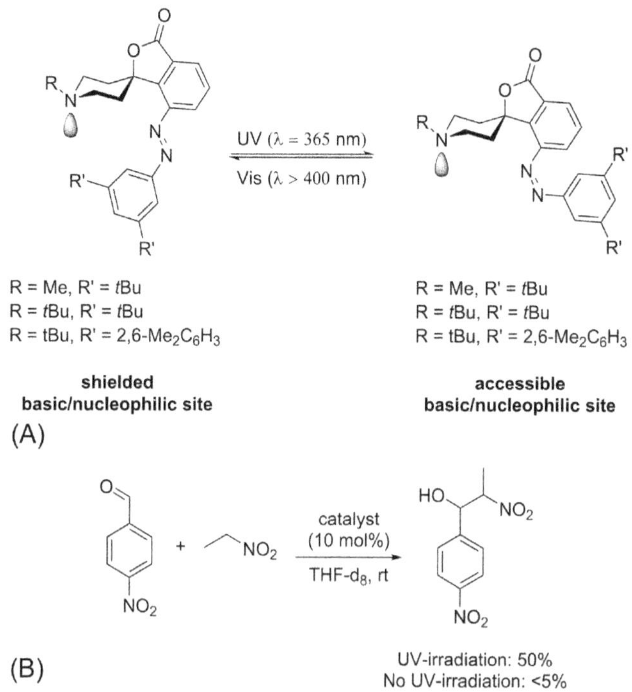

Fig. 15 (A) Design of an azobenzene-based photoswitchable base catalyst; (B) Light-controlled catalytic activity in Henry reaction.

azobenzene, which is ultimately responsible for the accessibility of the catalytic site by the reactants.

The synthesized molecules were shown to be photoactive with close to complete *trans*-to-*cis* isomerization; moreover, the *cis* forms showed remarkable thermal stability ($t_{1/2} = 286$ h at 20°C for R = R′ = *t*Bu, Fig. 15A). Titration experiments were performed to quantify the basicity of the shielded and deshielded bases and found that the basicity of the deshielded isomer is higher by an order of magnitude. This difference between the two forms was exploited in the base-catalyzed Henry reaction of 4-nitrobenzaldehyde and nitroethane, where the *trans* form of catalyst (R = *t*Bu, R′ = 2,6-$Me_2C_6H_3$) was practically inactive, while its *cis* form proved to be a more efficient catalyst (yield up to 50%). In a detailed mechanistic study,[41] the authors prepared the stilbene analogues of the azobenzene-based catalysts which turned out to be less efficient in Henry reaction. The longer C=C bond in stilbene compared to the N=N bond in azobenzene contributed to a less efficient shielding of the base, and hence, a lower ON/OFF ratio of the catalyst. Furthermore, the inefficient *cis*-to-*trans* isomerization in the case of the stilbene also lowers its potential as a photoswitchable catalyst.

The azobenzene-containing photoswitchable base was also successfully attached to silica particles and its photoactivity was confirmed in the surface bound system; however, no catalysis was demonstrated with it.[42]

Shibasaki and coworkers found that diamides tethered to azobenzenes can undergo reversible aggregation/dissolution cycles controlled by light via the isomerization of the azobenzene unit[43] (Fig. 16). The precipitated aggregates of the molecules having a *trans*-azobenzene moiety could be redissolved by UV-irradiation. Upon irradiation, the formation of the *cis*-azobenzene interrupted the H-bond network that was responsible for the aggregation and led to the dissolution of the molecules.

Based on this phenomenon, a system with incorporated catalytic function was developed[44] (Fig. 17A). The phenolic OH-groups of the previously reported structure (Fig. 16) were removed not to interfere with the target reactions and a nucleophilic pyridine moiety was introduced. The tailored structure, due to its lower solubility, was able to undergo aggregation/dissociation cycles upon UV and visible light stimuli on its own, without the presence of a molecule with opposite chirality. Irradiation of the photoswitchable diamide (Fig. 17A) with UV light led to a photostationary state composition of *trans*/*cis* = 4:96.

Nucleophilic catalysis by the switch was first tested in the reaction of 1-naphthol and Boc_2O (Fig. 17B). The *cis* form of the catalyst, which was soluble in the solvent mixture used, provided the product in about 70% yield after 180 min. In contrast, the *trans* form, due to its extensive aggregation, was found to be an inefficient catalyst giving a yield of about 10%. In $CHCl_3$,

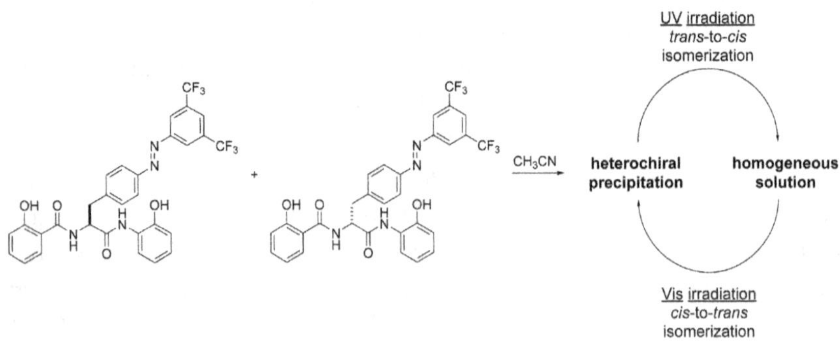

Fig. 16 Light-controlled reversible aggregation/dissolution cycles of diamides-containing photoisomerizable azobenzene units.

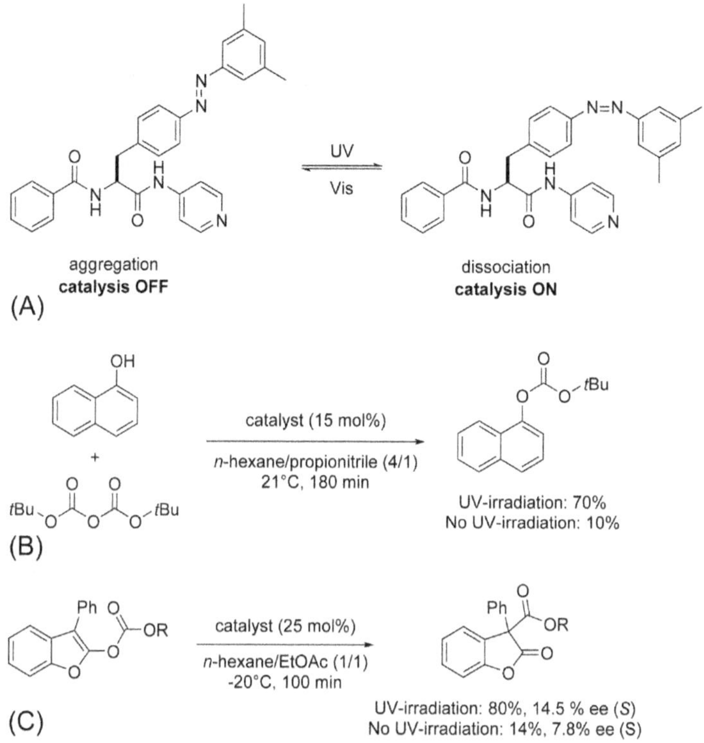

Fig. 17 Light-controlled catalysis by the reversible aggregation/dissolution of a diamide containing a photoisomerizable azobenzene unit.

in which both forms of the catalyst were soluble, no difference in the catalytic activity could be detected, which supported that the difference in catalytic activity was indeed due to the photoisomerization-controlled aggregation/dissociation of the catalyst. With this system, the in situ manipulation of the catalytic activity under reaction conditions could also be realized.

Control over the catalytic activity in the rearrangement of 2-acyloxybenzofuran to 3-acyl-2-benzofuranone could also be achieved by UV and visible light irradiation of catalyst (Fig. 17B). In this reaction, not only the yields could be controlled, but also the stereoselectivities were affected.

3.1.4 Photoswitchable N-heterocyclic carbene (NHC) catalysis

Bielawski and coworkers combined N-heterocyclic carbenes with photoswitchable dithienylethenes in order to control catalytic activity by light.[45] The key element of design was the alteration of the electron density on the NHC upon light-induced ring closure and ring opening (Fig. 18A). The

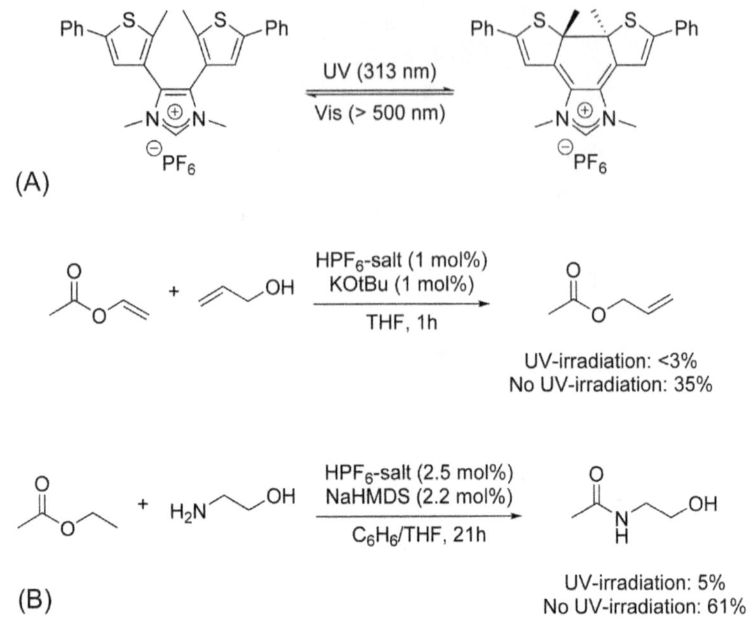

Fig. 18 (A) Design of a photoswitchable NHC catalyst; (B) Light-controlled catalytic activity in transesterification and amidation reactions.

success of the design was demonstrated in the catalysis of the transesterification of vinyl acetate with allyl alcohol and the amidation of ethyl acetate with 2-aminoethanol (Fig. 18B).

In both reactions, the UV-irradiation-induced ring closure resulted in the strong attenuation of the reaction rate, while the catalytic activity was restored by exposure to visible light. Photoswitchable NHC organocatalysts were also used to reversibly control the activation of ammonia[46] and the ring-opening polymerization of cyclic esters.[47]

3.2 Organometallic catalysis

The first example of controlling the activity of an organometallic catalyst by light was reported by Inoue and coworkers.[48] Their system consisted of an Al-porphyrin and a photoswitchable stilbazole derivative to reversibly control the chemical fixation of CO_2 (Fig. 19).

The key element of the design is that the Al-porphyrin requires the coordination of a base in order to catalyze the reaction of CO_2 and propylene oxide into propylene carbonate, while the stilbazole serves as a photoswitchable base. In its *trans* state, the coordination of the base to the Al-porphyrin was

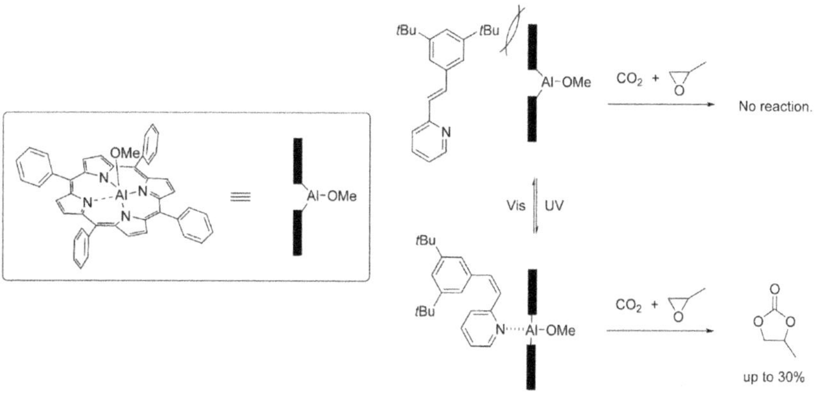

Fig. 19 Controlling chemical fixation of CO_2 with light by using an Al-porphyrin and a photoswitchable stilbazole base as catalysts.

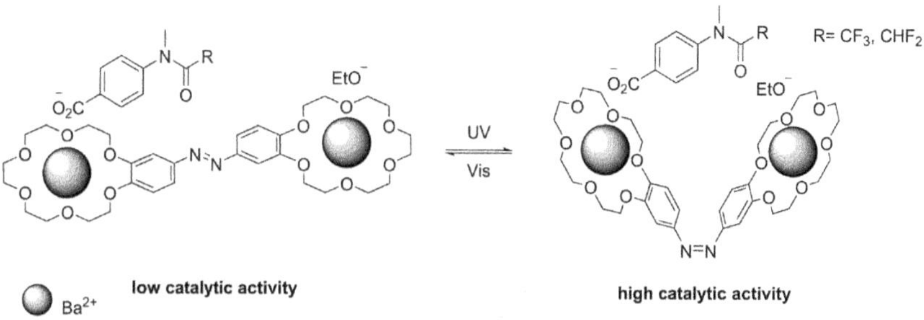

Fig. 20 Design of a Ba-ion-containing azobenzene-based crown ether for the light-controlled ethanolysis of anilides to anilines.

hindered by the *t*Bu groups; however, the UV-irradiation-generated *cis* form readily coordinated and activated the catalyst.

A Ba-ion complexing azobenzene-based crown ether was developed by Cacciapaglia and Mandolini (Fig. 20), which showed low catalytic efficiency in the ethanolysis (using $EtONMe_4$) of anilides to the corresponding anilines in its *trans* form, while the *cis* isomer facilitated the transformation by bringing the coordinating metal centers closer together, thus increasing the confinement of the reactant anilide and ethoxide anion.[49]

The *cis* form with its half-life of 7.9 h at rt. proved to be a stable catalyst and the back isomerization did not influence the outcome of the reaction considerably.

Several authors investigated the potential of switchable phosphine ligands in controlling transition metal catalysis.

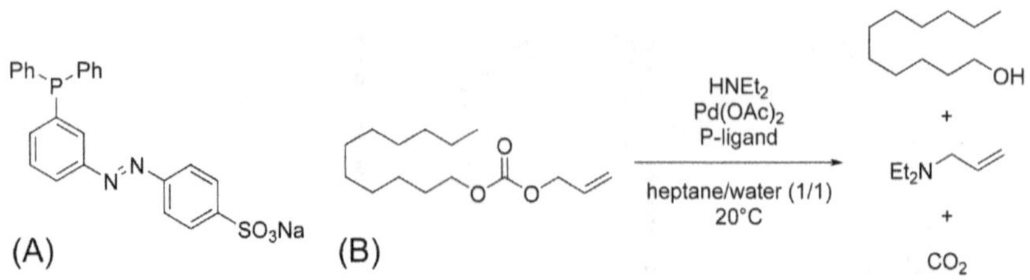

Fig. 21 (A) A water-soluble photoswitchable phosphane; (B) Pd-catalyzed cleavage of allyl-undecyl-carbonate in the presence of a water-soluble photoswitchable phosphane.

Monflier and coworkers reported the use of azobenzene-containing water-soluble phosphanes in the aqueous Pd-catalyzed cleavage of allyl-undecyl-carbonate[50] (Fig. 21).

A 2.5-fold increase in catalytic activity was obtained when the ligand was converted into its *cis* form by UV-irradiation compared to the activity in the *trans* form of the ligand. The observed activity difference between the two forms of the ligand was suggested to be due to the formation of different aggregates in water.

Shaw, Linehan, and coworkers synthesized photoresponsive azobenzene Rh(*bis*)diphosphine complexes and tested them in CO_2 hydrogenation reaction[51] (Fig. 22). Each metal complex was found active in the reaction that had to be performed in NMP due to the low solubility of the azobenzene-based complexes. Importantly, the ligand containing the alanine-based linker between the azobenzene and the diphosphine moieties (Fig. 22, $n = 1$) exhibited different reactivities based on the azobenzene configuration. In the *cis* state of the azobenzene, a 40% rate enhancement was measured compared to that in the *trans* state. On the other hand, the complex having the GABA linker (Fig. 22, $n = 2$) did not show any difference in activity in the two different configurations of the azobenzene.

The authors suggested that the reason for the different activity in the case of the shorter linker ($n = 1$) could be the change in electronics/ thermodynamics upon isomerization to the *cis* form, and/or in the *trans*

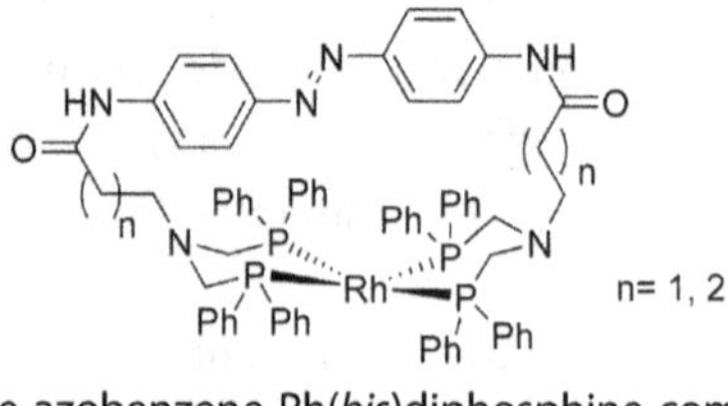

Fig. 22 Photoresponsive azobenzene Rh(*bis*)diphosphine complexes.

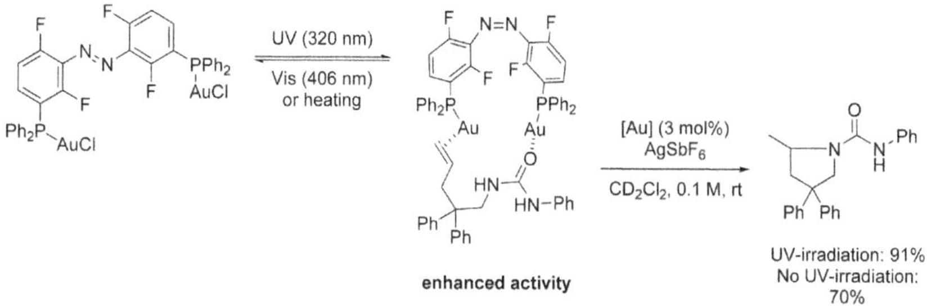

Fig. 23 Photoswitchable phosphine-based gold catalyst in light-controlled intramolecular hydroamination reaction.

form the azo group coordinates to the metal center that hinders the CO_2 hydrogenation reaction.

Photoswitchable phosphines have been combined with gold catalysis. Marinetti, Voituriez, and coworkers synthesized a photoresponsive gold complex that showed controllable catalytic activity in intramolecular hydroamination reaction of an *N*-alkenyl urea derivative[52] (Fig. 23).

The catalytic tests showed that the *cis* isomer had an enhanced activity compared to the *trans* form. The *cis* isomer provided 91% conversion after 44 min reaction compared to the 70% conversion under the same conditions obtained with the *trans* isomer. The authors suggested that the higher activity of the catalyst in the *cis* configuration is due to a cooperative effect between the metal centers that are in proximity in this state and are able to coordinate both the alkene and the carbonyl function within the molecule.

Freixa and coworkers explored the syntheses and use of photoisomerizable ligands including pyridines, bipyridines, and phosphines.[53] They found that triphenylphosphines having three pendant azobenzene units can be used to control the activity of a ruthenium catalyst in the solvolytic hydrogen generation from ammonia borane (Fig. 24).

In their *cis* form, the photoswitchable ligands provided enhanced activity (up to ~50%) for the catalyst, due to the change in the shape of the catalytic pocket, as suggested by the authors.

Several groups studied dithienylethene-based organometallic compounds.[54, 55] Branda and coworkers reported the control of stereoselectivity in the cyclopropanation of styrene using a dithienylethene-based photoswitchable chiral bis(oxazoline) ligand in combination with Cu-catalysis[56] (Fig. 25). In the open form of the ligand, a chiral environment was provided for the metal that resulted in enhanced stereoselectivity in the reaction.

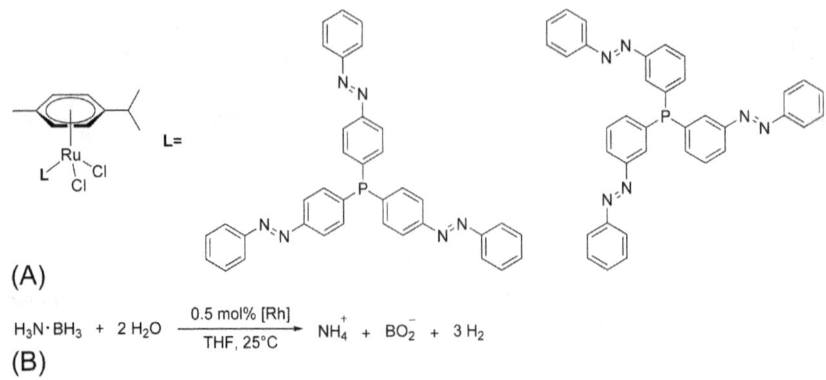

Fig. 24 (A) A ruthenium catalyst having photoswitchable triphenylphosphine ligands; (B) Solvolytic hydrogen generation from ammonia borane.

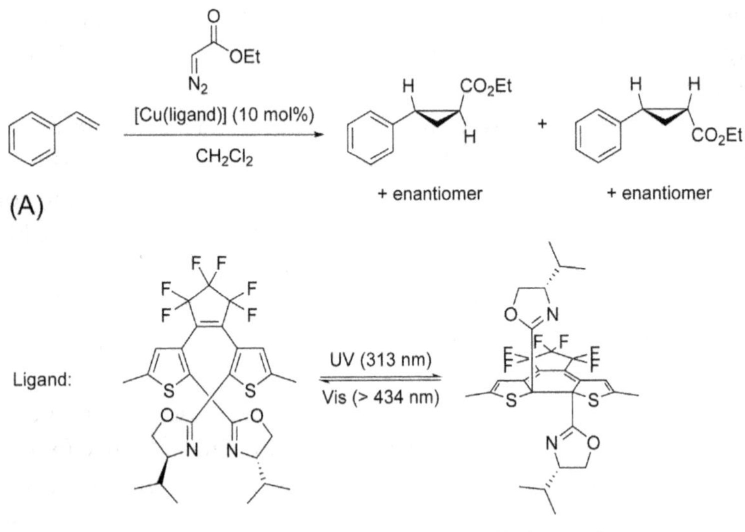

Fig. 25 Control of stereoselectivity in the cyclopropanation of styrene (A) using a dithienylethene-based photoswitchable chiral bis(oxazoline) ligand (B).

However, upon generating the closed form of the ligand by irradiation with UV light, the oxazoline rings were forced in a noncomplexing geometry that prevented stereoselective transformation.

Bielawski and coworkers also developed dithienylethene-based organometallic catalysis; however, in this case the different geometries of the two forms were not exploited, but the electronics around the metal was

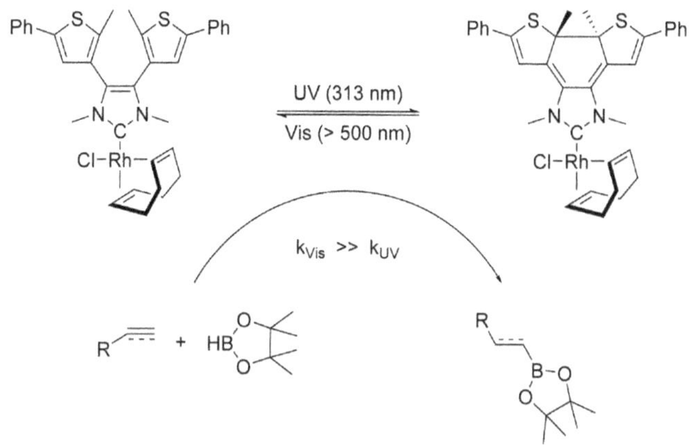

Fig. 26 Photoswitchable Rh-NHC catalyst for light-controlled hydroboration of alkynes and alkenes.

controlled.[57] The catalyst was composed of a dithienylethene switch and a fused N-heterocyclic carbene moiety that served as a ligand for Rh(I) (Fig. 26).

Under ambient light, in the open form of the ligand, the catalyst promoted the hydroboration of alkenes and alkynes; however, in the UV light-induced closed form the catalyst showed lower activity of up to an order of magnitude. The reason for the reduced activity was suggested to be due to the inhibition of the rate determining reductive elimination step, which originated from the decrease in electron-donating ability of the closed form of the NHC ligand.

Wolf and coworkers reported the use of photochromic dithienylethene-based bidentate phosphine ligands in the Cu(I)-catalyzed hydroboration reaction[58] (Fig. 27). UV-irradiation had a similar hindering effect on the reaction as described above.

4 Biphasic, heterogeneous, and supported catalysis

4.1 Biphasic and micellar reactions

In line with the aims of green chemistry, conditions to conduct organic reactions in water have been developed. Many of those apply different amphiphiles in order to solubilize the organic reactants in a microheterogeneous micellar system.[59]

Len and coworkers developed photoswitchable azobenzene-based surfactants that facilitate the reaction, the work-up, and the surfactant

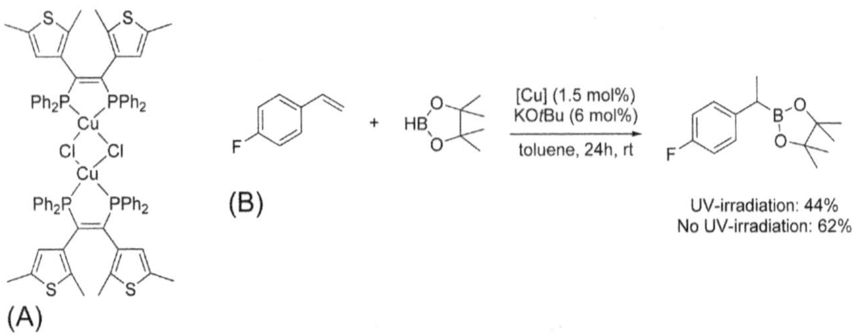

Fig. 27 Photochromic dithienylethene-based bidentate phosphine ligands (A) in the Cu (I)-catalyzed hydroboration reaction of alkenes (B).

recycling.[60] A sugar-azobenzene conjugate was used in the acylation of anilines, where the micellar system aided the transformation in the *trans* state of the surfactant, while upon irradiation to the more polar *cis* state, the product extraction and surfactant recycling were facilitated (Fig. 28A). The azobenzene-based surfactant overperformed the commercial sodium dodecyl sulfate (SDS) as it could be better recycled and reused due to the polarity control by isomerization up to 6 reaction cycles in the acylation of *p*-toluidine.

This concept was also extended to the aqueous Pd-catalyzed Tsuji-Trost reaction.[61] In this case, a PEGylated azobenzene served as the photoswitchable surfactant (Fig. 28B). The photoresponsive system overperformed commercially available surfactants (SDS, Tween 20, CTAB) and could be recycled after up to five consecutive reactions in the coupling of cinnamyl acetate and *p*TsNa.

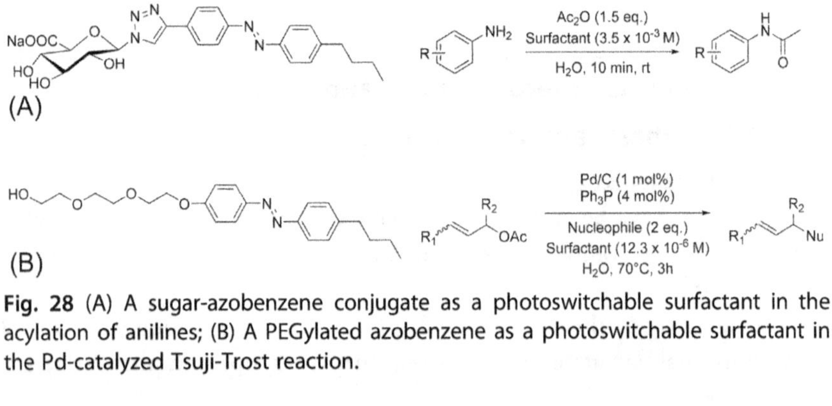

Fig. 28 (A) A sugar-azobenzene conjugate as a photoswitchable surfactant in the acylation of anilines; (B) A PEGylated azobenzene as a photoswitchable surfactant in the Pd-catalyzed Tsuji-Trost reaction.

Tan and coworkers used a diblock copolymer of N-isopropylacrylamide and azobenzene containing a chiral Ti(IV) salen complex to create photoresponsive metallomicelles for asymmetric sulfoxidation in water[62] (Fig. 29).

The azobenzene block was responsible for the photoresponsive nature of the system, while the N-isopropylacrylamide, being a thermoresponsive moiety, facilitated the temperature-controlled separation of the catalyst. At the reaction temperature (25°C), the polymer assembled into metallomicelles in water. The dark-adapted system (azobenzene in *trans* from) gave 82% conversion in the oxidation of phenyl methyl sulfide after 75 min reaction with 99% ee (*R* enantiomer in excess). Upon irradiation of the metallomicelles with UV light (30 or 60 s), an increased conversion of 99% was obtained. The authors rationalized this effect by morphological changes (based on TEM investigation) in the micellar system due to the isomerization of the *trans*-azobenzene into the more sterically demanding and dipolar *cis* form. The catalyst could be easily recovered and reused up to six times by precipitation at 32°C, which is the lower critical solution temperature of the polymer. Compared to the polymeric system, the neat Ti(salen) complex was not able to efficiently catalyze the reaction under the same condition, giving only 8% conversion in the oxidation of phenyl methyl sulfide.

As reversible phase transfer of nanoparticulate catalysts could be a practical means of catalyst removal and recycling, Wu, Li, and coworkers demonstrated an approach using azobenzenes as photoswitchable surfactants.[63] Polyoxometalates (POMs), discrete nano-sized metal oxide clusters, could be functionalized with cationic azobenzene-based surfactants through an ion-exchange reaction (Fig. 30). Unlike the free POM that is soluble in water, the surfactant-containing complex readily dissolved in organic solvents (toluene, THF, chloroform) due to the apolar backbone of the azobenzene molecules. Upon irradiation of the system, the *trans*-azobenzenes were converted to the more polar *cis* form, which rendered the core-shell system soluble in a strongly polar solvent system (H_2O/DMF 1:1).

This photoinduced reversible phase transfer was exploited in the oxidation of phenothiazine with cumene hydroperoxide catalyzed by the azobenzene surfactant-modified POM. The oxidation was found to be complete after 4 h at room temperature in toluene. Subsequently, DMF/H_2O mixture was added to the organic phase and the system was irradiated with UV light generating the more polar *cis* form of the surfactant, which led to the separation of the POM-surfactant catalyst complex from the organic reaction mixture. After separation of the aqueous phase and addition of

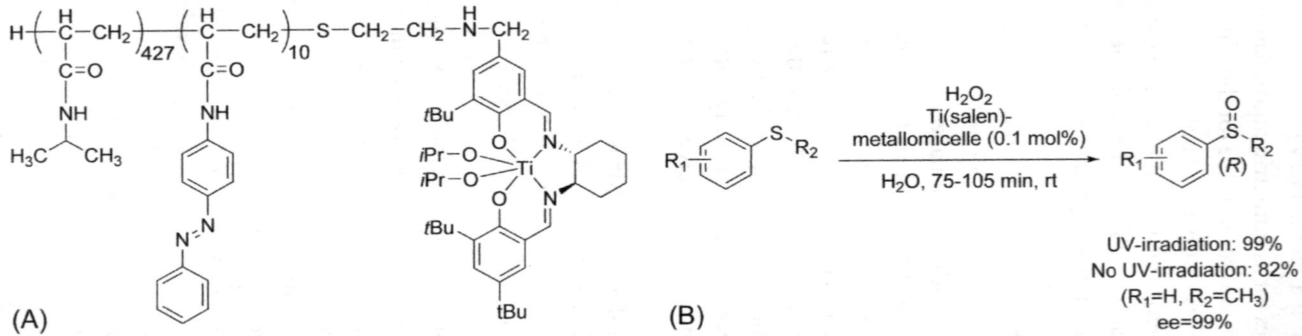

Fig. 29 (A) A diblock copolymer containing a photochromic azobenzene and a Ti(IV) salen complex; (B) Light-controlled asymmetric sulfoxidation in water.

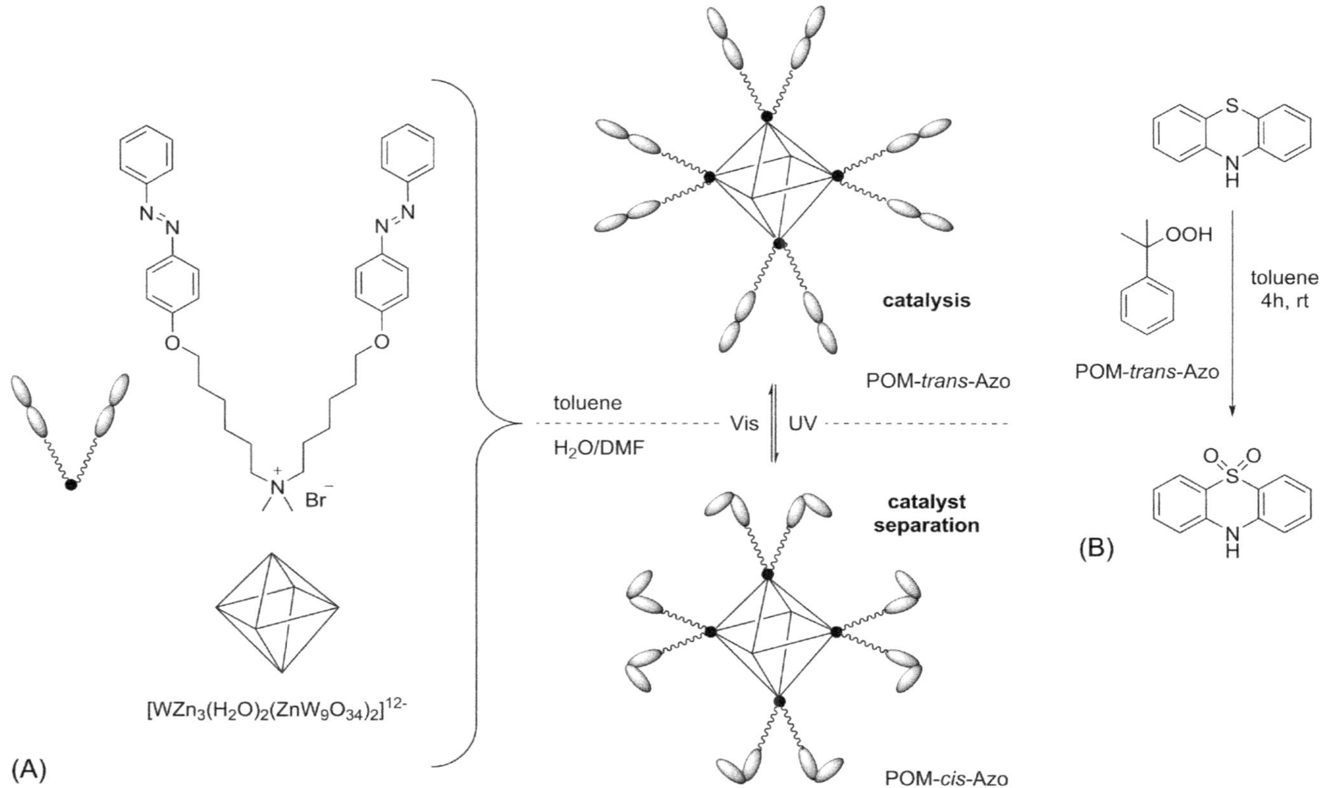

Fig. 30 (A) Combination of a polyoxometalate (POM) with an azobenzene surfactant to control reversible phase transfer of the catalyst; (B) Oxidation of phenothiazine with cumene hydroperoxide catalyzed by the azobenzene surfactant-modified POM.

further reactants in a toluene solution, the catalyst could be transferred back to the organic phase by irradiation with visible light for further oxidation catalysis.

4.2 Nanoparticle-based systems

4.2.1 Photoswitchable host-guest systems

The potential of reversible host-guest interactions between cyclodextrins and azobenzenes in controlling catalysis has been exploited in nanoparticle-based catalytic systems. Zhao and coworkers combined a Zn(II)-coordinated pyridyl-bis(cyclodextrin) complex with gold nanoparticles (~13 nm) that contain photoresponsive azobenzene surface ligands[64] (Fig. 31). To make the system water-miscible, additional hydrophilic surface ligands were added to the particle surface in a ratio of (azobenzene/hydrophilic ligand) 1:7. The Zn-complex itself has been shown to be an efficient supramolecular catalyst in ester hydrolysis.[65] When the Zn-complex and the particles were combined, the formation of a supramolecular interparticle host-guest complex occurred between the *trans*-azobenzenes and the cyclodextrin moieties of the catalyst. In this state, the Zn-centers were efficiently blocked from the substrate 4-nitrophenyl acetate and a low hydrolysis rate was observed. On the other hand, when the system was irradiated with UV light (365 nm) and the *trans*-azobenzenes were converted to the *cis* isomer, the reaction rate was close to that measured with the free Zn-complex. The reason for the rate enhancement is the weak interaction between the CD and the *cis*-azobenzene which resulted in the release of the catalyst that becomes available for the substrate. As the timescale of the hydrolysis reaction was longer than the photoswitching processes, the rate of the reaction could be enhanced and reduced during the course of the transformation.

Hu and coworkers also applied the host-guest chemistry between azobenzenes and CDs, but introduced an additional level of confinement.[66] In their system, mesoporous silica microspheres loaded with Ag nanoparticles were used as nanoreactors and the pore openings were controlled by the reversible binding between the silica-grafted azobenzene and α-CD (Fig. 32).

In the *trans* form of the azobenzenes, the formation of an inclusion complex with α-CD rendered the pores inaccessible for substrates. Irradiation with UV light and formation of the *cis*-azobenzene resulted in the disassembly of the complex, and hence, the pores were available for catalysis. Light-controlled switching of the catalytic activity of the system was demonstrated through the aqueous phase reduction of methylene blue dye with $NaBH_4$.

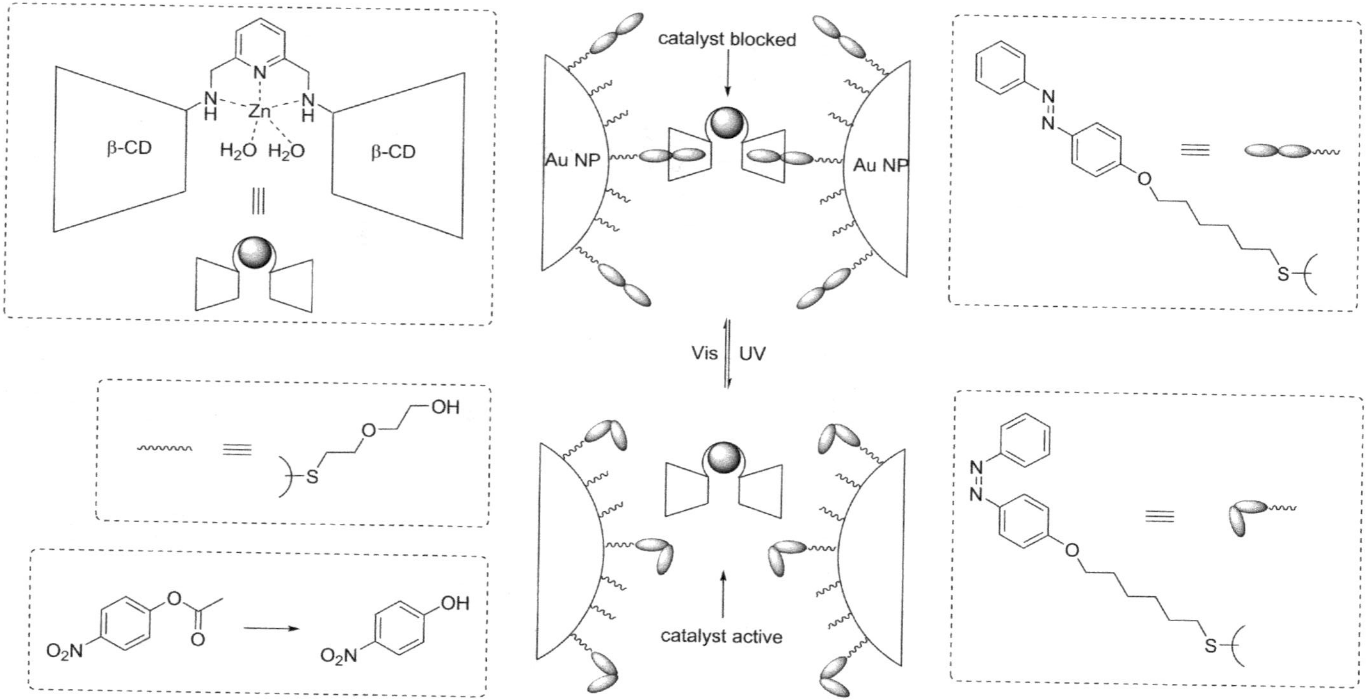

Fig. 31 A supramolecular catalyst-containing Zn(II)-coordinated pyridyl-bis(cyclodextrin) complex and gold nanoparticles (~13 nm) functionalized with photoresponsive azobenzene surface ligands.

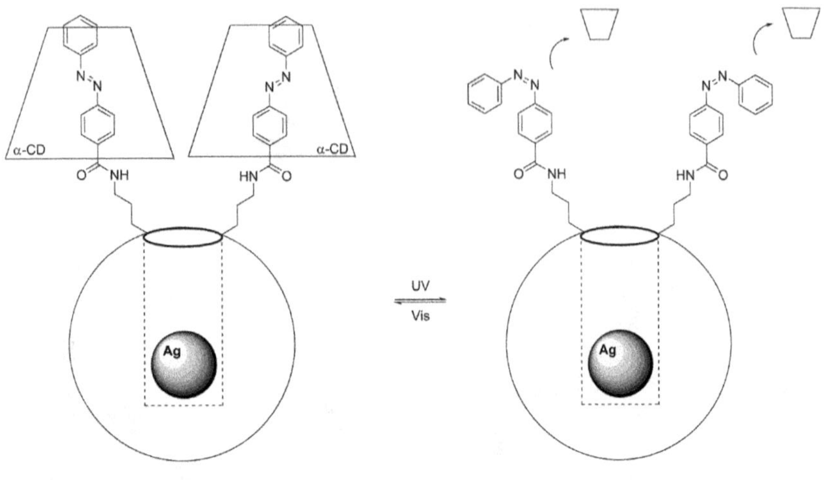

Fig. 32 Controlling the pore openings of mesoporous silica nanoreactors by light-controlled reversible binding between the silica-grafted azobenzene and α-CD.

A reverse approach in the design of the catalyst was followed by Chen, Tan, and coworkers, who immobilized *per*-6-thio-α-CD rings on the surface of gold nanoparticles (3.6 nm).[67] These α-CD-capped nanoparticles were used as hosts for azobenzenes. Similar to the previous example, the *trans* form of the azobenzene binds stronger to the CD than the *cis* form and so the association could be controlled by UV and visible light (Fig. 33A). An advantage of this design is that the solubility of the system could be controlled via the photochemically triggered association/dissociation phenomena, which aids the nanoparticle separation following catalysis.

In a two-phase water/toluene system, initially, the *trans*-azobenzene resides in the organic phase, while the hydrophilic α-CD-capped particles are dispersed in the aqueous phase (Fig. 33B). Upon agitation of the two-phase system, the trans-azobenzenes bind to the CD hosts, which results in a reverse micelle-like structure due to the long hydrophobic alkyl chains. Overall, the assembly leads to the transfer of the particles from water to toluene. Upon irradiating the inclusion complex, the *trans*-azobenzenes isomerize to the *cis* form that leads to disassembly followed by the separation of the components between the aqueous and organic phases. The phase transfer of the nanoparticles was shown to be reversible upon irradiation with UV and visible light. The photo-controlled solubility of the Au particles was exploited in the gold-catalyzed reduction of 4-nitrophenol (4-NP) to 4-aminophenol (4-AP) with $NaBH_4$. The aqueous phase reduction could be rapidly

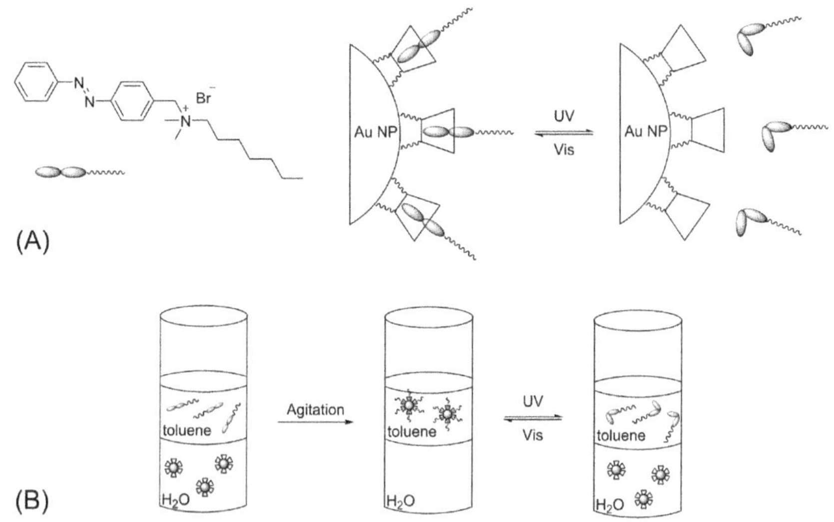

Fig. 33 (A) α-CD-capped gold nanoparticles as hosts for photoswitchable azobenzenes; (B) Controlling the solubility of α-CD-capped gold nanoparticles/azobenzene system via photochemically triggered association/dissociation phenomena.

quenched by the addition of a toluene solution of the *trans*-azobenzene derivative (Fig. 33A), due to the formation of the inclusion complex and its subsequent transfer to the organic phase. The photo-reversible phase transfer is also a means of recycling and reusing the gold catalyst. After the completion of the reduction, the Au particles could be removed from the aqueous phase by complexation with *trans*-azobenzene and could be transferred back to perform a subsequent catalytic transformation upon irradiation with UV light.

4.2.2 Light-controlled reversible aggregation

A nanoparticle property that can be reversibly controlled and exploited in catalytic applications is aggregation. Controlling catalytic activity of gold nanoparticles via their light-induced reversible aggregation was demonstrated by the group of Grzybowski.[68] In their system, gold nanoparticles were decorated with a mixed monolayer of an alkyl thiol-containing azobenzene and an alkyl amine (Fig. 34A).

In the *trans* form of the azobenzene ligands, the Au NPs could be suspended in toluene and were able to catalyze the hydrosilylation reaction of 4-methoxybenzaldehyde with diphenylsilane (Fig. 34B). When the NPs were irradiated with 365 nm UV light, the *trans*-to-*cis* isomerization took *place that generated a dipole within the molecules, which resulted in aggregation of*

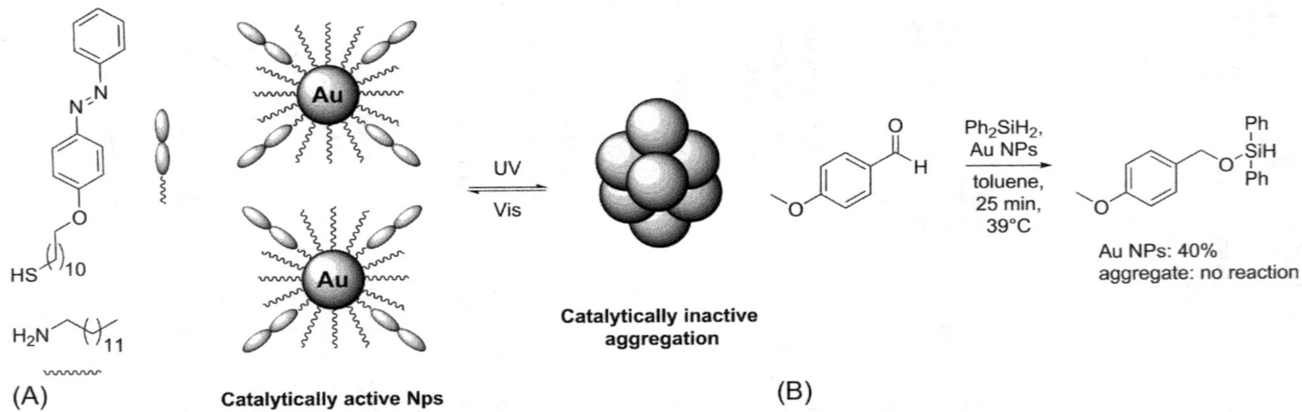

Fig. 34 (A) Controlling catalytic activity of gold nanoparticles via their light-induced reversible aggregation (B) in hydrosilylation reaction.

the particles in the nonpolar solvent (Fig. 34A). In the aggregated state, the particles were shown to be inactive catalysts in the hydrosilylation reaction. Due to the metastability of the aggregates, they could be disassembled upon irradiation with visible light that regenerated the *trans* isomer of the azobenzene ligand. The nature and composition of the mixed monolayer around the NPs were critical in the design of switchable NP aggregation. For a robust system, the combination of strongly binding azobenzene thiols that provide the reversibility of the system and loosely binding alkyl amines that facilitate catalysis was ideal. No catalysis took place when both the azobenzene and the alkyl components were thiol derivatives due to the poisoning of the gold catalyst. When both ligands were amines, the reversible aggregation did not take place because of the loose binding of the ligands. Furthermore, the fraction of the azobenzene component had to be between 0.23 and 0.33 in order to facilitate the light-driven aggregation/dispersion processes, but not to poison the catalyst. A limitation of this catalysis is that both the reactant and the product have affinity to gold surface that leads to the displacement of the azobenzene ligands, which eliminates the element of reversible control from the system.

A similar control was realized by using POMs decorated with azobenzene ligands identical as shown in Fig. 30.[69] In a toluene/THF medium, P_5W_{30} type POM with a *trans*-azobenzene ligand shell was well-dispersed and showed high activity in the oxidation of methyl phenyl sulfide with H_2O_2 (up to 90% conversion after 80 min at 30°C). When the POM-azobenzene catalyst was irradiated with UV light, upon which the *trans*-azobenzenes were converted to the *cis* form, the particles were aggregated and exhibited 15-times less catalytic activity in the oxidation reaction. This reversible aggregation behavior was also beneficial to recycle and reuse the catalyst by simple filtration following UV-irradiation. The POM-azobenzene system could be reused up to 5 times without considerable loss of activity.

While in the previous examples NP aggregation was used as a means to turn off catalytic activity, Klajn and coworkers demonstrated that the confinement effect associated with azobenzene isomerization-triggered aggregation can be an efficient tool for accelerating chemical reactions[70] (Fig. 35A).

They demonstrated that dimerization of anthracene derivatives (Fig. 35B) proceeds up to two orders of magnitude faster when occurs in the confinement of the NPs compared to the NP-free conditions. Furthermore, likely due to preorganizing interactions with the dipolar *cis*-azobenzenes, the product mixture was enriched in the kinetic *syn*-isomer, which is absent under NP-free conditions. This concept was also exploited

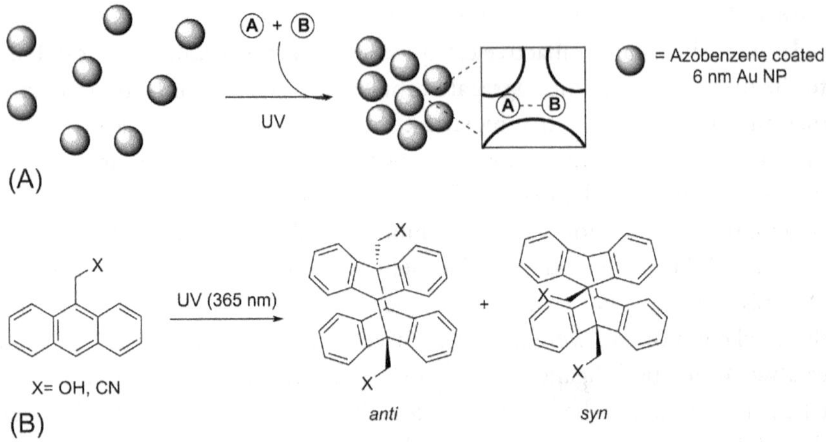

Fig. 35 (A) Azobenzene photoisomerization-triggered aggregation of gold NPs in the presence of reactants; (B) The dimerization of anthracene derivatives.

in the hydrolysis of acetals and, by using a mixed monolayer of azobenzene and a chiral ligand, enantioselective trapping was demonstrated as well. It has to be pronounced that this is one of the few examples of photoswitchable catalysis when not only the catalytic activity could be controlled, but also stereoselectivity can be affected. On the other hand, similarly to the previously discussed example, the limitation of this concept is the loss of photocontrol over the aggregation phenomena by ligand exchange of the azobenzene with the reaction substrates.

4.2.3 Controlling catalysis by surface ligand rearrangement

Metal nanoparticles applicable for catalytic transformations are mostly prepared by using a capping agent to stabilize them and prevent their aggregation. Encoding the ability of controllable rearrangement into the capping agent could provide a possibility to tune the catalytic activity of the NPs. Knecht and coworkers developed peptide-based capping agents that were able to stabilize Au NPs, while a covalently bound azobenzene photoswitch could affect the conformation of the peptide component on the surface of the NPs.[71, 72] The change in the structure of the capping agent was enough to alter the availability of the metal surface for catalytic reactions.[73] As a model reaction, 4-nitrophenol reduction with $NaBH_4$ was used. The effect of the azobenzene isomerization on the catalytic activity of the Au NP/peptide system was found to be dependent on the position of the azobenzene within the peptide sequence.[74] When the azobenzene was attached to the

N-terminus of the peptide, the activity of the biohybrid catalyst was greater if the azobenzene was in its *trans* state compared to that in the *cis* state. However, when the azobenzene was integrated as a mid-sequence into the peptide chain, the trend was the opposite. The greatest difference was found when the azobenzene moiety was linked to the C-terminus of the peptide. In this case, the catalyst having the azobenzene in *trans* configuration displayed very strong reduction capability, while in the *cis* configuration the reduction was inhibited in a broad temperature range.

4.2.4 Nanoparticle supported catalysis

Prins and coworkers used gold NPs (1.6 ± 0.4 nm) as a support for C_9 thiols with a 1,4,7-triazacyclononane Zn^{2+} complex headgroup, which has been shown to catalyze the aqueous transphosphorylation reaction of 2-hydroxypropyl-4-nitrophenylphosphate, a model substrate for RNA-hydrolysis.[75] To introduce an element of control, the catalytic NPs were combined with a photochromic azobenzene derivative, 4-(phenylazo)-benzoic acid, which could be exploited as light-sensitive cofactor in the catalytic process[76] (Fig. 36). The *trans* isomer of the azobenzene showed affinity to the monolayer on the gold NPs due to the favorable hydrophobic interaction with the apolar part of the alkyl thiols. This affinity was

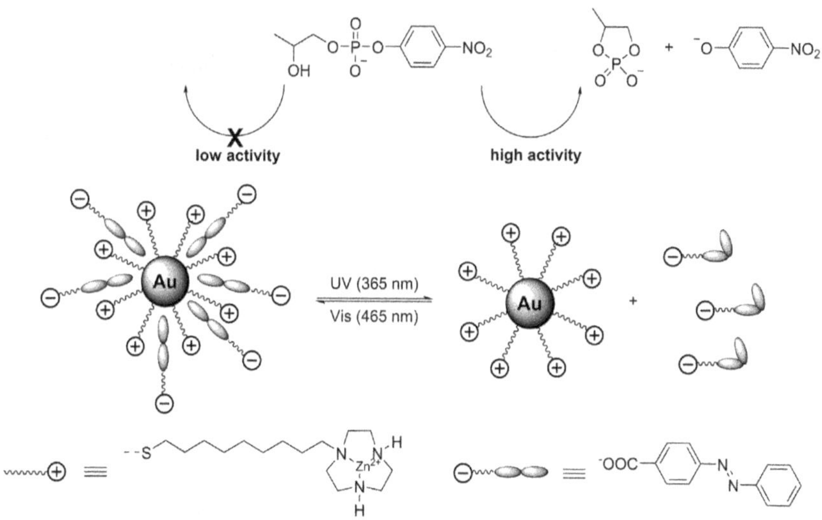

Fig. 36 Catalytic Au NPs combined with a photochromic azobenzene derivative as a light-sensitive cofactor to control transphosphorylation reaction.

about 2.2 times higher compared to that of the *cis* form with increased polarity, and hence, lower affinity to the monolayer.

This difference in affinity of the two forms of the azobenzene carboxylate to the NP system could be used to regulate the catalytic activity of the Zn-complex in an aqueous transphosphorylation reaction (Fig. 36). In its *trans* form, the azobenzene co-assembled with the catalytic monolayer on the NPs and hindered the reaction, while the more polar *cis* form was repelled from the surface, allowing higher catalytic activity. The regulation of catalysis could be repeated for several irradiation cycles, without the degradation of the catalytic NPs.

A photoswitchable gold NP (~3 nm)-supported organocatalyst was reported by Sashuk and coworkers.[77] The catalyst design involved a photoresponsive azobenzene unit, to which a proline organocatalyst was appended (Fig. 37A). In the *trans* state of the azobenzene, the proline unit was exposed to the reaction medium, while in the *cis* state of the switch, it was hidden from the reactants. The effect of this structural control was tested in the aldol

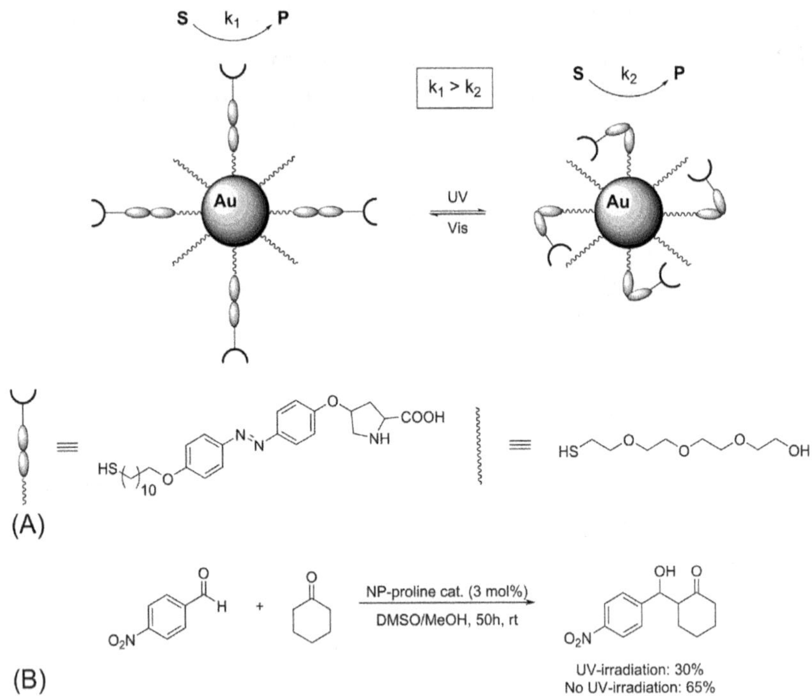

Fig. 37 (A) A photoswitchable, gold NP-supported azobenzene-proline organocatalyst; (B) Light-controlled catalysis in aldol reaction.

reaction of 4-nitrobenzaldehye with cyclohexanone (Fig. 37B). As expected, the two states of the catalyst showed different activity in the reaction, with the *trans* providing higher conversions. Importantly, the density of the azobenzene derivative on the NP surface had a strong impact on the efficiency of the system. If the density was too low (diluted with PEG), the *cis* form of the catalytic switch showed similar activity to the *trans* isomer, as it was easily accessible to the substrates. On the other hand, when the density was too high, the isomerization of the switches was hindered by steric interactions. The ideal composition of the surface layer was found to be azobenzene/PEG 1:1.

5 Conclusions and outlook

From the highlighted examples, it is clear that photochromic molecular switches are well-grounded and increasingly considered in the field of catalysis. The diversity of their applications is gradually expanding and departs from the original goal of activity control (ON/OFF, high/low) in homogeneous systems towards controlling selectivity, including stereoselectivity, controlling phase transfer, facilitating catalyst recycling and reuse, and the development of heterogeneous and supported systems. These latter examples are well in line with the principles and aims of green chemistry. On the other hand, parallel with the expansion of targeted applications, the number of photochromic scaffolds used in catalysis remains limited. The otherwise widely studied spiropyrans[78, 79] and the emerging donor-acceptor Stenhouse adducts[7] are rarely used, although the change in their polarity could be beneficial in catalytic applications as precedented in rather isolated examples.[80, 81] Another promising direction that has been moderately explored is the use of molecular capsules and cages that contain photochromic units as means of controlling and exploiting confinement effects in reaction rates and selectivities that are otherwise not easy to reach.[82, 83] New transformations and catalytic applications are expected from the combination of biological systems with photochromic control elements, which is a strongly developing area with already a number of interesting examples.[84–89]

Acknowledgments

Financial support from the National Research, Development and Innovation Office, Hungary [NKFIH Grant FK 123760], and through the Lendület Program of of the Hungarian Academy of Sciences is gratefully acknowledged.

References

1. Balzani, V.; Credi, A.; Venturi, M. *Molecular Devices and Machines: Concepts and Perspectives for the Nanoworld*; Wiley-VCH: Weinheim, Germany, 2008.
2. Browne, W. R.; Feringa, B. L., Eds. *Molecular Switches*; 2nd ed.; Wiley-VCH: Weinheim, Germany, 2011.
3. Sheldon, R. A.; Arends, I.; Hanefeld, U. *Green Chemistry and Catalysis*; Wiley-VCH: Weinheim, Germany, 2007.
4. Anastas, P.; Eghbali, N. Green Chemistry: Principles and Practice. *Chem. Soc. Rev.* **2010**, *39*, 301–312.
5. Török, B.; Dransfield, T., Eds. *Green Chemistry—An Inclusive Approach*; 1st ed.; Elsevier: Amsterdam, Netherlands, 2018.
6. van Dijken, D. J.; Kovaříček, P.; Ihrig, S. P.; Hecht, S. Acylhydrazones As Widely Tunable Photoswitches. *J. Am. Chem. Soc.* **2015**, *137*, 14982–14991.
7. Lerch, M. M.; Szymański, V.; Feringa, B. L. The (Photo)Chemistry of Stenhouse Photoswitches: Guiding Principles and System Design. *Chem. Soc. Rev.* **2018**, *47*, 1910–1937.
8. Petermayer, C.; Dube, H. Indigoid Photoswitches: Visible Light Responsive Molecular Tools. *Acc. Chem. Res.* **2018**, *51*, 1153–1163.
9. Stoll, R. S.; Hecht, S. Artificial Light-Gated Catalyst Systems. *Angew. Chem. Int. Ed.* **2010**, *49*, 5054–5075.
10. Neilson, B. M.; Bielawski, C. W. Illuminating Photoswitchable Catalysis. *ACS Catal.* **2013**, *3*, 1874–1885.
11. Göstl, R.; Senf, A.; Hecht, S. Remote-Controlling Chemical Reactions by Light: towards Chemistry with High Spatio-Temporal Resolution. *Chem. Soc. Rev.* **2014**, *43*, 1982–1996.
12. Vlatković, M.; Collins, B. S. L.; Feringa, B. L. Dynamic Responsive Systems for Catalytic Function. *Chem. A Eur. J.* **2016**, *22*, 17080–17111.
13. Dorel, R.; Feringa, B. L. Photoswitchable Catalysis Based on the Isomerisation of Double Bonds. *Chem. Commun.* **2019**, *55*, 6477–6486.
14. Bandara, H. M. D.; Burdette, S. C. Photoisomerization in Different Classes of Azobenzene. *Chem. Soc. Rev.* **2012**, *41*, 1809–1825.
15. Tian, H.; Yang, S. Recent Progresses on Diarylethene Based Photochromic Switches. *Chem. Soc. Rev.* **2004**, *33*, 85–97.
16. Irie, M.; Fukaminato, T.; Matsuda, K.; Kobatake, S. Photochromism of Diarylethene Molecules and Crystals: Memories, Switches, and Actuators. *Chem. Rev.* **2014**, *114*, 12174–12277.
17. Feringa, B. L. The Art of Building Small: from Molecular Switches to Motors (Nobel Lecture). *Angew. Chem. Int. Ed.* **2017**, *56*, 11060–11078.
18. van Leeuwen, T.; Lubbe, A. S.; Štacko, P.; Wezenberg, S. J.; Feringa, B. L. Dynamic Control of Function by Light-Driven Molecular Motors. *Nat. Rev. Chem.* **2017**, *1*, 0096.
19. Pollard, M. M.; Meetsma, A.; Feringa, B. L. A Redesign of Light-Driven Rotary Molecular Motors. *Org. Biomol. Chem.* **2008**, *6*, 507–512.
20. Ueno, A.; Takahashi, K.; Osa, T. Photoregulation of Catalytic Activity of β-Cyclodextrin by an Azo Inhibitor. *J. Chem. Soc. Chem. Commun.* **1980**, 837–838.
21. Ueno, A.; Takahashi, K.; Osa, T. Photocontrol of Catalytic Activity of Capped Cyclodextrin. *J. Chem. Soc. Chem. Commun.* **1981**, 94–96.
22. Lee, W.-S.; Ueno, A. Photocontrol of the Catalytic Activity of a b-Cyclodextrin Bearing Azobenzene and Histidine Moieties as a Pendant Group. *Macromol. Rapid Commun.* **2001**, *22*, 448–450.

23. Würthner, F.; Rebek, J., Jr. Light-Switchable Catalysis in Synthetic Receptors. *Angew. Chem. Int. Ed. Engl.* **1995**, *34*, 446–448.
24. Würthner, F.; Rebek, J., Jr. Photoresponsive Synthetic Receptors: Binding Properties and Photocontrol of Catalytic Activity. *J. Chem. Soc. Perkin Trans.* **1995**, *2*, 1727–1734.
25. Pieters, R. J.; Huc, I.; Rebek, J., Jr. Reciprocal Template Effects in a Replication Cycle. *Angew. Chem. Int. Ed. Engl.* **1994**, *33*, 1579–1581.
26. Dalko, P. I.; Moisan, L. In the Golden Age of Organocatalysis. *Angew. Chem. Int. Ed.* **2004**, *43*, 5138–5175.
27. MacMillan, D. W. C. The Advent and Development of Organocatalysis. *Nature* **2008**, *455*, 304–308.
28. Imahori, T.; Yamaguchi, R.; Kurihara, S. Azobenzene-Tethered Bis(Trityl Alcohol) as a Photoswitchable Cooperative Acid Catalyst for Morita–Baylis–Hillman Reactions. *Chem. A Eur. J.* **2012**, *18*, 10802–10807.
29. Unni, A. K.; Takenaka, N.; Yamamoto, H.; Rawal, V. H. Axially Chiral Biaryl Diols Catalyze Highly Enantioselective Hetero-Diels-Alder Reactions through Hydrogen Bonding. *J. Am. Chem. Soc.* **2005**, *127*, 1336–1337.
30. Connon, S. J. Organocatalysis Mediated by (Thio)Urea Derivatives. *Chem. A Eur. J.* **2006**, *12*, 5418–5427.
31. Doyle, A. G.; Jacobsen, E. N. Small-Molecule H-Bond Donors in Asymmetric Catalysis. *Chem. Rev.* **2007**, *107*, 5713–5743.
32. Osorio-Planes, L.; Rodríguez-Escrich, C.; Pericàs, M. A. Photoswitchable Thioureas for the External Manipulation of Catalytic Activity. *Org. Lett.* **2014**, *16*, 1704–1707.
33. Wang, J.; Feringa, B. L. Dynamic Control of Chiral Space in a Catalytic Asymmetric Reaction Using a Molecular Motor. *Science* **2011**, *331*, 1429–1432.
34. Vlatković, M.; Bernardi, L.; Otten, E.; Feringa, B. L. Dual Stereocontrol over the Henry Reaction Using a Light- and Heat-Triggered Organocatalyst. *Chem. Commun.* **2014**, *50*, 7773–7775.
35. Vlatković, M.; Volarić, J.; Collins, B. S. L.; Bernardi, L.; Feringa, B. L. Dynamic Control over Catalytic Function Using Responsive Bisthiourea Catalysts. *Org. Biomol. Chem.* **2017**, *15*, 8285–8294.
36. Pizzolato, S. F.; Collins, B. S. L.; van Leeuwen, T.; Feringa, B. L. Bifunctional Molecular Photoswitches Based on Overcrowded Alkenes for Dynamic Control of Catalytic Activity in Michael Addition Reactions. *Chem. A Eur. J.* **2017**, *23*, 6174–6184.
37. Eisenreich, F.; Kathan, M.; Dallmann, A.; Ihrig, S. P.; Schwaar, T.; Schmidt, B. M.; Hecht, S. A Photoswitchable Catalyst System for Remotecontrolled (Co)Polymerization In Situ. *Nature Catal.* **2018**, *1*, 516–522.
38. Viehmann, P.; Hecht, S. Design and Synthesis of a Photoswitchable Guanidine Catalyst. *Beilstein J. Org. Chem.* **2012**, *8*, 1825–1830.
39. Peters, M. V.; Stoll, R. S.; Kühn, A.; Hecht, S. Photoswitching of Basicity. *Angew. Chem. Int. Ed.* **2008**, *47*, 5968–5972.
40. Pescher, M. D.; van Wilderen, L. J. G. W.; Grützner, S.; Slavov, C.; Wachtveitl, J.; Hecht, S.; Bredenbeck, J. Ultrafast Light-Driven Substrate Expulsion from the Active Site of a Photoswitchable Catalyst. *Angew. Chem. Int. Ed.* **2017**, *56*, 12092–12096.
41. Stoll, R. S.; Peters, M. V.; Kuhn, A.; Heiles, S.; Goddard, R.; Bühl, M.; Thiele, C. M.; Hecht, S. Photoswitchable Catalysts: Correlating Structure and Conformational Dynamics with Reactivity by a Combined Experimental and Computational Approach. *J. Am. Chem. Soc.* **2009**, *131*, 357–367.
42. Stoll, R. S.; Hecht, S. Immobilization of a Photoswitchable Piperidine Base. *Org. Lett.* **2009**, *11*, 4790–4793.
43. Nojiri, A.; Kumagai, N.; Shibasaki, M. Reversible Heterochiral Aggregation/Dissociation of Bis(2-Hydroxyphenyl)Diamides Driven by UV/Vis Irradiation. *Angew. Chem. Int. Ed.* **2012**, *51*, 2137–2141.

44. Nojiri, A.; Kumagai, N.; Shibasaki, M. In Situ Manipulation of Catalyst Performance Via the Photocontrolled Aggregation/Dissociation State of the Catalyst. *Chem. Commun.* **2013**, *49*, 4628–4630.
45. Neilson, B. M.; Bielawski, C. W. Photoswitchable Organocatalysis: Using Light to Modulate the Catalytic Activities of N-Heterocyclic Carbenes. *J. Am. Chem. Soc.* **2012**, *134*, 12693–12699.
46. Teator, A. J.; Tian, Y.; Chen, M.; Lee, J. K.; Bielawski, C. W. An Isolable, Photoswitchable N-Heterocyclic Carbene: on-Demand Reversible Ammonia Activation. *Angew. Chem. Int. Ed.* **2015**, *54*, 11559–11563.
47. Neilson, B. M.; Bielawski, C. W. Photoswitchable NHC-Promoted Ring-Opening Polymerizations. *Chem. Commun.* **2013**, *49*, 5453–5455.
48. Sugimoto, H.; Kimura, T.; Inoue, S. Photoresponsive Molecular Switch to Control Chemical Fixation of CO_2. *J. Am. Chem. Soc.* **1999**, *121*, 2325–2326.
49. Cacciapaglia, R.; Di Stefano, S.; Mandolini, L. The Bis-Barium Complex of a Butterfly Crown Ether as a Phototunable Supramolecular Catalyst. *J. Am. Chem. Soc.* **2003**, *125*, 2224–2227.
50. Bricout, H.; Banaszak, E.; Len, C.; Hapiot, F.; Monflier, E. Amphiphilic Photo-Isomerisable Phosphanes for Aqueous Organometallic Catalysis. *Chem. Commun.* **2010**, *46*, 7813–7815.
51. Priyadarshani, N.; Ginovska, B.; Bays, J. T.; Linehan, J. C.; Shaw, W. J. Photoswitching a Molecular Catalyst to Regulate CO_2 Hydrogenation. *Dalton Trans.* **2015**, *44*, 14854–14864.
52. Arif, T.; Cazorla, C.; Bogliotti, N.; Saleh, N.; Blanchard, F.; Gandon, V.; Métivier, R.; Xie, J.; Voituriez, A.; Marinetti, A. Bimetallic Gold(I) Complexes of Photoswitchable Phosphines: Synthesis and Uses in Cooperative Catalysis. *Cat. Sci. Technol.* **2018**, *8*, 710–715.
53. Telleria, A.; van Leeuwen, P. W. N. M.; Freixa, Z. Azobenzene-Based Ruthenium(II) Catalysts for Light-Controlled Hydrogen Generation. *Dalton Trans.* **2017**, *46*, 3569–3578.
54. Samachetty, H. D.; Branda, N. R. Integrating Molecular Switching and Chemical Reactivity Using Photoresponsive Hexatrienes. *Pure Appl. Chem.* **2006**, *78*, 2351–2359.
55. Ko, C.-C.; Yam, V. W.-W. Coordination Compounds with Photochromic Ligands: Ready Tunability and Visible Light-Sensitized Photochromism. *Acc. Chem. Res.* **2018**, *51*, 149–159.
56. Sud, D.; Norsten, T. B.; Branda, N. R. Photoswitching of Stereoselectivity in Catalysis Using a Copper Dithienylethene Complex. *Angew. Chem. Int. Ed.* **2005**, *44*, 2019–2021.
57. Neilson, B. M.; Bielawski, C. W. Photoswitchable Metal-Mediated Catalysis: Remotely Tuned Alkene and Alkyne Hydroborations. *Organometallics* **2013**, *32*, 3121–3128.
58. Xu, Z.; Cao, Y.; Patrick, B. O.; Wolf, M. O. Photoswitching of Copper(I) Chromophores with Dithienylethene-Based Ligands. *Chem. A Eur. J.* **2018**, *24*, 10315–10319.
59. Dwars, T.; Paetzold, E.; Oehme, G. Reactions in Micellar Systems. *Angew. Chem. Int. Ed.* **2005**, *44*, 7174–7199.
60. Drillaud, N.; Banaszak-Léonard, E.; Pezron, I.; Len, C. Synthesis and Evaluation of a Photochromic Surfactant for Organic Reactions in Aqueous Media. *J. Org. Chem.* **2012**, 77, 9553–9561.
61. Billamboz, M.; Mangin, F.; Drillaud, N.; Chevrin-Villette, C.; Banaszak-Léonard, E.; Len, C. Micellar Catalysis Using a Photochromic Surfactant: Application to the Pd-Catalyzed Tsuji – Trost Reaction in Water. *J. Org. Chem.* **2014**, *79*, 493–500.
62. Gao, M.; Tan, R.; Hao, P.; Zhang, Y.; Deng, J.; Yin, D. Ultraviolet-Responsive Self-Assembled Metallomicelles for Photocontrollable Catalysis of Asymmetric Sulfoxidation in Water. *RSC Adv.* **2017**, 7, 54570–54580.

63. Yang, Y.; Zhang, B.; Wang, Y.; Yue, L.; Li, W.; Wu, L. A Photo-Driven Polyoxometalate Complex Shuttle and its Homogeneous Catalysis and Heterogeneous Separation. *J. Am. Chem. Soc.* **2013**, *135*, 14500–14503.
64. Zhu, L.; Yan, H.; Ang, C. Y.; Nguyen, K. T.; Li, M.; Zhao, Y. Photoswitchable Supramolecular Catalysis by Interparticle Host–Guest Competitive Binding. *Chem. A Eur. J.* **2012**, *18*, 13979–13983.
65. Tang, S.-P.; Zhou, Y.-H.; Chen, H.-Y.; Zhao, C.-Y.; Mao, Z.-W.; Ji, L.-N. Ester Hydrolysis by a Cyclodextrin Dimer Catalyst With a Tridentate N,N′,N″-Zinc Linking Group. *Chem. Asian J.* **2009**, *4*, 1354–1360.
66. Zhao, R.; Hu, J.; Niu, C.; Li, Y.; Hu, M.; Liu, R.; Li, S. A Smart Nanoreactor with Photo-Responsive Molecular Switches for Controlling Catalytic Reactions. *J. Mater. Chem. C* **2016**, *4*, 4748–4755.
67. Peng, L.; You, M.; Wu, C.; Han, D.; Öçsoy, I.; Chen, T.; Chen, Z.; Tan, W. Reversible Phase Transfer of Nanoparticles Based on Photoswitchable Host-Guest Chemistry. *ACS Nano* **2014**, *8*, 2555–2561.
68. Wei, Y.; Han, S.; Kim, J.; Soh, S.; Grzybowski, B. A. Photoswitchable Catalysis Mediated by Dynamic Aggregation of Nanoparticles. *J. Am. Chem. Soc.* **2010**, *132*, 11018–11020.
69. Yang, Y.; Yue, L.; Li, H.; Maher, E.; Li, Y.; Wang, Y.; Wu, L.; Yam, V. W.-W. Photo-Responsive Self-Assembly of an Azobenzene-Ended Surfactant-Encapsulated Polyoxometalate Complex for Modulating Catalytic Reactions. *Small* **2012**, *8*, 3105–3110.
70. Zhao, H.; Sen, S.; Udayabhaskararao, T.; Sawczyk, M.; Kučanda, K.; Manna, D.; Kundu, P. K.; Lee, J.-W.; Král, P.; Klajn, R. Reversible Trapping and Reaction Acceleration within Dynamically Self-Assembling Nanoflasks. *Nat. Nanotechnol.* **2016**, *11*, 82–88.
71. Tang, Z.; Lim, C.-K.; Palafox-Hernandez, J. P.; Drew, K. L. M.; Li, Y.; Swihart, M. T.; Prasad, P. N.; Walsh, T. R.; Knecht, M. R. Triggering Nanoparticle Surface Ligand Rearrangement via External Stimuli: Light-Based Actuation of Biointerfaces. *Nanoscale* **2015**, 7, 13638–13645.
72. Palafox-Hernandez, J. P.; Lim, C.-K.; Tang, Z.; Drew, K. L. M.; Hughes, Z. E.; Li, Y.; Swihart, M. T.; Prasad, P. N.; Knecht, M. R.; Walsh, T. R. Optical Actuation of Inorganic/Organic Interfaces: Comparing Peptide-Azobenzene Ligand Reconfiguration on Gold and Silver Nanoparticles. *ACS Appl. Mater. Interfaces* **2016**, *8*, 1050–1060.
73. Lawrence, R. L.; Scola, B.; Li, Y.; Lim, C.-K.; Liu, Y.; Prasad, P. N.; Swihart, M. T.; Knecht, M. R. Remote Optically Controlled Modulation of Catalytic Properties of Nanoparticles through Reconfiguration of the Inorganic/Organic Interface. *ACS Nano* **2016**, *10*, 9470–9477.
74. Lawrence, R. L.; Hughes, Z. E.; Cendan, V. J.; Liu, Y.; Lim, C.-K.; Prasad, P. N.; Swihart, M. T.; Walsh, T. R.; Knecht, M. R. Optical Control of Nanoparticle Catalysis Influenced by Photoswitch Positioning in Hybrid Peptide Capping Ligands. *ACS Appl. Mater. Interfaces* **2018**, *10*, 33640–33651.
75. Zaupa, G.; Mora, C.; Bonomi, R.; Prins, L. J.; Scrimin, P. Catalytic Self-Assembled Monolayers on au Nanoparticles: the Source of Catalysis of a Transphosphorylation Reaction. *Chem. A Eur. J.* **2011**, *17*, 4879–4889.
76. Neri, S.; Martin, S. G.; Pezzato, C.; Prins, L. J. Photoswitchable Catalysis by a Nanozyme Mediated by a Light-Sensitive Cofactor. *J. Am. Chem. Soc.* **2017**, *139*, 1794–1797.
77. Szewczyk, M.; Sobczak, G.; Sashuk, V. Photoswitchable Catalysis by a Small Swinging Molecule Confined on the Surface of a Colloidal Particle. *ACS Catal.* **2018**, *8*, 2810–2814.
78. Klajn, R. Spiropyran-Based Dynamic Materials. *Chem. Soc. Rev.* **2014**, *43*, 148–184.

79. Kortekaas, L.; Browne, W. R. The Evolution of Spiropyran: Fundamentals and Progress of an Extraordinarily Versatile Photochrome. *Chem. Soc. Rev.* **2019**, *48*, 3406–3424.
80. Niazov, T.; Shlyahovsky, B.; Willner, I. Photoswitchable Electrocatalysis and Catalyzed Chemiluminescence Using Photoisomerizable Monolayer-Functionalized Surfaces and Pt Nanoparticles. *J. Am. Chem. Soc.* **2007**, *129*, 6374–6375.
81. Rifaie-Graham, O.; Ulrich, S.; Galensowske, N. F. B.; Balog, S.; Chami, M.; Rentsch, D.; Hemmer, J. R.; de Alaniz, J. R.; Boesel, L. F.; Burns, N. Wavelength-Selective Light-Responsive DASA-Functionalized Polymersome Nanoreactors. *J. Am. Chem. Soc.* **2018**, *140*, 8027–8036.
82. Han, M.; Michel, R.; He, B.; Chen, Y.-S.; Stalke, D.; John, M.; Clever, G. H. Light-Triggered Guest Uptake and Release by a Photochromic Coordination Cage. *Angew. Chem. Int. Ed.* **2013**, *52*, 1319–1323.
83. Park, J.; Feng, D.; Yuan, S.; Zhou, H.-C. Photochromic Metal–Organic Frameworks: Reversible Control of Singlet Oxygen Generation. *Angew. Chem. Int. Ed.* **2015**, *54*, 430–435.
84. Agarwal, P. K.; Schultz, C.; Kalivretenos, A.; Ghosh, B.; Broedel, S. E., Jr. Engineering a Hyper-Catalytic Enzyme by Photoactivated Conformation Modulation. *J. Phys. Chem. Lett.* **2012**, *3*, 1142–1146.
85. Chen, Z.; Zhou, L.; Bing, W.; Zhang, Z.; Li, Z.; Ren, J.; Qu, X. Light Controlled Reversible Inversion of Nanophosphor-Stabilized Pickering Emulsions for Biphasic Enantioselective Biocatalysis. *J. Am. Chem. Soc.* **2014**, *136*, 7498–7504.
86. Kameta, N.; Manaka, Y.; Akiyama, H.; Shimizu, T. Bioreactors Based on Enzymes Encapsulated in Photoresponsive Transformable Nanotubes and Nanocoils End-Capped with Magnetic Nanoparticles. *Adv. Biosys.* **2018**, 1700214.
87. Wang, F.; Ju, E.; Guan, Y.; Ren, J.; Qu, X. Light-Mediated Reversible Modulation of ROS Level in Living Cells by Using an Activity-Controllable Nanozyme. *Small* **2017**, *13*, 1603051.
88. Wang, F.; Zhang, Y.; Du, Z.; Ren, J.; Qu, X. Designed Heterogeneous Palladium Catalysts for Reversible Light-Controlled Bioorthogonal Catalysis in Living Cells. *Nat. Commun.* **2018**, *9*, 1209.
89. Wang, S.; Yue, L.; Li, Z.-Y.; Zhang, J.; Tian, H.; Willner, I. Light-Induced Reversible Reconfiguration of DNA-Based Constitutional Dynamic Networks: Application to Switchable Catalysis. *Angew. Chem. Int. Ed.* **2018**, *57*, 8105–8109.

CHAPTER EIGHT

Photochemical rearrangements in organic synthesis and the concept of the photon as a traceless reagent

Corentin Lefebvre and Norbert Hoffmann
CNRS, Université de Reims Champagne-Ardenne, ICMR, Equipe de Photochimie, UFR Sciences, Reims, France

1 Introduction

In 1912 the famous Italian chemist Giacomo Ciamician (1857–1922) published his vision on a sustainable nonpolluting, a clean chemical industry in which chemical transformations are mainly carried out with light as it is done by green plants.[1] Four years before, he exposed even more precise ideas on sustainable chemistry. In a lecture before the French Chemical Society, he stated:

> *Mais, outre les ferments, il y a un autre agent qui est de la plus grande importance, pour les plantes du moins, et dont l'influence sur les processus organiques mérite une étude profonde: c'est la lumière.*[2]

In this way, he established a link between enzymatic (catalytic) and photochemical reactions. This event is considered as the beginning of green chemistry.[3] Almost in the same time, Emanuele Paternò also recognized the interest of reactions induced by light absorption for organic synthesis.[4] The interest of photochemical reactions in connection with sustainable chemistry and organic synthesis has then been neglected for a long time. However, since about 10 years in the academic research and since about 3 years in industrial research[5], we observe a bright renaissance of this research domain. Photocatalytic reactions with visible light play an important role in this development.[6]

Photochemical reactions of organic compounds are characterized by the fact that these compounds change their electronic configuration when they are excited by light absorption.[7,8] Consequently, their chemical reactivity is

Nontraditional Activation Methods in Green and Sustainable Applications
https://doi.org/10.1016/B978-0-12-819009-8.00008-6
© 2021 Elsevier Inc.
All rights reserved.

considerably modified, and compounds or compound families become available, which cannot or difficultly be synthesized using classical organic reactions.[9,10] Thus the photon has to be considered as a reagent that does not only increase the reactivity but also modifies it.[8] The traditional strong links between organic and physical photochemistry permit a profound understanding of such reactions that also facilitates their optimization.[11] Photochemical reactivity of a functional group is often complementary to the corresponding ground state reactivity.

In connection with sustainable chemistry, it is important to note that the photon is a traceless reagent.[12,13] Its application reduces the formation of side products and waste. The impact for green chemistry is particularly high when these reactions are carried out with sunlight as renewable energy resource.[13,14]

Photochemical reactions can also be induced using different kinds of photosensitization.[15] In such cases the light-absorbing compound transfers its excitation energy onto the substrate that reacts at its excited state. Other forms of sensitization such as processes involving electron or hydrogen atom transfer between the sensitizer and the substrate or reaction intermediates are also observed and frequently applied.

In this chapter, we mainly focus on photochemical rearrangements resulting from direct light absorption. Owing to their interest in many domains such as the synthesis of biologically active compounds or materials science, mainly transformations with heterocyclic compounds will be discussed. Nevertheless, it should be mentioned that a lot of photochemical rearrangements leading to complex isocyclic compounds have been and are currently studied.[16] As in the case of many photochemical reactions, the concept of the photon as traceless reaction is particularly well applied in such rearrangements.[12]

2 Photochemical position isomerization of heterocycles

Five-membered heterocyclic compounds are encountered in many biologically active compounds. Especially in the case of two or more heteroatoms, the position of these atoms and the substitution pattern significantly affect their properties. This can nicely be shown for the case of the heteroaromatic compounds such as imidazoles and pyrazoles possessing two nitrogen atoms. In the first case, both nitrogens are in positions 1 and 3, while in the second case, they are in positions 1 and 2. Typical examples of different biological activity of both position isomers are depicted in Fig. 1. The

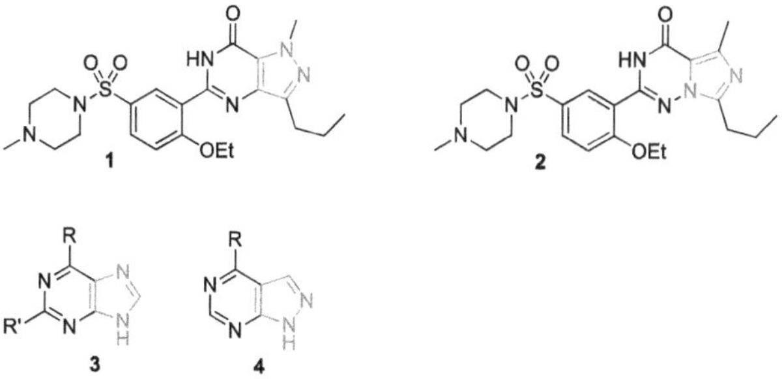

Fig. 1 The pyrazole derivative **1** possesses a higher activity than the corresponding imidazole derivative **2**. Similar effects are observed when replacing a purine compound **3** by its pyrazole position isomer **4**.

phosphodiesterase 5 (PDE) inhibition is increased when the pyrazole ring in **1** is replaced by an imidazole ring in **2**.[17] Different activities have been observed for purine (**3**) and corresponding pyrazole (**4**) derivatives as inhibitors of RNA or DNA glycosylase activity of Shiga toxin 1.[18]

Various synthesis methods have been developed to prepare imidazoles, pyrazoles, or similar heterocyclic compound families.[19] Of course the scope of each of these methods is limited. Photochemical rearrangements, in particular the transformation of one family into the other, could significantly extend the structural diversity. Pyrazole derivatives can be easily transformed into corresponding imidazoles (Scheme 1).[20] In a sequence of photochemical electrocyclization steps, bicyclic intermediates **8**, **9**, and **10** are generated from **5** involving $[_{\pi}2 + {}_{\sigma}2]$ electrocyclic reactions (blue arrows) and 1,3-sigmatopic shifts (red arrows). In a 1,3-sigmatropic shift, final compounds **6** and **7** are formed. The product ratio strongly depends on the reaction temperature. Thus the pyrazole **13** was transformed into compounds **14**, **15**, and **16** in different ratios (Scheme 2).[21] Intermediates such as **8**, **9**, or **10** (compare Scheme 1) are involved in the formation of **14** and **15**, while intermediates like **11** or **12** are involved in the formation **16**. More generally in such reactions, a competition between both mechanisms is discussed.[22] All these reactions are part of the larger family of circumambulatory rearrangements.[23]

Similar reactions have been carried out with benzopyrazoles such as **17** (Scheme 3).[24] Photochemical position isomerization yields the corresponding benzoimidazole derivatives **18**. Much better yields were obtained in the corresponding reaction of isomers **19** carrying an alkyl substituent on the nitrogen atom in 2 position. This compound can also be considered as being derived from ortho-benzoquinone. Compounds **19** were efficiently transformed into the benzoimidazoles **20**. In this reaction,

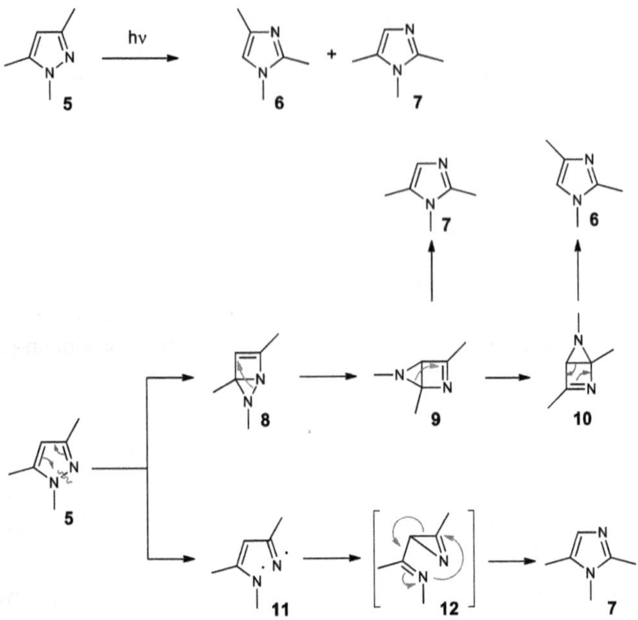

Scheme 1 Photochemical rearrangements of the pyrazole and imidazoles derivatives.

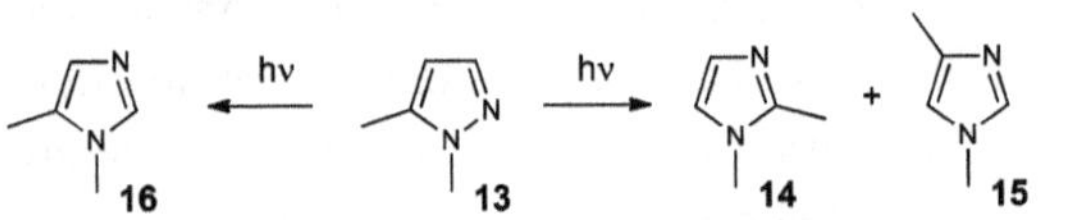

T(°C)	product ratio: **14/15**
30	1.6
10	3.4
0	6
-10	9
-30	23

Scheme 2 Temperature dependence of the product ratio **14/15**.

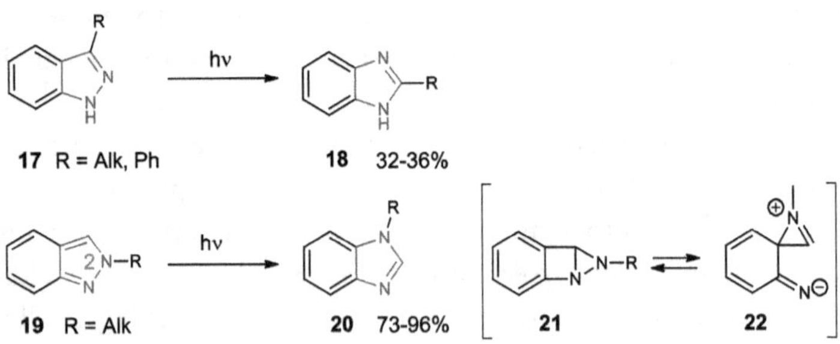

Scheme 3 Photochemical transformation of benzopyrazoles into corresponding benzoimidazoles. The irradiation has been carried out with high-pressure Hg-vapor lamps.

intermediates **21** and **22** are most probably involved.[25] In a reaction at low temperature (−60°C), the formation of compound **21** was observed. Upon warm-up, **21** yielded again the starting product **19**. For this reason the formation of an additional intermediate (**22**) was discussed.

Isoxazoles possess a similar structure than pyrazoles in which one nitrogen is replaced by an oxygen atom. Consequently, similar position isomerizations are observed with this compound family. Upon irradiation, compound **23** was transformed into compounds **24**, **25**, and **26** along with small amounts of benzoylacetone (Scheme 4).[26] In an extended mechanistic study, the azirine compound **24** was irradiated at $\lambda = 300$ nm, yielding the isoxazole **23**. When the same compound was irradiated at $\lambda = 254$ nm, oxazoles **25** and **26** were formed. It was suggested that in the first case, a zwitterion intermediate is involved, while in the second case, a diradical intermediate at the triplet state should be formed. This is an example that infringes the general accepted principle that a photochemical reaction always occurs from the lowest excited singlet or triplet state as it was formulated in analogy with the Kasha and Vavilov rules (see p. 40 in Ref. 7). Especially in cases of intramolecular reactions that are faster than interconversion processes, such infringes are observed. Similar reactions has been carried out with corresponding benz- or naphthisoxazoles.[27] Azirines are interesting synthesis intermediates. For example, in situ addition of β-dicarbonyl compounds leads to highly substituted pyrroles.[28]

Also, sulfur analogue isothiazoles were transformed in the same way. Upon irradiation, 4-methylisothiazole **27** yielded 4-methylthiazole **28** (Scheme 5).[29,30] In an acidic medium, protonation may occur at the nitrogen atom. Under these conditions the reaction is reversible. Under similar reaction conditions a migration of the phenyl substituent in the thiazole **29**

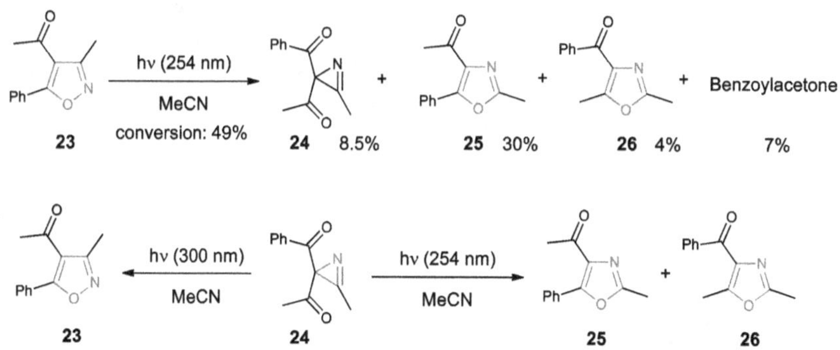

Scheme 4 Photochemical rearrangement of an acylisoxazole compound.

took place, leading to the regioisomer **30** along with small amounts of the isothiazole **31**.[31] As shown by a computational study, bicyclic intermediates (compare **8**, **9**, and **10** in Scheme 1) are involved in the main reaction.[32] Such migrations of substituents are also observed in photochemical rearrangements of oxazoles. Thus the positions of the aryl substituents in **32** have been exchanged, leading to the formation of **33** (Scheme 5).[33] In this case, substantial amounts of the isoxazole **34** were also obtained.

More complex isoxazolone compounds such as **35** also provide an access to oxazole derivatives. The irradiation ($\lambda = 300$ nm) of this compound efficiently yielded the oxazole **36** (Scheme 6).[34] The reaction can be carried out either in acetonitrile or acetone as solvent. In the latter case, triplet sensitization should be considered. After photochemical excitation, decarboxylation takes place, and the intermediate **37** is formed. Cyclization leads to the final product **36**. Similar transformations have been performed using pyrolysis conditions.

Oxazoles or thiazoles play a key role in many biologically active compounds. For example, they are encountered in cyclopeptides such as

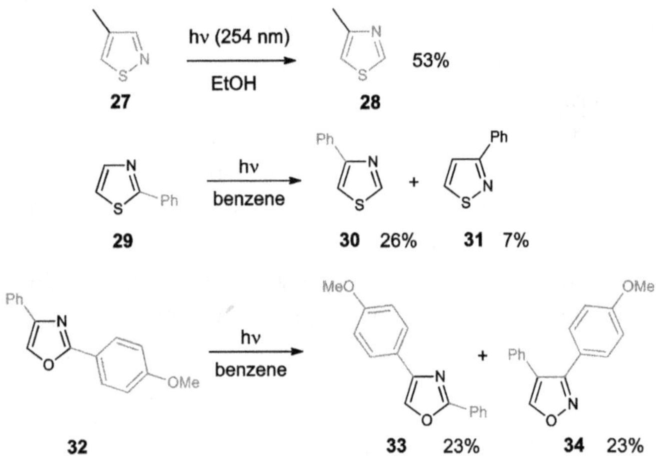

Scheme 5 Photochemical rearrangement of heteroatoms and migration of a substituent in isothiazole, thiazole, and oxazole derivatives.

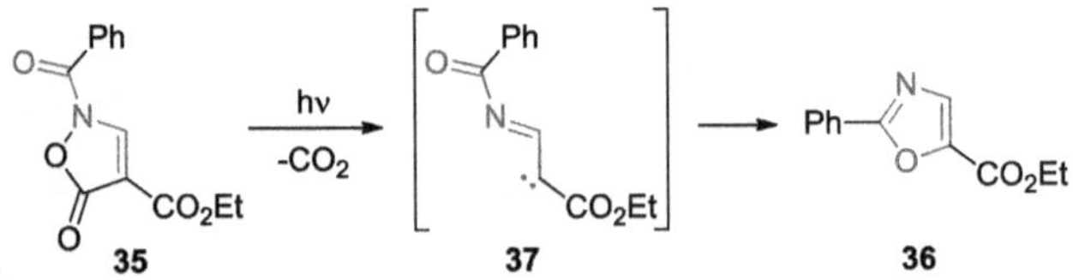

Scheme 6 Photolysis of an isoxazolone compound leading to an oxazole derivative.

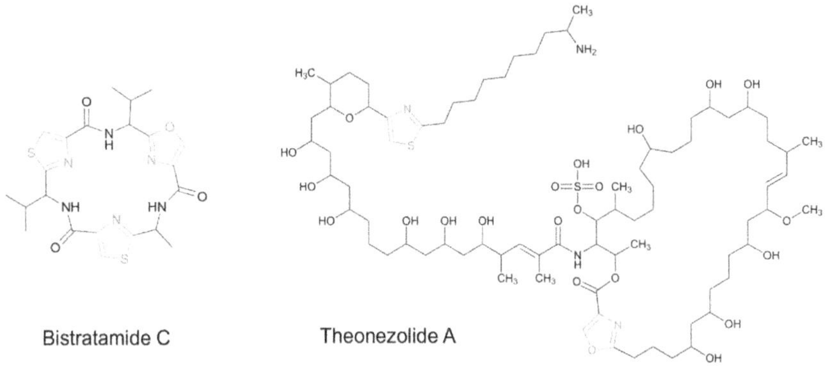

Fig. 2 Examples of natural products possessing oxazole and thiazole moieties.

bistratamide C or complex molecules such as theonezolide A (Fig. 2).[35,36] Reactions depicted in Scheme 5 may be applied to isomerization of partial structures in such compounds or to the synthesis of them or similar active compounds.[37]

Five-membered aromatics with three heteroatoms also play an important role in the domains of bioactive compounds or the preparation of new materials.[38] Photochemical reactions similar to those previously discussed have been performed with such compounds.[39] Often the cleavage of N—O bonds play a key role in such transformations.[40] The heteroatoms and the substituents change their places. Under photochemical conditions the heteroatoms in compound **38** rearrange, leading to the position isomer **39** (Scheme 7).[41] A urea function is transformed into a carbamate. When irradiated at longer wavelengths, the isoxadiazole derivative **40** was efficiently transformed into triazole derivative **41**.[42] Possessing a substituent at the middle nitrogen (position 2), it is a rather unusual triazole isomer. An attack of amine nitrogen atom of the side chain hydrazone on the nitrogen in 2 position (**42**) with formation of intermediate **43** was discussed as key step in this photochemical reaction.

Currently, photoredox catalysis and its application to organic synthesis considerably contribute to the renaissance of organic photochemistry.[6] Rearrangement reactions of heterocyclic compounds have also been carried out under these conditions.[43] When the isoxazolone derivative **44** is irradiated in the presence of the photoredox catalyst *fac*-Ir(ppy)$_3$, the oxazole derivative **45** is obtained in high yield (Scheme 8).[44] Also, in this case (compare Scheme 6), decarboxylation is involved. When the organic photoredox catalyst 4DPAIPN is used and in the presence of trimethylamine,

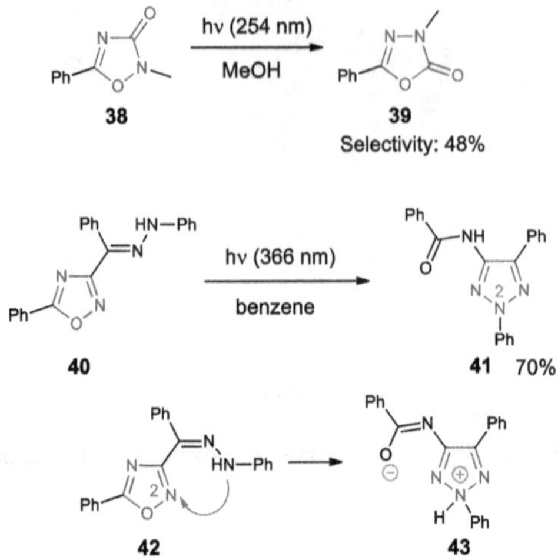

Scheme 7 Photochemical rearrangements with isoxadiazole derivatives.

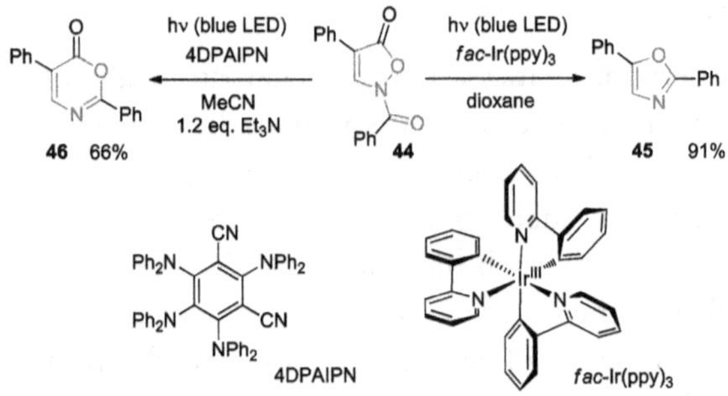

Scheme 8 Photoredox catalytic rearrangements of an isoxazolone derivative.

the ring-extension product **46** is obtained. In this case, it has been suggested that decarboxylation is avoided by electron transfer from the amine to a corresponding carboxyl radical intermediate. The reactions have been carried out with a large variety of substrates. Numerous corresponding thiocarbonyl derivatives have been transformed in the same way, however by direct light absorption.[45] No photocatalysis was necessary in this case.

3 Photochemical rearrangements induced by electron and hydrogen atom transfer (HAT)

Hydrogen transfer play a key role in many chemical[46] and biochemical processes.[47] Radical intermediates are generated, which can rearrange or react with other compounds leading to final products. Photochemical conditions provide sustainable methods to generate such intermediates. For example, utilization of toxic reagents still frequently used in radical chemistry is thus avoided.[48] In the context of organic photochemistry, often two mechanisms are often discussed for the hydrogen atom transfer (HAT) from a donor to an electronically excited acceptor: (1) The transfer occurs in one step, which means that the electron and the proton are transferred simultaneously. (2) The electron is transferred first and the proton follows.[49] In both cases, neutral radical intermediates are generated. In a more general context, these mechanisms are part of a large number of processes involving proton-coupled electron transfer (PCET).[50]

In connection with investigations of photochemical cycloadditions of aromatic compounds,[51] also aromatic ketones have been investigated. A one-step hydrogen atom transfer is involved in the photochemical transformation **47** (Scheme 9).[52] After photochemical excitation, hydrogen abstraction to the carbonyl function occurs, leading to the intermediate **48** possessing a ketyl radical and an allyl radical moiety. After a rearrangement via intermediate **49**, compound **50** is formed. Formation of the hemiacetal **51** and addition of the resulting hydroxyl function to the olefinic double bond lead to the final product **52** in high yield.

Many hydrogen abstraction reactions have been reported with electronically excited aromatic ketones.[53] A typical example is depicted in Scheme 10. Upon irradiation, the antiinflammatory drug ketoprofen **53** undergoes macrocyclization.[54] The efficiency of this reaction and the reaction of similar compounds depend on the relative configuration in the substrate. Such reactions may cause undesired side effects such as phototoxicity or photoallergy.

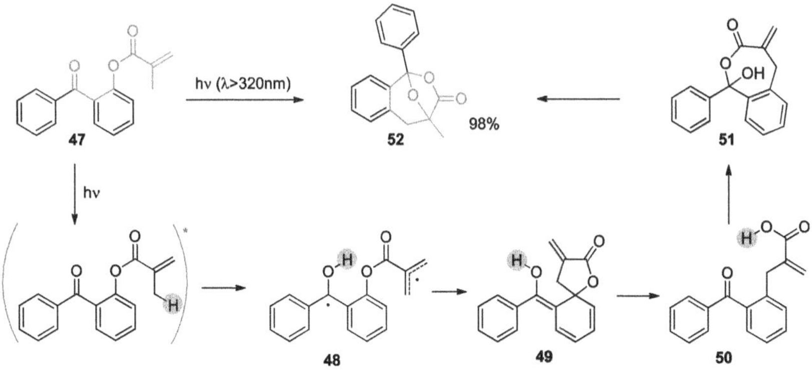

Scheme 9 A one-step hydrogen atom transfer (HAT) is involved in the photochemical reaction of the benzophenone derivative **47**.

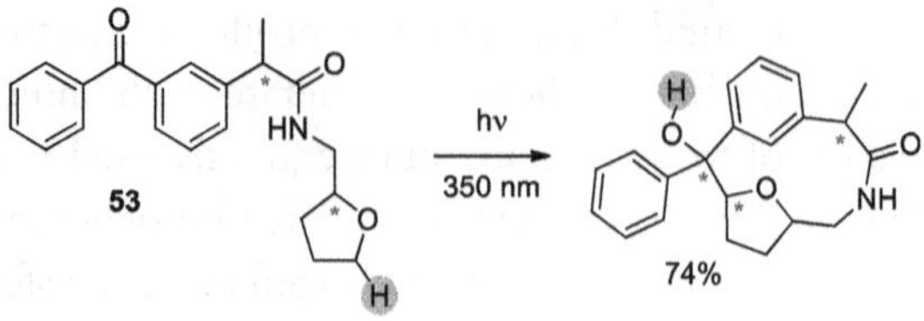

Scheme 10 Long-distance 1,12-hydrogen atom transfer leading to macrocyclization.

The present study focuses on the stereoselectivity and thus on a more profound understanding of the interaction between the drug and a biochemical structure, for example, an enzyme. Such interactions also significantly determine the extent of the adverse effects such as phototoxicity.

Also, α,β-unsaturated carbonyl compounds or α,β-unsaturated lactones are efficient hydrogen atom acceptors when excited to the triplet state. Such reactions are particular interesting for application to the synthesis of complex heterocyclic compounds.[55]

Various imides undergo similar hydrogen abstraction reactions.[56,57] When the anhydrogalactopyranose derivative **54** is irradiated, hydrogen abstraction occurs selectively in position 3, leading to the diradical intermediate **55** (Scheme 11).[58] The succinimide moiety is the hydrogen acceptor.

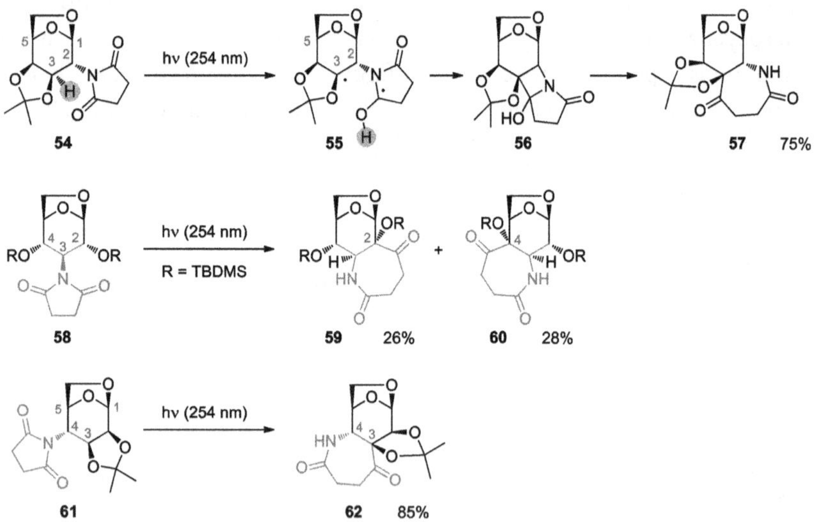

Scheme 11 Photochemical transformations of succinimide-substituted anhydropyranose compounds leading to annulated ketocaprolactams.

Radical combination yields the hydroxyl azetidine **56**, which rearranges to yield the final product **57**. This reaction sequence is a heterocyclic analogue of the Norrish-Yang reaction.[56] The reaction has also been carried out with anhydropyranose derivative carrying the succinimide substituent in other positions. In the case of compound **58**, hydrogen abstraction occurs in positions 2 and 4, leading to the caprolactams **59** and **60**, respectively.[59] In corresponding diastereoisomers, Norrish type II elimination products were also isolated. In the case of the isomer **61**, again a regioselective reaction was observed.[58] Compound **62** was obtained in high yields. Concerning the reactions of compounds **54** and **61**, it must be noted that in any case, hydrogen abstraction in position 1 or 5 does not occurred. C—H bond dissociation energies at such bridge points are high because particularly unstable σ-type radicals are formed. The same reaction has been carried out with corresponding glutarimide derivatives. A lot of such reactions have been carried out with phthalimides.[60,61]

The reaction has efficiently been applied to the synthesis of benzofuran or tetrahydrofuran compounds. Such moieties are found as core structure in many natural products. Upon UV irradiation of compound **63**, a hydrogen atom is transferred to the keto function from the benzotetrahydropyrane moiety from either the acetal position 3 (1,6-HAT) leading to compound **64** or the benzyl position 1 (1,8-HAT) leading to compound **65** (Scheme 12). The photochemical product **64** is a partial structure of the natural product crombenin.[62] Starting from the simple symmetrical trihydroxypropiophenone derivative **66**, aflatoxin M2 core structure **67** is obtained in only one step via photochemical HAT.[63] In the cyclic ketone **68**, hydrogen is abstracted from a benzyl position. Radical combination yields the furo[3,4-*c*]furan lignan paulownin.[64] It should be pointed out that this reaction is highly diastereoselective. The ketone function in α-keto esters are efficient hydrogen acceptors.[65] Photochemical HAT in compounds **69** yields **70** in high diastereoselectivity.[66] The photoadducts were transformed into the natural products coumestan and coumestrol. The reaction was also applied to the synthesis of alkaloids. In compound **71** the pyrrolidine amid moiety was added to the α-keto ester function.[67] The resulting bicyclic lactam **72** was transformed into the pyrrolizidine alkaloid isoretronecanol.

Benzoquinone or benzoquinoide structures play a key role in many syntheses (chemical or bio) of alkaloids possessing alkoxy benzene moieties such as benzo[d][1,3]dioxoles.[68,69] After light absorption by the quinone compound **73**, hydrogen is abstracted from the methoxy group

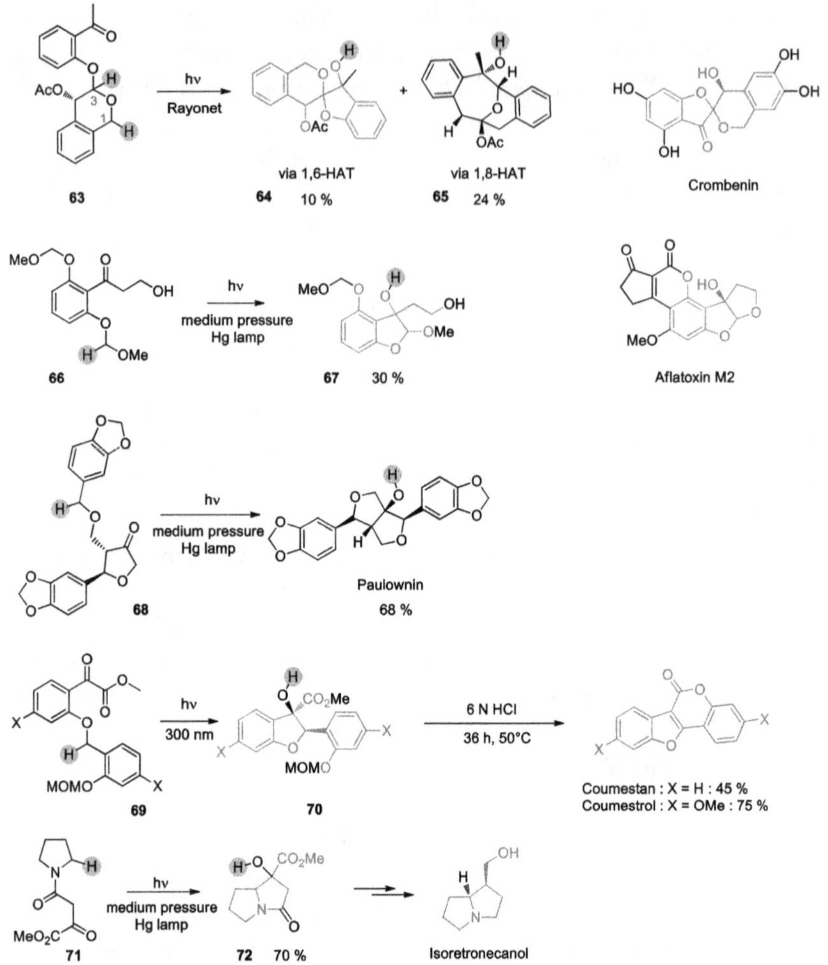

Scheme 12 Photochemical 1,6-HAT and its application natural product synthesis.

(Scheme 13).[70] Tautomerization and radical coupling steps lead to the formation of the benzo[*d*][1,3]dioxole structure **74**. Such tetrahydroisoquinolines are partial structures in the very big family of isoquinoline alkaloids. In particular, **74** is part of lurbinectedin. The same reaction was applied to the synthesis of tawaniaquinol A.[71] The transformation of **75** (tawaniaquinone F) can be carried out with sunlight.

HAT is also applied to the synthesis of complex nitrogen-containing polycyclic compounds in which more than one photochemical key step is involved. Upon irradiation the pyrrole derivative **76** first undergoes

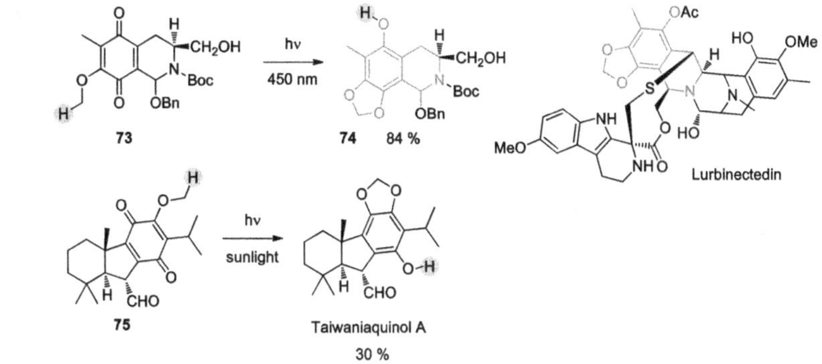

Scheme 13 Intramolecular HAT in quinone compounds and its application to the synthesis of natural products.

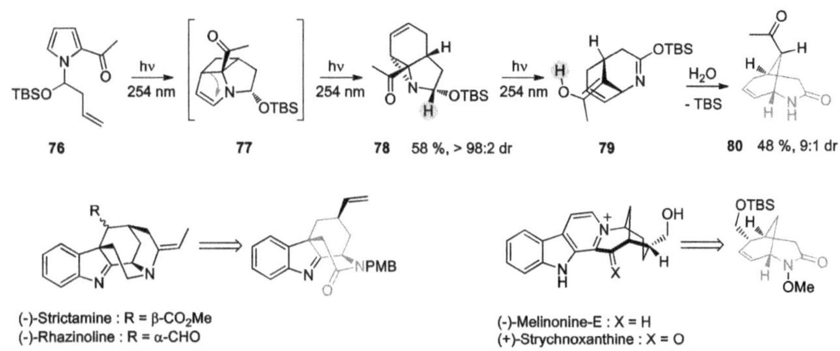

Scheme 14 Synthesis of bridged azocanone, a useful synthesis intermediate for alkaloids.

[2 +2] photocycloaddition (**77**) (Scheme 14).[72] In a 1,3-sigmatropic shift the azirine compound **78** is formed. This reaction occurs in a stereospecific way. Photochemical hydrogen abstraction from the hemiaminal position to the keto function leads to **79**. This transformation that is observed after a longer irradiation time resembles to a Norrish type II reaction. After tautomerization (enol) and hydrolysis, the bridged lactam **80** (azocanone) is obtained. It is a substructure of synthesis intermediates of alkaloids such as strictamine, rhazinoline, melinonine, or strychnoxanthine.[73]

Similar reactions have been carried out in flow reactors.[74] This technology is now often used for photochemical transformations.[75] It accounts for the particular challenges of chemical engineering in connection with the Lambert-Beer law. Furthermore, aspects of process safety and efficiency in connection with continuous production are addressed.

In some regards the photochemistry of imines resembles to that of carbonyl compounds. Under photochemical reaction conditions the oxazolone derivatives **81** abstract hydrogen from the acetal moiety, leading to the diradicals **82** (Scheme 15).[76] Radical combination yields the azaspirobicyclooctanedioxolane and azaspirobicyclooctanedioxane derivatives **83**. Especially in the case of the dioxolane compound ($n = 1$), thermal decarboxylation occurs, yielding the cyclic imines **84**. In the same case, compound **85** with a newly formed C—N bond is also obtained. It was shown that this product results from the same intermediate **82**, which was explained by the zwitterionic nature of the singlet state of **82**. Compound **85** possesses a core structure of molecules used for the treatment of diabetes.[77] Similar photochemical cyclization reactions have been carried out with α,β-unsaturated butyrolactone derivatives carrying an acetal or sugar substituent on the side chain.[55,78]

All previously discussed, photoinduced HATs are one-step processes, which means that the proton and the electron are transferred simultaneously. As mentioned previously in such photochemical reactions, also a two-step HAT is often discussed.[49,50] The one-step transfer takes place when the photochemical electron transfer is endergonic (Fig. 3A). However, the formation of the neural radical intermediates is exergonic. In the case of an electronically excited species, for example, the hydrogen acceptor, this step enables a dissipation of the excitation energy. When the photochemical electron transfer is also exergonic, a two-step process becomes possible, and the electron is transferred first and radical ion pair is formed (Fig. 3B). The proton follows and the same neural radical intermediates

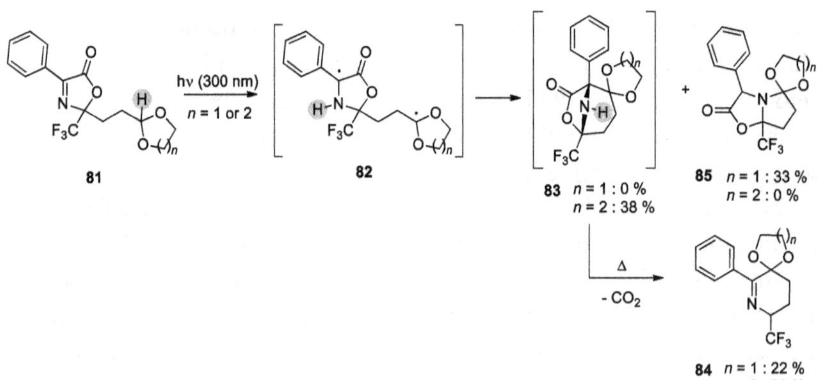

Scheme 15 Photochemical cyclization with oxazolone derivatives involving HAT.

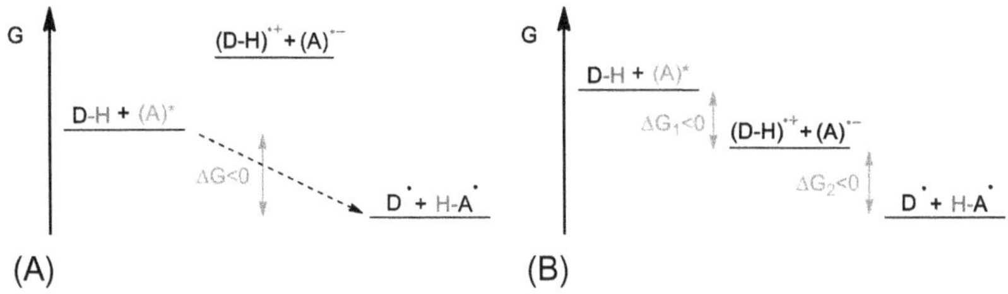

Fig. 3 Photochemical induced hydrogen transfer from a donor to an electronically excited acceptor. (A) One-step process and (B) two-step process.

are formed. The exergonicity of the photochemical electron transfer can be estimated with the Rehm-Weller relationship.[79] In the context of application to organic synthesis, it should be pointed out that the outcome of such reactions often significantly depends on the mechanism of the hydrogen atom transfer.

Amines are efficient electron donors and a two-step process often takes place. After photochemical excitation of the azetidinyl aryl ketone **86**, an electron transfer occurs from the azetitine nitrogen to the aromatic carbonyl group, leading to the intramolecular radical ion pair **87** (Scheme 16).[80] After proton transfer in the same direction, neural diradical **88a,b** is generated. Radical combination yields the bicyclic compound **89**. Dehydration and rearrangement yield the pyrrole derivative **90**. Photocatalysis (triplet sensitization) was also applied to perform similar pyrrole syntheses.[81]

As photochemical electron transfer is involved, photoredox catalysis can be used to perform such reactions. Photoinduced intramolecular addition of tertiary amines can also be carried out on numerous other unsaturated functional groups such as α,β-unsaturated lactams. In the case of quinolone compound **91**, however, the electron is not transferred to the electron-poor

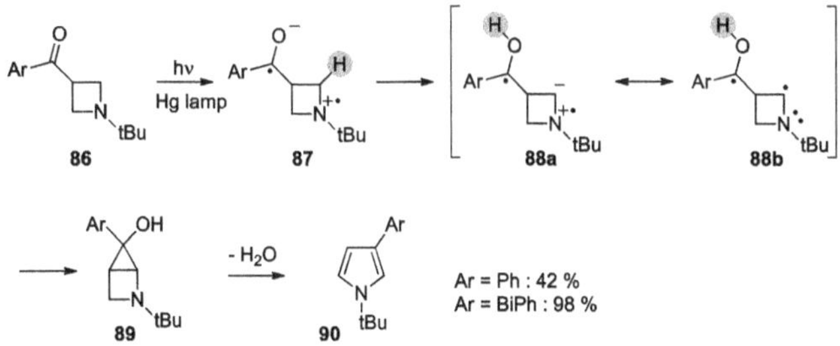

Scheme 16 Synthesis of pyrroles from azetidinyl aryl ketone using two-step HAT transfer.

alkene of the lactam function but on the photochemical excited sensitizer (Scheme 17).[82] The radical ion pair **93** is formed. The neutral radicals **94** are then formed by proton transfer. Intramolecular addition of the nucleophilic α-aminoalkyl radical to the electron-poor double bond occurs (**95**). In the last step a hydrogen atom is transferred from the ketyl radical to the oxoallyl radical, which leads to the final spirocyclic pyrrolizidine compound **96**. Details of the mechanism have also been discussed for the intermolecular addition of simple tertiary amines to α,β-unsaturated lactones.[83,84] For some time now, such reactions are also carried out using photoredox catalysis with visible light.[85] A further issue of the reaction depicted in Scheme 17 is asymmetric synthesis. In this reaction, chirality is induced by complexation of the substrate **92** with a chiral ligand, which carries the aromatic ketone as sensitizing group. Complexation occurs via two hydrogen bonds. In this regard the ketone also acts as shielding group to favor the approach of the α-aminoalkyl radical in one diastereotopic half space. The approach of the sensitizer to the substrates also facilitates the reaction steps of electron and hydrogen transfer. A lot of different photochemical reactions can be stereoselectively performed under these conditions.[86]

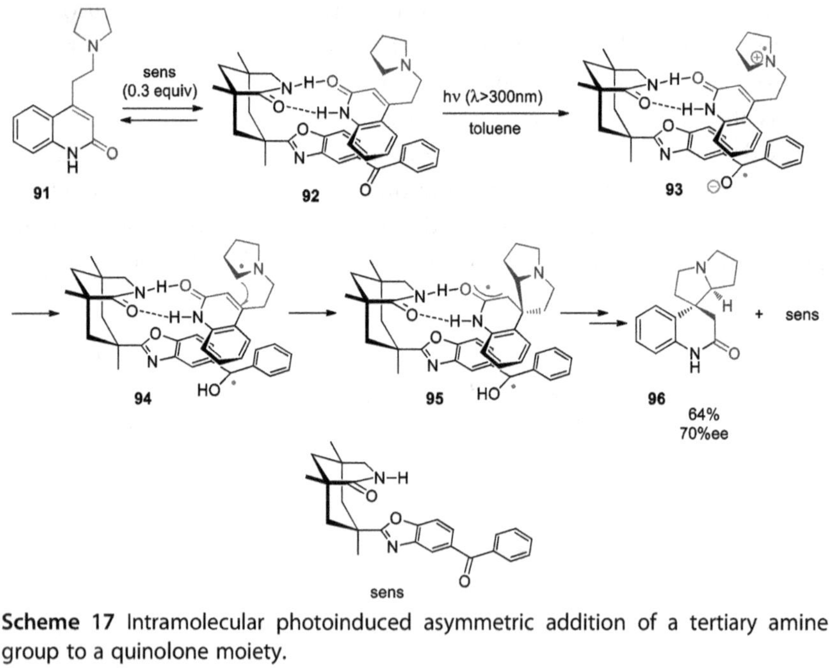

Scheme 17 Intramolecular photoinduced asymmetric addition of a tertiary amine group to a quinolone moiety.

Macrocyclizations have also been carried out successfully with these reaction.[60,61] The α-keto ester **97** carrying thioether containing side chains of different lengths undergoes easily macrocyclization (**98**) (Scheme 18).[87] Likewise tertiary amines, thioethers are efficient one-electron donors, and a two-step hydrogen transfer process is involved. The efficiency in macrocyclizations might be linked to the formation of radical ion pairs. In some cases, photopinacolization products are observed. Also, β-keto esters undergo similar reactions. Starting from **99**, various oxathiocanone derivatives **100** have been obtained.[88]

Intramolecular photochemical hydrogen atom transfer (HAT) is also observed with aromatic carbonyl compounds. In such reactions, hydroxylquinodimethane intermediates are formed. They can also be considered as phototautomers. A spectacular example is shown in Scheme 19.[89] After

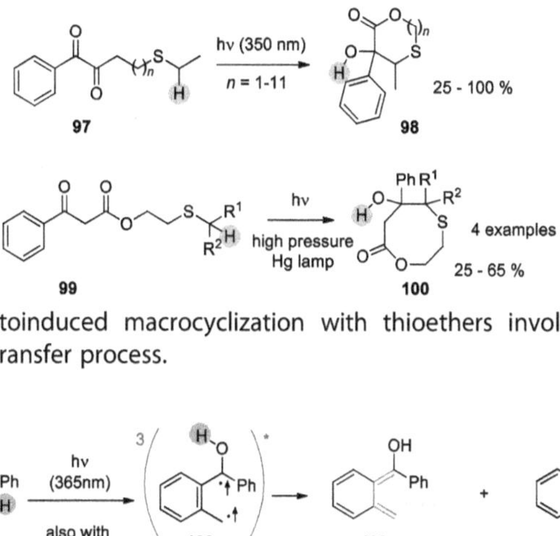

Scheme 18 Photoinduced macrocyclization with thioethers involving a two-step hydrogen atom transfer process.

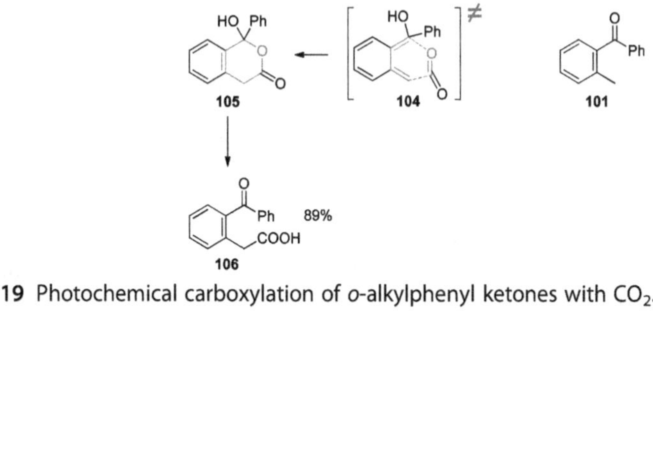

Scheme 19 Photochemical carboxylation of *o*-alkylphenyl ketones with CO_2.

photochemical excitation of **101** to the triplet state, hydrogen is transferred from the methyl group to the carbonyl oxygen (**102**). The corresponding singlet ground state **103** possesses a dienol moiety. Such intermediates can also be trapped by electron-poor alkenes in a Diels-Alder reaction.[90] In the present case, CO_2 was added, leading to the lactone intermediate **105**, which after ring opening yielded the keto carboxylic acid **106**. A theoretical investigation provided evidence for the formation of the typical transition state **104** of a [4 + 2] cycloaddition. Stereoisomers **103** and **107** of the dienol intermediate are formed in such reaction. While **103** undergoes cycloaddition, **107** gave the starting compound **101** by tautomerization.

Numerous applications of this reaction to the synthesis of natural products and heterocyclic compounds have been reported.[91] The *o*-hydroxymethylbenzaldehyde derivative **108** undergoes intramolecular hydrogen abstraction at the ortho side chain, yielding the dienol intermediate **109** (Scheme 20).[92] This step is favored by the presence of the *o*-methoxy group enabling the formation of a hydrogen bond. The dienol is trapped by an intramolecular Diels–Alder reaction leading to the final product **110**, which is a key intermediate in the synthesis of pleurotin. This compound was isolated from the mushroom *Pleurotus griseus* and possesses activity against Ehrlich ascites carcinoma, L-1210 lymphoid leukemia, and mammary tumors.

The Witkop reaction permits the synthesis of a variety of heterocyclic compounds possessing a lactam moiety.[93,94] It is mainly carried out with indole derivatives such as **111** (Scheme 21).[95] In a photochemical reaction of this compound, the product **112** is generated. Further intermolecular reactions lead to the natural product decursivine, which possesses antimalarial activity. After electronic excitation (**113**), an electron transfer from the indole moiety to the α,α-dichloroamide function occurs, leading to the intramolecular radical ion pair **114**. After release of a chloride ion, the diradical cation **115a,b** is formed. Cyclization occurs via radical

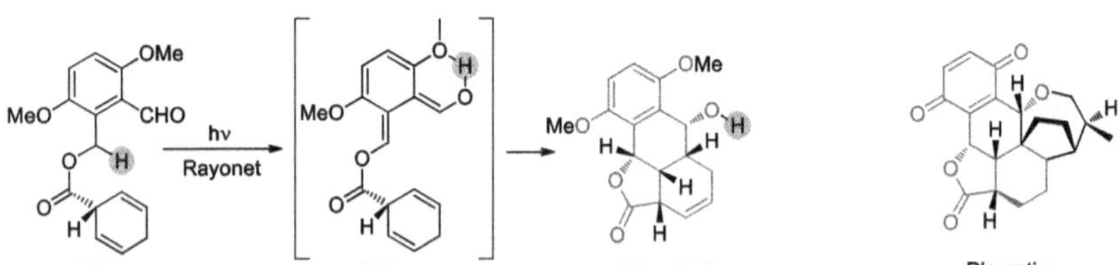

Scheme 20 Photoenolization of o-hydroxymethylbenzaldehyde derivative **108** via photochemical HAT. Synthesis of a core structure of pleurotin.

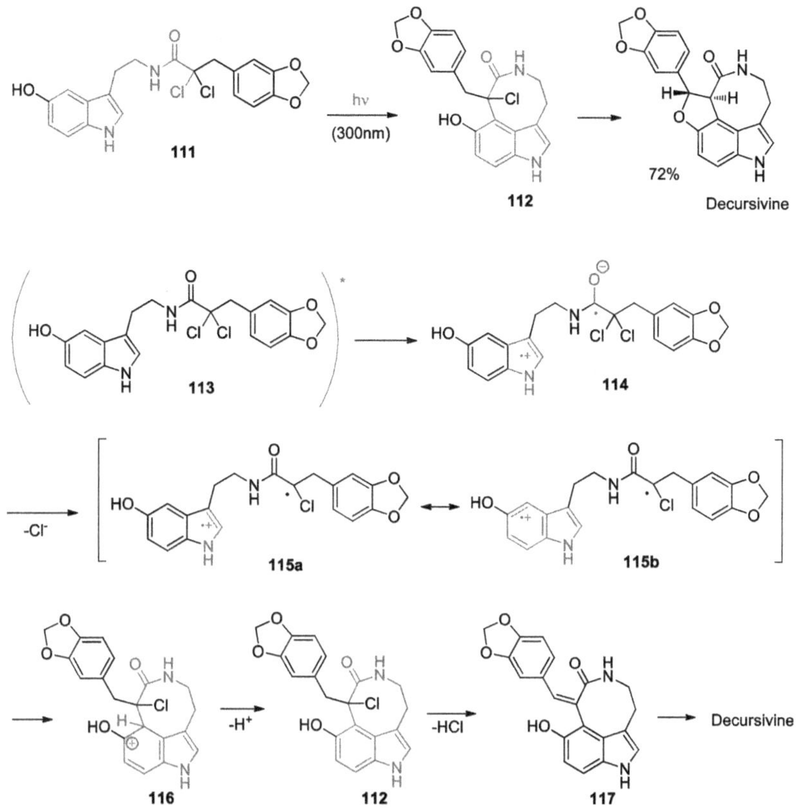

Scheme 21 Mechanism and application to the concise synthesis of decursivine of the Witkop reaction.

combination, and the intermediate **116** is generated, which is a typical Wheland intermediate of the electrophilic aromatic substitution. Consequently, deprotonation yields the final product **112** of the Witkop reaction. In the present case, elimination of HCl (**117**) and intramolecular Michael addition lead to decursivine in a very concise synthesis.

The Witkop reaction was also applied to the transformation of more complex compounds such as **118** in Scheme 22.[96] Only few more steps were necessary to obtain the synthesis target 2,7-dihydropleiocarpamine from the photochemical product **119**. The reliability of the reaction for the synthesis of particular isomers is nicely shown for the transformation of **120** into **121**[97] and in the corresponding transformation of the regioisomer **122** into **123**.[97,98] A large variety of catharanthine derivatives has thus become available by the same photochemical key step.

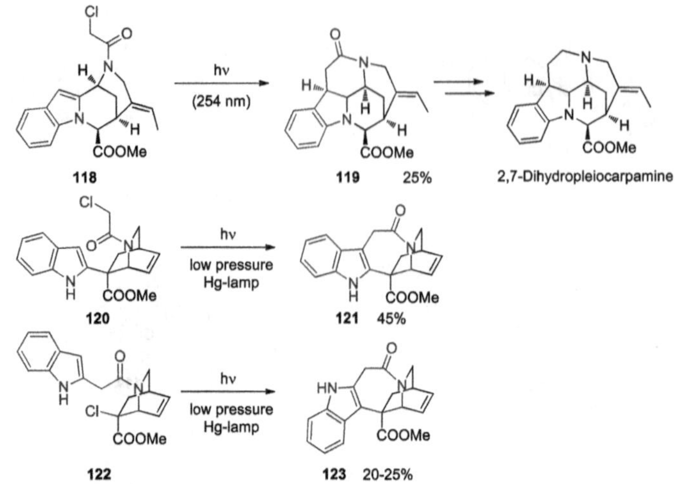

Scheme 22 Application of the Witkop reaction to the synthesis of 2,7-dihydropleiocarpamine and catharanthine derivatives.

A similar reaction mechanism is involved in the nucleophilic aromatic substitution under photochemical conditions (photochemical $S_{RN}1$ reaction).[99] Such an intramolecular reaction is observed in a basic reaction medium with the imide **124** carrying an arylbromide substituent (Scheme 23).[100] The following mechanism has been suggested. After photochemical excitation, intramolecular electron transfer occurs from the enolate moiety in **125** to the arylbromide substituent (**126**). Radical combination leads to the anion intermediate **127**. The final product **128** is obtained by release of bromide. A corresponding isoquinoline derivative **129** was also obtained. Due to the reduced reactivity of the corresponding substrate, a more basic reaction medium (3M NaOH) was necessary. A large variety of complex heterocyclic compounds have been synthesized with this reaction.[101]

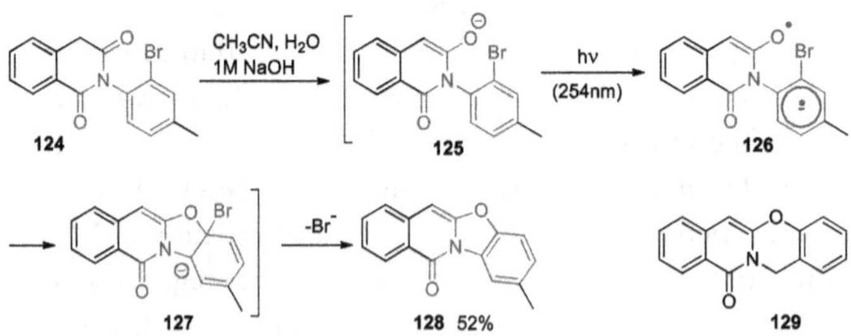

Scheme 23 Intramolecular photochemical $S_{RN}1$ reaction.

Photochemically induced hydrogen atom transfer also occurs from heteroatoms such as amide nitrogen to an electronically excited carbonyl function. A large variety of anthranilic acid derivatives such as **130** undergo such an intramolecular hydrogen atom transfer (Scheme 24).[102] Iminenol intermediates **133** are generated. It has been shown that such structures correspond to an excited singlet state S_1. A corresponding triplet ground state T_0 **134** may undergo further chemical transformation.[103] The particular reactive nitrogen-centered radical will add to the furan moiety, leading to the spirocyclic intermediate **135**. Radical combination can occur in two different positions, leading either to formal [2 + 4] adducts (compare **131**) via pathway a or to formal [4 + 4] adducts (compare **132**) via pathway b. Such reactions are particular interesting since they generate a high degree of molecular diversity and complexity, which is very much appreciated in the search of new biologically active compounds. In the present case, this diversity is further increased by the combination with the Suzuki reaction

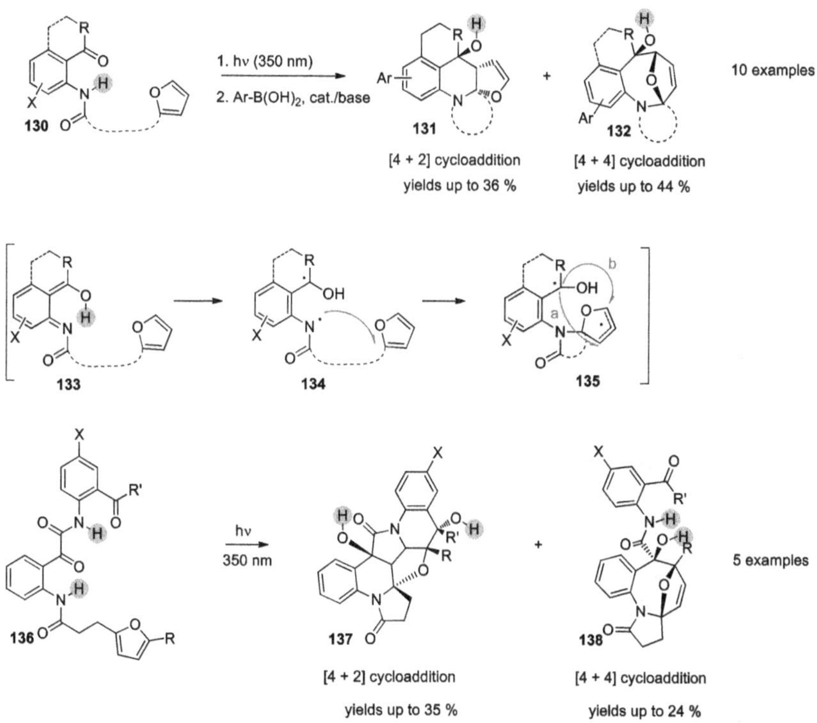

Scheme 24 Formation of complex polyheterocyclic compounds by photochemical transformation of anthranilic acid derivatives involving intramolecular hydrogen atom transfer.

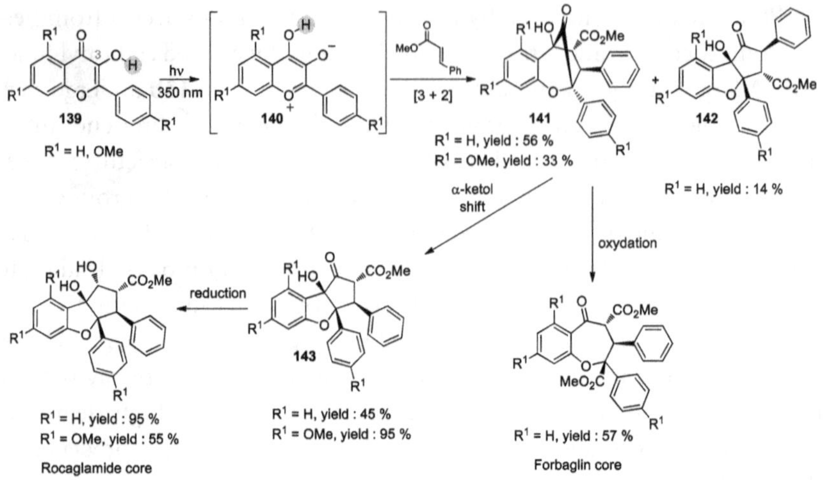

Scheme 25 ESIPT in photochemical transformations of 3-hydroxyl flavone compounds.

(**131** and **132**). Especially the molecular complexity can further be increased when the reaction is carried out with compounds possessing two formal anthranilic acid moieties such as in **136**.[104] In the case of an initial [4 + 2] cycloaddition, a second one takes place, and compound **137** is isolated. A corresponding reaction of the [4 + 4] adduct **138** is not possible. Such photochemical hydrogen atom or proton transfers between heteroatoms such as nitrogen or oxygen are also discussed in the context of excited state intramolecular proton transfer (ESIPT).[105]

An ESIPT is also involved in the photochemical transformations of 3-hydroxyl flavone compounds[106] such as **139** (Scheme 25). Upon irradiation in the presence of cinnamic esters, compounds **141** and **142** are obtained.[107] At the excited state an intramolecular proton transfer takes place, leading to the intermediate dipolar pyrylium structure **140**, which undergoes cycloaddition. Further transformations lead to the rocaglamide and forbaglin cores. Such compounds possess potent anticancer and antileukemic activity. In such reactions often a α-ketol shift is observed. In the present case, this reaction leads to the formation of compound **143**, which was transformed into a rocaglamide core.

4 Photochemical electrocyclic reactions

Numerous heterocyclic compounds undergo photochemically induced electrocyclic reactions, leading to constrained bicyclic moieties that are hardly accessible by conventional methods of chemical synthesis.[108,109] Pyridones such as **144** (Scheme 26) undergo photochemical $[_{\pi}2_s + _{\sigma}2_s]$ electrocyclization.[110] According to the Woodward–Hoffmann rules[111], this disrotatory cyclization is symmetry allowed at the excited state (Fig. 4). For reasons of molecular constraint, a corresponding conrotatory process (at the ground state) in the pyridone system is not possible. The photochemical reaction is therefore a privileged method for the preparation of bicyclic compounds such as **145** possessing a cyclobutene and a β-lactam moiety. The reaction was widely applied, and especially substituted cyclization products were stable and proved to be interesting synthesis intermediates.[112] An asymmetric version of the reaction has also been carried out (Scheme 26).[113] Thus the menthyloxy derivative **146** was transformed into **147a** and **147b**. Both diastereoisomers were separated by fractional crystallization. They were transformed into the corresponding enantiomeric monocyclic β-lactam compounds **148** and ent-**148**.

The electrocyclic reaction was also successfully performed with 4π systems containing an imine function such as in the pyrimidone derivative **149** (Scheme 27).[114] The bicyclic compound **150** thus obtained containing a

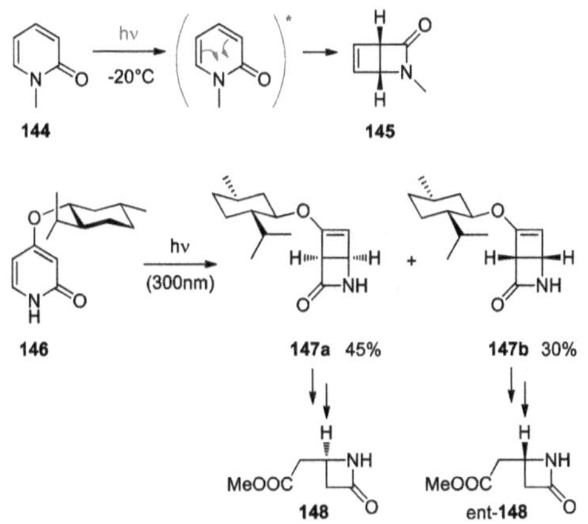

Scheme 26 Photochemical electrocyclization of pyrrolidone derivatives.

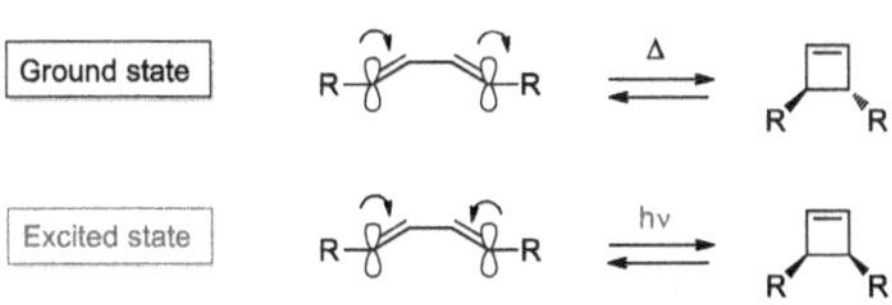

Fig. 4 Thermal conrotatory and photochemical disrotatory electrocyclization processes.

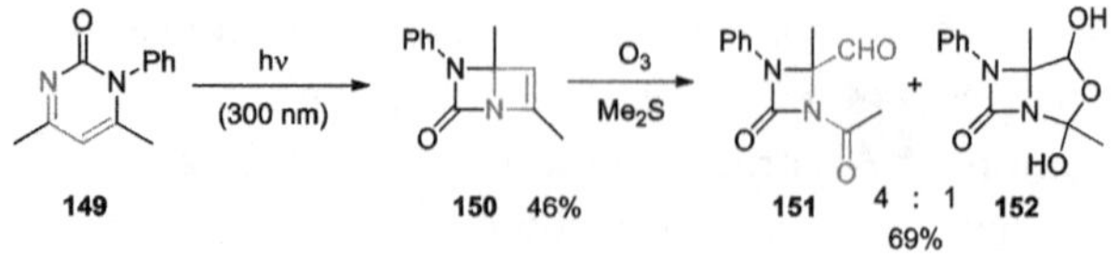

Scheme 27 Photochemical electrocyclization with a pyrimidone derivative.

four-membered cyclic urea function. An enamine function is incorporated in the second four-membered ring moiety. Such compounds are interesting synthesis intermediates. For example, ozonolysis yields compounds **151** and **152**. **152** is formed from **151** by addition of water.

The interest of this reaction for further application to organic synthesis of heterocyclic compounds has nicely been shown with the reaction of **153** (Scheme 28).[115] A variety of bicyclic photoproducts **154** were obtained in good yields. Epoxidation or reaction with ethoxycarbonylnitrene (generated from *N*-ethoxycarbonyl-*p*-nitrobenzyltriethylammonium bromide) of the cyclobutene moiety yielded the tricyclic compounds **155** and **156**, respectively. These compounds undergo thermal rearrangement, leading to 1,4-oxazepines **157** and 1,4-diazepines **158**. Such compounds are hardly available by conventional methods of organic chemistry. The overall synthesis consists in an insertion of either an oxygen or a nitrogen atom in the C—C bond between positions 4 and 5 of the initial pyridone substrate **153**.

The reaction has also been carried out with dihydropyrazidine such as **159** (Scheme 29).[116] The resulting 1,2-diazetidine **160** has been obtained in good yields. Gram-scale transformations have been successfully performed, and various transformations of the photoproducts have been carried out. Thus ring-opening metathesis with ethylene yielded the divinyl diazetidine **161** and RuO_2-catalyzed oxidation was used to synthesize tetracarboxylate derivatives **162**.

In the previously discussed transformations of six-membered heterocycles, the reaction centers are located in para positions. Similar reactions are

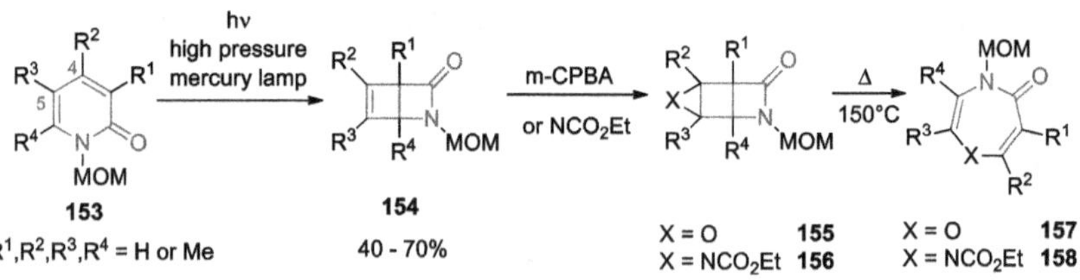

Scheme 28 Synthesis of 1,4-oxazepines **157** and 1,4-diazepines **158** starting from pyridines.

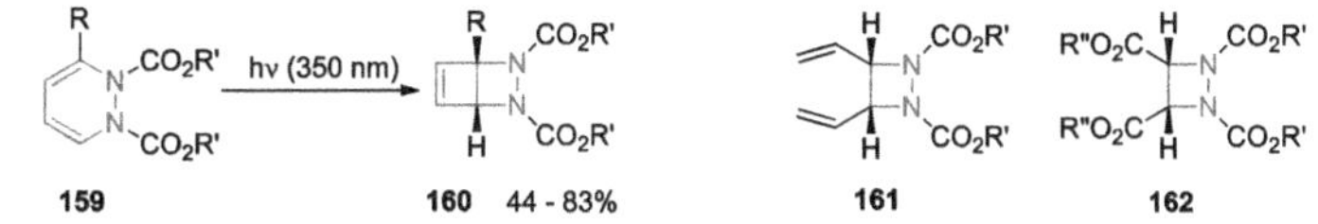

Scheme 29 Photocyclization of 1,2-dihydropyridazines.

also observed when the reaction centers are located in meta positions of the corresponding substrates. Topologically, both reactions are related to rearrangements of benzene (Scheme 30).[117,118] At the S_2 state (**163**), a C—C σ-bond is formed between positions 1 and 4, yielding Dewar benzene **164**. Under the reaction conditions, this compound undergoes intramolecular [2 + 2] photocycloaddition, yielding prismane **165**. In contrast, at the S_1 state, a C—C σ-bond is formed between positions 1 and 3, and the bicyclic intermediate **166** can be formed. Further C—C bond formation leads to benzvalene **167**. The by-product fulvene **168** is generated from **166** by fragmentation of a C—C bond of the cyclopropane moiety and hydrogen shift. The formation of bicyclo[3.1.0] intermediates like **166** is frequently discussed in photochemical reactions of pyridinium derivatives.[119] This reaction was frequently applied to the synthesis of biologically active compounds.[120]

Upon irradiation the pyridinium salt **169** undergoes photoisomerization by C—C bond formation in positions 6 and 2 (Scheme 31).[121,122] In the basic reaction medium, a hydroxyl function is added to the resulting allyl carbenium ion **170**. This reaction was carried out in almost quantitative yield. *cis*-Hydroxylation of the photochemical product **171** yielded mainly the *meso* isomer **172** in good yield along with minor amounts of the racemic compound **173**. Such polyhydroxylated cyclic amine compounds are

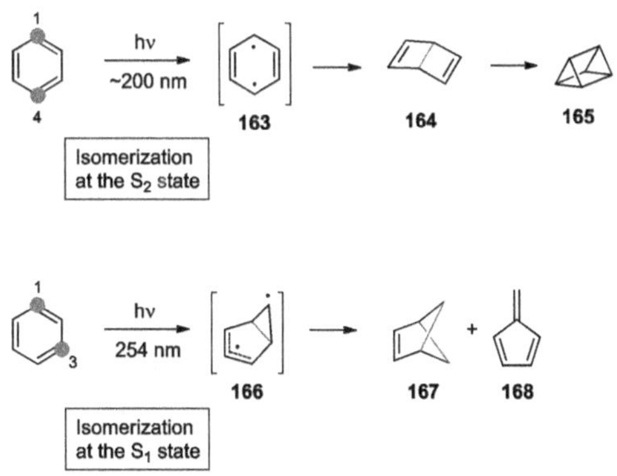

Scheme 30 Photochemical isomerizations of benzene.

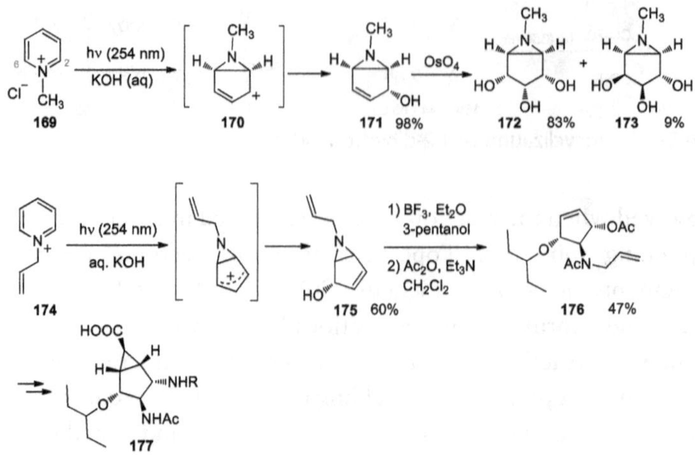

Scheme 31 Photochemical electrocyclization of pyridinium salts.

glycomimetics, and consequently, their glycosidase inhibitor activity is systematically investigated. The same photochemical reaction has been carried out with a corresponding N-allyl derivative **174**.[123] Treatment under acidic conditions and in the presence of 3-pentanol of the photochemical product led to the cyclopentenylamine derivative, which is acetylated (**176**). Further transformations yielded compound **177**. This scaffold was studied as influenza neuraminidase inhibitor. Considering an industrial application the photochemical transformation has also been carried out using a continuous flow photoreactor on multigram scale with a productivity of $3.7\ g\ L^{-1}\ h^{-1}$.[124] This technique is now widely applied to organic photochemistry to optimize these reactions,[75,125] also with a particular focus on heterocyclic chemistry.[126]

As in many cases of organic photochemical reactions, no particular protecting groups are necessary. This is nicely shown for the transformation of the glucosyl pyridinium salt **178** (Scheme 32).[127] The diastereoisomers **179a,b** of the azabicyclo[3.1.0]hexenol derivative has been obtained in high yield from the photochemical electrocyclization. The hydroxyl groups of the glucosyl moiety were not protected. Peracetylation after the photochemical reaction was used to separate the diastereoisomers by crystallization.

Due to its efficiency and large scope, the photochemical electrocyclization of pyridinium salts is an ideal key step in the synthesis of many biologically active compounds (in particular glycosidase inhibitors) and natural products, in particular cyclopentylamine derivatives of nitrogen-containing

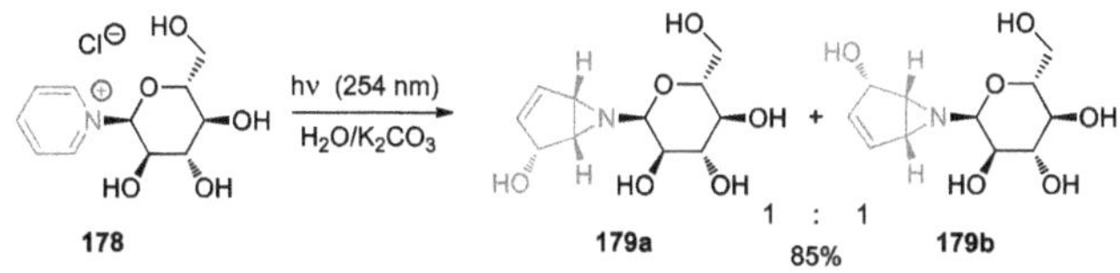

Scheme 32 Photochemical electrocyclization of a glucosyl pyridinium salt.

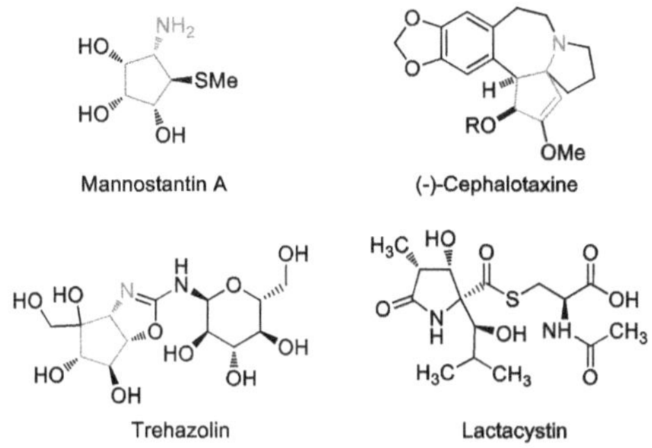

Fig. 5 Natural products that are accessible by photochemical electrocyclization of pyridinium salts.

heterocycles.[128] Some examples are shown in Fig. 5: mannostatin A,[129] trehazolin,[130] cephalotaxine,[131] or lactacystin.[132] Often complex carbon skeletons such as spirocyclic compounds are generated in the photochemical key step of these syntheses.

Photochemical electrocyclization of pyridine derivatives was also applied to interconversions of compounds belonging to a natural product family. Peroxidation of protoberberine yielded the zwitterion compound **180**, which is an ideal precursor for a photochemical electrocyclization (Scheme 33).[133] The photochemical reaction directly led to 8,14-cycloberberine with its cyclopentanone moiety. The photochemical reaction was carried out with a variety of substrates. Opening of the aziridine ring led to spirocyclic compounds that can be further transformed into corresponding alkaloids such as sibiricine. Stereoselective reduction of 8,14-cycloberberine with $NaBH_4$ yields mainly the *cis*-alcohol **181**, which is a precursor of dihydrofumagilin-1.[134]

Instead of protonation of the pyridine nitrogen, the photochemical cyclization can also be favored by N-oxidation. Such a reaction was applied to the synthesis of DNA topoisomerase I inhibitors as they are used for cancer therapy. This pharmaceutical activity was observed with camptothecin, a product from traditional Chinese medicine and isolated from *Camptotheca*

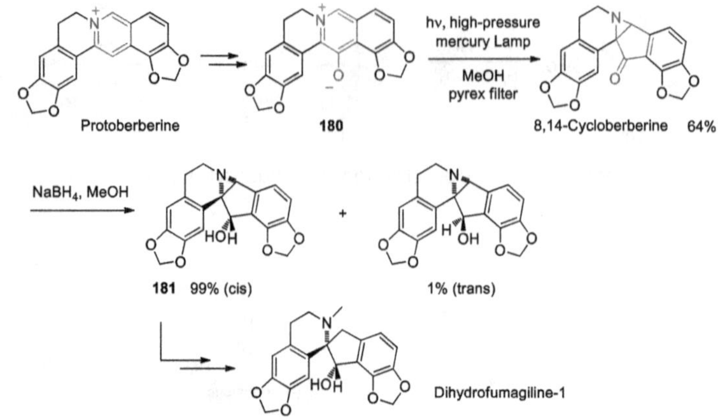

Scheme 33 Application of the photochemical electrocyclization to the chemistry of berberine alkaloids.

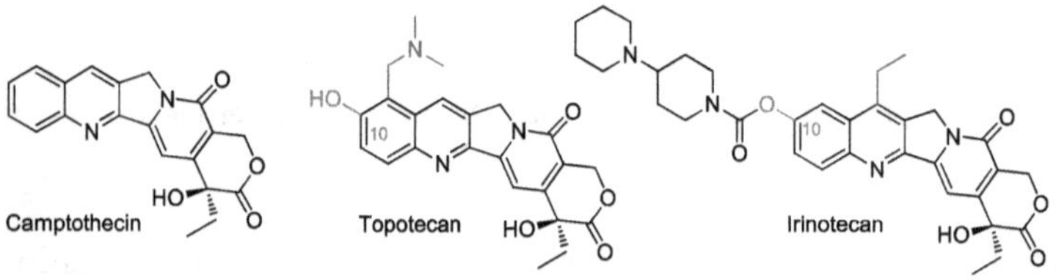

Fig. 6 DNA topoisomerase inhibitors. Derivatives of camptothecin.

acuminata (Fig. 6).[135] To increase the activity, derivatives such as topotecan or irinotecan had to be synthesized. An important step of many of these syntheses is the selective oxidation in position 10 of camptothecin. Camptothecin is easily oxidized at the pyridine nitrogen, leading to the N-oxide **182** (Scheme 34).[136] This oxidation state should be transferred into the 10 position. When **182** is irradiated in an acidic medium, electrocyclization in positions 2 and 13 takes place, leading to intermediate **183**. When compared with the previously discussed reactions of pyridinium salts, it can be stated that in the present case, the activation of the pyridine ring by N-protonation is replaced by an N-oxidation. Addition of a hydroxyl group is now favored in the 10 position, yielding the intermediate **184**. Acid-catalyzed dehydration and rearomatization in rings A and B lead to the final product **186**. The process has been optimized for the industrial production using a series of flat-bed reactors enabling a continuous flow production.[137] This technology permits a more economic and ecologically friendly production.[125] More precisely a higher concentrated solution of the substrate can be irradiated due to the reduced optical trajectory in flat-bed reactors. Thus the solvent

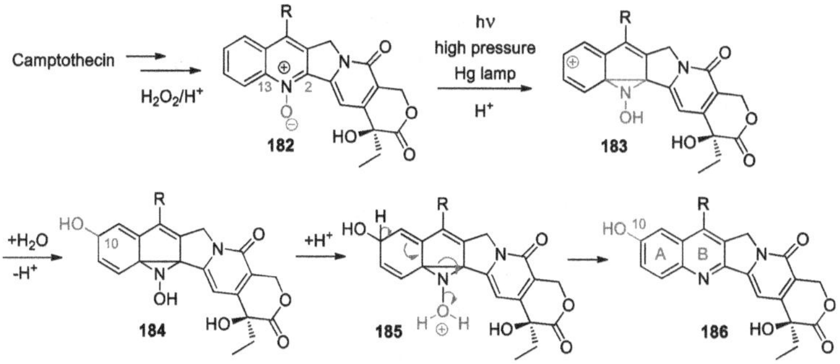

Scheme 34 Hydroxylation of camptothecin derivatives in position 10.

quantity is significantly reduced. Also, conversion and yields, especially the space-time yield (2 kg per day in 330 L in an industrial photoreactor system), are increased. Furthermore, processes conducted in microflow reactors are generally more safe.

In a similar way the pyridine system is activated by a *N*-carbamyl function for a photochemical rearrangement. Compounds **187** undergo rearrangement to the corresponding two regioisomeric **188a,b** (Scheme 35).[138] Further irradiation leads to bicyclic compounds such as **191**. The combination of this system with thermal rearrangements extends the molecular diversity. Thus compounds **190a,b** are obtained, which themselves undergo photochemical rearrangements to **192**, a regioisomer

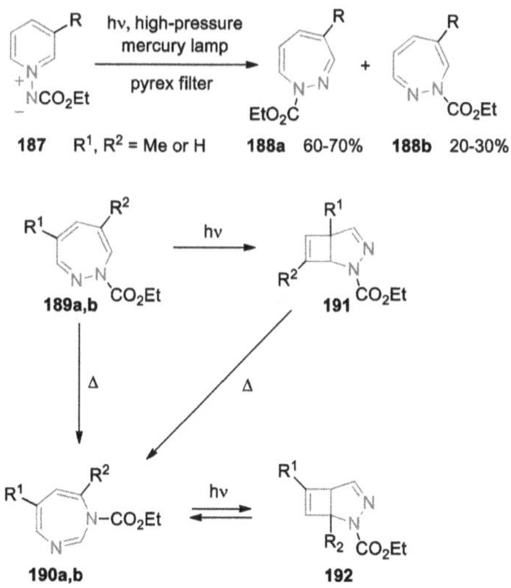

Scheme 35 Photochemical and thermal rearrangements of *N*-carbamylpyridines.

of **191**. It should be pointed out that these transformations can be performed under controlled conditions so that the formation of each compound family can selectively be addressed.

5 Photochemical cyclizations

Photochemical cyclizations involving at least two unsaturated moieties play an important role in heterocyclic chemistry.[108,139–141] These reactions have also been applied to the synthesis of photoswitches and photochromic compounds with various heterocyclic moieties.[142]

Such a reaction can be applied to the industrial production of (*S*)-indoline carboxylic acid **193** (Scheme 36).[143] The process starts with the synthesis of L-proline anilide **194**. In a very convenient step, the precursor **195** for the photochemical cyclization is obtained. This efficient step leads to **196** in a stereospecific way. After hydrolysis of the photochemical product, L-proline and the target compound **193** are isolated. They need to be separated. For the further production of (*S*)-indoline carboxylic acid **193**, the latter was used as template. Thus the corresponding precursor **197** is synthesized in the same way than **195**. The photocyclization of **197** is also highly efficient, and the photochemical product is obtained in a stereospecific way.

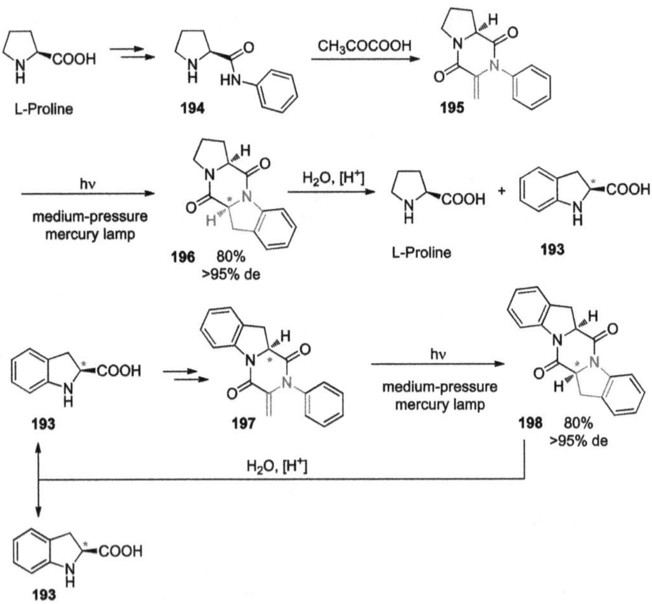

Scheme 36 Industrial process for the production of (*S*)-indoline carboxylic acid **193**.

After hydrolysis, two equivalents of (*S*)-indoline carboxylic acid **193** are obtained, and no separation is necessary. A part of the compound is recycled in the production process. In terms of chiral induction, the present process is a case of chiral amplification as it is frequently discussed in catalytic reactions.[144] Concerning chiral induction, many of such photochemical cyclizations have been carried out using asymmetric catalysis[86] or have been carried out in a chiral crystal matrix.[145]

Photochemical cyclization was used for the synthesis of carbazoles[146], dibenzofurans[147], and dibenzothiophenes[147] **199** (Scheme 37). In these reactions involving six electrons in the π-system, a dipolar intermediate such as **200** is generated. Such an intermediate has been trapped with 1,3-dipolar cycloaddition.[148] Further intermediates like **201** are formed by a tautomerization step. The final product in which aromaticity is reestablished is generated under mild oxidation conditions. The heteroatom between the two aromatic moieties can be replaced by an amide function such as in **203**. These functions possess two main mesomeric structures. In **204** a π-system with six electrons undergoes photocyclization.[149] According to the Woodward-Hoffmann rules, the cyclization occurs in a conrotatory way.[111] In the present case, this leads to the intermediate **205**. The final product **206** is obtained by elimination of methanol.

Compounds carrying α,β-unsaturated carboxyl or carbonyl groups and an aromatic substituent such as in **207** are also suitable substrates for photochemical cyclizations. Compounds **207** and **208** are in photostationary equilibrium (Scheme 38).[150] Both isomers undergo photocyclization, yielding **209** and **210**, respectively. Quinoline derivatives **211** carrying a

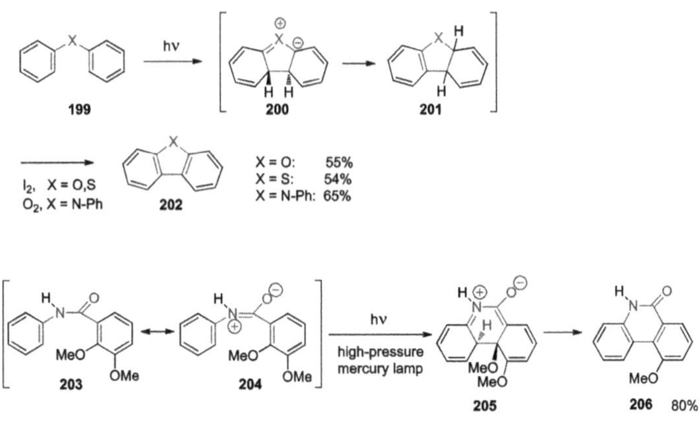

Scheme 37 Photocyclizations leading to carbazoles, dibenzofurans, dibenzothiophene, and dibenzopyridinones.

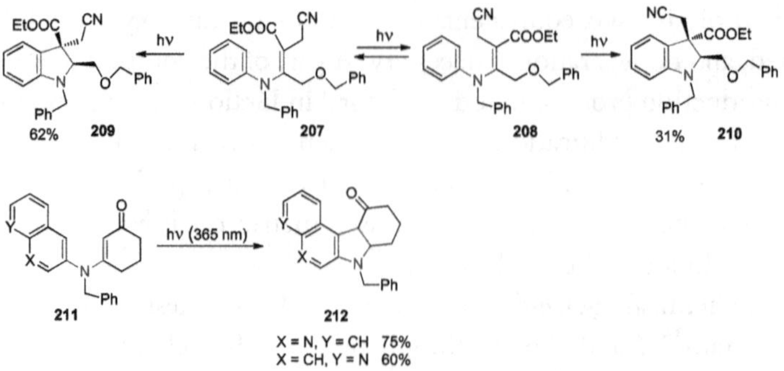

Scheme 38 Irradiation of compounds **207** and **208** (medium-pressure mercury-vapor lamp) yields indolines. Quinoline derivatives **212** have been obtained under similar conditions.

α,β-unsaturated ketone moiety have been transferred under similar conditions[151]. The corresponding photochemical products **212** have been obtained in good yields. In this case the irradiation was carried out with "black light" ($\lambda = 365$ nm). Irradiation with lower wavelength light led to fragmentation of the substrates.

As it has already been pointed out, nitrogen-containing heterocycles play an important role as biologically active compounds. Consequently the herein discussed reactions have been applied to the synthesis of a large number of such compounds. Two examples are shown in Scheme 39. Compound **213** undergoes photocyclization, yielding the tetracyclic compounds **214** and **215**.[152] This transformation is a nice example of photochemical

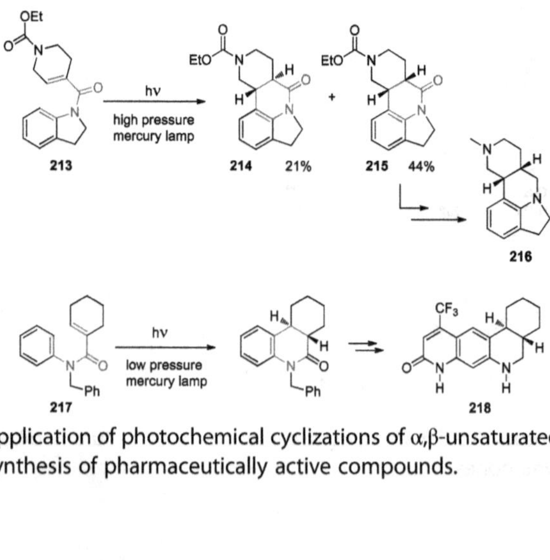

Scheme 39 Application of photochemical cyclizations of α,β-unsaturated anilids as key steps in the synthesis of pharmaceutically active compounds.

reactions that, starting from a relatively simple compound, in one step a complex compound is obtained. In the present case, especially the major photoproduct **215** contains already all structural elements of the target molecule **216**, which is a 5-$HT_{2C/2B}$ receptor antagonist. Such a pharmaceutical activity may be interesting for the therapy of diseases like eating disorders, migraine, obsessive–compulsive disorders, and anxiety. The photocyclization of the cyclohexene carbanilide derivative **217** was used as key step in the synthesis of **218**, which is a nonsteroidal androgen receptor agonist.[153]

The previously discussed reactions are a valuable tool for the synthesis of alkaloids. Some examples are depicted in Fig. 7. The cyclization of anilides was used for the synthesis of yohimbine[154] and xylopinine[155]. In the case of aspidospermidine[156], a α,β-unsaturated carbonyl moiety was cyclized. Lysergic acid[157] and a large variety of similar compounds such as chanoclavine-1[158] have been obtained from photocyclization products of the same furan carboxamide precursor. The photocyclization with more complex heterocyclic amides has been used to contribute to structure determination of pseudodistomins A and B.[159]

Photocyclization is a very suitable tool for the synthesis of helicenes.[160] A subfamily of these compounds contains heterocyclic moieties such as thiophenes. Heterohelicenes such as **219** (Scheme 40) have been obtained in this way.[161] A systematic synthesis strategy has been developed for the

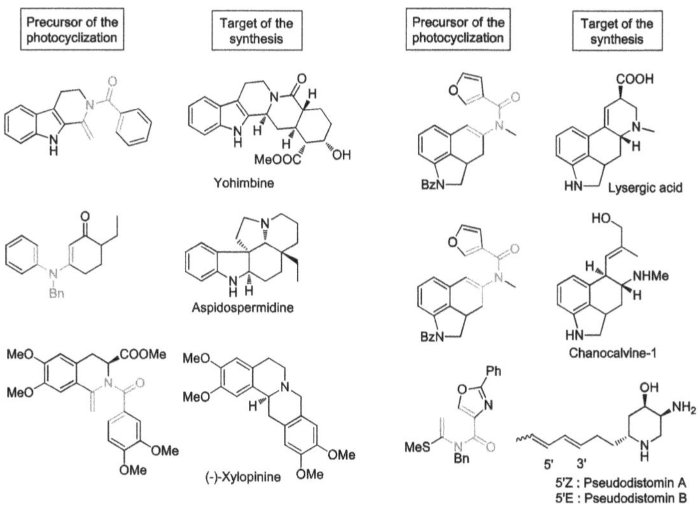

Fig. 7 Examples of the application of photocyclization reactions to the synthesis of alkaloids.

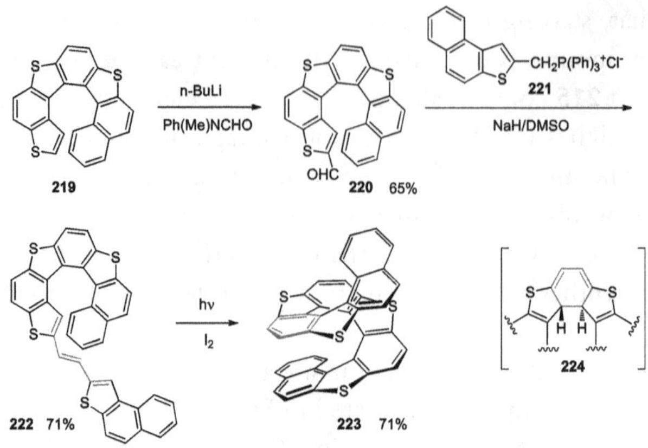

Scheme 40 Synthesis of heterohelicenes with thiophene moieties. Irradiations have been carried out with high-pressure mercury-vapor lamp.

preparation of homologues. Thus lithiation followed by carbonylation yielded **220**. Wittig reaction with **221** was used to prepare the substrate **222** for a photocyclization. The latter occurs in a conrotatory sense, leading to intermediate **224**. Under mild oxidation conditions, in the presence of iodine, oxidation (aromatization) takes place, and the final product **223** is obtained. A large variety of helicenes have been prepared using this strategy.

Complex polyheterocyclic systems can also been synthesized by highly efficient cyclization of carbodiimides such as **225** (Scheme 41).[162] After photochemical excitation the triplet species **226** is generated. Electronic excitation is achieved either by light absorption or by triplet sensitization of **225**. Radical addition to the alkyne function yields the diradical indole

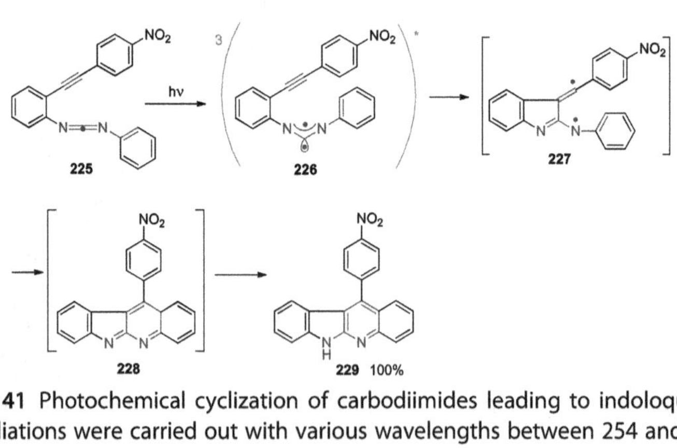

Scheme 41 Photochemical cyclization of carbodiimides leading to indoloquinolines. The irradiations were carried out with various wavelengths between 254 and 419 nm.

intermediate **227**. Radical combination occurs via addition to the phenyl substituent, leading to the formation of intermediate **228**. The final product **229** is an indoloquinoline system that is formed by tautomerization in the last step.

6 Conclusion

Photochemical rearrangements provide a convenient access to a large variety of organic compounds. Often photochemical products possess an unusual structure and can therefore be conceived as interesting synthesis intermediates. Although reported in the literature for a long time, the application of such photochemical rearrangements to the synthesis of heterocyclic compounds was rarely studied in a systematic way to determine their scope and limitations for application to organic synthesis in a more general context.

Heterocyclic compounds play a key role in many domains, for example, in the search of biologically active compounds for medicine or agriculture. Also in materials science, such compounds are systematically studied, for example, when they possess electronic semiconducting properties. It is therefore highly demanded to study photochemical reactions in connection with heterocyclic chemistry.

It must be particularly pointed out that most of the herein discussed reactions don't need any chemical activation. The photon is then the only reagent. It doesn't leave traces. Most of these photoreactions even don't need any photochemical sensitization. For many domains of chemistry, photochemical reactions represent an important tool of sustainability. In combination with other approaches such as catalysis or the use of products from renewable feedstock, numerous additional perspectives for sustainable chemistry can be developed.

References

1. Ciamician, G. The Photochemistry of the Future. *Science* **1912**, *36*, 385–394.
2. But besides the ferments, there is another reagent of high importance, at least for plants, and whose influence on organic reactions merits a profound investigation: it is light Ciamician, G. Sur les actions de la lumière. *Bull. Soc. Chim. Fr.* **1908**, *3*, i–xxvii.
3. (a) Albini, A.; Fagnoni, M. Giacomo Ciamician and the Concept of Green Chemistry. *ChemSusChem* **1908**, *2008* (1), 63–66; (b) Albini, A.; Fagnoni, M. Green Chemistry and Photochemistry Were Born at the Same Time. *Green Chem.* **2004**, *6*, 1–6.
4. Paternò, E. I nuovi orizzonti della sintesi in chimica organica. *Gazz. Chim. Ital.* **1909**, *39*, 213–219.

5. Bonfield, H. E.; Knauber, T.; Lévesque, F.; Moschetta, E. G.; Susanne, F.; Edwards, L. J. Photons as a 21st Century Reagent. *Nat. Commun.* **2020**, *11*. https://doi.org/10.1038/s41467-019-13988-4.
6. Stephenson, C. R. J.; Yoon, T. P.; MacMillan, D. W. C., Eds. *Visible Light Photocatalysis in Organic Chemistry*; Weinheim: Wiley-VCH, 2018.
7. Klán, P.; Wirz, J. *Photochemistry of Organic Compounds*; Wiley: Chichester, 2009.
8. Turro, N. J.; Schuster, G. Photochemical Reactions as a Tool in Organic Synthesis. *Science* **1975**, *187*, 303–312.
9. (a) Hoffmann, N. Photochemical Reactions as Key Steps in Organic Synthesis. *Chem. Rev.* **2008**, *108*, 1052–1103; (b) Bach, T.; Hehn, J. P. Photochemical Key Steps in Natural Product Synthesis. *Angew. Chem. Int. Ed.* **2011**, *50*, 1000–1045; (c) Liu, W.; Li, C.-J. Recent Synthetic Applications of Catalyst-Free Photochemistry. *Synlett* **2017**, *28*, 2714–2754; (d) Lefebvre, C.; Fortier, L.; Hoffmann, N. Photochemical Rearrangements in Heterocyclic Chemistry. *Eur. J. Org. Chem.* **2020**, *2020*, 1393–1404.
10. Beeler, A. B., Ed. Photochemistry in Organic Synthesis [Special Issue]. *Chem. Rev.* **2016**, *116* (17).
11. (a) Oelgemöller, M.; Hoffmann, N. Studies in Organic and Physical Photochemistry—An Interdisciplinary Approach. *Org. Biomol. Chem.* **2016**, *14*, 7392–7442; (b) Buzzetti, L.; Crisenza, G. E. M.; Melchiorre, P. Mechanistic Studies in Photocatalysis. *Angew. Chem. Int. Ed.* **2019**, *58*, 3730–3747.
12. Hoffmann, N. Photochemical Reactions of Aromatic Compounds and the Concept of the Photon as a Traceless Reagent. *Photochem. Photobiol. Sci.* **2012**, *11*, 1613–1641.
13. Oelgemöller, M.; Jung, C.; Mattay, J. Green Photochemistry: Production of Fine Chemicals with Sunlight. *Pure Appl. Chem.* **2007**, *79*, 1939–1947.
14. Oelgemöller, M. Solar Photochemical Synthesis: From the Beginnings of Organic Photochemistry to the Solar Manufacturing of Commodity Chemicals. *Chem. Rev.* **2016**, *116*, 9664–9682.
15. Michelin, C.; Hoffmann, N. Photosensitization and Photocatalysis—Perspectives in Organic Synthesis. *ACS Catal.* **2018**, *8*, 12046–12055.
16. See for example: Riguet, E.; Hoffmann, N. In *Comprehensive Organic Synthesis*; Knochel, P., Molander, G. A., Eds.; 2nd ed.; Vol. 5; Elsevier: Amsterdam, 2014; pp. 200–221.
17. Haning, H.; Niewöhner, U.; Schenke, T.; Es-Sayed, M.; Schmidt, G.; Lampe, T.; Bischoff, E. Imidazo[5,1-f][1,2,4]triazin-4(3H)-Ones, A New Class of Potent PDE 5 Inhibitors. *Bioorg. Med. Chem. Lett.* **2002**, *12*, 865–868.
18. Brigotti, M.; Carnicelli, D.; Accorsi, P.; Rizzi, S.; Montanaro, L.; Sperti, S. 4-Aminopyrazolo[3,4-d] as a Novel Inhibitor of the RNA and DNA Depurination Induced by Shiga Toxin 1. *Nucleic Acids Res.* **2000**, *28*, 2383–2388.
19. Yet, L. Rive-Membered Ring Systems: With More Than One N Atom. *Prog. Heterocycl. Chem.* **2016**, *28*, 275–315.
20. (a) Lablache-Combier, A. Photorearrangement of Nitrogen-Containing Arenes. In *Handbook of Photochemistry and Photobiology*; Horspool, W., Song, P.-S., Eds.; Boca Raton: CRC Press, 1995; pp. 1063–1120; (b) Pavlik, J. W. Photoisomerization of some nitrogen-containing hetero-aromatic compounds. In *Handbook of Photochemistry and Photobiology*; Horspool, W., Lenci, F., Eds.; 2nd ed.; Boca Raton: CRC Press, 2004; pp. 97/1–97/22.
21. Connors, R. E.; Burns, D. S.; Kurzweil, E. M.; Pavlik, J. W. Temperature-Dependent Phototransposition Chemistry of 1,5-Dimethylpyrazole and 1,2-Dimethylimidazole. *J. Org. Chem.* **1992**, *57*, 1937–1940.

22. Pavlik, J. W.; Martin, H. S.; Lambert, K. A.; Lowell, J. A.; Tsefrikas, V. M.; Eddins, C. K.; Kebede, N. Photochemistry of 4- and 5-Phenyl Substituted Isoxazoles. *J. Heterocyclic Chem.* **2005**, *42*, 273–281.
23. Childs, R. F. Circumambulatory Rearrangements. *Tetrahedron* **1982**, *38*, 567–608.
24. Tiefenthaler, H.; Dörscheln, W.; Göth, H.; Schmid, H. Photoisomerisierung von Pyrazolen und Indazolen zu Imiedazolen bzw. Benzimidazolen und 2-Aminobenzonitrilen. *Helv. Chim. Acta* **1967**, *50*, 2244–2258.
25. (a) Heinzelmann, W.; Märky, M.; Gilgen, P. Zum Mechanismus der photochemischen Umwandlung von 2-Alkyl-indazolen in 1-Alkyl-benzimidazole. I. Struktur und Reaktivität eines Zwischenproduktes. *Helv. Chim. Acta* **1976**, *59*, 1512–1527; (b) Heinzelmann, W.; Märky, M.; Gilgen, P. Zum Mechanismus der photochemischen Umwandlung von 2-Alkyl-indazolen in 1-Alkyl-benzimidazole. II. Photophysikalische und photochemische Primärprozesse. *Helv. Chim. Acta* **1976**, *59*, 1528–1546.
26. Sauers, R. R.; Hadel, L. M.; Scimone, A. A.; Stevenson, T. A. Photochemistry of 4-Acylisoxazoles. *J. Org. Chem.* **1990**, *55*, 4011–4019.
27. Grellmann, K. H.; Tauer, E. Photolysis of Benz- and Naphth-Isoxazoles: Evidence for the Intermediate Formation of Azirines. *J. Photochem.* **1977**, *6*, 365–370.
28. (a) Pusch, S.; Kowalczyk, D.; Opatz, T. A Photoinduced Cobalt-Catalyzed Synthesis of Pyrroles through In Situ-Generated Acylazirines. *J. Org. Chem.* **2016**, *81*, 4170–4178; (b) Pusch, S.; Schollmeyer; Opatz, T. A Light-Induced Vinylogous Nazarov-Type Cyclization. *Org. Lett.* **2016**, *18*, 3043–3045.
29. (a) Lablache-Combier, A.; Pollet, A. Etudes photochimiques-XIV: Photoréactions des méthylisothiazoles. *Tetrahedron* **1972**, *28*, 3141–3151; (b) Pavlik, J. W.; Pandit, C. R.; Samuel, C. J.; Day, A. C. Phototransposition Chemistry of Methylisothiazoles and Methylthiazoles. *J. Org. Chem.* **1993**, *58*, 3407–3410.
30. For a computational study see Su, M.-D. A Model Study on the Photochemical Isomerization of Isothiazoles and Thiazoles. *Phys. Chem. Chem. Phys.* **2014**, *16*, 17030–17042.
31. Pavlik, J. W.; Tongcharoensirikul, P.; Bird, N. P.; Day, A. C.; Barltrop, J. A. Phototransposition Chemistry of Phenylisothiazoles and Phenylthiazoles. 1. Interconversions in Benzene Solution. *J. Am. Chem. Soc.* **1994**, *116*, 2292–2300.
32. D'Auria, M. Ab Initio Study on the Photochemical Isomerization of Thiazole Derivatives. *Tetrahedron* **2002**, *58*, 8037–8042.
33. Maeda, M.; Kojima, M. Photorearrangements of Phenyloxazoles. *J. Chem. Soc., Perkin Trans. 1* **1977**, 239–247.
34. Prager, R. H.; Smith, J. A.; Weber, B.; Williams, C. M. Chemistry of 5-Oxadihydroisoxazoles. Part 18. Synthesis of Exazoles by the Photolysis and Pyrolysis of 2-acyl-5-oxo-2,5-Dihydroisoxazoles. *J. Chem. Soc., Perkin Trans. 1* **1997**, 2665–2672.
35. Jin, Z. Muscarine, Imidazole, Oxazole and Thiazole Alkaloids. *Nat. Prod. Rep.* **2016**, *33*, 1268–1317.
36. Wipf, P. Synthetic Studies of Biologically Active Marine Cyclopeptides. *Chem. Rev.* **1995**, *95*, 2115–2134.
37. (a) Hamdan, F.; Tahoori, F.; Balalaie, S. Synthesis of Novel Cyclopeptides Containing Heterocyclic Skeletons. *RSC Adv.* **2018**, *8*, 33893–33926; (b) Yeh, V. S. C. Recent Advances in the Total Syntheses of Oxazole-Containing Natural Products. *Tetrahedron* **2004**, *60*, 11995–12042; (c) Cox, M.; Prager, R. H.; Swensson, C. E. The Synthesis of some Chiral 2-Aminoalkyloxazole-4-Carboxylates from Isoxazol-5(2H)-Ones. *Aust. J. Chem.* **2003**, *56*, 887–896.
38. Pace, A.; Pierro, P. The New Era of 1,2,4-Oxadiazoles. *Org. Biomol. Chem.* **2009**, 7, 4337–4348.

39. Su, M.-D. Theoretical Investigation of the Photochemical Isomerizations of Indoxazene and Isoxazole. *J. Org. Chem.* **2009**, *74*, 6055–6063.
40. (a) Pace, A.; Pibiri, I.; Palumbo Piccionello, A.; Buscemi, S.; Vivona, N.; Barone, G. Experimental and DFT Studies on Competitive Heterocyclic Rearrangements. Part 2: A One-Atom Side-Chain versus the Classic Three-Atom Side-Chain (Boulton-Katritzky) Ring Rearrangement of 3-Acylamino-1,2,4-oxadiazoles. *J. Org. Chem.* **2007**, *72*, 2756–7888; (b) Pace, A.; Buscemi, S.; Vivona, N.; Silvestri, A.; Barone, G. Photochemistry of 1,2,4-Oxadiazoles. A DFT Study on Photoinduced Competitive Rearrangements of 3-Amino- and 3-*N*-Methylamino-5-Perfluoroalkyl-1,2,4-Oxadiazoles. *J. Org. Chem.* **2006**, *71*, 2740–2749.
41. Buscemi, S.; Cicero, M. C.; Vivona, N.; Caronna, T. Heterocyclic Photorearrangements. Photochemical Behavior of some 3,5-Disubstituted 1,2,4-Oxadiazoles in Methanol at 254 nm. *J. Heterocyclic Chem.* **1988**, *25*, 931–935.
42. D'Auria, M.; Frenna, V.; Marullo, S.; Racioppi, R.; Spinelli, D.; Viggiani, L. Photochemical Isomerization of Aryl Hydrazones of 1,2,4-Oxadiazol into the Corresponding Triazoles. *Photochem. Photobiol. Sci.* **2012**, *11*, 1383–1388.
43. Xuan, J.; Lu, L.-Q.; Chen, J.-R.; Xiao, W.-J. Visible-Light-Driven Photoredox Catalysis in the Construction of Carbocyclic and Heterocyclic Ring Systems. *Eur. J. Org. Chem.* **2013**, *2013*, 6755–6770.
44. Mei, M.; Anand, D.; Zhou, L. Divergent Conversion of N-Acyl-Isoxazol-5(2H)-Ones to Oxazoles and 1,3-Oxazin-6-Ones Using Photoredox Catalysis. *Org. Lett.* **2019**, *21*, 3548–3553.
45. Prager, R. H.; Taylor, M. R.; Williams, C. M. The Chemistry of 5-Oxodihydroisoxazoles. Part 19. The Synthesis and Photolysis of N-Thioaceulisoxazol-5(2H)-Ones. *J. Chem. Soc. Perkin Trans. 1* **1997**, 2673–2678.
46. Nechab, M.; Mondal, S.; Bertrand, M. P. 1,n-Hydrogen-Atom Transfer (HAT) Reactions in Which n≠5: An Updated Inventory. *Chem. A Eur. J.* **2014**, *20*, 16034–16059.
47. Nakano, Y.; Biegasiewicz, K. F.; Hyster, T. K. Biocatalytic Hydrogen Atom Transfer: An Invigorating Approach to Free-Radical Reactions. *Curr. Opin. Chem. Biol.* **2019**, *49*, 16–24.
48. See for example: (a) Baguley, P. A.; Walton, J. C. Flight from the Tyranny of Tin: The Quest for Practical Radical Sources Free From Metal Encumbrances. *Angew. Chem. Int. Ed.* **1998**, *37*, 3073–3082; (b) Povie, G.; Renaud, P. Lewis Acid-Water/Alcohol Complexes as Hydrogen Atom Donors in Radical Reactions. *Chimia* **2013**, *67*, 250–252; (c) Staveness, D.; Bosque, I.; Stephenson, C. R. J. Free Radical Chemistry Enabled by Visible Light-Induced Electron Transfer. *Acc. Chem. Res.* **2016**, *49*, 2295–2306.
49. (a) Hoffmann, N. Electron and Hydrogen Transfer in Organic Photochemical Reactions. *J. Phys. Org. Chem.* **2015**, *28*, 121–136; (b) Hoffmann, N. Photochemical Electron and Hydrogen Transfer in Organic Synthesis: The Control of Selectivity. *Synthesis* **2016**, *48*, 1782–7802.
50. See for example: (a) Miller, D. C.; Tarantino K. T.; Knowles, R. R. Proton-Coupled Electron Transfer in Organic Synthesis: Fundamentals, Applications, and Opportunities. *Top. Curr. Chem. (Z)* **2016**, *374*, 30; (b) Hoffmann, N. Proton-Coupled Electron Transfer in Photoredox Catalytic Reactions. *Eur. J. Org. Chem.* **2017**, *2017*, 1982–1992; (c) Ma, J.; Zhang X.; Phillips, D. L. Time-Resolved Spectroscopic Observation and Characterization of Water-Assisted Photoredox Reactions of Selected Aromatic Carbonyl Compounds. *Acc. Chem. Res.* **2019**, *52*, 726–737.
51. Streit, U.; Bochet, C. G. The Arene-Alkene Photocycloaddition. *Beilstein J. Org. Chem.* **2011**, 7, 525–542.
52. Nishio, T.; Sakurai, N.; Iba, K.; Hamano, Y.; Sakamoto, M. Intramolecular Photocyclization of 2-Acylphenyl Methacrylates: A Convenient Access to 4,5-Dihydro-

1,4-Epoxy-2-Benzoxepin-3(1*H*)-Ones (=Benzo[*c*]-6,8-Dioxabicyclo[3.2.1]Octan-7-Ones). *Helv. Chim. Acta* **2005**, *88*, 2603–2609.
53. (a) Breslow, R. Biomimetic Control of Chemical Selectivity. *Acc. Chem. Res.* **1980**, *13*, 170–177; (b) Wagner, P. J. Yang Photocyclization: Coupling of Biradicals Formed by Intramolecular Hydrogen Abstraction of Ketones. In *Handbook of Photochemistry and Photobiology*; Horspool, W., Lenci, F., Eds.; 2nd ed.; Boca Raton: CRC Press, 2004; pp. 58/1–58/70; (c) Nau, W. M.; Pischel, U. Photoreactivity of the n,π*-Excited Azoalkanes and Ketones. In *Organic Photochemistry and Photophysics*; Ramamurthy, V., Schanz, K., Eds.; Taylor & Francis: Boca Raton, 2006; pp. 75–129; (d) Rubin, M. B. Photoinduced Intermolecular Hydrogen Abstraction Reactions of Ketones. In *CRC Handbook of Organic Photochemistry and Photobiology*; Horspool, W. M., Song, P. S., Eds.; CRC Press: Boca Raton, 1995; pp. 430–436.
54. (a) Abad, S.; Boscá, F.; Domingo, L. R.; Gil, S.; Pischel, U.; Miranda, M. A. Triplet Reactivity and Regio-/Stereoselectivity in the Macrocyclization of Diastereomeric Ketoprofen-Quencher Conjugates Via Remote Hydrogen Abstraction. *J. Am. Chem. Soc.* **2007**, *129*, 7407–7420; (b) Pischel, U.; Abad, S.; Domingo, L. R.; Boscá, F.; Miranda, M. A. Diastereomeric Differentiation in the Quenching of Excited States by Hydrogen Donors. *Angew. Chem. Int. Ed.* **2003**, *42*, 2531–2534.
55. See for example: Jahjah, R.; Gassama, A.; Bulach, V.; Suzuki, C.; Abe, M.; Hoffmann, N.; Martinez, A.; Nuzillard, J.-M. Stereoselective Triplet-Sensitized Reactions of Furanone Derivatives. *Chem. Eur. J.* **2010**, *16*, 3341–3354.
56. Wessig, P.; Mühling, O. Abstraction of (γ ± n)-hydrogen by excited carbonyls. In *Synthetic Organic Photochemistry*; Griesbeck, A. G., Mattay, J., Eds.; New York: Marcel Dekker, 2005; pp. 41–87.
57. Kanaoka, Y.; Hatanaka, Y. Synthetic Photochemistry with the Imide System. Norrish Type II Cyclization of Alicyclic Imides. *J. Org. Chem.* **1976**, *41*, 400–401.
58. Thiering, S.; Sund, C.; Thiem, J.; Gieser, A.; Kopf, J. Synthesis of Imido-Substituted Glucosans and their Photocyclization towards Highly Functionalised Heterocycles. *Carbohydr. Res.* **2001**, *336*, 271–282.
59. Sund, C.; Thiering, S.; Thiem, J.; Kopf, J.; Stark, M. Photochemical Transformation of Levoglucosan Imides to Anhydrosugar-Annelated Azepanediones and Azocanediones. *Monatsh. Chem.* **2002**, *133*, 485–497.
60. (a) Kanaoka, Y. Photoreactions of Cyclic Imides. Examples of Synthetic Organic Photochemistry. *Acc. Chem. Res.* **1978**, *11*, 407–413; (b) Horvat, M.; Mlinarić-Majerski, K.; Basarić, N. Photochemistry of N-Aryl Substituted Phthalimides: H-Abstraction, Single Electron Transfer and Cycloadditions. *Croat. Chem. Acta* **2010**, *83*, 179–188; (c) Cindro, N.; Halasz, I.; Mlinarić-Majerski, K.; Basarić, N. Photoinduced H-Abstraction in Homo- and Protoadamantylphthalimide Derivatives in Solution and in Organized and Constrained Media. *Eur. J. Org. Chem.* **2013**, *2013*, 929–938; (d) Cindro, N.; Horvat, M.; Mlinarić-Majerski, K.; Griesbeck, A. G.; Basarić, N. Photoinduced Hemolytic C-H Activation in N-(4-Homoadamantyl)Phthalimide. *Beilstein J. Org. Chem.* **2011**, 7, 270–277.
61. (a) Oelgemöller, M.; Griesbeck, A. G. Photoinduced Electron Transfer Chemistry of Phthalimides: An Efficient Tool for C-C-Bond Formation. *J. Photochem. Photobiol., C* **2002**, *3*, 109–127; (b) Griesbeck, A. G.; Hoffmann, N.; Warzecha, K.-D. Photoinduced-Electron-Transfer Chemistry: From Studies on PET Processes to Application in Natural Product Synthesis. *Acc. Chem. Res.* **2001**, *40*, 128–140; (c) Yoon, U. C.; Mariano, P. S. The Synthetic Potential of Phthalimide SET Photochemistry. *Acc. Chem. Res.* **2001**, *34*, 523–533.
62. Cottet, F.; Cottier, L.; Descotes, G. Photocyclisations de cétoacétales aromatiques en vue d'une approche synthétique de la crombenine. *Can. J. Chem.* **1990**, *68*, 1251–1257.

63. Kraus, G. A.; Thomas, P. J.; Schwinden, M. D. An Approach to Aflatoxins Using Type II Photocyclization Reactions. *Tetrahedron Lett.* **1990**, *31*, 1819–1822.
64. Kraus, G. A.; Chen, L. A Total Synthesis of Racemic Paulownin Using a Type II Photocyclization Reaction. *J. Am. Chem. Soc.* **1990**, *112*, 3464–3466.
65. Pappas, S. P.; Pappas, B. C.; Blackwell, J. E., Jr. Photocyclization of Methyl-O-Benzyloxyphenylglyoxylate. *J. Org. Chem.* **1967**, *32*, 3066–3069.
66. Kraus, G. A.; Zhang, N. Hydrogen-Atom Abstraction/Cyclization in Synthesis. Direct Synthesis of Coumestan and Coumestrol. *J. Org. Chem.* **2000**, *65*, 5644–5646.
67. Gramain, J.-C.; Remuson, R.; Vallée, D. Intramolecular Photoreduction of α-Keto Esters. Total Synthesis of (±)-Isoretronecanol. *J. Org. Chem.* **1985**, *50*, 710–712.
68. Hesse, M. *Alkaloide Fluch und Segen der Natur?*; Verlag Helvetica Chimica Acta: Zürich, 2000.
69. (a) Micalizio, G. C.; Hale, S. B. Reaction Design, Discovery, and Development as a Foundation to Function-Oriented Synthesis. *Acc. Chem. Res.* **2015**, *48*, 663–673; (b) Bauman, V. T.; Shul'ts, E. E.; Kononchuk, V. V.; Bagryanskaya, I. Y.; Shakirov, M. M.; Tolstikov, G. A. Synthetic Transformations of Isoquinoline Alkaloids. 1-Alkynyl-3,6-dimethoxy-N-methyl-4,5α-epoxy-6,18-endoethenobenzo[i] isomorphinans and Their Transformations. *Russ. J. Org. Chem.* **2013**, *49*, 1502–1513; (c) Chuang, R. V.; Navarro, P.; Reisman, S. E. Benzoquinone-Derived Sulfinyl Imines as Versatile Intermediates for Alkaloid Synthesis: Total Synthesis of (−)-3-Demethoxyerythratidinone. *Chem. Sci.* **2011**, *2*, 1086–1089; (d) Martinez, E. J.; Corey, E. J. Enantioselective Synthesis of Saframycin A and Evaluation of Antitumor Activity Relative to Ecteinascidin/Saframycin Hybrids. *Org. Lett.* **1999**, *1*, 75–77.
70. He, W.; Zhang, Z.; Ma, D. A Scalable Total Synthesis of the Antitumor Agents Et-743 and Lurbinectedin. *Angew. Chem. Int. Ed.* **2019**, *58*, 3972–3975.
71. Thommen, D.; Jana, C. K.; Neuburger, M.; Gademann, K. Syntheses of Tawaniaquinone F and Tawaniaquinol A via An Unusual Remote C-H Functionalization. *Org. Lett.* **2013**, *15*, 1390–1393.
72. Koovits, P. J.; Knowles, J. P.; Booker-Milburn, K. I. Conformationally Driven Two- and Three-Photon Cascade Processes in the Stereoselective Photorearrangement of Pyrroles. *Org. Lett.* **2016**, *18*, 5608–5611.
73. (a) Li, W.; Chen, Z.; Yu, D.; Peng, X.; Wen, G.; Wang, S.; Xue, F.; Liu, X.-Y.; Qin, Y. Asymmetric Total Syntheses of the Akuammiline Alkaloids (−)-Strictamine and (−)-Rhazinoline. *Angew. Chem. Int. Ed.* **2019**, *58*, 6059–6063; (b) Xie, C.; Luo, J.; Zhang, Y.; Huang, S.-H.; Zhu, L.; Hong, R. Catalytic Aza-Wacker Annulation: Tuning Mechanism by the Activation Mode of Amide and Enantioselective Syntheses of Melinonine-E and Strychnoxanthine. *Org. Lett.* **2018**, *20*, 2386–2390.
74. Elliott, L. D.; Knowles, J. P.; Stacey, C. S.; Klauber, D. J.; Booker-Milburn, K. I. Using Batch Reactor Results to Calculate Optimal Flow Rates for the Scale-Up of UV Photochemical Reactions. *React. Chem. Eng.* **2018**, *3*, 86–93.
75. (a) Coyle, E. E.; Oelgemöller, M. Micro-Photochemistry: Photochemistry in Microstructured Reactors. The New Photochemistry of the Future? *Photochem. Photobiol. Sci.* **2008**, 7, 1313–1322; (b) Loubière, K.; Oelgemöller, M.; Aillet, T.; Dechy-Cabaret, O.; Prat, L. Continuous-Flow Photochemistry: A Need for Chemical Engineering. *Chem. Eng. Process.* **2016**, *104*, 120–132; (c) Michelin, C.; Lefebvre, C.; Hoffmann, N. Les reactions photochimiques à l'échelle industrielle. *Actual. Chim.* **2019**, *436*, 19–27; (d) Noël, T., Ed. *Photochemical Processes in Continuous-Flow Reactors*; World Scientific: London, 2017.
76. Lefebvre, C.; Michelin, C.; Martzel, T.; Djou'ou Mvondo, V.; Bulach, V.; Abe, M.; Hoffmann, N. Photochemically Induced Intramolecular Radical Cyclization Reactions with Imines. *J. Org. Chem.* **2018**, *83*, 1867–1875.

77. Tomoharu, I.; Yoshiyuki, S.; Teruyuki, N.; Makoto, B.; Junichi, N.; Toshio, N. *Banyu Pharma Co LTD*; JP 0052973, 2002.
78. (a) Lejeune, G.; Font, J.; Parella, T.; Alibés, R.; Figueredo, A. Intramolecular Photoreactions of (5S)-5-Oxymethyl-2(5H)-Furanones as a Tool for the Stereoselective Generation of Diverse Polycyclic Scaffolds. *J. Org. Chem.* **2015**, *80*, 9437–9445; (b) Xin, Y.; Rodríguez-Santiago, L.; Sodupe, M.; Álvarez-Larena, A.; Busqué, F.; Alibés, R. Intramolecular Photocycloaddition of 2(5H)-Furanones to Temporarily Tethered Terminal Alkenes as a Stereoselective Source of Enantiomerically Pure Polyfunctionalyzed Cyclobutanes. *J. Org. Chem.* **2018**, *83*, 3188–3199.
79. (a) Rehm, D.; Weller, A. Kinetics of Fluorescence Quenching by Electron and H-Atom Transfer. *Isr. J. Chem.* **1970**, *8*, 259–271; (b) Knibbe, K.; Rehm, D.; Weller, A. Intermediates and Kinetics of Fluorescence Quenching by Electron Transfer. *Ber. Bunsen. Phys. Chem* **1968**, *72*, 257–263.
80. Padwa, A.; Albrecht, F.; Singh, P.; Vega, E. Photochemical Transformations of Small Ring Carbonyl Compounds. XXX. Electron Transfer in the Photochemistry of Azetidinyl Ketones. *J. Am. Chem. Soc.* **1971**, *93*, 2928–2935.
81. See for example: Scholz, S. O.; Farney, E. P.; Kim, S.; Bates, D. M.; Yoon, T. P. Spin-Selective Generation of Triplet Nitrenes: Olefin Aziridination through Visible-Light Photosensitization of Azidoformates. *Angew. Chem. Int. Ed.* **2016**, *55*, 2239–2242.
82. Bauer, A.; Westkämper, F.; Grimme, S.; Bach, T. Catalytic Enantioselective Reactions Driven by Photoinduced Electron Transfer. *Nature* **2005**, *436*, 1139–1140.
83. (a) Bertrand, S.; Hoffmann, N.; Pete, J.-P. Highly Efficient and Stereoselective Radical Addition of Tertiary Amines to Electron-Deficient Alkenes—Application to the Enantioselective Synthesis of Necine Bases. *Eur. J. Org. Chem.* **2000**, *2000*, 2227–2238; (b) Bertrand, S.; Hoffmann, N.; Humbel, S.; Pete, J.-P. Diastereoselective Tandem Addition-Cyclization Reactions of Unsaturated Tertiary Amines Initiated by Photochemical Electron Transfer (PET). *J. Org. Chem.* **2000**, *65*, 8690–8703; (c) Hoffmann, N.; Görner, H. Photoinduced Electron Transfer from N-Methylpyrrolidine to Ketones and Radical Addition to an Electron-Deficient Alkene. *Chem. Phys. Lett.* **2004**, *383*, 451–455; (d) Harakat, D.; Pesch, J.; Marinković, S.; Hoffmann, N. Thiocarbonyl Compounds as Regulating Reagent in the Radical Addition of Tertiary Amines with Alkenes Using Photoelectron Transfer Conditions. *Org. Biomol. Chem.* **2006**, *4*, 1202–1205.
84. (a) Hoffmann, N.; Bertrand, S.; Marinković, S.; Pesch, J. Efficient Radical Addition of Tertiary Amines to Alkenes Using Photochemical Electron Transfer. *Pure Appl. Chem.* **2006**, *78*, 2227–2246; (b) Hoffmann, N. Efficient Photochemical Electron Transfer Sensitization of Homogeneous Organic Reactions. *J. Photochem. Photobiol., C* **2008**, *9*, 43–60; (c) Cossy, J.; Belotti, D. Generation of Ketyl Radical Anions by Photoinduced Electron Transfer (PET) Between Ketones and Amines. Synthetic Applications. *Tetrahedron* **2006**, *62*, 6459–6470.
85. Shi, L.; Xia, W. Photoredox Functionalization of C-N Bonds Adjacent to a Nitrogen Atom. *Chem. Soc. Rev.* **2012**, *41*, 7687–7697.
86. Brimioulle, R.; Lenhart, D.; Maturi, M. M. Bach Enantioselective Catalysis of Photochemical Reactions. *Angew. Chem. Int. Ed.* **2015**, *54*, 3872–3890.
87. Hu, S.; Neckers, D. C. Photochemical Reactions of Sulfide-Containing Alkyl Phenylglyoxylates. *Tetrahedron* **1997**, *53*, 7165–7180.
88. Yamazaki, Y.; Miyagawa, T.; Hasegawa, T. Photocyclization of 2-(Alkylthio)Ethyl Benzoylacetates Via Remote Proton Transfer Following Charge-Transfer Interaction between Excited Ketone Carbonyl and Thioether Chromophore. *J. Chem. Soc., Perkin Trans. 1* **1997**, 2979–2981.
89. Masuda, Y.; Ishida, N.; Murakami, M. Light-Driven Carboxylation of O-Alkylphenyl Ketones with CO_2. *J. Am. Chem. Soc.* **2015**, *137*, 14063–14066.

90. Nerdel, F.; Brodowski, W. UV-Lichtinduzierte Addition von Maleinsäureanhydrid an o-Methyl-Benzophenon. *Chem. Ber.* **1968**, *101*, 1398–1406.
91. Sammes, P. G. Photoenolisation. *Tetrahedron* **1976**, *32*, 405–422.
92. Kraus, G. A.; Chen, L. The Development of a Direct Synthetic Route to Pleurotin. *Synlett* **1981**, *2*, 89–90.
93. Gritsch, P. J.; Leitner, C.; Pfaffenbach, M.; Gaich, T. The Witkop Cyclization: A Photoinduced C-H Activation of the Indole System. *Angew. Chem. Int. Ed.* **2014**, *53*, 1208–1217.
94. Hoffmann, N.; Riguet, E. Aromatic Photochemical Reactions. In *Arene Chemistry*; Mortier, J., Ed.; Hoboken: Wiley, 2016; pp. 837–868.
95. (a) Mascal, M.; Modes, K. V.; Durmus, A. Concise Photochemical Synthesis of the Antimalarial Indole Alkaloid Decursivine. *Angew. Chem. Int. Ed.* **2011**, *50*, 4445–4446; (b) Qin, H.; Xu, Z.; Cui, Y.; Jia, Y. Total Synthesis of (±)-Decursivine and (±)-Serotobenine: A Witkop Photocyclization/Elimination/O-Michael Addition Cascade. *Angew. Chem. Int. Ed.* **2011**, *50*, 4447–4449.
96. Bennasar, M.-L.; Zulaica, E.; Jiménez, J.-M.; Bosch, J. Studies on the Synthesis of Mavacurine-Type Indole Alkaloids. First Total Synthesis of (±)-2,7-Dihydropleiocarpamine. *J. Org. Chem.* **1993**, *58*, 7756–7767.
97. Sundberg, R. J.; Hong, J.; Smith, S. Q.; Sabat, M.; Tabakovic, I. Synthesis and Oxidative Fragmentation of Catharanthine Analogs. Comparison to the Fragmentation – Coupling of Catharanthine and Vindoline. *Tetrahedron* **1998**, *54*, 6259–6292.
98. Szantay, C.; Bolcskei, H.; Gacs-Baitz, E. Synthesis of Vinca Alkaloids and Related Compounds XLVIII Synthesis of (+)-Catharanthine and (±)-Allocatharanthine. *Tetrahedron* **1990**, *46*, 1711–1732.
99. (a) Karapire, C.; Icli, S. Photochemical Aromatic Substitution. In *Handbook of Photochemistry and Photobiology*; Horspool, W., Lenci, F., Eds.; 2nd ed.; Boca Raton: CRC Press, 2004; pp. 37/1–37/14; (b) Schutt, L.; Bunce, N. J. Photodehalogenation of Aryl Halides. In *Handbook of Photochemistry and Photobiology*; Horspool, W., Lenci, F., Eds.; 2nd ed.; Boca Raton: CRC Press, 2004; pp. 38/1–38/18.
100. Senthilvelan, A.; Thirumalai, D.; Ramakrishnan, V. T. Photochemical Synthesis of Benzoxazolo[3,2-b]Isoquinolin-11-One and Isoquino[3,2-b][1,3]Benzoxazin-11-One under Basic Conditions. *Tetrahedron* **2005**, *61*, 4213–4220.
101. Aruna, Z.; Senthilvelan, A.; Thirumalai, D.; Muthusamy, S.; Ramakrishnan, V. T. Synthesis and Photocyclization of 1,2,4-Triazole-3-Thione. *Synthesis* **2006**, 3841–3848.
102. Cronk, W. C.; Mukhina, O. A.; Kutatetladze, A. G. Intramolecular Photoassisted Cycloadditions of Azaxylylenes and Postphotochemical Capstone Modifications Via Suzuki Coupling Provide Access to Complex Polyheterocyclic Biaryls. *J. Org. Chem.* **2014**, *79*, 1235–1246.
103. Mukhina, O. A.; Cronk, W. C.; Kumar, N. N. B.; Sekhar, M. C.; Samanta, A.; Kutateladze, A. G. Intramolecular Cycloadditions of Photogenerated Azaxylylenes: An Experimental and Theoretical Study. *J. Phys. Chem. A* **2014**, *118*, 10487–10496.
104. Cronk, W. C.; Mukhina, O. A.; Kutateladze, A. G. Photoinduced "Double Click" Cascade Offers Access to Complex Polyheterocycles from Readily Available Isatin-Based Photoprecursors. *Org. Lett.* **2016**, *18*, 3750–3753.
105. Kumpulainen, T.; Lang, B.; Rosspeintner, A.; Vauthrey, E. Ultrafast Elementary Photochemical Processes of Organic Molecules in Liquid Solution. *Chem. Rev.* **2017**, *117*, 10826–10939.
106. Duarte, L. G. T. A.; Germino, J. C.; de Ávila Braga, C.; Barboza, C. A.; Atvars, T. D. Z.; da Silva Santos, F.; Rodembusch, F. S. Photoacidity as a Tool to Rationalize Excited State Intramolecular Proton Transfer Reactivity in Flavonols. *Photochem. Photobiol. Sci.* **2018**, *17*, 231–238.

107. Gerard, B.; Jones, G., II; Porco, J. A., Jr. A Biomimetic Approach to the Rocaglamides Employing Photogeneration of Oxidopyryliums Derived from 3-Hydroxyflavones. *J. Am. Chem. Soc.* **2004**, *126*, 13620–13621.
108. Ninomiya, I.; Naito, T. *Photochemical Synthesis*; Academic Press: London, 1989; pp. 135–200.
109. Coote, S. C. 4-π-Photocyclization: Scope and Synthetic Applications. *Eur. J. Org. Chem.* **2020**, *2020*, 1405–1423.
110. Corey, E. J.; Streith, J. Internal Photoaddition Reactions of 2-Pyrone and N-Methyl-2-Pyridone: A New Synthetic Approach to Cyclobutadiene. *J. Am. Chem. Soc.* **1964**, *86*, 950–951.
111. Woodward, R. B.; Hoffmann, R. The Conservation of Orbital Symmetry. *Angew. Chem. Int. Ed.* **1969**, *8*, 781–932.
112. Furrer, H. Darstellung und Reaktionen von 3-Oxo-2-aza-bicyclo[2.2.0]hexenen-(5) und von 3-Oxo-1.2.4.6-tetramethyl-2.5-diazabicyclo[2.2.0]hexan. *Chem. Ber.* **1972**, *105*, 2780–2790.
113. Sato, M.; Katagiri, N.; Kaneko, C. Practicable Synthesis of (1R,4R)-5-(L-Menthoxy)-2-Azabicyclo[2.2.0]-hex-5-en-3-One and its Derivatives: New Building Blocks for Carbapenem Nuclei. *Tetrahedron Lett.* **1986**, *27*, 6091–6094.
114. Chandrakala, P. S.; Katz, A. K.; Carrell, H. L.; Sailaja, P. R.; Podile, A. R.; Nangia, A.; Desiraju, G. R. Synthesis, X-Ray Crystal Structures and Biological Evaluation of some Mono- and Bi-Cyclic 1,3-Diazetidin-2-Ones: Non-natural β-Lactam Analogues. *J. Chem. Soc. Perkin Trans. 1* **1998**, 2597–2608.
115. Kurita, J.; Yoneda, T.; Kakusawa, N.; Tsuchiya, T. Studies on Seven-Membered Heterocycles. XXXI. Synthesis of 1,4-Oxaepines and 1,4-Diazepines. *Chem. Pharm. Bull.* **1990**, *38*, 2911–2918.
116. Britten, T. K.; Kemmitt, P. D.; Halcovitch, N. R.; Coote, S. C. 4-P-Photocyclization of 1,2-Dihhydropyridazines: An Approach to Bicyclic 1,2-Diazetidines with Rich Synthetic Potential. *Org. Lett.* **2019**, *21*, 9232–9235.
117. Klán, P.; Wirz, J. *Photochemistry of Organic Compounds*; Chichester: Wiley, 2009; pp. 276–279.
118. (a) Gilbert, A. Ring Isomerization of Benzene and Naphthalene Derivatives. In *Handbook of Photochemistry and Photobiology*; Horspool, W., Song, P.-S., Eds.; Boca Raton: CRC Press, 1995; pp. 229–236; (b) Pincock, J. A. The Photochemistry of Substituted Benzenes: Phototranspositions and Photoadditions of Alcohols. In *Handbook of Photochemistry and Photobiology*; Horspool, W., Lenci, F., Eds.; 2nd ed.; Boca Raton: CRC Press, 2004; pp. 46/1–46/19.
119. (a) Mariano, P. S. A New Look at Pyridinium Salt Photochemistry. In *Handbook of Photochemistry and Photobiology*; Horspool, W., Lenci, F., Eds.; 2nd ed.; Boca Raton: CRC Press, 2004; pp. 100/1–100/10; (b) Zou, J.; Mariano, P. S. The Synthetic Potential of Pyridinium Salt Photochemistry. *Photochem. Photobiol. Sci.* **2008**, 7, 393–404.
120. Damiano, T.; Morton, D.; Nelson, A. Photochemical Transformation of Pyridinium Salts: Mechanistic Studies and Application in Synthesis. *Org. Biomol. Chem.* **2007**, *5*, 2735–2752.
121. Kaplan, L.; Pavlik, J. W.; Wilzbach, K. E. Photohydration of Pyridinium Ions. *J. Am. Chem. Soc.* **1972**, *94*, 3283–3284.
122. Schoenfeld, R. C.; Lumb, J.-P.; Ganem, B. Polyhydroxylated Aziridinylcyclopentanes as Glycomimetics: A New Competitive Inhibitor of a-Mannosidase. *Tetrahedron Lett.* **2001**, *42*, 6447–6449.
123. Colombo, C.; Pinto, B. M.; Bernardi, A.; Bennet, A. J. Synthesis and Evaluation of Influenza A Viral Neuraminidase Candidate Inhibitors Based on a Bicyclo[3.1.0]Hexane Scaffold. *Org. Biomol. Chem.* **2016**, *14*, 6539–6553.

124. Siopa, F.; Antonio, J. P. M.; Afonso, C. A. M. Flow-Assisted Synthesis of Bicyclic Aziridines Via Photochemical Transformation of Pyridinium Salts. *Org. Process Res. Dev.* **2018**, *22*, 551–556.
125. (a) Oelgemöller, M.; Goodine, T.; Malakar, P. Flow photochemistry—A green technology with a bright future. In *Sustainable Flow Chemistry: Methods and Applications*; Vaccaro, L., Ed.; 1st ed.; Weinheim: Wiley-VCH, 2017; pp. 1–24; (b) Mizuno, K.; Nishiyama, Y.; Ogaki, T.; Terao, K.; Ikeda, H.; Kakiuchi, K. Utilization of Microflow Reactors to Carry out Synthetically Useful Organic Photochemical Reactions. *J. Photochem. Photobiol., C* **2016**, *29*, 107–147.
126. Brandão, P.; Pineiro, M.; Pinho e Melo, T. M. V. D. Flow chemistry: Towards a More Sustainable Heterocyclic Synthesis. *Eur. J. Org. Chem.* **2019**, *2019*, 7188–7217.
127. Glarner, F.; Acar, B.; Etter, I.; Damiano, T.; Acar, E. A.; Bernardinelli, G.; Burger, U. The Photohydration of N-Glycosylpyridinium N,O-Acetals. *Tetrahedron* **2000**, *56*, 4311–4316.
128. Berecibar, A.; Grandjean, C.; Siriwardena, A. Synthesis and Biological Activity of Natural Aminocyclopentitol Glycosidase Inhibitors: Mannostatins, Trehazolin, Allosamidins, and their Analogues. *Chem. Rev.* **1999**, *99*, 779–844.
129. Ling, R.; Mariano, P. S. A Demonstration of the Synthetic Potential of Pyridinium Salt Photochemistry by its Application to the Stereocontrolled Synthesis of (+)-Mannostatin A. *J. Org. Chem.* **1998**, *63*, 6072–6076.
130. Feng, X.; Duesler, E. N.; Mariano, P. S. Pyridinium Salt Photochemistry in a Concise Route for the Synthesis of the Trehazolin Aminocyclitol, Trehazolamine. *J. Org. Chem.* **2005**, *70*, 5618–5623.
131. Zhao, Z.; Mariano, P. S. Application of the Photocyclization Reaction of 1,2-Cyclopenta-Fused Pyridinium Perchlorate to Formal Total Synthesis of (−)-Cephalotaxine. *Tetrahedron* **2006**, *62*, 7266–7273.
132. Zhou, J.; Gong, M.; Mariano, P. S.; Yoon, U. C. Observations Made in Exploring a Pyridinium Salt Photochemical Approach to the Synthesis of (+)-Lactacystin. *Bull. Kor. Chem. Soc.* **2008**, *29*, 89–93.
133. Hanaoka, M.; Mukai, C.; Nagami, K.; Okajima, K.; Yasuda, S. Chemical Transformation of Protoberberines. V. Photochemical Valence Isomerization of Berberinephenolbetaines to 8,14-Cycloberbines, Versatile Aziridine Derivatives for a Novel and Efficient Entry to Spirobenzylisoquinolines and Benzindenoazepines. *Chem. Pharm. Bull.* **1984**, *32*, 2230–2240.
134. Hanaoka, M.; Iwasaki, M.; Mukai, C. A Highly Stereoselective Synthesis of (±)-Dihydrofumariline-1. *Tetrahedron Lett.* **1985**, *26*, 917–920.
135. Cai, J.-C.; Hutchinson, C. R. Camptothecin. *The Alkaloids* **1983**, *21*, 101–137.
136. Sawada, S.; Okajima, S.; Aiyama, R.; Nokata, K.; Furuta, T.; Yokokura, T.; Sugino, E.; Yamaguchi, K.; Miyasaka, T. Synthesis and Antitumor Activity of 20(S)-Camptothecin Derivatives: Carbamate-Linked, Water-Soluble Derivatives of 7-Ethyl-10-Hydroxycamptothecin. *Chem. Pharm. Bull.* **1991**, *39*, 1446–1454.
137. Werner, S.; Seliger, R.; Rauter, H.; Wissmann, F. *(Heraeus GmbH) Quartz Glass Micro-Photoreactors and Synthesis of 10-Hydroxycamptothecin and 7-Alkyl-10-Hydroxycamptothecin*; EP 2 065 387 A2, 2009. US 2009/0143588 A1.
138. (a) Tsuchiya, T.; Kurita, J.; Kojima, H. Synthesis of Simple Monocyclic 1,3-Diazepines from 1,2-Diazepines. *J. Chem. Soc. Chem. Commun.* **1980**, 444–445; (b) Kurita, J.; Kojima, H.; Tsuchiya, T. Studies on Diazepines. XVI. Synthesis of Monocyclic 1,3-Diazepines. (1). Thermolysis of 1,2-Diazepines formed from Methylpyridine N-Imides. *Chem. Pharm. Bull.* **1981**, *29*, 3688–3695.
139. Hoffmann, N. Synthesis of Heterocycles by Photocyclization of Arenes. In *Handbook of Photochemistry and Photobiology*; Horspool, W., Lenci, F., Eds.; 2nd ed.; Boca Raton: CRC Press, 2004; pp. 34/1–34/20.

140. Gilbert, A. Cyclization of Stilbene and its Derivatives. In *Handbook of Photochemistry and Photobiology*; Horspool, W., Lenci, F., Eds.; 2nd ed.; Boca Raton: CRC Press, 2004; pp. 33/1–33/11.
141. Mallory, F. B.; Mallory, C. W. Photocyclization of Stilbenes and Related Molecules. *Org. React.* **1984**, *30*, 1–456.
142. (a) Irie, M. Discovery and Development of Photochromic Diarylethenes. *Pure Appl. Chem.* **2015**, *87*, 617–626; (b) Irie, M. Diarylethenes for Memories and Switches. *Chem. Rev.* **2000**, *100*, 1685–1716; (c) Irie, M.; Fukaminato, T.; Matsuda, K.; Kobatake, S. Photochromism of Diarylethene Molecules and Crystals: Memories, Switches, and Actuators. *Chem. Rev.* **2014**, *114*, 12174–12277.
143. (a) Roloff, A.; Meier, K.; Riediker, M. Synthetic and Metal Organic Photochemistry in Industry. *Pure Appl. Chem.* **1986**, *58*, 1267–1272; (b) Roloff, A.; Geschwind, H. W. *(Ciba-Geigy AG) Process for the Preparation of a Carboxylic Acid*; EP 018544 A1, 1986.
144. Soai, K.; Kawasaki, T.; Matsumoto, A. Asymmetric Autocatalysis of Pyrimidyl Alkanol and its Application to the Study of the Origin of Homochirality. *Acc. Chem. Res.* **2014**, *47*, 3643–3654.
145. Tanaka, K.; Toda, F. Solvent-Free Organic Synthesis. *Chem. Rev.* **2000**, *100*, 1025–1074.
146. (a) Grellmann, K.-H.; Sherman, G. M.; Linschitz, H. Photoconversion of Diphenylamines to Carbazoles and Accompanying Transient Species. *J. Am. Chem. Soc.* **1963**, *85*, 1881–1882; (b) Grellmann, K.-H.; Kühnle, W.; Weller, H.; Wolff, T. The Photochemical Formation of Dihydrocarbazoles from Diphenylamines and their Thermal Rearrangement and Disproportionation Reactions. *J. Am. Chem. Soc.* **1981**, *103*, 6889–6893.
147. (a) Zeller, K.-P.; Petersen, H. Photochemische Herstellung von Dibenzofuranen und Dibenzothiophenen. *Synthesis* **1975**, 532–533; (b) Zeller, K.-P.; Berger, S. Steric Hindrance in Substituted Dibenzofurans. *J. Chem. Soc., Perkin Trans. 2* **1977**, 54–58.
148. Schulz, A. G.; DeTar, M. B. Thiocarbonyl Ylides. Photogeneration, Rearrangement, and Cycloaddition Reactions. *J. Am. Chem. Soc.* **1976**, *98*, 3564–3572.
149. Kanaoka, Y.; Itoh, K. Photocyclization of Benzanilides to Phenanthridones with Elimination of the Ortho-Methoxy-Group. *Chem. Commun.* **1973**, 647–648.
150. Ibrahim-Ouali, M.; Sinibaldi, M.-E.; Troin, Y.; Guillaume, D.; Gramain, J.-C. Diastereoselective Photochemical Synthesis of 3,3′-Disubstuted Indolines. *Tetrahedron* **1997**, *53*, 16083–16096.
151. Blache, Y.; Benzech, V.; Chezal, J.-M.; Boule, P.; Viols, H.; Chavignon, O.; Teulade, J.-C.; Chapat, J.-P. Reactivity of Heterocyclic Enaminones: Regioselective Synthesis of Polyfused Indolones. *Heterocycles* **2000**, *53*, 905–916.
152. Nozulak, J.; Kalkman, H. O.; Floersheim, P.; Hoyer, D.; Schoeffter, P.; Buerki, H. R. (+)-cis-4,5,7a,8,9,10,11,11a-Octahydro-7*H*-10-methylindolo[1,7-*bc*][2,6]-naphthyridine: A 5-$HT_{2C/2B}$ Receptor Antagonist with Low 5-HT_{2A} Receptor Affinity. *J. Med. Chem.* **1995**, *38*, 28–33.
153. (a) Ninomiya, I.; Yamauchi, S.; Kiguchi, T.; Shinohara, A.; Naito, T. Photocyclization of Enamides. Part V. Photocyclization of α,β-Unsaturated Anilides. *J. Chem. Soc., Perkin Trans. 1* **1974**, 1747–1751; (b) Higuchi, R. I.; Edwards, J. P.; Caferro, T. R.; Ringgenberg, J. D.; Kong, J. W.; Hamann, L. G.; Arienti, K. L.; Marschke, K. B.; Davis, R. L.; Farmer, L. J.; Jones, T. K. 4-Alkyl- and 2,4-Dialkyl-1,2,3,4-Tetrahydro-8-Pyridono[4,6-g]Quinolones: Potent, Nonsteroidal Androgen Receptor Agonists. *Bioorg. Med. Chem. Lett.* **1999**, *9*, 1335–1340.
154. Naito, T.; Hirata, Y.; Miyata, O.; Ninomiya, I. Photocyclisation of Enamides. Part 27. Total Synthesis of Rac-Yohimbine, Rac-Alloyohimbine, and Rac-19,20-Didehydroyohimbines. *J. Chem. Soc., Perkin Trans. 1* **1988**, 2219–2225.

155. Kametani, T.; Takagi, N.; Toyota, M.; Honda, T.; Fukumoto, K. Total Synthesis of the Protoberberine Alkaloid (-)-Xylopinine by Photochemical 1,3-Asymmetric Induction. *J. Chem. Soc., Perkin Trans. 1* **1981**, 2830–2834.
156. Benchekroun-Mounir, V.; Dugat, D.; Gramin, J.-C.; Husson, H.-P. Stereocontrolled Formation of Octahydro-1H-Pyrrolo[2,3-d]Carbazoles by Reductive Cyclization: Total Synthesis of (±)-N-Benzylaspidospermidine. *J. Org. Chem.* **1993**, *58*, 6457–6465.
157. Ninomyia, I.; Hashimoto, C.; Kiguchi, T.; Naito, T. Photocyclisation of Enamides. Part 24. Total Synthesis of (±)-Isofumigaclavine B and (±)-Lysergic Acid. *J. Chem. Soc. Perkin Trans.* **1985**, *1*, 941–948.
158. Ninomiya, I.; Habe, N.; Kiguchi, T.; Naito, T. Photocyclization of Enamides. Part 35. New Total Syntheses of the Ergot Alkaloids (±)-Chanoclavine-1 and (±)-Isochanoclavine-1 Using a Fragmentation of 3-Amino Alcohols. *J. Chem. Soc. Perkin Trans.* **1991**, *1*, 3275–3285.
159. Naito, T.; Yuumoto, Y.; Kiguchi, T.; Ninomiya, I. Photocyclization of Amides. Part 39. General Strategy for the Synthesis of (±)-Tetrahydropseudodistomins: Synthesis of Triacetates of (±)-Tetrahydropseudodistomin and Proposed Structures of Pseudodistomins A and B. *J. Chem. Soc., Perkin Trans. 1* **1996**, 281–288.
160. Hoffmann, N. Photochemical Reactions Applied to the Synthesis of Helicenes and Helicene-Like Compounds. *J. Photochem. Photobiol., C* **2014**, *19*, 1–19.
161. Groen, M. B.; Schadenberg, H.; Wynberg, H. Synthesis and Resolution of Some Heterohelicenes. *J. Org. Chem.* **1971**, *36*, 2797–2809.
162. Schmittel, M.; Rodriguez, D.; Steffen, J.-P. A Highly Efficient Triplet Analogue of a Thermal Biradical Cyclization—The Photochemical C2-C6 Cyclization of Enyne-Heteroallenes. *Angew. Chem. Int. Ed.* **2000**, *39*, 2152–2155.

CHAPTER NINE

Electrochemical activation in organic synthesis applications

Bilal Nişancı
Department of Chemistry, Faculty of Science, Atatürk University, Erzurum, Turkey

1 Electrifying organic synthesis

Electrochemistry, which is called raw redox chemistry, is a very popular method used by synthetic chemists.[1] The application of the electric potential to initiate interaction between molecules and ensure the formation of new products is one of the important features of electrifying organic synthesis. Using electrons as a reagent also enables reactions with little or no waste formation. In addition, it is known that electrochemistry reduces the number of steps of many syntheses (step economic reactions),[2] creates milder conditions,[2] and enables green and sustainable technologies such as high atom economic reactions.[1,2]

Electrochemistry is one of the oldest synthetic methods discovered in the laboratory. Voltaic pile, the first battery capable of continuously supplying electricity to a circuit, was introduced by the Italian physicist Alessandro Volta in 1799.[1,3] In the 1830s Faraday added the terms electrolysis, anode, and cathode to the literature. At the same time the use of ionic salts as electrolytes was first introduced also by Faraday in this period.[1,3] Some years later, Kolbe observed the dimerization of radicals, the decarboxylation product of carboxylic acid by electrolysis (anodic oxidation) (Scheme 1).[1,4]

The first electrochemical dehalogenation method was carried out by Schoebein with cathodic reduction.[1] In 1898 Fritz Haber showed that if the potential at the cathode was kept constant, the expected reduction products could be generated by electrolytic processes. He performed the reduction of the nitrobenzene at the cathode.[5] The Tafel rearrangement is one of the

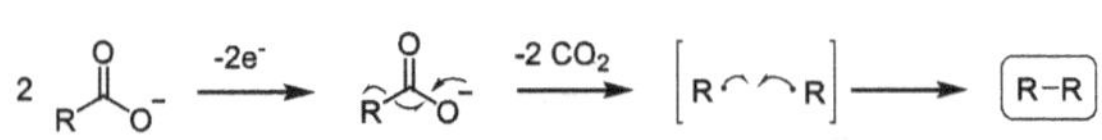

Scheme 1 Representative example of Kolbe's electrolysis.

https://doi.org/10.1016/B978-0-12-819009-8.00009-8
© 2021 Elsevier Inc. All rights reserved.

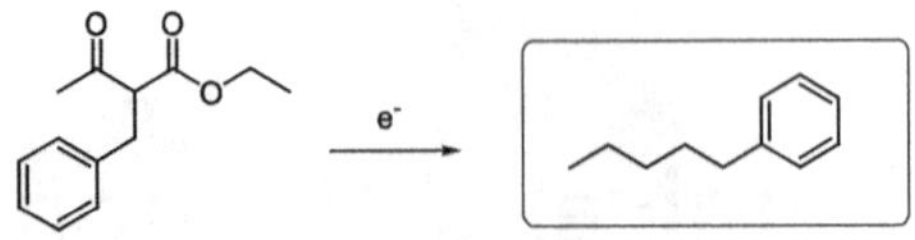

Scheme 2 Rearrangement of the alkyl group by the Tafel rearrangement.

most widely known electrochemical processes for the synthesis of hydrocarbons by cathodic reduction. The Tafel reaction was developed in 1907, and the reaction occurs via the rearrangement of the alkyl group (Scheme 2).[1,6]

The studies carried out in these pioneering times formed the basis of electrochemistry, and despite the passing of two centuries, it is still in the forefront of synthesis research. Especially, in recent years, there has been an extensive resurgence in the field of synthetic organic electrochemistry due to the growing interest in sustainable technologies.[1,7]

2 Electrochemical hardware

2.1 The electrochemical cell

Electrosynthesis requires a galvanic cell, a potentiostat, and two electrodes. There are two different cell types, undivided cells or divided cells. In divided cells the cathode and anode are separated with a semiporous membrane such as sintered glass, porous porcelain, polytetrafluoroethene, or polypropylene. Undivided cells, in which the anode and cathode are not separated, are more preferred because both oxidation and reduction can occur within the same region (Scheme 3).[8]

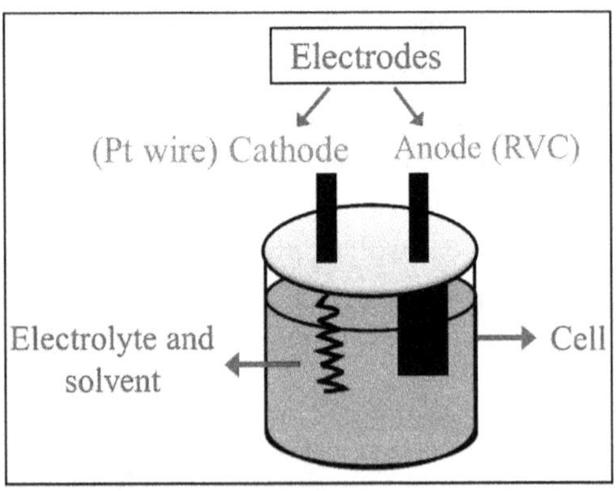

Scheme 3 Representative setup for an electrochemical cell.

2.2 The electrodes

The anode is defined as the electrode at which oxidation occurs, and the cathode is the electrode at which the reduction occurs. Conductive materials such as metals, semiconductors, graphite, and even conductive polymers are used as electrodes. Magnesium, platinum, and reticulated vitreous carbon (RVC) are the most commonly used electrodes in electroorganic reactions. In addition, at the end of the reaction, the electrodes can be easily removed from the reaction mixture and reused as long as they are stable to the applied reaction conditions.[9]

2.3 The electrolyte and the solvent

The appropriate solvent and electrolyte must be selected to provide electron transfer to the substrate with the lowest electrical resistance. The electrolytes are essential for the electrochemical reactions. Moreover, electrolytes can be capable to improve the conductivity of the reaction medium, and they should be stable under the electrochemical conditions. Commonly used electrolytes in organic synthesis are lithium perchlorate and tetraalkylammonium salts.[10]

The solvent selection depends on the reaction medium (protic or aprotic). For protic reactions, mixture of alcohol/water or dioxane/water is generally used as a solvent with the soluble salt as an electrolyte. In aprotic reactions, acetonitrile or dichloromethane is used as a solvent by the help of tetrabutylammonium salts as an electrolyte.[10] Regarding the solvent, its redox potential is also crucial. The solvent should be electrochemically stable at the applied redox potential.[10]

In 2017 a comprehensive review related with electrochemical activation in organic synthesis was published by Baran and coworkers.[1] The interest in electrochemical activation in organic synthesis by researchers has increased intensively over the last few years. For this reason, this chapter focuses on the electrifying organic synthesis performed in the last 3 years.

3 Electrochemical activation in organic synthesis

Transition metal-catalyzed C—H activation reactions in the last decade have attracted great attention from the researchers.[11] These reactions are generally carried out at elevated temperatures, with stoichiometric amounts of oxidants and expensive catalysts. The combination of C—H activation with electrocatalysis has emerged as a powerful new strategy

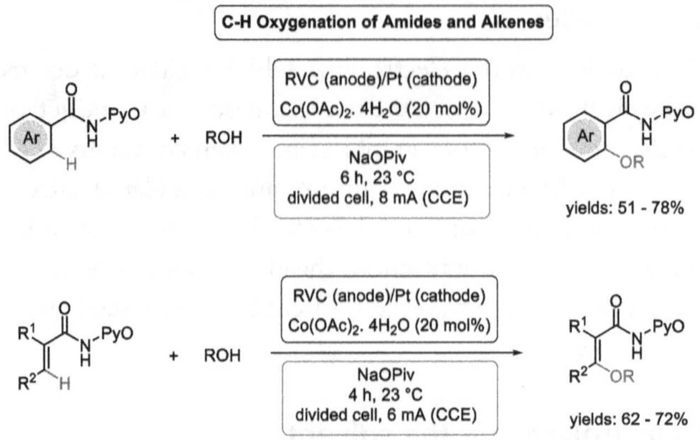

Scheme 4 C—H Oxygenation of amides and alkenes.

for molecular synthesis since it eliminates the use of toxic and expensive chemicals and requires only mild conditions.[12] C—H activation transformations were largely dominated by transition metal catalysts; herein, this chapter discussed some of them. In 2017 Sauermann et al. combined cobalt catalysis with electrochemistry to achieve C—H oxygenation reactions of amides and alkenes at room temperature (Scheme 4). The authors used a divided cell, constant current (8 mA), RVC (reticulated vitreous carbon) electrode (anode), and platinum electrode (cathode) with $Co(OAc)_2$. In addition, sodium trimethylacetate hydrate was used as an electrolyte and a base for this methodology.[13]

It was discovered by Tian et al. that cobalt-catalyzed electrochemical reactions were not limited to forming C—O bonds only. The authors have reported alkyne annulation by cobalt-catalyzed C—H activation in water/methanol mixture under room temperature conditions although the product yields were mostly moderate. The authors synthesized the annulation products in the absence of toxic metal oxidants (Scheme 5).[14]

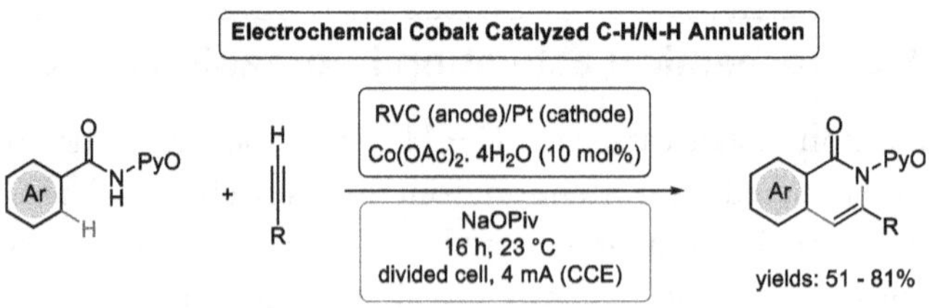

Scheme 5 Electrochemical cobalt-catalyzed C—H/N—H annulation.

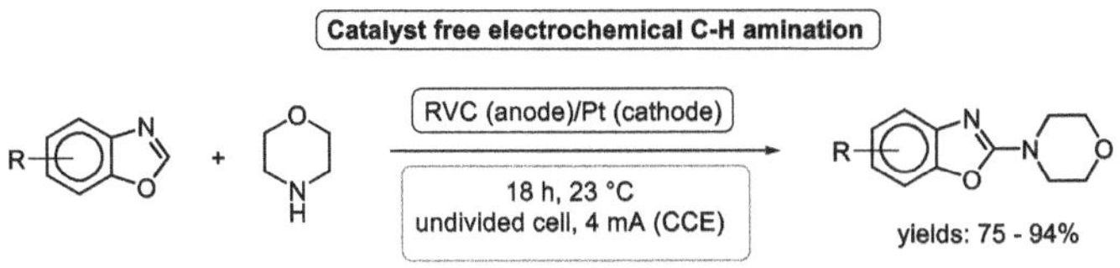

Scheme 6 Catalyst and oxidant-free electrochemical C—H amination of benzoxazole.

Taking advantage of the similar oxidation methodology, Qui et al. developed a new strategy for electrochemical azole C—H amination. The procedure proved feasible on azoles with reasonable yields under catalyst and reagent-free conditions. The authors used cyclic and acyclic aliphatic amines and anilines for the amination of benzoxazoles (Scheme 6). Electricity is the sole oxidant in this amination chemistry.[15]

Depending on the reactivity of the reagent used in the C—H amination reactions, small amounts of metal catalysts are also used in the reactions. Using nickel(II) acetate tetrahydrate (10 mol%) or nickel(II) chloride ethylene glycol dimethyl ether complex (20 mol%) as a catalyst with sodium trimethylacetate hydrate as an additive enables the electrochemical synthesis of C—H amination products from amides. In addition, nickel foam was placed as an anode in the undivided cell, while RVC worked as a cathode in this reaction. Corresponding amines were obtained in moderate yields. This study was the first nickel-catalyzed electrochemical C—H amination method in the literature (Scheme 7).[16]

Another reagent used for transition metal-catalyzed oxidative C—H reactions is carbon monoxide. Carbon monoxide is a cost-efficient and abundant chemical although its toxicity is a drawback in green synthetic methods. Moreover, C—H carbonylation with carbon monoxide is one of the most efficient methods for synthesizing carbonyl compounds. Zeng et al. developed a cobalt-catalyzed electrochemical reaction for oxidative

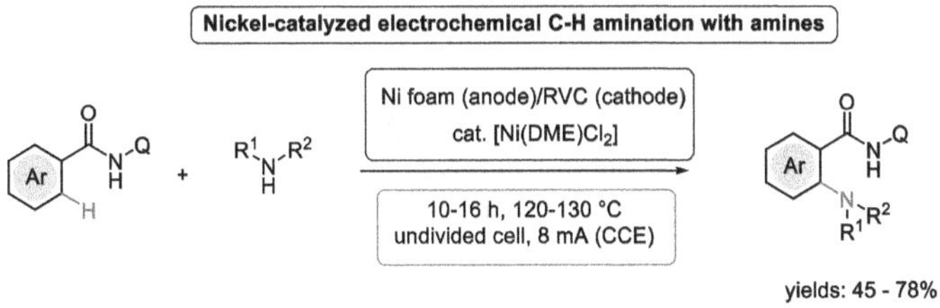

Scheme 7 Nickel-catalyzed electrochemical C—H amination with amines.

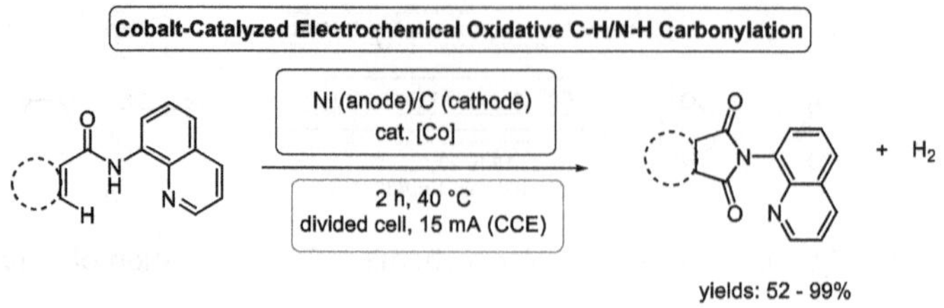

Scheme 8 Cobalt-catalyzed electrochemical oxidative C—H/N—H carbonylation.

C—H/N—H carbonylation by using carbon monoxide with the anodic oxidation. The cyclic products can be isolated with quantitative yields (Scheme 8).[17]

Due to extensive interest in electrooxidative C—H activation, many different syntheses have been developed with different types of catalyst for this methodology. The use of iridium as a catalyst led to the synthesis of electrooxidative C—H annulation products through weak O-coordination. In this topic, Qui et al. developed the first iridium-catalyzed electrooxidative C—H activation. Annulation products were synthesized and isolated in very good yields and excellent functional group tolerance including bromo- and iodoarenes. Moreover a gram-scale experiment was carried out and provided 84% (1.35 g) isolated yield from 1 g of acid (Scheme 9).[18]

Rhodium is also one of the commonly used metal catalysts in electrooxidative reactions. Ackermann reported the first electrooxidative

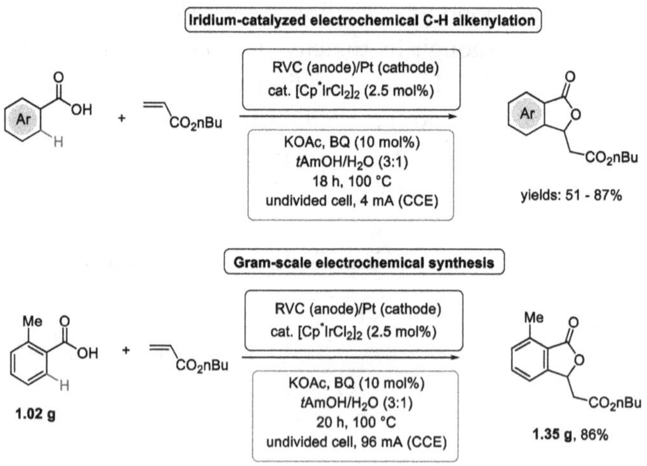

Scheme 9 Iridium-catalyzed C—H alkenylation and gram-scale synthesis.

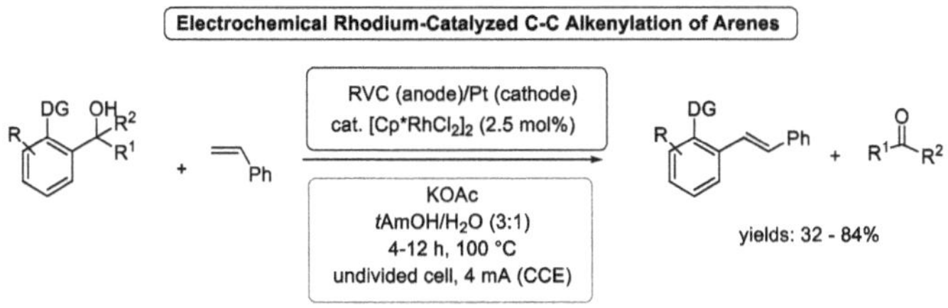

Scheme 10 Rhodium-catalyzed C—C alkenylation.

C—C activation by rhodium catalysis. The authors designed their experiments with pentamethylcyclopentadienylrhodium(III) chloride dimer ($[Cp^*RhCl_2]_2$) as the catalyst. Based on the optimization studies, it was concluded that the catalytic system provided the best performance by using the mixture of *tert*-amyl alcohol/H_2O as a solvent and KOAc as the additive in an undivided cell setup. Platinum plate was used as the cathode and RVC was used as the anode. The use of electrolyte was not required on this oxidative chemistry (Scheme 10).[19]

Hypervalent iodine compounds are chemicals that break the octet rule as iodine contains more than eight electrons in its valence shell. Hypervalent iodine derivatives have been shown to be an effective reagent in oxidation chemistry with many examples.[20–23] The use of hypervalent iodine with sustainable electricity contributes to the development of successful electro-oxidative methods.[20] By using iodobenzene as a mediator to generate phenyliodine bis(trifluoroacetate) (PIFA) in the reaction medium, an efficient C—H electrooxidation protocol was performed by applying Ru $(OAc)_2$(*p*-cymene). The electrooxidation was carried out in an undivided cell with platinum electrodes using weakly coordinating ketones as substrates (Scheme 11).[24]

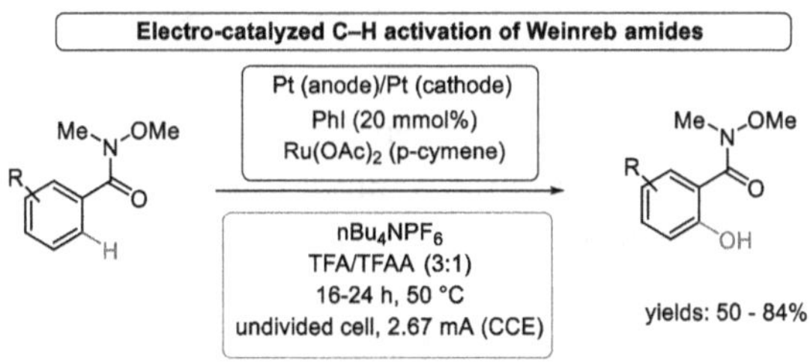

Scheme 11 Hypervalent iodine-catalyzed electrooxidation.

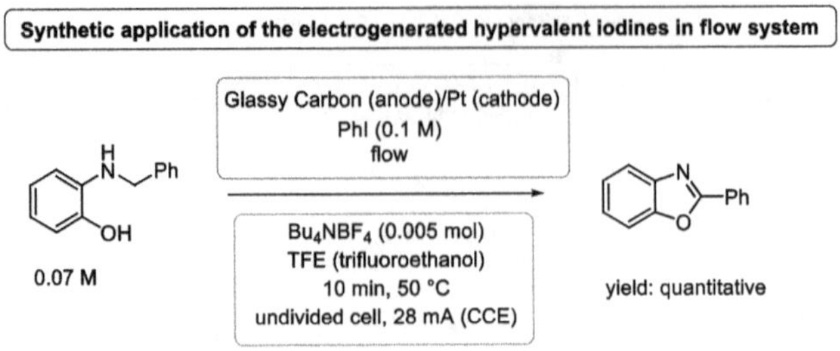

Scheme 12 Continuous flow electrochemical synthesis in flow system.

Another way to obtain hypervalent iodine compounds by electrochemical activation is using flow systems. Wirth developed an efficient electrochemical generator for hypervalent iodine derivatives. Moreover the use of electricity for this transformation helps to avoid the use of harmful and toxic oxidants. With this methodology the unstable hypervalent iodine compounds were easily synthesized and used in oxidation reactions. As a result, oxidation products were obtained with high yields in electrifying continuous flow system at room temperature (Scheme 12).[25]

Halogenated organic compounds play a very important role in synthetic chemistry, and they are essential in organic synthesis as starting materials and synthetic intermediates, designer molecules for materials science, industrial chemicals, and bioactive compounds.[26] Electrophilic halogenation in nature mainly occurs by oxidative halogenation through the catalyzed oxidation of the halide ion to form a halogenating reagent. In the laboratory, however, halogenation is usually carried out with hazardous, toxic, and corrosive molecular halogens and furthermore often performed in chlorinated solvents. A growing ecological awareness among chemists has coincided with an increased understanding of the importance of electrooxidative systems for halogenation reactions.[27]

Lei and coworkers prepared a series of halogenated compounds under catalyst- and oxidant-free conditions by electrochemical activation using an NaX system. This methodology has a broad substrate scope such as alkenes, heterocycles, alkynes, and even aliphatic hydrocarbons. In addition, all reactions can be scaled up to 200 mmol scale without losing efficiency or yields. The carbon rod was used as the anode and a platinum plate as the cathode within an undivided cell. Constant current was successfully employed for this clean and efficient halogenation protocol (Scheme 13).[27]

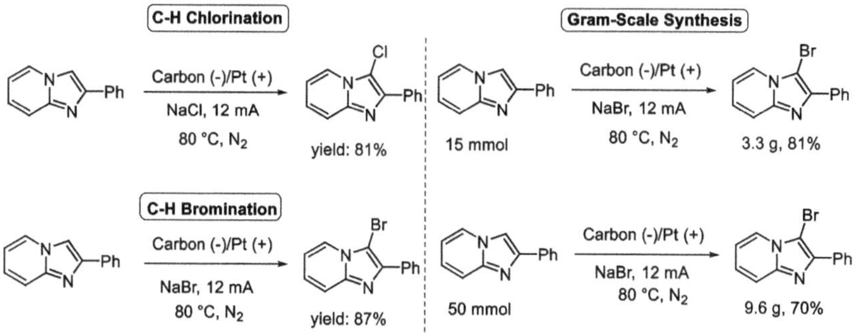

Scheme 13 Halogenation by electrochemical oxidation.

A reaction mechanism that was suggested for the C—H chlorination summarized in Scheme 13. According to the proposed mechanism, the reaction starts with anodic oxidation of chlorine and the formation of a chlorine radical. Then the chlorine radical attacks to the formed radical. Finally a one-electron oxidation and the following deprotonation occurs to the synthesis of chlorinated product (Scheme 14).[27]

Another attempt for the electrochemical halogenation has been carried out for the selective simultaneous N1-alkylation and C3-halogenation of indoles. The halogenated products were isolated with high yield using

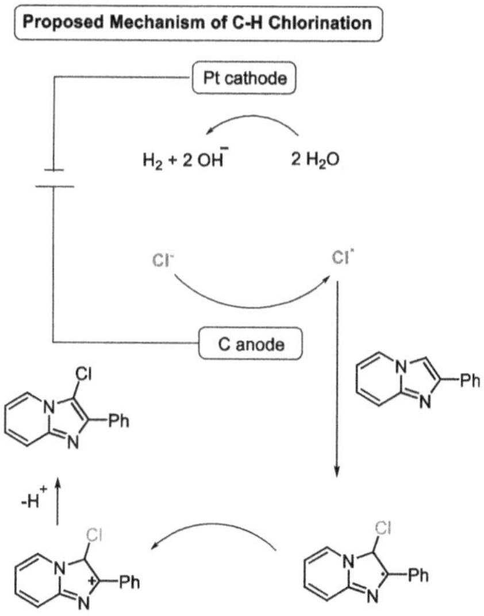

Scheme 14 Proposed mechanism for electrochemical C—H chlorination.

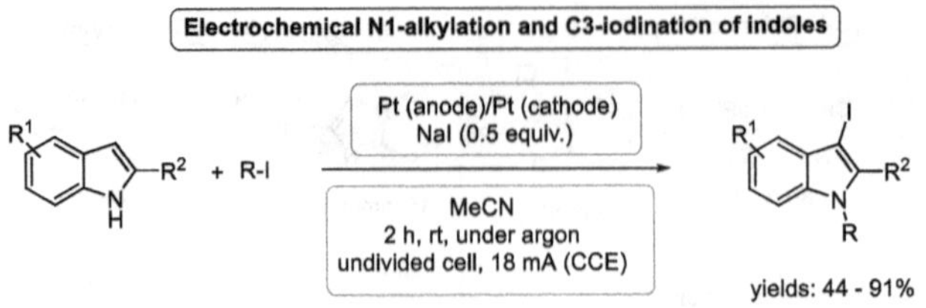

Scheme 15 Electrochemical N1-alkylation and C3-iodination of indole derivatives.

NaX as a halogen source at room temperature. This protocol proceeds in an undivided cell without using any additional oxidant or catalyst (Scheme 15).[28]

Baran and his group have added many new developments to electrifying organic synthesis. One of these works is related to electrochemical C(sp^3)-H fluorination with Selectfluor. They developed one of the most robust and simple methodologies for electrochemical halogenation chemistry. The reactions can be scaled up (100 g scale) easily and performed under catalyst-free conditions. Numerous types of functional groups (activating and deactivating) can be converted into their fluorinated analogues under short reaction times (Scheme 16).[29]

Polycyclic aromatic hydrocarbons (PAHs) are commonly used in optoelectronics, bioimaging, and catalysis. PAHs are described by their physical properties such as edge topology, electronic characteristics, and shape.[30,31] The need for simple and rapid access to polycyclic aromatic hydrocarbons (PAH) in materials chemistry attracted great interest. Ackermann et al. developed efficient and novel strategy for the synthesis of PAHs using the combination of electrocatalysis and [2 + 2 + 2] alkyne annulation by the help of electricity as a sustainable oxidant. The authors envisioned the electrochemical synthesis of PAHs from boronic acid derivates and biphenyl- or diarylacetylenes. The reaction occurs under rhodium electrocatalysis conditions (Scheme 17).[32]

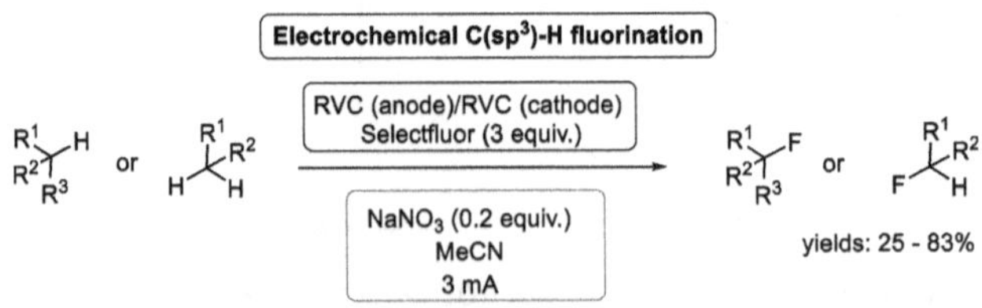

Scheme 16 Electrochemical C—H fluorination.

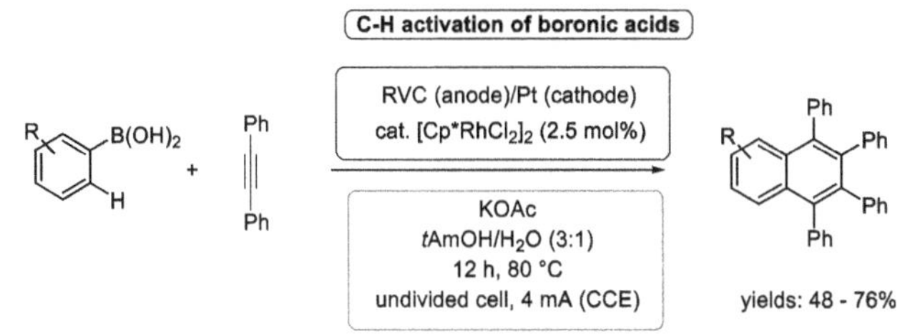

Scheme 17 Electrochemical synthesis of PAHs from boronic acid derivates.

Nitrogen-doped polycyclic aromatic hydrocarbons (aza-PAHs) were synthesized by the same group using similar methodology. Kong et al. performed feasible electrochemical synthesis of aza-PAHs via rhodium-catalyzed cascade C—H activation and domino annulations. Using imidamides and diphenylacetylene analogues enabled the synthesis of rhodaelectro-catalyzed cascade annulation products with broad substrate scope and excellent functional group tolerance. Some of the advantages of the developed cascade annulation reactions are (i) that they can be scaled up, (ii) occur under mild conditions, and (iii) have user-friendly setup (Scheme 18).[33]

Sulfonylhydrazines are molecules of high biological interest partially due to their antitumor activity. In addition, they can be used as a sulfonylation reagent.[34] Indole derivates are also known for their pharmaceutical and biological applications.[35,36] To eliminate the use of stoichiometric amounts of oxidants or transition metal catalysts in sulfonylation of indoles, Pan and coworkers developed an electrooxidation reaction between sulfonylhydrazines and indoles to synthesize sulfonylated hydrazine analogues (Scheme 19).[37]

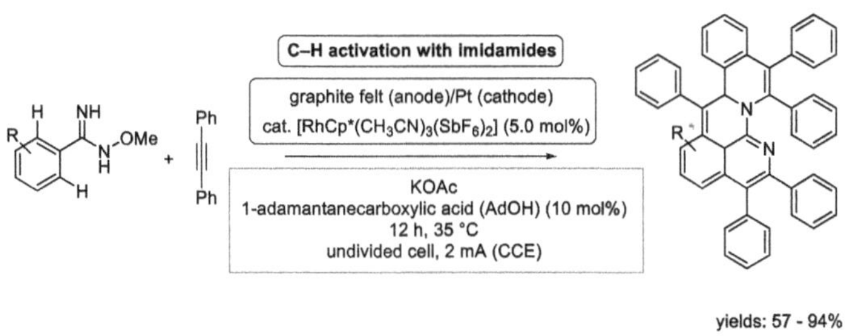

Scheme 18 Electrochemical access to aza-polycyclic aromatic hydrocarbons.

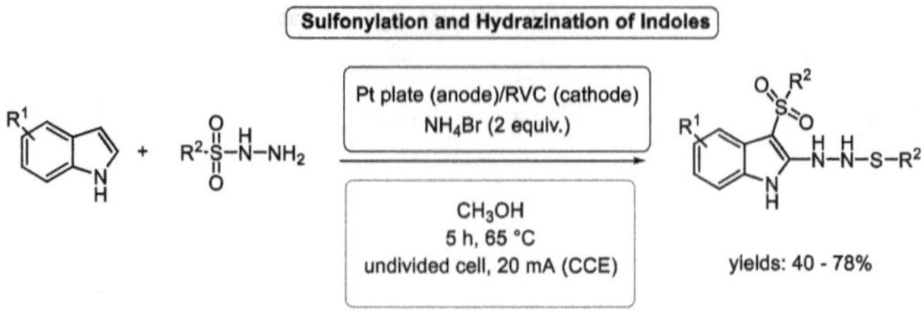

Scheme 19 Electrooxidative sulfonylation and hydrazination of indole derivates.

The use of limited or no additional oxidant in electroorganic chemistry makes electrochemistry attractive for researchers for the development of new protocols using electricity. In recent years, electrosynthetic application strategies are rapidly increasing. Ke et al. reported an electrochemical *N*-acylation protocol using carboxylic acids under mild reaction conditions at room temperature in water. Amides are obtained in good to excellent yields with very easy handle and simple methodology (Scheme 20).[38]

In another work, various enaminones were synthesized in high yields using decarboxylation coupling strategy. The enaminones were obtained from tolyl vinyl azides and α-keto acids under constant current conditions in an undivided cell. In addition, gram-scale reactions were also performed with high yield (81%). Both the electrolyte and the catalyst was n-Bu_4NI (Scheme 21).[39]

The synthesis of a broad variety of compounds that are widely used in the production of drugs, dyes, polymers, and plastics is enabled by hydrogenation reaction, which is one of the most essential reactions in the chemical industry and laboratory practice. Direct hydrogenation reactions are generally performed at high temperature and pressure using molecular hydrogen gas (H_2) supplied by pressurized tanks in a specific high-pressure reactors.[40–42] However, the requirement of special conditions

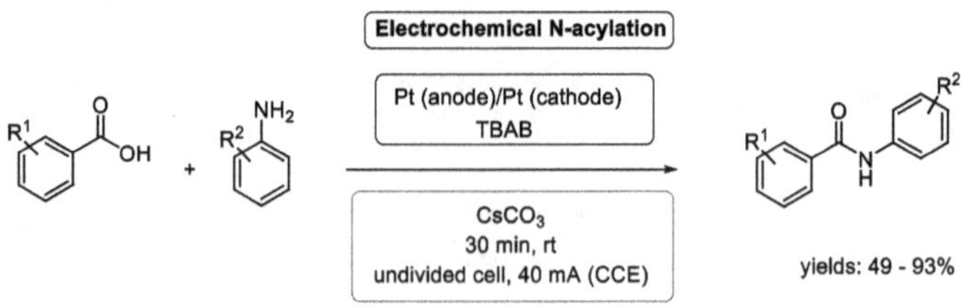

Scheme 20 Electrochemical acylation protocol from carboxylic acids.

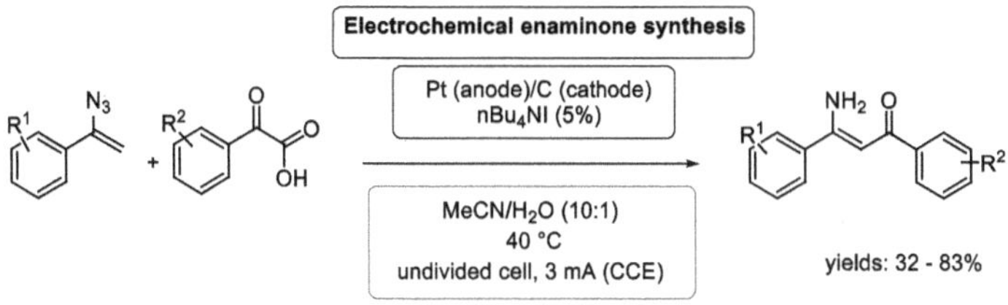

Scheme 21 Synthesis of enaminones using decarboxylation coupling strategy.

and the dangers associated by using these conditions make the development of hydrogenation methods providing high yields and product selectivity along with being easy operation, an important research topic. Hydrogenation carried out by electrochemical processes is an easy-to-handle, economic, and environmentally benign methodology to obtain a broad range of products.

It is known that there are many protocols related to electrochemical hydrogenation methodology in the literature.[1] In recent years, the advancement of these reduction methodologies has continued rapidly and increasingly. Egbert et al. developed the synthesis of corresponding hydrogenation products from a variety of substrates by using continuous electrocatalytic hydrogenation methodology. Using the flow electrolysis reactors, the authors obtained anilines, amines, and alcohols by the help of metal-supported graphitic carbon felt as an electrode (Scheme 22).[43]

Wei and coworkers demonstrated the selective reduction of cinnamaldehyde on RuO_2-SnO_2-TiO_2/Ti electrode. While low cinnamyl

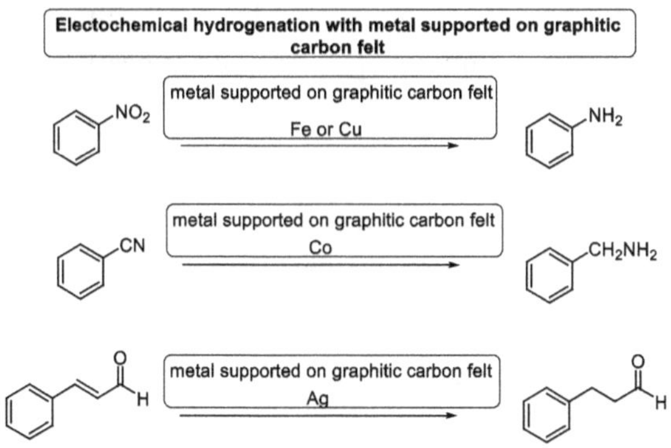

Scheme 22 Hydrogenation protocols under electrocatalytic conditions.

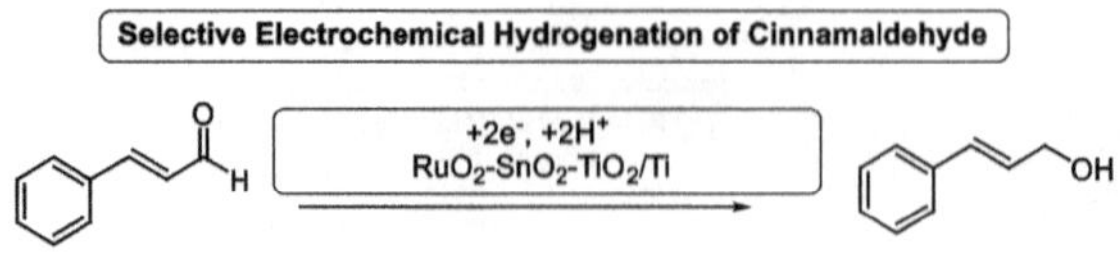

Scheme 23 Selective reduction of cinnamaldehyde.

alcohol selectivity was achieved with pure metal electrodes, the use of RuO_2-SnO_2-TiO_2/Ti electrode in acidic media increases the selectivity of cinnamyl alcohol up to 89% conversion (Scheme 23).[44]

The Birch reduction is a well-known process for the reduction of aromatic rings. The reduction is usually carried out with sodium or lithium metal in liquid ammonia and in the presence of an alcohol. Baran developed scalable reductive electrosynthetic method for the Birch reduction, which they called the e-Birch reaction. The authors used magnesium or aluminum as an electrode and dimethylurea as a water-soluble proton source. The protocol exhibits a wide-ranging substituent tolerance relative to classical electrochemical strategies and chemical dissolving metal reductions. Moreover the developed strategy can be applicable to other dissolving metal-type transformations such as the McMurry couplings, reductive ketone deoxygenations, and epoxide ring-opening reactions (Scheme 24).[45]

The use of ammonia as a hydrogen source is valuable because it is a high energy density hydrogen storage material. Li and coworkers published an electrochemical protocol for the hydrogenation of alkenes, alkynes, and ketones using ammonia as the hydrogen source on carbon electrodes. Variety of functional groups such as sulfide, benzyl, benzyl carbamate, and allyl carbamate were well tolerated in this methodology. In addition, in this process, ammonia gas was used at nearly atmospheric pressure (Scheme 25).[46]

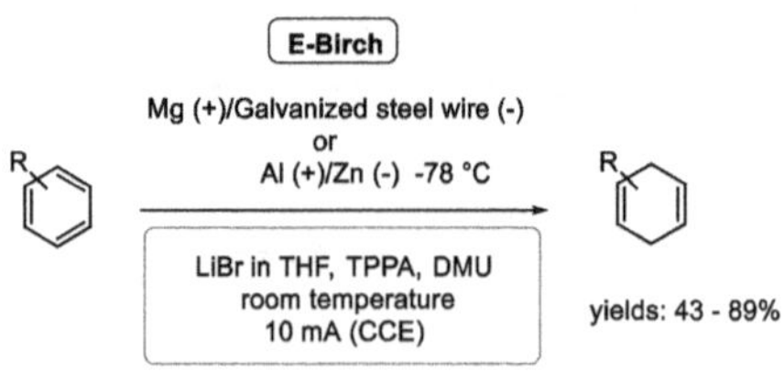

Scheme 24 Electrosynthetic method for the Birch reduction.

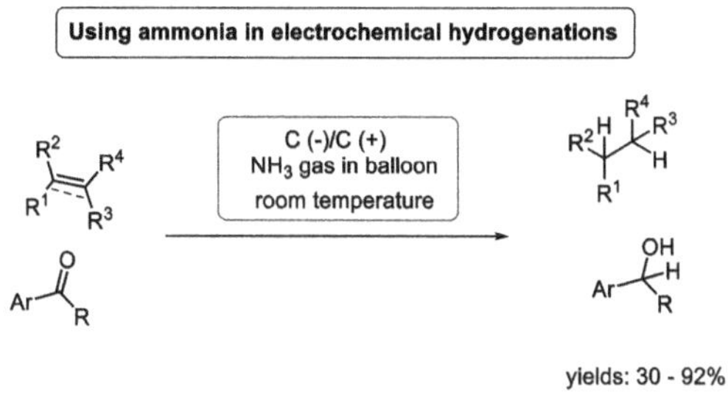

Scheme 25 Reduction of alkenes, alkynes, and ketones using ammonia as the hydrogen source on carbon electrodes.

Electrochemical dehydrogenation protocols are attracting great attention from the researchers similarly to the hydrogenation reactions. Lei et al. reported the electrochemical dehydrogenation of N-heterocycles by using TEMPO as the organoelectrocatalyst. This work is the first electrochemical acceptorless dehydrogenation process for the synthesis of N-heterocycles in the literature. The dehydrogenation of N-heterocycles takes place in the anode, and the release of H_2 takes place in cathode in undivided cell setup (Scheme 26).[47]

After this protocol, Wang developed a dehydrogenation methodology for the synthesis of N-heterocycles by applying metal-organic layer (MOL) as the catalyst using the same strategy (catalytic acceptorless dehydrogenation). The developed hybrid catalyst is efficient in the organic oxidation reactions and can be recycled and reused without losing its efficiency. The cell consists of RVC as the anode and Pt as the cathode, and the oxidation occurs at room temperature (Scheme 27).[48]

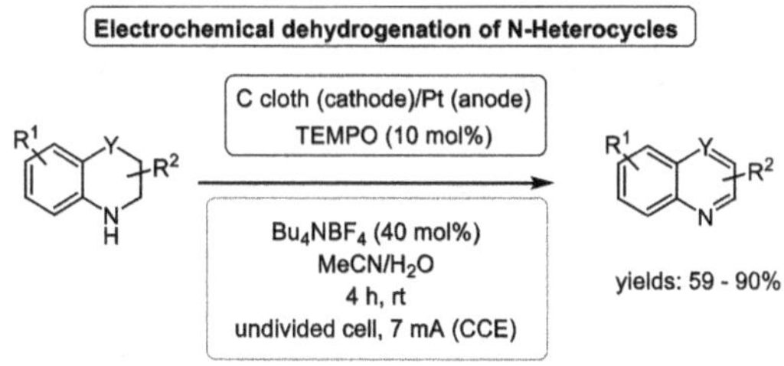

Scheme 26 The dehydrogenation of N-heterocycles with the electrochemical process.

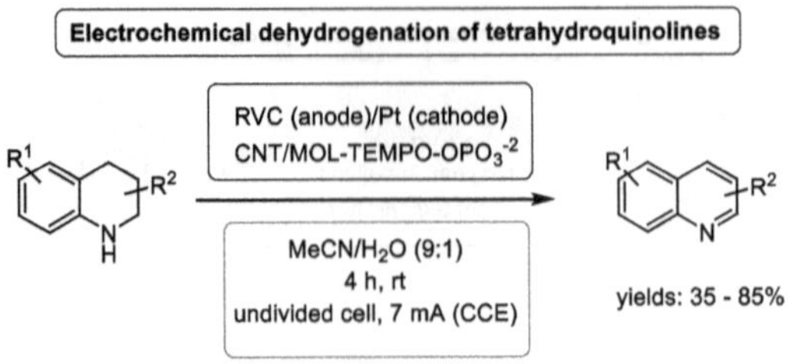

Scheme 27 Electrochemical dehydrogenation of tetrahydroquinolines.

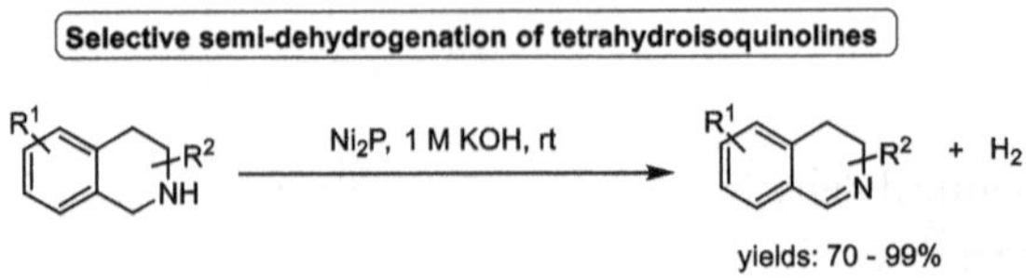

Scheme 28 Electrochemical selective semidehydrogenation of tetrahydroisoquinolines.

Selective electrochemical semidehydrogenation of tetrahydroisoquinolines is demonstrated by Zhang and coworkers. The authors used Ni_2P nanosheets as the electrode under aqueous conditions. Dihydroisoquinolines that are the semidehydrogenation products were obtained in high yields. In addition, the selective oxidation protocol can be applied to the gram-scale conditions as well (Scheme 28).[49]

4 Conclusions

Herein, we have demonstrated an overview of the recent developments in the field of electrochemical organic synthesis since 2017. Many different electrochemical methods such as C—H activation, halogenation, amination, hydrogenation, dehydrogenation, and addition reactions (4 + 2 annulation, sulfonylation, and acylation) have been summarized in this chapter.

Moreover, we have noted about an electrochemical setup and how an organic electrochemical synthesis takes place. Electrochemistry has many advantages such as scalability, the use of inexpensive and/or recyclable electrodes, the use of limited or no additional oxidant in synthesis, and high atom economical reactions. Electrochemical organic synthesis has been shown to be of great importance both in the field of green chemistry and in sustainable technology applications. Electroorganic synthesis, which has attracted great attention, will be widely applied in both research laboratories and industry in the coming years.

References

1. Yan, M.; Kawamata, Y.; Baran, P. Synthetic Organic Electrochemical Methods since 2000: On the Verge of a Renaissance. *Chem. Rev.* **2017**, *117*, 13230–13319.
2. Frontana-Uribe, B. A.; Little, R. D.; Ibanez, J. G.; Palma, A.; Vasquez-Medrano, R. Organic Electrosynthesis: A Promising Green Methodology in Organic Chemistry. *Green Chem.* **2010**, *12*, 2099–2119.
3. Yan, M.; Kawamata, Y.; Baran, P. S. Synthetic Organic Electrochemistry: Calling all Engineers. *Angew. Chem. Int. Ed.* **2018**, *57* (16), 4149–4155.
4. Kolbe, H. Beobachtungen über die oxydirende Wirkung des Sauerstoffs, wenn derselbe mit Hülfe einer elektrischen Säule entwickelt wird. *J. Prakt. Chem.* **1847**, *41*, 137–139.
5. Stock, J. Fritz Haber (1868-1934) and the Electroreduction of Nitrobenzene. *J. Chem. Educ.* **1988**, *65*, 337–338.
6. Tafel, J.; Hahl, H. Vollständige Reduktion Des Benzylacetessigesters. *Ber. Dtsch. Chem. Ges.* **1907**, *40*, 3312–3318.
7. Wiebe, A.; Gieshoff, T.; Möhle, S.; Rodrigo, E.; Zirbes, M.; Waldvogel, S. R. Electrifying Organic Synthesis. *Angew. Chem. Int. Ed.* **2018**, *57*, 5594–5619.
8. Bard, A. J.; Faulkner, L. R.; Leddy, J.; Zoski, C. G. *Electrochemical Methods: Fundamentals and Applications*, Vol. 2; Wiley: New York, 1980.
9. Kärkäs, M. D. Electrochemical Strategies for C–H Functionalization and C–N Bond Formation. *Chem. Soc. Rev.* **2018**, *47*, 5786–5865.
10. Grimshaw, J. *Electrochemical Reactions and Mechanisms in Organic Chemistry*; Elsevier, 2000.
11. Sauermann, N.; Meyer, T. H.; Qiu, Y.; Ackermann, L. Electrocatalytic C–H Activation. *ACS Catal.* **2018**, *8* (8), 7086–7103.
12. Qiu, Y.; Struwe, J.; Ackermann, L. Metallaelectro-Catalyzed C–H Activation by Weak Coordination. *Synlett* **2019**, *30* (10), 1164–1173.
13. Sauermann, N.; Meyer, T. H.; Tian, C.; Ackermann, L. Electrochemical Cobalt-Catalyzed C–H Oxygenation at Room Temperature. *J. Am. Chem. Soc.* **2017**, *139* (51), 18452–18455.
14. Tian, C.; Massignan, L.; Meyer, T. H.; Ackermann, L. Electrochemical C − H/N − H Activation by Water-Tolerant Cobalt Catalysis at Room Temperature. *Angew. Chem. Int. Ed.* **2018**, *57* (9), 2383–2387.
15. Qiu, Y.; Struwe, J.; Meyer, T. H.; Oliveira, J. C.; Ackermann, L. Catalyst- and Reagent-Free Electrochemical Azole C − H Amination. *Chem. Eur. J.* **2018**, *24* (49), 12784–12789.
16. Zhang, S. K.; Samanta, R. C.; Sauermann, N.; Ackermann, L. Nickel-Catalyzed Electrooxidative C − H Amination: Support for Nickel (IV). *Chem. Eur. J.* **2018**, *24* (72), 19166–19170.
17. Zeng, L.; Li, H.; Tang, S.; Gao, X.; Deng, Y.; Zhang, G.; Pao, C.-W.; Chen, J.-L.; Lee, J.-F.; Lei, A. Cobalt-Catalyzed Electrochemical Oxidative C–H/N–H Carbonylation with Hydrogen Evolution. *ACS Catal.* **2018**, *8* (6), 5448–5453.
18. Qiu, Y.; Stangier, M.; Meyer, T. H.; Oliveira, J. C.; Ackermann, L. Iridium-Catalyzed Electrooxidative C − H Activation by Chemoselective Redox-Catalyst Cooperation. *Angew. Chem.* **2018**, *130* (43), 14375–14379.
19. Qiu, Y.; Scheremetjew, A.; Ackermann, L. Electro-Oxidative C–C Alkenylation by Rhodium (III) Catalysis. *J. Am. Chem. Soc.* **2019**, *141* (6), 2731–2738.

20. Yang, F.; Ackermann, L. Ruthenium-Catalyzed C–H Oxygenation on Aryl Weinreb Amides. *Org. Lett.* **2013**, *15* (3), 718–720.
21. Amano, Y.; Nishiyama, S. Oxidative Synthesis of Azacyclic Derivatives through the Nitrenium Ion: Application of a Hypervalent Iodine Species Electrochemically Generated from Iodobenzene. *Tetrahedron Lett.* **2006**, *47* (37), 6505–6507.
22. Wang, K.; Fu, X.; Liu, J.; Liang, Y.; Dong, D. PIFA-Mediated Oxidative Cyclization of 1-Carbamoyl-1-Oximylcycloalkanes: Synthesis of Spiro-Fused Pyrazolin-5-One N-Oxides. *Org. Lett.* **2009**, *11* (4), 1015–1018.
23. Jiang, Q.; Zhao, A.; Xu, B.; Jia, J.; Liu, X.; Guo, C. PIFA-Mediated Esterification Reaction of Alkynes with Alcohols Via Oxidative Cleavage of Carbon Triple Bonds. *J. Org. Chem.* **2014**, *79* (6), 2709–2715.
24. Massignan, L.; Tan, X.; Meyer, T. H.; Kuniyil, R.; Messinis, A. M.; Ackermann, L. Electro-Catalyzed Hypervalent Iodine Orchestrated for Ruthenaelectro-Catalyzed C–H Oxygenation. *Angew. Chem.* **2020**, *132* (8), 3210–3215.
25. Elsherbini, M.; Winterson, B.; Alharbi, H.; Folgueiras-Amador, A. A.; Génot, C.; Wirth, T. Continuous-Flow Electrochemical Generator of Hypervalent Iodine Reagents: Synthetic Applications. *Angew. Chem. Int. Ed.* **2019**, *58*, 9811–9815.
26. Podgoršek, A.; Zupan, M.; Iskra, J. Oxidative Halogenation with "Green" Oxidants: Oxygen and Hydrogen Peroxide. *Angew. Chem. Int. Ed.* **2009**, *48* (45), 8424–8450.
27. Yuan, Y.; Yao, A.; Zheng, Y.; Gao, M.; Zhou, Z.; Qiao, J.; Hu, J.; Ye, B.; Zhao, J.; Wen, H.; Lei, A. Electrochemical Oxidative Clean Halogenation Using HX/NaX With Hydrogen Evolution. *iScience* **2019**, *12*, 293–303.
28. Sun, L.; Zhang, X.; Wang, C.; Teng, H.; Ma, J.; Li, Z.; Chen, H.; Jiang, H. J. Direct Electrosynthesis for N-Alkyl-C3-Halo-Indoles Using Alkyl Halide as Both Alkylating and Halogenating Building Blocks. *Green Chem.* **2019**, *21* (10), 2732–2738.
29. Takahira, Y.; Chen, M.; Kawamata, Y.; Mykhailiuk, P.; Nakamura, H.; Peters, B. K.; Reisberg, S. H.; Li, C.; Chen, L.; Hoshikawa, T.; Shibuguchi, T.; Baran, P. S. Electrochemical C (sp3)–H Fluorination. *Synlett* **2019**, *30* (10), 1178–1182.
30. Brasholz, M. "Super-Reducing" Photocatalysis: Consecutive Energy and Electron Transfers With Polycyclic Aromatic Hydrocarbons. *Angew. Chem. Int. Ed.* **2017**, *56* (35), 10280–10281.
31. Ito, H.; Segawa, Y.; Murakami, K.; Itami, K. J. Polycyclic Arene Synthesis by Annulative π-Extension. *J. Am. Chem. Soc.* **2018**, *141* (1), 3–10.
32. Kong, W. J.; Finger, L. H.; Oliveira, J. C.; Ackermann, L. Rhodaelectrocatalysis for Annulative C−H Activation: Polycyclic Aromatic Hydrocarbons through Versatile Double Electrocatalysis. *Angew. Chem. Int. Ed.* **2019**, *58* (19), 6342–6346.
33. Kong, W.-J.; Shen, Z.; Finger, L. H.; Ackermann, L. Electrochemical Access to Aza-Polycyclic Aromatic Hydrocarbons: Rhodaelectro-Catalyzed Domino Annulations. *Angew. Chem. Int. Ed.* **2020**,. https://doi.org/10.1002/anie.201914775.
34. Wei, W.; Liu, C.; Yang, D.; Wen, J.; You, J.; Suo, Y.; Wang, H. Copper-Catalyzed Direct Oxysulfonylation of Alkenes with Dioxygen and Sulfonylhydrazides Leading to β-Ketosulfones. *Chem. Commun.* **2013**, *49* (87), 10239–10241.
35. Kilic, H.; Bayindir, S.; Erdogan, E.; Saracoglu, N. Synthesis of Highly N-Substituted Indole Library Via Conjugate Additions of Indoline and Their Synthetic Tool Potentials. *Tetrahedron* **2012**, *68* (27–28), 5619–5630.
36. Kilic, H.; Aydin, O.; Bayindir, S.; Saracoglu, N. Condensation of Indoline With Some 1-, 2- and 1, 3-Diketones. *J. Heterocyclic Chem.* **2016**, *53* (6), 2096–2101.
37. Zhang, Y.-Z.; Mo, Z.-y.; Wang, H.; Wen, X.; Tang, H.-T.; Pan, Y.-M. Electrochemically Enabled Chemoselective Sulfonylation and Hydrazination of Indoles. *Green Chem.* **2019**, *21*, 3807–3811.
38. Ke, F.; Xu, Y.; Zhu, S.; Lin, X.; Lin, C.; Zhou, S.; Su, H. Electrochemical N-Acylation Synthesis of Amides under Aqueous Conditions. *Green Chem.* **2019**, *21* (16), 4329–4333.

39. Kong, X.; Liu, Y.; Lin, L.; Chen, Q.; Xu, B. Electrochemical Synthesis of Enaminones Via a Decarboxylative Coupling Reaction. *Green Chem.* **2019**, *21*, 3796–3801.
40. Eghbali, P.; Nişancı, B.; Metin, Ö. Graphene Hydrogel Supported Palladium Nanoparticles as an Efficient and Reusable Heterogeneous Catalysts in the Transfer Hydrogenation of Nitroarenes Using Ammonia Borane as a Hydrogen Source. *Pure Appl. Chem.* **2018**, *90* (2), 327–335.
41. Nişancı, B.; Dağalan, Z. A Facile and Highly Efficient Transfer Hydrogenation of Ketones and Aldehydes Catalyzed by Palladium Nanoparticles Supported on Mesoporous Graphitic Carbon Nitride. *J. Chem. Res.* **2019**, *44* (1–2), 14–19.
42. Ganjehyan, K.; Nişancı, B.; Sevim, M.; Daştan, A.; Metin, Ö. Monodisperse CuPt Alloy Nanoparticles Assembled on Reduced Graphene Oxide as Catalysts in the Transfer Hydrogenation of Various Functional Organic Groups. *Appl. Organomet. Chem.* **2019**, *33* (5), e4863.
43. Egbert, J. D.; Thomsen, E. C.; O'Neill-Slawecki, S. A.; Mans, D. M.; Leitch, D. C.; Edwards, L. J.; Wade, C. E.; Weber, R. S. Development and Scale-Up of Continuous Electrocatalytic Hydrogenation of Functionalized Nitro Arenes, Nitriles, and Unsaturated Aldehydes. *Org. Process. Res. Dev.* **2019**, *23* (9), 1803–1812.
44. Huang, X.; Zhang, L.; Li, C.; Tan, L.; Wei, Z. J. A. C. High Selective Electrochemical Hydrogenation of Cinnamaldehyde to Cinnamyl Alcohol on RuO_2–SnO_2–TiO_2/Ti Electrode. *ACS Catal.* **2019**, *9* (12), 11307–11316.
45. Peters, B. K.; Rodriguez, K. X.; Reisberg, S. H.; Beil, S. B.; Hickey, D. P.; Kawamata, Y.; Collins, M.; Starr, J.; Chen, L.; Udyavara, S.; Klunder, K.; Gorey, T. J.; Anderson, S. L.; Neurock, M.; Minteer, S. D.; Baran, P. S. Scalable and Safe Synthetic Organic Electroreduction Inspired by Li-Ion Battery Chemistry. *Science* **2019**, *363* (6429), 838–845.
46. Li, J.; He, L.; Liu, X.; Cheng, X.; Li, G. Electrochemical Hydrogenation With Gaseous Ammonia. *Angew. Chem. Int. Ed.* **2019**, *58* (6), 1759–1763.
47. Wu, Y.; Yi, H.; Lei, A. Electrochemical Acceptorless Dehydrogenation of N-Heterocycles Utilizing TEMPO as Organo-Electrocatalyst. *ACS Catal.* **2018**, *8* (2), 1192–1196.
48. Yang, L.; Ma, F. X.; Xu, F.; Li, D.; Su, L.; Xu, H. C.; Wang, C. J. C. A. A. J. Two-Dimensional Metal-Organic Layers for Electrochemical Acceptorless Dehydrogenation of N-Heterocycles. *Chem. Asian J.* **2019**, *14*, 3557–3560.
49. Huang, C.; Huang, Y.; Liu, C.; Yu, Y.; Zhang, B. Integrating Hydrogen Production With Aqueous Selective Semi-Dehydrogenation of Tetrahydroisoquinolines over a Ni_2P Bifunctional Electrode. *Angew. Chem. Int. Ed.* **2019**, *58*, 12014–12017.

CHAPTER TEN

Electrochemical anodic oxidative fluorination and trifluoroalkylation of organic compounds

Ricardo Belloso and Abid Shaikh
Department of Chemistry, Georgia Southern University, Statesboro, GA, United States

1 Introduction

Electrochemistry represents a powerful technique in the area of non-traditional activation methods in organic synthesis.[1] Discovery and applications of electrochemical industrial processes go back to late 1960s to produce large-scale chemical production.[2] Lately, electrochemical activation methods have reemerged as an important tool in mainstream organic synthesis.[3] This technique is usually referred to as "green" due to mild reaction conditions; electrolysis is performed at room temperature and at a low voltage with a high product chemoselectivity.[4] Also the use of electric energy to carry out chemical transformations significantly reduces the use of costly and harmful reagents, making the overall process clean and cost effective. Recent advancements in this field can also be related to commercially available electrochemical reactors that are easy to operate and have capability to perform reactions at variable batch sizes. Over the last decade, electrochemical synthesis has gone through a renaissance with its applications in organic synthesis that were not explored previously.[5]

1.1 Organofluorine chemistry

Fluorine is the most abundant halogen in the Earth's crust found in the form of mineral fluorspar.[6] Elemental fluorine was first isolated by a French chemist, Henri Moissan, in 1886 by electrochemical oxidation of potassium fluoride in hydrofluoric acid using platinum electrode.[7] Later, in 1906, Moissan received the Nobel Prize in Chemistry for his discovery of elemental fluorine (F_2). Surprisingly, there are only a few natural organofluorine compounds known, but the role of fluoride has been widely known to regulate

Nontraditional Activation Methods in Green and Sustainable Applications
https://doi.org/10.1016/B978-0-12-819009-8.00002-5
© 2021 Elsevier Inc.
All rights reserved.

the activity of various enzymes in physiological processes such as glycogen pathway, pyruvate dehydrogenase, and phosphatases.[8, 9] Fluorine is the most electronegative element in the periodic table, which brings unique chemical properties when bonded to carbon.[10] The C—F bond is highly polar and is even considered as ionic, which makes it the strongest bond compared with other halogens.[11] Organofluorine compounds have found many applications in industry due to the robust nature of the C—F bond, for example, polytetrafluoroethane (PFTE) possess high thermal stability, and fluorocarbons are used as refrigerants and fire retardants.[12] Over the past few decades, organofluorine compounds have emerged as potential therapeutics to treat a wide range of diseases.[13] Fluorine incorporation to organic molecules contributes to enhance biologically important properties such as an increase in lipophilicity of the overall molecule to facilitate drug delivery.[14] Also a relatively small van der Waals radius (1.47 Å) makes it a suitable candidate for the replacement of one or more hydrogen atoms in a molecule, resulting in a significant improvement in the overall molecular properties, which include hydrogen bonding, acidity, and solubility.[15] The introduction of fluorine into biologically active natural cortisone was first pursued in 1953 by Fried and Sabo.[16] The antiinflammatory activity of fluorinated cortisone was found to be 10 times greater than that of cortisone. In 1957 Heidelberger and coworkers reported that 5-fluorouracil had a remarkable antitumor activity.[17] These studies fueled the idea that fluorine plays an important role in the development of biologically active compounds. Recent studies have demonstrated that the replacement of a methyl with a trifluoromethyl group significantly alters the chemical reactivity and biological properties of the overall molecule.[18] In 1970 only a small number of fluorine-containing drugs were available in the market; however, their interesting pharmacological properties led to a tremendous development in fluorine-based drug design over the past two decades.[19, 20] Currently, more than 150 commercially available drugs contain one or more fluorine atoms in their structure.[21] Some of the most well-known organofluorine drugs are fluoxetine (antidepressant), ciprofloxacin (antibacterial), celecoxib (NSAID), atorvastatin (treatment of high triglycerides and cholesterol), and fluticasone propionate (antiinflammatory) (Fig. 1).

Recent applications of organofluorine compounds include potential therapeutics for HIV,[22] cancer[23] and Alzheimer's disease.[24] Fluorine is also used for radiolabeling compounds for positron emission tomography (PET) imaging in the form of radioactive ^{18}F isotope.[25] These groundbreaking studies now represent routine tools in rational drug design, and the search

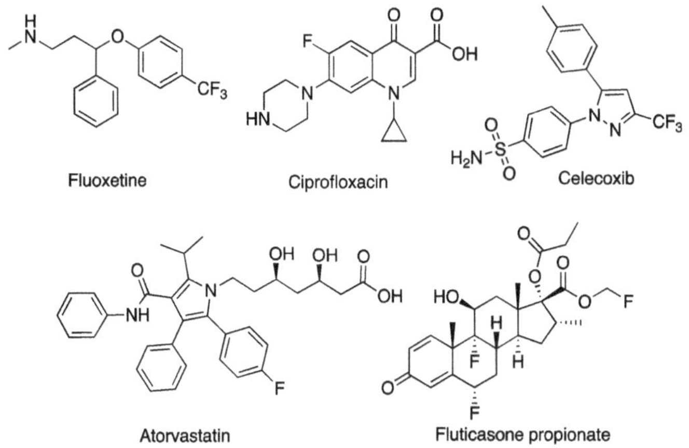

Fig. 1 Structures of some well-known fluorine-containing drugs.

for novel fluorine-based therapeutics is in the forefront of organic and medicinal chemistry research.[26]

1.2 Electrochemical fluorination

The first electrochemical fluorination (ECF) process for the direct introduction of fluorine into organic compounds was developed by Simons and Harland.[27] This process involves the use of nickel electrodes with HF as a reagent and an electrolyte at a high oxidation potential of 7 V in an undivided electrochemical cell. This method produced perfluoroalkylated products by replacing all the C—H bonds to C—F bonds including transformation of alkene carbons into saturated C—F bonds. Simons's process described a general approach for the introduction of fluorine into organic compounds, but a selective fluorination was still a challenge. Later in 1953 Schmidt and coworkers[28] reported a selective anodic oxidation of 1,1-diphenylethylene in KHF_2 and AgF under various solvent and electrolyte conditions. A straightforward and practical approach in ECF was developed by Knunyants and coworkers.[29] A selective benzylic fluorination of naphthalene using tetrabutylammonium fluoride trihydrofluoride ($Et_4NF{\cdot}3HF$) using a Calomel electrode with a potential of 1.8 V was carried out. This method resulted in regioselective monofluorination at C-5 and 5,8-difluorination of naphthalene in 27% and 70% product yields, respectively. Molten ammonium salts ($R_nNF{\cdot}mHF$) were later used extensively in ECF for a wide range of aromatic/heteroaromatic substrates and are still in current use.[30, 31] Electrolysis must be performed under anhydrous conditions to avoid the electrolysis of water into O_2 and H_2, which could possibly affect the outcome of the desired reaction.[32] Performing organic synthesis utilizing electrochemical activation has several advantages as compared with the traditional methods, because the reactions are fast, clean, and selective and eliminate the use of harmful redox reagents.

The basic electrochemical cell construction involves the use of a glass vessel; however, most of the fluorination methods involve the use of HF or formation of F_2 gas during the course of the reaction, and to avoid the damage, a steel cell is used.[33] There are two basic cell types that are commonly used, divided and undivided, where divided cell have electrodes that are separated and the electrons flow through a salt bridge. The electrodes' material employed is generally made of graphite, platinum, or steel. The electrolyte system contains HF salts that are stable in solvents such as acetonitrile, methanol, or dichloromethane. Commonly used electrolytes are HF: pyridine complexes (Olah's reagent) or similar tetraalkylammonium fluoride salts that can generate HF under electrolysis.[34]

Electrochemical reactions occur at the electrode surface and are initiated by a single electron transfer from the anode to the substrate or reagent by means of an electric current.[35] The current potential is controlled by a potentiostat with varying voltages. Once a radical species is generated at the anode, it further undergoes secondary reactions with the substrate to form the desired product. Over the past few decades, electrochemical methods have found tremendous applications in organic synthesis.[34] In this chapter, we provide a survey of literature specifically focusing on advancements in the area of electrochemical fluorination.

2 Electrochemical fluorination using N—F type reagents

Molten ammonium salts ($R_nNF{\cdot}m$ HF) have been widely used in selective partial fluorination of organic compounds and have attracted considerable interest over the past several decades as synthetic tools for preparing fluoroorganic compounds.[36] The first successful electrochemical partial fluorination was developed by Knunyants et al. utilizing tetraethylammonium hydrogen fluoride ($Et_3N{\cdot}3HF$) as a supporting electrolyte and a source of in situ fluoride ion.[37] Other variations such as $Et_4N{\cdot}4HF$, $Et_3N{\cdot}3HF$, and $Et_3N{\cdot}5HF$ have also been developed.[38] This discovery enabled performing selective fluorination under mild conditions with a diverse group of substrates.[39] Olah reported a novel pyridinium poly(hydrogen fluoride) as an inexpensive and a general fluorinating agent, which was later extensively used in fluorination of alkenes and arenes.[40] Molten salts or ionic liquids have many advantages over traditional solvents such as low vapor pressure, nonflammable, good electrical conductors due to presence of ions, and recyclable.[41] Over the past two decades, electrochemical fluorination has been used to achieve highly selective fluorinations by introduction of either single fluorine atom or small fluorinated groups into organic molecules of interest.[42] This section describes the remarkable contribution made by Laurent[43] and Fuchigami [42] on selective fluorination of organosulfur compounds and Yoneda[44] for aromatic partial fluorination.

2.1 Partial fluorination of benzenoid ring substrates

The groundbreaking work of Knunyants et al. on partial fluorination of naphthalene using molten salt of $Et_3N{\cdot}3HF$ as an electrolyte and acetonitrile as the solvent, provided monofluorinated product in 27% yield.[37] The fluorination occurred exclusively at the benzylic carbon with a trace amount of difluorinated product (Scheme 1A). This method was later utilized for a large number of substrates and is still very popular for regioselective fluorination.[42] A suggested mechanism was put forward by Rozhkov[45] based on applied reaction conditions such as the use of anhydrous aprotic solvent and standard platinum electrodes. Aromatic substrate undergoes anodic oxidation by losing an electron to form a radical cation, which undergoes a subsequent nucleophilic attack by a fluoride ion to form a radical species. Further oxidation forms a cation, which on deprotonation yields for the desired product (Scheme 1B). This mechanism had been widely accepted and is applicable for a broad range of substrates.

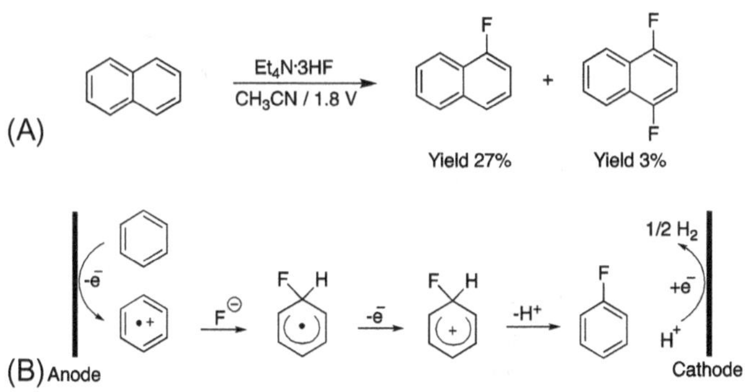

Scheme 1 (A) Rozhkov's partial fluorination of naphthalene. (B) A general mechanism for electrochemical fluorination under molten ammonium salts.

Electrochemical fluorination under solvent-free conditions using $Et_3N{\cdot}3HF$ as an electrolyte was reported by Meurs and Eilenberg.[46] This method was applicable to a range of substrates including aromatic ring heterocycles and alkenes; however, a strong oxidation potential provided a tetrafluoro product in case of phenanthroline replacing all the benzenoid hydrogens with fluorine atoms (Scheme 2A). A slight increase in the product yield was reported when solvent-free conditions were used. Later, this methodology was applied in the partial fluorination of 1,4-difluorobenzene to obtain 1,1,4,4-tetrafluorocyclohexadiene in 90% isolated yield (Scheme 2B).[47] At a high oxidation potential, quinoline provided 5,8-difluoroquinoline and 5,5,8,8-tetrafluoroquinoline products in 27% and 20% yields, respectively (Scheme 2C).[48]

Electrochemical fluorination of phenols has been studied extensively by Yoneda and coworkers.[49] In this study, $Et_3N{\cdot}5HF$ was used as an electrolyte to obtain 4,4-difluorocyclohexa-2,5-diene-1-one as the electrochemical outcome of the reaction, which, on further reduction with zinc, provided 4-fluorophenol in 80% yield (Scheme 3A). A similar approach was reported with fluorination of 1-naphthol derivatives using $Et_3N{\cdot}5HF$ and carbon fiber cloth anode to obtain 4,4-difluorocyclohexadienone derivatives in up to 89% yield (Scheme 3B).[50]

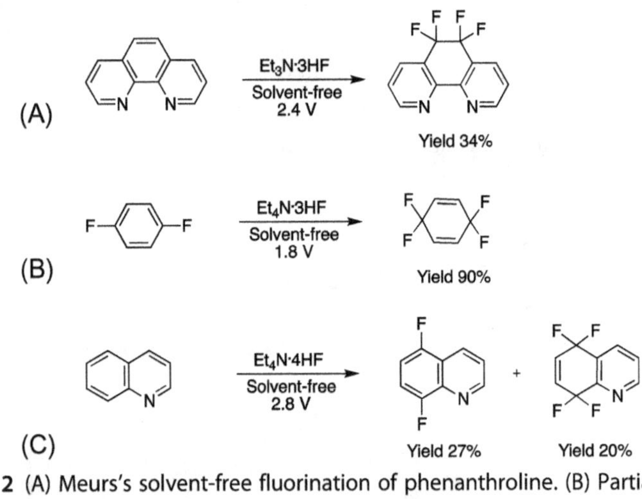

Scheme 2 (A) Meurs's solvent-free fluorination of phenanthroline. (B) Partial fluorination of 1,4-difluorobenzene to obtain 1,1,4,4-tetrafluorocyclohexadiene. (C) Electrochemical synthesis of 5,8-difluoroquinoline and 5,5,8,8-tetrafluoroquinoline.

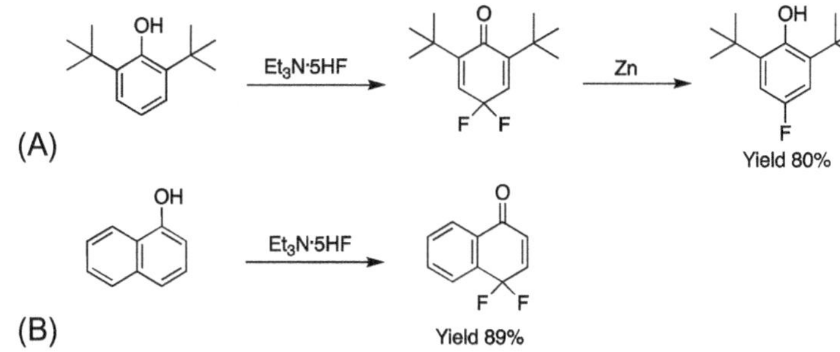

Scheme 3 (A) Yoneda's fluorination of phenols with $Et_3N·5HF$ as an electrolyte. (B) Fluorination of 1-naphthol derivatives using $Et_3N·5HF$ and carbon fiber cloth anode.

2.2 Electrochemical benzylic fluorination

The transformation of benzylic C—H bonds into C—F bonds has been widely pursued using traditional reagents[51, 52]; however, only few reports describe the electrochemical benzylic fluorination.[53] These reactions are initiated at the anode with the formation of a radical cation after losing an electron from the substrate. In this case a stable benzylic radical is generated, thus leading to the desired C—F bond formation. A systematic study was carried out by Middleton and coworkers[54] using 4-nitrotoluene as a model substrate for selective fluorination to obtain only monofluoro product when

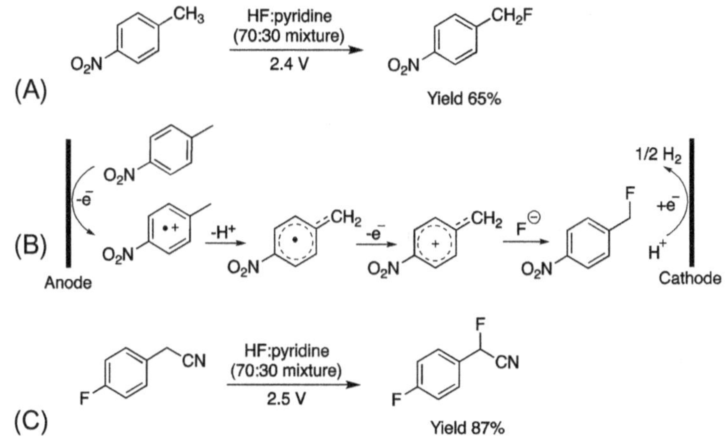

Scheme 4 (A) Electrolytic fluorination of 4-nitrotoulene using HF:pyridine (Olah's reagent). (B) Proposed mechanism of benzylic fluorination of 4-nitrotoluene. (C) Benzylic fluorination of 4-fluorophenylacetonitrile using HF:pyridine.

electrolyzed in HF:pyridine (70:30 mixture) at a potential of 2.4 V (Scheme 4A). The reaction pathway involved anodic oxidation to form the radical cation, which then loses hydrogen ion to obtain benzylic radical, which loses another electron at the anode to form benzylic carbocation. The combination of a fluoride ion with the benzylic cation forms the expected product (Scheme 4B). Similar work with 4-fluorophenylacetonitrile in acetonitrile gave 17% yield of the fluorinated product; however, performing reaction in presence of HF:pyridine (70:30 mixture) at 2.5 V on platinum surface improved the product yield to 87% (Scheme 4B).[55]

It was found that the presence of an electron-withdrawing group promotes mono- or difluorination at benzylic position (Scheme 5).[43] Also, benzylic electrochemical fluorination would occur if the phenyl group had an electron-donating group at the *para* position. It is necessary to have an electron-donating group present at the *para* position for the formation of a stable benzylic cation intermediate would react more readily with fluoride ions.

2.3 Electrochemical fluorination of alkenes

The partial fluorination of olefins is an important application of anodic oxidation as fluorine atoms can be directly incorporated into organic molecules. Vicinal difluorination of styrene was achieved using $Et_3N{\cdot}3HF$ and acetonitrile. A similar reaction with stilbene provided a vicinal difluoro product; however, due to the stability of the benzylic radical, a nucleophilic attack by acetonitrile is possible, leading to the formation of acetamide (Scheme 6A).[56] A proposed general alkene difluorination mechanism is depicted in Scheme 6B.

Lennox et al.[57] recently reported the deactivated olefins such as allylbenzene to readily undergo electrochemical difluorination in the presence of aryl iodide as a mediator. Electrolysis was performed using $Et_3N{\cdot}5.6HF$ or pyridine:5.6HF as an electrolyte and a mixture of dichloromethane:hexafluoroisopropanol (7:3) at a potential of 4 mA for 4 h to obtain the expected difluoro product in 81% yield (Scheme 7).

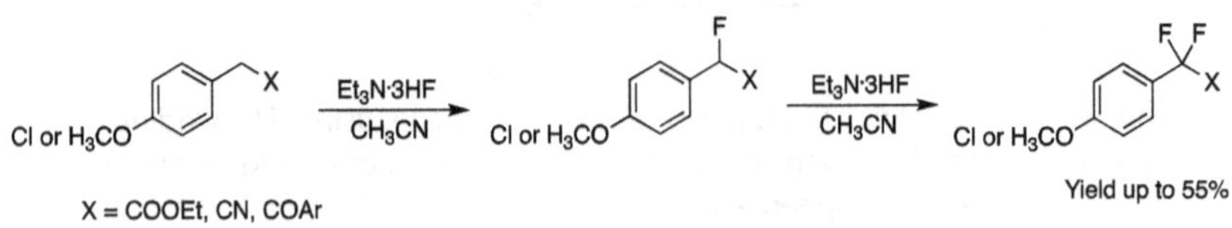

Scheme 5 Effect of electron-withdrawing groups at the benzylic electrochemical fluorination.

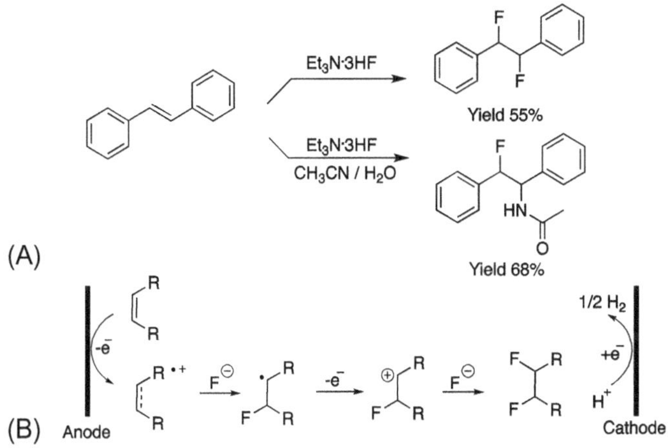

Scheme 6 (A) Difluorination of stilbene in presence of $Et_3N{\cdot}3HF$ and acetonitrile. (B) A proposed general mechanism of alkene vicinal difluorination.

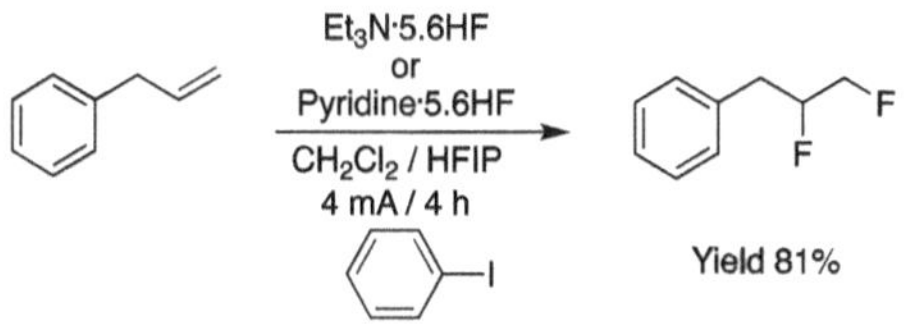

Scheme 7 Electrochemical difluorination of allylbenzene in presence of aryl iodide as a mediator.

Electrochemical fluorocyclization of *N*-allylcarboxamides for the synthesis of 5-fluoromethyl-2-oxazolines by using electrochemical oxidation was recently reported by Waldvogel and coworkers.[58] Anodic oxidation was carried out in the presence of $Et_3N{\cdot}5HF$ and iodoarene to form a hypervalent iodine species, which mediates the cyclization of *N*-allylcarboxamide to 5-fluoromethyl-2-oxazoline (Scheme 8). This method allowed application to various substrates furnishing the 2-oxazolines with yields up to 68%. This protocol offered a sustainable alternative over conventional reagent-based methods.

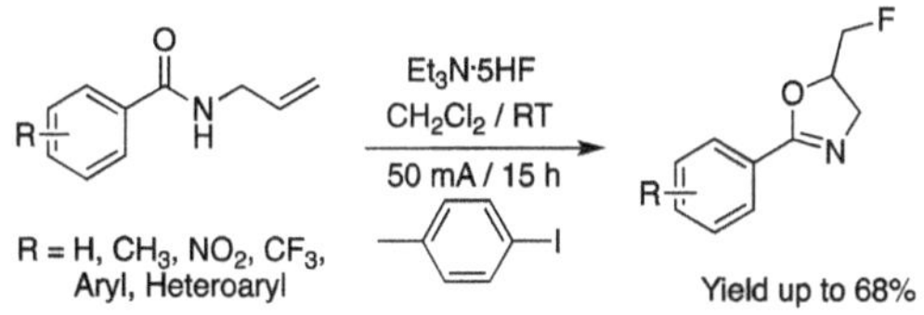

Scheme 8 Electrochemical fluorocyclization of *N*-allylcarboxamide to oxazoline.

2.4 Electrochemical fluorination of organosulfur compounds

Fluorination of an active methylene attached to an α-thio group has been studied extensively by Laurent et al.[43] and Fuchigami et al.[42] Ionic liquids $Et_3N{\cdot}3HF$ and HF:pyridine (Olah's reagent) were both tried, but $Et_3N{\cdot}3HF$ provided the best results. Organosulfur compound such as phenyl 2,2,2-trifluoroethyl sulfide was subjected to electrolysis in acetonitrile to obtain the product in 62% yield (Scheme 9A).[59] Fluorination only occurred at the side chain methylene carbon, and no fluorine substitution was observed on the phenyl group. The proposed mechanism by Fuchigami involves a Pummerer-type reaction with the formation of fluorosulfonium cation, which on further deprotonation and fluoride ion attack forms the expected product (Scheme 9B).

Later, Narizuka and Fuchigami investigated that the introduction of a heteroatom into organosulfur compounds influenced the product yields.[60] Synthesis of fluorinated β-lactams could be achieved in high yields using $Et_3N{\cdot}3HF$ and acetonitrile at a potential of 2 V (Scheme 10).

2.5 Electrochemical fluorination of heterocyclic compounds

A wide range of heterocyclic compounds has been explored for selective fluorination by Fuchigami and coworkers.[61] Anodic fluorination of

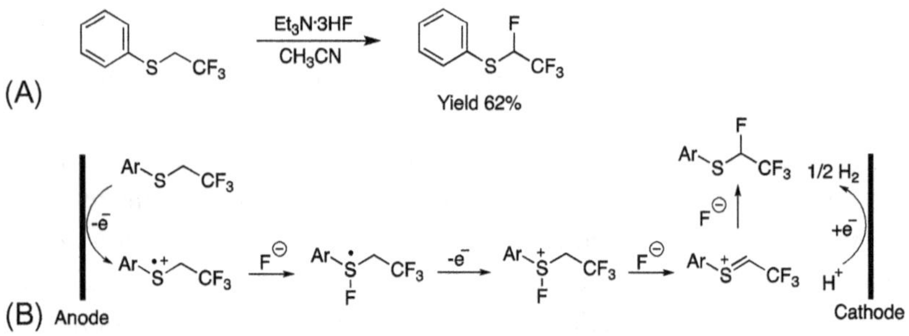

Scheme 9 (A) Electrochemical fluorination of methylene attached to α-thio group. (B) Mechanism of fluorination of methylene attached to α-thio group showing formation of fluorosulfonium cation.

Scheme 10 Synthesis of fluorinated β-lactams using electrochemical fluorination.

N-substituted pyrroles was performed in acetonitrile and $Et_3N\cdot 3HF$ under a constant current of 1.8 V (Scheme 11).[62] The fluorination with pyrrole having an electron-withdrawing cyano group proceeded smoothly to form the corresponding germinal difluoro product in 54% yield. The authors reported that fluorination of N-methylpyrrole without the cyano group did not occur as expected and a polymeric layer was formed at the platinum anode.

The anodic fluorination of flavanones was successfully carried out to provide the corresponding fluoro-substituted flavanones.[63] When $Et_4N\cdot 4HF$ was used, a 2,3-difluoro product was obtained in 68% yield. However, in the presence of $Et_3N\cdot 3HF$, a 3-fluoro-substituted product was obtained in up to 62% yield (Scheme 12). The presence of free-base Et_3N participated in dehydrofluorination of 2,3-difluoroflavone to form the 3-fluoroflavone.

3 Electrochemical trifluoromethylation

Introduction of fluorine-containing groups such as trifluoromethyl (CF_3) into organic molecules has attracted a significant attention in industry, academia, and materials science.[63] The incorporation of the CF_3 group can enhance pharmacokinetic properties such as increased in vivo metabolic stability, lipophilicity, and strong hydrogen binding ability to proteins.[64] Owing to its importance a great number of methods have been developed for the incorporation a CF_3 group into organic substrates.[65] The two major

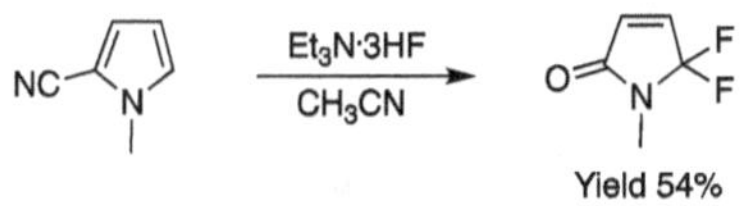

Scheme 11 The fluorination of N-methylpyrrole having an electron-withdrawing cyano group.

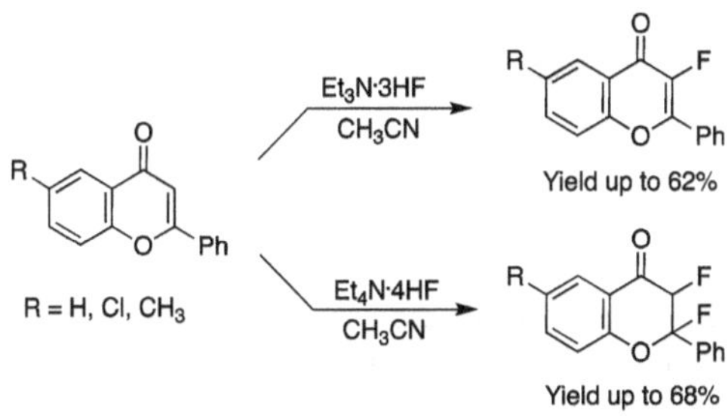

Scheme 12 Fluorination of flavanones in the presence of $Et_3N\cdot 3HF$ and $Et_4N\cdot 4HF$ electrolytes.

strategies involve late-stage trifluoromethylation using Langlois (CF_3SO_2Na), Togni (CF_3I), triflyl chloride (CF_3SO_2Cl), Baran ($(CF_3SO_2)_2Zn$), and Ruppert ($TMSCF_3$) reagents or using precursors such as ethyl trifluoropyruvate, ethyl trifluoroacetoacetate, and trifluoromethyl ketones.[66] Most of the traditional stoichiometric methods involve the use of metal-based catalyst or use of hazardous redox reagents.[65] The other strategy involves the use of an electrochemical process providing a cleaner, selective, and safer alternative that can be performed on a wide range of substrates under mild reaction conditions. A short literature overview of electrophilic trifluoromethylation is described in this section.

Recently, Zhang and coworkers[67] described a combined electrochemical trifluoromethylation/intermolecular cyclization of various *N*-alkyl-*N*-phenylmethacrylamides using the Langlois reagent as the source of CF_3 group. Transition metal-based redox catalyst ($MnBr_2$) and H_3PO_4 was used as oxidant without an electrolyte and platinum mesh as electrodes under electric current of 10 mA for 6 h in an undivided cell. The corresponding trifluoromethyloxindoles were obtained in 30%–77% yield (Scheme 13A). The scope of this methodology was further extended in the synthesis of

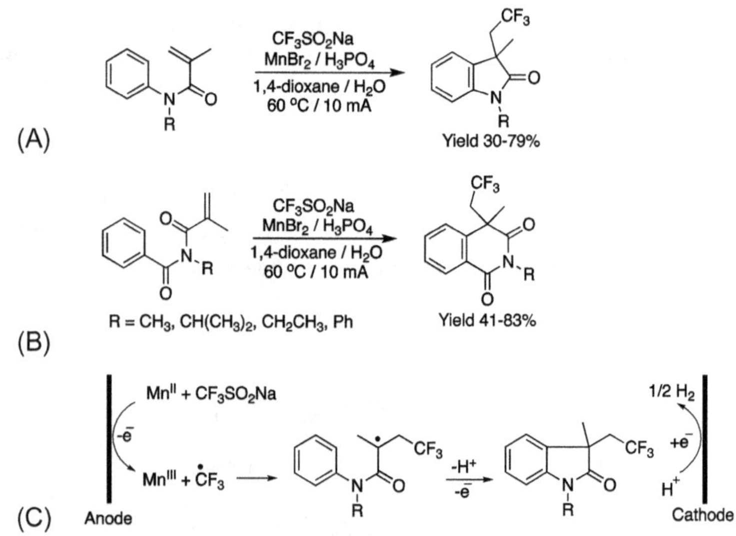

Scheme 13 (A) Electrochemical trifluoromethylation/intermolecular cyclization of *N*-alkyl-*N*-phenylmethacrylamides using the Langlois reagent. (B) Electrochemical trifluoromethylation/intermolecular cyclization of *N*-methacryloyl-*N*-methylarylamides using the Langlois reagent. (C) Proposed mechanism of tandem trifluoromethylation/intermolecular cyclization.

CF_3-containing hydroisoquinoline-1,3-dione derivatives staring from *N*-methacryloyl-*N*-methylarylamides under similar reaction conditions to obtain the product in 41%–83% yield (Scheme 13B).

A proposed mechanism for the trifluoromethylation/cyclization cascade is depicted in Scheme 13C. Anodic oxidation converts Mn^{II} to Mn^{III} along with the generation of CF_3 radical species, which then further reacts with the alkene to form a carbon radical. Addition of aromatic ring to the carbon radical accomplishes the cyclization, which after further dehydrogenation provides the final product.

Oxindoles and quinolinones represent a core motif for a variety of biologically active natural products and synthetic drugs.[68] Electrochemical oxidative trifluoromethylation/cyclization of C=C double bonds of *N*-arylacrylamides and *N*-heteroarylacrylamides was reported by Ruan et al.[69] The authors used the inexpensive Langlois reagent (CF_3SO_2Na) as a source of the trifluoromethyl group and performed the radical trifluoromethylation/cyclization cascade under catalyst-free, metal-free, and mild reaction conditions. Electrolysis was performed in an undivided cell using reticulated vitreous carbon (RVC) anode and platinum cathode and tetraethylammonium perchlorate (Et_4NClO_4) salt as electrolyte at an applied potential of 4 mA at 23°C for 16 h (Scheme 14A). This method provided the expected trifluoromethylated oxindole and quinolinone derivatives in up to 77% and 52% yields, respectively.

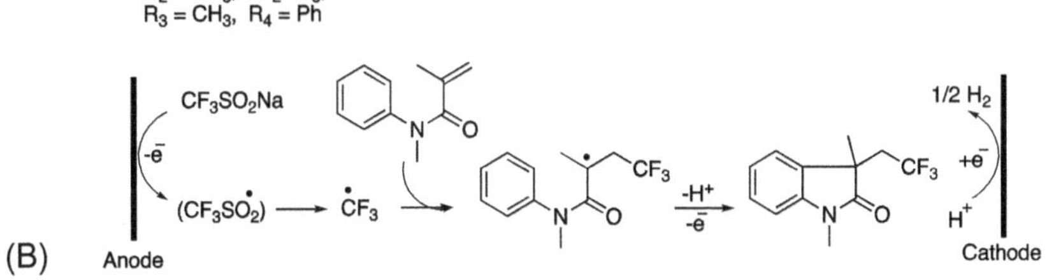

Scheme 14 (A) Electrochemical oxidative trifluoromethylation/cyclization of alkenes of *N*-arylacrylamides and *N*-heteroarylacrylamides using the Langlois reagent. (B) Proposed mechanism for the oxidative trifluoromethylation/cyclization of alkenes.

The anodic oxidation of CF_3SO_2Na produced a CF_3 radical via a single electron transfer (SET) mechanism. CF_3 radical further reacted with the alkene to provide the radical intermediate, which underwent intermolecular cyclization and aromatization to afford the expected product (Scheme 14B). Formation of radical species was confirmed by cyclic voltammetry (CV), and addition of 2,6-di-*tert*-butyl-4-methylphenol (BHT), a radical scavenging agent, suppressed the formation of expected product.

An interesting piece of work was reported by Hong and coworkers[70] to achieve decarboxylative trifluoromethylation of vinyl carbon starting from α,β-unsaturated carboxylic acids. Cinnamic acid was chosen as a substrate to react with the Langlois reagent (CF_3SO_2Na) under electrochemical oxidative decarboxylation. A series of electrolytes were employed under constant current of 7 mA in a DMSO/DCM solvent mixture and found that tetrabutylammonium hexafluorophosphate (Bu_4NPF_6) along with potassium *tert*-butoxide (KO*t*-Bu) provided the best results. An undivided cell was used in this study, equipped with graphite anode and platinum cathode. A series of substituted cinnamic acids were tried to obtain the required product in 40%–84% yield (Scheme 15A). The authors claimed the process to be a metal-free and oxidant-free approach with high product regioselectivities as compared with the traditional transition metal-catalyzed methods.

To investigate the mechanism of decarboxylative fluorination, a series of controlled experiments were conducted by adding radical scavengers such as 2,2,6,6-tetramethylpiperidine-1-oxyl (TEMPO) and 2,6-di-*tert*-butyl-4-methylphenol (BHT) to confirm the reaction involved in the formation of a CF_3 radical species. A proposed mechanism involved the formation

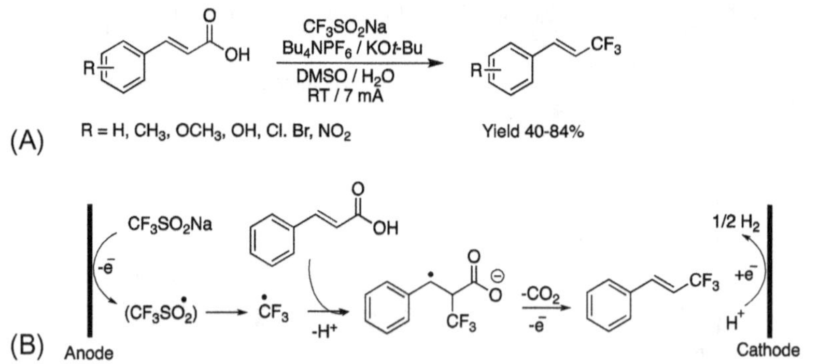

Scheme 15 (A) Decarboxylative trifluoromethylation of vinyl carbon of α,β-unsaturated carboxylic acids. (B) Proposed mechanism for the decarboxylative trifluoromethylation of cinnamic acid.

of CF_3 radical at the anode via a SET oxidation of CF_3SO_2Na. Subsequently, addition of a CF_3 radical to alkene forms an intermediate radical, which on further decarboxylation forms the desired product (Scheme 15B).

A recent report published by Guan and coworkers[71] described the electrochemical trifluoromethylation of alkenes along with 1,2-migration of alkyl group to form β-trifluoromethyl ketones utilizing CF_3SO_2Na as a radical source. 1,1-Diphenylprop-2-en-1-ol was used as model substrate and tetrabutylammonium fluoride (Bu_4NBF_4) as the electrolyte and acetonitrile as solvent in an undivided electrochemical cell connected to graphite anode and iron plate cathode. The expected product 4,4,4-trifluoro-1,2-diphenylbutan-1-one was obtained in 65% isolated yield with a graphite rod anode and iron plate cathode (Scheme 16A). Various substituted aromatic substrates provided the expected trifluoromethylated ketones in up to 65% isolated yield.

To gain more insight into the reaction mechanism, several radical trapping experiments were conducted with radical scavengers such as BHT and TEMPO. No desired product was detected as the CF_3 radical was trapped by scavengers. A single electron transfer at anode converted the CF_3SO_2Na to CF_3 radical, which abruptly reacted with alkene to form carbon radical that on further oxidation converted to a carbocation. Rearrangement of the carbocation took place via 1,2-migration of a phenyl group to provide the expected product (Scheme 16B).

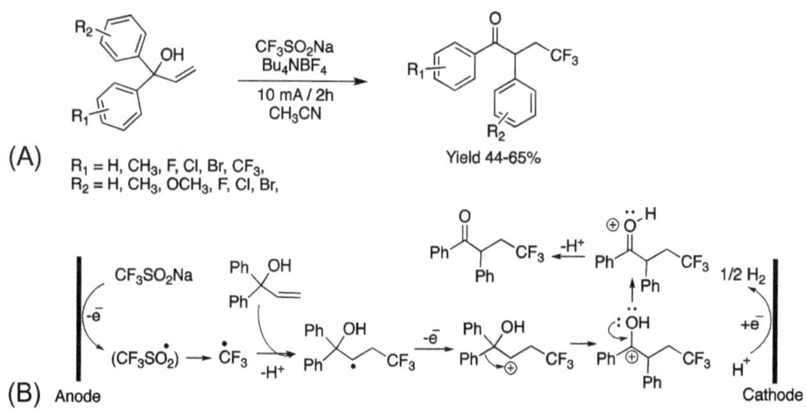

Scheme 16 (A) Electrochemical trifluoromethylation of alkenes along with 1,2-migration of alkyl group. (B) Proposed mechanism for the trifluoromethylation of alkene along with 1,2-migration of a phenyl group.

4 Conclusions and future outlook

Over the past two decades, selective fluorination has become a mainstream goal for organic synthetic chemists, and accordingly a large number of methods have been reported for the incorporation of a fluorine atom into organic molecules. The combination of ionic liquids with electrochemical activation has also proven to be very effective for the introduction of fluorine into a large number of substrates. Recent progress made in this area prompted the relevant industry to develop specifically designed, reliable, and controllable instruments. Electrochemical methods are fast, selective, clean, and energy efficient. Also the outcome of the electrochemical reaction depends on selection of the proper electrolyte, solvent, and substrate. A large number of chemical processes involve redox reactions; however, there are only a few industrial processes that constitute the use of electrochemical activation primarily due to the fact that electrochemical reactions are performed at a significantly smaller volume. The development of such suitable industrial processes is definitely an ideal goal for the sustainable chemical industry.

References

1. Wiebe, A.; Gieshoff, T.; Mohle, S.; Rodrigo, E.; Zirbes, D.; Waldvogel, S. R. Electrifying Organic Synthesis. *Angew. Chem. Int. Ed.* **2018**, *57*, 5594–5619.
2. Holm, E. J.; Rosen, B. R.; Baran, P. S. Synthetic Organic Electrochemistry: An Enabling and Innately Sustainable Method. *ACS Cent. Sci.* **2016**, *2*, 302–308.
3. Mohle, S.; Zirbes, M.; Rodrigo, E.; Gieshoff, T.; Wiebe, A.; Waldvogel, S. R. Modern Electrochemical Aspects for the Synthesis of Value-Added Organic Products. *Angew. Chem. Int. Ed.* **2018**, *57*, 2–23.
4. Frontana-Uribe, B. A.; Little, D. R.; Ibanez, J. G.; Palma, A.; Vasquez-Medrano, R. Organic Electrosynthesis: A Promising Green Methodology in Organic Chemistry. *Green Chem.* **2010**, *12*, 2099–2119.
5. Yan, M.; Kawamata, Y.; Baran, P. S., Synthetic Organic Electrochemical Methods Since 2000: On the Verge of a Renaissance. Chem. Rev. 2017, 117, 13230–13319.
6. Chambers, R. D. *Fluorine in Organic Chemistry*; Blackwell Publishing Ltd: Oxford, 2004.
7. Moissan, H. *Das fluor und seine verbindungen*, German Edition; Verlag von M. Krayn: Berlin, 1900.
8. Strunecka, A.; Patocka, J. Aluminofluoride Complexes: A Useful Tool in Laboratory Investigations, But a Hidden Danger for Living Organisms. In *Group 13 Chemistry, from Fundamentals to Application*; Shapiro, P., Atwood, D., Eds.; ACS Symposium Series 822: Washington, DC, 2002; pp. 271–282.
9. Strunecka, A.; Patocka, J. Aluminofluoride Complexes in the Etiology of Alzheimer's Disease. *Struct. Bond.* **2003**, *37*, 139–181.
10. Pauling, L. *The Nature of the Chemical Bond and the Structure of Molecules and Crystals: An Introduction to Modern Structural Chemistry*; Cornell University Press: Ithaca, New York, 1939.

11. O'Hagan, D. Understanding Organofluorine Chemistry. An Introduction to the C-F Bond. *Chem. Soc. Rev.* **2008**, *37*, 308–319.
12. Doz, P.; Kirsch, P. *Modern Fluoroorganic Chemistry: Synthesis, Reactivity, Applications*; Wiley-VCH Verlag GmbH: Weinheim, 2004; pp. 203–205.
13. Wang, J.; Sanchez-Rosello, M.; del Pozo, C.; Sorochinsky, A. E.; Fustero, S.; Soloshonok, V. A.; Liu, H. Fluorine in Pharmaceutical Industry: Fluorine-Containing Drugs Introduced to the Market in the Last Decade (2001–2011). *Chem. Rev.* **2014**, *114*, 2432–2506.
14. Silverman, R. B. *The Organic Chemistry of Drug Design and Drug Action*; Elsevier Academic Press: London, 2004; pp. 617–630.
15. Hagmann, W. K. The Many Roles for Fluorine in Medicinal Chemistry. *J. Med. Chem.* **2008**, *51*, 4359–4369.
16. Fried, J.; Sabo, E. F. Synthesis of 17 α-Hydroxycorticosterone and its 9 α-Halo Derivatives from 11-Epi-17 α-Hydrocorticosterone. *J. Am. Chem. Soc.* **1953**, *75*, 2273–2274.
17. Heidelberger, C.; Chaudhuri, N. K.; Danneberg, P.; Mooren, D.; Griesbach, L.; Duschinsky, R.; Schnitzer, R. J.; Pleven, E.; Scheiner, J. Fluorinated Pyrimidines, a New Class of Tumour-inhibitory Compounds. *Nature* **1957**, *179*, 663–666.
18. Furuya, T.; Kamlet, A. S.; Ritter, T. Catalysis for Fluorination and Trifluoromethylation. *Nature* **2011**, *473*, 470–477.
19. Ojima, I. *Fluorine in Medicinal Chemistry and Chemical Biology*; Wiley-Blackwell: Chichester, 2009.
20. Ojima, I. Exploration of Fluorine Chemistry at the Multidisciplinary Interface of Chemistry and Biology. *J. Org. Chem.* **2013**, *78*, 6358–6383.
21. Mei, H.; Fustero, S.; Medio-Simon, M.; Sedgwick, D. M.; Santi, C.; Ruzziconi, R.; Soloshonok, V. A. Fluorine-Containing Drugs Approved by the FDA in 2018. *Chem. A Eur. J.* **2019**, *25*, 11797–11819.
22. Frampton, J. E.; Perry, C. M. Emtricitabine: A Review of its Use in the Management of HIV Infection. *Drugs* **2005**, *65*, 1427–1448.
23. Jordan, V. C. Antiestrogens and Selective Estrogen Receptor Modulators as Multifunctional Medicines. *J. Med. Chem.* **2003**, *46*, 883–908.
24. Torok, M.; Abid, M.; Mhadgut, S. C.; Torok, B. Organofluorine Inhibitors of Amyloid Fibrillogenesis. *Biochemistry* **2006**, *45*, 5377–5383.
25. Miller, P. W.; Long, N. J.; Vilar, R.; Gee, A. D. Synthesis of ^{11}C, ^{18}F, ^{15}O, and ^{13}N Radiolabels for Positron Emission Tomography. *Angew. Chem. Int. Ed.* **2008**, *47*, 8998–9033.
26. Gillis, E. P.; Eastman, K. J.; Hill, M. D.; Donnelly, D. J.; Meanwell, N. A. Applications of Fluorine in Medicinal Chemistry. *J. Med. Chem.* **2015**, *58*, 8315–8359.
27. Simons, J. H.; Harland, W. J. The Electrochemical Process for the Production of Fluorocarbons. *J. Electrochem. Soc.* **1949**, *95*, 47–66.
28. Noel, M.; Suryanarayanan, V.; Chellammal, S. A Review of Recent Developments in the Selective Electrochemical Fluorination of Organic Compounds. *J. Fluorine Chem.* **1997**, *83*, 31–40.
29. Kuribayashi, S.; Kurioka, T.; Inagi, S.; Lu, H.-J.; Uang, B.-J.; Fuchigami, T. The Selective Electrochemical Fluorination of *S*-Alkyl Benzothioate and its Derivatives. *Beilstein J. Org. Chem.* **2018**, *14*, 389–396.
30. Fox, H. M.; Ruehlen, F. N.; Childs, W. V. Electrochemical Fluorination. *J. Electrochem. Soc.* **1971**, *118*, 1246–1249.
31. Fuchigami, T.; Tajima, T. Highly Selective Electrochemical Fluorination of Organic Compounds in Ionic Liquids. *J. Fluorine Chem.* **2005**, *126*, 181–187.
32. Shainyan, B.; Danilevich, Y. Electrochemical Fluorination of Aromatic Compounds in Anhydrous HF. *Russ. J. Org. Chem.* **2006**, *42*, 214–219.

33. Sawyer, D. T.; Roberts, J. L. *Experimental Electrochemistry for Chemists*, 1st ed.; John Wiley & Sons: New York, 1974; pp. 435–437.
34. Karkas, M. D. Electrochemical Strategies for C–H Functionalization and C–N Bond Formation. *Chem. Soc. Rev.* **2018**, *47*, 5786–5865.
35. Sartori, P.; Ignatev, N. The Actual State of our Knowledge about Mechanism of Electrochemical Fluorination in Anhydrous Hydrogen Fluoride (Simons Process). *J. Fluorine Chem.* **1998**, *87*, 157–162.
36. Fuchigami, T.; Inagi, S. Selective Electrochemical Fluorination of Organic Molecules and Macromolecules in Ionic Liquids. *Chem. Commun.* **2011**, *47*, 10211–10223.
37. Knunyants, I. L.; Rozhkov, I. N.; Bukhtiarov, A. V.; Goldin, M. M.; Kudryavtseu, R. V. New Method for Introducing Fluorine into the Aromatic Ring. *Izv. Akad. Nauk. SSSR, Ser. Khim* **1970**, 1207–1208.
38. Momota, K.; Kato, K.; Morita, M.; Matsuda, Y. Electrochemical Fluorination of Aromatic Compounds in Liquid $R_4NF{\cdot}mHF$-Part III. Fluorination of Trifluoromethylbenzenes. *Electrochim. Acta* **1994**, *39*, 681–690.
39. Fuchigami, T.; Tajima, T. *Selective Anodic Fluorination of Heterocyclic Compounds. In Fluorine-Containing Synthons*; American Chemical Society: Washington, DC, 2005; pp. 276–287.
40. Olah, G. A.; Welch, J. T.; Vankar, Y. D.; Nojima, M.; Kerekes, I.; Olah, J. A. Synthetic Methods and Reactions. 63. Pyridinium Poly(Hydrogen Fluoride) (30% Pyridine-70% Hydrogen Fluoride): A Convenient Reagent for Organic Fluorination Reactions. *J. Org. Chem.* **1979**, *44*, 3872–3881.
41. Tajima, T.; Nakajima, A.; Fuchigami, T. Electrochemical Partial Fluorination of Organic Compounds (76). Selective Anodic Fluorination of *N*-Protected Pyrroles towards Environmentally Friendly Process. *Electrochemistry* **2004**, *72*, 843–845.
42. Fuchigami, T.; Shimojo, M.; Konno, A.; Nakagawa, K. Electrolytic Partial Fluorination of Organic Compounds. 1. Regioselective Anodic Monofluorination of Organosulfur Compounds. *J. Org. Chem.* **1990**, *55*, 6074–6075.
43. Laurent, E. G.; Marquet, B.; Tardivel, R.; Thiebault, H. Nouvelle methode de preparation de cetones, ester et nitrile benzyliques α-fluoride OU α,α-difluores. *Tetrahedron Lett.* **1987**, *28*, 2359–2362.
44. Chen, S.-Q.; Hatakeyama, T.; Fukuhara, T.; Hara, S.; Yoneda, N. Electrochemical Fluorination of Aliphatic Aldehydes and Cyclic Ketones using $Et_3N{\cdot}5HF$ Electrolyte. *Electrochim. Acta* **1997**, *42*, 1951–1960.
45. Rozhkov, I. N. Radical-Cation Mechanism of the Anodic Fluorination of Organic Compounds. *Russ. Chem. Rev.* **1976**, *45*, 615–629.
46. Meurs, J. H. H.; Eilenberg, W. Oxidative Fluorination in Amine-hf Mixtures. *Tetrahedron* **1991**, *47*, 705–714.
47. Momota, K.; Yonezawa, T.; Hayakawa, Y.; Kato, K.; Morita, M.; Matsuda, Y. Synthesis of Fluoro-Cyclohexadienes by the Electrochemical Fluorination of *p*-Difluorobenzenes on a Preparative Scale. *J. Appl. Electrochem.* **1995**, *25*, 651–658.
48. Saeki, K.; Tomomitsu, M.; Kawazoe, Y.; Momota, K.; Kimoto, H. Fluorinated benzo [*h*] quinolines and benzo[*f*]quinolines. *Chem. Pharm. Bull.* **1996**, *44*, 2254–2258.
49. Fukuhara, T.; Akiyama, Y.; Yoneda, N.; Tada, T.; Hara, S. Effective Synthesis of Difluorocyclohexadienones by Electrochemical Oxidation of Phenols. *Tetrahedron Lett.* **2002**, *43*, 6583–6585.
50. Aoyama, M.; Fukuhara, T.; Hara, S. Selective Fluorination of Adamantanes by an Electrochemical Method. *J. Org. Chem.* **2008**, *73*, 4186–4189.
51. Bloom, S.; Pitts, C. R.; Woltornist, R.; Griswold, A.; Holl, M. G.; Lectka, T. Iron(II)-Catalyzed Benzylic Fluorination. *Org. Lett.* **2013**, *15*, 1722–1724.
52. Amaoka, Y.; Nagamoto, M.; Inoue, M. Metal-Free Fluorination of C(sp^3)-H Bonds Using a Catalytic *N*-Oxyl Radical. *Org. Lett.* **2013**, *15*, 2160–2163.

53. Zaikin, P. A.; Borodkin, G. I. *Electrophilic and Oxidative Fluorination of Aromatic Compounds: In Late-stage Fluorination of Bioactive Molecules and Biologically-relevant Substrates*; Elsevier, 2019; pp. 105–135.
54. Savett, S. C.; Lee, S. M.; Bradley, A. Z.; Kneizys, S. P.; LoBue, J. M.; Middleton, W. J. Microscale Electrocatalytic Fluorinations of 4-Nitrotoluene: Cell Construction, Computer Monitor and Control, and Chemistry. *Microchem. J.* **1993**, *48*, 192–199.
55. Lee, S. M.; Roseman, J. M.; Zartman, C. B.; Morrison, E. P.; Harrison, S. J.; Stankiewicz, C. A.; Middleton, W. J. Selective Electrolytic Fluorinations in 70% HF/30% Pyridine. *J. Fluorine Chem.* **1996**, 77, 65–70.
56. Bensadat, A.; Bodennec, G.; Laurent, E.; Tardivel, R. Étude de la mono et de la difluoration par oxydation électrochimique d'alcènes et d'alkylbenzènes. *Tetrahedron Lett.* **1977**, *18*, 3799–3802.
57. Doobary, S.; Sedikides, A.; Caidora, H.; Poole, D.; Lennox, A. Electrochemical Vicinal Difluorination of Alkenes: Sustainable, Scalable, and Amenable to Electron-Rich Substrates. *ChemRxiv* **2019**, 1–7.
58. Haupt, J. D.; Berger, M.; Waldvogel, S. R. Electrochemical Fluorocyclization of *N*-Allylcarboxamides to 2-Oxazolines by Hypervalent Iodine Mediator. *Org. Lett.* **2019**, *21*, 242–245.
59. Konno, A.; Nakagawa, K.; Fuchigami, T. New Mechanistic Aspects of Anodic Monofluorination of Halogenoalkyl and Alkyl Phenyl Sulphides. *J. Chem. Soc. Chem. Commun.* **1991**, 1027–1029.
60. Narizuka, S.; Fuchigami, T. Electrolytic Partial Fluorination of Organic Compounds. 8. Highly Regioselective Anodic Monofluorination of β-Lactams. *J. Org. Chem.* **1993**, *58*, 4200–4201.
61. Tajima, T.; Nakajima, A.; Fuchigami, T. Electrolytic Partial Fluorination of Organic Compounds. 83. Anodic Fluorination of *N*-substituted Pyrroles and Its Synthetic Applications to *Gem*-Difluorinated Heterocyclic Compounds. *J. Org. Chem.* **2006**, *71*, 1436–1441.
62. Hou, Y.; Higashiya, S.; Fuchigami, T. Electrolytic Partial Fluorination of Organic Compounds. 32. Regioselective Anodic Mono- and Difluorination of Flavones. *J. Org. Chem.* **1999**, *64*, 3346–3349.
63. Liang, T.; Neumann, C. N.; Ritter, T. Introduction to Fluorine and Fluorine-Containing Functional Groups. *Angew. Chem. Int. Ed.* **2013**, *52*, 8214–8264.
64. Nie, J.; Guo, H.-C.; Cahard, D.; Ma, J.-A. Asymmetric Construction of Stereogenic Carbon Centers Featuring a Trifluoromethyl Group from Prochiral Trifluoromethylated Substrates. *Chem. Rev.* **2011**, *111*, 455–529.
65. Chu, L.; Qing, F. L. Oxidative Trifluoromethylation and Trifluoromethylthiolation Reactions Using (Trifluoromethyl)Trimethylsilane as a Nucleophilic CF_3 Source. *Acc. Chem. Res.* **2014**, *47*, 513–522.
66. Ni, C.; Hu, M.; Hu, J. Good Partnership between Sulfur and Fluorine: Sulfur-based Fluorination and Fluoroalkylation Reagents for Organic Synthesis. *Chem. Rev.* **2015**, *115*, 765–825.
67. Zhang, Z.; Zhang, L.; Cao, Y.; Li, F.; Bai, G.; Liu, G.; Yang, Y.; Mo, F. Mn-Mediated Electrochemical Trifluoromethylation/C(sp^2)-H Functionalization Cascade for the Synthesis of Azaheterocycles. *Org. Lett.* **2019**, *21*, 762–766.
68. Gupta, A. K.; Bharadwaj, M.; Kumar, A.; Mehrotra, R. Spiro-Oxindoles as a Promising Class of Small Molecule Inhibitors of p53-MDM2 Interaction Useful in Targeted Cancer Therapy. *Top. Curr. Chem.* **2017**, *375*, 3.
69. Ruan, Z.; Huang, Z.; Xu, Z.; Mo, G.; Tian, X.; Yu, X.-Y.; Ackermann, L. Catalyst-free, Direct Electrochemical Tri- and Difluoroalkylation/Cyclization: Access to Functionalized Oxindoles and Quinolinones. *Org. Lett.* **2019**, *21*, 1237–1240.

70. Hong, H.; Li, Y.; Chen, L.; Li, B.; Zhu, Z.; Chen, X.; Chen, L.; Huang, Y. Electrochemical Synthesis Strategy for C_{Vinyl}-CF_3 Compounds through Decarboxylative Trifluoromethylation. *J. Org. Chem.* **2019**, *84*, 5980–5986.
71. Guan, Z.; Wang, H.; Huang, Y.; Wang, Y.; Wang, S.; Lei, A. Electrochemical Oxidative Aryl(Alkyl)Trifluoromethylation of Allyl Alcohols *via* 1,2-Migration. *Org. Lett.* **2019**, *21*, 4619–4622.

CHAPTER ELEVEN

Monitoring mechanochemical processes in situ and in real time

Adam A.L. Michalchuk, Anke Kabelitz, and Franziska Emmerling
BAM Federal Institute for Materials Research and Testing, Berlin, Germany

1 Mechanochemistry: The ex situ paradigm

The conversion of mechanical force into a chemical or physical transformation is arguably the oldest form of human-driven chemistry. From our prehistoric ancestors to the first written record of mechanically induced reactions by Theophrastus of Eresus in ca. 315 BCE through to the establishment of mechanochemistry in the late 19th century,[1–3] mechanochemical history is vast. Excellent reviews on the subject can be found elsewhere.[4, 5] In its early years, mechanochemical research focused almost exclusively on the study of inorganic materials and polymers. Since the early 2000s, however, global efforts have arisen devoted to the study of organic molecular solids.[6] The field of mechanochemistry now covers a broad scope of physical and chemical transformations. In its broadest definition as the application of mechanical energy to induce chemical or physical phenomena, this includes phenomena as diverse as high-pressure transformations,[7–9] the shock and impact induced initiation of energetic materials,[10] triboelectric effects,[11] and many more. Herein, we limit the scope of discussion, focusing solely on transformations conducted by ball mill grinding and related mechanical activators.

Many processes occur during ball mill grinding, including particle comminution, surface generation, formation of defects in the crystalline structures, phase transformations, and chemical reactions.[12] These processes were initially attributed to the heating associated with mechanical treatment. Examples are now known, however, where the thermal and mechanochemical reaction products differ.[2, 3] Moreover, it has been demonstrated that different effects can be observed in materials, depending on the type and rate of mechanical stressing.[13–15] Unraveling the behavior of molecules and materials exposed to different mechanical stresses has hence become a fascinating and diverse field of research. Much of this research has been

Nontraditional Activation Methods in Green and Sustainable Applications
https://doi.org/10.1016/B978-0-12-819009-8.00004-9
© 2021 Elsevier Inc.
All rights reserved.

conducted through the study of materials under extremes of pressure. These studies have included hydrostatic and nonhydrostatic compression of materials,[8, 16, 17] dynamic compression,[18, 19] and compression with controlled heating.[20, 21] Some studies have now directly linked the response of materials under controlled pressure experiments to mechanochemical reactivity.[22]

Traditional approaches for mechanochemical research involved the ex situ sampling of solid materials. In these approaches the mechanochemical reaction is stopped after variable lengths of time, and a representative sample is removed from the vessel for analysis. As mechanochemical research typically involves the study of solid reactants and product phases, it has been most conventional to investigate their processes using X-ray diffraction methods. Using laboratory techniques, for example, mechanochemical reactions have been successfully studied, including identifying intermediate phases,[23] following the formation of crystalline defects,[24, 25] identifying relative phase stability by competitive comilling,[26] and observing the rate of material doping.[27] The mechanochemical synthesis of the metal-organic framework $(H_2Im)[Bi(1,4\text{-}BDC)_2]$ (H_2Im = imidazolium cation, BDC = benzenedicarboxylate) was successfully monitored ex situ using a combination of powder X-ray diffraction (PXRD), Raman spectroscopy, and electron microscopy.[28] By collecting ex situ data at short time intervals (c. 10 s), a short-lived intermediate phase was identified. The data available by ex situ methods have been sufficient for detailed kinetic analyses in many cases.[29] Moreover, laboratory ex situ PXRD has also proven invaluable to explore the mechanistic details of mechanochemical transformations. In the mechanochemical synthesis of Ettringite, for example, ex situ diffraction studies demonstrated preferential amorphization of one reagent as a necessary reaction step.[30] Furthermore, by ex situ laboratory analysis, the effects of particle size[31] and surface structure[32, 33] have also been identified as driving forces for mechanochemical transformations in organic mechanosynthesis and polymorphism.

In a similar fashion, ex situ analysis has been performed by a variety of other methods. These techniques include Mössbauer spectroscopy,[34] thermal analysis,[24, 35] microscopy,[36] positron annihilation,[37] Raman[38] or infrared spectroscopy[39], THz time domain spectroscopy,[40] high-performance liquid chromatography (HPLC),[31, 41, 42] fluorescence spectroscopy,[43] and NMR spectroscopy.[41, 42, 44] HPLC methods have proven particularly important for ex situ monitoring of mechanochemical syntheses, as demonstrated by Oliveira et al.[45] By monitoring the rate of mechanochemical synthesis of 2,3-diphenylquinoxaline at various temperatures, it was possible to deduce the activation energy for the transformation. It was shown that the activation energy differed markedly above and below the eutectic temperature, being lower at lower temperatures. Similarly, HPLC methods have been used to investigate the effects of mechanochemical treatment on the dissolution profiles of pharmaceutical cocrystals, as in the case of pirfenidone.[46] The effects of mechanical treatment on the solubility and dissolution profile of carbamazepine cocrystals were similarly investigated by ex situ NMR.[47] In exceptional cases, ex situ analysis has been performed onsite at synchrotron sources. For example, the cocrystallization of caffeine with anthranilic acid by liquid-assisted grinding was followed in this way.[48] The use of synchrotron radiation allowed Emmerling and coworkers to reduce PXRD collection times from 10 s of minutes to ca. 30 s. Any aging of the ball-milled powder was thereby minimized.

Regardless of the speed of analysis, one must always disrupt the system to analyze ball-milled powder ex situ. Studies have shown that the interruption of the reaction process during ex situ investigations is not always a suitable approach as it can lead to different products compared with the continuous mechanical treatment.[14, 49–51] For instance, Cinčić and coworkers demonstrated how changes in the atmospheric composition can influence the mechanochemical reactivity of organic materials.[52] In other cases the development of powder rheology drives mechanochemical transformations, and its intermittent disruption can alter the reaction course.[53, 54] Furthermore, due to sampling, the ball to reactant ratio (BRR) is changed over the course of the reaction. Changes in the BRR can influence on the reaction rate.[55] Volatile, air-sensitive, or short-lived intermediates resulting from mechanical activation might be missed completely using this approach.[56, 57]

The previous examples highlight how ex situ analysis can influence mechanochemical reactions at the macroscopic level. However, the need for in situ and time-resolved studies is even more critical when considering these processes at the microscopic level. Collision theory underpins the statistical nature of mechanochemical processes.[58, 59] The reaction rate observed in such processes depends critically on (i) the reaction probability at a given mechanical energy impact per contact and (ii) the probability of contact formation at the moment of milling body impact (i.e., mixing).[60, 61] As each impact of the milling ball hits a different small fraction of the sample each time, the entire mechanochemical processes are discontinuous and stochastic. Each collision occurs on the time frame of 1 ms, whereas the transformation induced by mechanical stress often proceeds on the order of 10 s of nanoseconds or less.[62–64] This implies that the reaction is completed before the collision is finished and that in situ time-resolved analysis is required if any insights into the true mechanisms of these processes are to be obtained.

Time-resolved in situ monitoring of mechanochemical ball milling reactions dates to at least the 1970–80s.[65–67] During these early experiments, ball milling reactions were monitored by the evolution of gas pressure and temperature. In another example, both the pressure and temperature were monitored in real time during the mechanochemical synthesis of hydrogen storage material MgH_2 by high-energy reactive ball milling (Fig. 1).[68] Indeed, manometric measurements continue to find application in the organic mechanochemistry literature.[69] Optical cameras have also been used to identify the motions of milling balls within mechanochemical reactors under different milling frequencies.[70]

Disturbing the powdered material obtained during ball mill grinding has the potential to alter the reaction pathway.[14] Moreover, generating new powder mixtures can lead to stochastic variation in reaction progress, and

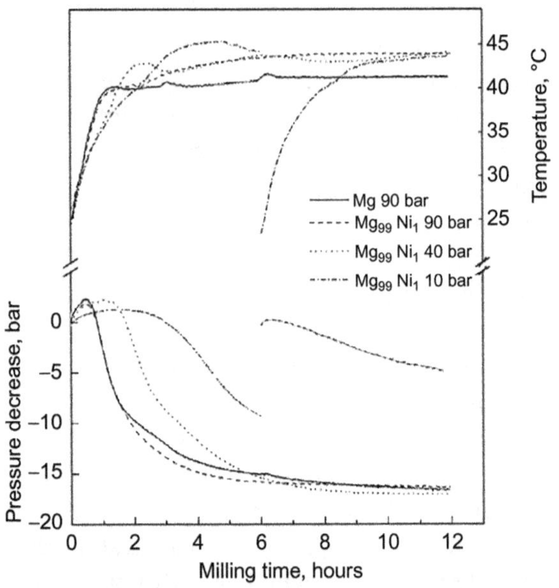

Fig. 1 Monitoring the mechanochemical synthesis of hydrogen storage material MgH_2 by time-resolved temperature and H_2 gas pressure measurements. *(Figure reproduced from Doppiu, S.; Schultz, L.; and Gutfleisch, O. In Situ Pressure and Temperature Monitoring during the Conversion of Mg into MgH2 by High-Pressure Reactive Ball Milling. J. Alloys Compd. 2007, 427 (1–2), 204–208. Copyright 2006 Elsevier B.V.)*

removing powder for ex situ analysis can have important consequence on the dynamics of the transformation.[48] In situ approaches (i.e., where the material is monitored in its nascent environment) are therefore of interest. Early work by James and coworkers[71] identified pseudofluid kinetics during ball mill synthesis of Zn-based zeolitic imidazole framework (ZIF) ZIF-8 by in situ Raman spectroscopy (Fig. 2A). In their experiments the ball milling reaction was stopped at different time intervals, and the Raman probe inserted into the vial; no powder was removed from the jar. Hence, these in situ (albeit not time-resolved) measurements allowed to monitor the kinetics of ball mill grinding of the same reaction mixture as a function of time. In a similar study the mechanochemical formation of pharmaceutical cocrystals was explored in situ.[72] Here, cocrystals of indomethacin with saccharin and theophylline with citric acid were prepared by ball mill grinding. As was performed by James and coworkers, the ball milling process was stopped at select time intervals, and a fiber-optic Raman probe was inserted directly into the powder material. By following the relative change in the intensities of characteristic vibrational bands, the kinetics of cocrystallization could be followed (Fig. 2B).

In 2013 Friščić and coworkers introduced an important step for time-resolved in situ monitoring of mechanochemical transformations. In their work, mechanochemical reactions in a shaker mill (MM200, Retsch) were followed using high-energy synchrotron PXRD and later by Raman spectroscopy.[73–75] The authors implemented a commercial mill equipped with an in-house built transparent cylindrical milling jar in their setup. From these initial proof-of-concept works, in situ investigations of mechanochemical reactions have developed rapidly into a new area of research.

2 Time-resolved in situ monitoring by powder X-ray diffraction

When considering solid-state transformations, X-ray diffraction (XRD) is among the most direct analytical approaches. For mechanochemical research, one typically considers powdered material and therefore powder X-ray diffraction (PXRD). In (P)XRD, photons are coherently scattered from the crystalline material. The angle of the scattered photons describes the structure of the crystalline lattice, while the intensity scattered photons at each angle is determined by the type and position of the atoms within the solid. Hence, (P)XRD provides a route to identify which crystalline phases are present in a solid mixture and can provide access to its

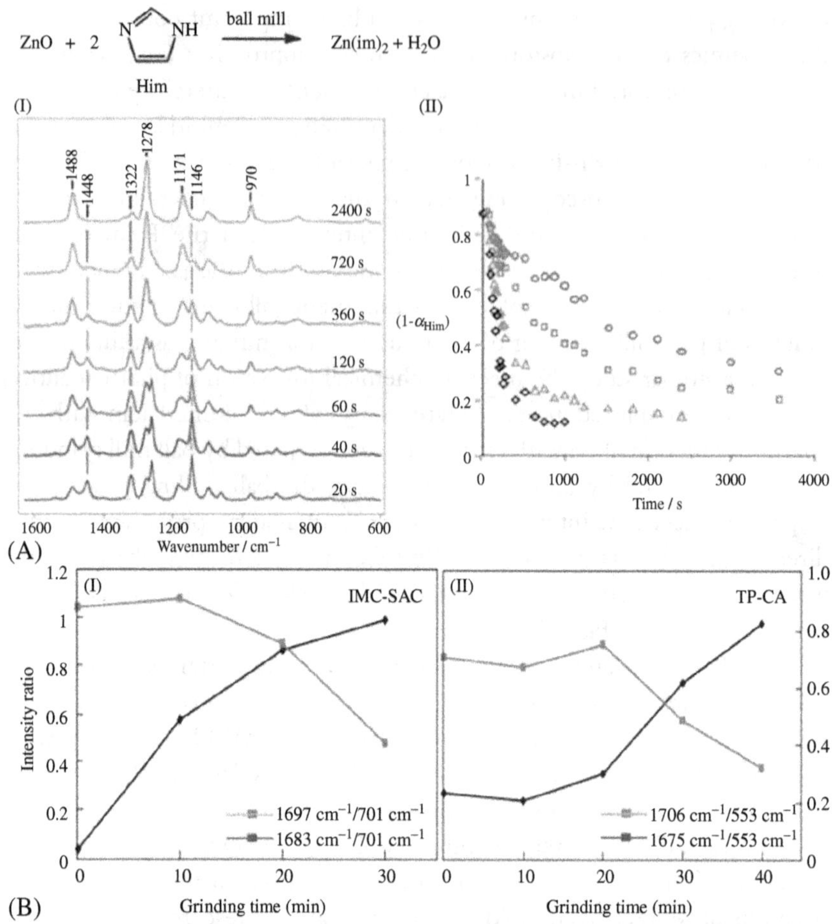

Fig. 2 In situ monitoring of mechanochemical transformations by Raman spectroscopy. (A) Mechanochemical formation of ZIF-8. (AI) Raman spectra collected in situ after stopping the ball milling process at select times. (AII) Kinetic profiles obtained by integrated peak intensities for milling at (black, diamond in print version) 25 Hz, (green, triangle in print version) 20 Hz, (blue, square in print version) 15 Hz, and (red, circle in print version) 10 Hz. (B) Mechanochemical preparation of (BI) indomethacin (IMC)-saccharine (SAC) and (BII) theophylline (TP)-caffeine (CA) cocrystals. *(Figures adapted from Ma, X.; Yuan, W.; Bell, S. E. J.; and James, S. L. Better Understanding of Mechanochemical Reactions: Raman Monitoring Reveals Surprisingly Simple "Pseudo-Fluid" Model for a Ball Milling Reaction. Chem. Commun. 2014, 50 (13), 1585–1587 with permission from The Royal Society of Chemistry and from Lin, H.; Zhang, G.; Hsu, P.; and Lin, S. A Portable Fiber-Optic Raman Analyzer for Fast Real-Time Screening and Identifying Cocrystal Formation of Drug-Coformer via Grinding Process. Microchem. J. 2013, 110, 15–20. Copyright 2013 Elsevier B.V.)*

atomic structure. The major requirement for in situ time-resolved (P)XRD is that the probing radiation must be incident on the sample and be free to leave the sample for subsequent detection. For this reason, nearly all in situ time-resolved mechanochemical investigations have so far used transparent Perspex (polymethylmethacrylate [PMMA]) vessels. Perspex is an amorphous polymer, thereby reducing any hindrance to the scattering by crystalline materials within the vessel.

The study of mechanochemical transformations by PXRD has been demonstrated for a wide range of material types, including inorganic solids, molecular solids, and framework materials. The first time-resolved PXRD monitoring study of mechanochemical synthesis was demonstrated for a series of ZIFs.[73] The mechanochemical ZIF synthesis was performed in a commercial Retsch MM200 vibratory ball mill under both neat and liquid-assisted grinding conditions. By monitoring characteristic Bragg peak intensities, formation of three ZIF compounds was investigated. Notably, these initial in situ time-resolved studies demonstrated the ability to identify short-lived intermediate phases with relative ease, as compared with the time-consuming approaches of ex situ analysis. The liquid-assisted formation of ZIF-8 from ZnO and imidazole was found to pass via formation of ZIF-4 (Fig. 3A). Similarly the product topology obtained by mechanochemical reaction of ZnO with 2-ethylimidazole depended on the quantity of liquid additive (here dimethylformamide [DMF]) (Fig. 3B).

Using time-resolved in situ PXRD monitoring, the presence of new framework topologies and framework formation mechanisms have been identified in a variety of mechanochemical syntheses. In ZIF-8, liquid-assisted grinding (LAG) with aqueous acetic acid led to apparent amorphization of the conventional ZIF-8 structure. Extended grinding of the amorphous phase drove the material to a previously unobserved topology.[76] The mechanism of mechanochemical formation of the metal-organic framework HKUST-1 was also found to depend on the synthesis conditions (Fig. 4).[77] Under neat grinding conditions, and during LAG with a variety of liquids, HKUST-1 was found to form directly from the reactant phases. In contrast, when LAG was performed with ethanol or DMF, an intermediate phase was first formed, followed by conversion to the known HKUST-1 form. Surprisingly, when this intermediate phase was investigated by ex situ monitoring, it was found to convert to a different intermediate phase. Hence the mechanochemical formation of HKUST-1 demonstrates where time-resolved in situ monitoring of ball mill grinding reactions can differ significantly from ex situ analysis.

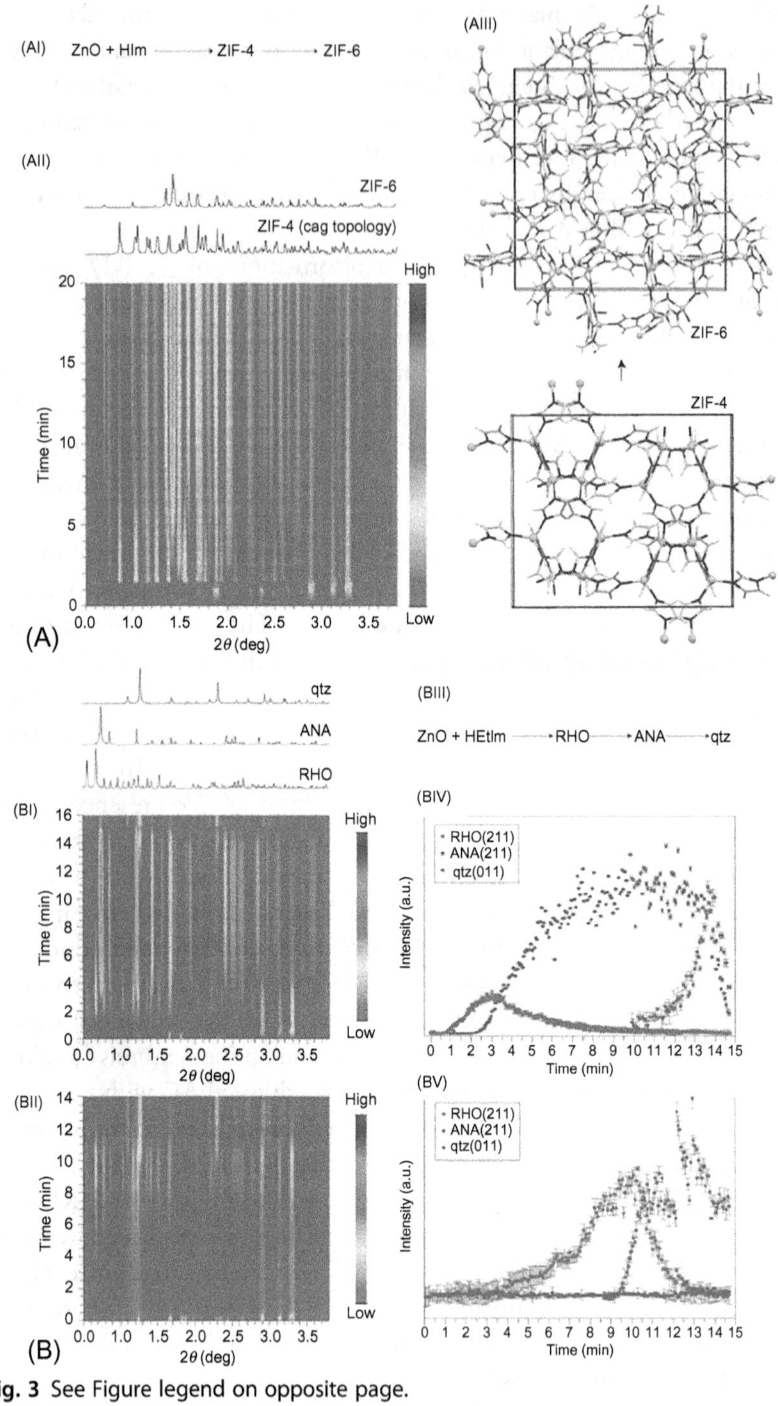

Fig. 3 See Figure legend on opposite page.

Understanding the mechanochemical formation of metal phosphonates has been an additional focus of time-resolved in situ PXRD studies. Cogrinding stoichiometric quantities of Cu(II) acetate and phenylphosphonate ($PhPO(OH)_2$) was shown to yield formation of the β-form of Cu $(O_3PC_6H_5)\cdot H_2O$ (β-CuPhPmH (2)) (Fig. 5B).[78] During the ball mill grinding process, two intermediate phases were identified: an unidentified phase and a known form of CuPhP2mH (4). Interestingly the latter could also be obtained via the same unidentified intermediate phase by cogrinding Cu(II) acetate with three molar equivalents of $PhPO(OH)_2$. In a similar study the mechanochemical formation of two Ce-based phosphonates was also investigated.[79] The in situ time-resolved PXRD analysis showed that the product phosphonates were formed within only two minutes of ball mill grinding and passed through an unknown intermediate phase. Similar studies have also explored the mechanosynthesis of manganese(II)-phosphonates.[80]

The capabilities of in situ time-resolved PXRD to monitor mechanochemical formation of inorganic materials sparked interest in its applications for the study of organic materials. Unlike the case of inorganic or metal-organic systems,[81] the poor scattering offered by organic materials has often rendered quantitative analysis difficult. However, it has been possible to extract qualitative trends, such as phase evolution profiles, and semiquantitative kinetic information. For example, the mechanochemical cocrystallization of carbamazepine and saccharine was followed by time-resolved PXRD.[82] Although full quantitative analysis was not possible, the in situ PXRD profiles showed that formation of the 2:1 cocrystal product passed via transient formation of the stoichiometric cocrystal phase.

Fig. 3 Time-resolved in situ monitoring of mechanochemical ZIF formation. (A) Liquid-assisted grinding formation of ZIF-8 from ZnO and imidazole (HIm). Dimethylformamide (DMF) was used as grinding liquid. (AI) Reaction scheme for the mechanochemical synthesis, (AII) time-resolved PXRD profiles obtained during ball mill grinding, and (AIII) pictorial representation of the reaction scheme demonstrating the structural differences between the intermediate product, ZIF-4, and the final product, ZIF-8. (B) Liquid-assisted grinding for mechanochemical ZIF formation between ZnO and 2-ethylimidazole (HEtIm). (BI, BIV) Time-resolved PXRD profiles obtained during ball mill grinding in the presence of 150 μL DMF and relative peak intensities of characteristic Bragg peaks. (BII, BV) Time-resolved PXRD profiles obtained during ball mill grinding in the presence of 25 μL DMF and relative peak intensities of characteristic Bragg peaks. RHO, ANA, and qtz denote ZIF topologies ρ, analcime, and β-quartz, respectively. *(Figures adapted from Friščić, T.; Halasz, I.; Beldon, P. J.; Belenguer, A. M.; Adams, F.; Kimber, S. A. J.; Honkimaki, V.; and Dinnebier, R. E. Real-Time and in Situ Monitoring of Mechanochemical Milling Reactions. Nat. Chem. 2013, 5, 66–73.)*

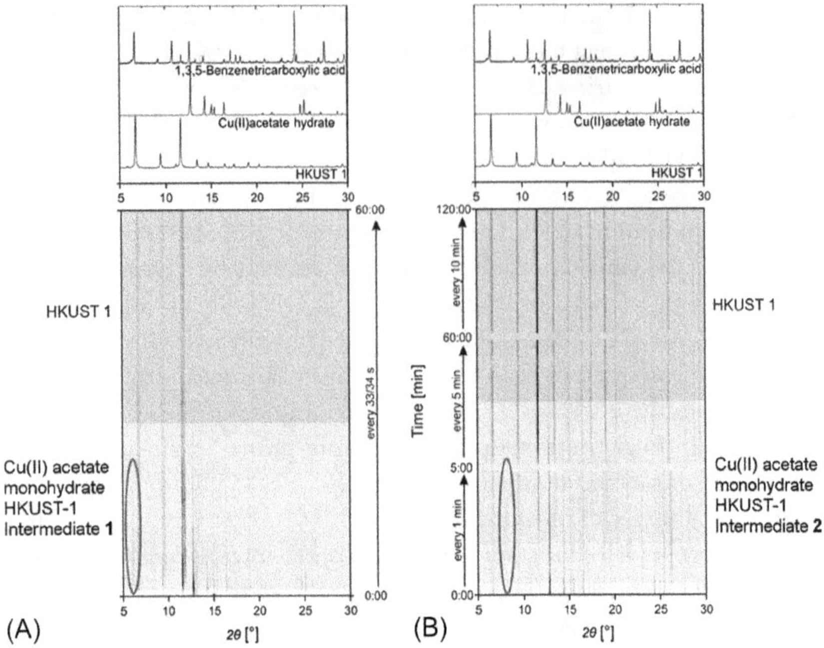

Fig. 4 Monitoring the ball mill grinding formation of HKUST-1 by PXRD. (A) Time-resolved, in situ monitoring of the reaction between copper(II) acetate hydrate and 1,3,5-benzenetricarboxylic acid with ethanol as the liquid. Intermediate 1 is visible in the first 15 min of ball mill grinding. (B) The same mechanochemical reaction followed by stepwise ex situ analyses. Ex situ analysis was performed by stopping the ball mill periodically and removing powder for PXRD sampling. *(Reprinted with permission from Stolar, T.; Batzdorf, L.; Lukin, S.; Žilić, D.; Motillo, C.; Friščić, T.; Emmerling, F.; Halasz, I.; and Užarević, K. In Situ Monitoring of the Mechanosynthesis of the Archetypal Metal-Organic Framework HKUST-1: Effect of Liquid Additives on the Milling Reactivity. Inorg. Chem. 2017, 56 (11), 6599–6608. Copyright 2017 American Chemical Society.)*

In other examples, time-resolved PXRD was used to investigate the role of liquid additives in directing cocrystal polymorphism during ball mill grinding.[83, 84] For the cocrystallization of benzamide with theophylline, ball mill grinding with nonpolar liquids (or under neat conditions) led to formation of the Form I cocrystal product.[84] Instead, in situ PXRD showed that with the addition of polar liquids, Form I acts as an intermediate phase toward the formation of cocrystal Form II. A substantially more complex series of cocrystal phases was later explored by LAG methods. Based on two racemic coformers, flurbiprofen (FLU) and proline (PRO), ball mill grinding facilitated the formation of 17 new cocrystal phases.[85] The time-resolved in situ PXRD studies showed that, in general, the 2:1 (FLU:PRO) cocrystal phases

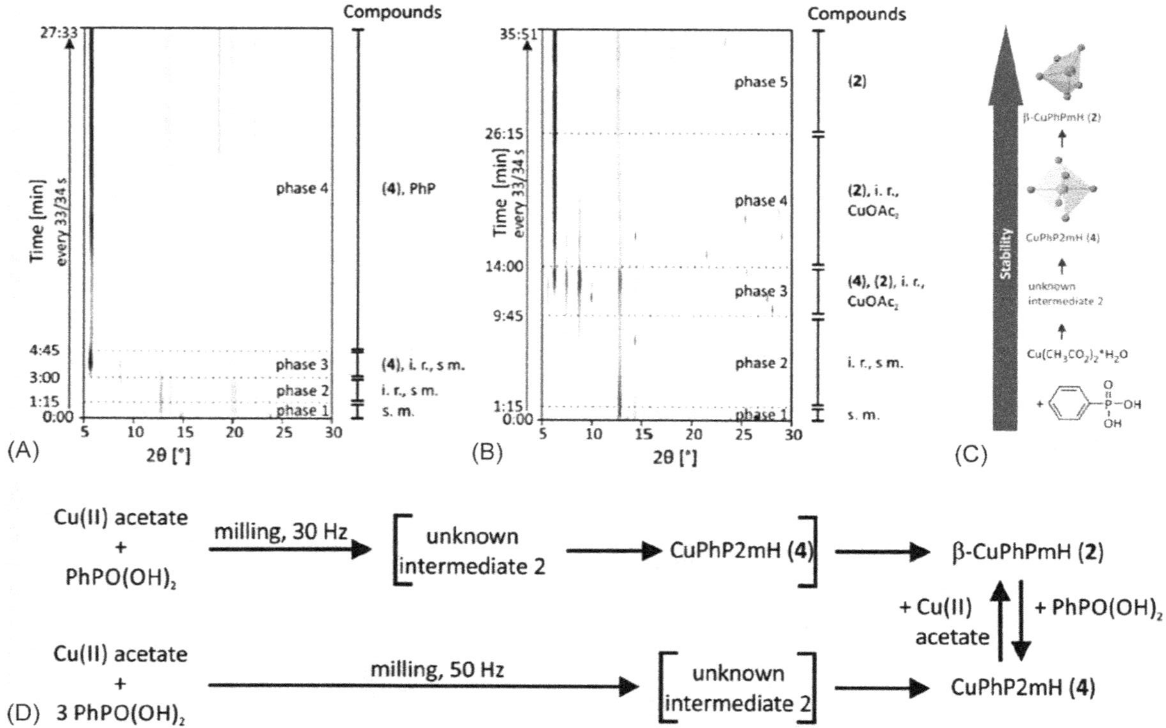

Fig. 5 In situ PXRD data for the mechanochemical synthesis of metal phosphonates. (A) synthesis of CuPhP2mH (**4**) at 50 Hz. (B) Synthesis of β-CuPhPmH (**2**) at 30 Hz. In each case the materials are described as i. r. = intermediate reflections of unknown compound, s. m. = starting materials, PhP = phenylphosphonic acid, $CuOAc_2$ = copper(II) acetate monohydrate. (C) Scheme of the mechanism for the copper phosphonates depending on the stability of the formed products. (D) Reaction scheme for the mechanochemical synthesis of copper(II) phenylphosphonates at 30 Hz and 50 Hz. The compounds in brackets symbolize the intermediates detected by in situ PXRD. *(Figure adapted from Wilke, M.; Kabelitz, A.; Gorelik, T. E.; Buzanich, A. G.; Reinholz, U.; Kolb, U.; Rademann, K.; and Emmerling, F. The Crystallisation of Copper(II) Phenylphosphonates. Dalton Trans. 2016, 45 (43), 17453–17463 with permission from The Royal Society of Chemistry.)*

were kinetically favored products, forming first during ball mill grinding. Prolonged grinding drove the systems instead toward stoichiometric cocrystal products. In the case of FLU + PRO mechanochemical cocrystallization, it was found that the choice of liquid used in the LAG had substantial influence on the rate of cocrystal formation and on the overall dynamics of the transformation. For example, when ground in the presence of ethanol, the 2:1 cocrystal product remained short-lived during the mechanochemical synthesis, while in the presence of methanol both the 2:1 and 1:1 cocrystal phases remained present until the end of the reaction (Fig. 6A).

The presence of intermediate products has been a useful result of in situ time-resolved PXRD investigations. Such information offers evidence (albeit inadvertently) of mechanochemical phase stability. To probe this in more explicit detail, competitive ball mill grinding transformations have been conducted and monitored in situ.[86] Using the theophylline-benzamide-benzoic acid ternary system, time-resolved in situ PXRD investigations were used to identify which binary cocrystal phases were favored during mechanical treatment. By observing the interconversion of binary phases, the authors proposed a stability ordering. The same was done using the theophylline-benzoic acid-isonicotinamide ternary system. Interestingly, time-resolved PXRD identified the presence of a 2:1 (benzoic acid:isonicotinamide) cocrystal as an intermediate toward formation of the mechanochemically stable benzoic acid:isonicotinamide binary cocrystal phase.

Monitoring the transformation of organic materials during ball mill grinding has not been restricted to multicomponent transformations. For example, in situ time-resolved PXRD was also shown to be instrumental in identifying the effects of polymorph seeding during mechanochemical processing in organic systems.[87] By seeding a system with traces of nicotinamide, the metastable form of benzamide was successfully obtained by ball mill grinding in the presence of ethanol. The seed to benzamide ratio was demonstrated to have marked influence on the rate of formation of the metastable form. This example highlights how time-resolved PXRD investigations are not restricted to formation of new materials but is also sensitive to changes in the structure of single-component systems.

Time-resolved in situ PXRD methods are not strictly necessary for identifying intermediate product phases. As discussed in Section 1, it is always possible to extract material for ex situ analysis after given reaction periods. Capturing the kinetics of a mechanochemical transformation by ex situ analysis, however, can be a great challenge. In most ex situ attempts, authors

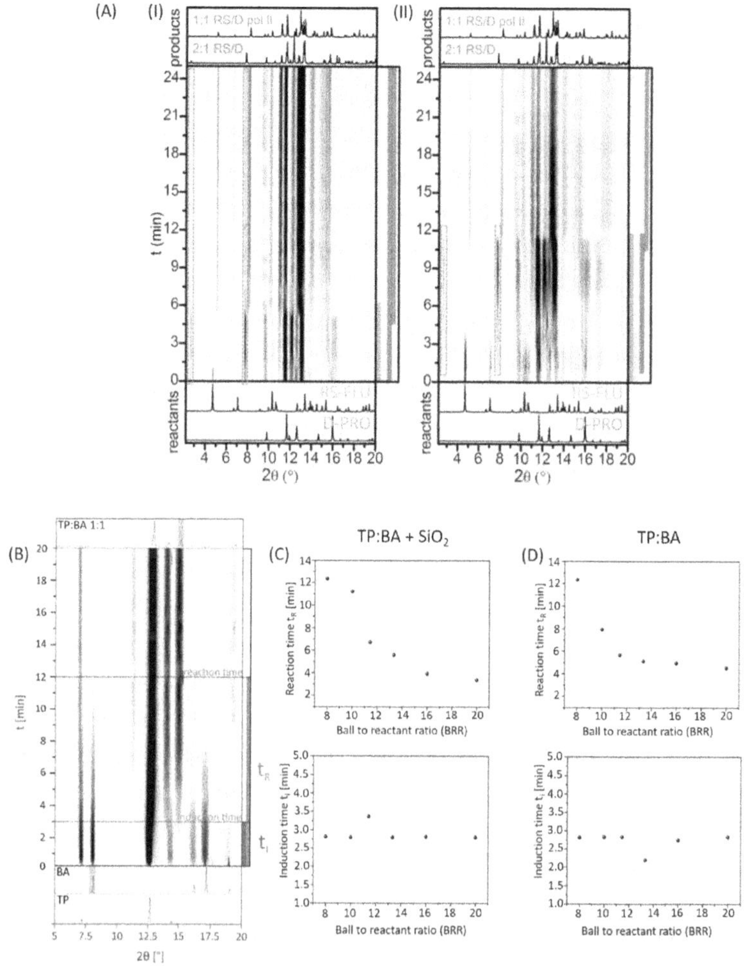

Fig. 6 Time-resolved in situ PXRD monitoring of multicomponent organic mechanochemical cocrystallization. (A) Formation of flurbiprofen (FLU) and proline (PRO) cocrystals in the presence of (I) methanol and (II) ethanol. (B) Cocrystal formation of theophylline (TP) and benzoic acid (BA) in molar ratio 1:1 under neat grinding conditions with a total mass of 1000 mg. The reaction time t_R (red, light gray in print version) shows the end of the reaction using the time when the reflection of the benzoic acid reflection (red) at $2\theta = 8°$ disappears. The induction time t_I (green, dark gray in print version) is the time when the first reflection (green) of the product at $2\theta = 15°$ appears. Reaction time t_R and induction time t_I for the formation of TP:BA (1:1) under neat grinding conditions as a function of the ball to reactant ratio (C) in the presence of SiO_2 and (D) without SiO_2. *(Part figure (A) reprinted with permission from Tumanova, N.; Tumanov, N.; Robeyns, K.; Fischer, F.; Fusaro, L.; Morelle, F.; Ban, V.; Filinchuk, Y.; Wouters, J.; and Leyssens, T.; et al. Opening Pandora's Box: Chirality, Polymorphism, and Stoichiometric Diversity in Flurbiprofen/Proline Cocrystals. Cryst. Growth Des. 2018, 18, 954–961. copyright 2018 American Chemical Society; Part figure (D) adapted with permission from Kulla, H.; Fischer, F.; Benemann, S.; Rademann, K.; and Emmerling, F. The Effect of the Ball to Reactant Ratio on Mechanochemical Reaction Times Studied by in Situ PXRD. CrstEngComm 2017, 19 (28), 3902–3907, with permission from The Royal Society of Chemistry.)*

have explored the kinetic profile by preparing new powder mixtures for each time point. This approach is met with the difficulty that each newly prepared powder mixture will be privy to unique mixing and reaction kinetics. Capturing the true kinetics of a mechanochemical transformation instead requires the same powder mixture to be followed in time. Unfortunately, intermittently disturbing the powder mixture can have detrimental effects on the apparent transformation mechanisms.[14, 53] Moreover, intermittent sampling can have more subtle effects, modifying the kinetic profile and hence its interpretation. By studying the known[88] mechanochemical cocrystallization of theophylline (TP) and benzoic acid (BA) (Fig. 6B), Kulla et al.[55] investigated the influence of the ball to reactant mass ratio (BRR) on the transformation kinetics. Under neat grinding conditions, stoichiometric mixtures of TP and BA were milled with two 10 mm stainless steel milling balls (total 8 g mass). The total quantity of reagent was varied from 400 mg through to 1000 mg, thereby modifying the BRR from 20 to 8. Time-resolved in situ PXRD showed that higher BRR led to markedly faster reaction rates (Fig. 6D). The increased reaction rates are presumably due to greater energy being transferred to the powder for each impact.[89] These studies clearly indicated that relatively small changes in BRR count have drastic influence on the apparent reaction rate. As such, intermittent sampling by ex situ analysis could introduce spurious effects on the kinetic profile. To explore the effects further, the authors conducted an additional set of experiments in which inert glassy SiO_2 was added to the reaction mixture. In these experiments the total BRR (i.e., reagent + SiO_2) was held constant, although the ball to reactive reagent ratio still varied (BRRR, i.e., ball to TP + BA ratio). In this case the same trends were observed for the BRRR and the BRR (Fig. 6C). In addition to the increase in impact energy, this suggested that a major influence of the BRR in this mechanochemical cocrystallization was the result of mixing.[90]

It has generally proved difficult to obtain time-resolved PXRD data for organic materials with sufficient quality for quantitative analysis. In a few cases, however, this has been achieved, thereby allowing reaction profiles to be examined in more detail. Cogrinding a stoichiometric mixture of γ-glycine (γGly) and oxalic acid dihydrate (OAD) led to formation of the stoichiometric cocrystal product.[91] However, the in situ time-resolved PXRD data showed formation of the 2:1 product as a transient intermediate phase. The improved quality of diffraction obtained for the cogrinding of γGly and OAD allowed for the detailed investigation of transformation kinetics as a function of milling frequency. Small changes in the milling frequency induced notable changes in the transformation kinetics. Moreover, analysis of the kinetic data revealed unexpected and abrupt changes in the kinetic profiles, believed to relate to rheological effects.[54] Similarly, time-resolved in situ PXRD data were used to investigate the transformation kinetics in two polymorphs of anhydrous caffeine.[90] In this case, kinetic analysis was performed as a function of milling ball size, whereby larger milling balls induced faster transformation.

Time-resolved in situ PXRD monitoring has not been restricted to the study of vibratory ball milling setups. Recently, mechanochemical cocrystallization was demonstrated with the Resonant Acoustic Mixer (RAM) technology.[92, 93] To investigate the RAM-induced mechanochemistry, the cocrystallization of pharmaceutically active carbamazepine (CBZ) and nicotinamide (NIC) was followed.[94] Starting from CBZ, stoichiometric quantities of reactant phases were mixed by RAM technology under liquid-assisted grinding conditions. The reaction progress was monitored in real time by PXRD, demonstrating the continuous growth of the cocrystal phase (Fig. 7). The smooth and continuous growth of product phase was unexpected on account of the absence of milling balls used in RAM. Moreover the high-quality PXRD data (Fig. 7D) allowed quantitative phase analysis. This phase analysis demonstrated how different mixing intensities could drive preferential formation of different product phases: CBZ dihydrate at low-intensity mixing and the cocrystal phase at higher mixing intensities. Through use of in situ time-resolved PXRD, this "ball-free" mechanochemical technology was demonstrated.

A number of cases have now been shown for inorganic/metal-organic materials for which time-resolved in situ PXRD data are obtained with sufficient quality for quantitative refinement.[81, 95] Even fewer cases have been demonstrated for organic materials.[90, 91, 94] While quantitative refinements have been possible in these few cases, structure solution of novel phases remains a great challenge. The need for high-quality data has led some authors to develop new ball milling vessels explicitly for in situ time-resolved data collection. Notable is the development by Casati et al.[96] In contrast to the Perspex (PMMA) jars used for most other in situ time-resolved PXRD studies, the setup proposed by Casati and coworkers implements a steel jar (Fig. 8A). The steel jar is surrounded by a small chamber composed of Kapton through which the X-ray beam is focused. As the vessel moves the powder falls into this chamber, thereby scattering the incident X-ray beam. The exceptional data quality obtained with this technique was initially demonstrated using inorganic samples including LaB_6, along with

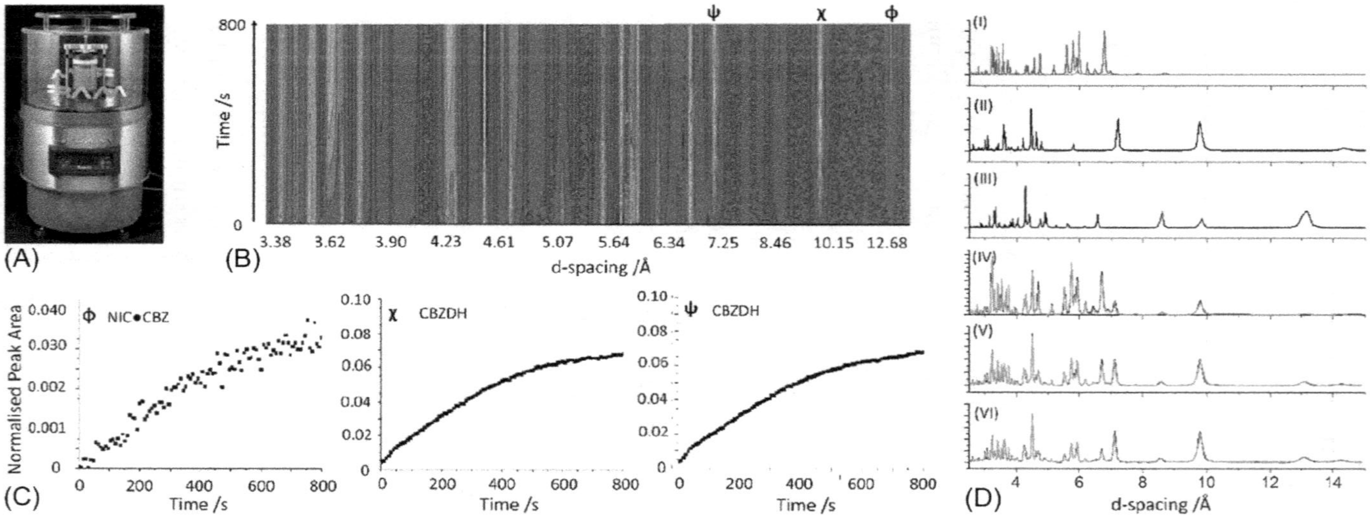

Fig. 7 Time-resolved in situ monitoring of mechanochemical cocrystal formation of carbamazepine (CBZ) and nicotinamide (NIC) by Resonant Acoustic Mixing. (A) Photograph of the LabRAM device. (B) Time-resolved PXRD profiles obtained during the mechanochemical synthesis. (C) Relative phase composition obtained by integration of characteristic Bragg reflections for the cocrystal (NIC•CBZ) and carbamazepine dihydrate (CBZDH). (D) Extracted powder diffraction profiles taken from (B). Simulated PXRD patterns for (DI) CBZ phase III (blue, light gray in print version) and NIC phase I (green, dark gray in print version), (DII) CBZDH, (DIII) CBZ-NIC, and Rietveld refined profiles at (DIV) 112 s, (DV) 400 s, and (DVI) 800 s. Experimental (black) and Rietveld refined profiles (red, light gray in print version) are shown. *(Figure adapted from Michalchuk, A. A. L.; Hope, K. S.; Kennedy, S. R.; Blanco, M. V.; Boldyreva, E. V.; and Pulham, C. R. Ball-Free Mechanochemistry: In Situ Real-Time Monitoring of Pharmaceutical Co-Crystal Formation by Resonant Acoustic Mixing. Chem. Commun. 2018, 54 (0), 4033–4036—Published by The Royal Society of Chemistry.)*

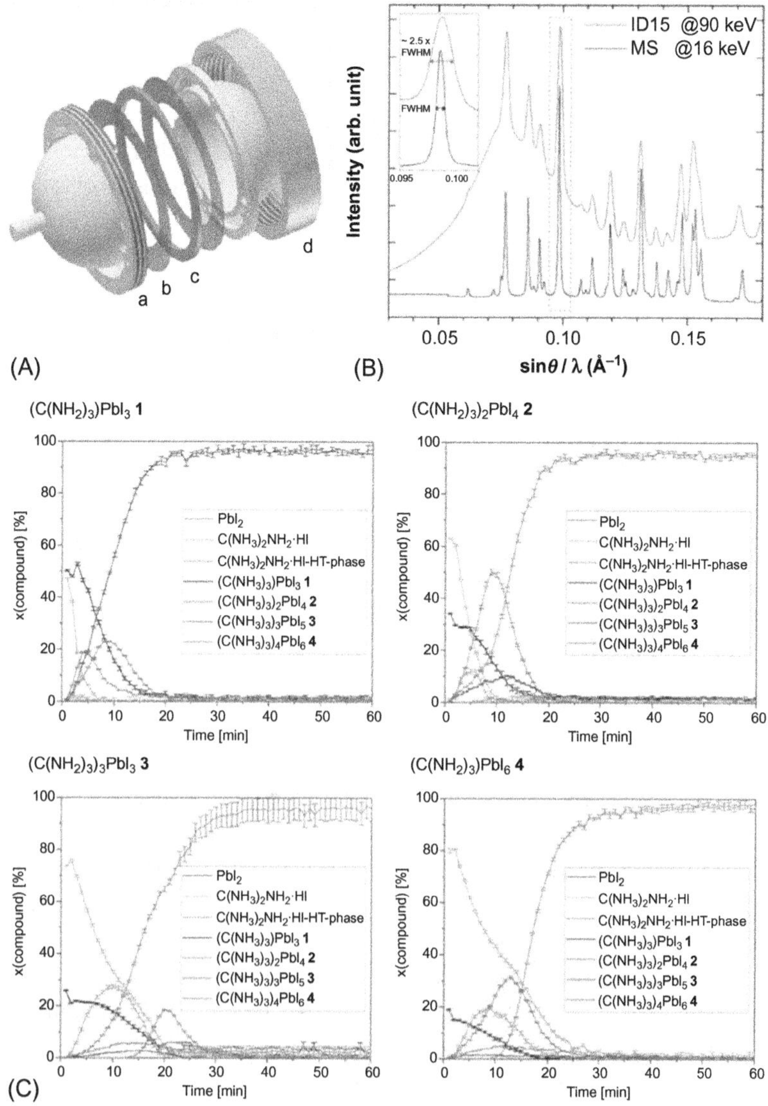

Fig. 8 Alternative setup for the in situ time-resolved monitoring of mechanochemical reactions. (A) Diagrammatic representation of the novel milling jar, showing (a) the stainless steel vessel, (b) Kapton sheets, (c) Teflon gasket, and (d) compression nut. (B) Comparison of in situ ball milling (red, upper line in print version) PXRD data obtained at the ESRF ID15 beamline (Grenoble, France) with that obtained in the new in situ setup (blue, lower line in print vesrion) at the SLS MS beamline (Villigen, Switzerland) for the mechanochemical reaction between $LiBH_4$ and $CsBH_4$. (C) Quantitative Rietveld phase analysis of the time-resolved PXRD data collected in the novel setup for the mechanochemical synthesis of (CI) $(C(NH_2)_3)PbI$, (CII) $(C(NH_2)_3PbI_4)_2$, (CIII) (C $(NH_2)_3PbI_5$, and (CIV) $(C(NH_2)_3PbI_6)_4$. *(Figures adapted with permission from Ban, V.; Sadikin, Y.; Lange, M.; Tumanov, N.; Filinchuk, Y.; Černý, R.; and Casati, N. Innovative in Situ Ball Mill for X-Ray Diffraction. Anal. Chem. 2017, 89 (24), 13176–13181 copyright 2017 American Chemical Society and from Wilke, M.; and Casati, N. Insight into the Mechanochemical Synthesis and Structural Evolution of Hybrid Organic–Inorganic Guanidinium Lead(II) Iodides. Chem. A Eur. J. 2018, 24 (67), 17701–17711. copyright 2018 John Wiley and Sons.)*

reactions between $LiBH_4$ and $CsBH_4$ (Fig. 8B) and between CsCl and KI. The potential offered by these new milling jars was further demonstrated with the mechanochemical synthesis of a series of four hybrid organic-inorganic lead(II) iodides, $(C(NH_2)_3)_nPbI_{2+n}$.[97] Depending on the initial reagent ratios, the four different materials could be prepared. In situ time-resolved analysis suggested the presence of the other phases as intermediates. Owing to the high-resolution PXRD data obtained using the new milling setup, accurate quantitative phase analysis could be captured for all seven phases over the course of these reactions (Fig. 8C).

In addition to the setup proposed by Casati et al., other attempts have been made to enhance the data quality of in situ time-resolved mechanochemical monitoring. Tumanov and colleagues,[98] for example, designed a set of 3D printed milling vessels based on polylactic acid (PLA). These 3D printed jars contained a small groove into which powder could fall for analysis. Importantly, these vessels are compatible with conventional mechanochemical reactors such in common use. As compared with the commonly used PMMA vessels, PLA shows reduced background for X-ray scattering, lower absorption at low energies, and higher angular resolution. Hence, these vessels offer an alternative approach to conducting in situ time-resolved PXRD studies based on conventional techniques for in situ investigations.

The development of new milling vessels is of great importance to facilitate the further development of time-resolved studies in mechanochemistry. Care must be taken, however, as it is also known where the type and rate of mechanical treatment can affect the outcome of the transformation.[14, 15] For example, the cocrystallization of succinic acid and piroxicam could be easily driven under impact intensive conditions. In contrast, under shear intensive conditions, the cocrystal dissociated back to its parent coformers.[99] Similarly, whereas the cocrystallization of glycine and malonic acid readily formed upon shear treatment, the product mixture was dominated by an unknown phase when exposed to heavy impact conditions.[53] Moreover, it has been recently demonstrated that even the milling jar material can influence on the mechanochemical transformation. This material dependence is demonstrated for the mechanochemical cocrystallization of pyrazinamide and pimelic acid.[83] The metastable cocrystal product obtained in a Perspex milling jar was found to be exceptionally more stable than the same product obtained by milling in a stainless steel milling jar.

3 Time-resolved in situ monitoring by vibrational spectroscopy

The most frequently used vibrational spectroscopy for monitoring mechanochemical transformation is Raman spectroscopy. In Raman spectroscopy, photons are inelastically scattered from a sample. The change in photon energy is equivalent to a vibrational frequency of the molecule or material. Hence, by monitoring the change in photon energy, Raman spectroscopy provides access to the vibrational spectrum of the material. As a laboratory-based technique, Raman spectroscopy has a distinct advantage over synchrotron-based PXRD methods. Moreover, Raman spectroscopy is not restricted to the study of crystalline materials. However, whereas PXRD is very sensitive to different crystalline forms of materials (i.e., polymorphism), changes in Raman Spectroscopy across solid-state transitions are often subtle. During the mechanochemical Knoevenagel condensation of *p*-fluorobenzaldehyde and malononitrile, formation of the condensation product was easily observed by Raman spectroscopy (Fig. 9).[100, 101] Unfortunately the two polymorphic forms have nearly identical vibrational spectra. Only through advanced statistical analysis could the two polymorphic forms be distinguished (Fig. 9B). In doing so, kinetic analysis could be undertaken for both the molecule synthesis and the polymorphic transformation from a laboratory-based Raman spectroscopy approach (Fig. 9C). The Knoevenagel condensation demonstrates the difficulty associated with monitoring polymorphic transformations by Raman spectroscopy. This is the general case except where low-frequency Raman spectroscopy is used.[102, 103]

Instead, Raman spectroscopy has been more amenable to systems in which new intra- and (strong) intermolecular interactions are formed. For this reason, Raman spectroscopy has found particular strength in the study of organic reactive chemistry by ball mill grinding. Ćurić and coworkers demonstrated the ball milling Pd-mediated C—H bond activation of azobenzenes.[104] By monitoring changes in the Raman spectra in real time, it was shown that the final product (a dicyclopalladate) passed through a transient intermediate phase (a monopalladate). In these reactions the formation of strong Pd—N interactions led to large shifting of the N—C and N—N vibrational bands. These shifts were unmistakable over the course of ball milling and hence could be used to investigate the rate of transformation under various conditions. Similar Pd-catalyzed C—H bond activations have

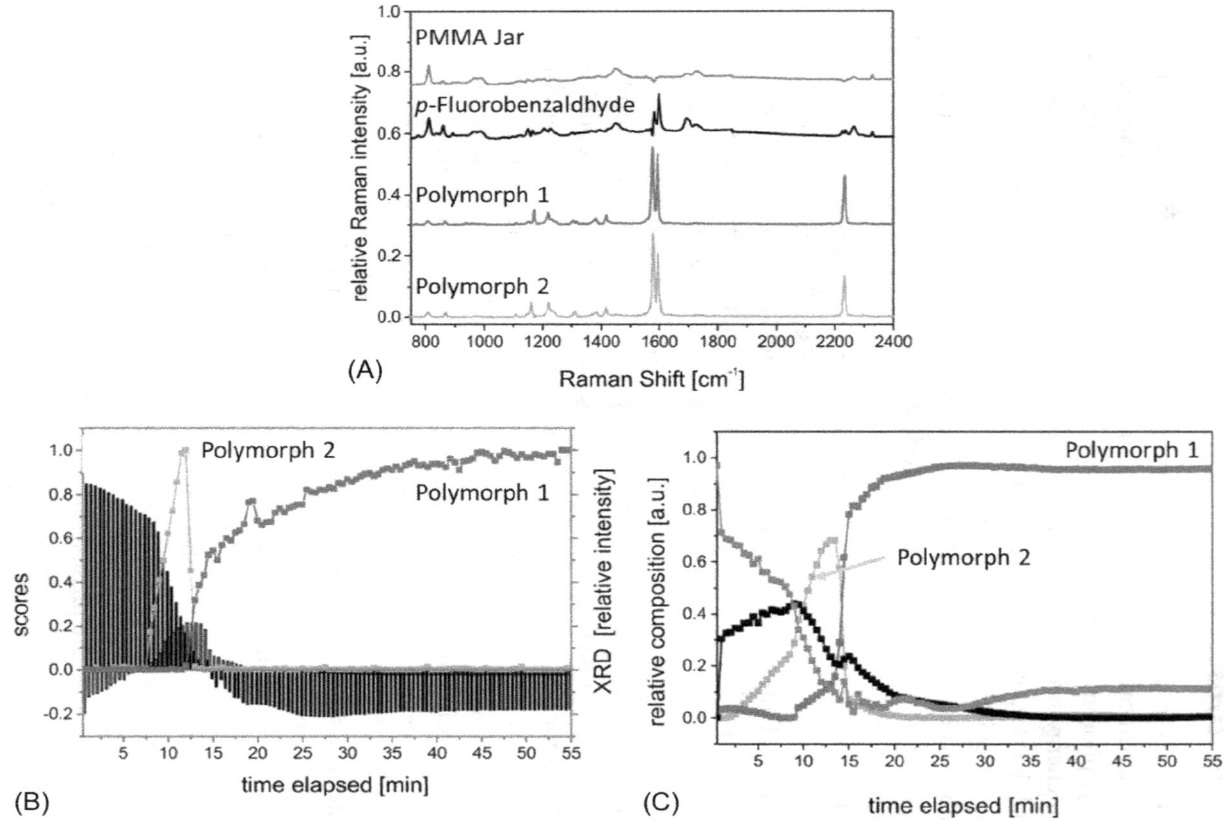

Fig. 9 Statistical analysis of time-resolved in situ Raman spectra to monitor the mechanochemical Knoevenagel condensation of *p*-fluorobenzaldehyde and malononitrile. (A) Raman spectra of the major reactant and milling jar (polymethylmethacrylate [PMMA]) and the two polymorphic forms of the product. (B) Principal component analysis of the Raman spectra (black and blue) and the integrated PXRD intensities for polymorphs 1 and 2 (red and green). (C) Composition of each phase: reactants (gray and black), polymorph 2 (green), and polymorph 1 (red). *(Figure adapted from Haferkamp, S.; Paul, A.; Michalchuk, A. A. L.; and Emmerling, F. Unexpected Polymorphism during a Catalyzed Mechanochemical Knoevenagel Condensation. Beilstein J. Org. Chem. 2019, 15, 1141–1148.)*

now been demonstrated under various liquid assisted and ionic liquid assisted grinding methods.[105]

The mechanochemical formation of coordination polymers has been also investigated by laboratory Raman spectroscopy. In the case of the reaction between cyanoguanidine (CNGE) and $CdCl_2$, time-resolved Raman spectra showed the formation of the coordination polymer by monitoring changes in the vibrational modes around the Cd coordination sphere.[75] With the aid of time-resolved spectroscopy, it was demonstrated how the stoichiometric and 2:1 CNGE-Cd complexes can be formed by ball mill grinding. Moreover, it was shown that the formation of the 2:1 product proceeded via the stoichiometric CNGE-Cd product only when a lighter milling ball was used. In contrast, heavier milling balls led to direct formation of the 2:1 product phase.

Laboratory-based Raman spectroscopy has also found use for monitoring a range of organic synthetic reactions. Strukil et al.[106] investigated the mechanochemical synthesis of aryl *N*-thiocarbamoylbenzotriazoles (Fig. 10). These compounds had been previously unattainable by solution-based processes. During the mechanosynthesis, however, the presence of unidentified vibrational bands in the Raman spectra indicated the transient formation of a highly speculated intermediate species. This allowed the authors to modify the mechanochemical synthesis and trap, for the first time, this new molecular species. The nucleophilic substitution of carbonyl groups was reported by Tireli et al.[107] Using as a model the reaction between 4-nitrobenzoyl azide[1] and 1,4-diaminobenzene (2) yielding N-(4-aminophenyl)-4-nitrobenzamide (3), time-resolved Raman spectroscopy allowed the authors to follow the kinetics of reaction as a function of liquid additive. It was found that the rate of reaction followed with the liquid basicity, thereby suggesting the liquid to act as a base catalyst for the transformation.

A key feature to the success of monitoring chemical transformations by Raman spectroscopy is the formation of new covalent bonds. In this light the formation of new cocrystals with strong hydrogen bonds have also been monitored successfully using time-resolved Raman spectroscopy. This was demonstrated by the mechanochemical formation of the pharmaceutical cocrystals between nicotinamide (NA) and suberic acid (SUB).[75] Depending on the ratio of initial reagents, either the stoichiometric or 2:1 NA:SUB products could be prepared. Owing to the similar Raman spectra of the 1:1 and 2:1 cocrystal phases, it proved difficult to unambiguously identify which cocrystal product was being formed based on Raman spectroscopy alone. However, very clear indications of reagent

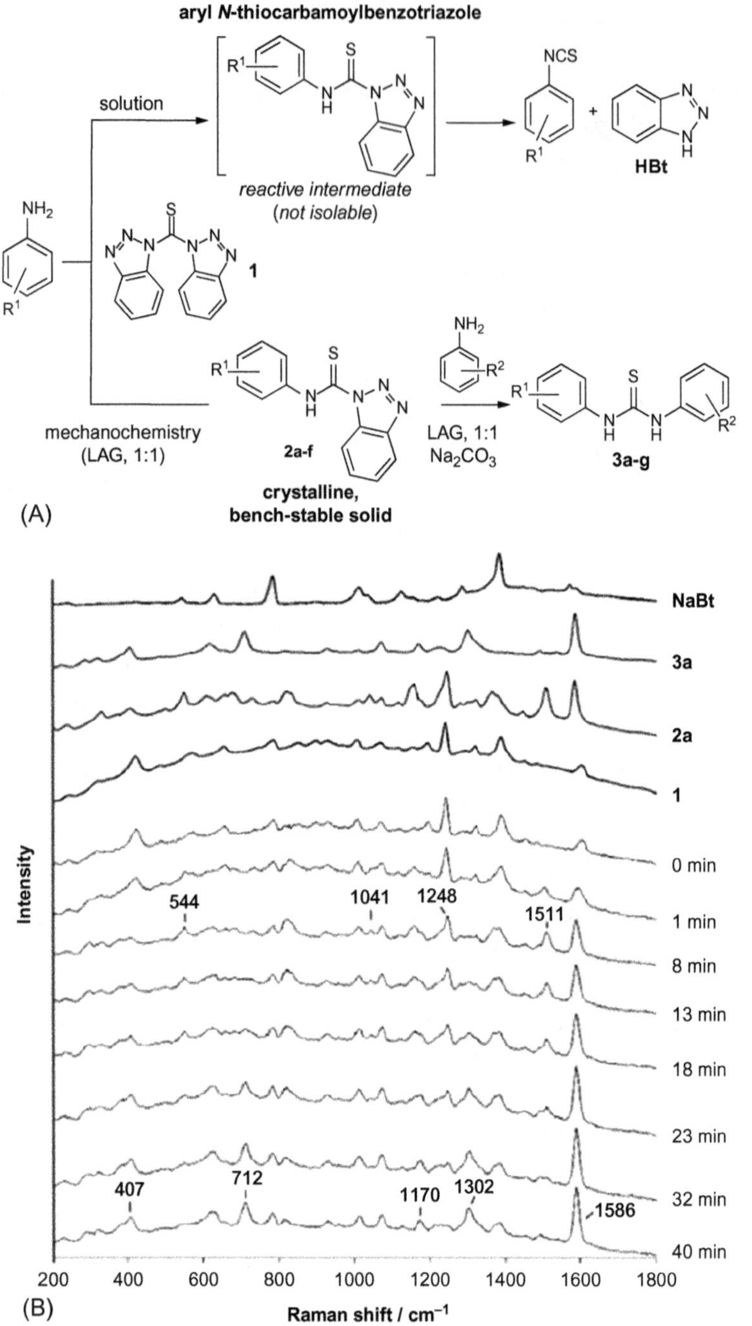

Fig. 10 Mechanochemical synthesis of aryl *N*-thiocarbamoylbenzotriazoles. (A) Reaction scheme comparing solution and mechanochemical synthetic routes. (2a) $R^1 = 4\text{-Br}$, (3a) $R^1 = R^2 = 4\text{-Br}$. (B) Time-resolved in situ Raman spectra collected during mechanochemical synthesis of (3a). *(Figure adapted from Štrukil, V.; Gracin, D.; Magdysyuk, O. V.; Dinnebier, R. E.; and Friščic, T. Trapping Reactive Intermediates by Mechanochemistry: Elusive Aryl N-Thiocarbamoylbenzotriazoles as Bench-Stable Reagents. Angew. Chem. Int. Ed. 2015, 54 (29), 8440–8443 copyright 2015 John Wiley and Sons.)*

consumption and product formation were visible. This allowed the authors to follow the effects of liquid additives on the rate of cocrystallization. Moreover, by analyzing the evolution of the Raman bands of the liquid additive (here acetonitrile), the authors were able to validate the direct participation of the liquid in the mechanochemical transformation.

Monitoring of mechanochemical transformations by Raman spectroscopy has not been limited to qualitative analyses. Using as a model system the liquid-assisted mechanochemical cocrystallization of ibuprofen and nicotinamide, the rate constants for the mechanochemical reaction were extracted from the time-resolved in situ Raman spectra (Fig. 11A).[109] The use of Raman spectroscopy was made largely possible by the formation of strong -NH...O hydrogen bonding during the cocrystallization associated with large shifts in the Raman signal. The rate constant was extracted across a range of temperatures. By fitting to the conventional Arrhenius equation, the variable temperature time-resolved Raman data allowed the apparent activation energy for the mechanochemical cocrystallization to be extracted. Despite yielding information on the bulk transformation (i.e., not at the elementary molecular scale), the extracted activation energy of 15 kJ mol^{-1} was obtained. Notably, this energy is on the order of hydrogen bonding and was therefore ascribed to the breaking of noncovalent intermolecular interactions in the reactant phases. Mechanochemical rate constants have been also extracted from ex situ analyses of covalent bond-forming reactions.[41] While aging of activated powders is not expected for covalent chemical reactions, liquid-assisted cocrystallizations are well known to progress without mechanical treatment. Quantitative assessment of ex situ rate constants for the latter therefore requires in situ time-resolved analysis. Using a second model cocrystallization, that of felodipine and imidazole, time-resolved in situ Raman spectroscopy was also used to extract mechanochemical rate constants as a function of milling ball size (Fig. 11B).[108] In this study the authors sought to elucidate whether the number of impacts is a rate-determining factor for the kinetics of the cocrystallization. The extracted rate constants suggest that the total ball mass is of little importance. Rather the kinetics obtained from time-resolved Raman spectroscopy showed that the reaction rate scales inversely with the number of milling balls.

A third milling parameter, the milling frequency, was additionally investigated by time-resolved in situ Raman spectroscopy. Rather than a mechanochemical cocrystallization, the synthesis of 2,3-diphenylquinoxaline from benzil and *o*-phenylenediamine was selected as a model system. The

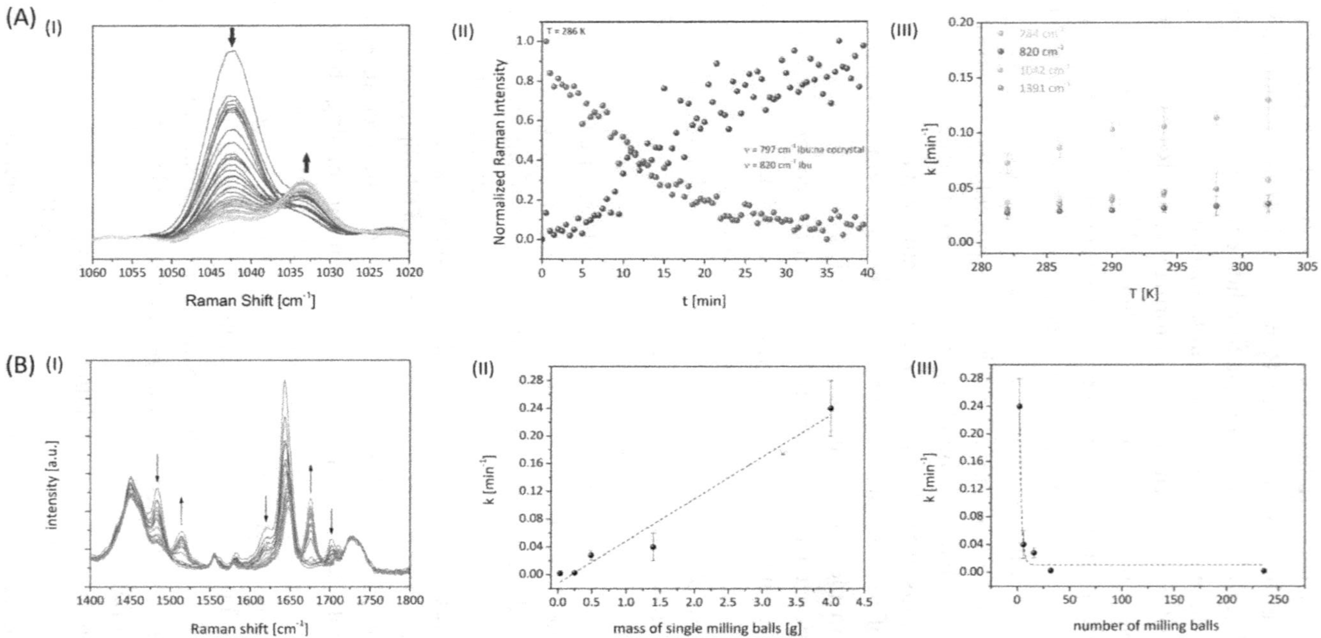

Fig. 11 Extrapolating mechanochemical rate constants by in situ time-resolved Raman spectroscopy. (A) Ball mill grinding cocrystallization of imidazole with felodipine.[108] (AI) Example time-resolved Raman spectra during ball mill grinding; (AII) Rate constants extracted for different milling ball mass. (AIII) Rate constants extracted for different number of milling balls. (B) Ball mill grinding cocrystallization of ibuprofen with nicotinamide.[109] (BI) Example time-resolved Raman spectra during ball mill grinding. (BII) Evolution of Raman bands with milling time at 298 K. (BIII) Rate constants extracted from Raman spectra as a function of milling temperature. *(Figure adapted from Fischer, F.; Fendel, N.; Greiser, S.; Rademann, K.; and Emmerling, F. Impact Is Important—Systematic Investigation of the Influence of Milling Balls in Mechanochemical Reactions. Org. Process Res. Dev. 2017, 21 (4), 655–659 and Fischer, F.; Wenzel, K.-J.; Rademann, K.; and Emmerling, F. Quantitative Determination of Activation Energies in Mechanochemical Reactions. PCCP 2016, 18, 23320–23325—Published by The Royal Society of Chemistry.)*

reaction rate was found to increase with milling frequency. At the lowest frequencies used (20 Hz), the transformation profile appeared almost linear, whereas the higher-frequency (30 Hz) profiles exhibited obvious sigmoidal structure. This qualitative difference in reaction profile was also noted during the ex situ monitoring of ZIF-8 formation by Raman spectroscopy.

In addition to elucidating the effects of external parameters on macroscopic mechanochemical transformations, Raman spectroscopy has been used to shed light on molecular-level mechanisms. In the study by Lukin and coworkers,[110] changes in the vibrational bands were followed during ball mill grinding to identify the presence and rate of hydrogen-deuterium isotopic exchange. As a model system, benzoic acid (BA) and paracetamol (PA) were selected (Fig. 12A). This system does not form a cocrystal product under ball milling conditions. If either of the reagents was deuterated prior to ball milling, traces of deuteration appeared in both phases upon milling. This suggested dynamic interactions between particle surfaces during ball milling, even in macroscopically noninteracting systems. The same dynamic transfer was also found in the reactive system, BA +2-pyridone (PYO) (Fig. 12B). In this case, ball mill grinding led to formation of a cocrystal product, accompanied by rapid isotopic exchange. In a similar study the same authors comilled BA with ^{13}C-labeled BA (BA*). Again, time-resolved Raman spectroscopy showed rapid formation of the heterodimer BABA*. Hence, by using time-resolved in situ Raman spectroscopy, it was demonstrated that ball mill grinding leads to rapid and dynamic exchange between solid particles. This process facilitates the diffusion conditions required for chemical reactivity.

Alternative to Raman spectroscopy, infrared (IR) spectroscopy has also been developed as a process analysis tool for time-resolved in situ monitoring of mechanochemical transformations. Unfortunately, as an absorption-based technique, IR spectroscopy is technically less amenable to solid-state process monitoring and has been therefore considerably less popular than Raman spectroscopy. An excellent example of how IR spectroscopy has been used is that by Tanaka and coworkers.[111] Using the Resonant Acoustic Mixing (RAM) technology, the authors explored the dynamics of powder mixing via near-infrared spectroscopy. As a model system, mixing was performed on dry powder mixtures of lactose, potato starch, and theophylline. Near-infrared spectra were collected in real time during the RAM mixing process to assess the evolution of powder homogeneity. With the help of advanced chemometric spectral analysis, the authors could quantify the presence of each phase within the mixture, thereby assessing the powder

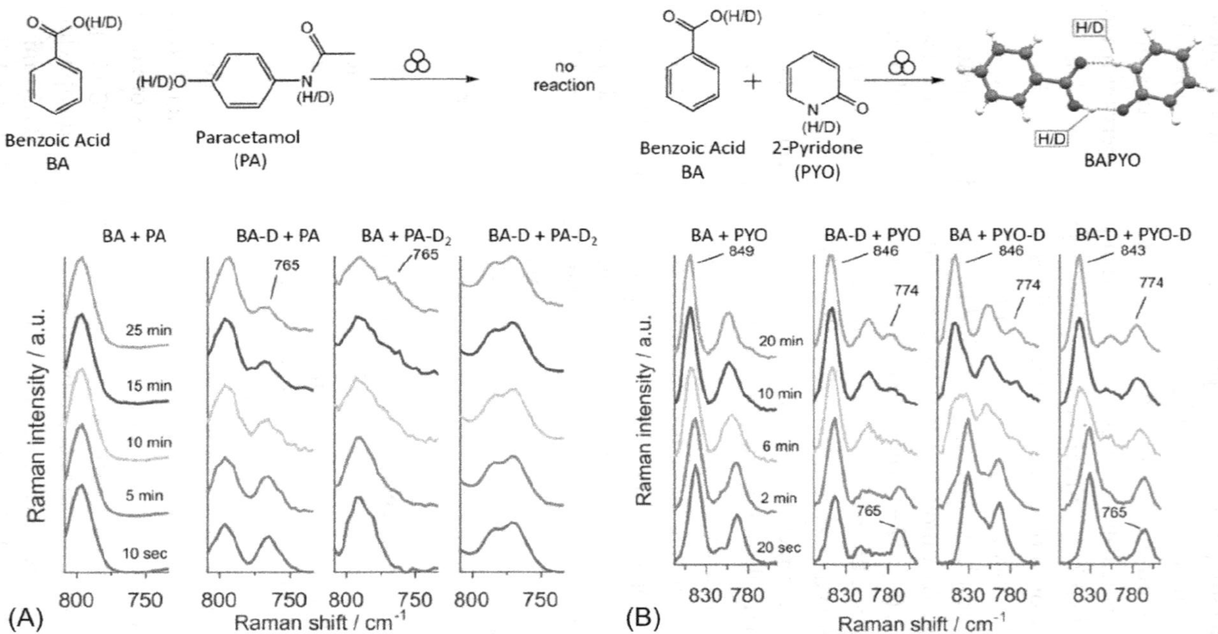

Fig. 12 Following isotopic exchange in organic materials using time-resolved in situ Raman spectroscopy. (A) Mechanochemical reaction scheme for failed attempt at mechanochemical cocrystallization of benzoic acid (BA) and paracetamol (PA), alongside the evolving Raman spectra that indicate an exchange of isotopically labeled protons. (B) Mechanochemical cocrystallization of benzoic acid (BA) and 2-pyridone (PYO) alongside Raman spectra which show exchange of isotopically labeled protons. *(Figure adapted from Lukin, S.; Tireli, M.; Stolar, T.; Barišić, D.; Blanco, M. V.; Di Michiel, M.; Užarević, K.; and Halasz, I. Isotope Labeling Reveals Fast Atomic and Molecular Exchange in Mechanochemical Milling Reactions. J. Am. Chem. Soc. 2019, 141 (3), 1212–1216. Copyright 2019 American Chemical Society.)*

composition. Remarkably, this allowed the authors to quantify the mixture to within 5% of the known composition.

4 Combining in situ time-resolved strategies: The tandem approach and beyond

Each analytical technique represents an additional dimension to process monitoring. X-ray diffraction, for example, is irreplaceable to monitor the evolution of crystalline phases, but does not provide insights into amorphous phases. Moreover, diffraction methods do not provide direct information on the evolution of temperature or pressure. In contrast, Raman spectroscopy detects the vibrational frequency of the molecule or material, thereby indicating changes in molecular interactions. Correspondingly, Raman spectroscopy is not restricted to the study of crystalline phases. Similarly, thermographic measurements are critical to monitor the evolution of temperature during mechanochemical transformations.

4.1 Combining PXRD and Raman spectroscopy

The importance of a multifaceted approach was recognized in the pioneering work by Emmerling and coworkers.[112] In the initial work the possibility to combine in situ time-resolved PXRD measurements with simultaneous Raman spectroscopy was demonstrated (Fig. 13). In a proof-of-principle study, four model reactions were investigated by the combined approach.[88, 113] In the mechanochemical synthesis of ZIF-8 (Fig. 13B), the characteristic (211) reflection of the cubic structure of ZIF-8 was observed almost immediately upon cogrinding the reagents ZnO and 2-methylimidazole under LAG conditions (in the presence of dimethylformamide and NH_4NO_3).[112] Time-resolved in situ PXRD measurements showed that the reaction completed within only two minutes of ball mill grinding. Whereas the PXRD data showed completion of ZIF-8 synthesis in under two minutes, the Raman spectra suggested slower relaxation of the 2-methylimidazole molecules into their equilibrium positions. Hence, PXRD alone is not capable of fully resolving the mechanochemical reactivity of the ZIF-8 synthesis. The additional information offered by combining PXRD and Raman spectroscopy is further demonstrated by the mechanochemical synthesis of a second framework material. In the mechanochemical synthesis of the metal-organic framework $(H_2Im)[Bi(BDC)_2]$ (H_2Im = imidazolium cation, BDC = benzenedicarboxylate) from bismuth(III) nitrate pentahydrate, terephthalic acid, and imidazole,

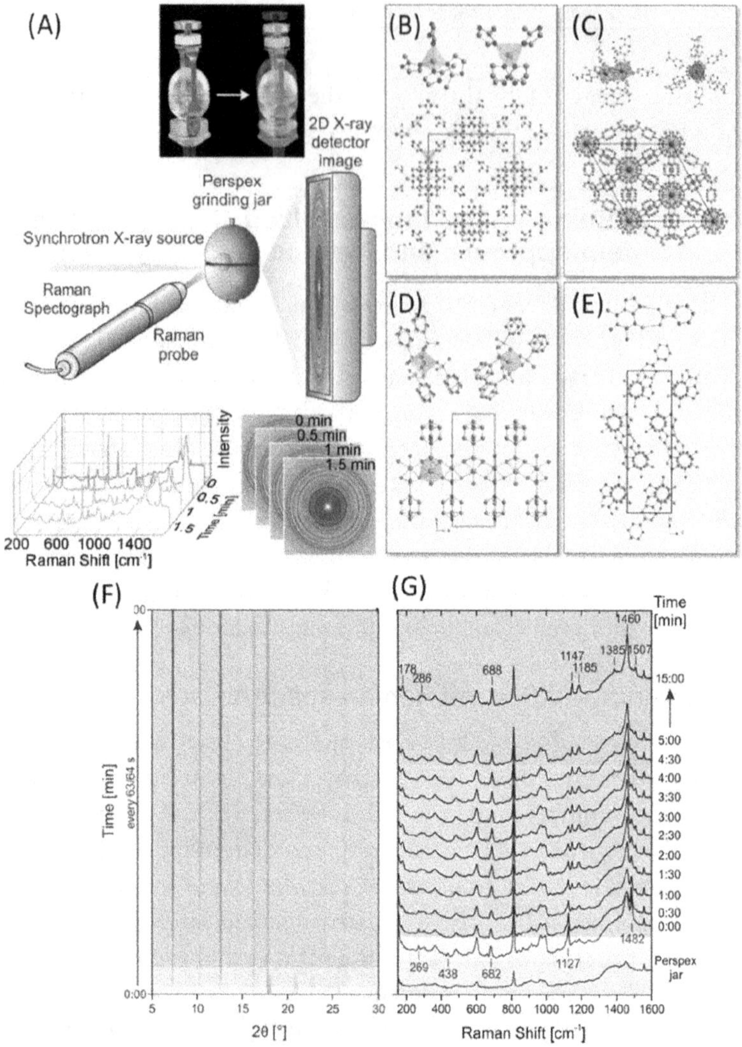

Fig. 13 Tandem approach to in situ time-resolved monitoring of mechanochemical transformations. (A) Schematic representation of the tandem approach combining synchrotron X-ray diffraction and Raman spectroscopy. (B) Coordination polyhedron of the Zn^{2+} cation (top) and unit cell of the metal-organic framework ZIF-8 along the crystallographic *b*-axis (bottom). (C) Coordination polyhedra of both Bi^{3+} cations in blue and violet (top) and unit cell of the metal-organic framework $(H_2Im)[Bi(BDC)_2]$ (H_2Im = imidazolium cation, BDC = benzenedicarboxylate) along the c-axis (bottom). (D) Coordination polyhedron of the Co^{2+} cation (top) and unit cell of the layered structure of the coordination polymer $CoPhPO_3$•H2O along the crystallographic c-axis (bottom). (E) Dimer building unit of theophylline and benzoic acid in the cocrystal theophylline: benzoic acid (top) and unit cell along the crystallographic *a*-axis (bottom). (F) In situ PXRD and (G) RS data for the mechanochemical synthesis of ZIF-8. The colors indicate the presence of reagents (yellow and orange) and ZIF-8 (red and blue) with imperfectly arranged 2-methylimidazolate molecules. At the bottom of the RS data, the spectrum for the empty PMMA jar (gray) is shown. *(Figure reproduced from Batzdorf, L.; Fischer, F.; Wilke, M.; Wenzel, K.-J. J.; and Emmerling, F. Direct in Situ Investigation of Milling Reactions Using Combined X-Ray Diffraction and Raman Spectroscopy. Angew. Chem. Int. Ed. 2015, 54 (6), 1799–1802. Copyright 2015 John Wiley and Sons.)*

time-resolved in situ data showed that the reaction begins within seconds (Fig. 13C). After less than one minute of milling, an unexpected intermediate phase was identified. The formation of the intermediate phase was further supported by the time-resolved Raman data, which suggested transient protonation of the imidazole molecule. Ex situ PXRD analysis confirmed these findings, showing the intermediate phase to be basic bismuth nitrate $Bi_6O_5(OH)_3(NO_3)_5(H_2O)_3$.[112]

The combination of PXRD and Raman spectroscopy was also demonstrated during the in situ investigation of the mechanochemical synthesis of a cobalt phosphonate (Fig. 13D).[112, 114] Cogrinding cobalt acetate with phenylphosphonic acid results in the final phosphonate structure via a stepwise process, as revealed by the in situ time-resolved investigation. An intermediate layered structure is formed after one minute of ball mill grinding, transforming to the final product after approximately three minutes. The PXRD data suggest that all phenylphosphonic acid is consumed with the formation of the layered intermediate phase. However, Raman data showed the presence of phenylphosphonic acid. It was assumed that this residual acid remains between the layers of the intermediate phase, facilitating conversion to the final product. The incorporation of phosphorylated glycinate as a linker in a 3D metal phosphonate framework, $Zn(O_3PCH_2NH_2CH_2CO_2).H_2O$, was also monitored by tandem time-resolved studies.[115] During the mechanosynthesis of this framework, both PXRD and Raman spectroscopy indicated the formation of a transient intermediate phase, which was identified as a 3:2 (Zn: linker) intermediate phase.

Owing to the considerable interest of the mechanochemical community in cocrystallization, the tandem monitoring approach has also been widely explored for these systems. For the synthesis of the cocrystal of theophylline and benzoic acid (Fig. 13E),[88] the time-resolved data from both PXRD and Raman spectroscopy indicated synthesis of the product without formation of intermediate phases.[112] The product reflections appeared after two minutes of milling time, with the transformation complete after six minutes of ball mill grinding. To date a number of cocrystallizations have been followed by combined in situ time-resolved PXRD and Raman spectroscopy.[84, 116, 117] While most studies have focused on the preparation of neutral cocrystals, the mechanism of formation of ionic cocrystals has also been investigated by tandem in situ time-resolved approaches. Linberg and coworkers[118] followed the mechanosynthesis of ionic cocrystals of glucose (Glc) and sodium salts, Glc_2NaX ($X = Cl$, Br, I). In all three cases, neat comilling a 2:1 mixture of the reactants led to formation of the respective salt, without any identifiable intermediate phase. In the case of the NaCl salt, however, mechanosynthesis favored formation of the hydrate, even under LAG with methanol. This was believed to result from inadvertent liquid-assisted grinding.[119]

The self-assembly of noncovalently bound materials is a remarkable phenomenon. This is particularly true under mechanochemical conditions where reactivity is often far from bulk equilibrium. As discussed earlier, much has been learned from in situ time-resolved monitoring about the mechanisms of binary cocrystal formation. Only recently the mechanochemical preparation of two ternary cocrystal phases was monitored by the combination of PXRD and Raman spectroscopy (Fig. 14).[117] In the case of the isonicotinamide (IA)-glutaric acid (GA)-pyrazinamide (PZA) ternary system, neat grinding at 40 Hz led initially to formation of the binary cocrystal phases, PZA:GA and IA:GA (Fig. 14C). Formation of both binary phases was also reported upon ball mill grinding. Continued grinding gradually converted the mixture to the X-ray phase pure ternary cocrystal product. In this system the Raman spectra of the binary and ternary cocrystal phases were not sufficiently unique. As such the Raman spectra could only be used to follow the conversion of reagent to product but could not distinguish the identity of the product phase. Surprisingly, formation of the binary phases could be suppressed by increasing the grinding frequency to 50 Hz. At this elevated frequency, no traces of either binary phases were observed, and only the ternary cocrystal product resulted (Fig. 14D). The LAG synthesis of ternary cocrystal with pyrazine-carboxylic acid (POA), IA:GA:POA, was also followed. Contrary to the typical effect of LAG, the formation rate of the ternary cocrystal product was decreased by LAG conditions. The time-resolved in situ PXRD studies showed that under LAG conditions, the binary intermediate phases were favored, hence decreasing the overall reaction rate. The initial evidence provided by the ternary cocrystal formation suggests not only that the overall rate of mechanochemical self-assembly will depend on milling parameters but that the apparent mechanism of mechanochemical self-assembly processes may as well. This is consistent with a similar phenomenon observed in the mechanochemical formation of Cd-based coordination polymers.[75]

The complementary strengths of Raman spectroscopy and PXRD make the combined approach well suited for the study of organic mechanosynthesis. In such cases the chemical transformations are followed at the molecular level by vibrational spectroscopy, whereas the

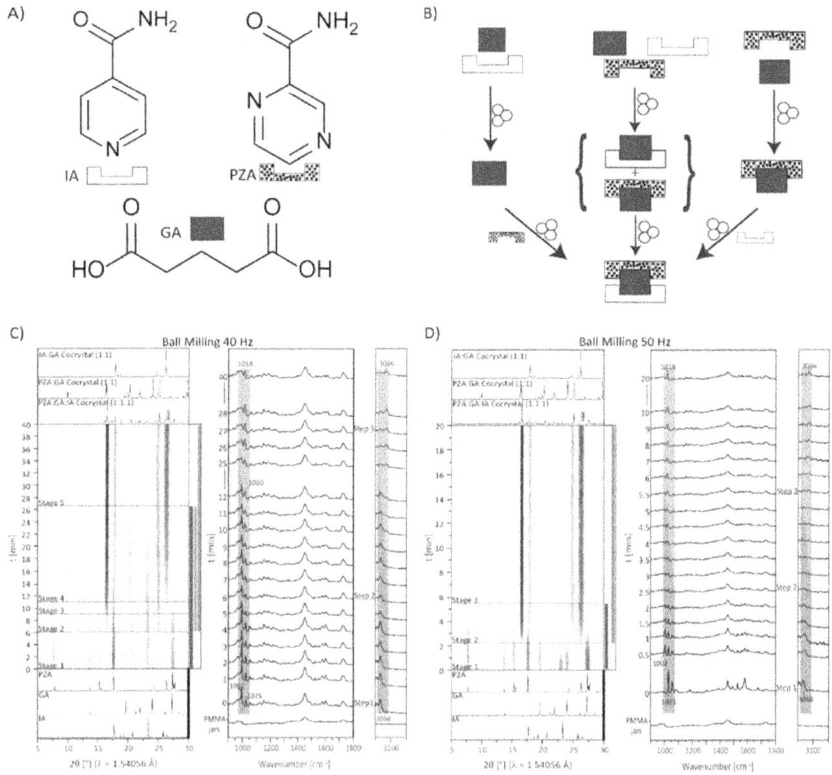

Fig. 14 Time-resolved in situ monitoring of the mechanochemical synthesis of ternary cocrystal pyrazinamide (PZA):glutaric acid (GA):isonicotinamide (IA). (A) Molecular structure of isonicotinamide, pyrazinamide, and glutaric acid. (B) Schematic representation of the reaction paths for the multicomponent cocrystal formation by ball mill grinding. Both binary phases are known to form upon grinding.[117] (C) In situ (left) PXRD profiles and (right) Raman spectra for the preparation of the ternary cocrystal by ball mill grinding at 40 Hz. (D) In situ (left) PXRD profiles and (right) Raman spectra for the preparation of the ternary cocrystal by ball milling at 50 Hz. For (C) and (D) coloring is used to indicate the presence of (blue, gray in print version) reactants, (green, dark gray in print version) binary phases, and (red, light gray in print version) the ternary cocrystal product. *(Figure adapted from Kulla, H.; Michalchuk, A. A. L.; and Emmerling, F. Manipulating the Dynamics of Mechanochemical Ternary Co-Crystal Formation. Chem. Commun. 2019, 1, 2–5 published by The Royal Society of Chemistry.)*

crystallization and polymorphism are identified by PXRD analysis. For the mechanochemical Knoevenagel synthesis between p-nitrobenzaldehyde with malononitrile,[120] the PXRD data suggested the initial stages of milling to reduce the reagent crystallinity. During this time, no chemical reaction was visible by either PXRD or Raman spectroscopy. Following this initial activation step, clear evolution of the Raman spectra was observed, followed by the growth of a new crystalline phase in the PXRD patterns. This lengthy induction period has also been noted for other chemical reactions, where it was suggested to manifest from the accumulation of mechanical energy during milling.[29] The example of the Knoevenagel condensation emphasizes the complementarity of the two methods. Whereas PXRD could not unambiguously discount a chemical reaction during the lengthy onset period, Raman spectroscopy could not clearly indicate the decrease in crystallinity over the initial stages of milling.

The power of combining Raman and X-ray scattering data has also been highlighted as a means to quantify the relative amounts of crystalline and noncrystalline phases during mechanochemical reactions.[121] Although X-ray-based methods offer a direct route to quantifying the phase composition of a mixture, the established techniques are only valid in cases where the structure of all phases in the mixture is known. Based on knowledge of the Raman spectra for each participating phase, it was possible to quantify the phase compositions by Raman spectroscopy. Comparison against quantitative analysis by X-ray methods showed important differences, which were ascribed to the fact that where X-ray methods are sensitive only to bulk crystalline phases, Raman spectroscopy methods provide insight into both crystalline and noncrystalline phases. Hence a comparison of the two approaches yields quantitation of the amorphous content of the system.

4.2 Combining PXRD and thermography

The importance of temperature in controlling mechanochemical transformations has been discussed for decades in the mechanochemical literature.[41, 42, 122] For example, Mack et al.[123] reported a correlation of the rate of a Diels-Alder reaction in solution to the rate of the same reaction under mechanochemical conditions. Comparing different starting materials and the corresponding calculated activation energies for each reaction, they were able to assign a maximum energy output for the used mill between 95.4 and 111.7 kJ mol^{-1}. The authors conclude that the conditions within the mill are comparable with those of the same reaction performed in solution: performing a reaction in solution at 90°C for 16 h led to the same yield as a mechanochemical reaction in a stainless steel vessel with one milling ball. Additional to external control of temperature—as is often the case—is the role of thermal feedback during mechanochemical processing. The mechanical work imparted by the mechanochemical reactor is in part converted to heat. The conversion to heat is greater at higher frequencies. The importance of monitoring temperature during mechanochemical processing has been reported previously, both during ball mill grinding and postgrinding. Takacs et al.[124] measured the temperature of milling jars and milling balls after mechanochemical reactions in different mills, a mixer mill and a planetary mill. In the mixer mill case, temperatures were recorded between 40°C and 80°C and for the planetary mill case up to 237°C was observed.

While thermocouple-based monitoring approaches are more common in solution systems, the intense mechanical force present during mechanochemical reactions often damages the thin thermocouple wires. Instead, successful monitoring of mechanochemical transformations has been made using infrared thermal cameras. Thermographic measurements performed using these cameras are restricted to monitoring the external temperature of the milling vessel. Recent studies using a series of thermocouples have, however, shown very little difference between the bulk temperature at the internal and external walls during mechanical treatment.[83]

The combination of PXRD and infrared thermography was combined to study the formation of a cement hydrate phase, Ettringite ($[Ca_3Al(OH)_6]_2\cdot(SO_4)_3\cdot 26H_2O$), during grinding.[125] The obtained in situ PXRD data of one-step process of stoichiometric amounts of $Ca(OH)_2$, $Al(OH)_3$, $CaSO_4\cdot 2H_2O$, and water are shown in Fig. 15A. Only 34% of Ettringite was formed after 2 h of grinding using the Rietveld method. To improve the yield of the product, the starting materials were milled without the addition of water (Fig. 15B). After 30 min of premilling the reflections of the starting materials disappear. After 90 min of neat grinding, a stoichiometric amount of water was added, and the reaction was prolonged for 30 min. Immediately after the addition of water, Bragg reflections corresponding to Ettringite appeared. At the end of the reaction, 94% of Ettringite was formed. Monitoring the temperature showed a steep decrease directly after the addition of water (Fig. 15C) followed by an increase in the temperature profile during the formation of Ettringite. The authors suggested that the exothermic reaction is possibly due to the activation of the starting materials by pregrinding.

The mechanochemical synthesis of manganese phosphonate from manganese(II) acetate tetrahydrate (MA) with *N,N*-bis(phosphonomethyl)glycine (BPG) (Fig. 16) offers another excellent example of the strength afforded by coupling PXRD and thermographic measurements.[126] Within less than a minute of ball mill grinding, all Bragg reflections of the starting phases were lost, thereby rendering PXRD analysis unhelpful. However,

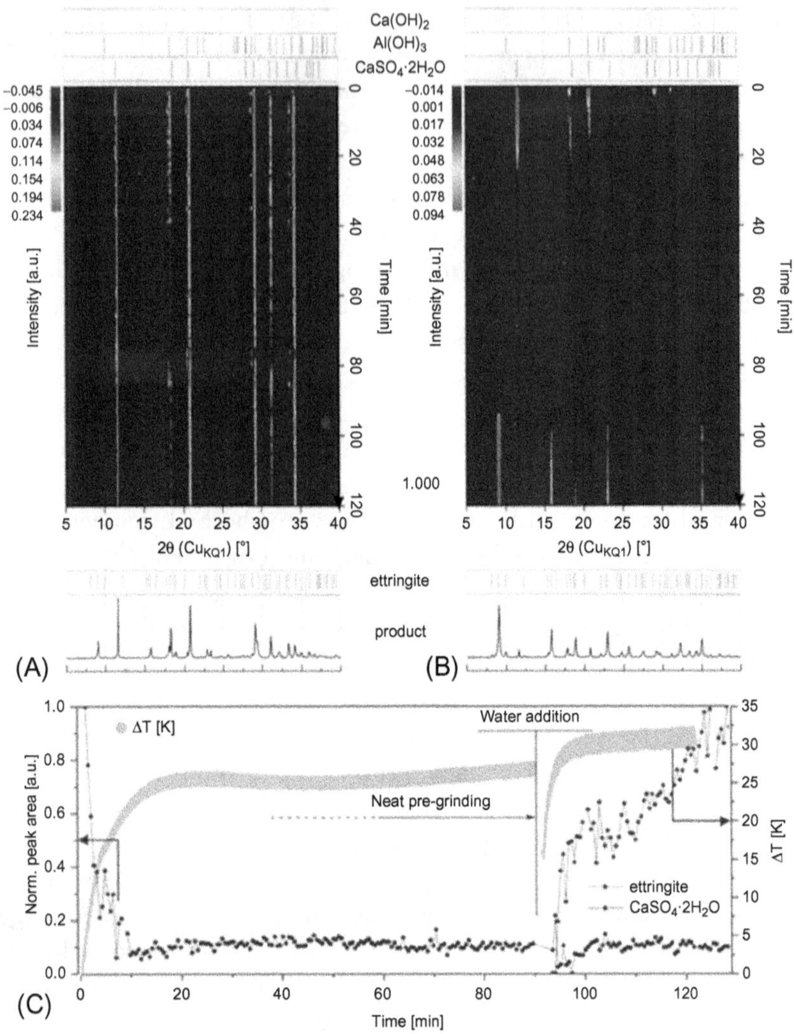

Fig. 15 In situ time-resolved monitoring of Ettringite formation during ball mill grinding. Time-resolved in situ PXRD profiles for (A) one-step mechanochemical synthesis with direct addition of water at the beginning and (B) two-step mechanochemical synthesis of Ettringite. The starting materials were pre milled for 90 min under neat grinding conditions. After that, water was added to the reaction mixture. (C) Left: progress of the integral Bragg peak intensity of $CaSO_4{\cdot}2H_2O$ (black) and Ettringite (blue, gray in print version) with time and the corresponding temperature profile (right, blue curve) for the two-step synthesis. *(Figure adapted from Stroh, J.; Ali, N. Z.; Maierhofer, C.; and Emmerling, F. Ettringite via Mechanochemistry: A Green and Rapid Approach for Industrial Application. ACS Omega 2019, 4 (1), 7734–7737. Copyright 2019 American Chemical Society.)*

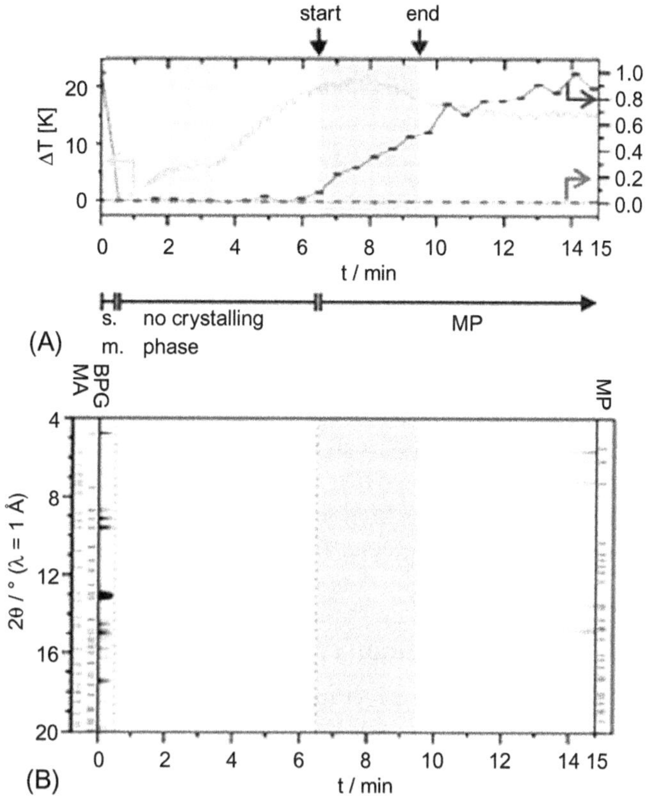

Fig. 16 Time-resolved in situ monitoring of manganese phosphonate synthesis by ball mill grinding. (A) Time-resolved thermographic measurements by infrared thermography. Gray line shows evolution of temperature, and integrated PXRD peak areas are indicated for (red, light gray in print version) manganese(II) acetate tetrahydrate (MA) and (blue, dark gray in print version) manganese phosphonate (MP). (B) Time-resolved in situ PXRD diffraction patterns. Color highlights indicate major events and are used to correlate data sets. *(Figure adapted from Kulla, H.; Haferkamp, S.; Akhmetova, I.; Rçllig, M.; Maierhofer, C.; Rademann, K.; and Emmerling, F. In Situ Investigations of Mechanochemical One-Pot Syntheses. Angew. Chem. Int. Ed. 2018, 57, 5930–5933. Copyright 2018 John Wiley and Sons.)*

with prolonged milling, there were clear indications of thermal effects. The initial rise in temperature was slow, suddenly becoming faster after about four minutes of grinding (orange in Fig. 16) This was suggested to result from the expulsion of water from the starting reagents, affecting both the heat capacity and rheology of the material. Slowly, as the thermographic measurements reached a plateau, indications of crystalline phases reappeared in the PXRD data. Hence, in the absence of crystalline material, the time-resolved in situ thermographic measurements were also able to provide mechanistic insights into the mechanochemical process.

The combination of thermography and PXRD also proved crucial to identifying the dominant origin of temperature increases during ball mill grinding.[127] During the mechanochemical synthesis of ZIF-8, the time-resolved PXRD data allowed the authors to correlate increase in measured temperature to known heats of formation for the evolving chemical system. Furthermore, by numerically modeling the rate of heat transfer out of the milling vessel, it became possible to identify the relative contribution of reaction heat and friction to the total thermal evolution of the milling vessel. Interestingly, it was found that the reaction heat generally contributed negligibly to the total heat evolution of the system. Instead, nearly the entire heating was due to friction between the sample and the milling assembly. Moreover the authors elucidated that changes in the sample properties could result in marked changes in the frictional heating of the system, thereby leading to unexpected temperature profiles.

4.3 Combining PXRD, Raman spectroscopy, and thermography

Following from the successful combinations of two process analytical methods, the number of tandem in situ time-resolved techniques was expanded to three.[126] The threefold tandem in situ time-resolved approach, combining Raman spectroscopy, PXRD, and thermography, is exemplified by the mechanochemical Knoevenagel reaction between p-nitrobenzaldehyde and malononitrile (Fig. 17).[126] Within only a minute of ball mill grinding, a sudden increase in temperature is observed (c. 20 K), followed by rapid loss in crystallinity of the reactant phases as observed by tandem PXRD. As the crystallinity continued to decrease, no change in the temperature was observed. Moreover, Raman spectroscopy did not suggest any chemical transformations were occurring during this time. After approximately 30 min ball mill grinding, all three techniques suggested a sudden and rapid reaction, with immediate crystallization of the product phase. A sharp temperature spike (blue in Fig. 17AI) was observed until the reaction finished and the temperature returned to its previous plateau temperature. Although the structural information is provided by the PXRD and Raman spectroscopic analysis, the thermal data confirm the rapid onset of this apparently exothermic reaction. Moreover, indication of the rise in temperature may be critical in understanding the rapid—and potentially self-accelerating—nature of this transformation. In the same

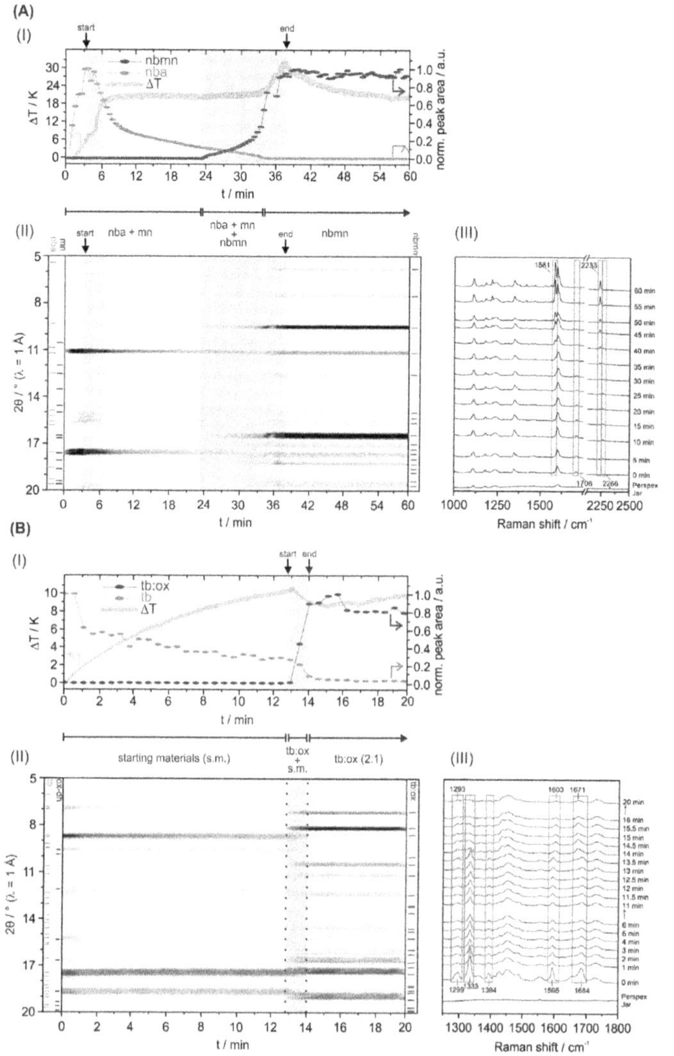

Fig. 17 Combining multiple techniques to following mechanochemical reactions in situ and in real time. (A) Mechanochemical Knoevenagel condensation, monitored by (I) thermography, (II) X-ray diffraction, and (III) Raman spectroscopy. (B) Mechanochemical cocrystallization of theobromine (TB) and oxalic acid dihydrate (OX-DH) to yield the 2:1 cocrystal product 2 TB:OX. (I) the temperature is monitored over the course of the reaction (gray) and compared against integral Bragg peak intensities (blue (dark gray in print version) and red (light gray in print version)). (II) The corresponding X-ray powder diffraction profile as a function of milling time. (III) Raman spectra. Colors are used to highlight points of interest between data sets. *(Reprinted with permission from Kulla, H.; Haferkamp, S.; Akhmetova, I.; Rçllig, M.; Maierhofer, C.; Rademann, K.; and Emmerling, F. In Situ Investigations of Mechanochemical One-Pot Syntheses. Angew. Chem. Int. Ed. 2018, 57, 5930–5933. Copyright 2018 John Wiley and Sons.)*

study the cocrystallization of organic molecules oxalic acid and theobromine was also monitored by the threefold tandem approach (Fig. 17B). Over the course of the milling, the temperature was found to rise slowly, as the crystallinity of theobromine was slowly lost. As the temperature rose to its maximum (ca. 10 K above room temperature), a sudden onset of cocrystallization was observed by all three methods. Indeed, all structural information is again attainable through the spectroscopic and diffraction data. However, it is interesting to again note the association of rapid product crystallization at the point of maximum temperature. This additional insight into the evolution of the macroscopically evolving system is crucial for determining the mechanism of action of mechanochemical transformations. All three techniques offer complementary information, mutually reinforcing and expanding on the interpretations obtained by each individual technique. As a relatively novel approach, few examples are so far present in the literature. However, the use of a larger number of tandem techniques is expected to become the new standard for time-resolved in situ monitoring of mechanochemical reactions.

4.4 Laboratory-based tandem approaches

Due to the limited access to synchrotron radiation facilities, further attempts have been made at detecting intermediates in situ by combining methods available in the laboratory. Thermography provides integral information on the temperature changes during the reaction without influencing the reaction conditions. Its combination with Raman spectroscopy to obtain structural information has rendered tandem laboratory analysis of mechanochemical transformations possible.

The thermal profile associated with mechanochemical transformations involves a convolution of many processes.[127] Most notably, these include the heating from the mechanical energy itself (e.g., from frictional heating and ball-wall impacts), particle comminution, and endo- or exothermic events occurring in the powder. To capture the effects of mechanical heating, the temperature profile of SiO_2 powder was measured during grinding using a thermographic camera, gray line in Fig. 18A. The bulk temperature profile for the milling experiment with SiO_2 powder exhibits a temperature increase $\Delta T = 7.5$ K over the course of the experiment.[128]

Using the thermographic infrared camera, coupled with Raman spectroscopy, the cocrystallizations of theobromine with oxalic acid (TB:OX) and pyrazinamide with oxalic acid (PZA:OX) were monitored in situ

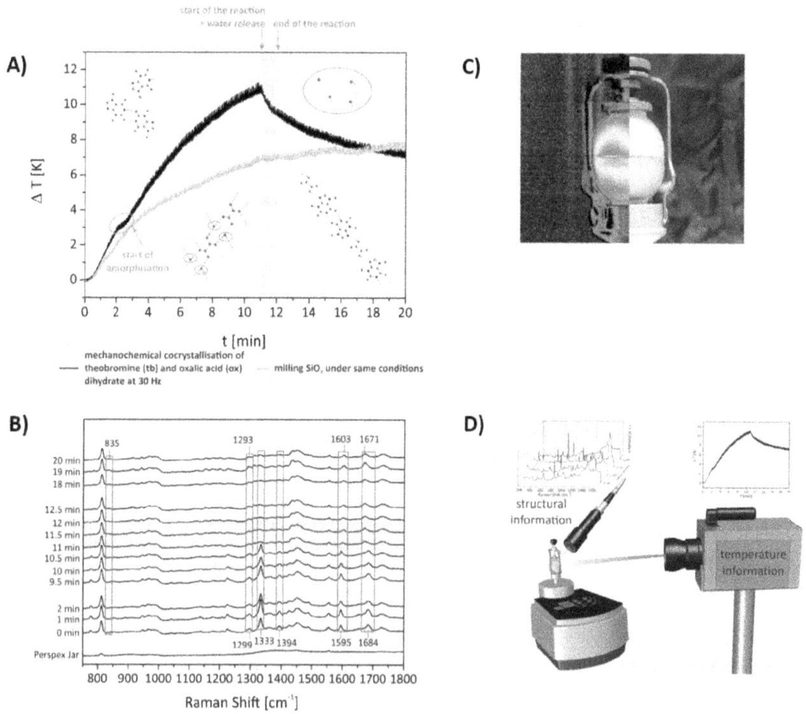

Fig. 18 Experimental setup used for measuring time-resolved Raman spectra in combination with thermograms during mechanochemical reactions. (A) Raman spectra were collected every 30 s; thermograms were recorded with a frequency of 50 Hz or 60 Hz. (B) Temperature development in the milling jar during the formation of the 2:1 cocrystal of theobromine and oxalic acid (OX) as a function of time (black line), compared the temperature curves obtained during milling pure SiO_2 powder (gray line) under the same conditions. The data are compared with simultaneously measured Raman spectra (C) and in situ synchrotron XRPD data (D) under the same milling conditions recorded in a separate experiment. The start of the reaction can be deduced from the appearance of the band at 835 cm^{-1} assigned to the C—C stretching of OX in the cocrystal. The end of the reaction can be determined from shifts of the bands at 1299, 1333, 1394, 1595, and 1684 cm^{-1}, which were attributed to the C—N and C=O stretching vibrations of pure TB: the start of the amorphization of reactants (orange, light gray in print version) and the reaction period (blue area, gray area in print version) are indicated. *(Adapted from reference Kulla, H.; Wilke, M.; Fischer, F.; Röllig, M.; Maierhofer, C.; and Emmerling, F. Warming up for Mechanosynthesis—Temperature Development in Ball Mills during Synthesis. Chem. Commun. 2017, 53 (10), 1664–1667 with permission from The Royal Society of Chemistry.)*

(Fig. 18A).[128] In the case of TB:OX, a 2:1 M ratio mixture of TB and OX dihydrate was ground for 20 min at 30 Hz. The time-resolved monitoring shows a kink in the thermal profile at approximately two minutes. Akin to other thermal-based methods, such features often indicate a change in the heat capacity of the material. Moreover, previous in situ PXRD studies on the TB:OX synthesis suggest amorphization at this time,[129] leading the authors to correlate this change in thermal profile to the onset of amorphization. Following this kink the temperature continues to rise until a maximum ($\Delta T = 11.3$ K) is reached. At this point a Raman band at 835 cm^{-1} appears (assigned to the C—C stretching of OX in the cocrystal) indicates the start of the short reaction period. The subsequent decrease in temperature of the jar was suggested to result from the cooling effect of the released crystallization water.

In a related example the same tandem approach was used to monitor the mechanochemical cocrystallization of the between cocrystal pyrazinamide and oxalic acid.[128] Using the mechanochemical approach the cocrystal was formed within 20 min ball mill grinding at 50 Hz. From the thermographic data collected in real time during the mechanochemical reaction (Fig. 19A), a steady increase in temperature was observed to a maximum temperature of $T = 36$ K. Tandem time-resolved in situ Raman spectra also indicated continuous formation of the cocrystal product in line with thermographic measurements, indicated by the growth of the vibrational band at 458 cm^{-1} and loss of the isolated pyrazinamide vibrational band at 415 cm^{-1}. Importantly, in the case of the cocrystallization of oxalic acid and pyrazinamide, the accumulated temperature is much larger for the cocrystallization process than when the pure component (here oxalic acid anhydrate) was milled alone. It was suggested that this additional heat relates to the exothermic effects of crystal formation. A small shoulder in the temperature curve around 1.5 min was also suggested to signify an onset of amorphization. The continuous increase in temperature observed for the cocrystallization of oxalic acid with pyrazinamide contrasts the thermal profiles observed the cocrystallization of oxalic acid with theobromine (Figs. 18A and 17AI). This suggests a potential difference in the driving force to cocrystallization in each system and reveals an important strength of thermographic measurements.

A particularly striking example of the strength of laboratory thermographic measurements is shown by the mechanochemical synthesis of Cd phosphonate $Cd(HO_3PPh)_2$ ($CdPhP_2$) (Fig. 19B).[128] Although not strictly a tandem approach, these laboratory thermographic measurements were

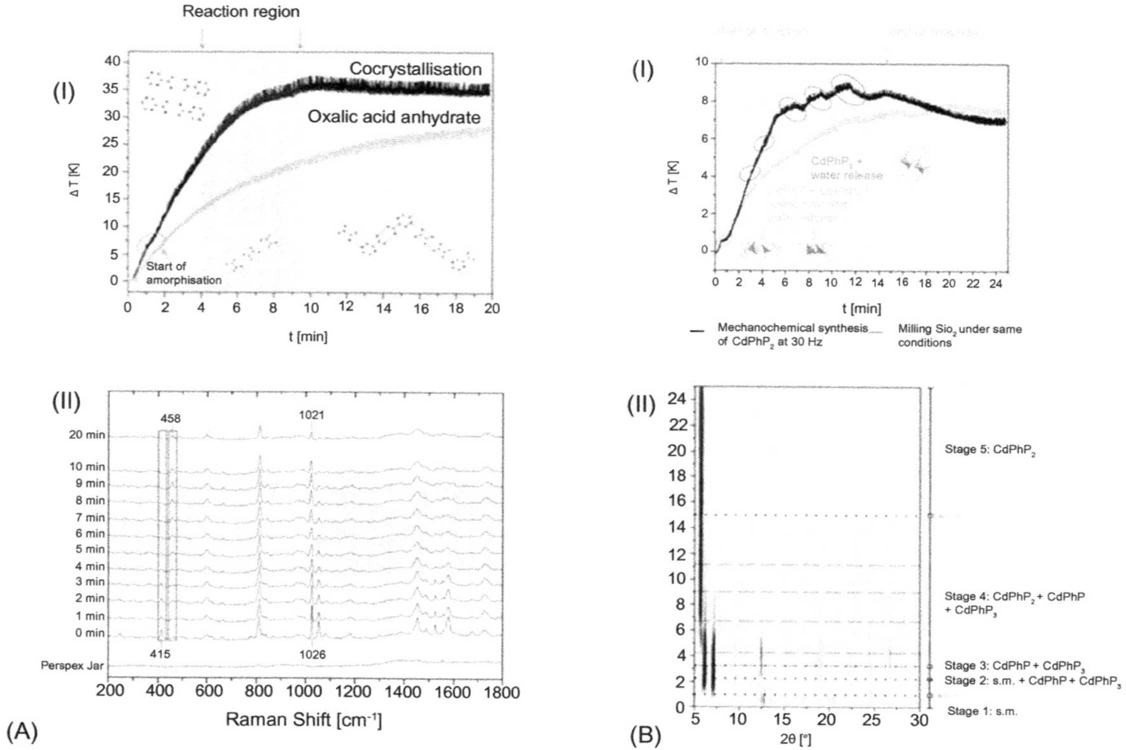

Fig. 19 (A) Monitoring the temperature evolution of mechanochemical cocrystallization of pyrazinamide-oxalic acid using a thermal camera. Curves are shown for the ball milling of a stoichiometric mixture of pyrazinamide + oxalic acid at 50 Hz (black) and for oxalic acid anhydrate at 50 Hz (gray). (B) Mechanochemical synthesis of Cd phosphonate. (I) In situ thermographic measurements during the mechanochemical synthesis of $Cd(HO_3PPh)_2$ ($CdPhP_2$) as a function of time (black line) in comparison to the data of SiO_2 powder under the same conditions (gray line). The changes during the milling process are shown in orange circles for the formation of the intermediates and the release of acetic acid and water. The red circles symbolize the product formation/ release of water. The reaction period is given in blue (gray in print version). (II) In situ 2D plot of the PXRD data of the formation of $Cd(HO_3PPh)_2$ ($CdPhP_2$) where the blue area shows the reaction period. *(Reproduced from Kulla, H.; Wilke, M.; Fischer, F.; Röllig, M.; Maierhofer, C.; and Emmerling, F. Warming up for Mechanosynthesis—Temperature Development in Ball Mills during Synthesis. Chem. Commun. 2017, 53 (10), 1664–1667 with permission from The Royal Society of Chemistry.)*

coupled to previous time-resolved in situ PXRD studies, which suggested a number of structural transformation to occur over the course of this reaction (Fig. 19BII).[130] By means of thermographic measurements, indications of the same structural transformations were also observed. Within only 30 s of ball milling, a notable kink was observed in the temperature profile, suggested to indicate the onset of the reaction and formation of the intermediates CdPhP and $CdPhP_3$. The release of acetic acid and water was suggested to be responsible for the slight cooling effect observed across this reaction. Two additional thermal evens at approximately 3 min ($\Delta T = 4.3$ K) and 4 min ($\Delta T = 5.8$ K) were correlated initial growth of the final product phase. The subsequent "bumps" (i.e., maximum followed by decrease in temperature) were ascribed to the release of further water and corresponded to decrease in the intermediate phase intensities in the PXRD study. This exceptional level of mechanistic detail from laboratory thermographic data exemplifies the true potential of this straightforward analytical approach. However, as evidenced from previous text, the underlying mechanisms are difficult to interpret without additional structural probes such as Raman spectroscopy or PXRD.

In addition to its use for mechanistic insight, laboratory thermographic measurements have also been used to tightly control mechanochemical bulk temperatures. This is exemplified in a recent example by Cindro and coworkers.[131] Using thermocouples the temperature was measured in real time and used in a feedback loop to tightly control the milling vessel temperature. In this way the authors were able to tightly control the bulk temperature (to within a couple of K) and explore the effects of temperature on the mechanochemical LAG synthesis of amide-urea compounds. Additional structural information was obtained by use of tandem in situ time-resolved Raman spectroscopy. These continued developments and novel applications of process monitoring promise to greatly extend our capability to explore and exploit mechanochemical methods. In doing so the use of time-resolved in situ techniques will open the door to understanding how these remarkable technologies work.

5 Summary

Mechanochemistry has gained enormous momentum as an alternative approach to synthesize a wide range of molecules and materials. Without need for solvent, these environmentally benign technologies promise to revolutionize both academic and industrial processing strategies. This exceptional promise led the International Union of Pure and Applied Chemistry to dub mechanochemistry as one of the *ten chemical innovations that will change our world*.[132] Unfortunately, despite the considerable interest in mechanochemical methods, their underlying mechanisms and guiding principles remain far from understood. As such, it is at present difficult to develop robust mechanochemical synthetic strategies a priori. Instead, immense trial-and-error efforts are required. Moreover, these efforts are not often transferrable between mechanochemical platforms. Hence the adoption of mechanochemistry to industry and academia remains slow.

Much of the difficulty with understanding mechanochemistry has surrounded the inability to follow transformations in real time. Instead, nearly all research to date has explored these solid-state transformations using stop-start ex situ analyses. Thus, by the time the reaction mixture is in fact measured, the transformation has long since finished or indeed has evolved away from the true mechanochemical product. Moreover, stop-start sampling strategies have been known to inadvertently affect the underlying mechanisms and kinetics, thereby leading researchers to erroneous conclusions.

With the goal of garnering control of mechanochemical transformations, researchers have sought to develop new strategies to follow these reactions in real time and in situ. Although early attempts were made, monitoring changes in temperature and pressure, very little structural information was obtained by these means. The development of time-resolved in situ X-ray diffraction and Raman spectroscopy methods have revolutionized mechanochemistry. Using these techniques, it is now possible to follow minute changes in crystallographic and molecular structure during the mechanochemical transformations. This has allowed unprecedented insights into the dynamics of mechanochemical reactions, has revealed the existence and role of intermediate and competitive phases in solid-state transformations, has hinted at the existence of intermediate amorphous phases, and has offered further knowledge as to how one can control mechanochemical reactions. Advances have been made in understanding mechanochemistry through the combined use of multiple in situ time-resolved techniques. This has included combinations of X-ray diffraction and Raman spectroscopy, Raman spectroscopy with thermography, X-ray diffraction with thermography, and a combination of all three approaches. By coupling techniques, one always obtains complementary and reinforced information regarding transformations and mechanisms. For example, while X-ray diffraction is critical to assess the evolution and transformation between crystalline phases, it offers little to no insight about noncrystalline phases. Hence, it is not straightforward to assess the onset of chemical transformations by X-ray diffraction. Instead, Raman spectroscopy is sensitive to changes in molecular structure and hence offers this missing information. Thermographic measurements provide additional detail regarding thermal driving forces in mechanochemical transformation and often real changes in the underlying reaction mechanism.

As time-resolved in situ methods continue to develop, new techniques will surely join the available toolbox. With refined data collection strategies and sampling procedures, an ever-increasing amount of information will be extracted from time-resolved in situ data. In this way the secrets underpinning mechanochemical reactions will become known, and the true potential of mechanochemical methods will be realized.

References

1. Takacs, L.; Carey-Lea, M. The Fater of Mechanochemistry. *Bull. Hist. Chem.* **2003**, *28* (1), 26–34.
2. Carey-Lea, M. Transformations of Mechanical into Chemical Energy. *Am. J. Sci.* **1894**, *47* (281), 377.
3. Carey-Lea, M. Researches on the Latent Image. *Br. J. Photogr.* **1866**, *13*, 84.
4. Takacs, L. The Historical Development of Mechanochemistry. *Chem. Soc. Rev.* **2013**, *42* (18), 7649–7659.
5. Boldyrev, V. V. Mechanochemistry in Siberia. *Her. Russ. Acad. Sci.* **2018**, *88* (2), 142–150.
6. Boldyreva, E. V. Mechanochemistry of Organic Solids: Where Are We Now? In *Frontiers in Mechanochemistry and Mechanical Alloying*; Kumar, R., Srikanth, S., Mehrotra, S. P., Eds.; CSIR-National Metallurgical Laboratory: Jamshedpur, India, 2011; pp. 17–30.
7. Fabbiani, F. P. A.; Pulham, C. R. High-Pressure Studies of Pharmaceutical Compounds and Energetic Materials. *Chem. Soc. Rev.* **2006**, *35* (10), 932–942.
8. Zakharov, B. A.; Boldyreva, E. V. High Pressure: A Complementary Tool for Probing Solid-State Processes. *CrstEngComm* **2019**, *21* (1), 10–22.
9. Tumanov, I. A.; Achkasov, A. F.; Boldyreva, E. V.; Boldyrev, V. V. Following the Products of Mechanochemical Synthesis Step by Step. *CrstEngComm* **2011**, *13* (7), 2213.
10. Avrami, L.; Hutchinson, R. The Sensitivity to Impact and Friction. In *Energetic Materials 2: Technology of the Inorganic Azides*; Fair, H., Walker, R. F., Eds.; Plenum Press: New York, 1977; pp. 111–159.
11. Pan, S.; Zhang, Z. Fundamental Theories and Basic Principles of Triboelectric Effect: A Review. *Friction* **2019**, 7 (1), 2–17.
12. Fernández-Bertran, J. Mechanochemistry : An Overview. *Pure Appl. Chem.* **1999**, *71* (4), 581–586.
13. Boldyrev, V. V. Mechanochemistry of Inorganic Solids. *Proc. Indian Natl. Sci. Acad.* **1986**, *52* (1), 400–417.
14. Michalchuk, A. A. L.; Tumanov, I. A.; Boldyreva, E. V. Complexities of Mechanochemistry: Elucidation of Processes Occurring in Mechanical Activators via Implementation of a Simple Organic System. *CrstEngComm* **2013**, *15* (32), 6403–6412.

15. Tumanov, I. A.; Achkasov, A. F.; Myz, S. A.; Boldyreva, E. V.; Boldyrev, V. V. Different Effect of Impact and Shear Mechanical Treatment on Mechanochemical Cocrystallization of Piroxicam and Succinic Acid. *Dokl. Chem.* **2014**, *457* (2), 154–159.
16. Konar, S.; Michalchuk, A. A. L.; Şen, N.; et al. A High-Pressure Study of Two Polymorphs of TNT Using Neutron Powder Diffraction and DFT Methods. *J. Phys. Chem. C* **2019**, *123*, 26095–26105.
17. Boldyrev, V.; Tkácová, K. Mechanochemistry of Solids : Past, Present, and Prospects. *J. Mater. Synth. Process.* **2000**, *8* (3/4), 121–132.
18. Jenei, Z.; Liermann, H. P.; Husband, R.; Méndez, A. S. J.; Pennicard, D.; Marquardt, H.; O'Bannon, E. F.; Pakhomova, A.; Konopkova, Z.; Glazyrin, K.; et al. New Dynamic Diamond Anvil Cells for Tera-Pascal per Second Fast Compression X-Ray Diffraction Experiments. *Rev. Sci. Instrum.* **2019**, *90* (6), 065114.
19. Coleman, A. L.; Gorman, M. G.; Briggs, R.; McWilliams, R. S.; McGonegle, D.; Bolme, C. A.; Gleason, A. E.; Fratanduono, D. E.; Smith, R. F.; Galtier, E.; et al. Identification of Phase Transitions and Metastability in Dynamically Compressed Antimony Using Ultrafast X-Ray Diffraction. *Phys. Rev. Lett.* **2019**, *122* (25), 255704.
20. Landers, A. G.; Brill, B. Pressure-Temperature Dependence of the B-D Polymorph Interconversion in HMX. *J. Phys. Chem.* **1980**, *84*, 3573–3577.
21. Bishop, M. M.; Chellappa, R. S.; Liu, Z.; Preston, D. N.; Sandstrom, M. M.; Dattelbaum, D. M.; Vohra, Y. K.; Velisavljevic, N. High Pressure-Temperature Polymorphism of 1,1-Diamino-2,2-Dinitroethylene. *J. Phys. Conf. Ser.* **2014**, *500* (5), 052005.
22. Sobczak, S.; Drożdż, W.; Lampronti, G. I.; Belenguer, A. M.; Katrusiak, A.; Stefankiewicz, A. R. Dynamic Covalent Chemistry under High-Pressure: A New Route to Disulfide Metathesis. *Chem. A Eur. J.* **2018**, *24* (35), 8769–8773.
23. Tumanov, I. A.; Achkasov, A. F.; Boldyreva, E. V.; Boldyrev, V. V. About the Possibilities to Detect Intermediate Stages in Mechanochemical Synthesis of Molecular Complexes. *Russ. J. Phys. Chem. A.* **2012**, *86* (6), 1014–1017.
24. Descamps, M.; Willart, J. Transformation of Pharmaceutical Compounds upon Milling and Comilling: The Role of Tg. *J. Pharm. Sci.* **2007**, *96* (5), 1398–1408.
25. Descamps, M.; Willart, J. F. Perspectives on the Amorphisation/Milling Relationship in Pharmaceutical Materials. *Adv. Drug Deliv. Rev.* **2016**, *100*, 51–66.
26. Caira, M. R.; Nassimbeni, L. R.; Wildervanck, A. F. Selective Formation of Hydrogen Bonded Cocrystals between a Sulfonamide and Aromatic Carboxylic Acids in the Solid State. *J. Chem. Soc. Perkin Trans.* **1995**, *2* (12), 2213–2216.
27. Avilés, M. A.; Córdoba, J. M.; Sayagués, M. J.; Gotor, F. J. Tailoring the Band Gap in the ZnS/ZnSe System: Solid Solutions by a Mechanically Induced Self-Sustaining Reaction. *Inorg. Chem.* **2019**, *58* (4), 2565–2575.
28. Tröbs, L.; Emmerling, F. Mechanochemical Synthesis and Characterisation of Cocrystals and Metal Organic Compounds. *Faraday Discuss.* **2014**, *170*, 109–119.
29. Belenguer, A. M.; Michalchuk, A. A. L.; Lampronti, G. I.; Sanders, J. K. M. Understanding the Unexpected Effect of Frequency on the Kinetics of a Covalent Reaction under Ball-Milling Conditions. *Beilstein J. Org. Chem.* **2019**, *15*, 1226–1235.
30. Stroh, J.; Feiler, T.; Zafar Ali, N.; Minas da Piedade, M. E.; Emmerling, F. Mechanistic Insights into a Sustainable Mechanochemical Synthesis of Ettringite. *ChemistryOpen* **2019**, *8* (7), 1012–1019.
31. Belenguer, A. M.; Friščić, T.; Day, G. M.; Sanders, J. K. M. Solid-State Dynamic Combinatorial Chemistry: Reversibility and Thermodynamic Product Selection in Covalent Mechanosynthesis. *Chem. Sci.* **2011**, *2* (4), 696–700.
32. Belenguer, A. M.; Lampronti, G. I.; Cruz-Cabeza, A. J.; Hunter, C. A.; Sanders, J. K. M. M. Solvation and Surface Effects on Polymorph Stabilities at the Nanoscale. *Chem. Sci.* **2016**, 7 (11), 6617–6627.

33. Belenguer, A. M.; Cruz-Cabeza, A. J.; Lampronti, G. I.; Sanders, J. K. M. On the Prevalence of Smooth Polymorphs at the Nanoscale: Implications for Pharmaceuticals. *CrstEngComm* **2019**, *21* (13), 2203–2211.
34. Šepelák, V.; Becker, K. D. Mössbauer Studies in the Mechanochemistry of Spinel Ferrites. *J. Mater. Synth. Process.* **2000**, *8* (3–4), 155–166.
35. Oliveira, P. F. M.; Willart, J.; Siepmann, J.; Siepmann, F.; Descamps, M. Using Milling to Explore Physical States : The Amorphous and Polymorphic Forms of Dexamethasone. *Cryst. Growth Des.* **2018**, *18*, 1748–1757.
36. Wagih, A.; Fathy, A.; Kabeel, A. M. Optimum Milling Parameters for Production of Highly Uniform Metal-Matrix Nanocomposites with Improved Mechanical Properties. *Adv. Powder Technol.* **2018**, *29* (10), 2527–2537.
37. Kajcsos, Z.; Marczis, L.; Tshakarov, C. G.; Gospodinov, G. G.; Horvath, D.; Vertes, A. Positron Annihilation Study of the Mechanochemical Reaction Between Zn + Se and Zn + S. *Nucl. Instrum. Methods* **1982**, *199*, 273–275.
38. Rehder, S.; Klukkert, M.; Löbmann, K. A. M.; Strachan, C. J.; Sakmann, A.; Gordon, K.; Rades, T.; Leopold, C. S. Investigation of the Formation Process of Two Piracetam Cocrystals during Grinding. *Pharmaceutics* **2011**, *3* (4), 706–722.
39. Osorio, J. G.; Hernández, E.; Romañach, R. J.; Muzzio, F. J. Characterization of Resonant Acoustic Mixing Using Near-Infrared Chemical Imaging. *Powder Technol.* **2016**, *297*, 349–356.
40. Nguyen, K. L.; Friscić, T.; Day, G. M.; Gladden, L. F.; Jones, W. Terahertz Time-Domain Spectroscopy and the Quantitative Monitoring of Mechanochemical Cocrystal Formation. *Nat. Mater.* **2007**, *6* (3), 206–209.
41. Andersen, J. M.; Mack, J. Decoupling the Arrhenius Equation via Mechanochemistry. *Chem. Sci.* **2017**, *8* (8), 5447–5453.
42. Andersen, J.; Brunemann, J.; Mack, J. Exploring Stable, Sub-Ambient Temperatures in Mechanochemistry *via* a Diverse Set of Enantioselective Reactions. *React. Chem. Eng.* **2019**, *4*, 1229–1236.
43. Frenette, M.; Cosa, G.; Friščić, T. Characterisation of Organic Solid Forms and Real-Time in Situ Monitoring of their Transformations Using Solid-State Fluorescence. *CrstEngComm* **2013**, *15* (25), 5100–5106.
44. Crawford, D. E.; Miskimmin, C. K. G.; Albadarin, A. B.; Walker, G.; James, S. L. Organic Synthesis by Twin Screw Extrusion (TSE): Continuous, Scalable and Solvent-Free. *Green Chem.* **2017**, *19* (6), 1507–1518.
45. Oliveira, P. F. M.; Baron, M.; Chamayou, A.; Baltas, M.; Guidetti, B.; Haruta, N.; Tanaka, K.; Sato, T. Lowering the Activation Energy under Mechanochemical Conditions: the Case of 2,3-Diphenylquinoxaline. *ChemistrySelect* **2016**, *1* (5), 984–988.
46. Kumari, N.; Bhattacharya, B.; Roy, P.; Michalchuk, A. A. L.; Emmerling, F.; Ghosh, A. Enhancing the Pharmaceutical Properties of Pirfenidone by Mechanochemical Cocrystallization. *Cryst. Growth Des.* **2019**, *19* (11), 6482–6492.
47. Batzdorf, L.; Zientek, N.; Rump, D.; Fischer, F.; Maiwald, M.; Emmerling, F. Make and Break—Facile Synthesis of Cocrystals and Comprehensive Dissolution Studies. *J. Mol. Struct.* **2017**, *1133*, 18–23.
48. Fischer, F.; Scholz, G.; Benemann, S.; Rademann, K.; Emmerling, F. Evaluation of the Formation Pathways of Cocrystal Polymorphs in Liquid-Assisted Syntheses. *CrstEngComm* **2014**, *16* (35), 8272–8278.
49. Takacs, L. Self-Sustaining Reactions Induced by Ball Milling. *Prog. Mater. Sci.* **2002**, *47* (4), 355–414.
50. Strukil, V.; Fábián, L.; Reid, D. G.; Duer, M. J.; Jackson, G. J.; Eckert-Maksić, M.; Friščić, T. Towards an Environmentally-Friendly Laboratory: Dimensionality and Reactivity in the Mechanosynthesis of Metal-Organic Compounds. *Chem. Commun.* **2010**, *46* (48), 9191–9193.

51. Halasz, I.; Friščić, T.; Kimber, S. A. J.; Užarević, K.; Puškarić, A.; Mottillo, C.; Julien, P.; Štrukil, V.; Honkimäki, V.; Dinnebier, R. E. Quantitative in Situ and Real-Time Monitoring of Mechanochemical Reactions. *Faraday Discuss.* **2014**, *170*, 203–221.
52. Cinčić, D.; Brekalo, I.; Kaitner, B. Effect of Atmosphere on Solid-State Amine-Aldehyde Condensations: Gas-Phase Catalysts for Solid-State Transformations. *Chem. Commun.* **2012**, *48* (95), 11683–11685.
53. Michalchuk, A. A. L.; Tumanov, I. A.; Drebushchak, V. A.; Boldyreva, E. V. Advances in Elucidating Mechanochemical Complexities via Implementation of a Simple Organic System. *Faraday Discuss.* **2014**, *170*, 311–335.
54. Hutchings, B. P.; Crawford, D. E.; Gao, L.; Hu, P.; James, S. L. Feedback Kinetics in Mechanochemistry: The Importance of Cohesive States. *Angew. Chem. Int. Ed.* **2017**, *56*, 1–6.
55. Kulla, H.; Fischer, F.; Benemann, S.; Rademann, K.; Emmerling, F. The Effect of the Ball to Reactant Ratio on Mechanochemical Reaction Times Studied by in Situ PXRD. *CrstEngComm* **2017**, *19* (28), 3902–3907.
56. Bowmaker, G. A.; Effendy, E.; Hanna, J. V.; Healy, P. C.; King, S. P.; Pettinari, C.; Skelton, B. W.; White, A. H. Solution and Mechanochemical Syntheses, and Spectroscopic and Structural Studies in the Silver(I) (Bi-)Carbonate: Triphenylphosphine System. *Dalton Trans.* **2011**, *40* (27), 7210–7218.
57. Braga, D.; Giaffreda, S. L.; Grepioni, F.; Polito, M. Mechanochemical and Solution Preparation of the Coordination Polymers Ag[N(CH2CH2)3N]2[CH 3 COO]· 5H2O and Zn[N(CH2CH 2)3N]Cl2. *CrstEngComm* **2004**, *6* (75), 458–462.
58. Avvakumov, E.; Senna, M.; Kosova, N. Introduction to Soft Mechanochemistry. In *Soft Mechanochemical Synthesis*; Avvakumov, E., Senna, M., Kosova, N., Eds.; Springer: Boston, MA, 2002; pp. 1–7.
59. Baláž, P.; Achimovicová, M.; Baláž, M.; Billik, P.; Zara, C. Z.; Criado, J. M.; Delogu, F.; Dutková, E.; Gaffet, E.; Gotor, F. J.; et al. Hallmarks of Mechanochemistry: From Nanoparticles to Technology. *Chem. Soc. Rev.* **2013**, *42* (18), 7571–7637.
60. Lapshin, O. V.; Smolyakov, V. K.; Boldyreva, E. V.; Boldyrev, V. V. Dynamics of the Homogenization of Binary Powder Mixtures. *Russ. J. Phys. Chem. A* **2018**, *92* (1), 66–69.
61. Lapshin, O. V.; Boldyrev, V. V.; Boldyreva, E. V. Mathematical Model of the Grinding and Mixing of Powder Binary Solids in a High-Energy Mill. *Russ. J. Phys. Chem. A* **2019**, *93* (8), 1592–1597.
62. Bellon, P.; Averback, R. S. Nonequilibrium Roughening of Interfaces in Crystals under Shear: Application to Ball Milling. *Phys. Rev. Lett.* **1995**, *74* (10), 1819–1822.
63. Chen, S.; Tolbert, W. A.; Dlott, D. D. Direct Measurement of Ultrafast Multiphonon Up-Pumping in High Explosives. *J. Phys. Chem.* **1994**, *98* (32), 7759–7766.
64. Holian, B. L.; Hammerberg, J. E.; Lomdahl, P. S. The Birth of Dislocations in Shock Waves and High-Speed Friction. *J. Computer-Aided Mater. Des.* **1998**, *5* (2–3), 207–224.
65. Butyagin, P. Y. Mechanochemical Reactions of Solids with Gases. *React. Solids* **1986**, *1*, 345–359.
66. Butyagin, P. Y.; Berestetskaya, I. V.; Kolbanov, I. V.; Pavlychev, I. K. Mechanochemical Hydrogenation of Graphite by Hydrogen. *Zh. Fiz. Khim.* **1986**, *60* (3), 579–584.
67. Arias, A. *Chemical Reactions of Metal Powders with Organic and Inorganic Liquids during Ball Milling*; National Aeronautics and Space Administration, 1975.
68. Doppiu, S.; Schultz, L.; Gutfleisch, O. In Situ Pressure and Temperature Monitoring during the Conversion of Mg into MgH2 by High-Pressure Reactive Ball Milling. *J. Alloys Compd.* **2007**, *427* (1–2), 204–208.
69. Brekalo, I.; Yuan, W.; Mottillo, C.; et al. Manometric Real-Time Studies of the Mechanochemical Synthesis of Zeolitic Imidazolate Frameworks. *Chem. Sci.* **2020**, *11*, 2141–2147.

70. Le Brun, P.; Froyen, L.; Delaey, L. The Modelling of the Mechanical Alloying Process in a Planetary Ball Mill: Comparison between Theory and in-Situ Observations. *Mater. Sci. Eng. A* **1993**, *161* (1), 75–82.
71. Ma, X.; Yuan, W.; Bell, S. E. J.; James, S. L. Better Understanding of Mechanochemical Reactions: Raman Monitoring Reveals Surprisingly Simple "Pseudo-Fluid" Model for a Ball Milling Reaction. *Chem. Commun.* **2014**, *50* (13), 1585–1587.
72. Lin, H.; Zhang, G.; Hsu, P.; Lin, S. A Portable Fiber-Optic Raman Analyzer for Fast Real-Time Screening and Identifying Cocrystal Formation of Drug-Coformer via Grinding Process. *Microchem. J.* **2013**, *110*, 15–20.
73. Friščić, T.; Halasz, I.; Beldon, P. J.; Belenguer, A. M.; Adams, F.; Kimber, S. A. J.; Honkimaki, V.; Dinnebier, R. E. Real-Time and in Situ Monitoring of Mechanochemical Milling Reactions. *Nat. Chem.* **2013**, *5*, 66–73.
74. Halasz, I.; Kimber, S. A. J.; Beldon, P. J.; Belenguer, A. M.; Adams, F.; Honkimäki, V.; Nightingale, R. C.; Dinnebier, R. E.; Friščić, T. In Situ and Real-Time Monitoring of Mechanochemical Milling Reactions Using Synchrotron X-Ray Diffraction. *Nat. Protoc.* **2013**, *8* (9), 1718–1729.
75. Gracin, D.; Strukil, V.; Friscic, T.; Halasz, I.; Uzarevic, K. Laboratory Real-Time and In Situ Monitoring of Mechanochemical Milling Reactions by Raman Spectroscopy. *Angew. Chem. Int. Ed.* **2014**, *53*, 6193–6197.
76. Katsenis, A. D.; Puskaric, A.; Strukil, V.; Mottillo, C.; Julien, P. A.; Uzarevic, K.; Pham, M.-H.; Do, T.-O.; Kimber, S. A. J.; Lazic, P.; et al. In Situ X-Ray Diffraction Monitoring of a Mechanochemical Reaction Reveals a Unique Topology Metal-Organic Framework. *Nat. Commun.* **2015**, *6*, 6662.
77. Stolar, T.; Batzdorf, L.; Lukin, S.; Žilić, D.; Motillo, C.; Friščić, T.; Emmerling, F.; Halasz, I.; Užarević, K. In Situ Monitoring of the Mechanosynthesis of the Archetypal Metal-Organic Framework HKUST-1: Effect of Liquid Additives on the Milling Reactivity. *Inorg. Chem.* **2017**, *56* (11), 6599–6608.
78. Wilke, M.; Kabelitz, A.; Gorelik, T. E.; Buzanich, A. G.; Reinholz, U.; Kolb, U.; Rademann, K.; Emmerling, F. The Crystallisation of Copper(II) Phenylphosphonates. *Dalton Trans.* **2016**, *45* (43), 17453–17463.
79. Wilke, M.; Akhmetova, I.; Rademann, K.; Emmerling, F. Mechanochemical Synthesis of Cerium(IV)-Phosphonates. *J. Mater. Sci.* **2018**, *53* (19), 13733–13741.
80. Akhmetova, I.; Schutjajew, K.; Wilke, M.; Buzanich, A.; Rademann, K.; Roth, C.; Emmerling, F. Synthesis, Characterization and in Situ Monitoring of the Mechanochemical Reaction Process of Two Manganese(II)-Phosphonates with N-Containing Ligands. *J. Mater. Sci.* **2018**, *53* (19), 13390–13399.
81. Užarević, K.; Štrukil, V.; Mottillo, C.; Julien, P. A.; Puškarić, A.; Friščić, T.; Halasz, I. Exploring the Effect of Temperature on a Mechanochemical Reaction by in Situ Synchrotron Powder X-Ray Diffraction. *Cryst. Growth Des.* **2016**, *16* (4), 2342–2347.
82. Halasz, I.; Puškarić, A.; Kimber, S. A. J.; et al. Real-Time in Situ Powder X-Ray Diffraction Monitoring of Mechanochemical Synthesis of Pharmaceutical Cocrystals. *Angew. Chem. Int. Ed.* **2013**, *52* (44), 11538–11541.
83. Kulla, H.; Becker, C.; Michalchuk, A. A. L.; Linberg, K.; Paulus, B.; Emmerling, F. Tuning the Apparent Stability of Polymorphic Cocrystals through Mechanochemistry. *Cryst. Growth Des.* **2019**, *19*, 7271–7279.
84. Fischer, F.; Heidrich, A.; Greiser, S.; Benemann, S.; Rademann, K.; Emmerling, F. Polymorphism of Mechanochemically Synthesized Cocrystals: A Case Study. *Cryst. Growth Des.* **2016**, *16* (3), 1701–1707.
85. Tumanova, N.; Tumanov, N.; Robeyns, K.; Fischer, F.; Fusaro, L.; Morelle, F.; Ban, V.; Filinchuk, Y.; Wouters, J.; Leyssens, T.; et al. Opening Pandora's Box: Chirality, Polymorphism, and Stoichiometric Diversity in Flurbiprofen/Proline Cocrystals. *Cryst. Growth Des.* **2018**, *18*, 954–961.

86. Fischer, F.; Lubjuhn, D.; Greiser, S.; Rademann, K.; Emmerling, F. Supply and Demand in the Ball Mill: Competitive Cocrystal Reactions. *Cryst. Growth Des.* **2016**, *16* (10), 5843–5851.
87. Fischer, F.; Greiser, S.; Pfeifer, D.; Jäger, C.; Rademann, K.; Emmerling, F. Mechanochemically Induced Conversion of Crystalline Benzamide Polymorphs by Seeding. *Angew. Chem. Int. Ed.* **2016**, *55* (46), 14281–14285.
88. Heiden, S.; Tröbs, L.; Wenzel, K. J.; Emmerling, F. Mechanochemical Synthesis and Structural Characterisation of a Theophylline-Benzoic Acid Cocrystal (1:1). *CrstEngComm* **2012**, *14* (16), 5128–5129.
89. Burmeister, C. F.; Kwade, A. Process Engineering with Planetary Ball Millss. *Chem. Soc. Rev.* **2013**, *42* (18), 7660–7667.
90. Michalchuk, A. A. L.; Tumanov, I. A.; Boldyreva, E. V. The Effect of Ball Mass on the Mechanochemical Transformation of a Single-Component Organic System: Anhydrous Caffeine. *J. Mater. Sci.* **2018**, *53* (19), 13380–13389.
91. Michalchuk, A. A. L.; Tumanov, I. A.; Konar, S.; Kimber, S. A. J.; Pulham, C. R.; Boldyreva, E. V. Challenges of Mechanochemistry: Is in Situ Real-Time Quantitative Phase Analysis Always Reliable? A Case Study of Organic Salt Formation. *Adv. Sci.* **2017**, *4*, 1700132.
92. Anderson, S. R.; Dubé, P.; Krawiec, M.; Salan, J. S.; Ende, D. J. A.; Samuels, P. Promising CL-20-Based Energetic Material by Cocrystallization. *Propellants, Explos., Pyrotech.* **2016**, *41* (5), 783–788.
93. Am Ende, D. J.; Anderson, S. R.; Salan, J. S. Development and Scale-Up of Cocrystals Using Resonant Acoustic Mixing. *Org. Process Res. Dev.* **2014**, *18* (2), 331–341.
94. Michalchuk, A. A. L.; Hope, K. S.; Kennedy, S. R.; Blanco, M. V.; Boldyreva, E. V.; Pulham, C. R. Ball-Free Mechanochemistry: In Situ Real-Time Monitoring of Pharmaceutical Co-Crystal Formation by Resonant Acoustic Mixing. *Chem. Commun.* **2018**, *54* (0), 4033–4036.
95. Užarević, K.; Halasz, I.; Friščić, T. Real-Time and in Situ Monitoring of Mechanochemical Reactions: A New Playground for all Chemists. *J. Phys. Chem. Lett.* **2015**, *6* (20), 4129–4140.
96. Ban, V.; Sadikin, Y.; Lange, M.; Tumanov, N.; Filinchuk, Y.; Černý, R.; Casati, N. Innovative in Situ Ball Mill for X-Ray Diffraction. *Anal. Chem.* **2017**, *89* (24), 13176–13181.
97. Wilke, M.; Casati, N. Insight into the Mechanochemical Synthesis and Structural Evolution of Hybrid Organic–Inorganic Guanidinium Lead(II) Iodides. *Chem. A Eur. J.* **2018**, *24* (67), 17701–17711.
98. Tumanov, N.; Ban, V.; Poulain, A.; Filinchuk, Y. 3D-Printed Jars for Ball-Milling Experiments Monitored in Situ by X-Ray Powder Diffraction. *J. Appl. Cryst.* **2017**, *50*, 994–999.
99. Tumanov, I. A.; Achkasov, A. F.; Myz, S. A.; Boldyreva, E. V.; Boldyrev, V. V. Different Effect of Impact and Shear Mechanical Treatment on Mechanochemical Cocrystallization of Piroxicam and Succinic Acid. *Dokl. Chem.* **2014**, *457* (2), 154–159.
100. Haferkamp, S.; Paul, A.; Michalchuk, A. A. L.; Emmerling, F. Unexpected Polymorphism during a Catalyzed Mechanochemical Knoevenagel Condensation. *Beilstein J. Org. Chem.* **2019**, *15*, 1141–1148.
101. Haferkamp, S.; Kraus, W.; Emmerling, F. Studies on the Mechanochemical Knoevenagel Condensation of Fluorinated Benzaldehyde Derivates. *J. Mater. Sci.* **2018**, *53* (19), 13713–13718.
102. Otaki, T.; Tanabe, Y.; Kojima, T.; Miura, M.; Ikeda, Y.; Koide, T.; Fukami, T. In Situ Monitoring of Cocrystals in Formulation Development Using Low-Frequency Raman Spectroscopy. *Int. J. Pharm.* **2018**, *542* (1–2), 56–65.

103. Roy, S.; Chamberlin, B.; Matzger, A. J. Polymorph Discrimination Using Low Wavenumber Raman Spectroscopy. *Org. Process Res. Dev.* **2013**, *17* (7), 976–980.
104. Juribasic, M.; Uzarevic, K.; Gracin, D.; Curic, M. Mechanochemical C–H Bond Activation: Rapid and Regioselective Double Cyclopalladation Monitored by in Situ Raman Spectroscopy. *Chem. Commun.* **2014**, *50*, 10287–10290.
105. Bjelopetrovic, A.; Lukin, S.; Halasz, I.; Uzarevic, K.; Dilovic, I.; Barisic, D.; Budimir, A.; Kulcsar, M. J.; Curic, M. Mechanism of Mechanochemical C-H Bond Activation in an Azobenzene Substrate by Pd II Catalysts. *Chem. A Eur. J.* **2018**, *24*, 10672–10682.
106. Štrukil, V.; Gracin, D.; Magdysyuk, O. V.; Dinnebier, R. E.; Friščic, T. Trapping Reactive Intermediates by Mechanochemistry: Elusive Aryl N-Thiocarbamoylbenzotriazoles as Bench-Stable Reagents. *Angew. Chem. Int. Ed.* **2015**, *54* (29), 8440–8443.
107. Tireli, M.; Juribašić Kulcsár, M.; Cindro, N.; Gracin, D.; Biliškov, N.; Borovina, M.; Ćurić, M.; Halasz, I.; Užarević, K. Mechanochemical Reactions Studied by in Situ Raman Spectroscopy: Base Catalysis in Liquid-Assisted Grinding. *Chem. Commun.* **2015**, *51* (38), 8058–8061.
108. Fischer, F.; Fendel, N.; Greiser, S.; Rademann, K.; Emmerling, F. Impact Is Important—Systematic Investigation of the Influence of Milling Balls in Mechanochemical Reactions. *Org. Process Res. Dev.* **2017**, *21* (4), 655–659.
109. Fischer, F.; Wenzel, K.-J.; Rademann, K.; Emmerling, F. Quantitative Determination of Activation Energies in Mechanochemical Reactions. *Phys. Chem. Chem. Phys.* **2016**, *18*, 23320–23325.
110. Lukin, S.; Tireli, M.; Stolar, T.; Barišić, D.; Blanco, M. V.; Di Michiel, M.; Užarević, K.; Halasz, I. Isotope Labeling Reveals Fast Atomic and Molecular Exchange in Mechanochemical Milling Reactions. *J. Am. Chem. Soc.* **2019**, *141* (3), 1212–1216.
111. Tanaka, R.; Takahashi, N.; Nakamura, Y.; Hattori, Y.; Ashizawa, K.; Otsuka, M. In-Line and Real-Time Monitoring of Resonant Acoustic Mixing by Near-Infrared Spectroscopy Combined with Chemometric Technology for Process Analytical Technology Applications in Pharmaceutical Powder Blending Systems. *Anal. Sci.* **2017**, *33* (1), 41–46.
112. Batzdorf, L.; Fischer, F.; Wilke, M.; Wenzel, K.-J. J.; Emmerling, F. Direct in Situ Investigation of Milling Reactions Using Combined X-Ray Diffraction and Raman Spectroscopy. *Angew. Chem. Int. Ed.* **2015**, *54* (6), 1799–1802.
113. Tröbs, L.; Emmerling, F. Mechanochemical Synthesis and Characterisation of Cocrystals and Metal Organic Compounds. *Faraday Discuss.* **2014**, *170*, 109–119.
114. Wilke, M.; Buzanich, A. G.; Reinholz, U.; Rademann, K.; Emmerling, F. The Structure and: In Situ Synthesis Investigation of Isomorphic Mononuclear Molecular Metal Phenylphosphonates. *Dalton Trans.* **2016**, *45* (23), 9460–9467.
115. Wilke, M.; Kabelitz, A.; Zimathies, A.; Rademann, K.; Emmerling, F. Crystal Structure and in Situ Investigation of a Mechanochemical Synthesized 3D Zinc N-(Phosphonomethyl)Glycinate. *J. Mater. Sci.* **2017**, *52* (20), 12013–12020.
116. Kulla, H.; Greiser, S.; Benemann, S.; Rademann, K.; Emmerling, F. In Situ Investigation of a Self-Accelerated Cocrystal Formation by Grinding Pyrazinamide with Oxalic Acid. *Molecules* **2016**, *21*, 917.
117. Kulla, H.; Michalchuk, A. A. L.; Emmerling, F. Manipulating the Dynamics of Mechanochemical Ternary Co-Crystal Formation. *Chem. Commun.* **2019**, *1*, 2–5.
118. Linberg, K.; Ali, N. Z.; Etter, M.; Michalchuk, A. A. L.; Rademann, K.; Emmerling, F. A Comparative Study of the Ionic Cocrystals NaX (α-d-Glucose)2 (X = Cl, Br, I). *Cryst. Growth Des.* **2019**, *19*, 4293–4299.
119. Tumanov, I. A.; Michalchuk, A. A. L.; Politov, A. A.; Boldyreva, E. V.; Boldyrev, V. V. Inadvertent Liquid Assisted Grinding: A Key to "Dry" Organic Mechano-Co-Crystallisation? *CrstEngComm* **2017**, *19* (21), 2830–2835.

120. Haferkamp, S.; Fischer, F.; Kraus, W.; Emmerling, F. Mechanochemical Knoevenagel Condensation Investigated in Situ. *Beilstein J. Org. Chem.* **2017**, *13*, 2010–2014.
121. Lukin, S.; Stolar, T.; Tireli, M.; Blanco, M. V.; Babić, D.; Friščić, T.; Užarević, K.; Halasz, I. Tandem in Situ Monitoring for Quantitative Assessment of Mechanochemical Reactions Involving Structurally Unknown Phases. *Chem. A Eur. J.* **2017**, *23* (56), 13941–13949.
122. Lee, C. H.; Mori, M.; Fukunaga, T.; Mizutani, U. Effect of Ambient Temperature on the Ma and Mg Processes in Ni–Zr Alloy System. *Jpn. J. Appl. Phys.* **1990**, *29* (3 R), 540–544.
123. McKissic, K. S.; Caruso, J. T.; Blair, R. G.; Mack, J. Comparison of Shaking versus Baking: Further Understanding the Energetics of a Mechanochemical Reaction. *Green Chem.* **2014**, *16* (3), 1628–1632.
124. Takacs, L.; McHenry, J. S. Temperature of the Milling Balls in Shaker and Planetary Mills. *J. Mater. Sci.* **2006**, *41* (16), 5246–5249.
125. Stroh, J.; Ali, N. Z.; Maierhofer, C.; Emmerling, F. Ettringite via Mechanochemistry : A Green and Rapid Approach for Industrial Application. *ACS Omega* **2019**, *4* (1), 7734–7737.
126. Kulla, H.; Haferkamp, S.; Akhmetova, I.; Rçllig, M.; Maierhofer, C.; Rademann, K.; Emmerling, F. In Situ Investigations of Mechanochemical One-Pot Syntheses. *Angew. Chem. Int. Ed.* **2018**, *57*, 5930–5933.
127. Užarević, K.; Ferdelji, N.; Mrla, T.; Julien, P. A.; Halasz, B.; Friščić, T.; Halasz, I. Enthalpy vs. Friction: Heat Flow Modelling of Unexpected Temperature Profiles in Mechanochemistry of Metal-Organic Frameworks. *Chem. Sci.* **2018**, *9* (9), 2525–2532.
128. Kulla, H.; Wilke, M.; Fischer, F.; Röllig, M.; Maierhofer, C.; Emmerling, F. Warming up for Mechanosynthesis—Temperature Development in Ball Mills during Synthesis. *Chem. Commun.* **2017**, *53* (10), 1664–1667.
129. Fischer, F.; Scholz, G.; Batzdorf, L.; Wilke, M.; Emmerling, F. Synthesis, Structure Determination, and Formation of a Theobromine: Oxalic Acid 2:1 Cocrystal. *CrstEngComm* **2015**, *17* (4), 824–829.
130. Wilke, M.; Batzdorf, L.; Fischer, F.; Rademann, K.; Emmerling, F. Cadmium Phenylphosphonates: Preparation, Characterisation and in Situ Investigation. *RSC Adv.* **2016**, *6* (42), 36011–36019.
131. Cindro, N.; Tireli, M.; Karadeniz, B.; Mrla, T.; Uzarevic, K. Investigations of Thermally Controlled Mechanochemical Milling Reactions. *ACS Sustain. Chem. Eng.* **2019**, 7, 16301–16309.
132. Gomollón-Bel, F. Ten Chemical Innovations that Will Change our World: IUPAC Identifies Emerging Technologies in Chemistry with Potential to Make our Planet more Sustainable. *Chem. Int.* **2019**, *41* (2), 12–17.

CHAPTER TWELVE

From solution-based nonconventional activation methods to mechanochemical procedures: The hydantoin case

Andrea Porcheddu[a,*], Clarence Charnay[b,*], Francesco Delogu[c], and Evelina Colacino[b]

[a]Dipartimento di Scienze Chimiche e Geologiche, Università degli Studi di Cagliari, Cittadella Universitaria, Monserrato, CA, Italy
[b]ICGM, Univ Montpellier, CNRS, ENSCM, Montpellier, France
[c]Dipartimento di Ingegneria Meccanica, Chimica, e dei Materiali, Università degli Studi di Cagliari, Cagliari, Italy

1 Introduction

Social responsibility, including environmental protection, is an indispensable principle to build a sustainable civilization. "Cultural barriers" and economic, technical, organizational, regulatory, and financial obstacles are significant impediments to realize a cultural shift toward an "eco-conscious" behavior. The demand for clean processes and environmentally friendly reaction conditions minimizing the hazard while increasing the performances of a chemical process have stimulated a growing interest in the scientific community. The culture of scientific discovery through the adoption of green chemistry practices[1] is an opportunity for competitiveness and innovation toward the design of new chemical (clean) processes. However, the broad adoption of green chemistry and the implementation of sustainable smart solutions in industry are still in infancy, and green chemistry-based regulations and incentives should be initiated and promoted by governments.

Safety, security, and more and more stringent regulatory requirements dictate the necessary improvements for a sustainable manufacturing,[2] the introduction of harmonized and tangible metrics,[3] and the adoption of newer and eco-efficient technologies. This approach fosters innovation, waste reduction (or no waste at all), and reduced costs while complying with the regulations and the need of performances of a company.

*These authors contributed equally to this work.

Nontraditional Activation Methods in Green and Sustainable Applications
https://doi.org/10.1016/B978-0-12-819009-8.00003-7
© 2021 Elsevier Inc.
All rights reserved.

The continuous search for new, rapid, environmentally benign alternatives to conventional chemical syntheses and transformations includes milder reaction conditions, shorter reaction times, and the development of energetically efficient procedures, especially in the perspective of large-scale processing.

The adoption of nonconventional activation methods for energy delivery[4] such as microwave irradiation,[5, 6] sonochemistry,[7, 8] electrochemistry,[9] photochemistry,[10] continuous flow,[11–14] and more recently mechanochemistry[15–19] represents a shade of green with proven effectiveness in diverse applications, compared with more conventional thermal activation techniques. The implementation of sustainable technologies minimizes scientific and technical difficulties and while complementing the REACH (Registration, Evaluation, Authorization and Restriction of Chemicals) legislation[20] can be useful to improve the ecological footprint of a process. However, especially on large-scale processing, a case-by-case analysis needs to be assessed before definitely affirming whether or not each of these processes is more convenient, especially from an energetic point of view.

One of the modern challenges that industrial and academic researchers are facing is to "think chemistry differently"[21] designing chemical processes "benign by design."[22]

In this context the preparation of active pharmaceutical ingredients (APIs) and more generally commodity chemicals for the industry is a social responsibility that starts with a change of mentality relying on a radically different approach to the organic synthesis, possibly accelerated if green chemistry becomes an integral part of chemistry education.[23]

2 Enabling Technologies at the service of hydantoin scaffold

The preparation of imidazolidine-2,4-diones, commonly known as hydantoins,[24–26] is a good example illustrating the breakthrough issued by the synergic association of nonconventional activation methods and sustainable approaches, also for the eco-friendly preparation of APIs and drug candidates (Fig. 1).

The modulation of biological activity and pharmacological properties of hydantoin derivatives is infinite, and it is intrinsically related to the high degree of possible combined substitutions and functionalization sites (FS). Moreover the presence of H-bond donor and acceptor positions (HBD and HBA, respectively) allows to modulate flexibility versus rigidity, hydrophilicity versus hydrophobicity, the ionic character, and chirality of the backbone (Fig. 1). The richness of hydantoin chemistry[24, 25] associated with the possibility to

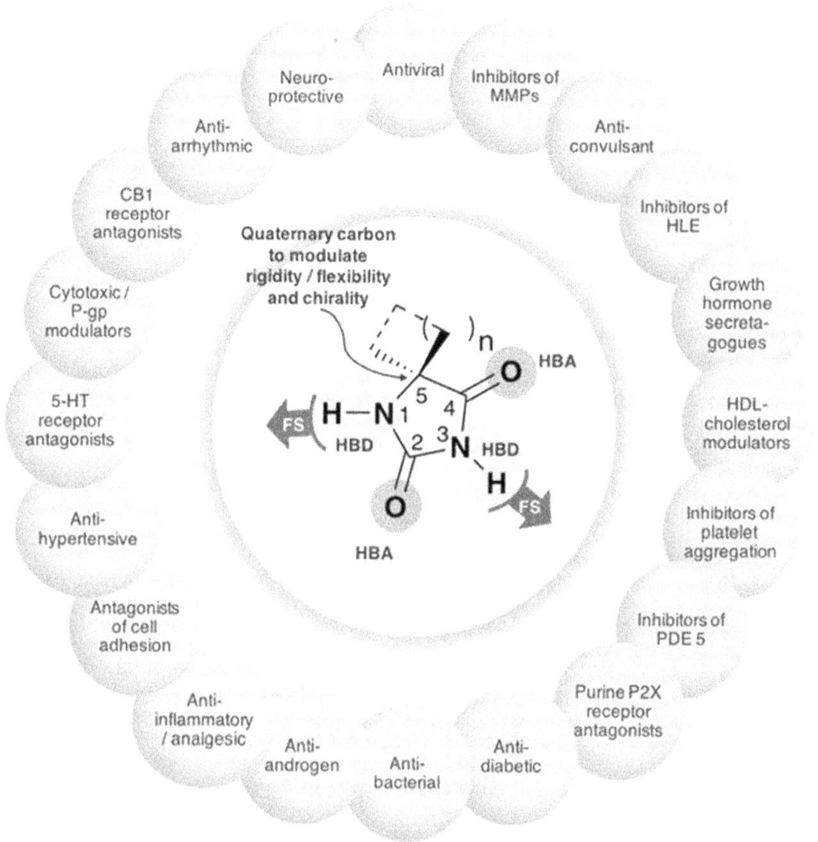

Fig. 1 Biological activities and structural features of hydantoin scaffold. *(Image reproduced with the permission of Royal Society of Chemistry from reference Colacino, E.; Porcheddu, A.; Charnay, C.; Delogu, F., From Enabling Technologies to Medicinal Mechanochemistry: An Eco-Friendly Access to Hydantoin-Based Active Pharmaceutical Ingredients. React. Chem. Eng.* ***2019****, 4, 1179–1188. https://doi.org/10.1039/c9re00069k.)*

target always new therapeutic areas is witnessed by an ever-increasing number of patents (1000)[27] and scientific publications (over 6000)[27] about their use and applications. It is worth of note that drugs containing hydantoin pharmacophores have entered in clinical trials[28, 29] and several pharmaceuticals have been already commercialized.[30] Therefore, it is of utmost importance to propose powerful, new, and sustainable synthetic methodologies and processes having a low environmental impact for their preparation.

The implementation of sustainable technologies and newer strategies (solvent-free) to improve the three historical synthetic methods, namely, Biltz,[31] Read,[32] and Bucherer-Bergs[33] methods to access hydantoins (Scheme 1 and Tables 1–3, respectively), already started, leading to a "new Renaissance" of old synthetic procedures.

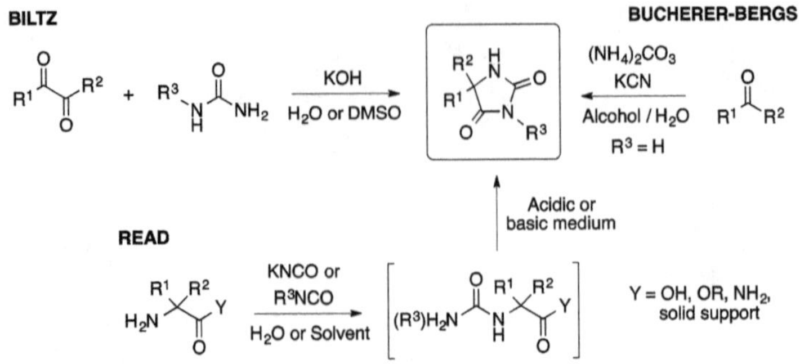

Scheme 1 Most common synthetic pathways to access hydantoins.

Table 1 Nonconventional activation methods for Biltz's reaction.

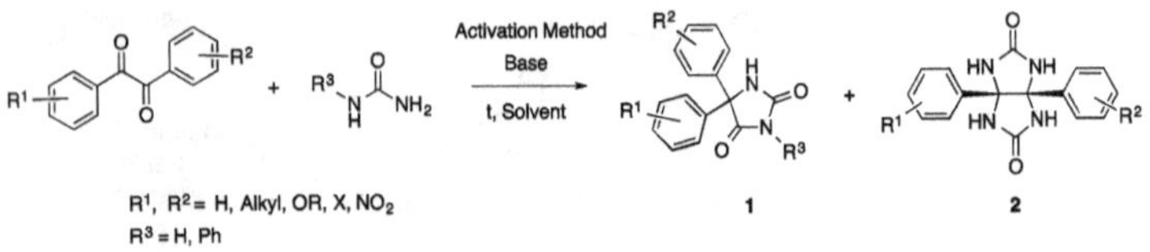

Entry	Activation method		Solvent/base	t (min)	1 (%)	2 (%)	#[a]
1[34]	MW[b]	100 W	Neat/NaOH	1.5	93–98	1–6	8
2[35]	MW[b,c]	300 W	Neat/PPE	3[c]	80–92	n.r.[d]	8
3[36]	MW[b,c]	750 W	DMSO:H_2O/KOH	30[c]	45–92	n.r.[d]	11
4[37]	US[e]	r.t.	EtOH/KOH	3–6	94–98	0	11
5[38]	US[e]	r.t.	DMSO:H_2O/KOH	3–5	97–99	0	6
6[39]	Grinding[f]	r.t.	HCl:EtOH/thiamine	10	63–74	n.r.[d]	2
7[40]	Milling[g]	r.t.[h]	Neat/NaOH	30–120	10–44	n.d.[d]	1

[a]# = number of prepared compounds.
[b]A domestic oven was used.
[c]Irradiation by intervals (pulsed method).
[d]n.r., not reported; n.d., not determined.
[e]Direct sonication using a probe.
[f]Manual grinding in a porcelain mortar with pestle.
[g]Both vibrating (at 30 Hz) and planetary (450 rpm) ball mills were used.
[h]Operational temperature of the jars during milling.

The most straightforward (and historical) method for the preparation of 5,5-disubstituted hydantoins is the Biltz synthesis,[31] involving a condensation reaction between the α-diketone benzil and urea in basic conditions (NaOH and KOH). The formation of 5,5-diphenylhydantoin (phenytoin **1**, $R^1 = R^2 = R^3 = H$; Table 1) proceeds via a benzylic rearrangement in which a phenyl group undergoes a 1,2-shift (similar to a pinacol rearrangement).[49] The stepwise mechanism[50] leads to the formation of the intermediate 4,5-diphenyl-4,5-dihydroxy-2-imidazolone, followed by a fast

Table 2 Nonconventional activation methods for Bucherer-Bergs reaction.

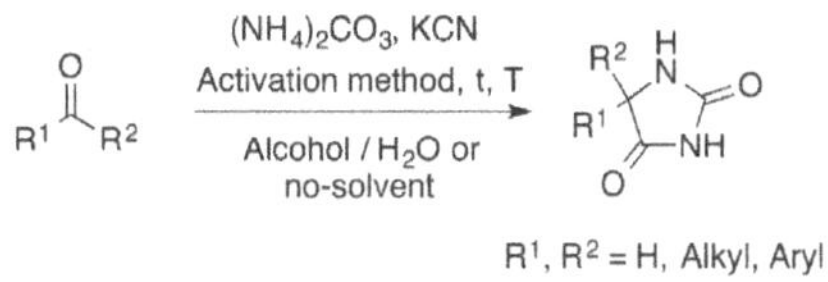

Entry	Activation method/*T* (°C)		*t* (min)	Additive	Yield (%)	#[a]
1[41]	MW	n.r.[b]	2–13	/	83–99	10
2[42]	MW	90	10	/	50–quant.	24
3[43]	US[c]	45	180–270	/	50–98	10
4[44]	US[d]	45	15–70	K-10[e]	85–98	15
5[45]	Milling[f]	r.t.[g]	25–50	ZnO	90–97	10
6[46]	Flow[h]	120 °C–20 bar	32	/	72–99	11
7[47]	Flow/$h\nu$[i]	80° to 6 bar	20	/	23–98[j]	9

[a]# = number of prepared compounds.
[b]n.r., not reported.
[c]Indirect sonication using an ultrasonic bath.
[d]Direct sonication using a probe.
[e]Montmorillonite K-10 was used as solid heterogeneous acid catalyst.
[f]A vibrating ball mill operating at 28 Hz was used.
[g]Operational temperature of the jars during milling.
[h]A biphasic system water/EtOAc was used.
[i]A mixture of Me-THF/MeOH was used as the solvent.
[j]Conversions by ^{1}H NMR (instead of yields) are here indicated.

Table 3 Preparation of phenytoin ($R^1 = R^2 = H$, from Table 1) by alternative energy inputs.

Entry	Activation method/*T* (°C)		*t* (min)	1 (%)	2 (%)	#[a]
1[34]	MW[b]	100 W	2	98	1.6	Biltz
2[36]	MW[b]	750 W	30[c]	74	n.q.[d]	Biltz
3[41]	MW[e]	n.r.[d]	10	84	/	Bucherer-Bergs
4[37,38]	US[f]	r.t.	3–4	98	0	Biltz
5[44]	US[f,g]	45 °C	72	86	/	Bucherer-Bergs
6[45]	Milling[h]	r.t.[i]	50	90	/	Bucherer-Bergs
7[40]	Milling[j]	r.t.[h]	30–120	10–44	n.q.[d]	Biltz
8[40]	Milling[k]	r.t.[h]	8 h	86	/	Read

[a]# = synthetic method.
[b]A domestic oven was used.
[c]Irradiation by intervals (pulsed method).
[d]n.r., not reported; n.q., not quantified.
[e]Experimental details were not reported.
[f]Direct sonication using a probe.
[g]Montmorillonite K-10 was used as solid heterogeneous acid catalyst.
[h]A vibrating ball mill operating at 25 Hz was used in the presence of ZnO nanoparticles.[48]
[i]Operational temperature of the jars during milling.
[j]Both vibrating (at 30 Hz) and planetary (450 rpm) ball mills were used.
[k]A planetary ball mill was used (450 rpm).

rearrangement.[49,50] This is also in agreement with the concomitant formation of the 3α,6α-diphenylglycoluril by-product (**2**, $R^1 = R^2 = R^3 = H$; Table 1), resulting in the condensation of the intermediate 4,5-diphenyl-4,5-dihydroxy-2-imidazolone with a second equivalent of urea.

Since then, several reports described the use of Biltz's method to access 5,5-diphenylhydantoin derivatives from unsymmetrical benzils ($R^1 \neq R^2$), leading to N^3-substituted hydantoins from substituted ureas (R^3 = alkyl, aryl). The use of nonconventional activation methods applied to Biltz's synthesis are summarized in Table 1. Although yields are generally good, these studies aimed to increase the productivity of 5,5-diaryl hydantoins **1** over the time and to find the optimum conditions avoiding the concomitant formation of 3α,6α-diarylglycoluril by-products **2**. In this regard and on the basis of the available data, "direct" sonication (entries 4 and 5, Table 1)[37,38] outperformed for achieving both goals, and waste was reduced compared with the traditional methods, due to a drastic reduction of the amount of base to promote the rearrangement (from 9 to 1.8 mmol). Strikethrough, enhanced reaction rates were obtained under microwave activation, in solution,[36] or in neat conditions[34,35] with NaOH or in the presence of a large excess of polyphosphate esters (PPE) (entries 1–3). However, in this last case, even if "solvent economy" is attained, the process suffers from poor "atom economy" and waste generation, dramatically impacting on the sustainability of the whole process.

The water-soluble 3α,6α-diarylglycoluril by-products **2**, if present in the reaction crude, could be easily removed by the precipitation with water of the final products **1**.

Considering their simplicity, strategies based on unconventional means of mixing to delivery energy to the system and on the use of solvent-free approaches were also investigated. Manual grinding in a porcelain mortal with pestle, using vitamin B1 as a base (entry 6),[39] and the use of ball-milling devices equipped with stainless steel jars and balls (entry 7)[40] were not efficient in terms of productivity. In particular, manual grinding led to comparable yields as for the thermal heating but displayed enhanced kinetics, while by mechanochemistry the conversion of the substrates was never complete, and 3α,6α-diphenylglycoluril **2** was always present.[40]

A systematic investigation of Read's synthesis by mechanochemistry was also reported: starting from a large variety of α-amino acid methyl ester hydrochlorides, two general approaches (involving similar reaction intermediates and steps) were disclosed using potassium cyanate[40] or isocyanates[51] (Scheme 2).

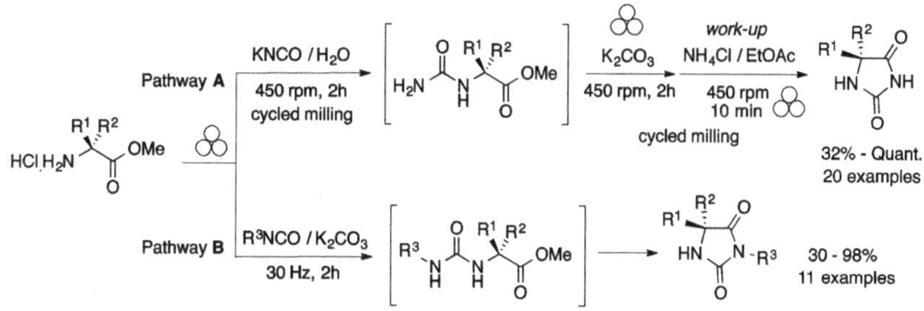

Scheme 2 Mechanochemical Read's preparation of substituted hydantoins. Hanusa's formalism to represent the reaction activated by mechanochemical energy[52] was used.

The mechanochemical *one-pot*/two-step procedure using a 2:1 ratio of KNCO/H_2O led to the corresponding ureido derivatives without need of adding any extra base (*reagent economy*) to generate the free nucleophilic amine responsible of the reaction on the cyanate, the base (KOH) being generated in situ by the spontaneous decomposition of KNCO in the presence of water (pathway A; Scheme 2).[40] The intramolecular cyclization step occurred in basic conditions affording pure hydantoins without need of any purification step. The procedure presents the advantage of completely eliminating the organic solvents during the synthesis and limiting their amount during the work-up procedure. Indeed, milling an excess of ammonium chloride suspended in EtOAc led to the precipitation of the inorganic salts into the milling jars, while the pure hydantoins were recovered in the organic phase after filtration/evaporation steps.

Compared with solution synthesis, the reaction conditions are simpler, the yields are better, and the work-up is easier; the mechanochemical method is general and particularly suitable to react sterically hindered quaternary α-amino acid methyl esters, leading to 5,5-disusbstituted hydantoins with a threefold increase in yield compared with solution synthesis. The mechanochemical reactivity of α-amino acid methyl esters was also investigated with alkyl or arylisocyanates, leading to *N*-3-substituted hydantoins in *high atom economy* (pathway B, Scheme 2).[51]

Microwave-activated hydantoins using Read's method was also reported on solid[53,54] support or liquid phases such as poly(ethylene glycol)s (PEGs)[55,56] and fluorous tags.[57,58] In addition to the well-assessed benefits coming by the use of microwave activation, the advantage of using supported synthesis relays in an efficient purification of the reaction mixtures.

In this regard the reaction of isocyanates with preloaded resins with amino acids, followed by a *one-pot* two step base-mediated intramolecular

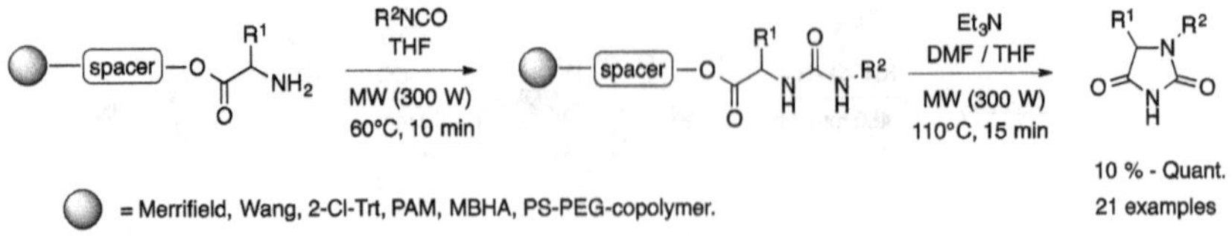

Scheme 3 Solid-phase synthesis of hydantoins.[53]

cyclization/cleavage, afforded 3,5-disubstituted hydantoins in high purity and yields[53] (Scheme 3). As expected, microwave activation increased the kinetics of the reaction (few minutes instead of few hours) compared with thermal heating under reflux,[59] and the supported ureido derivatives not able to cyclize remained attached to the solid support.

The nature of the R^1 and R^2 substituents on the amino acid side chains and isocyanates and the nucleofuge properties of the resin determine the rate of cyclative cleavage. In the same fashion, highly functionalized hydantoins were successfully prepared under microwave activation by liquid-phase synthesis using soluble polymer supports such as PEGs (Scheme 4).[60,61]

Because of their permanent dipole, high boiling point, and thermal stability, PEG-based reaction media are particularly indicated to be used in organic transformation involving microwave activation. Indeed, their good adsorption of microwave energy[6,55] favors efficient heating. However, it is questionable whether, in the case of liquid-phase synthesis, additional solvent is really needed at each step. At room temperature, PEGs with a molecular weight above 900 Da are waxy solids that become liquid reaction media upon melting (45–60°C). Based on their intrinsic characteristics, reactions involving PEG polymers can be performed in neat conditions upon heating, reducing the environmental impact of the syntheses (*solvent economy*) providing waste minimization. Moreover, upon completion of the reaction, the polymer support is removed by a precipitation/filtration work-up, allowing the recycling of the polymer and the recovery of the final products in the homogeneous solution in high yields and shorter reaction times compared with preheated oil bath (from several days to several minutes).

Liquid-phase preparation of tricyclic hydantoins via a Read-like method was also investigated under microwave activation starting from perfluoroalkylated bicyclic proline derivatives obtained *one-pot* [3 + 2] cycloaddition reaction between a fluorous amino ester (limiting reagent), a substituted benzaldehyde (generating in situ azomethine ylides), and an *N*-alkyl maleimide[57] (Scheme 5).

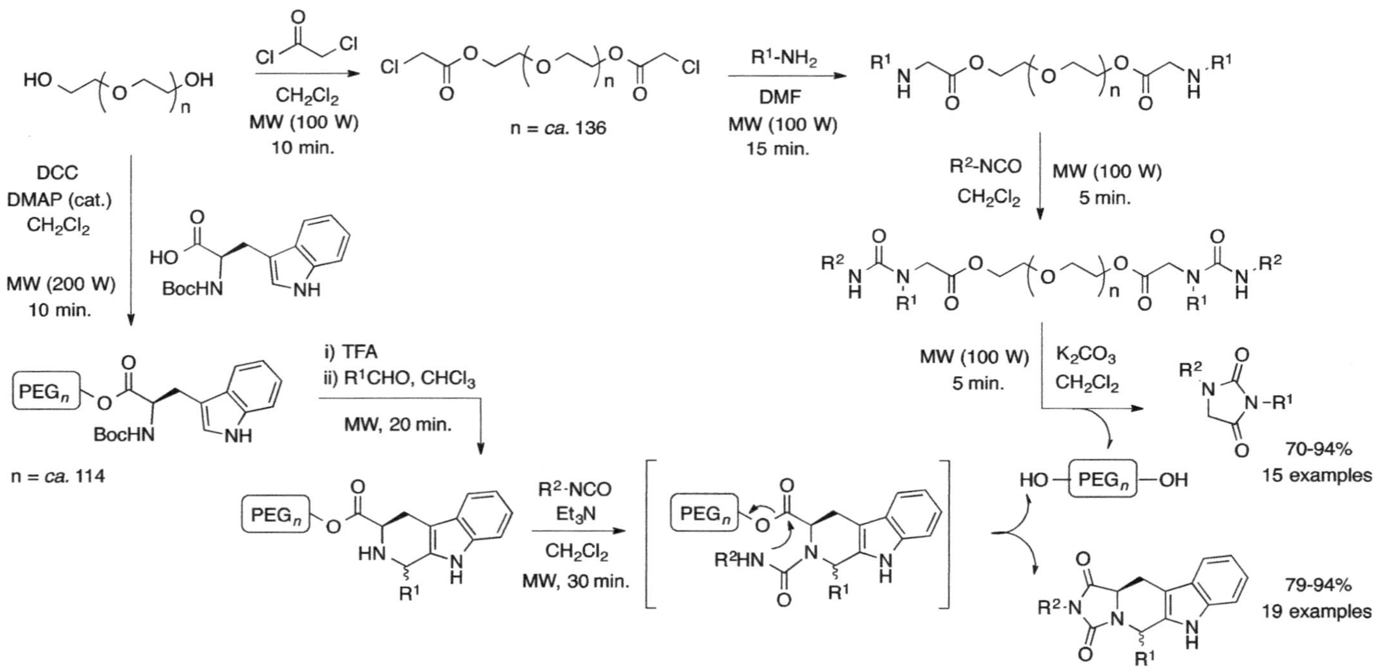

Scheme 4 Microwave-assisted liquid-phase synthesis of hydantoins on PEG support.[60,61]

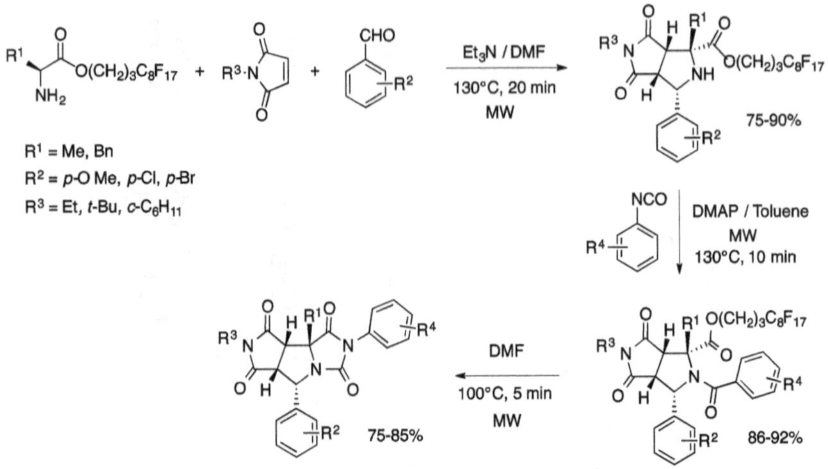

Scheme 5 "Fluorous tag" for the preparation of fused hydantoin under microwave activation.[57]

Fluorous solid-phase extraction (F-SFE) allowed a simple purification during each step. Indeed the unreacted nonfluorous components were eluted out with a fluorophobic solvent, while the fluorous-tagged product is eluted out from the cartridge using a fluorophilic solvent such as MeOH.

Alternative energy sources were also explored for the improvement of the Bucherer-Bergs reaction (Scheme 1 and Table 2). The reaction occurs *one-pot* and starts after the thermal decomposition of ammonium carbonate, generating in situ the reactive gases such as ammonia and carbon dioxide. The addition of nitrile anion on the imine generated by the condensation reaction between the carbonyl compound is followed by a carbamoylation reaction in the presence of carbon dioxide leading to the unstable α-amino nitrile intermediate (Scheme 1). Hydantoin is formed during a cyclization/rearrangement cascade reaction.

To achieve acceptable yields, high temperature is usually required, and the use of a large excess of cyanide leads to the concomitant formation of polymer by-products and the release of toxic hydrogen cyanide during the cyclization step. The Bucherer-Bergs reaction conditions are intrinsically unsafe due to (i) elevated risks of intoxication, (ii) overpressure generated during reaction performed in a sealed tube (by generation in situ of volatile gas), or (iii) the need to use a large excess of ammonium carbonate to compensate the escape of the reactive gases in the headspace of the flask when using traditional batch techniques.

In comparison with the conventional activation methods, reaction rates were strongly enhanced, and heat transfer and mass transport improved when using microwave and ultrasound activation (entries 1–4, Table 2), providing cleaner reaction profiles and consequently higher yields, overcoming some of the disadvantages of the Bucherer-Bergs reaction conditions.

Acidic heterogeneous catalysts are performed similarly under sonochemical activation[44] or in solvent-free mechanochemical procedures[45] (entries 4 and 5). This is not surprising: the activation effects, based on unusual and extreme reaction conditions generating high local pressure and temperature, are similar and both allow to reach otherwise forbidden activation barriers,[62–64] sometimes promoting a switch in the selectivity of a reaction.[7,65,66]

A step forward to overcome the intrinsic drawbacks of Bucherer-Bergs reaction and its practical limitations in conventional batch methods relies on the use of biphasic liquid-liquid (water/EtOAc) continuous flow technologies[67] (entry 6).[46] In a modified version of Bucherer-Bergs reaction, commercially available amines were used as starting materials, and the reaction was performed under combined photochemical multistep gas-liquid continuous flow[14] using the biomass-based Me-THF, CO_2[68] instead of ammonium carbonate, and O_2[69] as cheap, abundant, and renewable oxidant (entry 7).[47]

Indeed, differently from conventional batch methods, continuous flow methods allow (i) an accurate control of the reaction parameters; (ii) to increase gas solubility in solvents and a more efficient heat and mass transfer; (iii) faster reactions, higher yields, and selectivity; (iv) increased productivity within scalability; (v) the possibility to use gases and solvents in pressurized conditions and at high temperature; and (vi) to operate in safer conditions when handling hazardous, toxic, or highly reactive reagents and intermediates, rapidly consumed and no more isolated.

3 Miscellaneous methods

3.1 CDI-mediated synthesis and cyclization of unsymmetrical ureas

An appealing and sustainable approach for the preparation of 3,5-disubstituted hydantoins, including 5-substituted-3-(alkoxycarbonyl)-alkyl-hydantoins, relies on the 1,1′-dicarbonyl imidazole (CDI)-mechanochemical-mediated activation of *N*-terminal position of α-amino esters or dipeptides. The *one-pot*/two-step cyclization reaction involves an unsymmetrical urea (formed

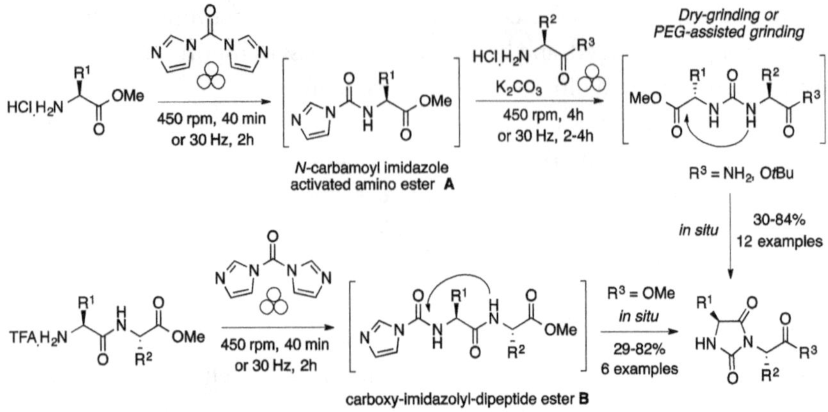

Scheme 6 Mechanochemical preparation of 3,5-disubstituted hydantoins from α-amino esters or dipeptides with or without PEG as grinding additive.[70,71]

from an *N*-carbamoyl imidazole-activated amino ester **A**) or carboxy-imidazolyl-dipeptide ester **B** (Scheme 6).[70,71]

The use of air-stable, cheap, and easy-to-handle CDI as acyl transfer reagent is the nontoxic alternative to generally toxic isocyanates (and not broadly commercialized), producing relatively innocuous waste by-products easy to remove (imidazole and gaseous CO_2). It is worth of notice that by mechanochemistry no base was needed to generate in situ the nucleophilic *N*-terminal amine.[21,70] Indeed the strong activation provided by mechanochemistry allowed the direct reaction of a nonnucleophilic dipeptide salt with CDI. It was proposed that the driving force of the reaction consisted of a sequence of acid-base reactions generating intermediates **C** and **D** having the distal nitrogen of the imidazole nucleus activated by protonation (Scheme 7).[70]

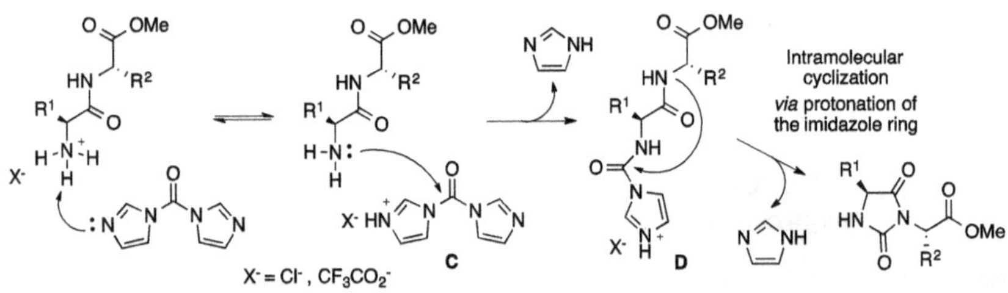

Scheme 7 Proposed mechanism for the mechanochemical CDI-mediated synthesis of 3,5-disubstituted hydantoins from dipeptide salts. *(Scheme adapted from reference Konnert, L.; Gonnet, L.; Halasz, I.; Suppo, J.-S.; de Figueiredo, R. M.; Campagne, J.-M.; Lamaty, F.; Martinez, J.; Colacino, E., Mechanochemical Preparation of 3,5-Disubstituted Hydantoins from Dipeptides and Unsymmetrical Ureas of Amino Acid Derivatives. J. Org. Chem.* ***2016****, 81, 9802–9809. https://doi.org/10.1021/acs.joc.6b01832.)*

Dry grinding experiments were complemented by experiments involving poly(ethylene) glycols (PEGs) as safe grinding assisting agents (liquid-assisted grinding [LAG]), leading to cleaner reaction profiles. Indeed the addition of PEGs in the second step of the reaction hampered the formation of the symmetrical urea generated by reaction of the residual starting amino ester with the newly formed *N*-carbamoyl imidazole amino ester intermediate **A** (Scheme 6).[71]

Similarly the synergic combination of CDI-mediated synthesis and mechanochemical processes proved to be successful also for the preparation of hydantoin-based bis-silylated derivatives, precursors of biohybrid bridged silsesquioxanes having uniform and narrow size (150 nm) distribution and morphology, obtained via a one-pot/two-step *mechanochemical assisted sol-gel procedure* (Scheme 8).[72]

Mechanochemistry provided an inedited approach to access organosilicon compounds in shorter reaction times compared with solution

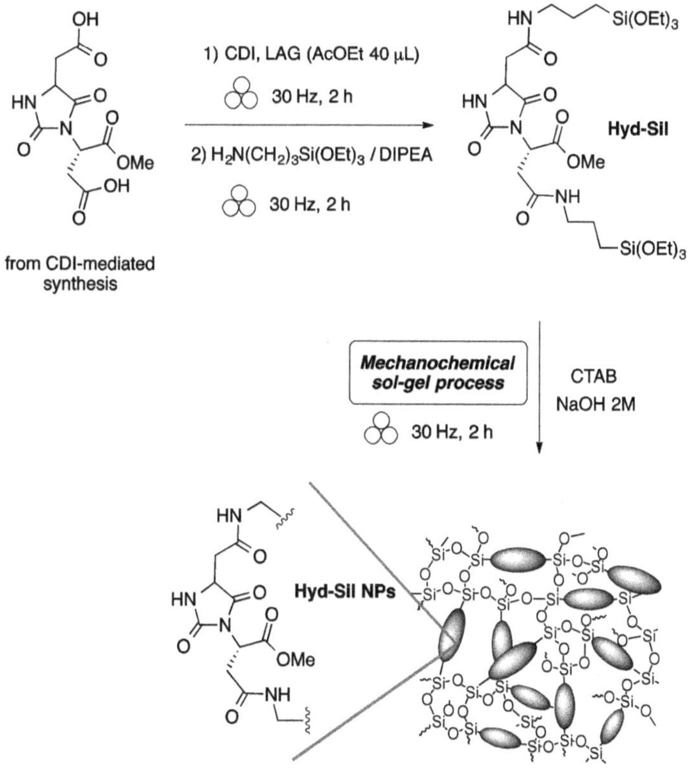

Scheme 8 Mechanochemical sol-gel process to hydantoin-based bridged polysilsesquioxanes[72] (CTAB, cetyltrimethylammonium bromide).

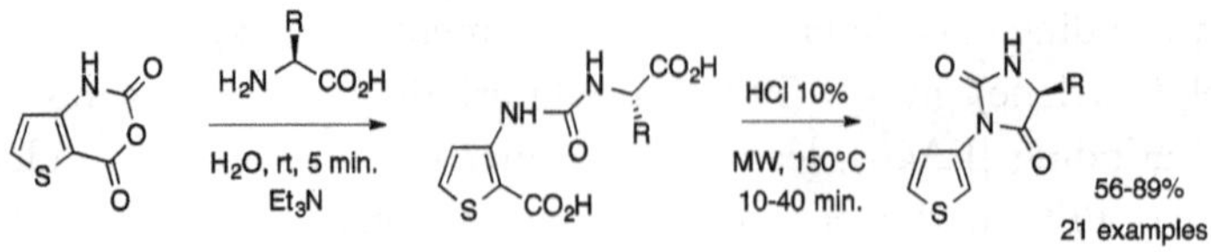

Scheme 9 Acid-mediated cyclocondensation/decarboxylation of 2-thienoic acid derivatives under microwave activation.

procedures under conventional stirring (4 h instead of 12 h), avoiding the need of excess of reagents and dry conditions and preventing the tricky and technically complex recovery of the bis-silylated organosilica precursors. Additionally the preparation of these nanohybrid systems, incorporating complex functionalities covalently bound into the silica framework and endowed with potential biological activity,[25] was not successful under commonly used Stöber-like sol-gel procedures.[73]

The preparation of hydantoins via a base-[54,74] or acid-mediated[75] intramolecular cyclization reaction of unsymmetrical ureido derivatives obtained from amino esters was also investigated under microwave activation. Microwaves were effective to access stereoselectively, a series of polycyclic hydantoins by a two-step reaction, from azomethine ylide cycloaddition/$Ba(OH)_2$-mediated cyclization reaction of carbanilide[54,74] (similarly to Scheme 4). Several 3-(thien-3-yl)imidazolidine-2,4-dione analogues[75] were also prepared from thiaisatoic anhydride via an acid-mediated cyclocondensation/decarboxylation of the corresponding 2-thienoic acids (Scheme 9).[75]

3.2 Multicomponent reactions (MCR)

Merging synthetically efficient methodologies such as MCR, able to generate several bonds in a single operation, with nontraditional activation methods is a very promising approach to the sustainable preparation of complex heterocycles, including highly functionalized hydantoins. Ugi-cascade reactions[76] were recently disclosed under microwave activation[77–79] (Scheme 10).

5-Hydroxyl hydantoins could be obtained via the *one-pot* two-step Ugi five-component condensation reaction/cyclization reaction in moderate yields (pathway A,[77] Scheme 10). A similar approach involving propionic acid was also investigated (pathway B)[79], but the formation of the corresponding 5-hydroxyl hydantoins was never observed. The authors proposed a reaction mechanism claiming that the driving force of the

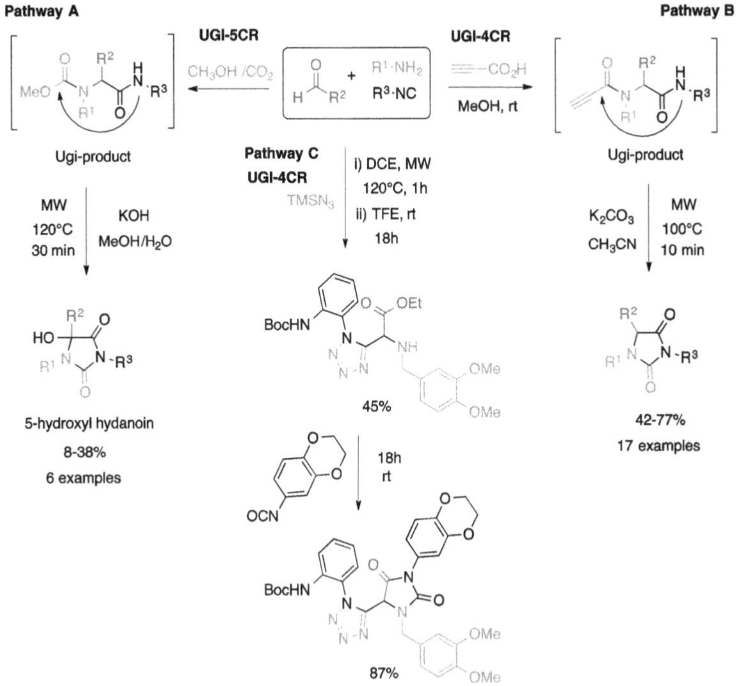

Scheme 10 Access to hydantoin derivatives by Ugi-MCR/cascade reactions under microwave activation.

reaction is related to the departure of acetylene, favored under microwave activation.[79] Highly functionalized 1,5-disubstituted tetrazolo-hydantoins were also prepared in acid media (pathway C)[78] via a sequence involving a TMS-N_3-Ugi MCR/isocyanate addition/cyclisation reaction, with microwave activation used only during the first step, even if the generality of the method cannot be fully assessed because only one example for the preparation of the hydantoin scaffold was reported.

3.3 Photochemical rearrangements

The application of light-induced transformations to the hydantoins scaffold is very scarce, despite the importance of photochemistry as eco-friendly synthetic tools to access organic molecules.[10] Intrinsically green and despite its operational simplicity and the mild reaction conditions requiring the absorption of a photon (a green reagent not generating any waste) to activate the reaction, scaling up at industrial level remains a challenge.

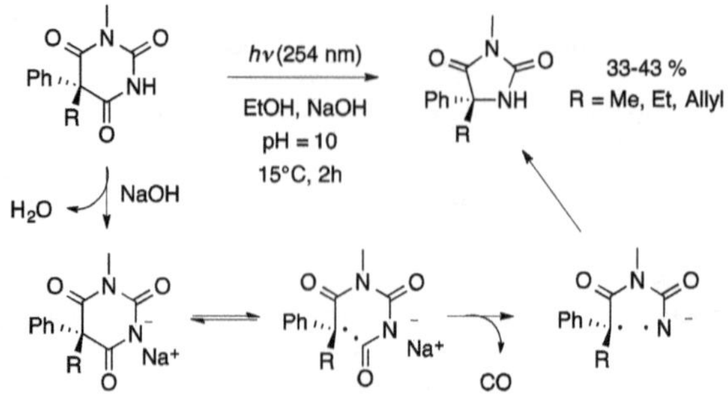

Scheme 11 Photoinduced contraction of barbituric acid derivatives.[80–82]

The only examples reported address the obtention of hydantoins by photochemically induced hydrolysis of barbituric acid derivatives,[80–82] the reaction occurring stereospecifically in aqueous alkaline medium. The conversion occurs through the homolytic scission of C-4/C-5 bond, followed by carbon monoxide elimination and stereospecific ring contraction of the pyrimidine ring without a change in the relative position of the substituents at C-5 carbon (Scheme 11).

More recently a 100% atom economy reaction involving α-ketoacids and carbodiimides was discovered adventitiously. The visible light-induced domino reaction led to 5-hydroxyhydantoins via an *O*-acyl urea intermediate. The strong nucleophilic character of the ketocarbonyl group allowed the intramolecular nucleophilic addition of the *N*-substituted urea to generate an iminooxazolidin-5-one derivative, rearranging spontaneously to the corresponding 5-hydroxyhydantoin (Scheme 12).[83]

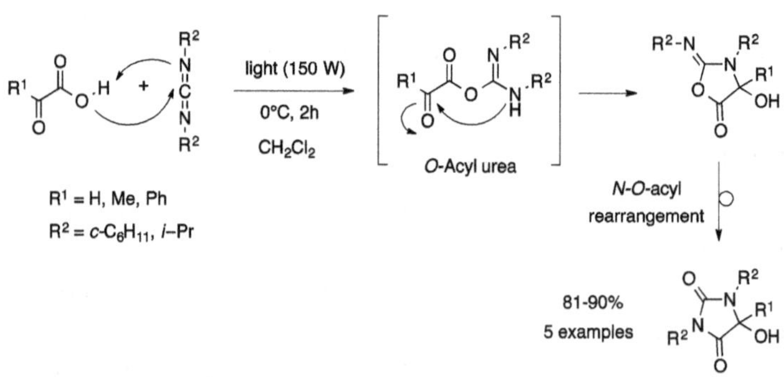

Scheme 12 Visible light-induced domino reaction to 5-hydroxyhydantoins.[83]

3.4 Preparation of active pharmaceutical ingredients and commodity chemicals

The preparation of commodity chemicals or value-added product for the industry involving alternative energy inputs has been rarely exploited. However, mechanochemistry remains the more sustainable process among the diverse nontraditional activation methods to access hydantoin-based active pharmaceutical ingredients (APIs),[26,40,48,51,84] enzymes inhibitors,[70] commodity chemicals,[51,71] and advanced silicon-based nanomaterials for biomedical applications.[72] Mechanochemistry is a technology "benign by design"[85] and constitutes a novel and attractive avenue fostering innovation, paving the way to an active area of investigation referred to as "medicinal mechanochemistry,[86] aiming to disclose new sustainable alternatives to solution-based procedures for the preparation of APIs. Operationally simple and energetically efficient (also at a manufacturing scale through twin screw extrusion processes),[87] the mechanochemical approach highly reduces the environmental impact for the production of biomolecules. Indeed, cleaner reaction profiles and simplified work-up procedures associated with avoidance of solvents and high temperatures (sometimes responsible for API degradation) allow reagent and waste economy. The notable features described so far are shown for the mechanochemical preparation of marketed drugs, commodity chemicals, and hybrid nanoparticles, avoiding harmful reaction conditions and compared with chemistry in solution when possible.

3.5 Antiepileptic drug phenytoin

More recently, nonconventional activation methods were explored for the preparation of the marketed antiepileptic drug phenytoin (5,5-diphenylhydantoin) (Table 3). Historically prepared by Biltz's reaction,[31] activation by microwaves (entries 1 and 2),[34,36] ultrasound (entry 4),[37,38] or mechanochemically (entry 7) [40] led to phenytoin mostly in high yields, with concomitant formation of variable amount of 3α,6α-diphenylglycoluril by-product **2**. In the case of Bucherer-Bergs reaction and despite the limitations usually encountered and already highlighted, the preparation of phenytoin resulted straightforward whatever was the activation method used (entries 3, 5, and 6).

As for sonochemically enhanced reaction, mechanochemical procedures are characterized by extreme reaction conditions (high local temperature and pressure).[62–64] In the case of phenytoin, noteworthy is the selectivity switch observed in the case of the mechanochemical Biltz reaction (entry 7), compared with solution-based procedures (including ultrasound-activated processes), where the favorite pathway was the 1,2-shift rearrangement of a phenyl group.[49] Independently on the mechanochemical process conditions investigated (both in vibrating and planetary mills),[40] phenytoin was always recovered in a moderate yield (10%–44%), being the 3α,6α-diphenylglycoluril by-product **2** and 1,3-bis(2-oxo-1,2-diphenylethylidene)urea mainly formed, the last one never observed in solution previously (Scheme 13). In this regard, one of the notable features of mechanochemistry relies on the possibility to alter the selectivity of a reaction,[66] allowing activation barriers and sometimes forbidden pathways, which give access to new products otherwise impossible to be obtained in solution.

A modified Read method[32] was the only possible alternative to Biltz's reaction (still remaining the principal pathway to access phenytoin in solution, also at large scale) or to the intrinsically toxic Bucherer-Bergs reaction to access phenytoin by mechanochemistry (Scheme 13).

The *one-pot*/two-step preparation of phenytoin involving 5,5-diphenylglycine methyl ester hydrochloride and trimethylsilyl isocyanate (TMS-NCO) generated an intermediate urea that can be either isolated or cyclized in situ in basic conditions (Scheme 13)[40] As previously highlighted, also in this case, no base was required to generate the nucleophilic free amino group from the quaternary amino acid derivative before its addition on the isocyanate. Indeed the base is generated in situ thanks to a "cooperative effect" between the substrate hydrochloride salt and the water added. In the presence of water, TMS-NCO is decomposed in situ generating gaseous ammonia NH_3 and hexamethyldisilazane (HMDS), the last one needing a source of HCl, provided by the amino ester derivative hydrochloride salt. Moreover, water played the additional task to remove in situ the *N*-TMS-protecting group from the transient trimethylsilyl ureido derivative, leading to the monosubstituted urea, cyclized in the second step. The successful symbiosis of mechanochemical procedures (also involving gaseous reactants)[88] and the suitable choice of the reactants (and their stoichiometry) provided an "all-in-one" approach leading to "reagent economy" and simplified work-up procedures (also performed by milling). The addition of EtOAc, in which phenytoin is soluble, allowed the precipitation of the inorganic salts (CsCl), while the other wastes, consisting of easily removable gases (CO_2 and NH_3) or volatile liquids (MeOH and Me_3SiOH), were eliminated by evaporation.

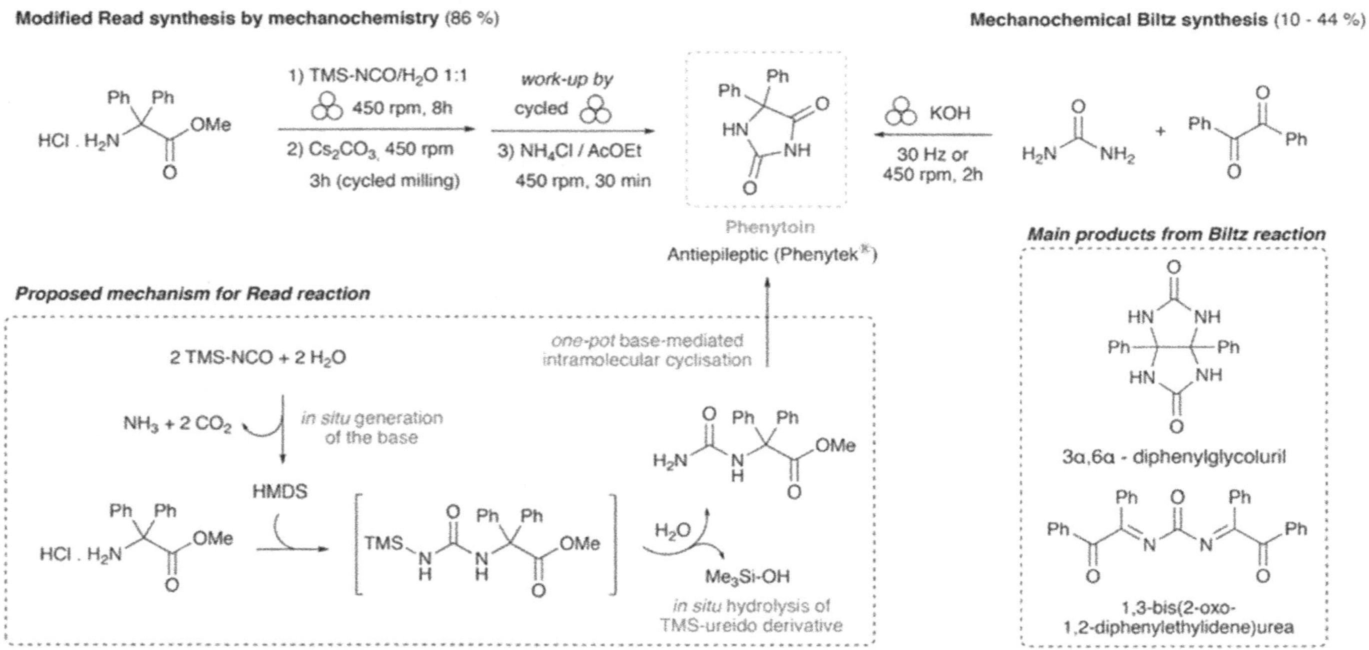

Scheme 13 Mechanochemical Biltz and Read methods for the preparation of the antiepileptic drug phenytoin.[40] *(Image reproduced with the permission of Royal Society of Chemistry from reference Colacino, E.; Porcheddu, A.; Charnay, C.; Delogu, F., From Enabling Technologies to Medicinal Mechanochemistry: An Eco-Friendly Access to Hydantoin-Based Active Pharmaceutical Ingredients. React. Chem. Eng.* ***2019****, 4, 1179–1188. https://doi.org/10.1039/c9re00069k.)*

3.6 Anticonvulsant drug ethotoin

Two solution-based procedures were reported for the preparation of ethotoin, leading to similar average yields (65%) after chromatographic purification and based on a cyclization reaction of 5-phenylhydantoic acid[89] or a rearrangement of ethyl *N*-Boc-phenylglycinamide with triflic anhydride.[90] However, with the exception of ball-milling procedures,[51,84] alternative energy inputs were never explored so far to access the anticonvulsant drug ethotoin. Mechanochemistry provided safer procedures and a straightforward preparation of ethotoin by four different methods[51,84] involving commercially available and cheap substrates (methods **A–D**, Scheme 14).

Methods **A–D** share some common features:[92] (i) 5-phenylglycine methyl ester hydrochloride was used as reactant at some point of the process; (ii) substrates were activated by a CDI (methods **B**[51] and **D**[93]/**D′**[84]), alternatively (iii) (iso)cyanates reactants (Read method **A**[51] and **A′**[40] to **C**[51]) were used from commercially available suppliers or generated in situ by Lossen rearrangement (method **D**[84]), (iv) ethotoin is always obtained via a base-mediated cyclization step from unsymmetrical urea generated in situ, and (v) work-up is simplified because of the cleaner reaction profiles compared with solution synthesis, the final product being recovered as a precipitate.

As mentioned so far for the preparation of phenytoin, the use of a KNCO/H_2O mixture (method **A′**[40]) prevent to add a base ("reagent economy"). Together with the advantages already mentioned earlier on the use of eco-friendly carbonyl equivalent CDI, it is worth mentioning that CDI is also a convenient reagent adopted in the pharmaceutical industry for the large-scale production of APIs.[94, 95] A sustainable alternative to the use of hazardous ethyl isocyanate was to generate it in situ during the fourfold mechanochemical CDI-mediated cascade Lossen transposition from the corresponding hydroxamic acid (method **D′**[84]). CDI played a double role: it activated the hydroxamic acid while releasing imidazole, which promoted the self-sustaining transposition reaction leading to the cyclic carbonate **Q**, which decomposed to ethyl isocyanate and CO_2, similarly to solution-based procedures.[96] Indeed, once ethyl isocyanate is generated in situ, it is immediately reacted, with no polymerization and no formation of ethylamine from the corresponding propyl hydroxamic acid precursor, thus avoiding the formation of symmetrical diethyl urea by-product by reaction with CDI, a side reaction usually encountered in solution. The use of mechanochemistry to activate Lossen transposition is also advantageous to overcome the drawbacks related to the low solubility of hydroxamic acids in solution, a

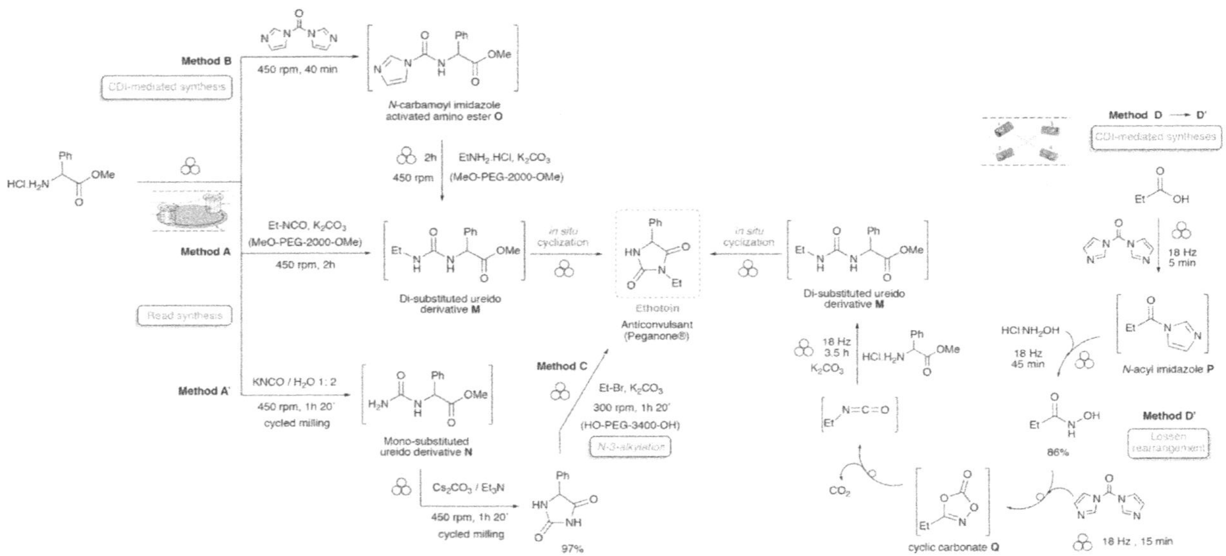

Method	Yields (%)		Waste
	Without PEG	With PEG	
A (Read's by EtNCO)[43]	35	63	2 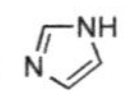 H_2O KCl(Br) MeOH CO_2
B (CDI / $EtNH_2 \cdot HCl$)[43]	35	65	
C (*N*-3-alkylation by Et-Br)[43]	35	45	
D (Lossen's by CDI)[85]	52	n.r.	

Scheme 14 Mechanochemical preparation of anticonvulsant drug ethotoin.[51,84] (n.p. = not performed. The scheme is partially readapted from Refs. [26,91] with the permission of Royal Society of Chemistry.)

real limitation to fully exploit the potential of this rearrangement in solution. It is worth highlighting that mechanochemical Lossen transposition (methods **D/D′**)[84, 93] outperformed also in comparison with the three previously reported mechanochemical methods, namely, **A**, **B**, or **A′/C**,[40,51] in terms of yields (ethotoin was obtained in 52% instead of 35% yield) and circumvented the use of an excess of grinding additives such as PEGs, able to increase the mechanochemical productivity of ethotoin, obtained with cleaner reaction profiles (Scheme 14).[51,71,97]

PEG solvents as grinding agents were already used for the preparation of cocrystals (POLAG effect)[98] and in metal-mediated tribochemical transformations.[99, 100] Among the intrinsic advantages of PEG polymers, extensively described in the previous sections, PEG matrix is a valid, safe, and eco-friendly alternative to classic solvents (e.g., CH_3NO_2 or EtOAc) used in liquid-assisted grinding (LAG)[101] procedures, being nontoxic, nonvolatile, and not flammable, reducing the risks of explosions or overpressure potentially encountered on large-scale LAG procedures.

3.7 APIs containing an *N*-acylhydrazone: Nitrofurantoin and dantrolene

Mechanochemistry proved its efficiency in the stereoselective gram-scale preparation of several hydantoins containing the *N*-acylhydrazone bond, including two APIs such as nitrofurantoin (an antibacterial agent) and dantrolene (used for the treatment of malignant hyperthermia) (Scheme 15).[48]

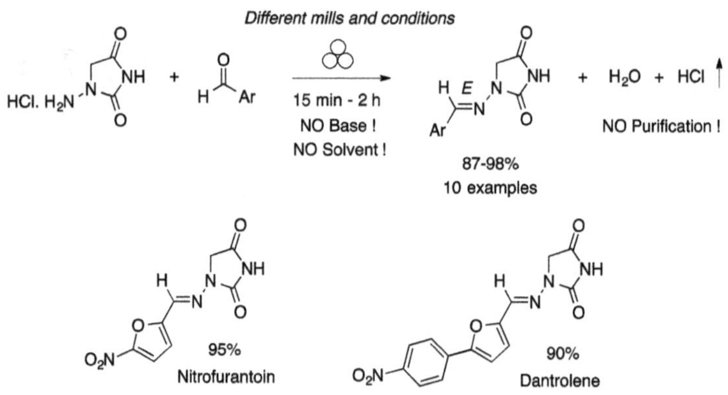

Scheme 15 Mechanochemical preparation of *N*-acylhydrazone APIs: nitrofurantoin and dantrolene. *(The scheme is partially readapted from referernce Colacino, E.; Porcheddu, A.; Charnay, C.; Delogu, F., From Enabling Technologies to Medicinal Mechanochemistry: An Eco-Friendly Access to Hydantoin-Based Active Pharmaceutical Ingredients. React. Chem. Eng.* ***2019****, 4, 1179–1188. https://doi.org/10.1039/c9re00069k with the permission of Royal Society of Chemistry.)*

The reaction conditions are eco-compatible and with low environmental impact: as previously illustrated for other mechanochemical transformations, also in this case the activation provided by mechanical shocks, avoided the use of any base (*reagent economy*). The reaction is practically waste-free (*waste economy*) and even not a drop of solvent was necessary during all the process (work-up included). Indeed, only water and gaseous HCl were formed as by-products, and the final products (crystalline powders) were obtained pure and without any postreaction work-up, just by scratching the powders out of the jars. Worth of notice is the stability of the *E*-hydrazones in the presence of HCl and water: no hydrolysis was observed, with the water most probably confined in the crystals and thus not available to promote any isomerization/hydrolysis. The reaction is general, and it was investigated using different milling devices and jar materials. The API nitrofurantoin was also analyzed by PXRD: the polymorph obtained by milling was identical to the FDA-approved drug, prefiguring mechanochemistry as an alternative process for drug manufacturing of these drugs. For both solution and ball-milling preparation, the green metrics and costs of the syntheses were compared, clearly showing the advantages of the mechanochemical process (Table 4).

The mechanochemical preparation of nitrofurantoin and dantrolene presents several additional advantages compared with solution-based procedures: (i) elimination of (harmful) solvents (DMF, ACN, and EtOH); (ii) simplification of the process by elimination of large amounts of concentrated aqueous solutions of strong bases and acids (NaOH, HCl 30%, and HNO_3-H_2SO_4) and avoiding corrosion problems and hydrazone hydrolysis; (iii) reduction of the synthetic steps; (iv) reduction of the energetic costs of the process, avoiding heating (between 45°C and 90°C) or cooling (between −2°C and 4°C) of the reaction mixtures; (v) improved

Table 4 Comparative green metrics for nitrofurantoin and dantrolene.[100]

Nitrofurantoin	In solution/by milling	Dantrolene
8/0.25	Reaction time (h)	1/2
95/95	Yield (%)	95/90
16/0.29	*E*-factor	239/0.30
81	Atom economy	85
n.d./40.7	Costs (€/g)[a]	133.9/54.7
5	Costs (€/g) from commercial sources	500

[a]Energetic costs are not included.

throughput/hour and environmental footprint of the process; and (vi) reduced costs for the preparation of 1 g of final product.

3.8 Value-added *N*-1 chloramines

The solvent-based preparation of *N*-1 chlorinated hydantoins, finding applications as antimicrobial agents for polymer textiles,[102] was extensively described in the presence of gaseous chlorine, *N*-chlorosuccinimide (NCS), trichloroisocyanuric acid (TCCA), *tert*-butyl, and unstable sodium hypochlorite. However, the selectivity for *N*-1/*N*-3-chlorination strictly depended on the pH variation and adjustments. Surprisingly, only the use of mechanochemistry was reported as nonconventional activation methods for the *N*-chlorination of hydantoins. The two-step sequence relies on the preparation of *N*-3-allyl (ADMH) and *N*-3-ethyl (EDMH) 5,5-dimethylhydantoins via an intramolecular cyclization reaction of the corresponding unsymmetrical ureido derivatives obtained from amino esters (via a CDI-mediated or Read method), followed by the *N*-1 chlorination reaction (Scheme 16).[51]

The *N*-1 chlorination of *N*-3-allyl (ADMH) and *N*-3-ethyl (EDMH) 5,5-dimethylhydantoins was achieved in the presence of solid and stable $Ca(ClO)_2$, and the final products were recovered after precipitation of the calcium salts in cyclohexane. The use of solid $Ca(ClO)_2$ in mechanochemical procedures was a safe and easy-to-handle alternative to the unstable NaClO and outperformed in comparison with the swimming pool bactericide TTCA, with improved kinetics and yields and better waste and atom economy.

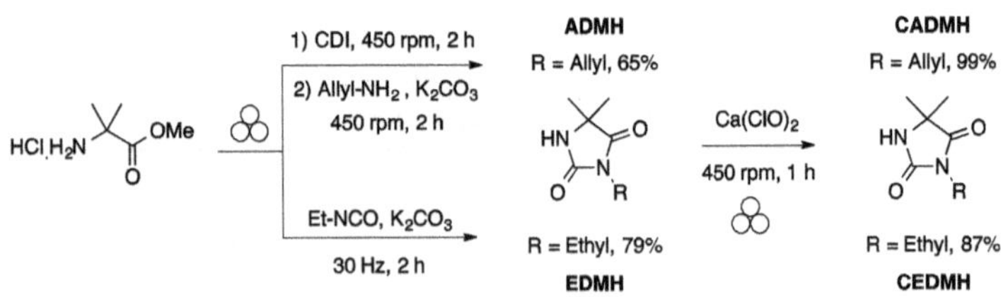

Scheme 16 Mechanochemical preparation of bactericides: *N*-1 chlorinated hydantoins.[51] *(The scheme is partially readapted from reference Colacino, E.; Porcheddu, A.; Charnay, C.; Delogu, F., From Enabling Technologies to Medicinal Mechanochemistry: an Eco-Friendly Access to Hydantoin-Based Active Pharmaceutical Ingredients. React. Chem. Eng.* ***2019****, 4, 1179–1188. https://doi.org/10.1039/c9re00069k with the permission of Royal Society of Chemistry.)*

4 Conclusions

Compared with the classical thermal methods, the association of enabling technologies and new strategies in solution or solvent-free allowed the access to a large panel of diversely substituted hydantoins and nanoparticles for advanced applications. Harmful reagents and reaction conditions could be avoided; kinetics (*time economy*), yields, and selectivity are improved (or switched); reaction profiles were cleaner; reaction conditions were safer (easily controlled) and milder; the work-up procedures are simplified. These advantages translate into a reduced environmental footprint (*reagent and waste economy*) of the whole process. In some cases, new compounds otherwise not easily accessible (or not accessible at all) could be obtained by a careful selection of the most suitable technology for solution-based or solvent-free procedures (dry media or milling).

However, the benefits of unconventional approaches go beyond, and considerations related to the scalability of each technology, heat, mass transfer, and product separation need to be taken into account and studied in cooperation with chemical engineers to disclose case by case the most highly desirable and energy-efficient process for a specific (industrial) application.

Acknowledgments

E.C. is grateful to J.W. for support and dedicates this manuscript to her father. Part of the text was readapted from Ref. [26] by kind permission of the Royal Society of Chemistry. This chapter is based upon work from COST Action CA18112,[103–105] supported by COST (European Cooperation in Science and Technology).[106].

References

1. Erythropel, H. C. E.; Zimmerman, J. B.; de Winter, T. M.; Petitjean, L.; Melnikov, F.; Lam, C. H.; Lounsbury, A. W.; Mellor, K. E.; Janković, N. Z.; Tu, Q.; Pincus, L. N.; Falinski, M. M.; Shi, W.; Coish, P.; Plata, D. L.; Anastas, P. T. The Green ChemisTREE: 20 Years after Taking Root with the 12 Principles. *Green Chem.* **2018**, *20*, 1929–1961. https://doi.org/10.1039/c8gc00482j.
2. Roschangar, F.; Colberg, J.; Dunn, P. J.; Gallou, F.; Hayler, J. D.; Koenig, S. G.; Kopach, M. E.; Leahy, D. K.; Mergelsberg, I.; Tucker, J. L.; Sheldon, R. A.; Senanayake, C. H. A Deeper Shade of Green: Inspiring Sustainable Drug Manufacturing. *Green Chem.* **2017**, *19*, 281–285. https://doi.org/10.1039/C6GC02901A.
3. Roschangar, R.; Sheldon, R. A.; Senanayake, C. H. Overcoming Barriers to Green Chemistry in the Pharmaceutical Industry—The Green Aspiration LevelTM Concept. *Green Chem.* **2015**, *17*, 752–768. https://doi.org/10.1039/C4GC01563K.
4. Nasir Baig, R. B.; Varma, R. S. Alternative Energy Input: Mechanochemical, Microwave and Ultrasound-Assisted Organic Synthesis. *Chem. Soc. Rev.* **2012**, *41*, 1559–1584. https://doi.org/10.1039/c1cs15204a.

5. Kappe, C. O.; Stadler, A.; Dallinger, D. *Microwaves in Organic and Medicinal Chemistry*, 2nd ed, Vol. 52; Wiley-VCH, 2012; p. 683. Completely Revised and Enlarged Edition. ISBN: 978-3-527-33185-7.
6. Kappe, C. O. Controlled Microwave Heating in Modern Organic Synthesis. *Angew. Chem. Int. Ed.* **2004**, *43*, 6250–6284. https://doi.org/10.1002/anie.200400655.
7. Cravotto, G.; Cintas, P. Forcing and Controlling Chemical Reactions with Ultrasound. *Angew. Chem. Int. Ed.* **2007**, *46*, 5476–5478. https://doi.org/10.1002/anie.200701567.
8. Lupacchini, M.; Mascitti, A.; Giachi, G.; Tonucci, L.; d'Alessandro, N.; Martinez, J.; Colacino, E. Sonochemistry in Non-Conventional, Green Solvents or Solvent-Free Reactions. *Tetrahedron* **2017**, *73*, 609–653. https://doi.org/10.1016/j.tet.2016.12.014.
9. Yan, M.; Kawamata, Y.; Baran, P. S. Synthetic Organic Electrochemistry: Calling All Engineers. *Angew. Chem. Int. Ed.* **2008**, *57*, 4149–4155. https://doi.org/10.1002/anie.201707584.
10. Kärkäs, M. D.; Porco, J. A., Jr.; Stephenson, C. R. J. Photochemical Approaches to Complex Chemotypes: Applications in Natural Product Synthesis. *Chem. Rev.* **2016**, *116*, 9683–9747. https://doi.org/10.1021/acs.chemrev.5b00760.
11. Akwi, F. M.; Watts, P. Continuous Flow Chemistry: where Are we Now? Recent Applications, Challenges and Limitations. *Chem. Commun.* **2018**, *54*, 13894–13928. https://doi.org/10.1039/c8cc07427e.
12. Bogdan, A. R.; Dombrowski, A. W. Emerging Trends in Flow Chemistry and Applications to the Pharmaceutical Industry. *J. Med. Chem.* **2019**, *62*, 6422–6468. https://doi.org/10.1021/acs.jmedchem.8b01760.
13. Plutschack, M. B.; Pieber, B.; Gilmore, K.; Seeberger, P. H. The Hitchhiker's Guide to Flow Chemistry. *Chem. Rev.* **2017**, *117*, 11796–11893. https://doi.org/10.1021/acs.chemrev.7b00183.
14. Britton, J.; Raston, C. B. Multi-Step Continuous-Flow Synthesis. *Chem. Soc. Rev.* **2017**, *46*, 1250–1271. https://doi.org/10.1039/c6cs00830e.
15. Takacs, L. The Historical Development of Mechanochemistry. *Chem. Soc. Rev.* **2013**, *42*, 7649–7659. https://doi.org/10.1039/c2cs35442j.
16. Stuart, J. L.; Adams, C. J.; Bolm, C.; Braga, D.; Collier, P.; Friscic, T.; Grepioni, F.; Harris, K. D. M.; Hyett, G.; Jones, W.; Krebs, A.; Mack, J.; Maini, L.; Orpen, A. G.; Parkin, I. P.; Shearouse, W. C.; Steed, J. W.; Waddell, D. C. Mechanochemistry: Opportunities for New and Cleaner Synthesis. *Chem. Soc. Rev.* **2012**, *41*, 413–447. https://doi.org/10.1039/C1CS15171A.
17. Stolle, A.; Ranu, B., Eds. *Green Chemistry Series*; *Ball Milling Towards Green Synthesis: Applications, Projects, Challenges*; RSC, 2015.
18. *Mechanochemistry: From Functional Solids to Single Molecule, Faraday Discuss*, Vol. 170; RSC: Cambridge, UK, 2014.
19. Margetic, D.; Strukil, V. *Mechanochemical Organic Synthesis*; Elsevier: Amsterdam, NL, 2016; p. 386.
20. Regulation (EC) No 1907/2006 of the European Parliament and of the Council of 1918 December 2006 Concerning the Registration, Evaluation, Authorisation and Restriction of Chemicals (REACH), Establishing a European Chemicals Agency, http://ec.europa.eu/enter.
21. This expression was used for the first time in: Lanzillotto, M.; Konnert, L.; Lamaty, F.; Martinez, J.; Colacino, E. Mechanochemical 1,1′-Carbonyldiimidazole-Mediated Synthesis of Carbamates. *ACS Sustain. Chem. Eng.* **2015**, *3*, 2882–2889. https://doi.org/10.1021/acssuschemeng.5b00819.
22. Anastas, P. T.; Farris, C. A. Benign by Design: Alternative Synthetic Design for Pollution Prevention. *ACS Symposium Series 557*; American Chemical Society: Washington DC, 1994.

23. Colacino, E.; Dayaker, G.; Morère, A.; Friščić, T. Introducing Students to Mechanochemistry via Environmentally Friendly Organic Synthesis Using a Solvent-Free Mechanochemical Preparation of the Antidiabetic Drug Tolbutamide. *J. Chem. Educ.* **2019**, *96*, 766–771. https://doi.org/10.1021/acs.jchemed.8b00459.
24. Konnert, L.; Lamaty, F.; Martinez, J.; Colacino, E. Recent Advances in the Synthesis of Hydantoins: The State of the Art of a Valuable Scaffold. *Chem. Rev.* **2017**, *117*, 13757–13809. https://doi.org/10.1021/acs.chemrev.7b00067.
25. Meusel, M.; Gütschow, M. Recent Developments in hydantoin Chemistry. A review. *Org. Prep. Proced. Int.* **2009**, *36*, 391–443. https://doi.org/10.1080/00304940409356627.
26. Colacino, E.; Porcheddu, A.; Charnay, C.; Delogu, F. From Enabling Technologies to Medicinal Mechanochemistry: an Eco-Friendly Access to Hydantoin-Based Active Pharmaceutical Ingredients. *React. Chem. Eng.* **2019**, *4*, 1179–1188. https://doi.org/10.1039/c9re00069k.
27. Data extrapolated by Chemical abstract data base (Scifinder, w. s. o., accessed September 23, 2019 *Data extrapolated by Chemical abstract data base.* Scifinder http://www.scifinder.org/. accessed 23 September 2019.
28. https://clinicaltrials.gov/. accessed 23 September 2019.
29. https://www.clinicaltrialsregister.eu/. accessed 23 September 2019.
30. U.S. Food and Drug Administration. http://www.accessdata.fda.gov/scripts/cder/drugsatfda/. accessed 23 September 2019.
31. Biltz, H. Constitution of the Products of the Interaction of Substituted Carbamides on Benzil and Certain New Methods for the Preparation of 5,5-Diphenylhydantoin. *Ber. Dtsch. Chem. Ges.* **1908**, *41*, 1379–1393.
32. Read, W. T. Researches on Hydantoins. Synthesis of the Soporific 4,4-Phenylethyl-Hydantoin (Nirvanol). *J. Am. Chem. Soc.* **1922**, *44*, 1746–1755. https://doi.org/10.1021/ja01429a017.
33. Bucherer, H. T.; Lieb, V. A. Syntheses of Hydantoins. II. Formation of Substituted Hydantoins from Aldehydes and Ketones. *J. Prakt. Chem.* **1934**, *141*, 5–43.
34. Safari, J.; Naeimi, H.; Ghanbari, M. M.; Sabzi Fini, O. Preparation of Phenytoin Derivatives under Solvent-Free Conditions Using Microwave Irradiation. *Russ. J. Org. Chem.* **2009**, *45*, 477–479. https://doi.org/10.1134/S1070428009030270.
35. Paul, S.; Gupta, M.; Gupta, R.; Loupy, A. Microwave Assisted Synthesis of 1,5-Disubstituted Hydantoins and Thiohydantoins in Solvent-Free Conditions. *Synthesis* **2002**, *1*, 75–78. https://doi.org/10.1055/s-2002-19308.
36. Muccioli, G. G.; Poupaert, J. H.; Wouters, J.; Norberg, B.; Poppitz, W.; Scriba, G. K. E.; Lambert, D. M. A Rapid and Efficient Microwave-Assisted Synthesis of Hydantoins and Thiohydantoins. *Tetrahedron* **2003**, *59*, 1301–1307. https://doi.org/10.1016/S0040-4020(03)00033-4.
37. Safari, J.; Moshtael Arani, N.; Ramezan Isfahani, A. Ultrasound-Enhanced Green Synthesis of 5,5-Diphenylhydantoin Derivatives Using Symmetrical or Unsymmetrical Benzils. *Chin. J. Chem.* **2010**, *28*, 255–258. https://doi.org/10.1002/cjoc.201090062.
38. Safari, J.; Arani, N. M. A Rapid and Efficient Ultrasound-Assisted Synthesis of 5,5-Diphenylhydantoins and 5,5-Diphenyl-2-Thiohydantoins. *Ultrason. Sonochem.* **2011**, *18*, 640–643. https://doi.org/10.1016/j.ultsonch.2010.09.001.
39. Fahmy, A. F. M.; Rizk, S. A.; Hemdan, M. M.; El-Sayed, A. A.; Hassaballah, A. I. Efficient Green Synthesis and Computational Chemical Study of Some Interesting Heterocyclic Derivatives as Insecticidal Agents. *J. Heterocyclic Chem.* **2018**, *55*, 2545–2555. https://doi.org/10.1002/jhet.3308.
40. Konnert, L.; Reneaud, B.; de Figueiredo, R. M.; Campagne, J.-M.; Lamaty, F.; Martinez, J.; Colacino, E. Mechanochemical Preparation of Hydantoins from Amino

Esters: Application to the Synthesis of the Antiepileptic Drug Phenytoin. *J. Org. Chem.* **2014**, *79*, 10132–10142. https://doi.org/10.1021/jo5017629.

41. Safari, J.; Gandomi-Ravandi, S.; Javadian, L. Microwave-Promoted Facile and Rapid Synthesis Procedure for the Efficient Synthesis of 5,5-Disubstituted Hydantoins. *Synthetic Commun.* **2013**, *43*, 3115–3120. https://doi.org/10.1080/00397911.2012.730647.
42. Prevet, H.; Flipo, M.; Roussel, P.; Deprez, B.; Willand, N. Microwave-Assisted Synthesis of Functionalized Spirohydantoins as 3-D Privileged Fragments for Scouting the Chemical Space. *Tetrahedron Lett.* **2016**, *57*, 2888–2894. https://doi.org/10.1016/j.tetlet.2016.05.065.
43. Li, J.; Li, L.; Li, T.; Li, H.; Liu, J. An Efficient and Convenient Procedure for the Synthesis of 5,5-Disubstituted Hydantoins under Ultrasound. *Ultrason. Sonochem.* **1996**, *3*, S141–S143. https://doi.org/10.1016/1350-1477(96)00011-2.
44. Safari, J.; Javadian, L. Montmorillonite K-10 as a Catalyst in the Synthesis of 5, 5-Disubstituted Hydantoins under Ultrasound Irradiation. *J. Chem. Sci.* **2013**, *125*, 981–987. https://doi.org/10.1007/s12039-013-0427-5.
45. Maddah, B. Highly Efficient and Rapid Synthesis of Diverse Hydantoin Derivatives Using Nano-Ordered ZnO Catalyst under Mechanochemical Ball Milling. *Iran. Chem. Soc.* **2017**, *5*, 58–66.
46. Monteiro, J. L.; Piebera, B.; Corrêa, A. J.; Kappe, C. O. Continuous Synthesis of Hydantoins: Intensifying the Bucherer–Bergs Reaction. *Synlett* **2016**, *27*, 83–87. https://doi.org/10.1055/s-0035-1560317.
47. Vukelic´, S.; Koksch, B.; Seeberger, P. H.; Gilmore, K. A Sustainable, Semi-Continuous Flow Synthesis of Hydantoins. *Chem. Eur. J.* **2016**, *22*, 13451–13454. https://doi.org/10.1002/chem.201602609.
48. Colacino, E.; Porcheddu, A.; Halasz, I.; Charnay, C.; Delogu, F.; Guerra, R.; Fullenwarth, J. Mechanochemistry for "No Solvent, No Base" Preparation of Hydantoin-Based Active Pharmaceutical Ingredients: Nitrofurantoin and Dantrolene. *Green Chem.* **2018**, *20*, 2973–2977. https://doi.org/10.1039/c8gc01345d.
49. Dunnavant, W. R.; James, F. L. Molecular Rearrangements. I. The Base-catalyzed Condensation of Benzil with Urea. *J. Am. Chem. Soc.* **1956**, *78*, 2740–2743. https://doi.org/10.1021/ja01593a025.
50. Butler, A. R.; Leitch, E. Mechanistic Studies in the Chemistry of Urea. Part 2. Reaction with Benzil, 4,4′-Dimethylbenzil, and 4,4′-Dimethoxybenzil. *J. Chem. Soc. Perkin Trans. 2* **1977**, *14*, 1972–1976. https://doi.org/10.1039/P29770001972.
51. Konnert, L.; Dimassi, M.; Gonnet, L.; Lamaty, F.; Martinez, J.; Colacino, E. Poly(Ethylene) Glycols and Mechanochemistry for the Preparation of Bioactive 3,5-Disubstituted Hydantoins. *RSC Adv.* **2016**, *6*, 36978–36986. https://doi.org/10.1039/c6ra03222b.
52. Rightmire, N. R.; Hanusa, T. P. Advances in Organometallic Synthesis with Mechanochemical Methods. *Dalton Trans.* **2016**, *45*, 2352–2362. https://doi.org/10.1039/C5DT03866A. The formalism for mechanochemically activated reactions was recently proposed by this group.
53. Colacino, E.; Lamaty, F.; Martinez, J.; Parrot, I. Microwave-Assisted Solid-Phase Synthesis of Hydantoin Derivatives. *Tetrahedron Lett.* **2007**, *48*, 5317–5320. https://doi.org/10.1016/j.tetlet.2007.05.084.
54. Gong, Y.-D.; Sohn, H.-Y.; Kurth, M. J. Microwave-Mediated Intramolecular Carbanilide Cyclization to Hydantoins Employing Barium Hydroxide Catalysis. *J. Org. Chem.* **1998**, *63*, 4854–4856. https://doi.org/10.1021/jo980419w.
55. Colacino, E.; Martinez, J.; Lamaty, F.; Patrikeevaa, L. S.; Khemchyan, L. L.; Ananikov, V. P.; Beletskaya, I. P. PEG as an Alternative Reaction Medium in Metal-Mediated Transformations. *Coord. Chem. Rev.* **2012**, *256*, 2893–2920. https://doi.org/10.1016/j.ccr.2012.05.027.

56. Chen, J.; Spear, S. K.; Huddleston, J. G.; Rogers, R. D. Polyethylene Glycol and Solutions of Polyethylene Glycol as Green Reaction Media. *Green Chem.* **2005**, 7, 64–82. https://doi.org/10.1039/B413546F.
57. Zhang, W.; Lu, Y.; Hiu-Tung Chen, C.; Curran, D. P.; Geib, S. Fluorous Synthesis of Hydantoin-, Piperazinedione-, and Benzodiazepinedione-Fused Tricyclic and Tetracyclic Ring Systems. *Eur. J. Org. Chem.* **2006**, *2059*, 2055. https://doi.org/10.1002/ejoc.200600077.
58. Zhang, W.; Nagashima, T.; Lu, Y.; Hiu-Tung Chen, C. A Traceless Perfluorooctylsulfonyl Tag for Deoxygenation of Phenols Under Microwave Irradiation. *Tetrahedron Lett.* **2004**, *45*, 4611–4613. https://doi.org/10.1016/j.tetlet.2004.04.115.
59. Kim, S. W.; Ahn, S. Y.; Koh, J. S.; Lee, J. H.; Ro, S.; Cho, H. Y. Solid Phase Synthesis of Hydantoin Library Using a Novel Cyclization and Traceless Cleavage Step. *Tetrahedron Lett.* **1997**, *38*, 4603–4606. https://doi.org/10.1016/S0040-4039(97)00986-6.
60. Lee, M.-J.; Sun, C.-M. Traceless Synthesis of Hydantoin by Focused Microwave Irradiation. *Tetrahedron Lett.* **2004**, *45*, 437–440. https://doi.org/10.1016/j.tetlet.2003.10.123.
61. Yeh, W.-P.; Chan, W.-J.; Sun, M.-L.; Sun, C.-M. Microwave-Assisted Traceless Synthesis of Hydantion-Fused b-Carboline Scaffold. *Tetrahedron* **2007**, *63*, 11809–11816. https://doi.org/10.1016/j.tet.2007.09.054.
62. Boldirev, V. V. Mechanochemistry and Sonochemistry. *Ultrason. Sonochem.* **1995**, *2*, S143–S145. https://doi.org/10.1016/1350-4177(95)00019-3.
63. Suslick, K. S. Mechanochemistry and Sonochemistry: Concluding Remarks. *Faraday Discuss.* **2014**, *170*, 411–422. https://doi.org/10.1039/c4fd00148f.
64. Cravotto, G.; Cintas, P. Harnessing Mechanochemical Effects with Ultrasound-Induced Reactions. *Chem. Sci.* **2012**, *3*, 295–307. https://doi.org/10.1039/c1sc00740h.
65. Ando, T.; Sumi, S.; Kawate, T.; Ichihara, J.; Hanafusa, T. Sonochemical Switching of Reaction Pathways in Solid-Liquid Two-phase Reactions. *J. Chem. Soc. Chem. Commun.* **1984**, 439–440. https://doi.org/10.1039/C39840000439.
66. Hernandez, J.; Bolm, C. Altering Product Selectivity by Mechanochemistry. *J. Org. Chem.* **2017**, *82*, 4007–4019.
67. Weeranoppanant, N. Enabling Tools for Continuous-Flow Biphasic Liquid–Liquid Reaction. *Reac. Chem. Eng.* **2019**, *4*, 235–243. https://doi.org/10.1039/c8re00230d.
68. Seo, H.; Nguyen, L. V.; Jamison, T. F. Using Carbon Dioxide as a Building Block in Continuous Flow Synthesis. *Adv. Synth. Catal.* **2019**, *361*, 247–264. https://doi.org/10.1002/adsc.201801228.
69. Hone, C.; Kappe, C. O. The Use of Molecular Oxygen for Liquid Phase Aerobic Oxidations in Continuous Flow. *Top. Curr. Chem.* **2019**, *377*, 1–44. https://doi.org/10.1007/s41061-018-0226-z.
70. Konnert, L.; Gonnet, L.; Halasz, I.; Suppo, J.-S.; de Figueiredo, R. M.; Campagne, J.-M.; Lamaty, F.; Martinez, J.; Colacino, E. Mechanochemical Preparation of 3,5-Disubstituted Hydantoins from Dipeptides and Unsymmetrical Ureas of Amino Acid Derivatives. *J. Org. Chem.* **2016**, *81*, 9802–9809. https://doi.org/10.1021/acs.joc.6b01832.
71. Mascitti, A.; Lupacchini, M.; Guerra, R.; Taydakov, I.; Tonucci, L.; d'Alessandro, N.; Lamaty, F.; Martinez, J.; Colacino, E. Poly(Ethylene Glycol)S as Grinding Additives in the Mechanochemical Preparation of Highly Functionalized 3,5-Disubstituted Hydantoins. *Beilstein J. Org. Chem.* **2017**, *13*, 19–25. https://doi.org/10.3762/bjoc.13.3.
72. Lupacchini, M.; Mascitti, A.; Tonucci, L.; d'Alessandro, N.; Colacino, E.; Charnay, C. From Molecules to Silicon-Based Biohybrid Materials by Ball Milling. *ACS Sustain. Chem. Eng.* **2018**, *6*, 511–518. https://doi.org/10.1021/acssuschemeng.7b02782.

73. Hu, L.-C.; Khiterer, M.; Huang, S.-J.; Chan, J. C. C.; Davey, J. R.; Shea, K. J. Uniform, Spherical Bridged Polysilsesquioxane Nano- and Microparticles by a Non-emulsion Method. *Chem. Mater.* **2010**, *22*, 5244–5250. https://doi.org/10.1021/cm101243m.
74. Gong, Y.-D.; Kurth, M. J. Microwave-Promoted Reactions: Intramolecular Carbanilide Cyclization to Hydantoins Employing Barium Hydroxide Catalyst. *Tetrahedron Lett.* **1998**, *39*, 3379–3382. https://doi.org/10.1016/S0040-4039(98)00497-3.
75. Brouillette, Y.; Lisowski, V.; Guillon, J.; Massip, S.; Martinez, J. Efficient One-Pot Microwave-Assisted Synthesis of 3-(Thien-3-Yl)Imidazolidine-2,4-Dione Analogs. *Tetrahedron* **2007**, *63*, 7538–7544. https://doi.org/10.1016/j.tet.2007.05.057.
76. Ugi, I.; Steinbrueckner, C.; Isonitrile, I. X. α-Addition von Immonium-Ionen und Carbonsäure-Anionen an Isonitrile. *Chem. Ber.* **1961**, *94*, 2802–2814. https://doi.org/10.1002/cber.19610941032.
77. Huang, A.; Wo, K.; Pierson, A.; Lee, S. Y. C.; Yang, E.; Johnson, C. 5-Hydroxyl Hydantoins via One-Pot Microwave-Assisted Air Oxidation of Ugi Products. *J. Heterocyclic Chem.* **2016**, *53*, 1499–1504. https://doi.org/10.1002/jhet.2454.
78. Medda, F.; Martinez-Ariza, G.; Hulme, C. A Facile and Concise Route toward the Synthesis of Novel Imidazotetrazolodiazepinones Via Post-Condensation Modifications of the Ugi-Azide Adduct. *Tetrahedron Lett.* **2015**, *56*, 5295–5298. https://doi.org/10.1016/j.tetlet.2015.07.083.
79. Xu, Z.-G.; Ding, Y.; Menga, J.-P.; Tanga, D.-Y.; Li, Y.; Lei, J.; Xu, C.; Chen, Z.-Z. Facile Construction of Hydantoin Scaffolds via a Post-Ugi Cascade Reaction. *Synlett* **2018**, *29*, 2199–2202. https://doi.org/10.1055/s-0037-1610234.
80. Barton, H.; Bojarski, J.; Mokrosz, J. Photochemical Ring Opening of Barbital. *Tetrahedron Lett.* **1982**, *23*, 2133–2134. https://doi.org/10.1016/S0040-4039(00)87280-9.
81. Barton, H.; Bojarski, J.; Zurowska, A.; Ekiert, L. Photoinduced Stereospecific Formation of Substituted Hydantoin from Hexobarbital. *J. Photochem. Photobiol. A Chem.* **1990**, *54*, 187–196. https://doi.org/10.1016/1010-6030(90)85031-Q.
82. Barton, H.; Bojarski, J.; Zurowska, A. Stereospecificity of the Photoinduced Conversion of Methylphenobarbital to Mephenytoin. *Arch. Pharm.* **1986**, *319*, 457–461. https://doi.org/10.1002/ardp.19863190514.
83. Meza-León, R. L.; Bernès, S.; Cortés-López, G. N.; Mastranzo, V. M.; Sosa-Rivadeneyra, M.; Sartillo-Piscil, F. Synthesis of 5-Hydroxy Hydantoins via a Tandem Process. *Tetrahedron Lett.* **2016**, *57*, 4232–4234. https://doi.org/10.1016/j.tetlet.2016.08.022.
84. Porcheddu, A.; Delogu, F.; De Luca, L.; Colacino, E. From Lossen Transposition to Solventless "Medicinal Mechanochemistry". *ACS Sustain. Chem. Eng.* **2019**, *7*, 12044–12051. https://doi.org/10.1021/acssuschemeng.9b00709.
85. Anastas, P.; Farris, C. A., Eds. *The expression "Benign by design" was previously used in other contexts. See: Benign by Design: Alternative Synthetic Design for Pollution Prevention*; American Chemical Society: Washington, DC, 1994.
86. Tan, D.; Loots, L.; Friscic, T. Towards Medicinal Mechanochemistry: Evolution of Milling from Pharmaceutical Solid Form Screening to the Synthesis of Active Pharmaceutical Ingredients (APIs). *Chem. Commun.* **2016**, *52*, 7760–7781. https://doi.org/10.1039/C6CC02015A.
87. For one of the first examples on the use of Twin Screw Extrusion process to scale-up mechanochemical procedures see: Crawford, D.; Casaban, J.; Haydin, R.; Giri, N.; McNally, T.; James, S. L. Synthesis by extrusion: continuous, large-scale preparation of MOFs using little or no solvent. *Chem. Sci.* **2015**, *6*, 1645–1649. https://doi.org/10.1039/C4SC03217A.

88. Bolm, C.; Hernandez, J. Mechanochemistry of Gaseous Reactants. *Angew. Chem. Int. Ed.* **2019**, *58*, 2–17. https://doi.org/10.1002/anie.201810902.
89. Dudley, K. H.; Bius, D. L. The Synthesis of Optically Active 5-Phenylhydantoins. *J. Heterocyclic Chem.* **1973**, *10*, 139–278. https://doi.org/10.1002/jhet.5570100208.
90. Liu, H.; Yang, Z.; Pan, Z. Synthesis of Highly Substituted Imidazolidine-2,4-Dione (Hydantoin) through Tf_2O-Mediated Dual Activation of Boc-Protected Dipeptidyl Compounds. *Org. Lett.* **2014**, *16*, 5902–5905. https://doi.org/10.1021/ol502900j.
91. Andersen, J.; Mack, J. Mechanochemistry and Organic Synthesis: from Mystical to Practical. *Green Chem.* **2018**, *20*, 1435–1443. https://doi.org/10.1039/C7GC03797J.
92. A Detailed Description of Each Method is Given in the Original Article Herein Cited.
93. Mocci, R.; De Luca, L.; Delogu, F.; Porcheddu, A. An Environmentally Sustainable Mechanochemical Route to Hydroxamic Acid Derivatives. *Adv. Synth. Catal.* **2016**, *358*, 3135–3144. https://doi.org/10.1002/adsc.201600350.
94. Dale, D. J.; Draper, J.; Dunn, P. J.; Hughes, M. L.; Hussain, F.; Levett, P. C.; Ward, G.; Wood, A. S. The Process Development of a Scaleable Route to the PDE5 Inhibitor UK-357,903. *Org. Process Res. Dev.* **2002**, *6*, 767–772. https://doi.org/10.1021/op9900683.
95. Dale, D. J.; Dunn, P. J.; Golightly, C.; Hughes, M. L.; Levett, P. C.; Pearce, A. K.; Searle, P. M.; Ward, G.; Wood, A. S. The Chemical Development of the Commercial Route to Sildenafil: A Case History. *Org. Process Res. Dev.* **2000**, *4*, 17–22. https://doi.org/10.1021/op9900683.
96. Dubé, P.; Fine Nathel, N. F.; Vetelino, M.; Couturier, M.; Aboussafy, C. L.; Pichette, S.; Jorgensen, M. L.; Hardink, M. Carbonyldiimidazole-Mediated Lossen Rearrangement. *Org. Lett.* **2009**, *11*, 5622–5625. https://doi.org/10.1021/ol9023387.
97. For a first report involving the use of PEG as phase transfer agent for API synthesis (phenytoin) see: Poupaert, J. H.; Do Keyser, J.-L.; Vandervorst, D.; Dumont, D. Phase-Transfer Catalysis by Poly(ethylene glycol) 600 in the Biltz Synthesis of Phenytoin. *Bull. Soc. Chim. Belg.* **1984**, *93*, 493–495. https://doi.org/10.1002/bscb.19840930612.
98. Hasa, D.; Rauber, G. S.; Voinovich, D.; Jones, W. Cocrystal Formation through Mechanochemistry: from Neat and Liquid-Assisted Grinding to Polymer-Assisted Grinding. *Angew. Chem. Int. Ed.* **2015**, *54*, 7371–7375. https://doi.org/10.1002/anie.201501638.
99. Giachi, G.; Frediani, M.; Oberhauser, W.; Lamaty, F.; Martinez, J.; Colacino, E. Ultrasounds in Melted Poly(Ethylene Glycol) Promote Copper-Catalyzed Cyanation of Aryl Halides with $K_4[Fe(CN)_6]$. *ChemSusChem* **2014**, 7, 919–924. https://doi.org/10.1002/cssc.201300997.
100. Declerck, V.; Colacino, E.; Bantreil, X.; Martinez, J.; Lamaty, F. Poly(Ethylene Glycol) as Reaction Medium for Mild Mizoroki–Heck Reaction in a Ball-Mill. *Chem. Commun.* **2012**, *48*, 11778–11780. https://doi.org/10.1039/c2cc36286d.
101. Friscic, T. New Opportunities for Materials Synthesis Using Mechanochemistry. *J. Mater. Chem.* **2010**, *20*, 7599–7605. https://doi.org/10.1039/C0JM00872A.
102. Chen, Z.; Sun, Y. N-Halamine-Based Antimicrobial Additives for Polymers: Preparation, Characterization, and Antimicrobial Activity. *Ind. Eng. Chem. Res.* **2006**, *45*, 2634–2640. https://doi.org/10.1021/ie060088a.
103. For more information on COST Action CA18112 "Mechanochemistry for Sustainable Industry" (MechSustInd), see http://www.mechsustind.eu/ (Accessed 14 October 2020).

104. Bala´zˇ, M.; Vella-Zarb, L.; Hernandez, J.; et al. A Disruptive Innovation for the Industry of the Future. *Chem. Today* **2019**, *37*, 32–34.
105. Herna´ndez, J. G.; Halasz, I.; Crawford, D. E.; et al. European Research in Focus: Mechanochemistry for Sustainable Industry (COST Action MechSustInd). *Eur. J. Org. Chem.* **2020**, 8–9.
106. For more information on COST Association, see http://www.cost.eu/ (Accessed 14 October 2020).

CHAPTER THIRTEEN

Application of mechanochemical activation in synthetic organic chemistry

Gábor Varga, Pál Sipos, and István Pálinkó
University of Szeged, Szeged, Hungary

1 Introductory remarks

The famous philosopher Aristotle's statement "No Coopora nisi Fluida" means "no reaction is possible in the absence of solvent" was invalidated by the pioneering works of Toda at the end the previous century showing that many organic reactions of solution chemistry would be reproducible in solid state too.[1, 2] Many important named reactions such as Morita–Baylis–Hillman, Michael, Wittig, Ritter, Heck, Suzuki–Miyaura, and Sonogashira couplings; Ugi, Passerini, and Biginelli multicomponent reactions; and Paal–Knorr and Hantzsch reactions for the synthesis of heterocyclic compounds and others like aldol dimerization, copper-catalyzed cross-dehydrogenative coupling, click reactions, and C—O, C–X, and C—N bond formation are included, and many additional examples are listed in a recent book[3] and review article.[4]

In the solid-state reaction, the ingredients are mixed intimately in the form of fine powder, and the energy transfer provided by grinding/milling facilitates chemical reaction by mechanochemical activation. The IUPAC definition of a mechanochemical reaction is "a chemical reaction that is induced by the direct absorption of mechanical energy."[5]

Mechanochemistry is regularly used in the area of crystal engineering, nanoscience, or minerals engineering.[6] Recently, mechanochemical activation became a frequently applied method in organic syntheses as well.[3, 4, 7, 8] The technique has several green aspects, since it does not use solvent or reduces reaction times. Nevertheless, the most interesting transformations are those, which exhibit different reactivity to conventional solution-based reactions.[6, 9] It is expected and indeed was found that chemical reactions,

https://doi.org/10.1016/B978-0-12-819009-8.00001-3
© 2021 Elsevier Inc.
All rights reserved.

initiated by this form of energy input, lead to different products and provide access to reactions, which are inaccessible by other more conventional activation methods.

2 Basics of mechanochemistry

2.1 Tools and scientific background of mechanochemical treatment[8]

Mechanochemical treatment most frequently means grinding or milling. Grinding can be performed using a mortar and pestle, and can be a mild way of energy transfer. However, its reproducibility is low, since the extent of energy transfer highly depends on the experimentalist, especially when a reaction requiring longer mechanical activation has to be performed. Therefore, mostly, electronic milling devices are applied. The mixer mill is one type of ball milling machine, which uses the movement of ball bearings to apply mechanical force to the reagents. In this case, reagents are loaded into jars, and one or more ball bearings are added. The jars are then mounted horizontally and shaken at the desired frequency. The main mechanical energy applied to the reagents is impact force. The other most common type of ball mill is a planetary mill. In this case the reagents and ball (s) are loaded as before, but the motion is different. In this instance, the jars spin counterdirectionally to the spinning disc they are mounted on. The central spinning disc is called the "sun wheel," and the jars orbit around the central spinning axis. In both cases, the material and size of the jars and balls can be altered. The main type of force applied in planetary milling is shear force. Stirred media ball mills are also available, in which the reagents and balls are stirred together.

For mechanochemistry, each type of milling device can be used to achieve different scales. The mixer mill is effective at gram scales, which are suitable for laboratory investigations. However, for anything larger, other types of mill must be used, and different sizes of planetary mills are available. For pilot and manufacturing scales, stirred media ball mills are used.

If the process is modified from batch to continuous, extruders must be used instead of mills. In organic synthesis, twin-screw extrusion (TSE; mechanochemistry in flow) was used for Knoevenagel condensation of various aldehydes and barbituric acid (Scheme 1).[10]

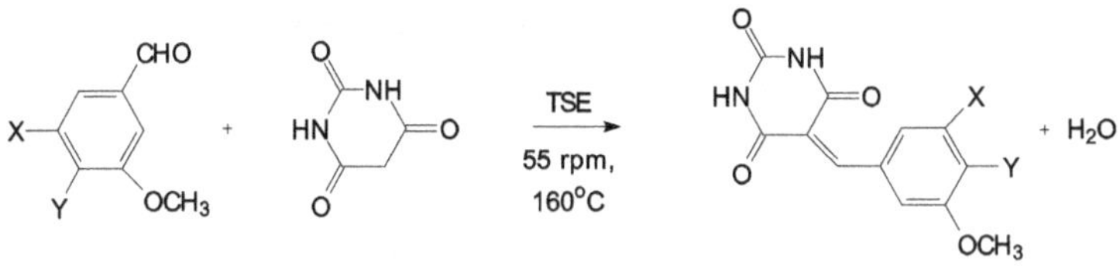

Scheme 1 Reaction of aldehydes with barbituric acid in the solid state by twin-screw extrusion (TSE).

The three main variables affecting mechanochemical reactions are the kinetic energy of the ball(s) prior to collision, the mode of energy transfer to the reagents and the frequency of collisions. These should be controlled; however, it is not an easy exercise.

The amount of kinetic energy the ball(s) possesses prior to collision is the maximum amount of energy, which can be transferred to the reagents per collision. The mode of energy transfer can be by direct impact, under which the material is locally compressed, or shear force, in which a reactive face is exposed. It has been shown that these different types of energy absorption can lead to different outcomes.[11] Different types of ball mills achieve different ratios of impact and shear forces.

Differences in mixing or mass transfer can affect the outcome of any reaction. In milled reactions, this can be a difficult variable to control, and may have a dramatic effect on the outcome of a reaction. To help homogeneity and mixing, grinding agents can be used. When one of the reactants is liquid, for example, in a two-step milling, when dry milling is performed first to achieve complete mixing, then wet milling (often called as liquid-assisted grinding, abbreviated as LAG) has to be performed, the reaction mixture can become sticky and resemble a paste or gum, which prevents efficient mass and energy transfer.[12] To overcome these difficulties, it is common to use an auxiliary material such as silica, alumina, talc, or inorganic salts as a milling agent/auxiliary or adsorbent.[13, 14] It is important that such additives are chemically inert toward the desired reaction; otherwise, they might interfere with the reactivity.

Finally, those variables, which are usually altered in any standard reaction development are also assessed during the optimization of a mechanochemical reaction like stoichiometry, reaction time, and temperature. Nevertheless, temperature cannot be controlled easily; the reaction vessel heats up due to the collisions, and is dependent on the filling degree volumes of grinding balls, sample and vessel size as well as oscillating frequency.

2.2 Reactions proceeding differently on mechanochemical treatment than under wet chemical conditions

As it has been mentioned before, the most interesting reactions are those, which proceed differently (with altered rate and/or reaction pathway, i.e., selectivity) under mechanical activation than in conventional wet chemistry environment. Reactions having significant green aspects include solvent-free multistep reactions or special setups in which the authors combine mechanochemical activation with other methods such as irradiation with light or ultrasound. Representative examples of these reaction types are collected, and those are detailed, which were published after the appearance of the already referenced reviews[1, 2, 4, 6] and the perspective[9], that is, from 2016 to 2017 to date.

2.2.1 Reactions accelerated by mechanochemical activation

A tutorial review lists a number of reactions (and gives the corresponding schemes and the original references) in which considerable acceleration in reaction rates resulted in increased product yields.[8] These examples include the synthesis of Cu–NHC complexes (NHC is N-heterocyclic carbene, and planetary mill was applied), which are used as organometallic catalysts in a variety of reactions: palladacycle synthesis in a liquid-assisted milling process from asymmetrically substituted azobenzene and $Pd(OAc)_2$; iridium(III)-catalyzed C—H bond amidation of benzamides with sulfonyl azides, where the active cationic Ir(III) catalyst was formed in situ in the mixer mill; oxidative C–H/C–H coupling, where biaryl products could be obtained in both high selectivity and yield in one hour from electron-deficient oximes and arenes in the presence of a palladium catalyst and oxidant; solvent-accelerated palladium-catalyzed Suzuki–Miyaura cross-coupling of aryl chlorides with boronic acids under ball milling conditions; mono- and difluorination of 1,3-diketones with Selectfluor in liquid-assisted mechanochemical reactions; enantioselective fluorination of β-keto esters using N-fluorobenzenesulfonimide (NFSI) under ball milling conditions; and (*S*)-proline-catalyzed enantioselective aldol reaction of 4-nitrobenzaldehyde with tetrahydrothiopyran-4-one and N-alkylation of glyoxal-bridged bisaminal derivatives of cyclam (1,4,8,11-tetraazacyclotetradecane) and cyclen (1,4,7,10-tetraazacyclododecane).

A recent research article describes an experimental work in which primary and secondary alcohols were oxidized to the corresponding aldehydes and ketones in an eco-friendly and cost-efficient way by mechanical

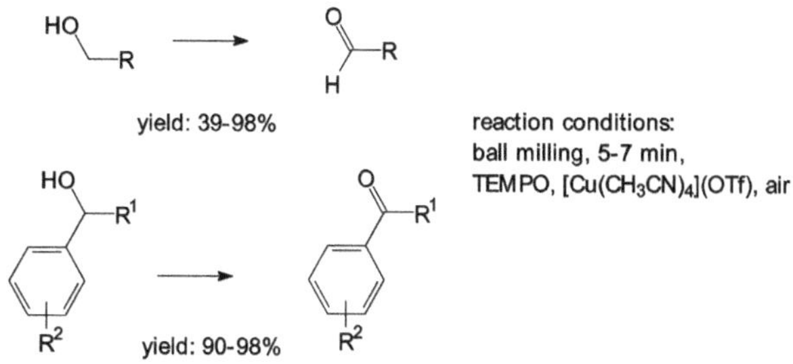

Scheme 2 Oxidation of primary and secondary alcohols to aldehydes (13 examples) and ketones (8 examples), respectively, in the solid state under ball milling conditions.

processing in air (Scheme 2).[15] Ball milling was shown to promote the quantitative conversion of a broad set of alcohols into carbonyl compounds with no trace of an overoxidation to carboxylic acids. The mechanochemical reaction exhibited higher yields and rates than the classical, homogeneous, TEMPO-based oxidation.

Copper-catalyzed mechanochemical click reactions using Cu(II), Cu(I), and Cu(0) catalysts, respectively, have been successfully implemented to provide novel 6-phenyl-2-(trifluoromethyl)quinolines with a phenyl-1,2,3-triazole moiety at O-4 of the quinoline core (Scheme 3).[16] Milling procedures proved to be significantly more efficient than the corresponding solution-based reactions, with up to a 15-fold gain in yield. Efficiency of both solution and milling procedures depended on the *p*-substituent in the azide reactant, resulting in H < Cl < Br < I reactivity sequence.

The solvent-free mechanochemical polymerization of trimethylene carbonate using 1,8-diazabicyclo[5.4.0]undec-7-ene (DBU) and

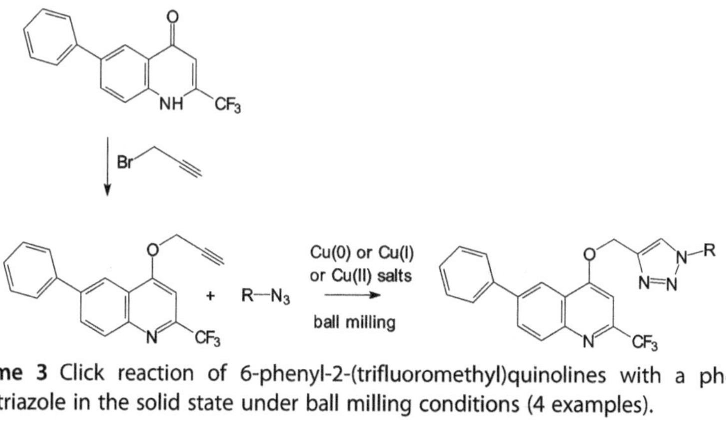

Scheme 3 Click reaction of 6-phenyl-2-(trifluoromethyl)quinolines with a phenyl-1,2,3-triazole in the solid state under ball milling conditions (4 examples).

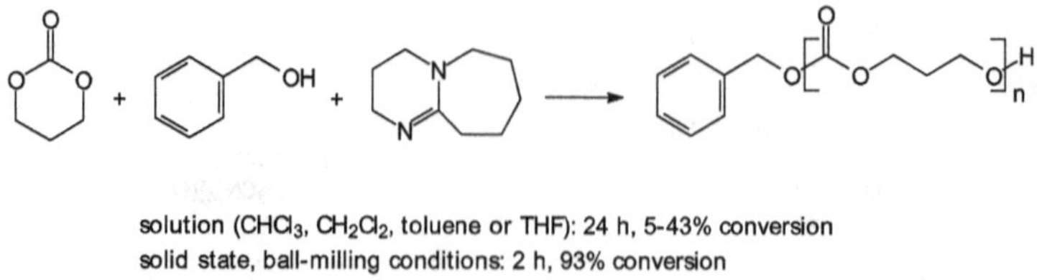

Scheme 4 Polymerization of trimethylene carbonate in solution or in the solid state under ball milling conditions.

1,5,7-triazabicyclo[4.4.0]dec-5-ene (TBD) organocatalysts were performed under ball milling conditions exhibiting significant rate enhancements compared with polymerizations in solution (Scheme 4).[17] The use of TBD enhanced the efficiency, and all polymerizations reached completion within 5 min despite physical mixing limitations. However, the mechanochemical polymerization was accompanied by degradation processes, which limited the molecular weight. Liquid-assisted grinding did not show any protective effect.

2.2.2 Alternative or new reaction pathways not possible in solution

These reaction types are significantly scarcer than those with activity enhancement; therefore the publication time limitation is omitted in this section.

In solution, anilines react with bis(benzotriazolyl)methanethiones and form their corresponding isothiocyanates and benzotriazoles through intermediate N-(thiocarbamoyl)benzotriazoles (Scheme 5). The intermediate could not be isolated from a solution-based reaction; however, it was bench-stable in the solid state, and could be isolated and used for the synthesis of both symmetrical and nonsymmetrical thioureas.[18]

Simple solvent-free milling or LAG to combine *p*-toluenesulfonamide and *N,N′*-dicyclohexylcarbodiimide (DCC) for two hours was unsuccessful, as was the overnight reflux in dichloromethane or acetone (Scheme 6). However, using 5 mol% of CuCl as a catalyst in the reaction mixture, the desired sulfonyl guanidine product was isolated in 81% yield, which was further boosted to 98% by LAG using nitromethane (Scheme 6).[19, 20]

Attempts to expand this reactivity to sulfonimides led to an unexpected C—N insertion, through which saccharin was expanded into 1,2,4-benzothiadiazepines in a single catalytic step (Scheme 7).[21] The scope of the reaction could be extended to aliphatic, aromatic, symmetrical and nonsymmetrical carbodiimide reactants and to linear sulfonimides leading to a two-atom chain extension.

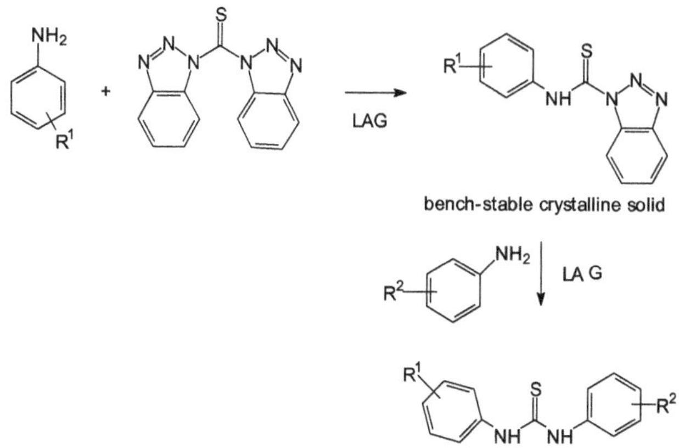

Scheme 5 The *N*-thiocarbamoylbenzotriazole intermediates are isolable bench-stable compounds if they are produced mechanochemically under LAG conditions (6 examples).

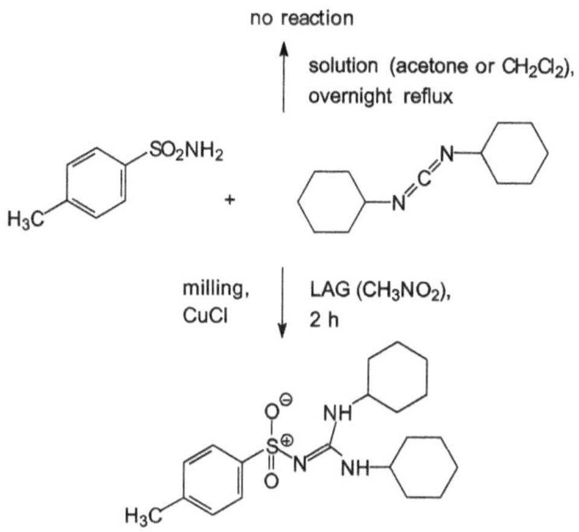

Scheme 6 There was no reaction between *p*-toluenesulfonamide and *N,N*′-dicyclohexylcarbodiimide in solution, but 98% yield was obtained in the solid state under LAG conditions.

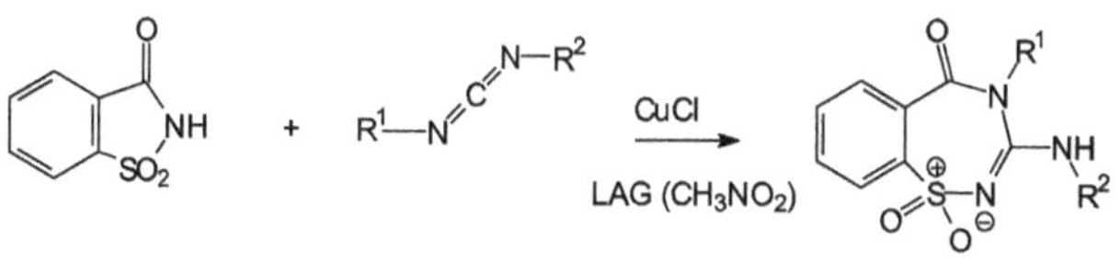

Scheme 7 C—N insertion reaction in the ring expansion reaction of saccharin to 1,2,4-benzothiadiazepines under LAG conditions (4 examples, 78%–95% yield).

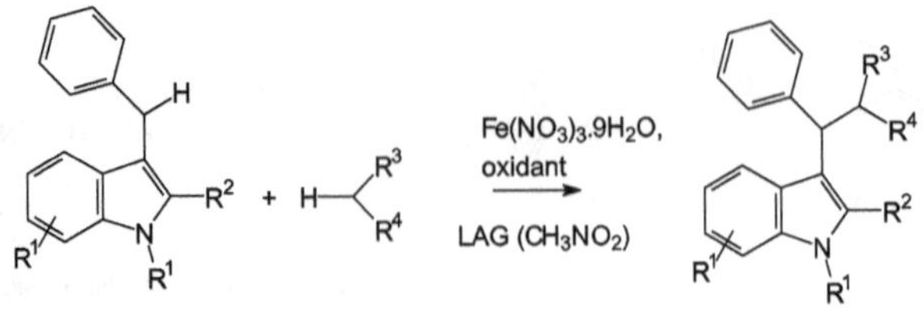

Scheme 8 An Fe(III)-catalyzed cross-dehydrogenative coupling of 3-benzyl-indoles with methylenes leading to bisindoles (24 examples).

An Fe(III)-catalyzed cross-dehydrogenative coupling (CDC) of 3-benzyl-indoles produced bisindoles under ball milling conditions (Scheme 8) in good yields. In contrast, the solution-based reaction in dichloroethane solvent only produced the product in traces.[22]

Highly functionalized iptycenes could be synthesized by iterative Diels–Alder/aromatization reactions under solvent-free conditions.[23] Good yield (87%) of adduct could be achieved in a ball mill using $ZnCl_2$ as a Lewis acid. In contrast, only 5% of this material was obtained when the reaction was carried out in solution in the presence of $ZnCl_2$ at 80°C after 24 h.

2.2.3 Solvent-free multistep reactions

A one-pot two-step mechanochemical synthesis of fluorinated pyrazolones was developed, in which heterocycle formation and subsequent fluorination were successfully applied across a range of substrates resulting in 12 difluorinated pyrazolones in moderate to excellent yields (Scheme 9).[24]

The pseudo-five-component reaction between β-dicarbonyl compounds (two molecules), diamines, and α-iodoketones (two molecules), prepared in situ from aryl ketones, was performed efficiently under mechanochemical conditions involving high-speed vibration milling with a single zirconium oxide ball (Scheme 10).[25] This reaction afforded symmetrical

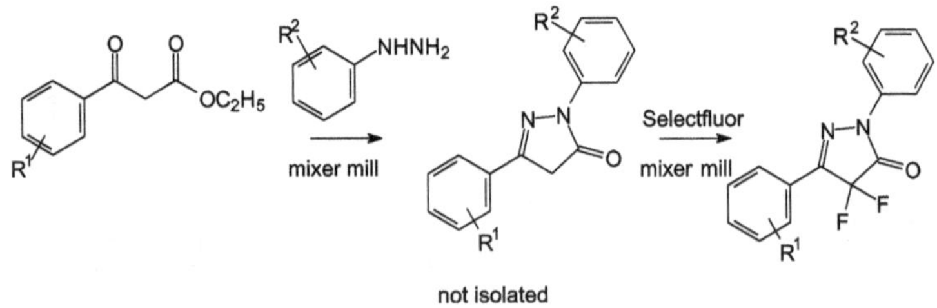

Scheme 9 Mechanochemical synthesis of pyrazolone derivatives (12 examples, yields: 37%–83%) in a mixer mill without the isolating the product of the first reaction.

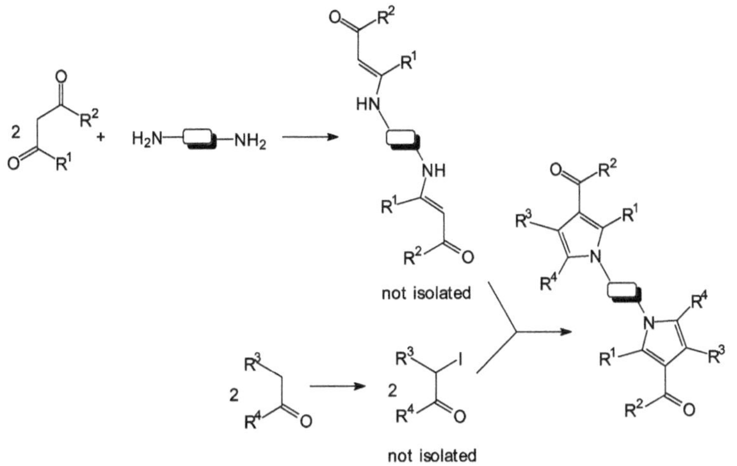

Scheme 10 Pseudo-five-component reaction in a high-speed mixer mill affording symmetrical bispyrrole derivatives joined by a spacer (15 examples, yields: 45%–74%).

frameworks containing two pyrrole or fused pyrrole units joined by a spacer. The method was also extended to the synthesis of a special product containing three identical pyrrole fragments via a pseudo-seven-component reaction (Scheme 11).[25]

A multistep synthetic route to access mono- and diacylglycerol derivatives was performed including the synthesis of diacylglycerol–coumarin conjugates.[26] Specifically, the synthesis of mono- and diacylglycerols required first, the application of solvent-free functional group protection chemistry in ball mills; second, the implementation of metal-catalyzed epoxide ring opening; and third, the development of solvent-free ester formation (Scheme 12).

2.2.4 Synthesis of supramolecules

In the literature, the mechanochemical synthesis of classical small organic molecules has been well explored including even multistep syntheses. However, the concept of supramolecular chemistry under mechanomilling conditions only has limited number of examples. The original works have been reviewed recently.[27] The reader is referred to this review; here, we only show two typical examples.

In 2019, a solvent-free mechanochemical synthesis of pillar[5]arene-containing [2]rotaxanes was reported.[28] Mixing a 2:1 ratio of pillar[5]arene (wheel) with dodecanedioyl dichloride (axle) in $CHCl_3$ resulted in the formation of pseudorotaxane, which was further treated with

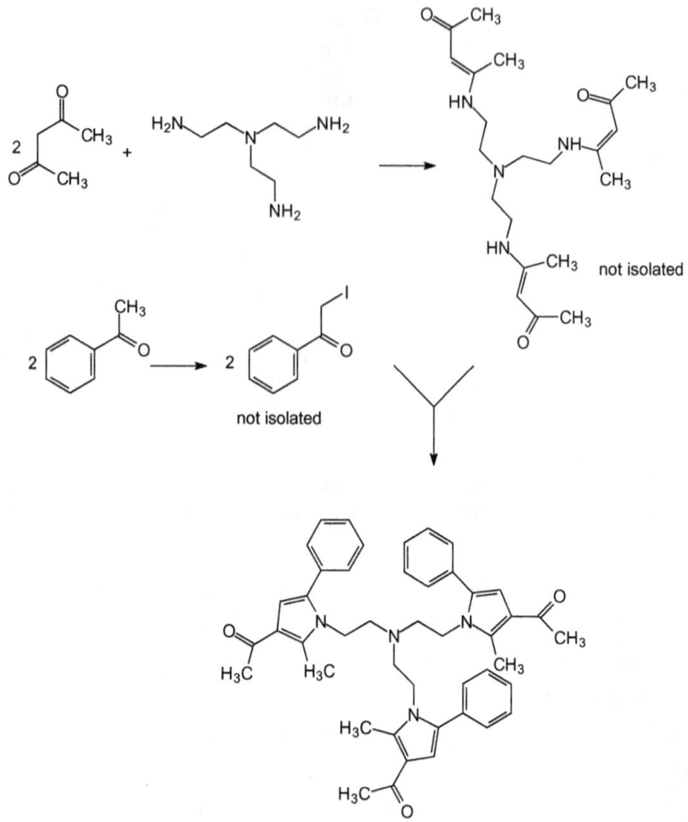

Scheme 11 Pseudo-seven-component reaction in a high-speed mixer mill affording a symmetrical trispyrrole derivative with 35% yield joined by a functionalized spacer.

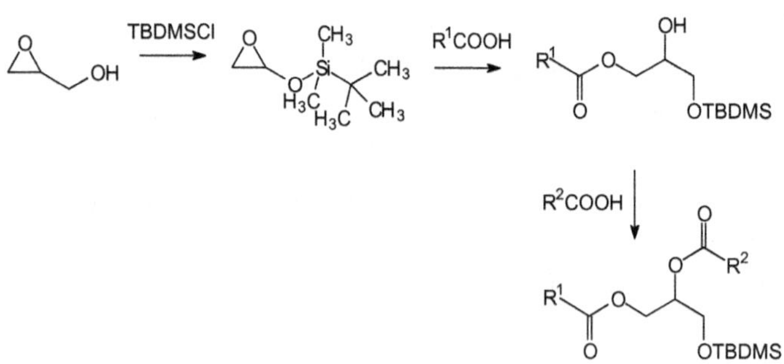

Scheme 12 Synthetic reaction steps, performed in a ball mill, to obtain diacylglycerol derivatives.

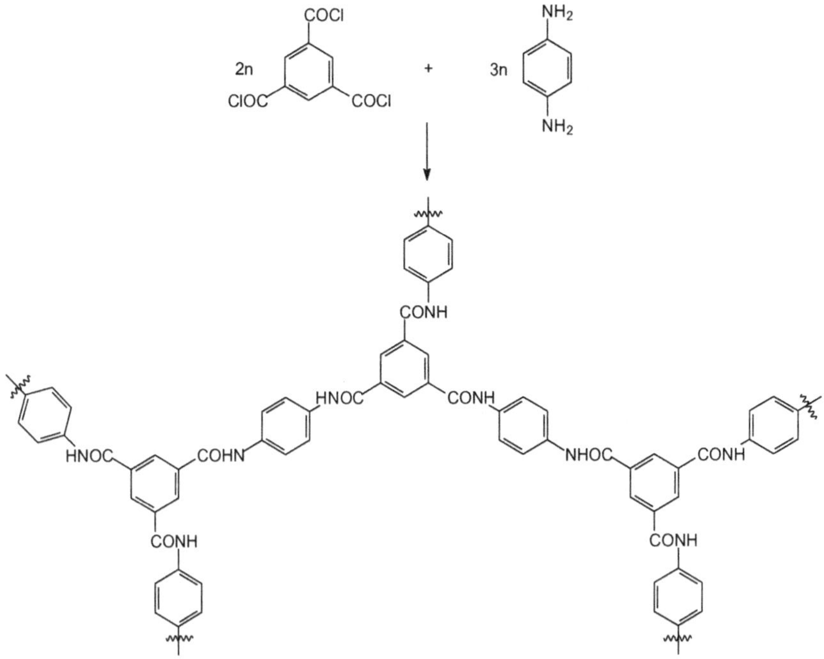

Scheme 13 2D polymerization between 1,3,5-benzenetricarbonyl chloride and 1,4-phenylenediamine under ball milling conditions.

different amines (stopper) in a stainless steel jar with four steel balls under milling conditions (30 Hz for 1–2 h). When, for example, *N*-methyl-1,1-diphenylmethanamine was used as one of the stoppers, diamido[2] rotaxane was obtained with high yield (c. 87%).

The synthesis of two-dimensional aromatic polyamides (2DAPAs), under solvent-free ball milling conditions, was performed by reacting 1,3,5-benzenetricarbonyl chloride and 1,4-phenylenediamine in a ball mill affording 75% of the desired 2D polymer within 15 min (Scheme 13).[29] When using 4,4′-diaminobiphenyl in place of 1,4-phenylenediamine, the analogous two-dimensional structure comprising biphenylene units was obtained within the same time, but with lower yield (≈65%).

2.2.5 *Mechanochemical activation coupled to other activation methods*

Dual activation of an organic system by light and ball milling techniques permitted studying the role of eosin Y in the borylation of aryldiazonium salts in a ball mill under catalyst-free conditions (Scheme 14).[30] Simultaneous neat grinding/irradiation of the reactants led to the formation of boronates in a

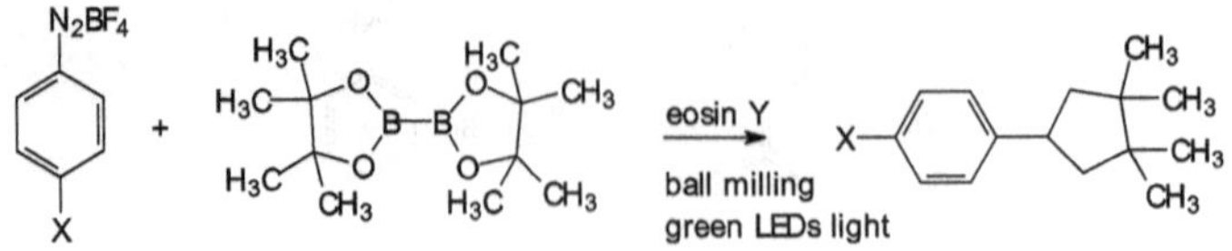

Scheme 14 Simultaneous activation by light and ball milling: borylation of aryldiazonium salts (5 examples, yield: 41%–68%).

molten state. The catalyst-free liquid-assisted grinding/irradiation reaction is a direct photolysis pathway facilitated by substrate–solvent charge-transfer complex formation.

A recently emerged technique is Resonant Acoustic Mixing (RAM), which uses low-frequency, high-amplitude acoustic resonance to agitate powder samples.[31–33] Sample mixing is achieved through longitudinal vibrations and vortexes generated by an oscillating plate at the bottom of a reaction vessel. Because no milling media are present, there is significantly less frictional heating during this mechanochemical process. Cocrystals of carbamazepine and nicotinamide were prepared using RAM in the presence of a small amount of liquid.[34] The technique exhibited good scalability.

3 Conclusion and outlook

This survey of recent reviews and original works indicate that mechanochemical activation in organic synthesis (as well as in many other areas of chemistry[35, 36]) is very useful, and it is in the forefront of current research efforts. It is reasonable to suggest that mechanochemistry is going to become a major activation method in the toolbox of chemists and materials scientists in the future as well. Finally, let us give a surely not complete list of earlier reviews, some of which give a historic view of mechanochemistry[37], others works[38, 39] survey reactions in more detail that can be found in the most recent and already cited monograph[3]. Arguably, however, the tutorial reviews[40, 41] are the most useful, especially for the newcomers to the field. Moreover, papers with purely educational focus[42, 43] are also available, in which mechanochemical syntheses are described for organic laboratory practicals giving a brief account of the preceding research.

References

1. Toda, F. Solid State Organic Chemistry: Efficient Reactions, Remarkable Yields, and Stereoselectivity. *Acc. Chem. Res.* **1995**, *28*, 480–486.
2. Tanaka, K.; Toda, F. Solvent-free Organic Synthesis. *Chem. Rev.* **2000**, *100*, 1025–1074.
3. Margetić, D.; Štrukil, V. *Mechanochemical Organic Synthesis*; Elsevier, 2016.

4. Achar, T. K.; Bose, A.; Mal, P. Mechanochemical Synthesis of Small Organic Molecules. *Beilstein J. Org. Chem.* **2017**, *13*, 1907–1931.
5. Nič, M.; Jirát, J.; Košata, B.; Jenkins, A.; McNaught, A., Eds. *IUPAC Compendium of Chemical Terminology*; IUPAC, 2009.
6. Baláž, P. *Mechanochemistry in Nanoscience and Minerals Engineering*; Springer, 2008.
7. Tan, D.; Friščić, T. Mechanochemistry for Organic Chemists: An Update. *Eur. J. Org. Chem.* **2018**, 18–33.
8. Howard, J. L.; Cao, Q.; Browne, D. L. Mechanochemistry as an Emerging Tool for Molecular Synthesis: What Can it Offer? *Chem. Sci.* **2018**, *9*, 3080–3094.
9. Hernández, J. G.; Bolm, C. J. *Org. Chem.* **2017**, *82*, 4007–4019.
10. Crawford, D. E.; Miskimmin, C. K. G.; Albadarin, A. B.; Walker, G.; James, S. L. Organic Synthesis by Twin Screw Extrusion (TSE): Continuous, Scalable and Solvent-free. *Green Chem.* **2017**, *19*, 1507–1518.
11. Michalchuk, A. A. L.; Tumanov, I. A.; Drebushchak, V. A.; Boldyreva, E. V. Advances in Elucidating Mechanochemical Complexities *via* Implementation of a Simple Organic System. *Faraday Discuss.* **2014**, *170*, 311–335.
12. Ferencz, Z.; Szabados, M.; Ádok-Sipiczki, M.; Kukovecz, Á.; Kónya, Z.; Sipos, P.; Pálinkó, I. Mechanochemically Assisted Synthesis of Pristine Ca(II)Sn(IV)-Layered Double Hydroxides and their Amino Acid Intercalated Nanocomposites. *J. Mater. Sci.* **2014**, *49*, 8478–8486.
13. Kumar, N.; Biswas, K. Fabrication of Novel Cryomill for Synthesis of High Purity Metallic Nanoparticles. *Rev. Sci. Instrum.* **2015**, *86*, 83903.
14. Do, J. L.; Mottillo, C.; Tan, D.; Štrukil, V.; Friscić, T. Mechanochemical Ruthenium-catalysed Olefin Metathesis. *J. Am. Chem. Soc.* **2015**, *137*, 2476–2479.
15. Porcheddu, A.; Colacino, E.; Cravotto, G.; Delogu, F.; de Luca, L. Mechanically Induced Oxidation of Alcohols to Aldehydes and Ketones in Ambient Air: Revisiting TEMPO-assisted Oxidations. *Beilstein J. Org. Chem.* **2017**, *13*, 2049–2055.
16. Tireli, M.; Maračić, S.; Lukin, S.; Kulcsár, M. J.; Žilić, D.; Cetina, M.; Halasz, I.; Raić-Malić, S.; Užarević, K. Solvent-free Copper-catalyzed Click Chemistry for the Synthesis of *N*-Heterocyclic Hybrids Based on Quinoline and 1,2,3-Triazole. *Beilstein J. Org. Chem.* **2017**, *13*, 2352–2363.
17. Park, S.; Kim, J. G. Mechanochemical Synthesis of Poly(Trimethylene Carbonate)S: An Example of Rate Acceleration. *Beilstein J. Org. Chem.* **2019**, *15*, 963–970.
18. Štrukil, V.; Gracin, D.; Magdysyuk, O. V.; Dinnebier, R. E.; Friščić, T. Trapping Reactive Intermediates by Mechanochemistry: Elusive Aryl N-Thiocarbamoyl-benzotriazoles as Bench-stable Reagents. *Angew. Chem. Int. Ed.* **2015**, *54*, 8440–8443.
19. Tan, D.; Mottillo, C.; Katsenis, A. D.; Štrukil, V.; Friščić, T. Development of C–N Coupling Using Mechanochemistry: Catalytic Coupling of Arylsulfonamides and Carbodiimides. *Angew. Chem. Int. Ed.* **2014**, *53*, 9321–9324.
20. Štrukil, V. Mechanochemical Organic Synthesis: The Art of Making Chemistry Green. *Synlett* **2018**, *29*, 1281–1288.
21. Tan, D.; Friščić, T. Carbodiimide Insertion into Sulfonimides: One-step Route to Azepine Derivatives by a Two-atom Saccharin Ring Expansion. *Chem. Commun.* **2017**, *53*, 901–904.
22. Yu, J.-B.; Zhang, Y.; Jiang, Z.-J.; Su, W.-K. Mechanically Induced Fe(III) Catalysis at Room Temperature: Solvent-free Cross-dehydrogenative Coupling of 3-Benzylic Indoles with Methylenes/Indoles. *J. Org. Chem.* **2016**, *81*, 11514–11520.
23. Zhao, Y.; Rocha, S. V.; Swager, T. M. Mechanochemical Synthesis of Extended Iptycenes. *J. Am. Chem. Soc.* **2016**, *138*, 13834–13837.
24. Howard, J. L.; Nicholson, W.; Sagatov, Y.; Browne, D. L. One-pot Multistep Mechanochemical Synthesis of Fluorinated Pyrazolones. *Beilstein J. Org. Chem.* **2017**, *13*, 1950–1956.

25. Leonardi, M.; Villacampa, M.; Menéndez, J. C. High-speed Vibration-milling-promoted Synthesis of Symmetrical Frameworks Containing Two or Three Pyrrole Units. *Beilstein J. Org. Chem.* **2017**, *13*, 1957–1962.
26. Ardila-Fierro, K. J.; Pich, A.; Spehr, M.; Hernández, J. G.; Bolm, C. Synthesis of Acylglycerol Derivatives by Mechanochemistry. *Beilstein J. Org. Chem* **2019**, *15*, 811–817.
27. Bose, A.; Mal, P. Mechanochemistry of Supramolecules. *Beilstein J. Org. Chem.* **2019**, *15*, 881–900.
28. Holler, M.; Stoerkler, T.; Louis, A.; Fischer, F.; Nierengarten, J.-F. Mechanochemical Solvent-free Conditions for the Synthesis of Pillar[5]Arene-containing [2]Rotaxanes. *Eur. J. Org. Chem.* **2019**, 3401–3405.
29. Yang, Y.; Bu, F.; Liu, J.; Shakir, I.; Xu, Y. Mechanochemical Synthesis of Two-dimensional Aromatic Polyamides. *Chem. Commun.* **2017**, *53*, 7481–7484.
30. Hernández, J. G. Mechanochemical Borylation of Aryldiazonium Salts; Merging Light and Ball Milling. *Beilstein J. Org. Chem.* **2017**, *13*, 1463–1469.
31. Tanaka, R.; Takahashi, N.; Nakamura, Y.; Hattori, Y.; Ashizawa, K.; Otsuka, M. Verification of the Mixing Processes of Active Pharmaceutical Ingredient, Excipient and Lubricant in a Pharmaceutical Formulation Using a Resonant Acoustic Mixing Technology. *RSC Adv.* **2016**, *6*, 87049–87057.
32. Osorio, J. G.; Muzzio, F. J. Evaluation of Resonant Acoustic Mixing Performance. *Powder Technol.* **2015**, *278*, 46–56.
33. Osorio, J. G.; Hernández, E.; Romañach, R. J.; Muzzio, F. J. Characterization of Resonant Acoustic Mixing Using near-infrared Chemical Imaging. *Powder Technol.* **2016**, *297*, 349–356.
34. am Ende, D. J.; Anderson, S. R.; Salan, J. S. Development and Scale-up of Cocrystals Using Resonant Acoustic Mixing. *Org. Process Res. Dev.* **2014**, *18*, 331–341.
35. Ralphs, K.; Hardacre, C.; James, S. L. Application of Heterogeneous Catalysts Prepared by Mechanochemical Synthesis. *Chem. Soc. Rev.* **2013**, *42*, 7701–7718.
36. Boldyreva, E. Mechanochemistry of Inorganic and Organic Systems: What Is Similar, What Is Different? *Chem. Soc. Rev.* **2013**, *42*, 7719–7738.
37. Takacs, L. The Historical Development of Mechanochemistry. *Chem. Soc. Rev.* **2013**, *42*, 7649–7659.
38. Wang, G.-W. Mechanochemical Organic Synthesis. *Chem. Soc. Rev.* **2013**, *42*, 7668–7700.
39. Lyakhov, N. Z.; Grigorieva, T. F.; Barinova, A. P.; Vorsina, I. A. Mechanochemical Synthesis of Organic Compounds and Composites with their Participation. *Russ. Chem. Rev.* **2010**, *79*, 189–203.
40. Tan, D.; García, F. Main Group Mechanochemistry: from Curiosity to Established Protocols. *Chem. Soc. Rev.* **2019**, *48*, 2274–2292.
41. Andersen, J.; Mack, J. Mechanochemistry and Organic Synthesis: from Mystical to Practical. *Green Chem.* **2018**, *20*, 1435–1443.
42. Colacino, E.; Dayaker, G.; Morère, A.; Friščić, T. Introducing Students to Mechanochemistry Via Environmentally Friendly Organic Synthesis Using a Solvent-free Mechanochemical Preparation of the Antidiabetic Drug Tolbutamide. *J. Chem. Educ.* **2019**, *96*, 766–771.
43. Wixtrom, A.; Buhler, J.; Abdel-Fattah, T. Mechanochemical Synthesis of Two Polymorphs of the Tetrathiafulvalene-chloranil Charge Transfer Salt: An Experiment for Organic Chemistry. *J. Chem. Educ.* **2014**, *91*, 1232–1235.

CHAPTER FOURTEEN

Sonomechanochemistry

Giancarlo Cravotto[a], Katia Martina[a], Maria Jesus Moran[a], and Pedro Cintas[b]

[a]Department of Drug Science and Technology, and NIS-Centre for Nanostructured Interfaces and Surfaces, University of Turin, Turin, Italy
[b]Department of Organic and Inorganic Chemistry, Faculty of Sciences and IACYS-Green Chemistry and Sustainable Development Unit, University of Extremadura, Badajoz, Spain

1 Introduction

Understanding how sound waves convey mechanical energy to a given system, ranging from materials to chemical reactions as well as biological cells and tissues, represents a current challenge of a broad and far-reaching field, which can appropriately be denoted as *sonomechanochemistry*. It may be surprising to most scientists not familiarized with sonochemistry that sound is actually mechanical energy. Waves move through an elastic medium by means of consecutive compression and rarefaction phases. Acoustic waves can be characterized by numerous physical parameters, albeit intensity/amplitude and frequency are the most usual ones. Ranging from a few hertz (Hz) in the case of infrasound to some megahertz (a zone employed for medical imaging and diagnosis), audible sound corresponds to a narrow window between 10 Hz and c. 20,000 Hz (20 kHz). For comparative and pedagogical purposes, Fig. 1 shows an eye-catching cartoon of sound frequencies. Two well-defined windows are of particular interest in sonochemistry: (1) the so-called *power ultrasound* (US), from 20 kHz (even if frequencies above 16–18 kHz are inaudible for most adults) to 100 kHz, where physical effects are prevalent, and (2) high-frequency ultrasound, from 200 to 2000 kHz (2 MHz), where effects of chemical nature, including an enhancement of radical production, are much more relevant.

Like in other irradiative processes (e.g., photochemistry or microwaves), an accurate control of external parameters and conditions (vide infra) is a must, even more compulsory in sonochemistry than in other techniques if one may say so. It is not unusual to read sonochemical protocols, in high-profile journals too, which cannot be reproduced as working conditions are not specified, namely, frequency, amplitude, temperature, and type of ultrasonic generator or piezoelectric transducer. Nor did peers recognize

Nontraditional Activation Methods in Green and Sustainable Applications
https://doi.org/10.1016/B978-0-12-819009-8.00005-0
© 2021 Elsevier Inc.
All rights reserved.

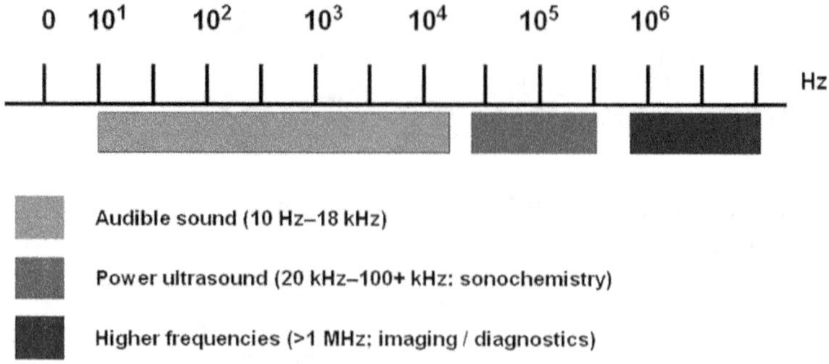

Fig. 1 The spectrum of sound frequencies.

the importance of these measurements. The absence of this information can lead to false positives or overestimations of a spotted sonochemical reaction. The systematic use of well-established "dosimeters," by which either physical or chemical effects of ultrasound can be determined with confidence, enables useful and absolute comparison of results obtained with different equipment.

It is pertinent to emphasize that sound, unlike electromagnetic radiation, lacks quantum character, and as a result, there is no direct correlation between frequency and energy, or in other words, similar effects can be attained within a broadband of frequencies. Moreover the wavelengths of sound are usually much longer than those of photons carrying less energy, which excludes any interaction between sound and matter at atomic/molecular levels.

Most effects detected in sonochemical reactions arise from a unique phenomenon—*cavitation*—that releases enough kinetic energy, capable of inducing bond-forming and bond-breaking processes and driving reactions to completion. Acoustic cavitation in liquids takes place when a pressure wave of sufficient intensity disrupts the cohesive forces, thereby creating micrometer-sized bubbles that collapse violently to restore the bulk liquid. Extreme conditions of temperature and pressure can be reached within collapsing bubbles (typical figures are 4500–5000 K and 1500–2000 bar), which are considerably attenuated at the liquid-gas interface. High cooling rates (up to 10^{10} K/s) are consistent with a quasi-adiabatic process, as minor thermal variations are observed in the bulk solution. Volatile substrates and solvents may then be trapped in the bubbles and undergo homolytic cleavage and/or conversion to excited species. This primary sonochemistry can subsequently be followed by secondary effects as free radicals released to the bulk solution will react with molecules present at the vicinity of the collapsing bubble or recombine to form stable species. Bubble implosion is accompanied by shock waves and shear stress that propagate through the liquid, and they collectively represent the main physical effects associated with cavitation (Fig. 2). In heterogeneous reactions, where liquid-solid interfaces are involved, such forces are often accompanied by high-velocity flow induced by the pressure gradient, *acoustic streaming*, which enhance both mass and heat transfer. Generation of microjets, hammering, and interparticle collisions account for the well-known mechanical action of acoustic cavitation like erosion, surface cleaning by removing passivating coatings, and micromixing leading to rapid emulsification.

A rigorous treatment of acoustic cavitation lies beyond the scope of this article. Also, mathematical modeling is complex indeed as cavitation exhibits a nonlinear behavior that cannot be described by linear acoustics. From a practical standpoint, it suffices to have in mind, in line with the aforementioned description, that bubble collapse and hence sonochemical activity will largely be dependent on physicochemical properties of solvents, presence of dissolved gases, and characteristics of the incident acoustic wave. Both comprehensive and concise monographs provide good overviews of cavitational chemistry and its applications, with preliminary chapters on parameters influencing cavitation and methods to quantify the resulting cavitation activity.[1–5]

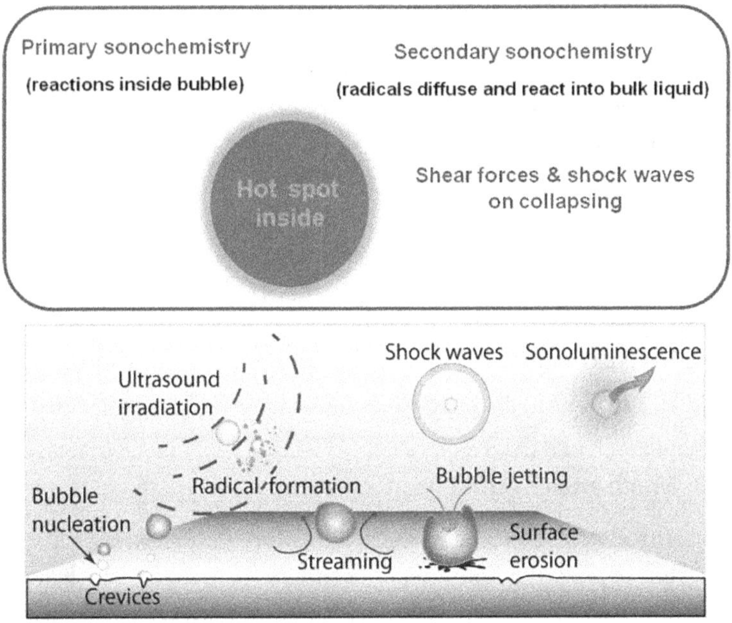

Fig. 2 Chemical and mechanical effects caused by acoustic cavitation.

The current impact of sonochemistry in numerous disciplines is primarily related to the concept of sustainability aided by enabling technologies, which chiefly include photochemistry, electrochemistry, microwave irradiation, and mechanochemistry. All of them have the potential to influence dramatically numerous reactions and physicochemical processes by reducing reaction times, enhancing yields, and avoiding the formation of unwanted side products or competing pathways, thus favoring selectivity and product isolation. Unfortunately the lure of green chemistry has often engendered both hype and misconceptions in modern literature. Chemical and process metrics categories have been introduced to assess the greenness of a global transformation, *from craddle to the grave*, in terms of efficiency, selectivity, mass productivity, wastes, inherent hazards, renewability, recyclability, product quality, and cost.[6–10] Clearly, sonochemical methods should strongly adhere to these considerations if green protocols are to be claimed.[11, 12]

Although this chapter concentrates on mechanical effects of sonication, the chemical and physical pathways of cavitation cannot easily be disentangled. The distinctive point is the fact that while most mechanochemical applications involve solid phases or slurries containing little amounts of solvents, sonochemical transformations do require a liquid phase to induce cavitation and propagate sound waves. The cavitational collapse and, as a result, the forces created at the bubble-liquid interface can be fined-tuned by means of external parameters, frequency, and power in particular. Furthermore, mechanical action may still occur in the absence of cavitation, thus preventing structural damage or uncontrolled side reactions. The pulses of sonochemistry lie in multiple effects, among others: acceleration of chemical reactions; the use of less forcing experimental conditions; reduction of synthetic steps; initiation of stubborn reactions; elimination of induction periods; or enhancement of catalytic activity and radical production. Such effects, as mentioned, are especially noticeable in heterogeneous sonochemistry, that is, on and at the vicinity of solid surfaces. Along with improved mass and heat transfer caused by ultrasonic agitation, species resulting from cavitation will also be able to interact with adsorbed reagents and catalytic sites of solid materials.

2 Mechanochemistry in context

Intuitively the concept of mechanical force implies physical contact, unlike other stimuli like light, heat, electricity, or gravity. Mechanochemistry was likely the first science performed by our ancestors at the dawn of civilizations in the form of rubbing sticks together for fire, as well as milling, sliding, grinding, drilling, or stretching to induce unconsciously chemical transformations. The term was introduced by Ostwald more than 1 century ago, to denote simply the coupling of mechanical energy and chemical energy.[13, 14] This definition is broad enough to be in practice ambiguous. The modern IUPAC definition indicates that a mechanochemical reaction involves a chemical reaction that is induced by the direct absorption of mechanical energy.[15] Mechanochemistry can also be accurately described as the coupling between macroscopic motion or mechanical loads and chemical reactivity.[16, 17] Yet, mechanochemistry entails two different conceptual frameworks: on one hand mechanical energy for making and breaking bonds and on the other effects that result in conformational changes or structural rearrangements. The latter would lie rather in the field of mechanophysics.[18] Accordingly, even if not usually employed, it is convenient for clarity to speak of *covalent* and *noncovalent* mechanochemistry.[19–21] In the covalent domain, mechanical actions cause molecular-level fragmentations yielding either a new compound or material with different properties. These changes are usually irreversible, and the initial state is not recovered when the force is removed. Noncovalent mechanochemistry, called *soft mechanochemistry* as well, involves reversible changes that do not cause any rupture of chemical bonds but rather conformational variations and structural reorganization of molecular or supramolecular entities, often leading to further physicochemical responses. These applications take strong inspiration from nature, as mechanotransduction pervades all living organisms. At the cellular level, any mechanical deformation induced by flow, elongation, or pressure may trigger diverse signaling pathways. Cell motion; ion-channel opening in membranes, thereby altering electrolyte concentrations; or left-right asymmetry in embryonic development represent a few examples of key mechanisms governed by mechanical inputs.[22–26] As we shall discuss, sonication has found applicability in both covalent (bond-breaking reactions) and noncovalent (aggregation, nucleation, self-assembly, or disassembly) mechanochemistry.

In less than one decade, mechanochemistry has moved from a peripheral activity exploited largely by solid-state scientists and some synthetic chemists with variable success, to a multidisciplinary research field that engages people from disparate disciplines such as polymers, nanomaterials, catalysis, drug synthesis, or biomechanical applications.[27–32] The attractiveness is twofold: firstly, mechanochemical transformations generally involve solid reagents, thus overcoming the use of toxic volatile solvents and resulting in environmentally friendly processes. Likewise the operational protocols are simple and easy to make using inexpensive mortar or yet relatively cheap ball mills. According to green chemistry principles, however, solvent-free mechanochemical transformations should be assessed with caution as post-synthetic operations and workup may not be green at all, that is, involving hazardous protocols and organic solvents (e.g., liquid-liquid extraction or chromatography).[6, 7, 12]

3 The force of sound

As noted earlier, sound can be produced by a number of phenomena that, to large extent, carry mechanical forces or pressures in objects and media, such as vibrating bodies, changing airflow and supersonic flow, and time-dependent heat sources. Human and animals use sound to communicate. Our inner ear behaves actually as a musical instrument that transforms vibrations into sound. The hearing organ, called *cochlea*, possesses a spiral structure composed of sensory cells that respond to different frequencies of sound. Some mammals, in particular bats, dolphins, and whales, have developed biosonars capable of spotting frequencies beyond the human hearing (>20 kHz). This evolutionary mechanism, *echolocation*, lets them see the world via the echo of high-frequency vibrations.[33]

As mentioned too, in the absence of cavitation, sound waves carry little energy. Like other radiative processes, working intensity and frequency should be specified. Even if the range of audible frequencies is narrow (from ca. 10 Hz to 18–20 kHz), the range of sound intensity to which the human ear responds is indeed remarkable. The ratio between painful sound/noise and the weakest signal we can hear is larger than 10.[12] On the contrary the ratio of audible frequencies is nearly 10^3 (or more than nine octaves in musical terms). The least amount of sound energy humans can hear is ca. 10^{-20} J (for comparative purposes one quantum of light in the middle of the visible spectrum is $\sim 4 \times 10^{-19}$ J).[34] As frequency rises, a higher threshold intensity is required for inducing cavitation in liquids. It is no surprise that medical imaging exploits the megahertz band, where high resolution can be attained while cavitation will hardly occur at low intensity. Chemical and mechanical effects exhibit an inverse dependence on acoustic frequency.[35] At frequencies lower than 100 kHz, the mechanical effect threshold, determined by erosion measurements in aluminum foil, nearly equals the threshold for cavitation to occur. However, the threshold for mechanical effect is much higher than the cavitation threshold at high frequency.[36] The shear forces generated from collapsing bubbles are strongest at low frequencies (in the typical working band between 20 and 40 kHz) and drop off at higher frequencies. This trend is related to the larger bubble radius before collapse at lower frequencies, which results in a larger empty volume that the surrounding bulk liquid will immediately replace after collapse.[37] Alternatively the rate of radical production increases with frequency up to a maximum located around 500 kHz (a Gaussian-like curve is observed between 200 and 800 kHz) and then decreasing near megahertz frequencies as mentioned earlier.

A priori the connection between mechanochemistry and sonochemistry is not obvious. A tribochemical interpretation was first suggested in the 1990s, where stimulation of chemical reactions involving solids takes place through friction at boundary surfaces.[38] Cavitation, by virtue of its nonlinearity, leads to distortion of the sinusoidal waveform, which become nearly triangular as the shock wave forms. Bubble oscillations exhibit chaotic behavior, and local disturbances may extend over large distances in the liquid.[39, 40] Bubbles, with shapes far from ideal sphericity, also deform near surfaces upon collapse, which decreases the internal heating within the cavity as inferred from sonoluminescence studies but imparts sufficient mechanical energy that erodes the surface and increases its reactivity.[41]

In mechanochemistry, different mechanisms account for the reactivity in the solid state, although interparticle collisions, diffusion, and even the generation of a eutectic phase upon mixing may provide a plausible rationale.[19, 28, 42–44] In solution, however, molecular motion averages out the orientation of the molecules and directional effects vanish quickly. Unlike heterogeneous reactions, in homogeneous systems, mechanical effects, if any, can be negligible for small molecules, but macromolecular entities, namely, polymers, could still undergo plastic deformations. Even if large molecules, owing to their low volatility, will not undergo cleavage inside the bubbles, shear forces may fragment homolytically polymer chains, or the sonolysis of solvent or small monomers will induce further reactions. In fact, chain-growth radical polymerization can efficiently be performed under controlled sonication.[45, 46]

Mechanochemical phenomena induced under sonication can be related to the effects caused by transient elongational flows when bubbles generated by propagating ultrasound waves collapse.[29, 47–49] In such flow fields the strain rate depends on the time from the onset of the collapse and hence the radius of the imploding cavity and on the distance to the bubble. Because the solvent flow rate in close vicinity to a collapsing bubble decreases quickly with distance from the bubble surface, the two termini of a polymer chain located at different distances from that surface will experience very different flow rates, and the chain becomes stretched.[29] With respect to other types of flow, ultrasonic fields can provide high strain rates (up to $10^7\ s^{-1}$), while average masses required for chain scission are reduced to the range of 10^4–10^5 Da.[47] Long chains serve then as tweezers capable of propagating a vibrating force and focusing its energy on reactive or weak bonds. This conceptual framework represents the core of polymer sonochemistry tailored to mechanochemical sensors and actuators, which has resulted in a series of innovative applications, disclosing in addition new mechanistic insights and challenges.[29, 48, 50–54]

4 Mechanophore activation by sonication

A mechanophore is essentially a molecular unit, fragment, or functional group that undergoes chemical activation under mechanical stress. As noted in previous remarks, the mechanical action can be achieved by numerous methods, both in solution and solid state, leading to either conformational changes or bond-breaking/bond-forming reactions (soft and covalent mechanochemistry, respectively). Force-induced structural changes can be monitored at the single-molecule level with a force-transducer device, (AFM tip, for instance), where the restoring force is measured as a function of the extension of the molecule along its entire length.[21] In covalent mechanochemistry, we are interested in measuring the collective response of molecules to external mechanical stimuli, which translate into changes in physicochemical properties as a result of bond reorganization.

It is quite obvious that mechanophores implicitly involve mechanically labile bonds, such as strained molecules, and/or weak bonds that experience facile fragmentations with shock waves. This effect can be augmented in molecules adsorbed on surfaces (heterogeneous sonochemistry) or embedded into polymer chains in solution. The latter emerges from the pioneering work of Melville and Murray in the 1950s, who carried out ultrasonication of solutions of a styrene-containing poly(methacrylate), which led to polymer degradation and repolymerization.[55] This result was consistent with homolytic bond cleavage arising from radical species as initiators. Uncontrolled fragmentation in polymers would however be of little synthetic value. Fortunately, further studies demonstrated that cavitation-induced cleavage in polymers is not purely stochastic and chains tend to break near the middle of large chains. By placing mechanophores at that position, specific cleavage can be accomplished under sonication, as proven by Moore and coworkers with a labile azo group near the center of poly (ethyleneglycol).[56] The mechanophore concept is globally shown in Fig. 3, where bond rupture occurs at chemical functionalities placed near the middle point of polymer chains. In agreement with the large tensile forces transmitted by acoustic agitation to vibrating chains, such an activation fails in mechanophore end-capped polymers, that is, if chains are not attached at both termini of the functional group.

In addition to its role to convey force, the chemical nature of polymer chains may be a secondary parameter capable of augmenting sonochemical activity, although such effects have been scarcely scrutinized in mechanochemical studies. Most polymers in organic media may behave as nonionic surfactants, which influence bubble instability and collapse, as well documented in aqueous sonochemistry.[57] In particular, surfactants may reduce bubble coalescence, which leads to larger bubbles and clustering that attenuate the sound field.

As illustrated in Fig. 3, mechanophore activation does not necessarily imply bond cleavage, but productive structural changes that cause physicochemical changes.[50–53] These may include isomerization, light emission (mechanoluminescence), color changes, delivery of small molecules, or activation of latent catalysts, all useful as chemical probes or analytical diagnosis.

It is fair to say that mechanochemistry with sound waves received considerable attention, even skepticism, in the scientific community after a 2007 *Nature* paper by Moore and associates reporting on the apparent violation of Woodward-Hoffmann rules with a benzocyclobutene mechanophore when

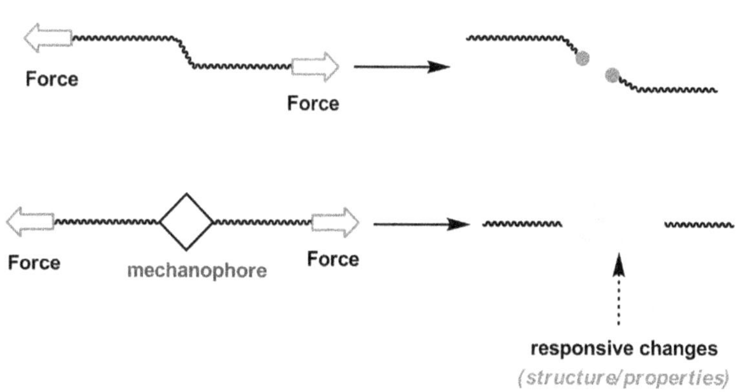

Fig. 3 Mechanical activation of polymer chains under sonication, either causing bond cleavage *(top)* or structural changes *(bottom)*.

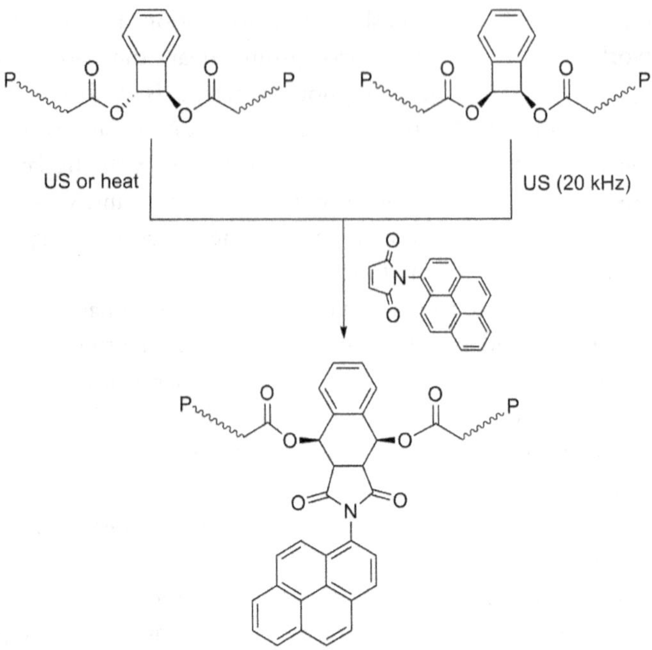

Fig. 4 Altering orbital symmetry conservation rules by sonication. P denotes a polymer chain of α-methoxy-ω-amino-poly(ethylene glycol).

activated by sonication relative to thermal conditions.[58] As shown in Fig. 4, both *cis*- and *trans*-benzocyclobutenes within polymer chains undergo ring opening yielding an *E,E*-configured diene, a fact that could be confirmed by derivatization with an aromatic chromophore. Clearly, ultrasound induces a forbidden ring opening that cannot be predicted by symmetry rules. Under thermal activation the *trans* isomer would lead to the *E,E*-diene and the *cis* isomer to the *E,Z*-diene only. Photochemical irradiation would reverse that stereochemical bias.

Key applications of mechanophore activation in ultrasound fields are related to chemical monitoring or design of selective actuators. A well-exploited mechanophore to this end is the spiropyran moiety, which undergoes C—O bond cleavage and isomerization to produce a conjugated heterodiene system. As a result a noticeable color change, usually from yellow to pink, is visible to the naked eye.[59] In a similar way the four-membered 1,2-dioxetane ring is a suitable mechanophore that experiences facile ring opening when the adjacent polymer chains are subjected to ultrasonic agitation.[60] Dioxetane fragmentation is accompanied by intense chemiluminescence, again a visual change of significance in the preparation of mechanoresponsive polymeric materials. As light is emitted from the adamantanone-excited state that arises from ring opening, further color tuning could be achieved by energy transfer to appropriate aromatic acceptors (Fig. 5).

More synthetic applications are collected in Fig. 6 through two representative examples. A *gem*-dichlorocyclopropyl indene carbocycle provides another strained mechanophore sensitive to vibrating force. When that framework is glued on polymeric chains and irradiated with ultrasound, liberation of hydrogen chloride (HCl) takes place by a sequential transformation involving bond cleavage, rearrangement, and elimination-aromatization.[61] Precatalysts or latent catalysts can be generated in situ from mechanophores embedded in polymer chains.[62] Bond cleavage, often labile metal-carbon bonds, produces reactive species that participate subsequently in organic or organometallic reactions.

Extension of these mechanochemical strategies from synthesis to materials with specific properties has been recently illustrated by the preparation of plastic semiconductors based on ladderenes. Such strained organics represent a new type of sensitive mechanophores, which consist of five-fused cyclobutyl rings adopting a staircase-like conformation. Due to ring strain, they undergo facile ring fragmentation under mechanical force. A terminal cyclobutene ring of the chlorosubstituted [5]ladderene monomer fragments through ring-opening metathesis polymerization (ROMP) with a Grubbs ruthenium carbene catalyst. Subsequent dehydrohalogenation gives rise to hydrocarbon homopolymers with high loading of ladderene mechanophores along the entire backbone, which are then irradiated with ultrasound in an inert atmosphere (Fig. 7).[63] The mechanical action causes formation of a polyacetylene, a typical organic semiconductor, which self-aggregates after prolonged sonication. The effect is visible as acoustic radiation converts a soluble, nearly colorless material into an insoluble polymer having an intense blue color.

In the search for conjugated homo- and heteropolymers beyond polyladderene, the same group of Xia and coworkers has applied this strategy to ladderene-norbornene block copolymers[64] and benzoladderenes.[65] The initially formed nonconjugated polyladderenes were mechanochemicall transformed into conjugated systems in moderate yields by sonication, often at lower temperatures and more controlled timing than ladderenes.

Polymers incorporating more than one mechanophore units, that is, multimechanophore architectures, have also been generated.[66] The advantages of such highly functionalized systems are quite obvious,

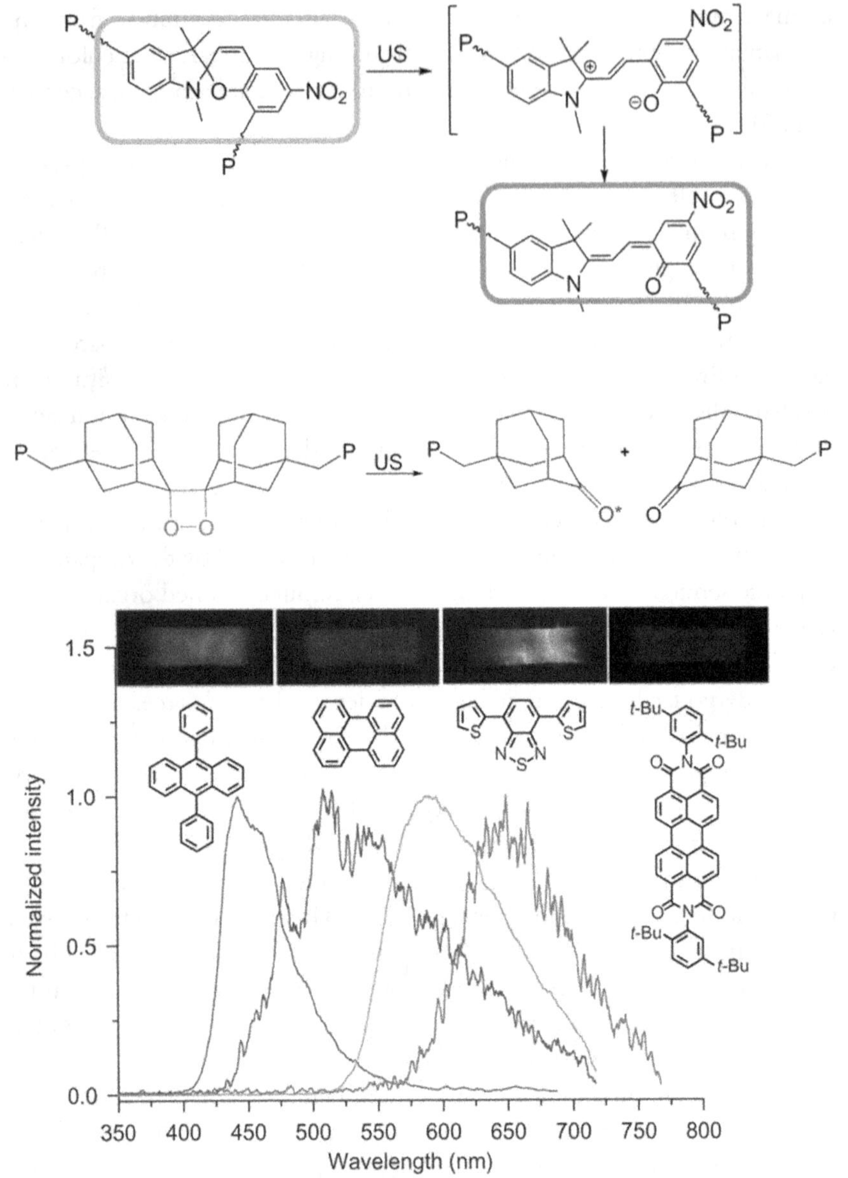

Fig. 5 Ultrasound-induced activation of covalent bonds in strained mechanophores leading to color change or chemiluminescence. P denotes a poly(methyl acrylate) side chain. *The image at the bottom is reproduced from Chen, Y. L.; Spiering, A. J. H.; Karthikeyan, S.; Peters, G. W. M.; Meijer, E. W.; Sijbesma, R. P. Mechanically Induced Chemiluminescence from Polymers Incorporating a 1,2-Dioxetane Unit in the Main Chain. Nat. Chem. 2012, 4, 559–562 with permission. Copyright 2012 MacMillan Publishers Ltd.*

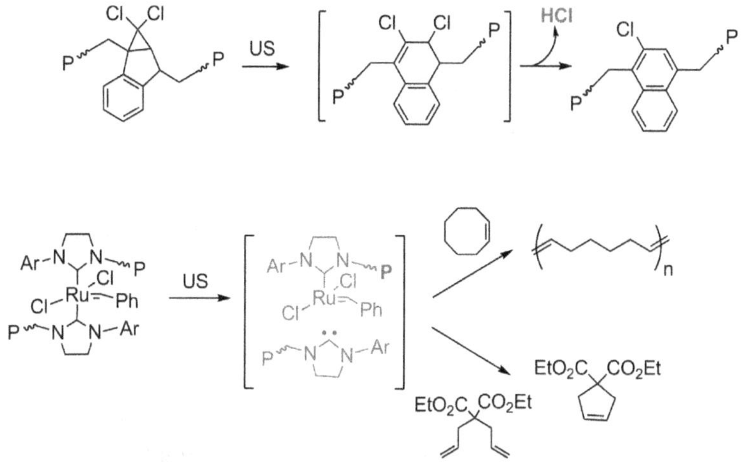

Fig. 6 Liberation of reactive species from mechanophore-containing polymers, which may be used in synthetic organic chemistry.

as they enable varied reactivity and construction of multiresponsive materials. In a recent example, the groups of Craig and Tang and their associates have synthesized ruthenocene polymers using well-known protocols of atom-transfer radical polymerization and ROMP and then subjected to ultrasonication, which produces preferential scission at the metallocene unit along the polymer backbone. Compared with ferrocene, having a Fe-cyclopentadienyl (Cp) bond distance of 1.66 Å, ruthenocene has a longer

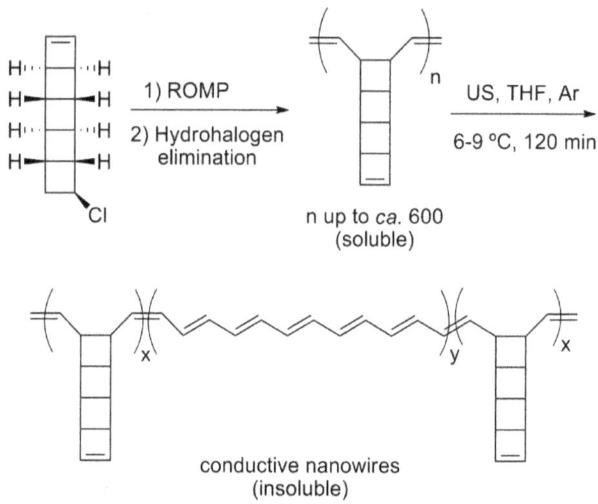

Fig. 7 Polyladderene-polyacetylene copolymers form semiconductive nanowires under the mechanical action of sonication.

Ru-Cp bond length (1.82 Å) and hence a higher bond dissociation energy for both homolytic and heterolytic fragmentations. Despite this thermodynamic stability, ruthenocene is also a labile mechanophore for selective scission in a manner similar to previous studies with ferrocene mechanophores. The mechanical strength of ruthenocene under ultrasound-based force field could be quantified by competition between bond scission at the metallocene and the nonscissile ring-opening of gem-dichlorocyclopropane mechanophore within the same polymer.[67]

Counterintuitively, in a nonconventional use of mechanochemistry, mild sonication can stabilize, through stretching, compounds that would otherwise be unstable in the absence of external force, as demonstrated recently by single-molecule force spectroscopy of maleimide-thiol adducts.[68] Such adducts are widely employed in bioconjugation and materials chemistry, but they are likewise prone to undergo retro-Michael reaction and hence thiol exchange in vivo that diminishes the therapeutic effect of bioconjugates. Cao and coworkers show that pulsed sonication in aqueous buffer (pH 7.4) of a polyethylene glycol (PEG)-maleimide reagent modified with the clinically used antibody trastuzumab accelerates adduct hydrolysis, thus stabilizing the resulting conjugates and reducing substantially the competing retro-Michael reaction (Fig. 8). Moreover, ultrasound does not alter the binding of the antibody to its target ligand, because

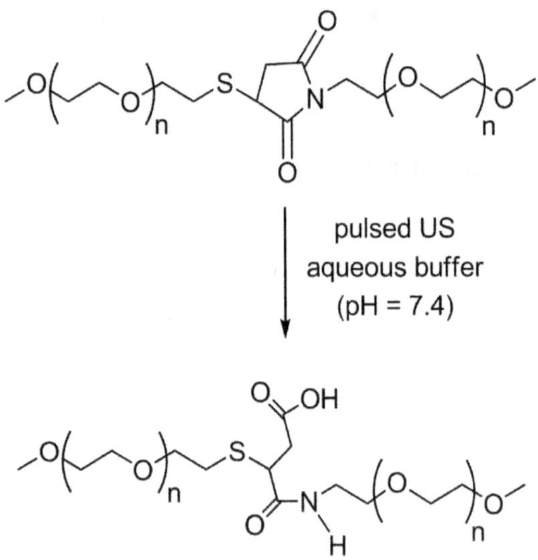

Fig. 8 Mechanochemical stabilization and hydrolytic acceleration of maleimide-thiol adducts under pulsed ultrasonication.

antibody-PEG conjugates formed under sonication showed similar ligand-binding activity to those of nonirradiated treatment in an enzyme-linked immunosorbent assay.

In another promising application targeted to trigger cellular responses and biomonitoring, mechanochemical transduction in synthetic elastomers has been accomplished by means of high-intensity focused ultrasound (HIFU), which employs, for biomedical applications, frequencies much higher than those of conventional power ultrasound. Two types of mechanophores embedded in a cross-linked polydimethylsiloxane (PDMS) network could be activated under HIFU irradiation. One reaction involves the reversible isomerization of a naphthopyran unit leading to a visible color change at the focal spot depending upon irradiation exposure. In a second example, HIFU is also capable of fragmenting a dioxetane mechanophore resulting in visible blue light at the focal point (Fig. 9).[69] The experimental setup works at 1 MHz with average intensities varied from 39.4 to 376 W/cm^2 (Fig. 10). Control experiments ruled out thermal

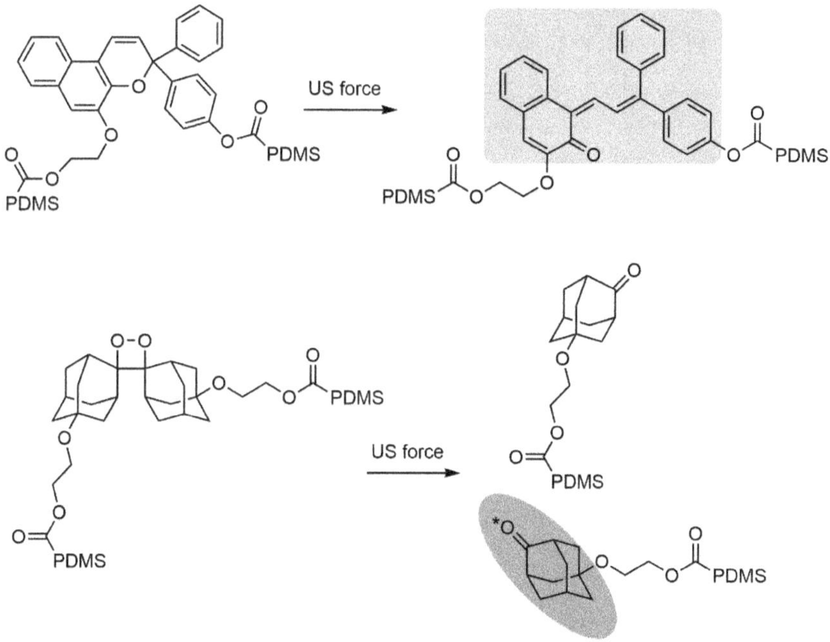

Fig. 9 Synthesis of mechanochromic and mechanoluminescent polydimethylsiloxane (PDMS) elastomers, whose mechanophores are locally activated under high-intensity focused ultrasound (HIFU).

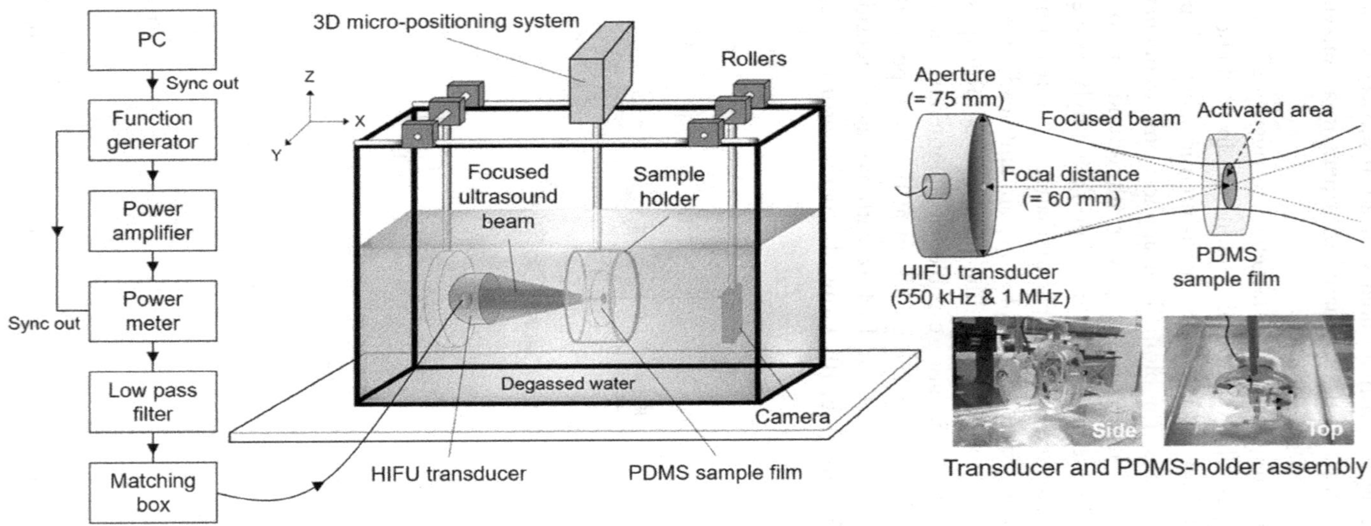

Fig. 10 Schematic diagram of the experimental setup employed for HIFU-induced mechanochromic activation of PDMS films. *Reproduced with permission from Kim, G.; Lau, V. M.; Halmes, A. J.; Oelze, M. L.; Moore, J. S.; Li, K. C. High-Intensity Focused Ultrasound-Induced Mechanochemical Transduction in Synthetic Elastomers. Proc. Natl. Acad. Sci. U. S. A. 2019, 116, 10214–10222. Copyright 2019 American Association for the Advancement of Science.*

energy to cause mechanical responses; only minor increases in temperature (<6°C) were observed at intensities below 354 W/cm^2, although above that threshold, thermal variations can be significant and thermal ablation would then be a dominant mechanism masking any mechanochemical effect. The device paves the way to noninvasive ultrasonic energy triggering mechanical stimuli compatible with biological systems.

4.1 Structure-reactivity relationships in sonomechanochemical reactions

Ultrasound-assisted mechanochemistry is clearly circumscribed to polymer mechanochemistry, although how sound waves induce mechanical transduction requires further examination of some structural parameters that affect polymer scission to a large extent. As mentioned, sonication and in general shear forces on polymers lead to preferential chain scission in the middle of the macromolecule.[47, 55] A more recent study on covalently bonded Diels-Alder step-growth polymers, assessed by theoretical modeling and high-temperature NMR spectroscopy, shows that the observed selectivity is a consequence of entropic factors.[70] Nonetheless, there are some discrepancies between kinetic and thermodynamic parameters measured for polymeric species depending on either chain length or average molecular weight. Early work conducted in the late 1950s on the ultrasonic degradation of polymers concluded that chain length is the decisive parameter governing the rate of chain scission.[71] Mechanophore-aided reinvestigation of this issue has been performed by Moore and associates in an attempt to decouple the effects of molecular weight and degree of polymerization (i.e., polymer chain length). This group analyzed the sonochemical ring opening of a spiropyran mechanophore and its conversion into the merocyanine structure (Fig. 5), embedded in acrylate polymers, for which both repeating units and side chain constitution were varied.[72] Ultrasonic irradiation (10.7 W/cm^2 at 3–5°C) in 2-butanone, where all of the polymers were soluble, was implemented in a continuous flow setup coupled with real-time UV-Vis monitoring for obtaining kinetic information. Plots of kinetic constants against degree of polymerization, for which all data showed a single linear relationship, evidenced that chain length is the main parameter accounting for mechanochemical activation, irrespective of molecular weight and side chain constitution. On the contrary, plots of rate constants versus molecular-weight afforded different linear regressions depending on the pendant group (Fig. 11).

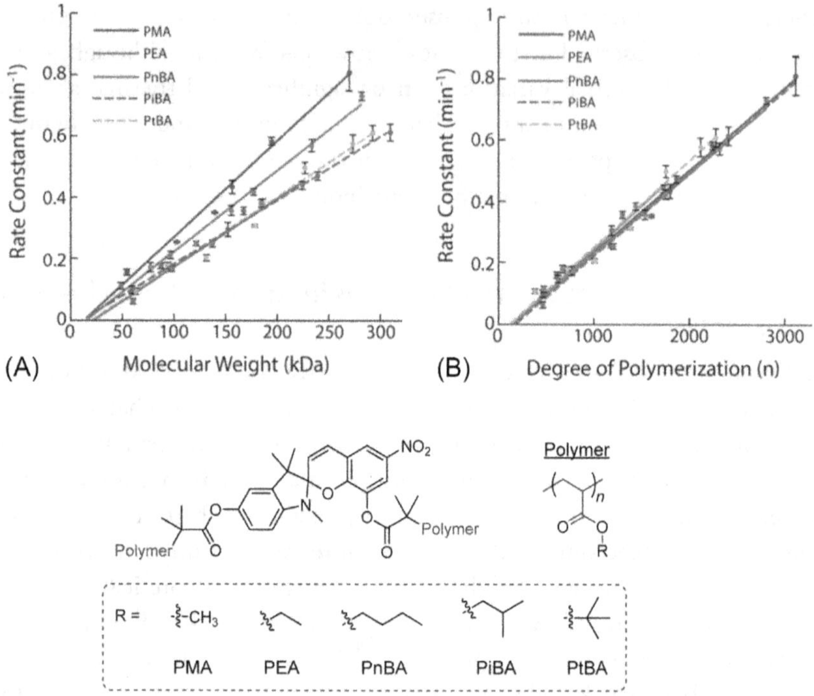

Fig. 11 Plots showing activation rates of ultrasound-induced spiropyran scission as a function of (A) number average molecular weight and (B) degree of polymerization. Each point represents the average of two measurements: error bars denoting the range of two values. *Reproduced with permission from May, P. A.; Munaretto, N. F.; Hamoy, M. B.; Robb, M. J.; Moore, J. S. Is Molecular Weight or Degree of Polymerization a Better Descriptor of Ultrasound-Induced Mechanochemical Transduction? ACS Macro Lett. 2016, 5, 177–180. Copyright 2016 The American Chemical Society.*

Other structural features present in polymer architectures, namely, branching, cross-linking, hydrogen bonding, and folding, influence the fate of mechanical action through solvodynamic shear forces. In another recent study on ultrasound-induced chain fragmentation of polyphenylene-based dendritic polymers, Choi et al. measured degradation rate constants across variable degrees of polymerization (~100–600) and found that larger side chains led to increased degradation rates. Results were also consistent with the fact that polymers having elongated and rigid conformations degrade more rapidly.[73] In addition, experimental evidence and computational modeling suggest that polymer chains undergo multiple scissions during ultrasonic activation. This multiscission had been previously reported for linear polymers, being a molecular weight-dependent phenomenon.[47] The study of Choi et al. suggests instead that multiscission occurs at lower degrees of polymerization (for polymers with large side chains), simply because larger side chains also increase the total mass.

The influence of acoustic forces on cross-linking can be better modeled with single-chain polymer nanoparticles in solution. To this end, sonication has been applied to dilute samples of linear and cross-linked polymers, for which variations in molecular weight, hydrodynamic radius, and viscosity were evaluated.[74] An increase in cross-linking enhances the structural stability of the polymer, and thus polymers containing 10 and 15 mol% cross-linked motifs remained practically unaffected after ~2-h sonication, with average masses changing from 132 and 136 kDa to 117 and 127 kDa, respectively. This contrasts with a linear polymer whose average molecular weight decreased from 99 to 64 kDa after a similar sonication exposure. It should be mentioned that mechanochemical reactions are faster by increasing cross-linking density, although the scission occurs at sacrificial bonds that do not involve chain fragmentation. These features are much noticeable in low-molecular weight polymers, which undergo less tensile forces than their higher homologs, and in addition, bond cleavage takes place above a threshold mass. A small 100-kDa polymer containing more than 10 mol% intramolecular cross-linking becomes essentially insensitive to shear forces in solution. In an interesting extrapolation to intramolecular ligand-metal bonds, the incorporation of rhodium-π bonds into the double bonds of a linear polybutadiene backbone leads to weak intramolecular cross-links. Mechanical stress induced by sonication is absorbed at the weak Rh-π bonds, rather than on mechanophores within the linear chain, thereby causing some sort of chain unfolding,[75] in a similar way to the aforementioned ruthenocene mechanophores.[67] Interestingly, as the Rh center lies into close contact with the alkene moiety, thermal refolding occurs spontaneously, thus providing a new strategy for intramolecular self-healing in polymers. More sophisticated structural variations have been recently introduced to check the proof of concept of mechanical bonds as efficient force actuators of mechanical energy. The rate of a retro-Diels-Alder reaction under tension (thermal or ultrasound) is decreased when the cycloadduct mechanophore is embedded in the axis of a rotaxane, because the latter represents a high-stress region that competes with the mechanical activation at the scissile bond.[76]

All mechanochemical scissions induced by cavitational collapse do invariably involve some sort of thermal ingredients. While polymer scientists may overlook this issue, sonochemistry still enables ways of disentangling both effects by judiciously controlling bubble collapse. Sijbesma et al. have addressed this concern by activating latent carbene catalysts after mechanical scission in solutions saturated with methane (CH_4) instead of argon.[77] Polyatomic gases dissolved in liquids attenuate, often suppress, cavitation with respect to monoatomic gases like noble gases.[78] Inhibition of secondary sonochemical reactions, such as radical production or solvent pyrolysis, increases catalyst lifetime and hence catalyst performance. Furthermore, CH_4 is more soluble in organic solvents than Ar, and the greater solubility leads to less violent bubble collapse and as a result to slower rates of polymer scission.

Quantitative estimations of mechanical force, which can be inferred from theoretical models aided by ab initio and DFT calculations, are required to establish unequivocally structure-reactivity relationships.[19, 21, 79] A convenient and popular approach is known as *Constrained Geometries Simulate External Force (COGEF)*, where the distance between two atoms is gradually increased while the energy is concomitantly minimized for each step.[80] Moore and associates were able to determine structure-mechanochemical activity correlations when cyclobutane-containing polymers with variable substitution degree and configuration at the strained ring were ultrasonically cleaved in a formal [2 + 2] cycloaddition leading to alkenes.[81] Methyl esters derived from six *cis*- and *trans*-cyclobutanes having two, one, or no cyano substituents (DCC/DCT, MCC/MCT, and NCC/NCT pairs, respectively) were interrogated using the COGEF method (Fig. 12).

The theoretical study unveiled two key conclusions: (1) The *cis*-configured cyclobutane is more reactive than the *trans* isomer; in other words, less force is needed for bond cleavage, and (2) the reactivity of mechanophores increases as cyclobutane substitution increases. The latter can be attributed to enhanced ring strain caused by steric interactions. These predictions could be verified when such mechanophores were covalently linked to poly(methyl acrylate) (PMA) polymers with molecular weights from 35 to 125 kDa, which were grown by single-electron transfer living radical polymerization starting from the corresponding α-bromoester derivatives as initiators (Fig. 12). Ultrasonic activation was then carried out on all the PMA-based polymers along with two control polymers: one where the cyclobutane ring was replaced by an alkane moiety and the other a PMA homopolymer having neither a center-based mechanophore nor the alkane. The results were fitted to linear relationships of k (rate constant of polymer cleavage) versus M_i (initial average molecular weight). As predicted by the computational analysis, *cis* cyclobutanes were more reactive than their *trans*

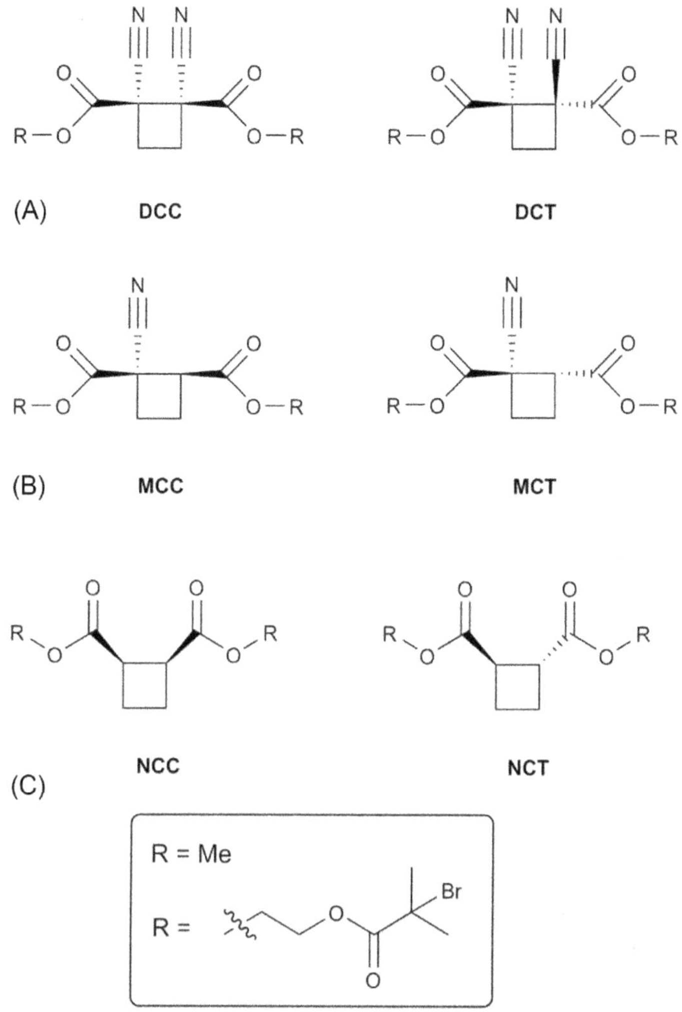

Fig. 12 Chemical structures of (A) *cis* and *trans* dicyano-substituted cyclobutanes (DCC and DCT, respectively); (B) *cis* and *trans* cyano-substituted cyclobutanes (MCC and MCT, respectively); and (C) *cis* and *trans* configured cyclobutanes lacking cyano groups (NCC and NCT, respectively). All these mechanophores were derivatized with methyl esters for theoretical calculations together with α-bromoisobutyryl groups as initiators for polymerization.

derivatives. Moreover the more highly substituted cyclobutanes underwent faster mechanical scission than the less highly substituted ones (Fig. 13). The control polymers were less reactive under sonication than cyclobutane-containing polymers, with the exception of the NCT mechanophore.

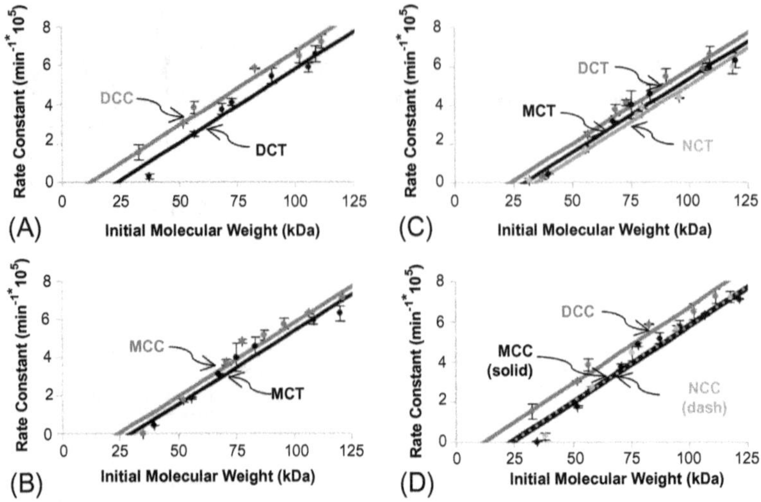

Fig. 13 Plots of linear relationships between ultrasound-mediated polymer scission rate constants and initial polymer molecular weights for cyclobutane-containing polymers: (A) fragmentation of DCC ($R^2 = 0.974$) and DCT ($R^2 = 0.962$); (B) cleavage of MCC ($R^2 = 0.958$) and MCT ($R^2 = 0.961$); (C) cleavage of DCT, MCT, and NCT; and (D) scission of DCC, MCC, and NCC polymers. *Adapted and reproduced from Kryger, M. J.; Munaretto, A. M.; Moore, J. S. Structure-Mechanochemical Activity Relationships for Cyclobutane Mechanophores. J. Am. Chem. Soc. 2011, 133, 18992–18998. With permission. Copyright 2011 American Chemical Society.*

4.2 Mechanochemical polymerization initiated by ultrasound

Sonochemical degradation of synthetic and natural polymers has long been known, and as mentioned, it harnesses both shear forces and reactive radicals generated through bubble collapse. Conversely, radicals generated ultrasonically, like hydroxy radicals from a Fenton reaction, for example, can also sustain a green catalyst-free chain-growth polymerization. Unlike the aforementioned mechanophore activations conducted usually at low frequencies (20 kHz), these radical polymerizations have been achieved at high frequencies (c. 400 kHz), which produce little mechanical forces, thereby bypassing chain-breaking reactions.[82] This smart protocol is however conceptually different from recent *ultrasonically induced polymerizations* based on mechanical activation, such as mechanically controlled atom transfer radical polymerization (mechano-ATRP).[83]

Like other stimulus-mediated electron transfer protocols, mechano-ATRP is dependent on Cu(II) complexes as catalysts, whereas piezoelectric materials (e.g., $BaTiO_3$ or ZnO nanoparticles) are employed as *chemical*

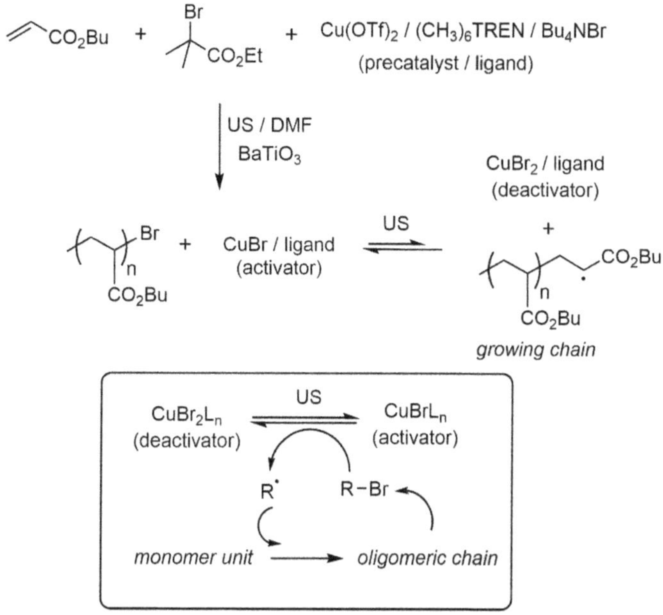

Fig. 14 Ultrasound-induced mechanoradical polymerization of *n*-butyl acrylate using a Cu(II) precatalyst and piezoelectric nanoparticles of $BaTiO_3$. Me_6TREN = *N,N,N′,N′,N′′,N′′*-hexamethyl[tris(aminomethyl)amine]. The bottom shows a mechanistic simplification for the sonomechanical activation.

transducers. Accordingly, electron transfer from piezoelectric particles to the Cu(II) complex leads to Cu(I) species that activate an alkyl halide reagent, whereas sonication serves as mechanical input to control polymer growth. Fig. 14 shows the radical polymerization of *n*-butyl acrylate, where a suspension of $BaTiO_3$ nanoparticles was sonicated in a typical Suslick cell containing n-butyl acrylate, ethyl α-bromoisobutyrate as initiator, and equimolar amounts of $Cu(OTf)_2$, M_6TREN, and Bu_4NBr as the catalyst precursor.[84] As depicted in Fig. 14 as well, the role of ultrasound appears to be twofold. Firstly, it induces piezoelectric reduction of the Cu(II) complex yielding activated Cu(I) species as polymerization activator. The latter promotes the formation of alkyl radical that adds a monomer onto the growing chain. At the same time, ultrasound produces alkyl radicals by homolytic fragmentation, thus propagating the radical polymerization.

The versatility of this step-growth polymerization can also be illustrated by the well-known copper-catalyzed azide-alkyne click cycloaddition affording a linear polytriazole. Using an appropriate trialkyne a linear polymer could be cross-linked mechanochemically to produce a gel that

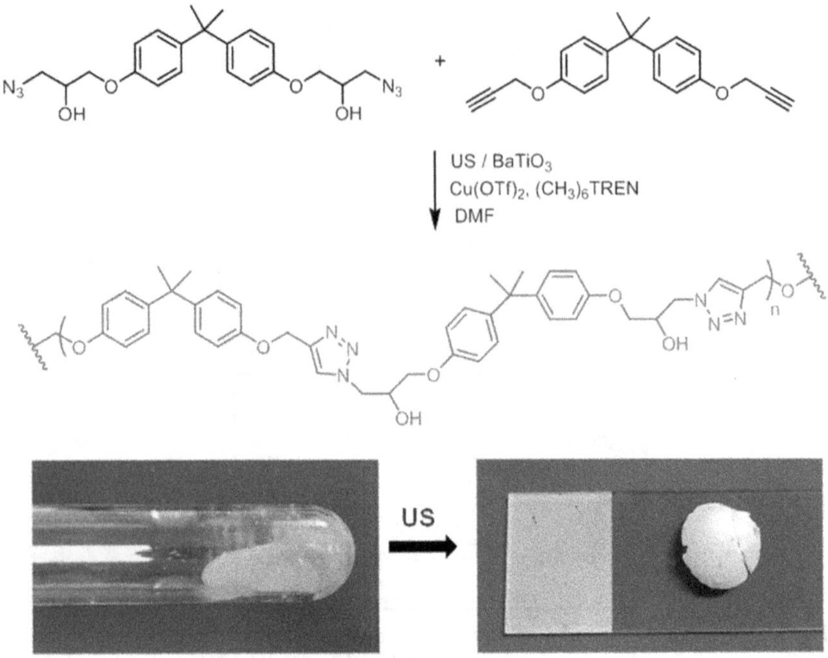

Fig. 15 Schematic representation of click polymerization using activated Cu(I) catalyst generated by ultrasonic reduction of Cu(II) species with $BaTiO_3$ particles. The process can be applied to cross-linking of a polyurethane-trialkyne mixture giving rise to a solid gel after sonication. *Adapted and reproduced from Mohapatra, H.; Ayarza, J.; Sanders, E. C.; Scheuermann, A. M.; Griffin, P. J.; Esser-Kahn, A. P. Ultrasound Promoted Step-Growth Polymerization and Polymer Crosslinking Via Copper Catalyzed Azide-Alkyne "Click" Reaction. Angew. Chem. Int. Ed. 2018, 57, 11208–11212. With permission. Copyright 2018 Wiley-VCH Verlag GmbH & Co.*

solidified after prolonged sonication (Fig. 15),[85] the latter being a current application of sound waves toward soft matter as we shall see in the next section.

It is worth mentioning that theoretical simulations suggest that triazole unclicking may be feasible, although mechanical stretching alone appears to be insufficient to explain the effect of sonication. Mechanophore cleavage might also stem from radical-initiated reactions and/or solvent-assisted ring opening.[86] The issue became controversial as experiments reporting triazole uncliking under sonication proved to be a fiasco and the paper in question was retracted. Nevertheless, further simulations of an external pulling force employing the COGEF approach again have revealed that mechanical reversibility is indeed possible in strain-promoted azide-alkyne cycloadditions.[87] The key parameter for cycloreversion to occur is the direction of the applied force acting on the triazole ring. Such a control might therefore be difficult to achieve in a bulk acoustic field, while selective directionality of force via atomic force microscopy for instance could eventually show the potential of strained cycloalkynes as mechanophores in mechanoresponsive polymers.

5 Molecular assembly and disassembly

The effects of ultrasound to induce molecular aggregation in liquids by enhancing, among others, hydrophobic effects[88] and subsequently disrupting noncovalent interactions after prolonged exposures are phenomena that may be tailored for diverse applications. Sonication-aided immobilization of solvent molecules in supramolecular networks, which leads to gels or hydrogels, constitutes a paradigmatic case. Formation and evolution of sonogels in inorganic materials, especially metal oxides generated by sol-gel processes, have been known for several decades.[89] Application to organic molecules has emerged in recent years as a promising field in sonochemistry and materials science.[90, 91]

Gelation of a liquid submitted to the fast "agitation" of ultrasound waves seems to be counterintuitive and hard to believe. The watershed was presumably the 2005 paper by Naota and Koori, who reported almost instant gelation of various organic solvents using a dinuclear Pd complex as gelator, which is stabilized by intramolecular π-stacking interactions. Ultrasound promotes gelation by disrupting such interactions and favoring intermolecular ones through a planar interlocked structure.[92] By the same time, Wang et al. also described the rapid gelation of an electroactive thiafulvalene derivative, where ultrasound-assisted gelation is accompanied by charge-transfer interaction.[93]

From a mechanistic standpoint, it seems obvious that gelation in an acoustic field depends on inherent variables of the system in question, that is, nature and extent of intermolecular interactions, number of hydrogen donor and acceptor atoms, solubility, and viscosity of the fluid, together with working parameters of the ultrasonic setup. Given the fact that most experiments are conducted in cleaning baths with poor control of power, frequency, or temperature, the successful discovery of gelation under sonication, often at lower concentration of gelator than those typical for classical gelation, demonstrates the robustness of the ultrasonic method. Among a range of effects to be taken into account, ultrasound-induced self-assembly involves both competition and cooperation of attractive forces in solution.

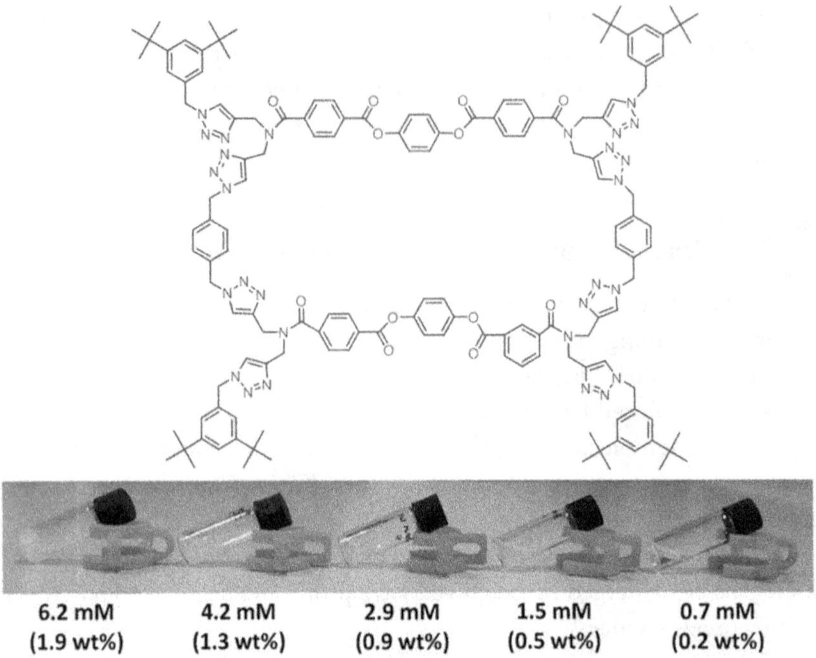

Fig. 16 Chemical structure of a 68-membered macrocycle and visual representation of the ultrasound-induced gelation: gels formed upon sonication in the same solvent at different concentrations *(bottom). Adapted and reproduced from Núñez-Villanueva, D.; Jinks, M. A.; Magenti, J. G.; Hunter, C. A. Ultrasound-Induced Gelation of a Giant Macrocycle. Chem. Commun. 2018, 54, 10874–10877. With permission. Copyright 2018 The Royal Society of Chemistry.*

In particular, sonication transforms intramolecular interactions into intermolecular ones, as the energy released by bubble collapse and shear forces helps to overcome the activation barrier of otherwise kinetically disfavored self-aggregation pathways.[94] This can easily be appreciated in a recent example where a giant macrocycle gelifies under sonication in acetonitrile (Fig. 16).[95] The resulting sonogel exhibit notable thermostability and shows a typical fibrillar morphology. Sonication appears to favor stacking interactions among aromatic groups that drive the self-assembly process, while H-bonding should not be responsible for gelation in the absence of strong H-bond donor/acceptor atoms.

Along with the key role exerted by a given low-molecular mass gelator, the solvent itself may not be completely innocent. Inverse-freezing materials, which solidify upon heating, and non-Newtonian liquids may open the door to novel opportunities in sonochemistry. Aqueous methylcellulose

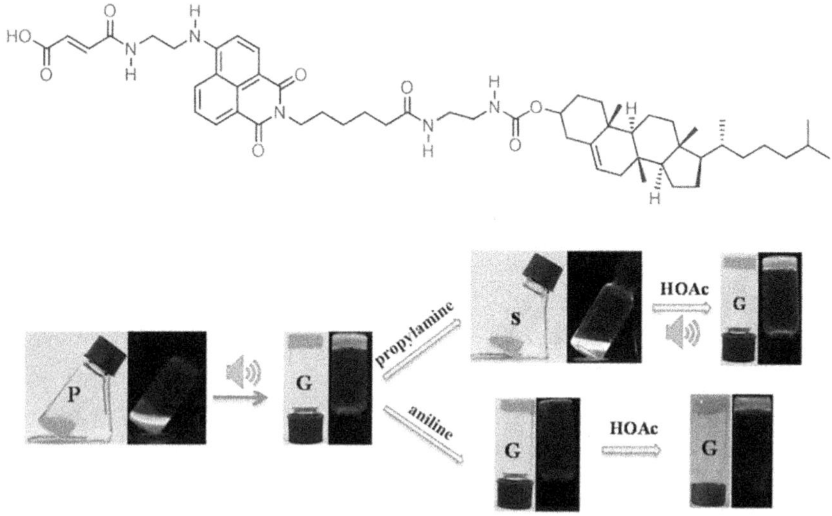

Fig. 17 Low-molecular mass naphthalimide derivative containing a cholesteryl group and schematic diagram of the ultrasound-accelerated gelation and visual recognition of aliphatic and aromatic amines. *Reproduced with permission from Pang, X.; Yu, X.; Lan, H.; Ge, X.; Li, Y.; Zhen, X.; Yi, T. Visual Recognition of Aliphatic and Aromatic Amines Using a Fluorescent Gel: Application of a Sonication-Triggered Organogel. ACS Appl. Mater. Interfaces 2015, 7, 13569–13577. Copyright 2015. The American Chemical Society.*

solutions undergo gelation upon heating. Shock waves caused by mechanical impact produce almost instant gelation. The process is endothermic, and the energy uptake enables large shock wave attenuation, thereby showing potential in materials tailored for energy dissipation and shock mitigation.[96]

Sonogels have become increasingly important as responsive materials and sensors. Yu and coworkers have developed to that end self-healing gels with color-tunable emission for visual recognition.[91, 97–99] Most examples are based on cholesteryl-appended 1,8-naphthalimides, which are sensitive to ultrasound. As shown in Fig. 17, this kind of amphiphilic material is able to gelate alcohols (e.g., propanol) via a precipitate-to-gel transformation under sonication for less than 2 min. The resulting S-gel enables a facile visual recognition of aliphatic and aromatic amines. Addition of *n*-propylamine to that gel is accompanied by a strong enhancement of the fluorescence intensity along with gel-to-sol conversion. In stark contrast, addition of aniline causes fluorescence quenching, and the gel state remains unaffected.[97] The sensing mechanisms, as inferred from spectroscopic measurements, depend on the analyte. Aliphatic amines trigger *trans*-to-*cis* isomerization of the alkenyl group of organogelator with a concomitant gel-sol transition, while aromatic amines tend to insert into the supramolecular network via H-bonding and charge-transfer interactions that do not alter the gel state.

Probably the most salient applications of this mechanical action is portrayed by exfoliation and intercalation processes en route to 2D materials, including both inorganic layered structures and the most promising 2D-organic crystals, which benefit from mild sonication, often in pulsed mode to avoid layer's damage and leading to fine dispersions of few-layer materials.[48, 100, 101]

Graphene continues to be the compound of choice, and the number of exfoliating agents used in its preparation has increased significantly. In general the mechanism of exfoliation and stabilization of graphene involves the use of surfactants and planar molecules that disrupt the van der Waals interactions between consecutive graphite layers and intercalate through strong π-π interactions with the basal planes of graphene layers (Fig. 18). Chemical

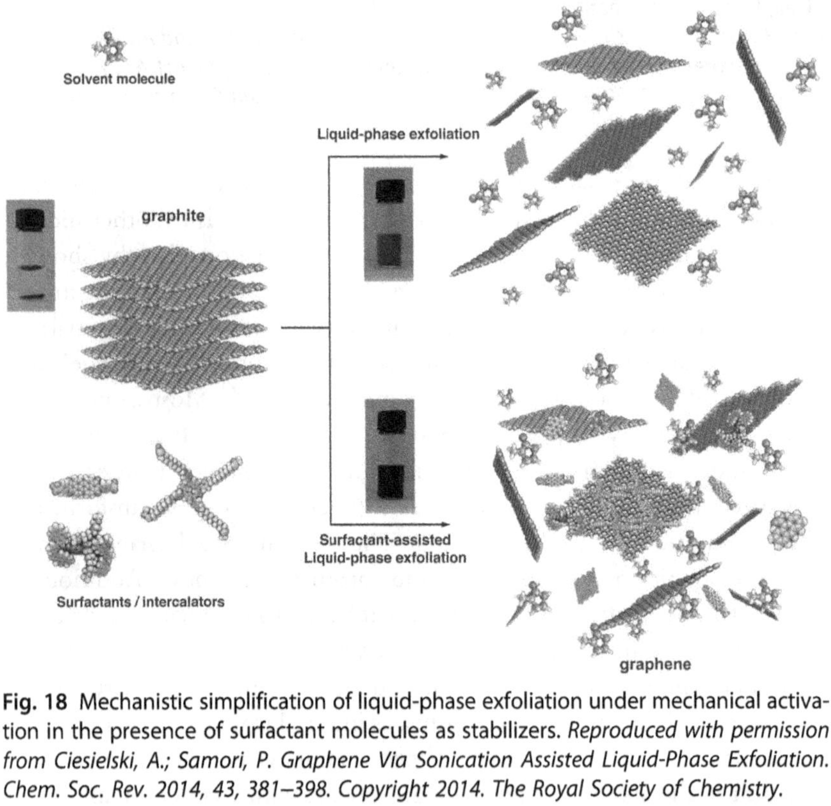

Fig. 18 Mechanistic simplification of liquid-phase exfoliation under mechanical activation in the presence of surfactant molecules as stabilizers. *Reproduced with permission from Ciesielski, A.; Samori, P. Graphene Via Sonication Assisted Liquid-Phase Exfoliation. Chem. Soc. Rev. 2014, 43, 381–398. Copyright 2014. The Royal Society of Chemistry.*

and mechanical exfoliation can be conducted by a plethora of methods, including sonication itself, and the workup process is followed by ultrasonication again to remove the attached solid and producing a fine dispersion. Exfoliation is usually conducted in high-boiling solvents like amides or haloaromatics, which in turn provide higher cavitational energy than volatile derivatives after bubble collapse. Greener protocols have been developed in aqueous solutions in ultrasonic baths (on-off pulsating sequence) and temperatures below 30°C.[102]

Easily prepared graphene oxide (GO), from cheap graphite, can be further activated and chemically modified by ultrasonication. Thus, in a recent example, Cu-, Fe-, and Cu/Fe-doped GO nanocomposites have been obtained with a submicron size smaller than that of pristine GO. In line with previous and well-established intercalation protocols under sonication, such GO nanocomposites may undergo intercalation with drugs, for instance, than can subsequently be released in aqueous solution.[103]

6 Sonocrystallization

If there is a physical process that illustrates the mechanical action of ultrasound in an outstanding fashion, it should surely be crystallization assisted by power ultrasound. Crystallization operations are of utmost importance in various large-scale industries (chemical, pharmaceutical, and food) and parameters like purity, yield, crystal structure, or size distribution must be carried out with care and accuracy. The basic mechanisms of crystallization involve nucleation and growth, which may be influenced by sonication. Contrary to common wisdom where agitation is employed for dissolving solids and preventing precipitation, ultrasound is generally reported to accelerate nucleation and crystal growth, by narrowing the metastable zone (vide infra) and leading to crystals with smaller size and more homogeneous size distributions than conventional methods. Sonocrystallization is not a new subject although a renewed interest took place from the mid-2000s onward.[104, 105] The ultrasound effect lies clearly in the shear forces in liquids associated to the presence of cavitational bubbles. However, the physical mechanisms at the bubble scale have not yet been fully clarified; nonetheless, recent work addresses the theoretical bases of sonocrystallization to explain the thermodynamic and kinetic aspects surrounding nucleation and growth.[106] Herein, we will highlight a few representative cases along with recent mechanistic insights and applications of crystallization enhanced and tuned by sound waves.

The effects of sonication capable of modifying habits and size in crystals occur through multiple cavitation cycles. Forces after collapse break one crystal into two or more fragments along its longest dimension. This reduces the length of the crystal or the growth rate in that dimension, and as a result, crystals grow thicker over time. As mentioned earlier an immediate conclusion is that ultrasonic irradiation enhances nucleation by narrowing the metastable zone (MZ), thereby reaching supersaturation levels faster.[107] The lower concentration limit of the MZ is defined by the equilibrium solubility of the compound in a given solvent, whereas the upper limit is the supersaturation line below which nucleation is slow and above which nucleation will be occurring spontaneously.[105] Nucleation can be observed in the vicinity of cavitating bubbles, and early studies pointed out a power threshold for nucleation to occur in devices operating at low frequency (usually from 20 to 40 kHz),[108] and the average crystal size decreases as the ultrasonic power increases.[109]

A well-known model case for mechanistic insights involves ice crystallization, as controlling ice nucleation has enormous industrial importance, especially freezing of fresh foodstuffs.[106] Chow and associated investigated the effects of sonication on the primary and secondary nucleation of water in sucrose solutions. The team showed that primary nucleation of ice can be achieved at higher nucleation temperatures under irradiation and that the nucleation temperature increases with increasing the acoustic power.[110, 111] Preexisting ice crystals were fragmented in the vicinity of stable oscillating bubbles, which generates nucleation sites and modifies the overall size distribution (Fig. 19).

Further investigation by Zeiger and Suslick has showed further insights into sonofragmentation mechanisms of molecular crystals, which can be different from those of metal powders.[112] Using aspirin slurries irradiated with a titanium horn (20 kHz, 10 W), the dominant mechanism appears to be the direct interaction of shock waves with crystalline particles (Fig. 20), whereas neither particle-horn collisions nor particle-wall collisions constitute the source of primary breakage. Instead, interparticle collisions favored by bubble collapse are the main effect on the erosion and damage of metal particles. A more recent study by Van Gerven et al. unveils the distinctive effects of *stable* and *transient* cavitation on particle fragmentation using paracetamol crystals.[113] As expected, both frequency and power exerted a significant influence on particle breakage. The latter was notable at low frequencies (from 30 kHz to 166 kHz), while at frequencies of 850 kHz and above, no breakage occurred. The zone in between the two previous ranges

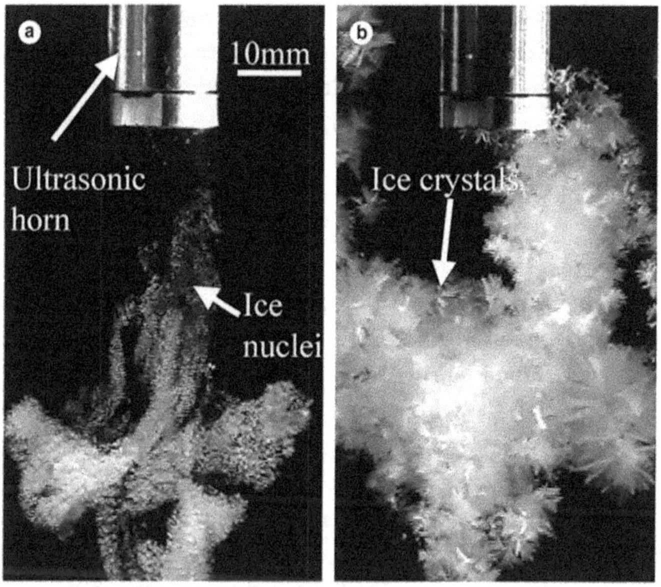

Fig. 19 Pictures of ice nucleation in a sucrose solution (15 wt%) at 3.4°C using an ultrasonic probe: (A) ice nuclei observed just after an ultrasonic pulse and (B) ice crystals photographed 5 s later. *Reproduced with permission from Chow, R.; Blindt, R.; Chivers, R.; Povey, M. A Study on the Primary and Secondary Nucleation of Ice by Power Ultrasound. Ultrasonics 2005, 43, 227–230. Copyright 2005. Elsevier B.V.*

enhanced the fragmentation of the smaller particles by bubble implosion, albeit the large crystals remained largely unaffected. An average size of 75 μm was found to be independent of the applied power, so long as stable cavitation bubbles are generated. A size threshold of 35 μm could not be

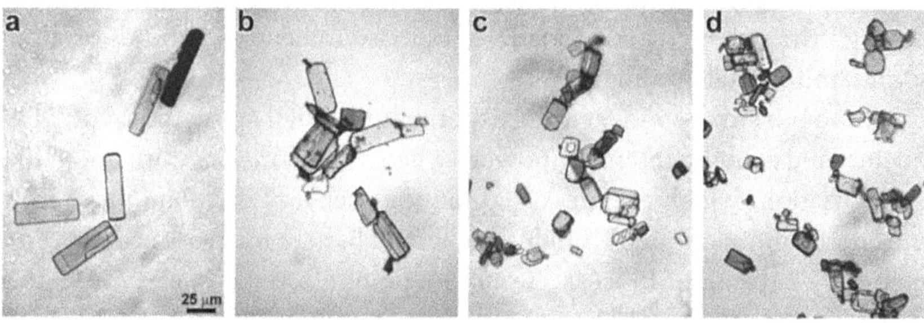

Fig. 20 Acoustic fragmentation of aspirin crystals viewed under polarizing microscopy: (A) before sonication, (B) 1-min irradiation, (C) 3-min irradiation, and (D) 10-min irradiation. *Reproduced with permission from Zeiger, B. W.; Suslick, K. S. Sonofragmentation of Molecular Crystals. J. Am. Chem. Soc. 2011, 133, 14530–14533. Copyright 2011. The American Chemical Society.*

reduced under such conditions regardless of the frequency of power. On the contrary, transient cavitation, as deduced from sonoluminescence measurements, causes higher breakage rates that depend strongly on the acoustic power. At high powers (up to 200 W), very small particles could be obtained without threshold size. Remarkably, crystal abrasion occurred with transient cavitation bubbles as well, and that effect is more pronounced at higher powers.

In the context of *crystal abrasion* and/or *crystal attrition*, ultrasound has proven to be an efficient strategy to that end, relative to other common protocols using glass or ceramic beads. Enhanced crystal attrition has become an innovative and thought-provoking method of chiral crystallization, as pioneered by Viedma,[114] which however requires the presence of solid conglomeratic phases, that is, each crystal containing homochiral molecules, like in classical Pasteur's resolution. *Ultrasound-enhanced deracemization* has been successfully applied to both organic molecules and achiral substances forming chiral phases in the solid state.[115–117] The latter is exemplified by the iconic case of sodium chlorate ($NaClO_3$); this inorganic salt crystallizes as two enantiomorphous crystals that can be resolved by Viedma ripening. Working at 41.2 kHz and c. 35 W of absorbed power, Stefanidis et al. have showed that grinding of enantiomorphous $NaClO_3$ crystals leads to smaller crystal size and narrower size distributions than glass bead-enhanced grinding (Fig. 21).[117] As a result the initial deracemization, governed by the Gibbs-Thomson rule and Ostwald ripening (dissolution of smaller crystals feeding the larger ones), is faster with sonication, although prolonged ultrasonication hinders a further increase of the enantiomeric excess, presumably due to the absence of solubility gradients induced by crystal size. This limitation can be overcome nevertheless, through the synergetic combination of ultrasound and glass beads or ultrasound and crystal seeding.

Although either sonochemical experiment is chiefly characterized by frequency and acoustic intensity (power), a heterogeneous transformation, like crystallization, which produces a continuous delivery of a solid phase from the energy input of bubble implosion, would benefit from quantitative estimation of cavitation. In fact, case studies point to crystal nucleation, rather than fragmentation, as the major influence of sonication, with a rapid equilibrium reached at the initial stages of irradiation by precipitating most of the crystals.[118] Sonoluminescence measurements constitute the suitable method to estimate the cavitational action, as the onset of bubble activity increased the nucleation rate and diminished concomitantly crystal size. Lee and

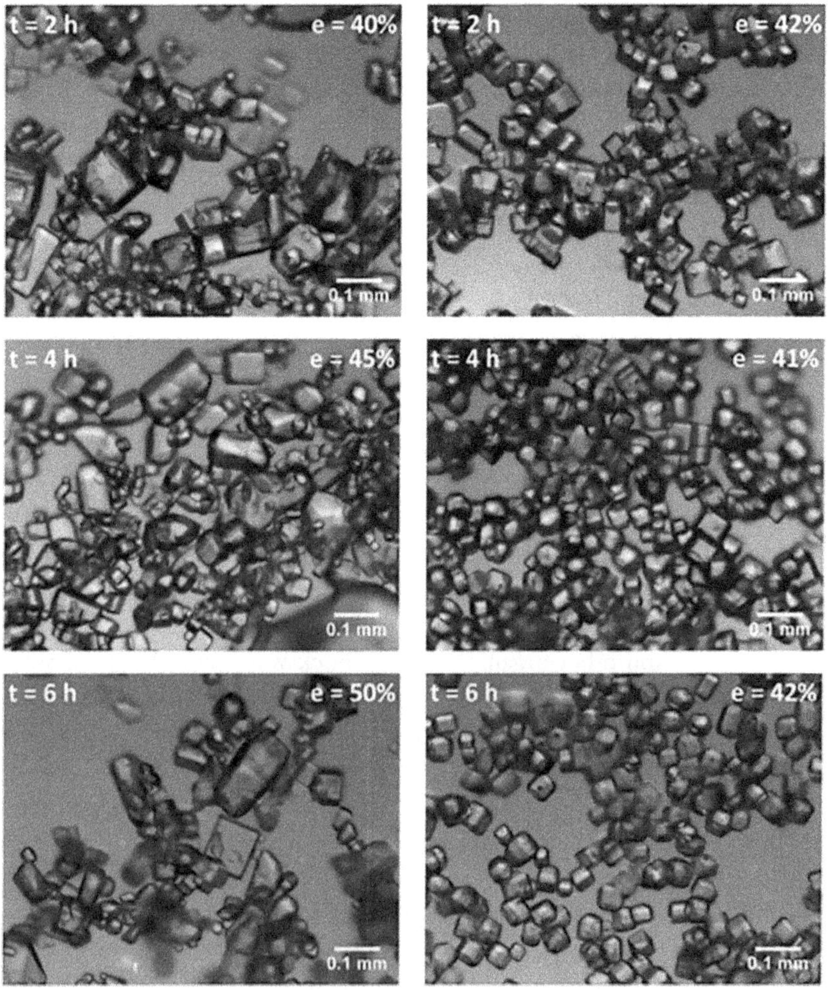

Fig. 21 Comparative images of crystal sizes at different times during the grinding of $NaClO_3$ using glass beads at 700 rpm *(left)* and sonication at 41.2 kHz *(right)*. *Reproduced with permission from Xiouras, C.; Van Aeken, J.; Panis, J.; Ter Horst, J. H.; Van Gerven, T., Stefanidis, G. D. Attrition-Enhanced Deracemization of NaClO3: Comparison Between Ultrasonic and Abrasive Grinding. Cryst. Growth Des. 2015, 15, 5476–5484. Copyright 2015. The American Chemical Society.*

coworkers evaluated the antisolvent crystallization of NaCl within a broad frequency range (22–1080 kHz) and acoustic calorimetric power (0–30 W), showing that a power threshold has to be overcome before a decrease in crystal size and that power coincided with the power threshold for sonoluminescence emission from bubbles.[119]

Since cavitation in liquids triggers shock waves at the bubble interface, which control mechanical performance to a substantial extent, comparisons with the action of neat shock wave energy may provide new working guidelines. The fast generation of gels in liquids under shock waves has been mentioned earlier,[96] and shock-induced mechanical reactions in polymers[120] and crystalline metal-organic frameworks (MOFs)[121, 122] have also been evaluated. These substances, usually decorated by zeolite-type structures, can trap small molecules and store energy, and shock energy attenuation measurements show their potential application as protective materials. Under shock compression, MOFs undergo plastic deformations by different mechanisms that resemble the effects of high pressure, such as powder compaction, nanopore collapse, and endothermic bond scission.[122]

For certain solids the mechanical action exerted by grinding, scratching, or thermal shock may lead to light emission, a well-known phenomenon called triboluminescence or mechanoluminescence.[123, 124] Acoustic cavitation can elicit mechanoluminescence as well using high-intensity ultrasound in liquid slurries of saccharose (whose triboluminescence was reported in the 17th century) and other organic crystals. Shock waves after bubble implosion can accelerate the small solid particles suspended in the liquid phase, thus giving rise to intense interparticle collisions.[125–127] Eddingsaas and Suslick reported different mechanoluminescence (ML) spectra for the sonication of a slurry of resorcinol in hydrocarbon and perdeuterated hydrocarbon solvents depending on the sparged gas (N_2, He, or O_2). Such ML spectra not only are more intense than those attained by manual grinding but also exhibit new emission lines consistent with gas-phase chemical reactions during the mechanoluminescent event (Fig. 22).[126]

As we have emphasized throughout this chapter, the mechanical action of ultrasound waves is produced in a liquid medium, and sonocrystallization follows this routine. Cocrystallization, which affords new solid phases and is especially relevant in the pharmaceutical industry, is conducted by numerous solvent-free procedures or involving little amount of solvent. In a recent study, Crawford and coworkers went against the tenet of cavitation in liquids, by performing successful solventless cocrystal syntheses with paracetamol or aspirin in an empty ultrasonic bath.[128] Quantitative conversions, as monitored by powder x-ray diffraction, were obtained within 20–60 min, although liquid-assisted irradiation improved substantially the outcome. Even if a rationale was not established, cavitation should clearly be discarded as the primary effect. Probably the role of standing waves capable of inducing

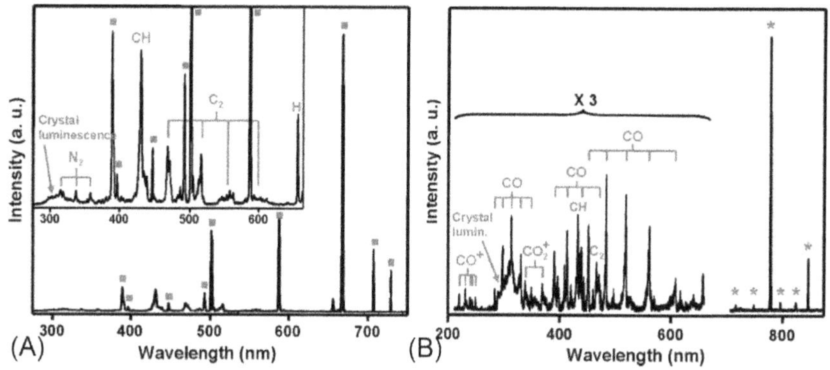

Fig. 22 (A) ML spectra of a sonicated resorcinol slurry (150–250 μm) in hexadecane (10% *w*/w) sparged with He, showing very strong emission from He-excited states (squares). The inset (10-fold intensity enlargement) displays extensive C_2 and H α line emission. No sonoluminescence was observed. (B) ML from sonication of resorcinol in hexadecane sparged with O_2, showing both O atom emission (*) and extensive molecular emission. *Reproduced with permission from Eddingsaas, N. C.; Suslick, K. S. Intense Mechanoluminescence and Gas Phase Reactions from the Sonication of an Organic Slurry. J. Am. Chem. Soc. 2007, 129, 6718–6719. Copyright 2007. The American Chemical Society.*

agglomeration of solids at nodal points[50, 90] needs to be reinvestigated in this context.

For biological and medicinal purposes, crystallization of large biomolecules is a major challenge, as they cannot easily be accommodated in crystal lattices. Techniques to facilitate biomolecular crystallization have long been established and require a careful control of multiple variables (precipitant, temperature, buffer, metal ions, additives, etc.).[129] For proteins, ultrasonication has demonstrated similar effects to those obtained with small molecules, notably faster nucleation leading to crystals that grow at low supersaturation levels.[130, 131]

Interestingly, external vibrations, noise, and environmental sound have occasionally been reported to be beneficial for crystallization of proteins and other biomolecules like tRNA.[129] These observations may hardly be reproducible, although in contrast to the long-held principle that the mother liquor should not be disturbed in any way, "agitation" does sometimes induce a positive effect. Thus oscillatory flow increases nucleation and crystal yield of insulin,[132] and low-frequency mechanical vibrations have also been applied to protein crystallization.[133] Mechanical vibrations, however, may have variable effects and depend on parameters like direction, speed, temperature, and frequency range (4–200 Hz). A typical vibration wave at 30 Hz, which is common in planetary balls mills too, lies in the lower zone of audible sound. Yet, such a vibration helps to grow better faceted crystals, though the likelihood of obtaining crystals at low supersaturation decreases. On the contrary, vibration helps to obtain crystals in highly supersaturated solutions in which either amorphous precipitates or tiny crystals normally appear.[133]

The effect of audible sound on protein crystallization has been systematically evaluated using 15 commercially available proteins and a computer-driven loudspeaker as source of sound.[134] In most cases a frequency of 5000 Hz (5 kHz) was found to produce optimal effects and gave better results than sound at 1000 Hz. In further screening in droplet experiments under frequencies from 500 to 17,500 Hz, crystals appeared in all of the droplets irrespective of the frequency employed, although the number of crystals varied with the frequency (maximum at 5000 Hz). Even if devoid of cavitation, sound irradiation induces phenomena that collectively affected crystal formation and evolution, such as solvent evaporation, bubble formation, or increase of temperature. Such effects are gathered in Fig. 23 using lysozyme as protein model. Small bubbles formed when crystallization droplets were exposed to audible sound, which did not disappear during

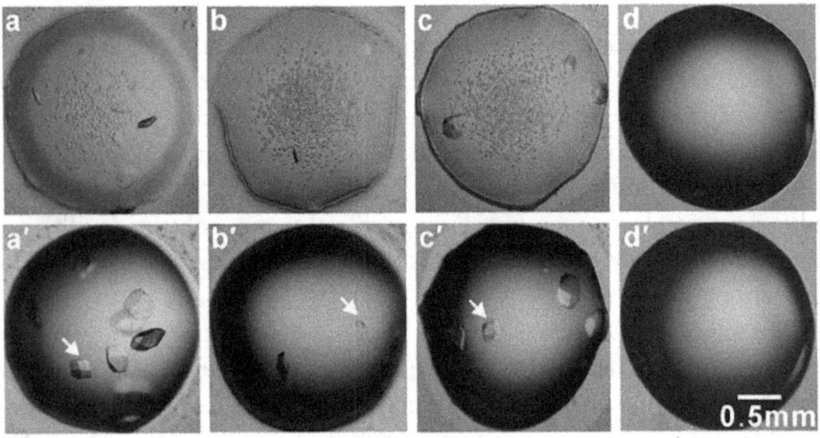

Fig. 23 Formation of bubbles and crystallization of lysozyme (20 mg/mL) in saline solution (concentration of NaCl = 50 mg/mL) under audible sound (5000 Hz). Bubbles appeared after 24-h irradiation (*A–C*) and disappeared after stopping sound irradiation for 24 h (*A′–C′*). Control experiments without sound for 24 h (*D*) and 48 h (*D′*). Crystallization was performed using the hanging-drop method. *Reproduced with permission from Zhang, C.-Y.; Wang, Y.; Schubert, R.; Liu, Y.; Wang, M.-Y.; Chen, D.; Guo, Y.-Z.; Dong, C.; Lu, H.-M.; Liu, Y.-M.; Wu, Z.-Q.; Betzel, C.; Yin, D.-C. Effect of Audible Sound on Protein Crystallization. Cryst. Growth Des. 2016, 16, 705–713. Copyright 2016. The American Chemical Society.*

irradiation, but they vanished when the sound was stopped. Control experiments showed no detectable bubbles, thereby pointing to sound as bubble source. Most crystals developed in the vicinity of bubbles, which served as nucleation sites for new crystals to appear after irradiation. This study not only discloses new insights into the role of audible frequencies present in the environment but also paves the way for processing other complex materials. Certainly a new sonochemical science is waiting for exploration at sub-ultrasonic frequencies.[135]

7 Outlook

The examples discussed earlier represent a narrow spectrum of sonochemical applications, yet demonstrating the capabilities of high-energy vibrations in liquids in an impressive manner. The interplay between purely chemical reactions inside the bubble and mechanical impacts derived from bubble collapse has been known for several decades. Now, with increasing progress in allied solvent-free mechanochemistry, the complementarity of sound-based techniques has been revealed in greater detail. The research is still less matured than we could think of, and there is plenty room to start new explorations.

The research on sonomechanochemistry remains partly empirical, often overlooking the control of acoustical field parameters and mechanistic interpretation at atomic and molecular level. This makes clear that there is not only a single rationale by which sound waves make things happen, and forces in liquids may stimulate chemical transformations depending on the starting conditions and substrates. Supramolecular systems, crystal nucleation, phase transitions, dynamic covalent and noncovalent linkages, and others, may all benefit from the mild "bulk" activation fueled by sound waves, even in the largely ignored audible range. Chemistry and more importantly biochemical and biomedical applications will be enhanced in the near future with a touch of sound.

References

1. Mason, T. J.; Lorimer, J. P. *Applied Sonochemistry. The Uses of Power Ultrasound in Chemistry and Processing*; Wiley-VCH: Weinheim, 2002.
2. Chen, D.; Sharma, S. K.; Mudhoo, A. *Hanbook on Applications of Ultrasound. Sonochemistry for Sustainability*; CRC Press-Taylor & Francis Group: Boca Raton, FL, 2011.
3. Gallego-Juárez, J. A.; Graff, K. F. *Power Ultrasonics. Applications of High-Intensity Ultrasound*; Elsevier: Amsterdam, 2015.

4. Chatel, G. *Sonochemistry. New Opportunities for Green Chemistry*; World Scientific Publishing: London, 2017.
5. Lévêque, J.-M.; Cravotto, G.; Delattre, F.; Cintas, P. *Organic Sonochemistry. Challenges and Perspectives for the 21st Century*; Springer, 2018.
6. Constable, D. J. C.; Curzons, A. D.; Cunningham, V. L. Metrics to Green Chemistry: Which Are the Best? *Green Chem.* **2002**, *4*, 521–527.
7. Jiménez-González, C.; Constable, D. J. C.; Ponder, C. S. Evaluating the Greennes of Chemical Processes and Products in the Pharmaceutical Industry-A Green Chemistry Primer. *Chem. Soc. Rev.* **2012**, *41*, 1485–1498.
8. Sheldon, R. A. Metrics of Green Chemistry and Sustainability: Past, Present, and Future. *ACS Sustain. Chem. Eng.* **2018**, *6*, 32–48.
9. Andraos, J. Useful Tools for the Next Quarter Century of Green Chemistry Practice: A Dictionary of Terms and a Data Set of Parameters for High Value Industrial Commodity Chemicals. *ACS Sustain. Chem. Eng.* **2018**, *6*, 3206–3214.
10. Andraos, J. *The Algebra of Organic Synthesis. Green Metrics, Design Strategy, Route Selection, and Optimization*; CRC Press-Taylor & Francis Group: Boca Raton, FL, 2012.
11. Fegade, S. L.; Tremblay, J. P. Misinterpretation of Green Chemistry. *Ultrason. Sonochem.* **2017**, *37*, 686–687.
12. Cintas, P. Ultrasound and Green Chemistry: Further Comments. *Ultrason. Sonochem.* **2016**, *28*, 257–258.
13. Ostwald, W.; Drucker, C. *Handbuch der Allgemeinen Chemie*; Akademische Verlagsgesellschaft mbH: Leipzig, 1919; pp. 70–77.
14. Reis, A. Handbuch der Allgemeinen Chemie. Herausgegeben von Wilhelm Ostwald und Carl Drucker Bd. I: Die Chemische Literatur und Die Organisation Der Wissenschaft. *Z. Elektrochem. Angew. Phys. Chem.* **1919**, *25*, 371–372.
15. Fernández-Bertan, J. F. Mechanochemistry: An Overview. *Pure Appl. Chem.* **1999**, *71*, 581–586.
16. Boulatov, R. Demonstrated Leverage. *Nat. Chem.* **2013**, *5*, 84–86.
17. Boulatov, R. Polymer Mechanochemistry. *Topics in Current Chemistry*, Vol. 369; Springer: New York, 2015; pp. v–xiii. and references therein.
18. Kaupp, G. Mechanochemistry: The Varied Applications of Mechanical Bond-Breaking. *CrstEngComm* **2009**, *11*, 388–403.
19. Ribas-Arino, J.; Marx, D. Covalent Mechanochemistry: Theoretical Concepts and Computational Tools with Applications to Molecular Nanomechanics. *Chem. Rev.* **2012**, *112*, 5412–5487.
20. Lavalle, P.; Boulmedais, F.; Schaaf, P.; Jierry, L. Soft Mechanochemistry: Mechanochemistry Inspired by Nature. *Langmuir* **2016**, *32*, 7265–7276.
21. Garcia-Manyes, S.; Beedle, A. E. M. Steering Chemical Reactions with Force. *Nat. Rev. Chem.* **2017**, *1*, 0083.
22. Chidanguro, T.; Weng, W.; Simon, Y. C. Mechanochemistry: Inspiration From Biology. In *Mechanochemistry in Materials*; Simon, Y. C., Craig, S. L., Eds.; Royal Society of Chemistry: Cambridge, 2018; pp. 1–35. Chapter 1.
23. Chalfie, M. Neurosensory Mechanotransduction. *Nat. Rev. Mol. Cell Biol.* **2009**, *10*, 44–52.
24. Majkut, S.; Dingal, P. C.; Discher, D. E. Stress Sensitivity and Mechanotransduction During Heart Development. *Curr. Biol.* **2014**, *24*, R495–R501.
25. Levin, M. Left-Right Asymmetry in Embryonic Development: A Comprehensive Review. *Mech. Dev.* **2005**, *122*, 3–25.
26. Yuan, S.; Zhao, L.; Brueckner, M.; Sun, Z. Intraciliary Calcium Oscillations Initiate Vertebrate Left-Right Asymmetry. *Curr. Biol.* **2015**, *25*, 556–567.
27. Nasir Baig, R. B.; Varma, R. S. Alternative Energy Input: Mechanochemical, Microwave and Ultrasound-Assisted Organic Synthesis. *Chem. Soc. Rev.* **2012**, *41*, 1559–1584.

28. Do, J.-L.; Friščić, T. Chemistry 2.0: Developing a New, Solvent-Free System of Chemical Synthesis Based on Mechanochemistry. *Synlett* **2017**, *28*, 2066–2092.
29. Akbulatov, S.; Boulatov, R. Experimental Polymer Mechanochemistry and its Interpretational Frameworks. *ChemPhysChem* **2017**, *18*, 1422–1450.
30. Muñoz-Batista, M. J.; Rodriguez-Padron, D.; Puente-Santiago, A. R.; Luque, R. Mechanochemistry: Toward Sustainable Design of Advanced Nanomaterials for Electrochemical Energy Storage and Catalytic Applications. *ACS Sustain. Chem. Eng.* **2018**, *6*, 9530–9544.
31. Howard, J. L.; Cao, Q.; Browne, D. L. Mechanochemistry as an Emerging Tool for Molecular Synthesis: What Can It Offer? *Chem. Sci.* **2018**, *9*, 3080–3094.
32. Leonardi, M.; Villacampa, M.; Menéndez, J. C. Multicomponent Mechanochemical Synthesis. *Chem. Sci.* **2018**, *9*, 2042–2064.
33. Fletcher, N. H. *Acoustic Systems in Biology*; Oxford University Press: New York, 1992.
34. Rossing, T. D. Introduction to Acoustics. In *Springer Handbook of Acoustics*; Rossing, T. D., Ed.; Springer: New York, 2007; pp. 1–6. Chapter 1.
35. Mason, T. J.; Cobley, A. J.; Graves, J. E.; Morgan, D. New Evidence for the Inverse Dependence of Mechanical and Chemical Effects on the Frequency of Ultrasound. *Ultrason. Sonochem.* **2011**, *18*, 226–230.
36. Nguyen, T. T.; Asakura, Y.; Koda, S.; Yasuda, K. Dependence of Cavitation, Chemical Effect and Mechanical Effect Thresholds on Ultrasonic Frequency. *Ultrason. Sonochem.* **2017**, *39*, 301–306.
37. Ashokkumar, M.; Lee, J.; Iida, Y.; Yasui, K.; Kozuka, T.; Tuziuti, T.; Towata, A. Spatial Distribution of Acoustic Cavitation Bubbles at Different Ultrasound Frequencies. *ChemPhysChem* **2010**, *11*, 1680–1684.
38. Boldyrev, V. V. Mechanochemistry and Sonochemistry. *Ultrason. Sonochem.* **1995**, *2*, S143–S145.
39. Lauterborn, W.; Kurz, T. Physics of Bubble Oscillations. *Rep. Prog. Phys.* **2010**, *73*, 106501.
40. Lauterborn, W.; Kurz, T.; Akhatov, I. Nonlinear Acoustics in Fluids. In *Springer Handbook of Acoustics*; Rossing, T. D., Ed.; Springer: New York, 2007; pp. 257–297. Chapter 8.
41. Suslick, K. S.; Eddingsaas, N. C.; Flannigan, D. J.; Hopkins, S. D.; Xu, H. The Chemical History of a Bubble. *Acc. Chem. Res.* **2018**, *51*, 2169–2178.
42. Beyer, M. K. Coupling of Mechanical and Chemical Energy: Proton Affinity as a Function of External Force. *Angew. Chem. Int. Ed.* **2003**, *42*, 4913–4915.
43. Rothenberg, G.; Downie, A. P.; Raston, C. L.; Scott, J. L. Understanding Solid/Solid Organic Reactions. *J. Am. Chem. Soc.* **2001**, *123*, 8701–8708.
44. Bowmaker, G. A. Solvent-Assisted Mechanochemistry. *Chem. Commun.* **2013**, *49*, 334–348.
45. McKenzie, T. G.; Karimi, F.; Ashokkumar, M.; Qiao, G. G. Ultrasound and Sonochemistry for Radical Polymerization: Sound Synthesis. *Chem. A Eur. J.* **2019**, *25*, 5372–5388.
46. Teo, B. M.; Grieser, F.; Ashokkumar, M. Applications of Ultrasound to Polymer Synthesis. In *Hanbook on Applications of Ultrasound. Sonochemistry for Sustainability*; Chen, D., Sharma, S. K., Mudhoo, A., Eds.; CRC Press-Taylor & Francis Group: Boca Raton, FL, 2011; pp. 475–500. Chapter 19.
47. Nguyen, T. Q.; Liang, O. Z.; Kausch, H. H. Kinetics of Ultrasonic and Transient Elongational Flow Degradation: A Comparative Study. *Polymer* **1997**, *38*, 3783–3793.
48. Martina, K. Interplay Between Mechanochemistry and Sonochemistry. In *Polymer Mechanochemistry*; Cintas, P., Cravotto, G., Barge, A., Boulatov, R., Eds.; Topics in Current Chemistry, Vol. 369; Springer: New York, 2015; pp. 239–284.

49. Weng, W. Mechanochemistry of Topological Complex Polymer Systems. In *Polymer Mechanochemistry*; Zhang, H., Lin, Y., Xu, Y., Boulatov, R., Eds.; Topics in Current Chemistry, Vol. 369; Springer: New York, 2015; pp. 135–208.
50. Cravotto, G.; Gaudino, E. C.; Cintas, P. On the Mechanochemical Activation by Ultrasound. *Chem. Soc. Rev.* **2013**, *42*, 7521–7534.
51. Li, J.; Nagamani, C.; Moore, J. S. Polymer Mechanochemistry: From Destructive to Productive. *Acc. Chem. Res.* **2015**, *48*, 2181–2190.
52. Lévêque, J.-M.; Cravotto, G.; Delattre, F.; Cintas, P. Ultrasound as Mechanical Force. In *Organic Sonochemistry. Challenges and Perspectives for the 21st Century*; Springer, 2018; pp. 87–98. Chapter 6.
53. Sijbesma, R. P. Mechanochemical Reactions Reporting and Repairing Bond Scission in Polymers. In *Polymer Mechanochemistry*; Clough, J. M., Balan, A., Boulatov, R., Eds.; Topics in Current Chemistry, Vol. 369; Springer: New York, 2015; pp. 209–238.
54. Weder, C. Mechanochemistry in Polymers with Supramolecular Mechanophores. In *Polymer Mechanochemistry*; Haehnel, A. P., Sagara, Y., Simon, Y. C., Boulatov, R., Eds.; Topics in Current Chemistry, Vol. 369; Springer: New York, 2015; pp. 345–376.
55. Melville, H. W.; Murray, A. J. R. The Ultrasonic Degradation of Polymers. *Trans. Faraday Soc.* **1950**, *46*, 996–1009.
56. Berkowski, K. L.; Potisek, S. L.; Hickenboth, C. R.; Moore, J. S. Ultrasound-Induced Site-Specific Cleavage of Azo-Functionalized Poly(Ethyleneglycol). *Macromolecules* **2005**, *38*, 8975–8978.
57. Wood, R. J.; Lee, J.; Bussemaker, M. J. A Parametric Review of Sonochemistry: Control and Augmentation of Sonochemical Activity in Aqueous Solutions. *Ultrason. Sonochem.* **2017**, *38*, 351–370.
58. Hickenboth, C. R.; Moore, J. S.; White, S. R.; Sottos, N. R.; Baudry, J.; Wilson, S. R. Biasing Reaction Pathways with Mechanical Force. *Nature* **2007**, *446*, 423–427.
59. Davis, D. A.; Hamilton, A.; Yang, J.; Cremar, L. D.; Gough, D. V.; Potisek, S. L.; Ong, M. T.; Braun, P. V.; Martinez, T. J.; White, S. R.; Moore, J. S.; Sottos, N. R. Force-Induced Activation of Covalent Bonds in Mechanoresponsive Polymeric Materials. *Nature* **2009**, *459*, 68–72.
60. Chen, Y. L.; Spiering, A. J. H.; Karthikeyan, S.; Peters, G. W. M.; Meijer, E. W.; Sijbesma, R. P. Mechanically Induced Chemiluminescence from Polymers Incorporating a 1,2-Dioxetane Unit in the Main Chain. *Nat. Chem.* **2012**, *4*, 559–562.
61. Diesendruck, C.; Steinberg, B. D.; Sugai, N.; Silberstein, M. N.; Sottos, N. R.; White, S. R.; Braun, P. V.; Moore, J. S. Proton-Coupled Mecanochemical Transduction: A Mechanogenerated Acid. *J. Am. Chem. Soc.* **2012**, *134*, 12446–12449.
62. Piermattei, A.; Karthikeyan, S.; Sijbesma, R. P. Activating Catalysts With Mechanical Force. *Nat. Chem.* **2009**, *1*, 133–137.
63. Chen, Z. X.; Mercer, J. A. M.; Zhu, X. L.; Romaniuk, J. A. H.; Pfattner, R.; Cegelski, L.; Martinez, T. J.; Burns, N. Z.; Xia, Y. Mechanochemical Unzipping of Insulating Polyladderene to Semiconducting Polyacetylene. *Science* **2017**, *357*, 475–478.
64. Su, J. K.; Feist, J. D.; Yang, J.; Mercer, J. A. M.; Romaniuk, J. A. H.; Chen, Z.; Cegelski, L.; Burns, N. Z.; Xia, Y. Synthesis and Mechanochemical Activation of Ladderene-Norbornene Block Copolymers. *J. Am. Chem. Soc.* **2018**, *140*, 12388–12391.
65. Yang, J.; Horst, M.; Romaniuk, J. A. H.; Jin, Z.; Cegelski, L.; Xia, Y. Benzoladderene Mechanophores: Synthesis, Polymerization, and Mechanochemical Transformation. *J. Am. Chem. Soc.* **2019**, *141*, 6479–6483.
66. Browser, B. H.; Craig, S. L. Empowering Mechanochemistry With Multi-Mechanophore Polymer Architectures. *Polym. Chem.* **2018**, *9*, 3583–3593.

67. Sha, Y.; Zhang, Y.; Xu, E.; McAlister, C. W.; Zhu, T.; Craig, S. L.; Tang, C. Generalizing Metallocene Mechanochemistry to Ruthenocene Mechanophores. *Chem. Sci.* **2019**, *10*, 4959–4965.
68. Huang, W.; Wu, X.; Gao, X.; Yu, Y.; Lei, H.; Zhu, Z.; Shi, Y.; Chen, Y.; Qin, M.; Wang, W.; Cao, Y. Maleimide-Thiol Adducts Stabilized Through Stretching. *Nat. Chem.* **2019**, *11*, 310–319.
69. Kim, G.; Lau, V. M.; Halmes, A. J.; Oelze, M. L.; Moore, J. S.; Li, K. C. High-Intensity Focused Ultrasound-Induced Mechanochemical Transduction in Synthetic Elastomers. *Proc. Natl. Acad. Sci. U. S. A.* **2019**, *116*, 10214–10222.
70. Pahnke, K.; Brandt, J.; Gryn'ova, G.; Lin, C. Y.; Altintas, O.; Schmidt, F. G.; Lederer, A.; Coote, M. L.; Barner-Kowollik, C. Entropy-Driven Selectivity for Chain Scission: Where Macromolecules Cleave. *Angew. Chem. Int. Ed.* **2016**, *55*, 1514–1518.
71. Thomas, J. R. Sonic Degradation of High Polymers in Solution. *J. Phys. Chem.* **1959**, *63*, 1725–1729.
72. May, P. A.; Munaretto, N. F.; Hamoy, M. B.; Robb, M. J.; Moore, J. S. Is Molecular Weight or Degree of Polymerization a Better Descriptor of Ultrasound-Induced Mechanochemical Transduction? *ACS Macro Lett.* **2016**, *5*, 177–180.
73. Peterson, G. I.; Bang, K.-T.; Choi, T.-L. Mechanochemical Degradation of Denpols: Synthesis and Ultrasound-Induced Chain Scission of Polyphenylene-Based Dendronized Polymers. *J. Am. Chem. Soc.* **2018**, *140*, 8599–8608.
74. Levy, A.; Wang, F.; Lang, A.; Galant, O.; Diesendruck, C. E. Intramolecular Cross-Linking: Addressing Mechanochemistry With a Bioinspired Approach. *Angew. Chem. Int. Ed.* **2017**, *56*, 6431–6434.
75. Levy, A.; Feinstein, R.; Diesendruck, C. E. Mechanical Unfolding and Thermal Refolding of Single-Chain Nanoparticles Using Ligand-Metal Bonds. *J. Am. Chem. Soc.* **2019**, *141*, 7256–7260.
76. Zhang, M.; De Bo, G. Impact of a Mechanical Bond on the Activation of a Mechanophore. *J. Am. Chem. Soc.* **2018**, *140*, 12724–12727.
77. Groote, R.; Jakobs, R. T. M.; Sijbesma, R. P. Performance of Mechanochemically Activated Catalysts Is Enhanced by Suppression of the Thermal Effects of Ultrasound. *ACS Macro Lett.* **2012**, *1*, 1012–1015.
78. Mason, T. J. Practical Sonochemistry. In *User's Guide to Applications in Chemistry and Chemical Engineering*; Ellis Horwood Ltd.: Chichester, 1991; p. 23. Chapter 1.
79. Makarov, D. E. Perspective: Mechanochemistry of Biological and Synthetic Molecules. *J. Chem. Phys.* **2016**, *144*, 030901.
80. Beyer, M. K. The Mechanical Strength of a Covalent Bond Calculated by Density Functional Theory. *J. Chem. Phys.* **2000**, *112*, 7307–7312.
81. Kryger, M. J.; Munaretto, A. M.; Moore, J. S. Structure-Mechanochemical Activity Relationships for Cyclobutane Mechanophores. *J. Am. Chem. Soc.* **2011**, *133*, 18992–18998.
82. McKenzie, T. G.; Colombo, E.; Fu, Q.; Ashokkumar, M.; Qiao, G. G. Sono-RAFT Polymerization in Aqueous Medium. *Angew. Chem. Int. Ed.* **2017**, *56*, 12302–12306.
83. Pan, X.; Fantin, M.; Yuan, F.; Matyjaszewski, K. Externally Controlled Atom Transfer Radical Polymerization. *Chem. Soc. Rev.* **2018**, *47*, 5457–5490.
84. Mohapatra, H.; Kleiman, M.; Esser-Kahn, A. P. Mechanically Controlled Radical Polymerization Initiated by Ultrasound. *Nat. Chem.* **2017**, *9*, 135–139.
85. Mohapatra, H.; Ayarza, J.; Sanders, E. C.; Scheuermann, A. M.; Griffin, P. J.; Esser-Kahn, A. P. Ultrasound Promoted Step-Growth Polymerization and Polymer Crosslinking Via Copper Catalyzed Azide-Alkyne "Click" Reaction. *Angew. Chem. Int. Ed.* **2018**, *57*, 11208–11212.

86. Smalø, H. S.; Uggerud, E. Rigid Opening vs. Direct Bond Scission of the Chain in Polymeric Triazoles Under the Influence of an External Force. *Chem. Commun.* **2012**, *48*, 10443–10445.
87. Jacobs, M. J.; Schneider, G.; Blank, K. G. Mechanical Reversibility of Strain-Promoted Azide-Alkyne Cycloaddition Reactions. *Angew. Chem. Int. Ed.* **2016**, *55*, 2899–2902.
88. Tuulmets, A.; Hagu, H.; Salmar, S.; Cravotto, G.; Järv, J. Ultrasonic Evidence of Hydrophobic Interactions. Effect of Ultrasound on Benzoin Condensation and Some Other Reactions in Aqueous Ethanol. *J. Phys. Chem. B* **2007**, *111*, 3133–3138.
89. Blanco, E.; Esquivias, L.; Litrán, R.; Piñero, M.; Ramírez-del-Solar, M.; De la Rosa-Fox, N. Sonogels and Derived Materials. *Appl. Organomet. Chem.* **1999**, *13*, 399–418.
90. Cravotto, G.; Cintas, P. Molecular Self-Assembly and Patterning Induced by Sound Waves. The Case of Gelation. *Chem. Soc. Rev.* **2009**, *38*, 2684–2697.
91. Yu, X.; Chen, L.; Zhang, M.; Yi, T. Low-Molecular-Mass Gels Responding to Ultrasound and Mechanical Stress: Towards Self-Healing Materials. *Chem. Soc. Rev.* **2014**, *43*, 5346–5371.
92. Naota, T.; Koori, H. Molecules that Assemble by Sound: An Application to the Instant Gelation of Stable Organic Fluids. *J. Am. Chem. Soc.* **2005**, *127*, 9324–9325.
93. Wang, C.; Zhang, D.; Zhu, D. A Low-Molecular-Mass Gelator With an Electroactive Tetrathiafulvalene Group: Tuning the Gel Formation by Charge-Transfer Interaction and Oxidation. *J. Am. Chem. Soc.* **2005**, *127*, 16372–16373.
94. Bardelang, D. Ultrasound Induced Gelation: A Paradigm Shift. *Soft Matter* **2009**, *5*, 1969–1971.
95. Núñez-Villanueva, D.; Jinks, M. A.; Magenti, J. G.; Hunter, C. A. Ultrasound-Induced Gelation of a Giant Macrocycle. *Chem. Commun.* **2018**, *54*, 10874–10877.
96. Parvari, G.; Rotbaum, Y.; Eichen, Y.; Rittel, D. Impact-Induced Gelation in Aqueous Methylcellulose Solutions. *Chem. Commun.* **2018**, *54*, 12578–12581.
97. Pang, X.; Yu, X.; Lan, H.; Ge, X.; Li, Y.; Zhen, X.; Yi, T. Visual Recognition of Aliphatic and Aromatic Amines Using a Fluorescent Gel: Application of a Sonication-Triggered Organogel. *ACS Appl. Mater. Interfaces* **2015**, 7, 13569–13577.
98. Yu, X.; Ge, X.; Geng, L.; Lai, H.; Ren, J.; Li, Y.; Yi, T. Cyclodextrin-Assisted Two-Component Sonogel for Visual Humidity Sensing. *Langmuir* **2017**, *33*, 1090–1096.
99. Guo, J.; Yu, X.; Zhang, Z.; Li, Y. Self-Healing Gels Triggered by Ultrasound with Color-Tunable Emission Based on Ion Recognition. *J. Colloid Interface Sci.* **2019**, *540*, 134–141.
100. Cravotto, G.; Cintas, P. Sonication-Assisted Fabrication and Post-Synthetic Modifications of Graphene-Like Materials. *Chem. A Eur. J.* **2010**, *16*, 5246–5259.
101. Ciesielski, A.; Samori, P. Graphene Via Sonication Assisted Liquid-Phase Exfoliation. *Chem. Soc. Rev.* **2014**, *43*, 381–398.
102. González-Domínguez, J. M.; León, V.; Lucio, M. I.; Prato, M.; Vázquez, E. Production of Ready-to-Use Few-Layer Graphene in Aqueous Suspensions. *Nat. Protoc.* **2018**, *13*, 495–506.
103. Radziuk, D.; Mikhnavets, L.; Vorokhta, M.; Matolín, V.; Tabulina, L.; Labunov, V. Sonochemical Formation of Copper/Iron-Modified Graphene Oxide Nanocomposites for Ketorolac Delivery. *Chem. Eur. J.* **2019**, *25*, 6233–6245.
104. Ruecroft, G.; Hipkiss, D.; Ly, T.; Maxted, N.; Cains, P. W. Sonocrystallization: The Use of Ultrasound for Improved Industrial Crystallization. *Org. Process Res. Dev.* **2005**, *9*, 923–932.
105. Tung, H.-H.; Paul, E. L.; Midler, M.; McCauley, J. A. *Crystallization of Organic Compounds. An Industrial Perspective*; John Wiley & Sons, Inc.: Hoboken, NJ, 2009; pp. 154–166. 264–268.
106. Baillon, F.; Espitalier, F.; Cogné, C.; Peczalski, R.; Louisnard, O. Crystallization and Freezing Processes Assisted by Power Ultrasound. In *Power Ultrasonics. Applications of*

High-Intensity Ultrasound; Gallego-Juárez, J. A., Graff, K. F., Eds.; Elsevier: Amsterdam, 2015; pp. 845–874. Chapter 28.
107. Ramisetty, K. A.; Rasmuson, A. C. Controlling the Product Crystal Size Distribution by Strategic Application of Ultrasonication. *Cryst. Growth Des.* **2018**, *18*, 1697–1709.
108. Acton, E.; Morris, G. J. *Method and Apparatus for the Control of Solidification in Liquids*; US Patent WO 99/20420, 1992; pp. 1–49.
109. Amara, N.; Ratsimba, B.; Wilhelm, A.-M.; Delmas, H. Crystallization of Potash Alum: Effect of Power Ultrasound. *Ultrason. Sonochem.* **2001**, *8*, 265–270.
110. Chow, R.; Blindt, R.; Chivers, R.; Povey, M. The Sonocrystallisation of Ice in Sucrose Solution: Primary and Secondary Nucleation. *Ultrasonics* **2003**, *41*, 595–604.
111. Chow, R.; Blindt, R.; Chivers, R.; Povey, M. A Study on the Primary and Secondary Nucleation of Ice by Power Ultrasound. *Ultrasonics* **2005**, *43*, 227–230.
112. Zeiger, B. W.; Suslick, K. S. Sonofragmentation of Molecular Crystals. *J. Am. Chem. Soc.* **2011**, *133*, 14530–14533.
113. Jordens, J.; Appermont, T.; Gielen, B.; Van Gerven, T.; Braeken, L. Sonofragmentation: Effect of Ultrasound Frequency and Power on Particle Breakage. *Cryst. Growth Des.* **2016**, *16*, 6167–6177.
114. Viedma, C. Chiral Symmetry Breaking During Crystallization: Complete Chiral Purity Induced by Nonlinear Autocatalysis and Recycling. *Phys. Rev. Lett.* **2005**, *94*, 065504.
115. Xiouras, C.; Fytopoulos, A.; Jordens, J.; Boudouvis, A. G.; Van Gerven, T.; Stefanidis, G. D. Applications of Ultrasound to Chiral Crystallization, Resolution and Deracemization. *Ultrason. Sonochem.* **2018**, *43*, 184–192.
116. Rougeot, C.; Guillen, F.; Plaquevent, J.-C.; Coquerel, G. Ultrasound-Enhanced Deracemization: Toward the Existence of Agonist Effects in the Interpretation of Spontaneous Symmetry Breaking. *Cryst. Growth Des.* **2015**, *15*, 2151–2155.
117. Xiouras, C.; Van Aeken, J.; Panis, J.; Ter Horst, J. H.; Van Gerven, T.; Stefanidis, G. D. Attrition-Enhanced Deracemization of $NaClO_3$: Comparison Between Ultrasonic and Abrasive Grinding. *Cryst. Growth Des.* **2015**, *15*, 5476–5484.
118. Nalesso, S.; Bussemaker, M. J.; Sear, R. P.; Hodnett, M.; Lee, J. Development of Sodium Chloride Crystal Size during Antisolvent Crystallization Under Different Sonication Modes. *Cryst. Growth Des.* **2019**, *19*, 141–149.
119. Lee, J.; Yasui, K.; Ashokkumar, M.; Kentish, S. E. Quantification of Cavitation Activity by Sonoluminescence to Study the Sonocrystallization Process Under Different Ultrasound Parameters. *Cryst. Growth Des.* **2018**, *18*, 5108–5115.
120. Ren, Y.; Banishev, A. A.; Suslick, K. S.; Moore, J. S.; Dlott, D. D. Ultrafast Proton Transfer in Polymer Blends Triggered by Shock Waves. *J. Am. Chem. Soc.* **2017**, *139*, 3974–3977.
121. Su, Z.; Shaw, W. L.; Miao, Y.-R.; You, S.; Dlott, D. D.; Suslick, K. S. Shock Wave Chemistry in a Metal-Organic Framework. *J. Am. Chem. Soc.* **2017**, *139*, 4619–4622.
122. Zhou, X.; Miao, Y.-R.; Shaw, W. L.; Suslick, K. S.; Dlott, D. D. Shock Wave Energy Absorption in Metal-Organic Framework. *J. Am. Chem. Soc.* **2019**, *141*, 2220–2223.
123. Ubba, E.; Tao, Y.; Yang, Z.; Zhao, J.; Wang, L.; Chi, Z. Organic Mechanoluminescence with Aggregation-Induced Emission. *Chem. Asian J.* **2018**, *13*, 3106–3121.
124. Mukherjee, S.; Thilagar, P. Renaissance of Organic Triboluminescent Materials. *Angew. Chem. Int. Ed.* **2019**, *58*, 7922–7932.
125. Eddingsaas, N. C.; Suslick, K. S. Light from Sonication of Crystal Slurries. *Nature* **2006**, *444*, 163.
126. Eddingsaas, N. C.; Suslick, K. S. Intense Mechanoluminescence and Gas Phase Reactions from the Sonication of an Organic Slurry. *J. Am. Chem. Soc.* **2007**, *129*, 6718–6719.

127. Eddingsaas, N. C. Mechanoluminescence Induced by Acoustic Cavitation. In *Triboluminescence. Theory, Synthesis, and Applications*; Olawale, D. O., Okoli, O. O. I., Fontenot, R. S., Hollerman, W. A., Eds.; Springer: Cham, 2016; pp. 237–271.
128. Roy, D.; James, S. L.; Crawford, D. E. Solvent-Free Sonochemistry as a Route to Pharmaceutical Co-Crystals. *Chem. Commun.* **2019**, *55*, 5463–5466.
129. McPherson, A. *Crystallization of Biological Macromolecules*; Cold Spring Harbor Laboratory Press: New York, 1999; pp. 215–269. Chapter 6.
130. Villaseñor, A. G.; Wong, A.; Shao, A.; Garg, A.; Kuglstatter, A.; Harris, S. F. Acoustic Matrix Microseeding: Improving Protein Crystal Growth With Minimal Chemical Bias. *Acta Crystallogr.* **2010**, *D66*, 568–576.
131. Crespo, R.; Martins, P. M.; Gales, L.; Rocha, F.; Damas, A. M. Potential Use of Ultrasound to Promote Protein Crystallization. *J. Appl. Cryst.* **2010**, *43*, 1419–1425.
132. Parambil, J. V.; Schaepertoens, M.; Williams, D. R.; Heng, J. Y. Y. Effects of Oscillatory Flow on the Nucleation and Crystallization of Insulin. *Cryst. Growth Des.* **2011**, *11*, 4353–4359.
133. Lu, Q.-Q.; Yin, D.-C.; Liu, Y.-M.; Wang, X.-K.; Yang, P.-F.; Liu, Z.-T.; Shang, P. Effect of Mechanical Vibration on Protein Crystallization. *J. Appl. Cryst.* **2010**, *43*, 473–482.
134. Zhang, C.-Y.; Wang, Y.; Schubert, R.; Liu, Y.; Wang, M.-Y.; Chen, D.; Guo, Y.-Z.; Dong, C.; Lu, H.-M.; Liu, Y.-M.; Wu, Z.-Q.; Betzel, C.; Yin, D.-C. Effect of Audible Sound on Protein Crystallization. *Cryst. Growth Des.* **2016**, *16*, 705–713.
135. Vinatoru, M.; Mason, T. J. Can Sonochemistry Take Place in the Absence of Cavitation?—A Complementary View of How Ultrasound Can Interact With Materials. *Ultrason. Sonochem.* **2019**, *52*, 2–5.

CHAPTER FIFTEEN

Diastereo- and enantioselective synthesis by nontraditional activation methods: Ultrasonic, microwave, electro-, photo- and mechanochemically activated reactions

Manisha Mishra and Béla Török
Department of Chemistry, University of Massachusetts Boston, Boston, MA, United States

1 Introduction

Chiral chemistry was established by Louis Pasteur, a French chemist and microbiologist, when he separated the two isomers of sodium ammonium tartrate by hand in 1848.[1] However, it took about a century more to reveal that the chirality plays a key role not only in the life of plants and animals but also in pharmaceutical, agricultural, and other chemical industries. All proteins, enzymes, amino acids, carbohydrates, nucleosides, and a number of alkaloids and hormones are chiral compounds. In pharmaceutical industries, 56% of the drugs currently in use are chiral products, and 88% of these compounds are marketed as racemates.[2] Although enantiomers have the same general chemical structure, most of them exhibit marked differences in their biological activities such as pharmacology, toxicology, pharmacokinetics, and metabolism. For chiral drugs, it is often seen that only one enantiomer is effective against a particular disease while the other enantiomer has different pharmacological activity or even toxicity.[3] For example, (*R*)-flurbiprofen can modulate γ-secretase and has the potential to treat the symptoms of Alzheimer's disease, while (*S*)-flurbiprofen is toxic because it can inhibit cyclooxygenase directly.[4, 5] The mechanisms of the mode of action of many chiral drugs within biological environment are now explained. Thus it is important to promote the chiral separation and analysis of racemic drugs in both pharmaceutical industry and clinic to eliminate the

Nontraditional Activation Methods in Green and Sustainable Applications
https://doi.org/10.1016/B978-0-12-819009-8.00012-8
© 2021 Elsevier Inc.
All rights reserved.

unwanted isomer from the preparation and to find an optimal treatment and a right therapeutic control for the patient. Therefore chirality is now a top-class subject for academic research and for pharmaceutical development.[6] Developing new approaches toward the synthesis of enantiopure compounds, which have remarkable significance in pharmaceutical and agrochemical industries, has been considered as one of the most challenging tasks by chemists for a long time. Furthermore, development of resource-efficient and sustainable chemical methodologies and processes is also one of the most important goals of synthetic organic chemistry in the 21st century. Various attempts were undertaken to minimize the adverse environmental impact while maximizing the efficiency of organic synthesis. One such important development for the sustainable synthesis of asymmetric molecules is the use of nontraditional activation method that began several decades ago. This chapter will include the use of ultrasounds, microwaves, photochemistry, electrochemistry, and mechanochemistry in asymmetric synthesis. Given the extended history and earlier reviews on the topic, the focus will be on the recent developments achieved in asymmetric synthesis of chiral compounds. Thus the literature review will be limited to the last two decades.

2 Asymmetric reactions in Sonochemistry

Ultrasonic irradiation is a very popular and promising technique for the activation of organic synthesis reactions.[7, 8] It has also been a prominent contributor to green chemical technology from synthesis to environmental studies. Syntheses associated with ultrasounds often save energy over conventional heating, increase the product yield in shorter reaction times, and decrease the use of volatile and toxic organic solvents.[9–13] The use of ultrasonic radiation in the field of chiral synthesis has also been extensively explored.[14] In this chapter the new developments achieved between 2000 and 2020 will be summarized. In the examples that are cited here, nontraditional activation method is involved in the key steps, forming chiral centers, or they are applied in one of the steps during the total synthesis of a target.

2.1 Hydrogenation

The application of ultrasonic irradiation has been found beneficial in asymmetric catalytic hydrogenations.[15] One of the most frequently used catalysts is the Raney Ni that is a well-known skeletal metal catalyst and is frequently

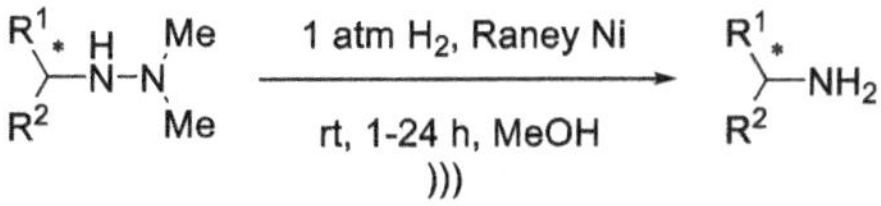

Scheme 1 Ultrasound-assisted cleavage of nitrogen-nitrogen bonds in hydrazines by Raney nickel-catalyzed hydrogenolysis.

used in hydrogenation reactions.[16] Thus it is not a surprise that this popular catalyst was explored in asymmetric hydrogenation reactions under sonication. Chiral primary amines were synthesized by reductive N—N cleavage of N-substituted *N,N*-dimethylhydrazines by Raney Ni-catalyzed hydrogenations reaction under sonication[17] (Scheme 1). Although the starting material was already chiral, during this process, the original stereochemistry or the optical purity did not change, and no racemization was observed.

The Raney Ni was also successfully applied in one of the steps during the total synthesis of the marine natural product (+)-ptilocaulin. The step involved desulfurization hydrogenation catalyzed by Raney Ni under ultrasonic conditions.[18]

Tartaric acid-modified Raney Ni catalysts were found to be highly effective in the hydrogenation of β-ketoesters and β-diketones.[19] The catalyst was prepared under sonochemical conditions and was used to hydrogenate a number of the aforementioned substrates,[20] including methyl 3-cyclopropyl-3-oxopropanoate, which reaction produced the corresponding alcohol in 98% ee (Scheme 2).[21, 22]

One of the most significant developments in transfer hydrogenation was reported by Noyori et al. with the use of chlororuthenium(II) arene precursors and chiral monoarylsulfonylated-1,2-diamines or β-amino alcohols as ligands.[23–26] Ultrasound-promoted asymmetric transfer hydrogenation of ketones using Ru(II)(arene)/amino alcohol catalyst system was achieved successfully by R.V. Chaudhary and his group[27] to synthesize chiral benzyl

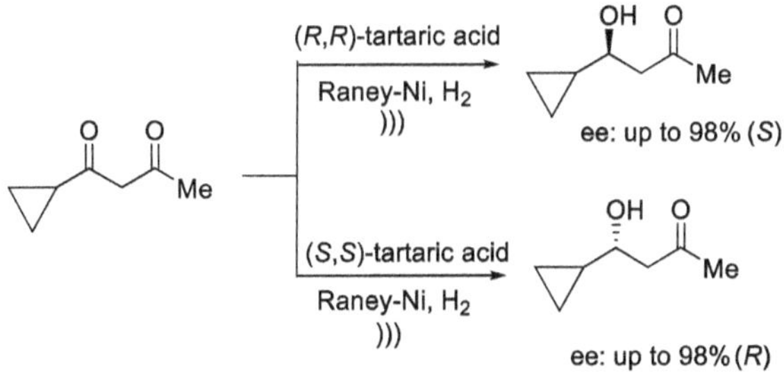

Scheme 2 Sonochemical enantioselective hydrogenation of a β-ketoester.

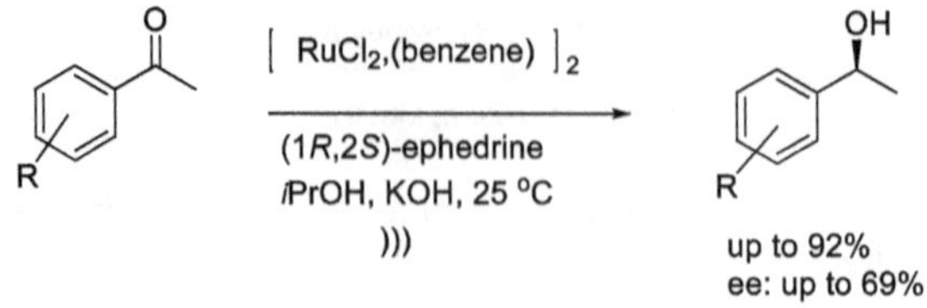

Scheme 3 Ruthenium complex-catalyzed sonochemical reduction of substituted acetophenones.

alcohols (Scheme 3) as useful synthons.[28] The results obtained clearly showed several folds of enhancement in the activity of the catalyst without negatively affecting the enantioselectivity.

High enantioselectivities were achieved in ultrasound-promoted asymmetric transfer hydrogenation of aromatic ketones by Li et al.,[29] utilizing a cationic rhodium complex $(Cp^{*}RhTsDPEN)^{+}(CF_3SO_3)^{-}$ entrapped within Me-SBA-15 and Me-SBA-16 as depicted in Scheme 4.

The aforementioned ion-pair immobilization strategy was found to be beneficial because it can anchor different types of chiral catalysts for various asymmetric syntheses and the catalysts can be recovered easily. The heterogeneous cationic rhodium catalyst shown in Scheme 4 was investigated and applied in the ultrasound-promoted asymmetric transfer hydrogenation of different aromatic ketones in aqueous medium. The resulting reaction showed excellent catalytic activities and high enantioselectivities (Scheme 5).

The sonochemical asymmetric hydrogenation of isophorone (3,3,5-trimethyl-2-cyclohexenone) by proline-modified Pd/Al_2O_3 catalysts was described by Mhadgut et al.[30] (Scheme 6). It was observed that presonication of a commercial Pd/Al_2O_3-proline catalytic system resulted in highly enhanced enantioselectivities (up to 85% ee) and produced hydrogenated isophorone.

Another heterogeneous catalytic system where the ultrasonic treatment resulted in significant benefits is the cinchona alkaloid-modified Pt and Pd catalyzed hydrogenations. The cinchona-Pt/Al_2O_3 catalytic system was

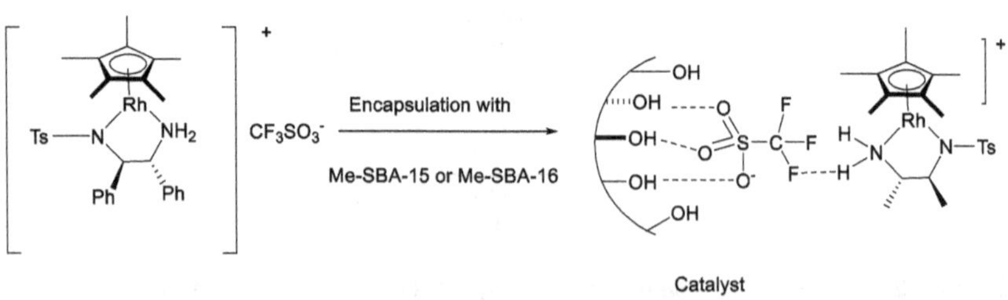

Scheme 4 Preparation of a cationic rhodium complex.

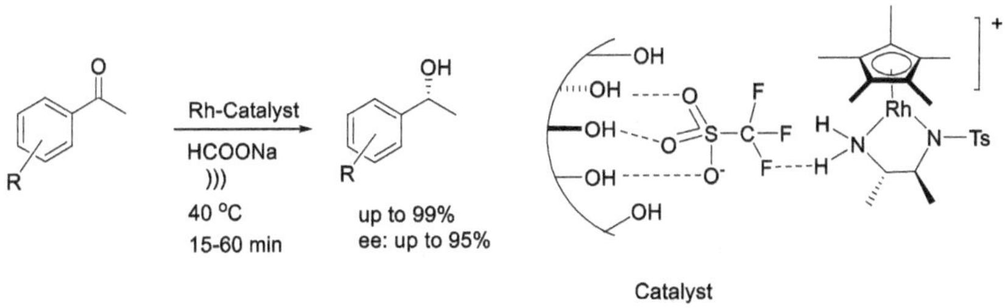

Scheme 5 Cationic rhodium complex-catalyzed reduction of acetophenone derivatives.

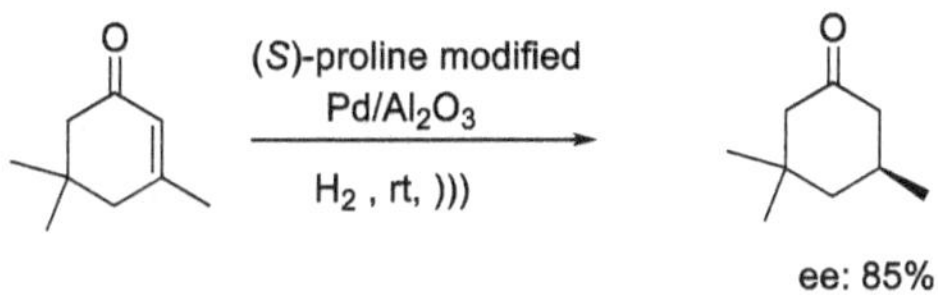

Scheme 6 Asymmetric hydrogenation of C=C of isophorone on a Pd/Al_2O_3 catalyst using (*S*)-proline as a chiral modifier.

studied in the hydrogenation of α-ketoesters[31] and activated α-carbonyl moieties[32] under ultrasonic irradiation[33] (Scheme 7). The reactions showed that the sonochemical pretreatment of the catalyst-modifier system resulted in enhanced selectivity in the hydrogenation with excellent ee values (up to 98% ee) under mild and widely varied experimental conditions.[34] The improved enantiodifferentiation was explained by the synergistic effect of the surface cleaning and size optimization of the metal particles and enhanced cinchona adsorption initiated by sonication.

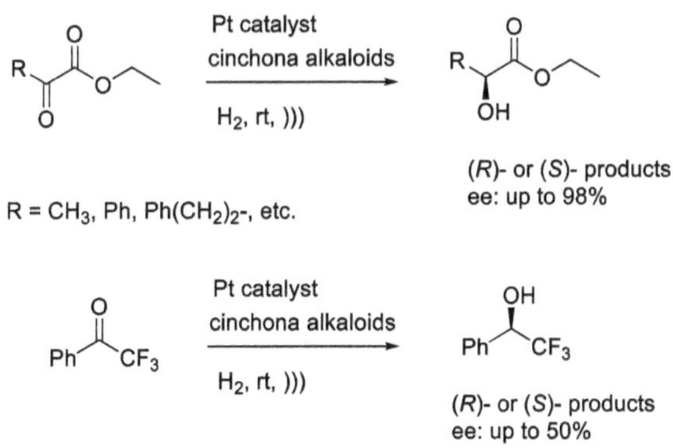

Scheme 7 Cinchona modified-Pt catalyzed enantioselective hydrogenation of activated carbonyl groups.

2.2 Michael additions with phase transfer catalysis

Phase transfer catalysis (PTC) is one of the most important and efficient techniques for various organic transformations due to its operational simplicity, mild reaction conditions, use of safe and inexpensive reagents and solvents, and possibility of conducting reactions on a large scale. The role of chiral phase transfer catalysts (CPTC) is to transfer reagents from the aqueous phase into the organic phase, thus providing the organic substrate and the required anion to form the corresponding product in the organic phase.[35–41] In one of the first enantioselective applications, the asymmetric Michael addition of diethyl *N*-acetyl-aminomalonate to chalcone was studied where the chirality was induced by a chiral phase transfer catalyst, namely, (−)-*N*-benzyl-*N*-methyl-ephedrinium bromide[42] (Scheme 8). It was observed that sonication increased the yield significantly. However, the enantiomeric excesses were found to be same as those obtained in the standard silent procedure.

To increase the chemical yield and enantiomeric excesses, most of the researchers concentrated their efforts on the development of chiral multisite phase transfer catalysts (CMPTCs)[43] to improve the catalytic actions. Siva et al. reported[44] that the catalytic enantioselective Michael reaction of diethyl malonate with a chalcone derivative can be effectively promoted by quaternary ammonium salts from cinchonine as phase transfer catalysts (Scheme 9). The efficacy of the asymmetric Michael addition reactions of malonate and chalcone was enhanced under ultrasonic irradiation conditions.

The multisite containing chiral quaternary ammonium ion-based catalysts derived from mesitylene also provided very good chemical yields and enantiomeric excesses (up to 98% ee) for performing the Michael addition reaction under mild base and ultrasonic irradiation conditions (Scheme 10).[45]

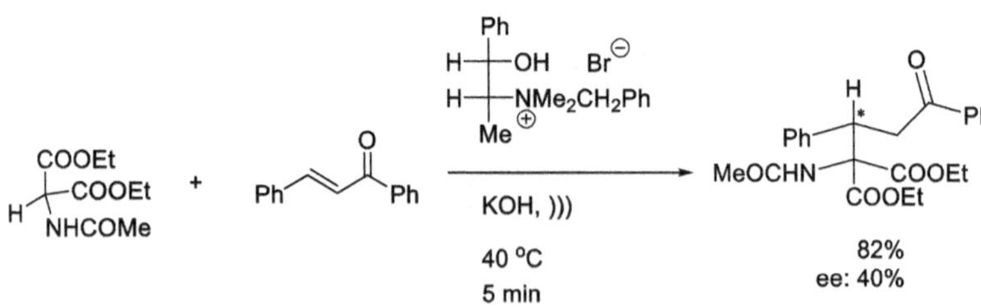

Scheme 8 An asymmetric sonochemical Michael addition of chalcone.

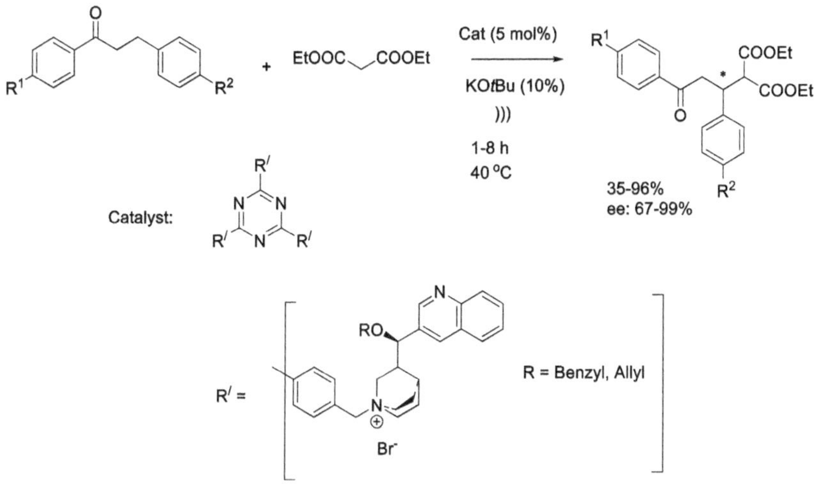

Scheme 9 Enantioselective sonochemical Michael addition of diethylmalonate using chiral multisite phase transfer catalysts (CMPTC) in aqueous media.

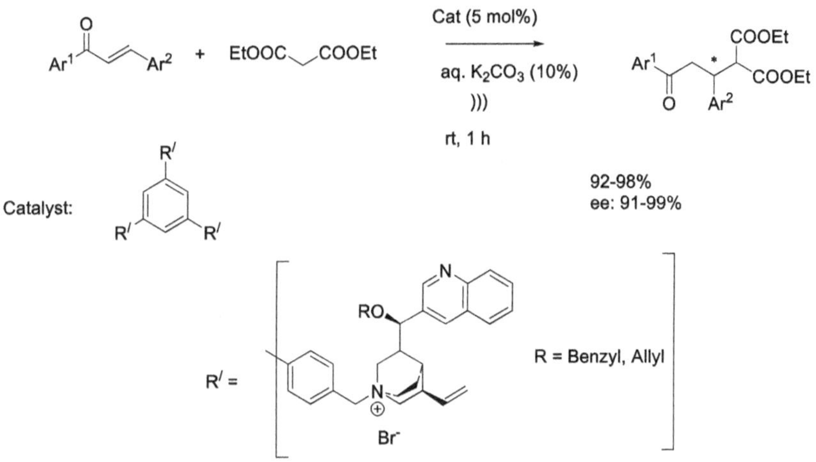

Scheme 10 Enantioselective sonochemical Michael addition of diethyl malonate to chalcone.

2.3 Organometallic chemistry

The application of ultrasounds in organometallic chemistry is one of the most popular fields in sonochemistry. Ultrasounds can effectively activate metals, thus accelerating their reactions. For example, the sonochemical activation was used by Baldoli and coworkers to carry out the enantioselective synthesis of β-aminoesters and β-lactams by the Reformatsky condensation (Scheme 11).[46]

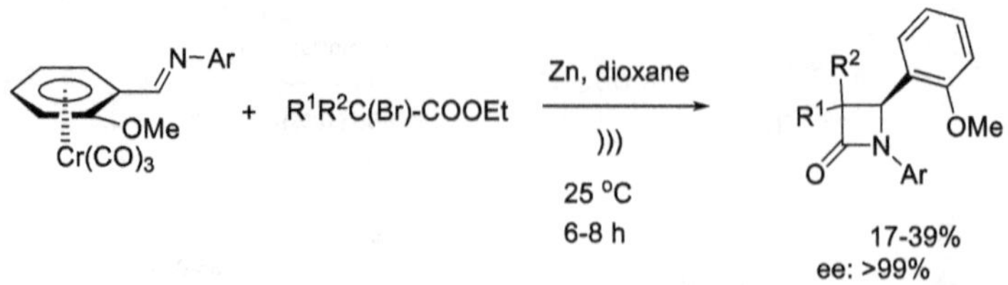

Scheme 11 Enantioselective Reformatsky condensation by ultrasonic activation.

An enantioselective synthesis of new C2-symmetry ferrocenylalkylamines was developed by Woltersdof and his group under ultrasonic irradiation in an amination reaction of 1-ferrocenylalkyl acetates (Scheme 12).[47] *Trans* 1-ferrocenylalkyl acetate produced the pure trans-diastereomer besides only very small amounts of *meso* compound. It was observed that the reaction occurred with a significantly greater rate under sonochemical activation than in the comparative silent method. However, the diastereomeric excess was same in both cases.

Hunt et al. described the ultrasound accelerated cleavage of diastereomeric hydroxylamines to chiral amines in Zn/AcOH system (Scheme 13).[48] The N—O bond in hydroxylamines was readily cleaved using the zinc-acetic acid-ultrasound medium, and the resulting amine was protected without purification as either its *tert*-butyl or benzyl carbamates. Products were obtained with high optical purity.

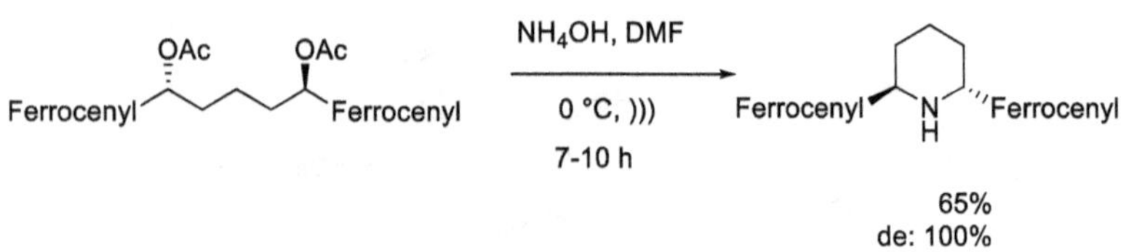

Scheme 12 Sonochemical amination and cyclization of 1-ferrocenylalkyl acetates.

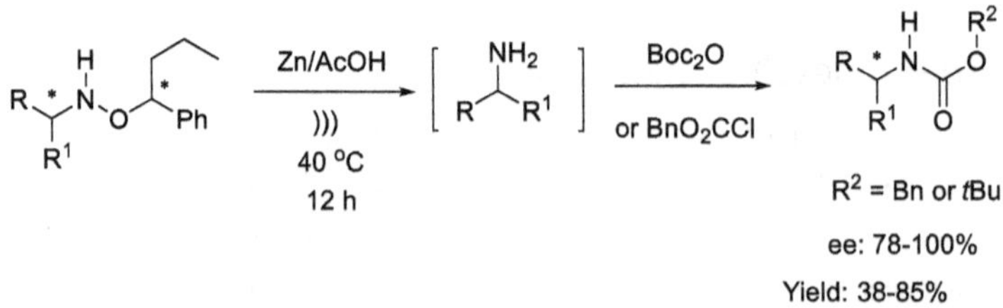

Scheme 13 Cleavage of hydroxylamines and simultaneous protection of the resulting amines.

2.4 Synthesis of asymmetric Schiff bases

The synthesis of enantiomeric Schiff bases has gained considerable significance in recent years due to their various utilities in the medical field with particular attention to the environmentally friendly and economical applications. A novel method for the synthesis of asymmetric Schiff bases from α-amino acids and aldehydes under ultrasonic conditions in aqueous medium was developed by Sendil et al.[49] The desired products were obtained in good yields (Scheme 14) in a very short time period (5 min). No racemization was observed during the reaction.

In addition to imines the chiral *tert*-butylsulfinylimines developed by Ellman et al.[50] have been used as substrates in organic synthesis. They have proved to be as effective auxiliaries in several reports for making chiral amine-containing precursors and biologically significant complex molecules.[51, 52] Moreover, (*S*)- and (*R*)-*tert*-butylsulfinamides and (*S*)- and (*R*)-α-methylbenzylamines are widely accepted chiral ammonia equivalents among all the chiral amino substrates.[53] A versatile, simple sonochemical method was reported[54] for the construction of *tert*-butylsulfinylimines of aldehydes under solvent-free and aerobic conditions. This method utilizes silica supported *p*-toluenesulfonic acid (*p*-TSA-SiO_2) as an efficient, safe, and inexpensive catalyst under sonochemical condition (Scheme 15). The product retained the original stereochemistry.

3 Microwave-assisted catalytic asymmetric synthesis

Microwave-assisted organic synthesis (MAOS)[55] uses the region of the electromagnetic spectrum that lies between IR radiation and radio frequencies and corresponds to wavelengths of 1 cm–1 m (frequencies of 30 GHz–300 MHz, respectively). Heating in microwave instruments relies upon the ability of some liquids and solids to transform electromagnetic energy into heat, thereby activating chemical reactions. Since the observed thermal

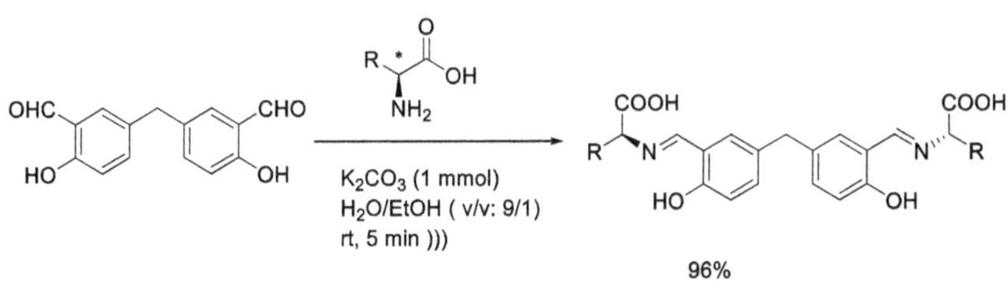

Scheme 14 Synthesis of asymmetric Schiff bases from α-amino acids and aldehydes.

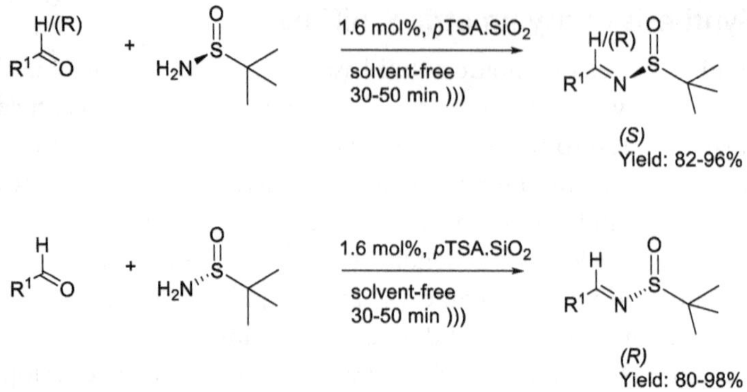

Scheme 15 Synthesis of chiral imines under solvent-free conditions via ultrasonic activation.

effect depends on the properties of material exposed to microwave irradiation (materials for chemical reactions do not interact equally with the commonly used microwave frequencies), it is possible to change the selectivity of a reaction and obtain an alternative product distribution in comparison with the reaction carried out under conventional heating conditions.[56] Due to its clear advantages, microwave activation has attracted the attention of synthetic organic chemists, who have started to apply this nonconventional heating technique routinely in their own work.[55, 57–59] The reduced timescales and parallel improved yields in the case of a great number of chemical reactions and the potential environmentally friendly factor with reduced energy consumption[60] under microwave conditions are the main reasons for using microwave techniques over traditional heating. Herein, microwave-assisted synthesis of several asymmetric molecules under various catalytic conditions will be presented.

3.1 Asymmetric allylic alkylation

Palladium- and molybdenum-catalyzed allylic substitution reactions were studied intensively especially in the synthesis of chiral compounds.[61, 62] Among the Pd-catalyzed transformations, the Pd-catalyzed asymmetric allylic alkylations (commonly called as Tsuji-Trost reaction) of active methylene compounds are one of the most powerful tools for C—C and C-X bond formations.[63, 64] Palladium-phosphine-oxazoline catalytic system was applied by Kaiser et al.[65] on *racemic* 1,3-diphenyl-2-propenyl acetate to perform the allylic substitution reaction with diethylmalonate under microwave-assisted conditions (Scheme 16). The reaction produced very

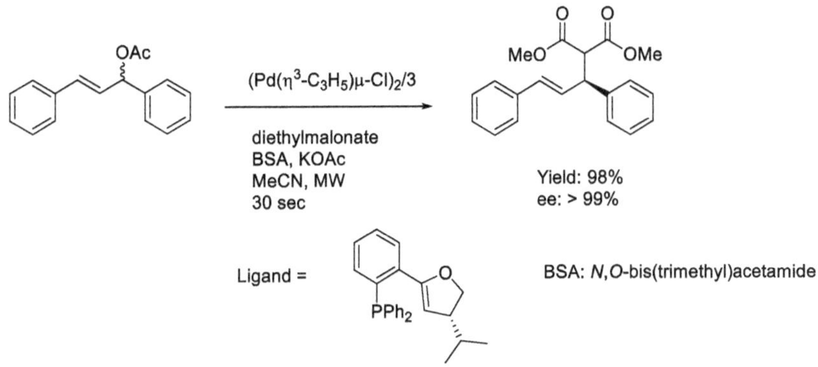

Scheme 16 Microwave-assisted Pd-catalyzed asymmetric allylic substitution.

high yield of the product with high ee's in a very short reaction time (in 30 s). In the absence of microwave irradiation, the reaction rate of the asymmetric alkylation was very low.

Braga et al. also developed a series of cysteine-derived oxazolidine and thiazolidine ligands and conducted the aforementioned reaction to test the potential of a ligand in the asymmetric palladium catalyzed allylation using *racemic* 1,3-diphenyl-2-propenyl acetate and dimethyl malonate as a nucleophile.[66] The product was isolated in high yields and good enantioselectivity with microwave flash heating (Scheme 17). Use of chiral β-selenoamides in the same reaction accomplished higher enantioselectivity.[67]

The allylic substitution reaction was also carried out with hindered allylic substrate using a phosphinooxazoline as the chiral ligand with complete conversion under microwave irradiation[68] (Scheme 18). The reaction time was

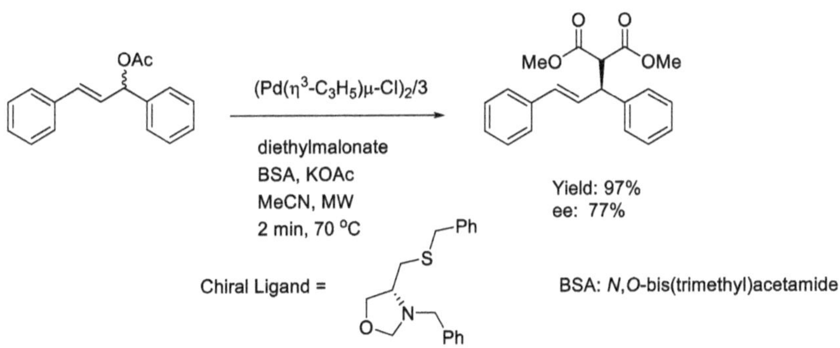

Scheme 17 Asymmetric allylic substitution in presence of cysteine-derived oxazolidine catalyst.

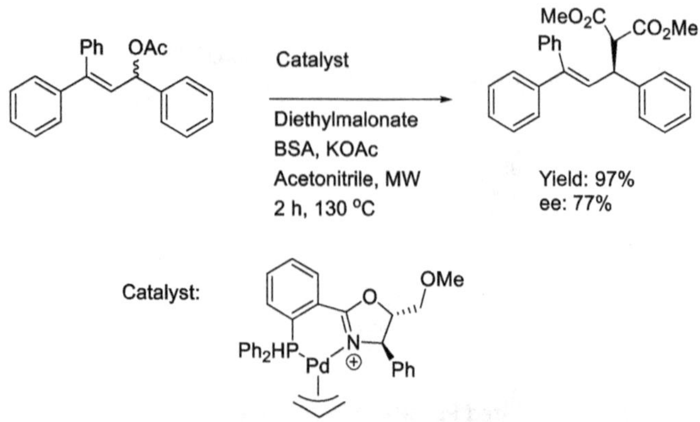

Scheme 18 Asymmetric allylic substitution catalyzed by a Pd-phosphinooxazoline complex.

dramatically shortened without any loss of enantioselectivity compared with the conventionally heated reaction.

An asymmetric allylic amination was investigated in a continuous-flow system with resin anchored phoshinooxazoline-Pd catalyst (Scheme 19).[69] The polymer-supported catalyst was derived via the strategy of click chemistry (Scheme 20) and efficiently applied in allylic amination, especially when the reactions are promoted by microwaves.

Molybdenum-catalyzed allylic substitution reactions are preferred due to the lower cost of molybdenum compared with iridium and platinum and the robustness of the in situ forming catalytic system.[70] Bis(pyridine-2-carboxamides) are the most efficient ligands for the Mo-catalyzed process

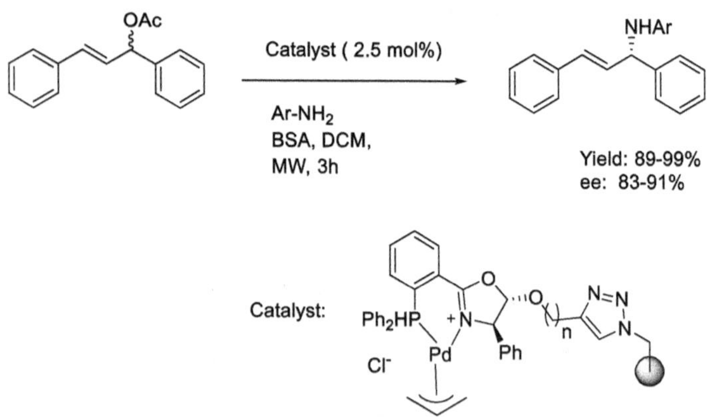

Scheme 19 Asymmetric allylic amination by a polymer supported phosphinooxazoline-Pd catalyst.

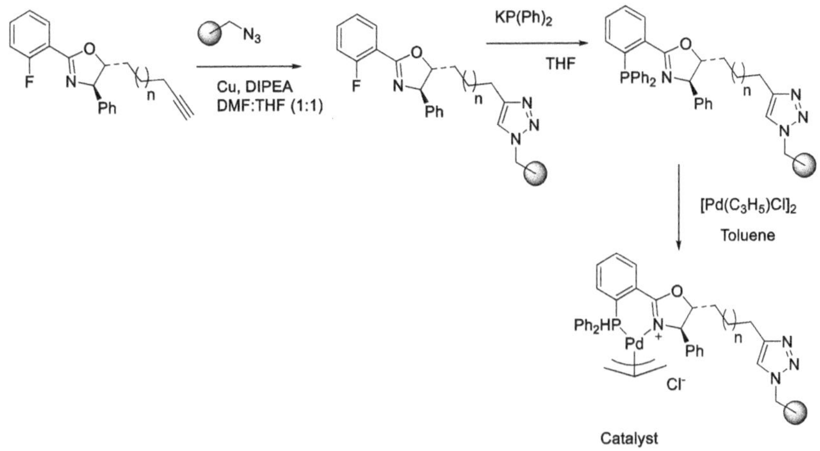

Scheme 20 Synthesis of a polymer supported phosphinooxazoline-Pd catalyst.

and are easily prepared from a picolinic acid derivative and a chiral enantioenriched 1,2-diamine.[71] This efficient ligand was used in Mo-catalyzed microwave-mediated asymmetric allylic alkylation reactions by Litto et al. (Scheme 21).[72] The reaction produced high enantioselectivity with 99% yield in reactions with branched racemic substrates.

3.2 Asymmetric Heck reaction

Palladium-catalyzed coupling of an aryl or alkyl halides or triflates with an alkene, a variation of the Heck reaction, is an important and a powerful route for the formation of carbon-carbon bonds. Pd-catalyzed intermolecular

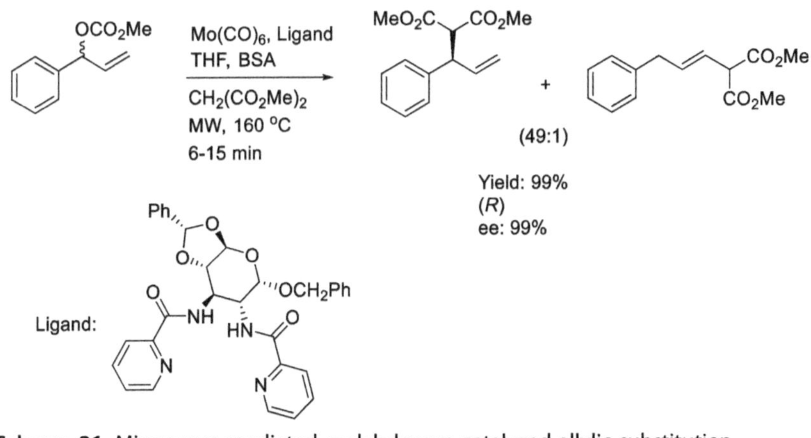

Scheme 21 Microwave-mediated molybdenum-catalyzed allylic substitution.

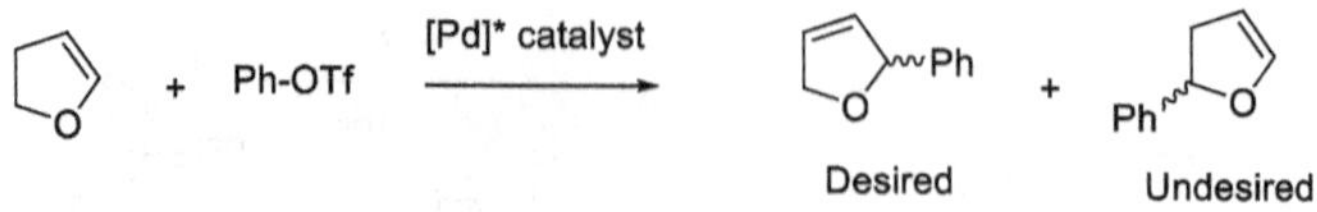

Scheme 22 A generic scheme of Pd-catalyzed intermolecular Heck reaction.

Heck reaction of 2,3-dihydrofuran with phenyl triflate produced the desired coupling product along with an undesired product, which was obtained due to an isomerization in the reaction process as described in Scheme 22.[73]

Several literature reports described[74–76] the successful application of phosphanyloxazoline (PHOX) ligands in intermolecular Heck reactions. The long reaction time for full conversion is the only drawback; however, it is limited to the application of N and P ligands. This issue in asymmetric Heck reactions has been resolved by applying microwave irradiation. The reaction of 2,3-dihydrofuran, cyclopentene, and 4,7-dihydro-1,3-dioxepin with phenyl or cyclohexyl triflate was carried out in presence of the Pd-complexed ligands (Fig. 1) producing the corresponding products in excellent yield with high regio- and enantioselectivities (Scheme 23).[77–79]

Intramolecular asymmetric Heck reaction under microwave conditions was carried out as a key intermediate step in the total synthesis of (+)-minfiensin (Scheme 24).[80]

3.3 Organocatalysts in asymmetric synthesis

The Mannich reactions were intensively studied under proline-catalyzed conditions, which required high catalyst loading and had a long reaction time.[81, 82] A direct reaction of cyclohexanone, formaldehyde, and various anilines when performed under microwave irradiation required only 0.5 mol% catalyst ((*S*)-proline) loading and afforded the product with high

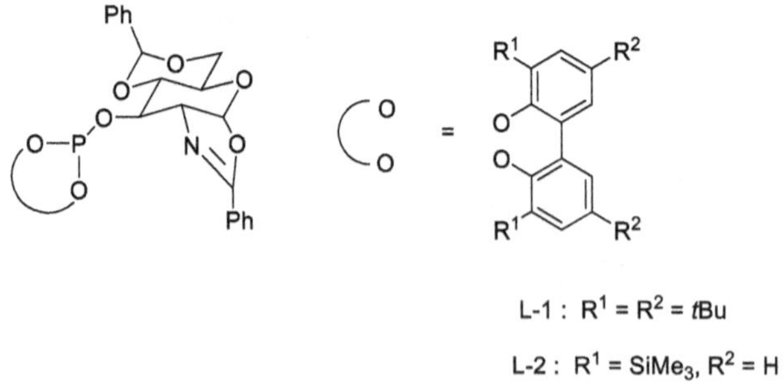

Fig. 1 Phosphite-oxazoline ligand library.

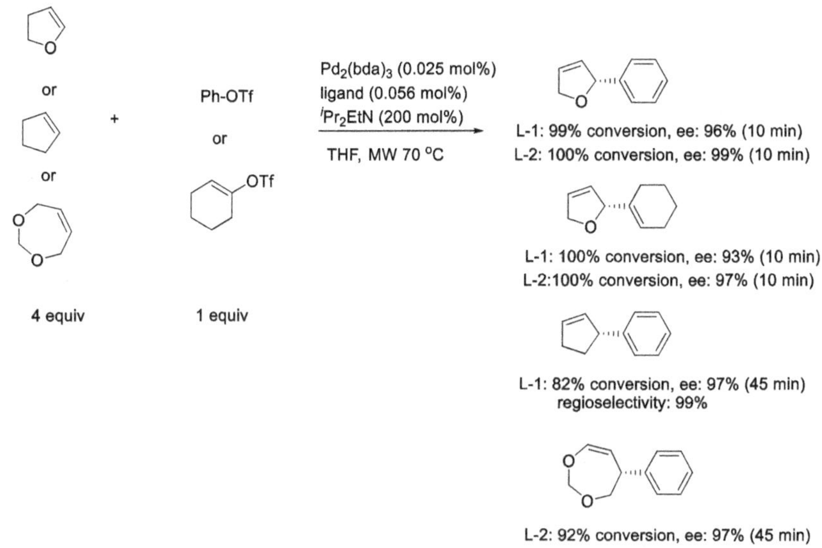

Scheme 23 Enantioselective Heck reactions using Pfaltz's PHOX and BINAP ligands.

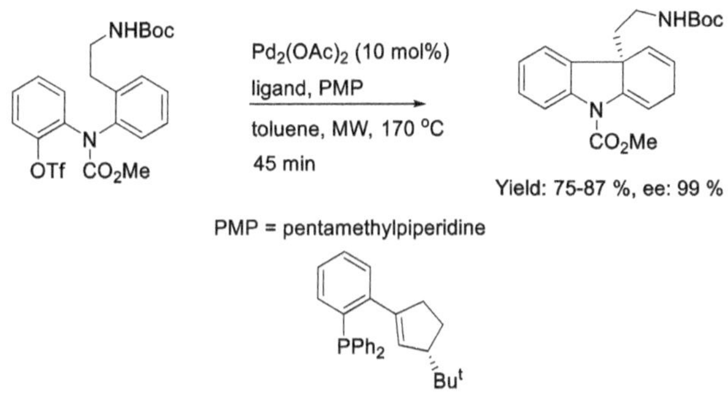

Scheme 24 Intramolecular Heck reaction under microwave conditions.

yield and high enantioselectivity.[83] In the last step the carbonyl group was transformed into the corresponding alcohol under these reaction conditions (Scheme 25).

The intramolecular enantioselective aldol condensations catalyzed by organocatalysts such as proline (1 mol%) under microwave conditions have been explored by Pan et al. with the goal to prepare the (*S*)-Wieland-Miescher ketone in a rapid reaction (Scheme 26).[84] The enantioselectivity and the conversion ratio were very similar to that of the reaction performed under conventional condition. The reaction time, however, was successfully

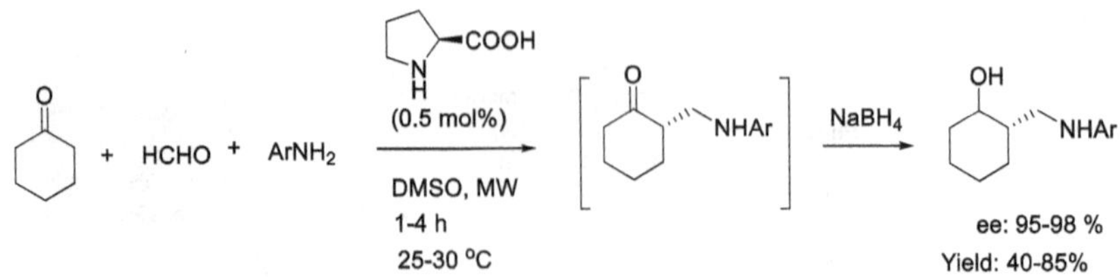

Scheme 25 Proline-catalyzed microwave-assisted enantioselective Mannich reaction.

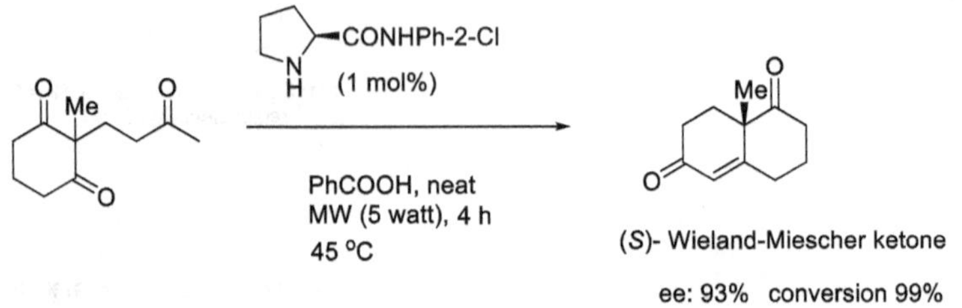

Scheme 26 Proline-catalyzed microwave-assisted intramolecular aldol condensation.

reduced from 5 days to less than 4 h. In the mechanism of microwave-assisted asymmetric cyclization, it is possible that microwave irradiation enhanced the molecular collision and promoted the formation of enamine/iminium intermediate. The use of excessive microwave power decreased the selectivity significantly.

Use of organocatalysts under microwave activation in aldol reaction, conjugate addition, and Diels-Alder reaction was studied by Alexakis and Mosse (Scheme 27).[85] In all cases the reaction time was dramatically shortened without loss of selectivity. Most importantly the catalyst loading in the conjugate addition was decreased from 15 to 5 mol% while maintaining good reactivity. A specific microwave effect was suggested to be involved in the Michael addition where a nonpolar solvent transparent to MW is used but not in aldol and Diels-Alder reactions where thermal effects should predominate as a result of the polar nature of the solvent: DMSO and MeOH/H_2O. One, however, must not forget that even if the solvent is microwave transparent in the Michael reaction, the reactants are all at least somewhat polar and thus would absorb the microwaves themselves.

Another important example of using organocatalyst is the synthesis of the six-membered nitrogen-containing ring of the morphine scaffold, a very familiar building block in natural products.[86] They were formed by an intramolecular aldol process of an aza-tethered dicarbonyl compound in the

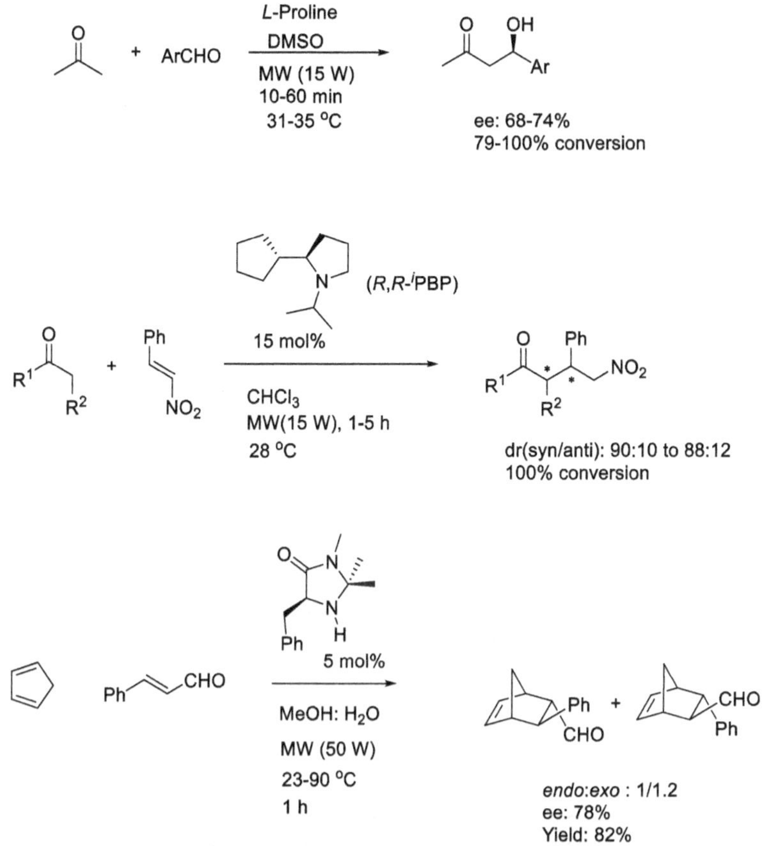

Scheme 27 Application of organocatalyst in aldol condensation, conjugate addition, and Diels-Alder reaction.

presence of a proline derivative under microwave conditions treating a ketoaldehyde using CH_3CN as a solvent in the presence of water producing 70% overall yield and 72% ee (Scheme 28).[87]

A highly enantioselective proline-catalyzed addition of ethyl trifluoropyruvate to ketones was successfully achieved by using a special microwave technique, the so-called subzero temperature microwave activation.[88] Part of the reaction was carried out at −25°C, while the sample was irradiated with microwaves. This was possible by using a special apparatus: a double jacketed vessel with a microwave transparent cooling liquid circulated in the jacket. This continuous cooling kept the reaction mixture at the desired temperature despite the continuous microwave irradiation providing elevated enantioselectivity in the reaction (Scheme 29).

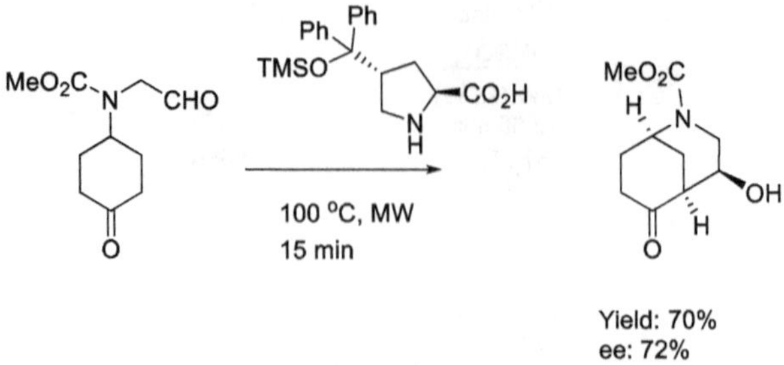

Scheme 28 Intramolecular aldol condensation toward nitrogen heterocycles.

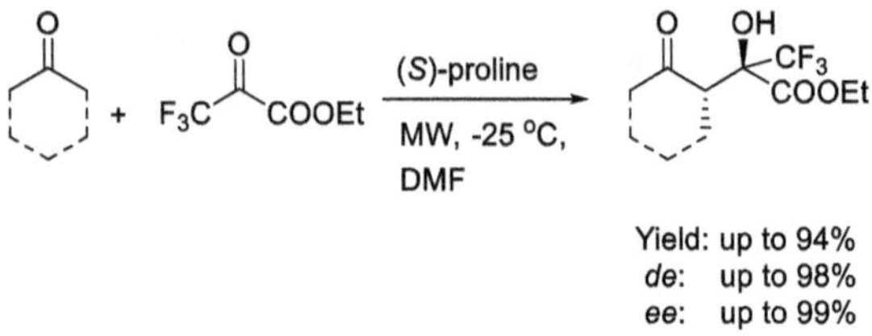

Scheme 29 Enantioselective proline-catalyzed addition of ethyl trifluoropyruvate to ketones with microwave activation at −25°C.

3.4 Asymmetric reductions

Asymmetric transfer hydrogenation (ATH) is a highly valuable and versatile method for the reduction of carbon-heteroatom double bonds, especially in ketones.[89, 90] Microwave-assisted enantioselective reduction was first reported by Lutsenko and Moberg in 2001 via a ruthenium-catalyzed hydrogen transfer reaction of acetophenones.[91] In this study, Noyori's monotosylated (*R*,*R*)-diphenylethylenediamine was employed[23] as the chiral ligand and propan-2-ol as the hydrogen source. Under standard conditions the reaction required 15 h to reach completion, affording (*R*)-1-phenylethanol in 95% yield and 97% ee. Utilizing microwave heating the authors could produce the same product in 90% yield and 82% ee after only 9 min of irradiation at 60 W (Scheme 30).

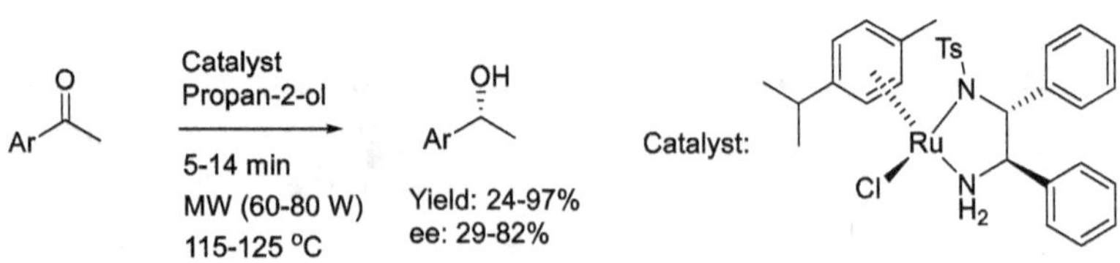

Scheme 30 Ruthenium-catalyzed asymmetric hydrogen transfer for the reduction of acetophenones.

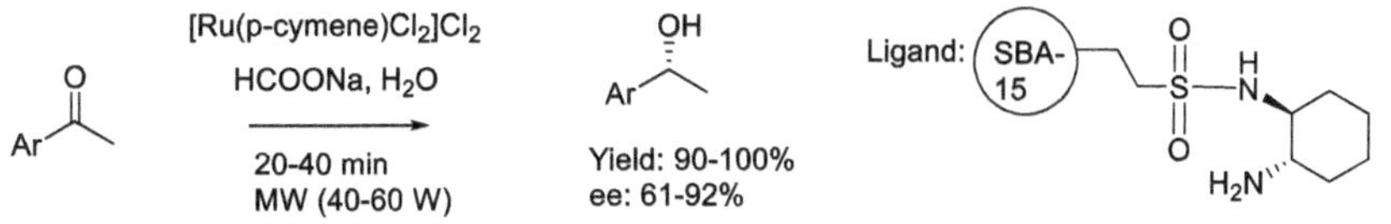

Scheme 31 SBA-15-anchored asymmetric ligand-complexed ruthenium catalyst and the reduction of acetophenones (SBA-Santa Barbara Amorphous silica).

The aforementioned reduction was also carried out by Jin and coworkers[92] by using a SBA-15-supported ligand derived from (1*R*,2*R*)-diaminocyclohexane. An excellent conversion (>90%) and good enantioselectivities (61%–92%) were obtained using this catalytic system after only 20–40 min of microwave irradiation (Scheme 31).

The first report of a simple microwave-assisted one-step cyclization/hydrogenation process for the production of menthols was described by Balu et al. (Scheme 32).[93] The selective conversion of citronellal to menthols was effectively carried out in a one-step reaction under microwave conditions catalyzed by supported nanoparticles on a mesoporous material providing (−)-menthol with good diastereoselectivities in the case of (+)-citronellal as starting material. Two percent Pt/Ga-MCM-41 was found to be the optimum catalyst for the reaction, with a selectivity for menthols above 85%, in short reaction times (Scheme 32). When enantiomerically pure (+)-citronellal was employed as starting material instead of the racemic compound, a 55% maximum selectivity to (−)-menthol was observed. The addition of small quantities of a chiral modifier such as cinchonidine increased the selectivity to (−)-menthol up to a 75% ee. Supported nanoparticles on such mesoporous materials were also proved to be highly stable and reusable under the investigated conditions, offering a potentially interesting alternative to the conventional production of these important natural products.

Several examples on the use of microwaves to promote the ATH of ketones can be found in literature. However, the reports of microwave-assisted

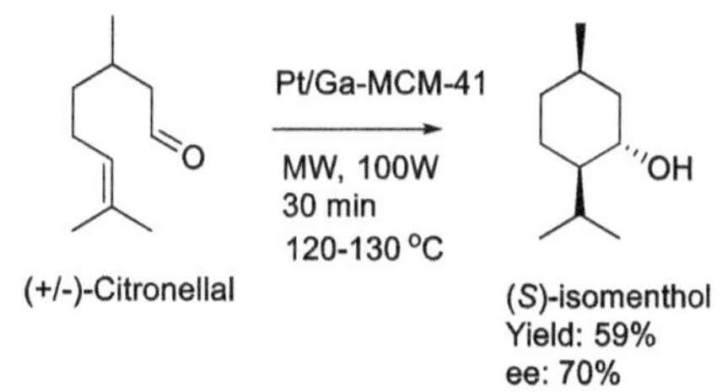

Scheme 32 Application of Pt supported on MCM-41 in an asymmetric hydrogen transfer reaction.

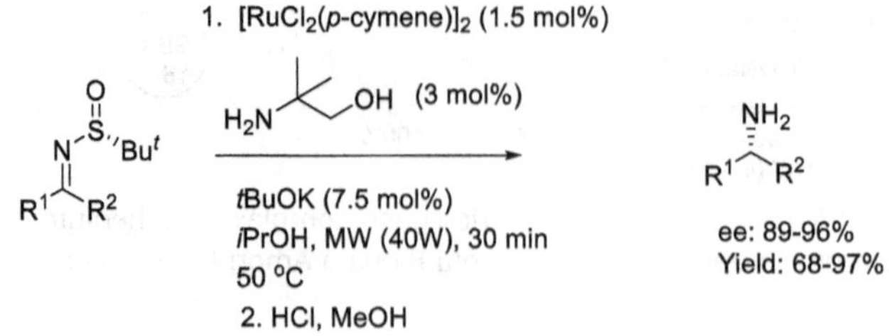

Scheme 33 Synthesis of chiral amines from *N*-(*tert*-butylsulfinyl)ketimines using ruthenium catalyst.

ATH of the C=N bond are very few. A very efficient procedure for the asymmetric transfer hydrogenation of *N*-(*tert*-butylsulfinyl)imines promoted by microwave irradiation was developed by Pablo et al. (Scheme 33).[94] This methodology achieved the reduction of a variety of aromatic, heteroaromatic, and aliphatic ketimines in very short reaction times, leading to the expected α-branched primary amines in excellent yields and with very high enantiomeric purities after desulfinylation. Both enantiomers of the amine are readily accessible by using the same ruthenium catalyst and changing the absolute configuration of the sulfinyl group bonded to the nitrogen atom of the imine.

A microwave-assisted asymmetric hydrogenation in the synthesis of a *C*-glycoside analogue of β-D-galactosyl hydroxylysine was achieved by using Burk's catalyst with exceptional selectivity (>99% de) and yields (95%).[95] The reaction was carried out in 2-propanol as solvent under microwave heating at 70°C for 40 min (Scheme 34). Interestingly, increasing the reaction temperature by 10°C resulted in a threefold loss in selectivity, highlighting the difficulty in obtaining both rapid reaction rates and high stereoselectivities at elevated temperatures.

A new highly stereoselective hydrogenation protocol for enol-acetate carboxylates was developed under microwave heating condition using

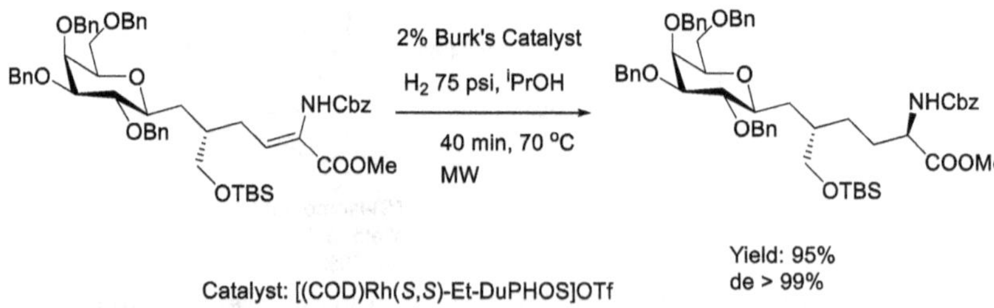

Scheme 34 Use of Burk's catalyst in an asymmetric hydrogen transfer.

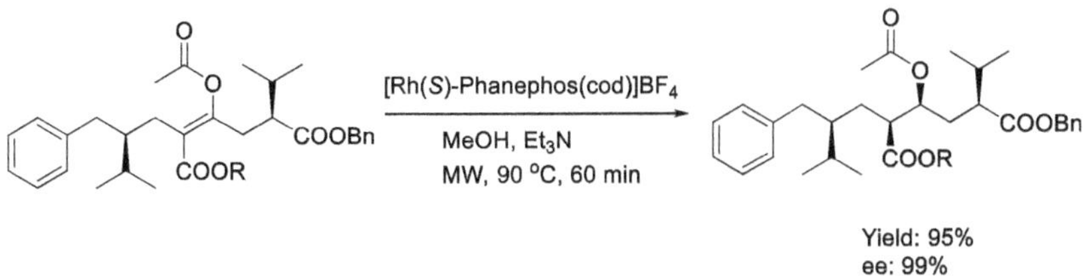

Scheme 35 Reduction of enolacetate carboxylate using rhodium catalyst.

[Rh-(*S*)-PhanoPhos(cod)]BF_4 as the catalyst (Scheme 35).[96] Microwave irradiation yielded the corresponding product in 95% yields and 99:1 enantioselectivity. The scaled up reaction (6 g of starting material) also produced the product in high yield, thus confirming the robustness of the catalytic process.

3.5 Miscellaneous microwave-assisted transformations

Tetrahydropyrans are useful synthons for biologically important compounds.[97] A very simple yet useful method for preparing tetrahydropyrans was developed by Guiry and Singh[98] where $ZrCl_4$ was used as a catalyst in deprotection of 1,3-dioxane/dioxalane. The reaction produced 6-methoxy-substituted tetrahydropyrans under microwave irradiation in good yield (Scheme 36).

[3,3]-Sigmatropic rearrangement (Cope rearrangement) of α-quaternary butenolides offered a straightforward access to γ-tertiary or γ-quaternary furanones when heated at 180°C under microwave irradiation.[99] The reaction produced the new stereogenic center with no erosion of the optical purity (Scheme 37).

The asymmetric ring opening of epoxides with nitrogen nucleophiles is a powerful tool for the preparation of chiral β-amino alcohols. Dioos and Jacobs reported[100] the first asymmetric epoxide ring opening under microwave heating catalyzed by a commercially available chiral (salen)Cr(III)

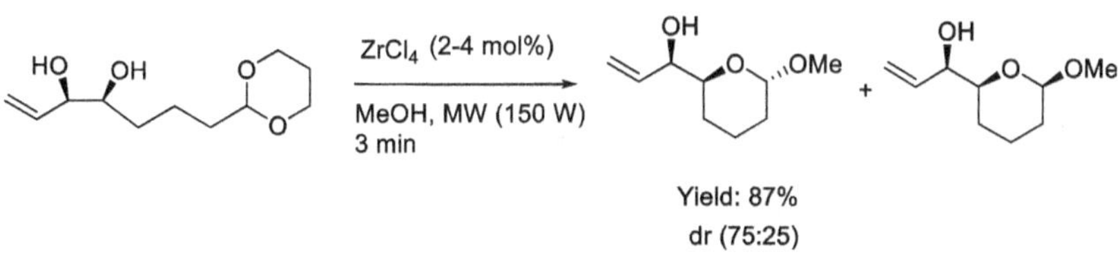

Scheme 36 $ZrCl_4$-catalyzed deprotection of 1,3-dioxane/dioxalane and the simultaneous formation of 6-methoxy-substituted diastereomeric tetrahydropyrans.

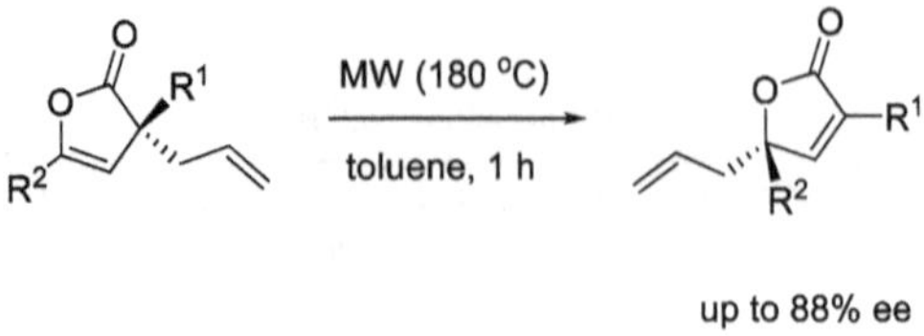

Scheme 37 A microwave-assisted Cope rearrangement.

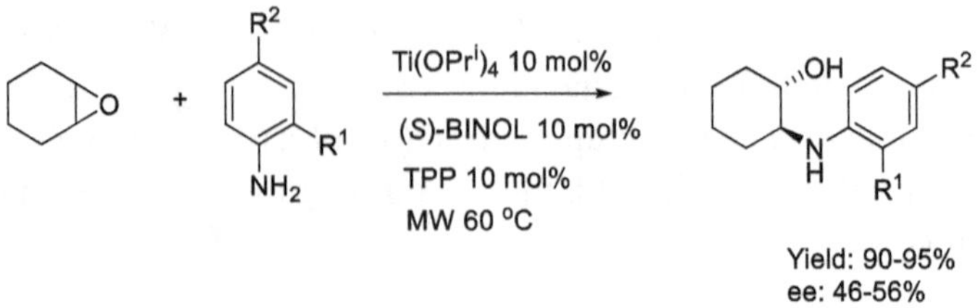

Scheme 38 Asymmetric ring opening of cyclohexene oxide with anilines catalyzed by a chiral Ti complex.

complex. Later, Kureshy et al. reported[101] the asymmetric ring opening of *meso*-epoxides with various anilines, catalyzed by a Ti-(*S*)-BINOL (Ti-(*S*)-1,1′-bi-2-naphthol). Excellent yields (90%–95%) and reasonable enantioselectivities (46%–56% ee) were obtained after only 30 s of microwave irradiation at 60°C (Scheme 38). The microwave-assisted reaction was 10 times and 235 times faster than the classical oil bath heating and room temperature reactions, respectively. The reaction was also enabled to produce the corresponding product with no significant decrease in enantiopurity. Interestingly a sharp decrease in enantioselectivity was observed when the reaction temperature was increased to 70°C, suggesting the decomposition of the Ti-(*S*)-BINOL complex at this temperature.

Polymer-immobilized diarylprolinol (α,α-bis[bis-3,5-(trifluoromethyl) phenyl]prolinol silyl ether anchored onto polystyrene) was synthesized by Beletskaya et al.,[102] and its catalytic activity was studied in the addition of aliphatic aldehydes to azodicarboxylates (Scheme 39). Reaction products were obtained in high yields (up to 96%) and with excellent enantioselectivity (up to 99%). The catalyst can be repeatedly used at least in three cycles without loss of enantioselectivity.

A rapid and highly efficient method for the Mannich reaction was developed by Ghomia and Zahedi.[103] The coupling reactions of cyclohexanone, various substituted aromatic amines, and aromatic aldehydes proceeded well under ultrasonic irradiation and microwave irradiation with Fe_3O_4-L-proline

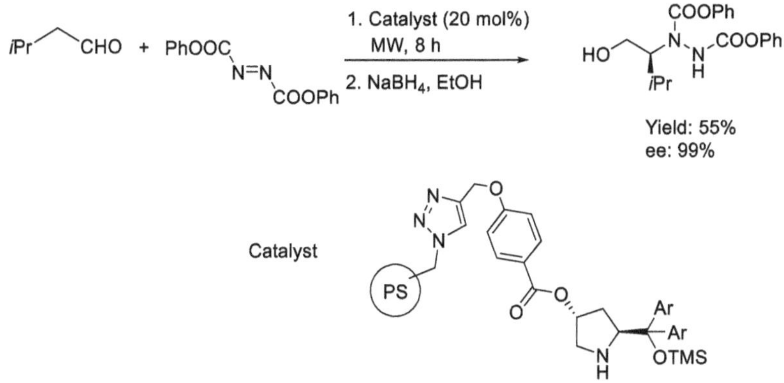

Scheme 39 A microwave-assisted addition of aliphatic aldehydes to azodicarboxylates catalyzed by polystyrene-supported proline derivatives (PS-polystyrene).

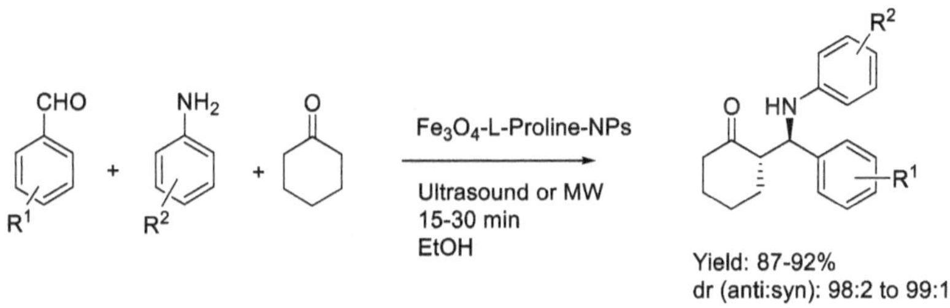

Scheme 40 Application of Fe_3O_4-L-proline nanoparticles (NPs) as chiral catalyst in a Mannich reaction.

nanoparticles as chiral catalyst in EtOH (Scheme 40). Reusability, eco-friendliness, inexpensive catalyst, short reaction times, high yields, and easy workup are advantages of this protocol.

4 Asymmetric reactions by electrochemical activation

Preparative electrolysis is commonly used in industrial processes to prepare bulk chemicals on ton-scale. Classic examples include the chloralkali process,[104] wherein aqueous sodium chloride is electrolyzed to give chlorine gas and sodium hydroxide, and the Hall–Héroult process, which provides elemental aluminum by electrolysis of Al_2O_3.[105] These profoundly useful applications of electrochemistry are producing millions of metric tons of these valuable chemicals and materials. Yet, examples of large scale electrochemistry for use in organic synthesis and the fine chemicals industry remain scarce. This is perplexing given the fact that this technique generally features relatively mild conditions, good functional group tolerance, and high chemoselectivity. Furthermore the ease with which many electrochemical reactions can be scaled up, as well as the intrinsic "greenness" of the reactions (the electric current is used in place of stoichiometric oxidants or reductants), makes this chemistry attractive in a process chemistry setting. These advantages are now being adopted in synthesizing wide range of compounds including chiral organic molecules as well, which will be discussed in this section.

Electrodes are crucial to organic electrochemical synthesis. Substrate molecules undergo electron transfers on the surface of the electrode. Such heterogeneous processes can impose mass transport limitations and a high kinetic barrier. Furthermore, reactions in the space near the electrode surface (known as the double layer) lead to accumulation of high energy species (e.g., radical cations and anions). Some of these reactive intermediates can diffuse back into the bulk solution for downstream functionalization, while others decompose to trigger electrode deactivation (known as passivation), performing further reactions. To circumvent these problems a mediator (or redox catalyst) can be used that undergoes heterogeneous electron transfer on the electrode surface to form a stabilized intermediate. These reactive species can then oxidize or reduce a substrate molecule homogeneously in an indirect electrolytic process. Few common powerful organic redox mediators such as triaryl amines[106] and nitroxyl radicals[107] were popularized during the 1970s and 1980s. Transition metal complexes and ionic halides represent two other types of common mediators; the latter can undergo anodic oxidation to generate molecular halogen, hypohalite, or cationic halogen species. Synthetic applications of electrolysis-generated base[108] and acid109 are studied as mediators to effect proton transfer in an environmentally friendly fashion. Miller's invention of the chiral electrode in 1975 pointed to a new direction in asymmetric catalysis, although it has not resulted in many practical applications to date.[110] Here the illustrative examples in the context of the synthesis of asymmetric molecules will be highlighted showing the intrinsic benefits of using the electrochemical technique. It is worth mentioning that while the electrochemical methods can readily produce products with high diastereoselectivity, achieving high enantioselectivity is still a challenge in many reactions.

4.1 Anodic oxidation

4.1.1 Amidation

Zhu and coworkers[111] developed an unprecedented electrocatalytic method in which olefin-bearing amidyl radicals can be efficiently prepared through the use of ferrocene (Fc) as the redox catalyst and subjected to intramolecular hydroamidation to synthesize valuable diastereomeric lactams and cyclic carbamates (Scheme 41). The success of this transformation relies on the use of a nonpolar solvent to lower the oxidation potential of anilides relative to that of Fc, which facilitates the electron transfer between the substrate and Fc^+.

4.1.2 Oxidative coupling

A metal- and oxidant-free method for the catalytic asymmetric oxidative coupling of tertiary amines with simple ketones was achieved via the combination of electrochemical oxidation in presence of chiral primary amine catalyst.[112] Simple ketones such as cyclopentanone, cyclohexanone, and cycloheptanone, as well as an acyclic ketone, can be applied in the developed catalytic system, allowing the rapid construction of diverse, optically active C1-alkylated tetrahydroisoquinoline derivatives with good yields and high stereoselectivities (Scheme 42).

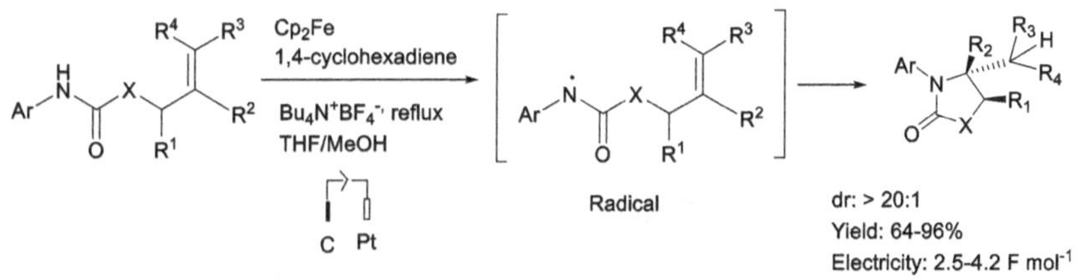

Scheme 41 Intramolecular olefin hydroamidation using amidyl radicals.

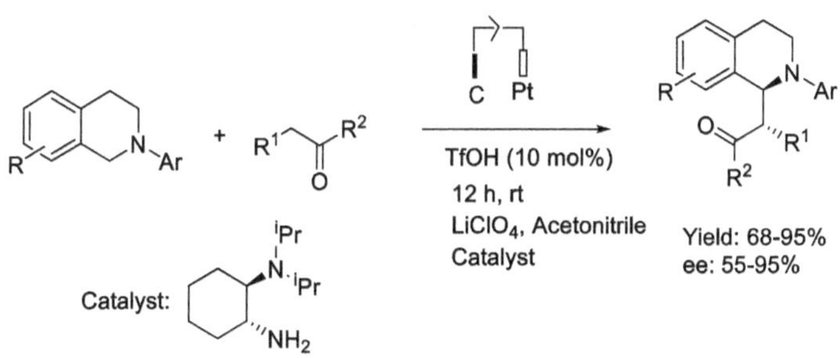

Scheme 42 Asymmetric cross-dehydrogenative coupling of tertiary amines with simple ketones.

4.1.3 Olefin oxidation

Anodic oxidation of enol ethers gives rise to radical cations with considerable radical character. These reactive species can undergo facile cyclization with an electron rich olefin. Commonly denoted as anodic olefin couplings, these processes enable umpolung disconnections that are broadly applicable in the synthesis of complex molecules. This chemistry has been rigorously examined by Tang et al.[113] The anodic oxidation of the starting olefin led to a 75% isolated yield of a bicyclic product containing five consecutive stereogenic atoms, one of which was a quaternary carbon (Scheme 43). The product was generated as a 6:1 ratio of stereoisomers at C4 (the major isomer is shown).

Okada and coworkers[114] established the novel electrocatalytic formally [2 + 2] cycloaddition reactions between anodically activated aliphatic enol ethers and olefins possessing an alkoxyphenyl group. The alkoxyphenyl group is necessary for the reaction intermediates to complete the formation of the cyclobutane ring through intramolecular electron transfer between the cyclobutane radical cation and the alkoxyphenyl group. Anodic oxidation of 3,4-dihydro-2H-pyran in the presence of an excess amount (20 equiv.) of 2-allylanisole produced the desired cycloadduct in excellent yield as a mixture of diastereomers (Scheme 44).

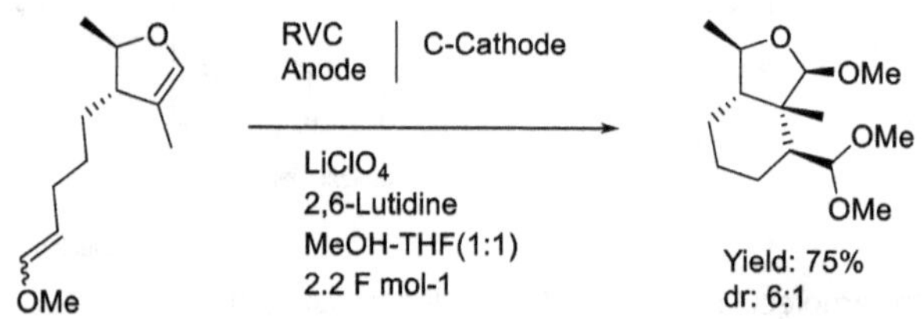

RVC Anode = Reticulated vitreous carbon anode

Scheme 43 Cyclization through anodic olefin coupling.

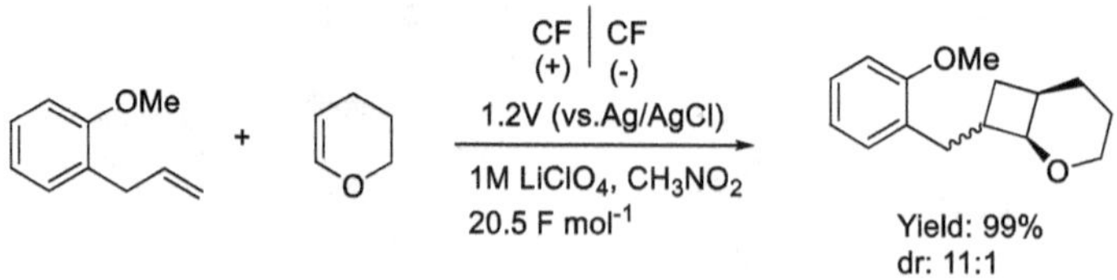

Scheme 44 [2 + 2] Cycloadditions between anodically activated aliphatic enolethers and unactivated olefins possessing an alkoxyphenyl group.

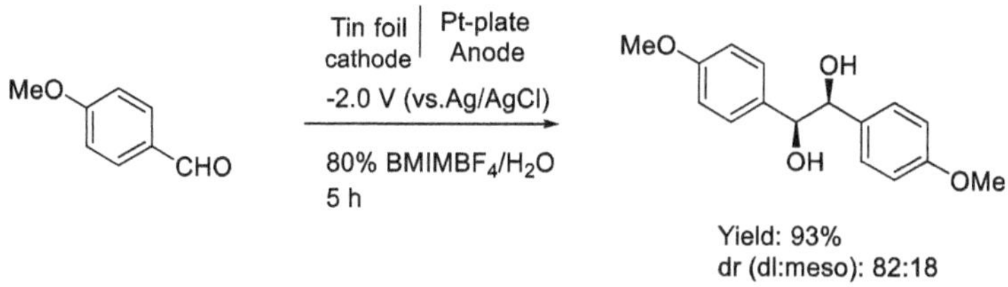

Scheme 45 Electrochemical pinacol coupling of benzaldehydes.

4.2 Cathodic reduction

Electrochemical pinacol coupling of aromatic aldehydes and ketones can be selectively performed using 80% [BMIM][BF_4]-H_2O as the electrolytic medium.[115] The resulting products were obtained in high yields with moderate diastereoselectivity (Scheme 45). The developed electrochemical method avoids the use of catalyst-cocatalyst systems for reductive coupling, thereby preventing the formation of any metallic byproducts. This method provides a simple synthetic route to 1,2-diols. The electrolyte can be recycled and reused at least up to five cycles.

5 Asymmetric synthesis by photochemical activation

The utility of ultraviolet and visible light in organic reactions has been attracting great attention, especially in green chemical synthesis. Visible light is particularly important for green chemistry; it is readily available, renewable, and abundant. Moreover the production of the reactive species, triggered by visible light through energy or electron transfer in a radical form, could lead to novel chemical transformations affording new and unusual products.[116, 117] Generally, this approach relies on the ability of metal complexes and organic dyes to engage in single-electron transfer (SET) processes with organic substrates upon photoexcitation with visible light. Many of the most commonly employed visible light photocatalysts are polypyridyl complexes of ruthenium and iridium and are typified by the complex tris(2,2′-bipyridine) ruthenium(II) or $Ru(bpy)_3^{2+}$. These complexes absorb light in the visible region of the electromagnetic spectrum to give stable, long-lived photoexcited states. The catalysts, redox-active species, upon irradiation with simple household light bulbs, transform into a remarkably chemoselective trigger to induce unique and valuable catalytic processes. In the past decade, various visible light-induced reactions have been designed and have emerged as practical methods for organic synthesis.[118, 119] Despite these significant achievements in classical asymmetric synthesis, surprisingly, the stereocontrol of the photocatalyzed reactions remains a challenging task. Following are the examples of the development of the visible light-induced asymmetric reactions in synthesis of chiral molecules.

5.1 Single bifunctional catalysis strategy

Photocatalyzed asymmetric reactions, involving a single catalyst, are common and attractive owing to their efficiency and new activation scenarios of synergistic photoexcitation and asymmetric catalysis. A chiral iridium complex was used by Meggers et al. to catalyze enantioselective trichloromethylation of 2-acyl imidazoles and 2-acyl pyridines under visible light irradiation (Scheme 46).[120] The desired products were obtained in moderate to excellent yield (62%–96%) with ≥99% ee. The chiral iridium enolate, generated via the coordination between an iridium complex and an acyl substrate in situ, is a key intermediate for stereo control and serves as a visible light-initiated photosensitizer for inducing electron transfer catalysis.

A chiral Rh complex, which resembles to the structure of the above Ir complex (shown in Scheme 46), was used for a condensation reaction between 2-acyl imidazoles or 2-acylpyridines with α-ketoesters, activated by visible light (Scheme 47). The desired products were obtained in high yield (up to 99%) and with excellent stereoselectivity (up to >20:1 dr and up to >99% ee).[121]

Asymmetric [2 + 2]-photocycloadditions were carried out by the same research group using the same chiral Ir-catalyst described earlier (Scheme 44)

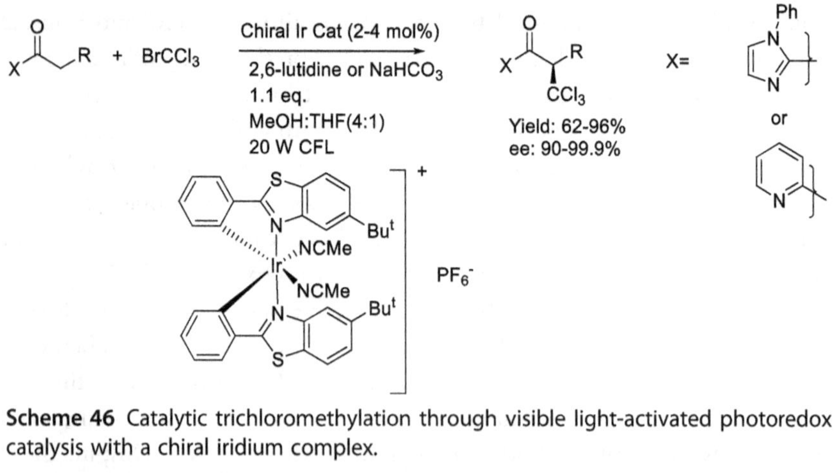

Scheme 46 Catalytic trichloromethylation through visible light-activated photoredox catalysis with a chiral iridium complex.

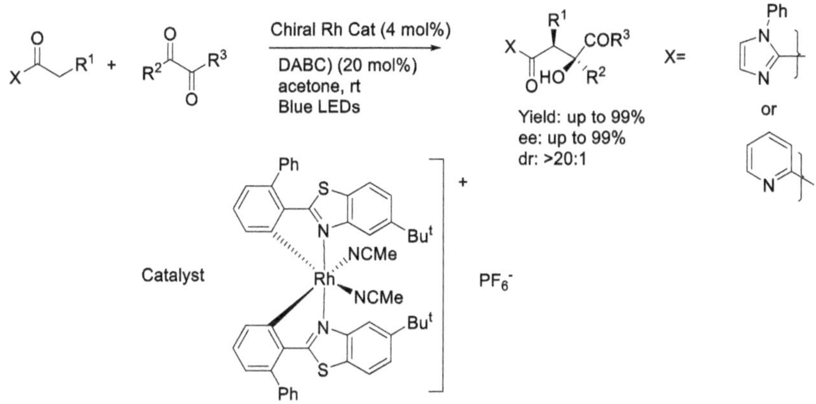

Scheme 47 Visible light-activated condensation reaction using a chiral rhodium catalyst.

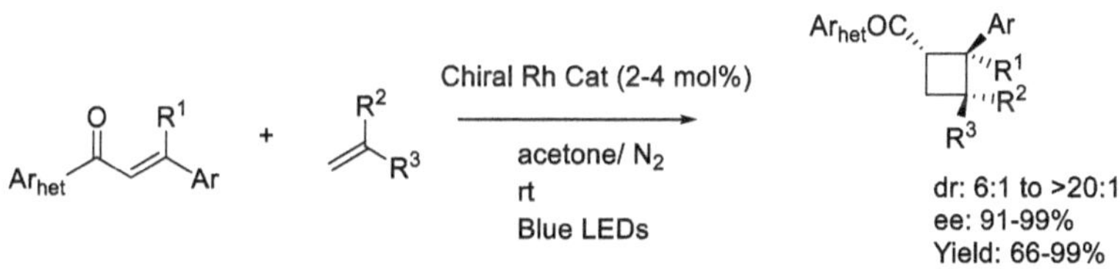

Scheme 48 Intermolecular [2 + 2] photocycloadditions catalyzed by a visible light-excited asymmetric Lewis acid.

furnishing chiral cyclopentanes, pyrrolidines, and cyclopentenes, respectively, with moderate to excellent yield and excellent enantioselectivities (Scheme 48).[122]

Another enantioselective excited-state [2 + 2] photocycloaddition of 3-alkoxyquinolones was studied by Skubi et al. with the application of chiral-iridium complexes as triplet energy transfer photocatalysts (Scheme 49).[123] The ligands containing pyrazole N—H moieties serve as hydrogen bond donors, which control the enantiomeric ratio. Although the reaction fulfills the green requirements in most aspects, it must be noted that the solvent mixture used is not desirable in environmentally benign applications.

5.1.1 Dual catalysis strategy

Some chemical transformations that cannot be catalyzed by a single catalyst are carried out in the presence of two catalyst-containing systems. In this strategy, organic molecules can generate a diverse array of reactive intermediates with photocatalysts, while a chiral cocatalyst introduces chirality via

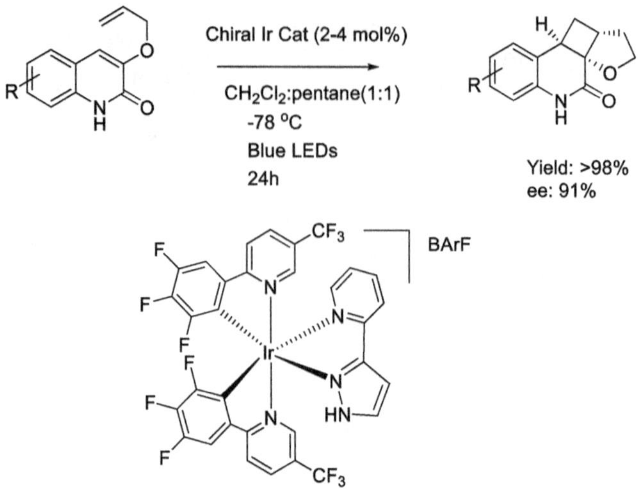

Scheme 49 Intramolecular [2 + 2] photocycloaddition of 3-alkoxyquinolones.

different activation modes simultaneously. Following are some examples of asymmetric transformations using the dual catalysis strategy. The combination of a photocatalyst with organocatalysts such as chiral amines was successfully used for photochemical reactions. Since its introduction by MacMillan and coworkers in 2008,[124] the group has developed a series of asymmetric transformations based on merging photoredox catalysis and amine catalysis. Recently the asymmetric alkylation of aldehydes was developed by MacMillan and his group using α-bromocyanoalkyls to generate optically-pure β-cyanoaldehydes, which is an important building block of various pharmacophore motifs[125] (Scheme 50).

A novel reaction platform by MacMillan et al. produced α-alkyl carbonyl adducts enantioselectively via coupling of simple aldehyde substrates with a

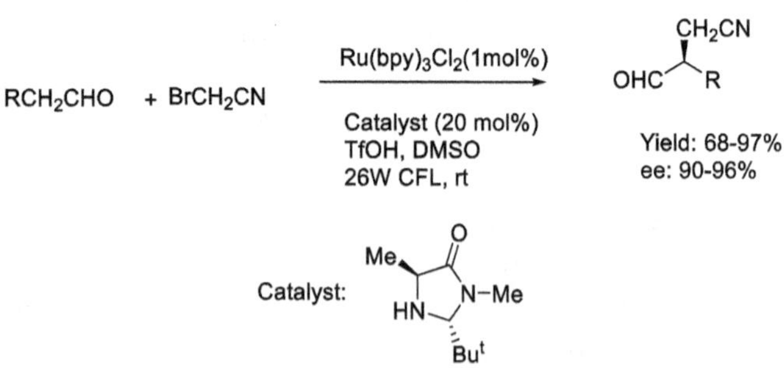

Scheme 50 Enantioselective α-alkylation of aldehydes by photoredox organocatalysis.

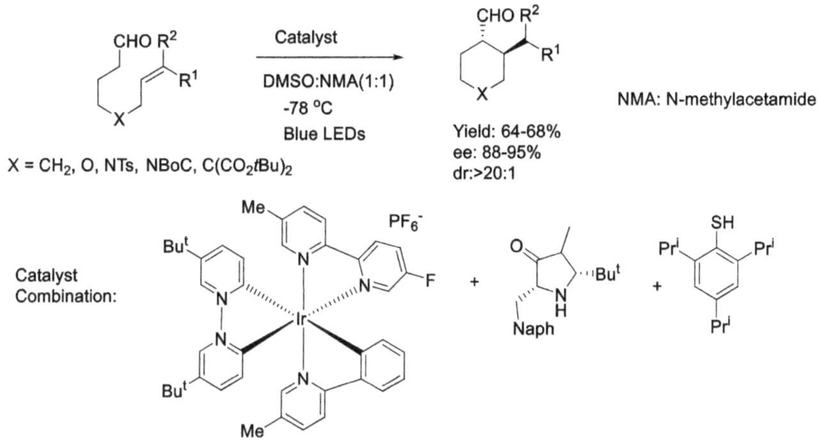

Scheme 51 Visible light-mediated enantioselective α-alkylation of aldehydes using simple olefins.

wide variety of olefins (Scheme 51).[126] This visible light-mediated transformation was accomplished via a hydrogen atom and electron borrowing mechanism that involves three discrete catalytic systems working simultaneously. This multicatalytic process enabled both intra- and intermolecular coupling of aldehydes to construct both cyclic and acyclic products, respectively.

In 2016 Meggers' group reported an asymmetric α-amination and α-alkylation catalyzed by a chiral-at-metal rhodium-based Lewis acid in combination with a photoredox sensitizer.[127] Electron acceptor-substituted aryl azides and α-diazo carboxylic esters were used as substrates to react with 2-acyl imidazoles in the visible light-activated synthesis. This novel proton- and redox-neutral method provided yields of up to 99% and excellent enantioselectivities of up to >99% ee with broad functional group compatibility (Scheme 52). Mechanistic investigations suggest that an intermediate rhodium enolate complex acts as a reductive quencher to initiate a radical process with the aryl azides and α-diazo carboxylic esters serving as precursors for nitrogen and carbon-centered radicals, respectively. This is the first report on using aryl azides and α-diazo carboxylic esters as substrates for asymmetric catalysis under photoredox conditions. These reagents have the advantage that molecular nitrogen is the leaving group and sole byproduct in this reaction.

Jiang and coworkers adopted a cooperative catalyst strategy by combining the dicyanopyrazine-derived chromophore (DPZ) photocatalyst, the

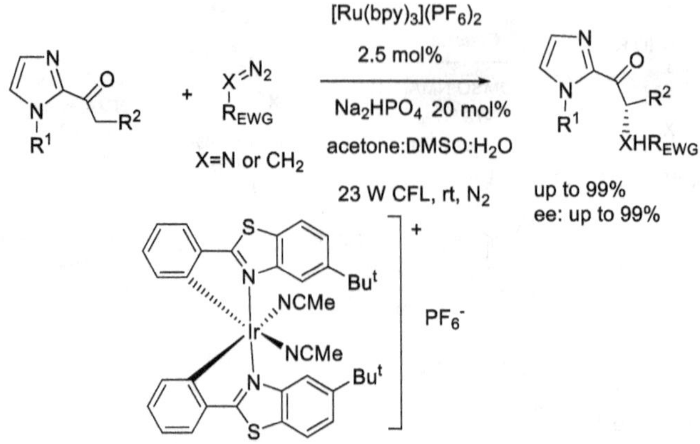

Scheme 52 Visible light-activated asymmetric α-alkylation/amination of ketones.

Lewis base catalyst β-isocinchonine (β-ICD) and the inorganic salt cocatalyst (NaBArF) for the enantioselective aerobic oxidative C(sp^3)-H olefination of tetrahydro-β-carbolines (THCs) and tetrahydroisoquinolines (THIQs).[128] The dicyanopyrazine-derived chromophore is an excellent metal-free photoredox catalyst that facilitates the excitation of the nonactivated substrates with reactive radical species with low catalyst loading. With this highly efficient triple-catalyst system, the olefination products were synthesized in high yields with excellent regio- and enantioselectivity (up to 95% ee, Scheme 53).

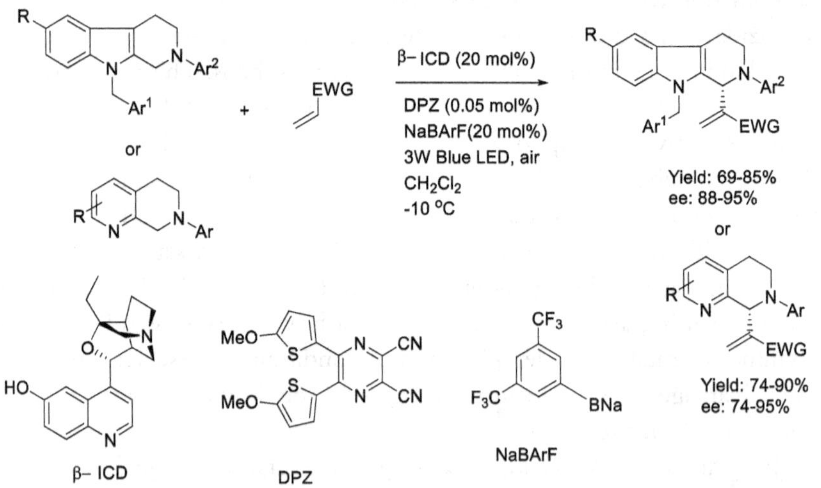

Scheme 53 Enantioselective oxidative C(sp^3)-H olefination of amines using photoredox and asymmetric catalysis.

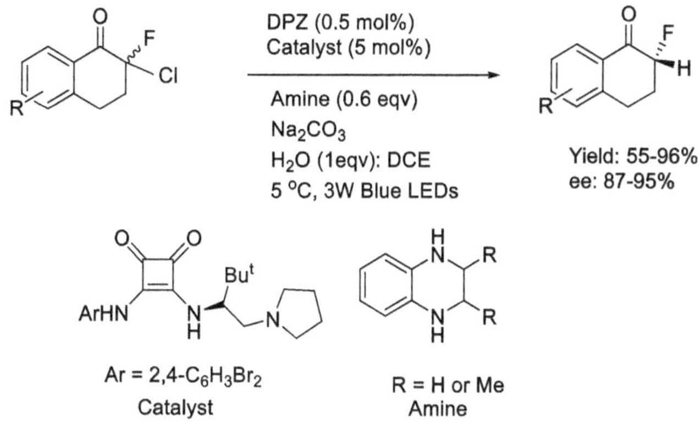

Scheme 54 Visible light-driven photoredox catalyzed dehalogenation-enantioselective protonation reaction.

A visible light-driven photoredox catalyzed dehalogenation-enantioselective protonation reaction was reported very recently by the same group.[129] A cooperative catalytic system consisting of a dicyanopyrazine-derived chromophore (DPZ) and a chiral H-bonding catalyst with secondary amine reductants was developed to initiate this transformation. A wide range of α-halogen-substituted cyclic and acyclic ketones were obtained in high yield with good to excellent enantioselectivity (Scheme 54).

Enzymes have been given serious consideration as catalyst candidates for organic synthesis as in their native environment they possess the ability to provide excellent selectivity for unique transformations.[130, 131] Enzyme catalysis can be combined with other catalytic modes, including photocatalysis. In 2018 Hyster's group demonstrated a new strategy for combining photoredox catalysis with enzyme catalysis. They proposed that the redox potential of the substrate can be attenuated upon its binding to the enzyme active site to enable selective electron transfer. As a result, they attempted to use the nicotinamide-dependent double-bond reductases (DBR) in photoredox catalysis to achieve the enantioselective deacetoxyl protonation of ketones (Scheme 55).[132]

6 Mechanochemical activation: Ball milling in asymmetric synthesis

The term mechanochemistry has been introduced in periodicals recently.[133, 134] According to IUPAC a mechanochemical reaction is defined as "a chemical reaction that is induced by the direct absorption of mechanical energy."[135] Ball milling is a mechanical method broadly used to granulate

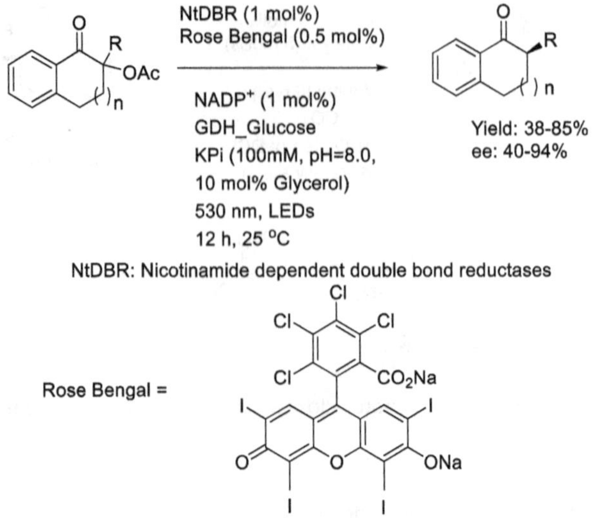

Scheme 55 Photoredox catalysis in nicotinamide-dependent oxidoreductases in enantioselective deacetoxyl protonation of ketones.

minerals into very fine particles for the preparation or alteration of inorganic solids. Ball milling is a simple, rapid, cost-effective green technology with enormous potential. In general the advantages of mechanochemical approaches are that they do not require a solvent to perform the reaction (solvent-free reaction) and produce considerably pure products. The mechanochemical technique has been explored in organic synthesis[136, 137] mainly in metal-mediated cross coupling reactions[138–140] organocatalysis [141, 142] and in miscellaneous applications[143] in medicinal and pharmaceutical chemistry. Reports on asymmetric syntheses carried out under ball-milling conditions are still limited. Examples of most commonly applied reactions for the synthesis of chiral molecules are listed in the succeeding text.

6.1 Condensation

The aldol condensation is one of the most powerful strategies in synthetic organic chemistry since it allows the formation of new C—C bonds that facilitates the construction of larger molecules from smaller units. Hernandez and Juaristi demonstrated that dipeptide (*S*)-proline-(*S*)-phenylalanine is a more efficient organocatalyst in the asymmetric aldol reaction between cyclohexanone or cyclopentanone and various aldehydes under solvent-free conditions using ball-milling activation compared with the reaction in solution (Scheme 56).[144] This is a significant result and suggests that noncovalent interactions, which are enhanced in the absence of solvents, are crucial for

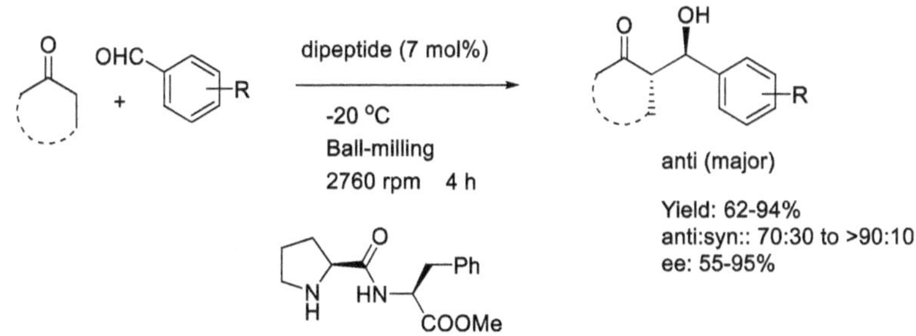

Scheme 56 Asymmetric aldol reaction organocatalyzed by (*S*)-proline-containing dipeptides under ball-milling activation.

the course of the reaction. The reaction proceeds efficiently, affording the antialdol products with good diastereo- and enantioselectivities in a relatively short time.

Chiral amine macrocycle metal(II) complexes are excellent catalysts for the asymmetric aldol reaction. Tanaka et al. used chiral amine macrocycle complexed with $CoBr_2$ as a catalyst in aldol reaction between cyclic ketones and substituted aromatic aldehydes under solvent-free conditions using a planetary ball mill (Scheme 57).[145] The antialdol adduct was obtained with higher enantioselectivity (up to 90% ee), compared with the same reaction in solution or in neat liquid with traditional magnetic stirring.

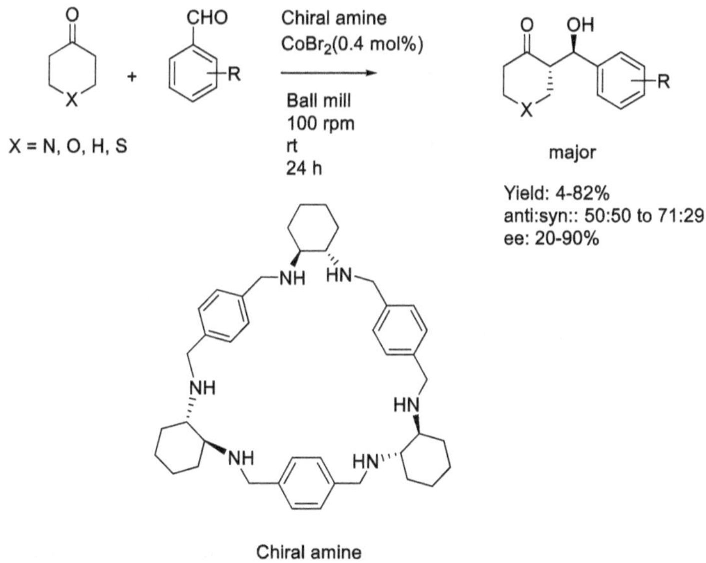

Scheme 57 Asymmetric aldol reactions catalyzed by chiral amine macrocycle-metal(II) complexes.

6.2 Michael addition reaction

Wang et al. reported a highly enantioselective organocatalytic Michael reaction of 1,2-carbonyl compounds with nitroolefins under solvent-free conditions using a planetary ball mill in the presence of cinchona-derived chiral squaramide.[146] Various substituted nitroalkenes underwent addition in a short reaction time (5–30 min) under ball milling (400 rpm) to provide an easy access to Michael adducts with good to excellent yields (63%–95%) and excellent enantioselectivity (91%–99% ee) (Scheme 58).

The ball milling-induced hydrogen bond-mediated organocatalytic addition of α-nitrocyclohexanone to aryl nitroalkenes under solvent-free conditions was developed to provide the corresponding addition products with high yields and excellent stereoselectivities (Scheme 59).[147]

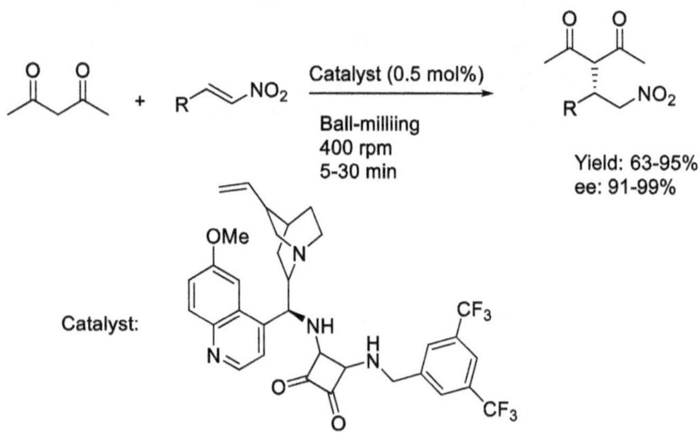

Scheme 58 Asymmetric Michael addition by a mechanochemical activation.

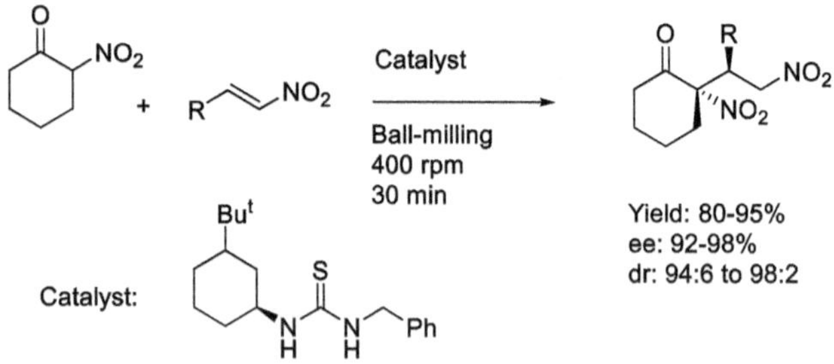

Scheme 59 Asymmetric Michael addition of α-nitrocyclohexanone to aryl nitroalkenes in a ball mill.

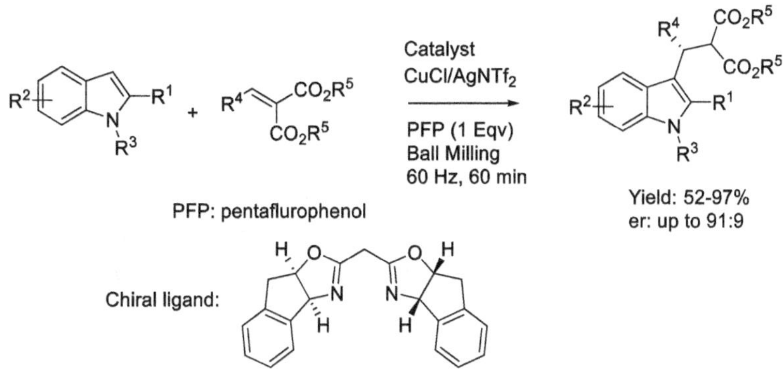

Scheme 60 Copper-catalyzed asymmetric Michael-type Friedel-Crafts alkylation of indoles.

Recently a mechanochemical version of the asymmetric Michael-type Friedel-Crafts alkylation of indoles was carried out with arylidenemalonates. The reaction proceeds under ambient conditions using a chiral bis (oxazoline)-copper catalyst in a ball mill. The product 3-substituted indole derivatives were obtained in good to excellent yields in short reaction times and with good enantioselectivities[148] (Scheme 60).

6.3 C—C coupling reactions

Solvent-free cross-dehydrogenative coupling (CDC) reactions were reported in literature under high energy ball-milling activation.[149] Solvent-free three-component coupling of aldehydes, alkynes, and amines (A^3 coupling) was achieved by Li et al.[150] where the desired adduct was produced with high ee's in short reaction times. Air-stable $Cu(OTf)_2$ and readily available PyBox were used as catalytic system in the reaction (Scheme 61). The chiral catalytic system was recovered and reused five times maintaining high performance without any additional replenishment of ligands.

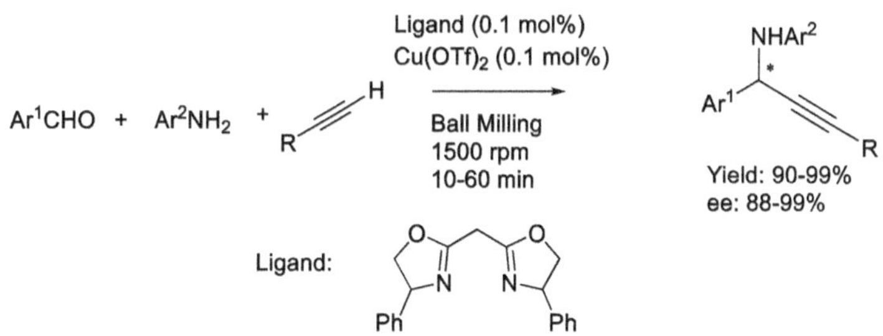

Scheme 61 Cross-dehydrogenative coupling reactions in a ball mill.

7 Conclusion

In recent years, there has been a growing awareness of the importance of the environmentally friendly design of chemical products and processes. The concept of sustainability has strongly impacted the chemical community, which is focusing more on minimizing the use of hazardous substances and adopting green synthetic strategies from renewable and sustainable resources for both starting materials and activation techniques. Stereoselective synthesis, including the preparation of pure enantiomers of organic molecules following green protocols, is still a challenge to synthetic chemists. Application of various nontraditional activation methods has led scientists to make advances in this area. Some syntheses under ultrasonic influence produce specific enantiomer rapidly, whereas other reactions show a keen preference for furnishing asymmetric centers under microwave condition. Electrochemistry has many innate advantages, including the ease of scalability, the use of inexpensive and/or recyclable electrodes, and potential fine-tuning of the reactions. It is a direct method for interacting with organic molecules that in many cases permits one to "dial in" to selectivity by choosing the right potential of the electrodes. Visible light-induced asymmetric syntheses are significantly empowered routes for chiral compounds in a green way. On the other hand, mechanochemically activated synthesis has proved to be an efficient tool for constructing various types of organic compounds, including potentially biologically active chiral scaffolds by environmentally friendly synthetic protocols. Despite the advantages of aforementioned methodologies in asymmetric synthesis, some problems still exist that should be overcome in the future. Two of the most important parameters in synthetic chemistry—temperature and pressure—are difficult to control and regulate inherently in microwave heating and ball milling methods, respectively. Another challenge to solve is the harmonization of reaction variables. Thus choice of ideal activation techniques with an ideal activating partners can provide excellent chiral environment as desired. However, in comparison with traditional heating processes, the methods described earlier offer several advantages such as (i) use of less solvent or no solvent, (ii) easy workup process, (iii) shorter reaction times, and (iv) green synthetic route. These achievements will inspire researchers to explore more in this field, making nontraditional activation techniques valuable tools for organic synthesis.

References

1. Drayer, D. E. The Early History of Stereochemistry. In *Drug Stereochemistry. Analytical Methods and Pharmacology*; Wainer, I. W., Ed.; 2nd ed.; Marcel Dekker Publisher: New York, 1993; pp. 1–24.
2. Katzung, B. G. Basic and Clinical Pharmacology. In *The Nature of Drugs*; 9th ed.; Lange Medical Books/McGraw Hill: New York, 2004; pp. 3–5.
3. Shen, Q.; Wang, L.; Zhou, H.; Jiang, H.; Yu, L.; Zeng, S. Stereoselective Binding of Chiral Drugs to Plasma Proteins. *Acta Pharmacol. Sin.* **2013**, *34*, 998–1006.
4. Wolfe, M. S. Inhibition and Modulation of γ-Secretase for Alzheimer's Disease. *Neurotherapeutics* **2008**, *5*, 391–398.
5. Thomas, K.; Sonya, P.; Eriksen, J. L.; Vallie, H.; Paul, M. M.; Koo, E. H.; Golde, T. E.; Michelle, N. Chronic Administration of *R*-Flurbiprofen Attenuates Learning Impairments in Transgenic Amyloid Precursor Protein Mice. *BMC Neurosci.* **2007**, *8*, 54–65.
6. Bartók, M.; Wittman, G. In *Stereochemistry of Heterogeneous Metal Catalysis*; Bartók, M., Ed.; Wiley: Chichester, 1985; pp. 163–164. Chapter 11.
7. Boudjouk, P. In *Ultrasounds, Its Chemical, Physical, and Biological Effects*; Suslick, K. S., Ed.; VCH: Weinheim, 1988.
8. Suslick, K. S. Sonocatalysis. In *Handbook of Heterogeneous Catalysis*; Ertl, G., Knozinger, H., Weitkamp, J., Eds.; Vol. 3; Wiley-VCH: New York, 1997.
9. Pinjari, D. V.; Pandit, A. B. Room Temperature Synthesis of Crystalline CeO_2 Nanopowder: Advantage of Sonochemical Method over Conventional Method. *Ultrason. Sonochem.* **2011**, *18*, 1118–1123.
10. Chatel, G. How Sonochemistry Contributes to Green Chemistry? *Ultrason. Sonochem.* **2018**, *40*, 117–122.
11. Draye, M.; Kardos, N. Advances in Green Organic Sonochemistry. *Top. Curr. Chem.* **2016**, *374*, 1–29.
12. Reddy, S. S.; Reddy, B. N.; Reddy, P. V. G.; Reddy, G. V.; Sarma, L. S. Mild and Efficient Synthesis of 5-(2,2-Difluoro-1-Phenyl Cyclopropyl)-N-Substituted-1,3,4-Oxadiazol-2-Amines Via Graphene Oxide as Catalyst under Ultrasonic Irradiation Conditions. *ChemistrySelect* **2017**, *2*, 356–363.
13. Ahankar, H.; Ramazani, A.; Slepokura, K.; Lis, T.; Joo, S. W. Synthesis of Pyrrolidinone Derivatives From Aniline, an Aldehyde and Diethyl Acetylenedicarboxylate in an Ethanolic Citric Acid Solution Under Ultrasound Irradiation. *Green Chem.* **2016**, *18*, 3582–3593.
14. Török, B.; Balázsik, K.; Török, M.; Felföldi, K.; Bartók, M. Asymmetric Reactions in Sonochemistry. *Ultrason. Sonochem.* **2001**, *8*, 191–200.
15. Török, B.; Balázsik, K.; Török, M.; Felföldi, K.; Bartók, M. Heterogeneous Asymmetric Reactions 20. Effect of Ultrasonic Variables on the Enantiodifferentiation in Cinchona-Modified Platinum-Catalyzed Sonochemical Hydrogenations. *Catal. Lett.* **2002**, *81*, 55–62.
16. Smith, G. B.; Notheisz, F. *Heterogeneous Catalysis in Organic Chemistry*; Academic Presss: San Diego, 1999.
17. Alexakis, A.; Lensen, N.; Mangeney, P. Ultrasound-Assisted Cleavage of Nitrogen-Nitrogen Bonds in Hydrazines by Raney Nickel. *Synlett* **1991**, 625–626.
18. Schellhaas, K.; Schmaiz, H. G.; Bats, J. W. Chiral [η6-Arene-Cr$(CO)_3$] Complexes as Synthetic Building Blocks: a Short Enantioselective Total Synthesis of (+)-Ptilocaulin. *Chem. A Eur. J.* **1998**, *4*, 57–66.
19. Izumi, A. Modified Raney Nickel (MRNi) Catalyst: Heterogeneous Enantiodifferentiating (Asymmetric) Catalyst. *Adv. Catal.* **1983**, *32*, 215–271.
20. Choliq, A. A.; Nakae, R.; Watanabe, M.; Misaki, T.; Fujita, M.; Okamoto, Y.; Sugimura, T. Enhanced Enantioselectivity Achieved at Low Hydrogen Pressure for

the Asymmetric Hydrogenation of Methyl Acetoacetate over a Tartaric Acid NaBr-Modified Raney Nickel Catalyst: a Kinetic Study. *Bull. Chem. Soc. Jpn.* **2019**, *92*, 1175–1180.
21. Nakagawa, S.; Sugimura, T.; Tai, A. Almost Perfect Enantio-Differentiating Hydrogenation of Methyl 3-Cyclopropyl-3-Oxopropanoate over Tartaric Acid Modified Raney Nickel Catalyst. *Chem. Lett.* **1997**, *9*, 859–860.
22. Sugimura, T. Recent Progress in Tartaric Acid-Modified Raney Nickel System for Enantio-Differentiating Hydrogenation. *Catal. Surv. Jpn.* **1999**, *3*, 37–42.
23. Hashiguchi, S.; Fujii, A.; Takehara, J.; Ikariya, T.; Noyori, R. Asymmetric Transfer Hydrogenation of Aromatic Ketones Catalyzed by Chiral Ruthenium(II) Complexes. *J. Am. Chem. Soc.* **1995**, *117*, 7562–7563.
24. Fujii, A.; Hahsiguchi, S.; Uematsu, N.; Ikariya, T.; Noyori, R. Ruthenium(II)-Catalyzed Asymmetric Transfer Hydrogenation of Ketones Using a Formic Acid-Triethylamine Mixture. *J. Am. Chem. Soc.* **1996**, *118*, 2521–2522.
25. Haack, K. J.; Hashiguchi, S.; Fujii, A.; Ikariya, T.; Noyori, R. The Catalyst Precursor, Catalyst and Intermediate in the Ru(II)-Promoted Asymmetric Hydrogen Transfer between Alcohols and Ketones. *Angew. Chem. Int. Ed.* **1997**, *36*, 285–288.
26. Takehara, J.; Hashiguchi, S.; Fujii, A.; Inoue, S.; Ikariya, T.; Noyori, R. Amino Alcohol Effects on the Ruthenium(II)-Catalyzed Asymmetric Transfer Hydrogenation of Ketones in Propan-2-Ol. *Chem. Commun.* **1996**, *2*, 233–234.
27. Shingote, S. K.; Kelkar, A. A.; Borole, Y. L.; Joshi, P. D.; Chaudhari, R. V. Ultrasound Promoted Asymmetric Transfer Hydrogenation of Ketones Using Ru(II)Arene/Amino Alcohol Catalyst System. *Ultrason. Sonochem.* **2008**, *15*, 289–293.
28. Török, B.; Balázsik, K.; Szöllösi, G.; Felföldi, K.; Bartók, M. Ultrasonics in Asymmetric Syntheses. Sonochemical Enantioselective Hydrogenation of Prochiral C=O Groups over Platinum Catalysts. *Chirality* **1999**, *11*, 470–474.
29. Xu, Y.; Cheng, T.; Long, J.; Liu, K.; Qian, Q.; Gao, F.; Liu, G.; Lia, H. An Ion-Pair Immobilization Strategy in Rhodium-Catalyzed Asymmetric Transfer Hydrogenation of Aromatic Ketones. *Adv. Synth. Catal.* **2012**, *354*, 3250–3258.
30. Mhadgut, S. C.; Bucsi, I.; Török, M.; Török, B. Sonochemical Asymmetric Hydrogenation of Isophorone on Proline Modified Pd/Al_2O_3 Catalysts. *Chem. Commun.* **2004**, *9*, 84–985.
31. Török, B.; Balázsik, K.; Török, M.; Szöllösi, G.; Bartók, M. Asymmetric Sonochemical Hydrogenation of α-Ketoesters Over Cinchona-Modified Platinum Catalysts. *Ultrason. Sonochem.* **2000**, 7, 151–158.
32. Balázsik, K.; Szöri, K.; Felföldi, K.; Török, B.; Bartók, M. Asymmetric Synthesis of Alkyl 5-Oxo-Tetrahydrofuran-2-Carboxylates by Enantioselective Hydrogenation of Dialkyl 2-Oxoglutarates Over Cinchona Modified Pt/Al_2O_3 Catalysts. *Chem. Commun.* **2000**, 555–556.
33. Török, B.; Balázsik, K.; Szakonyi, G.; Felföldi, K.; Bartók, M. Enantiodifferentiation in Asymmetric Sonochemical Hydrogenations. *Catal. Lett.* **1998**, *52*, 81–84.
34. Török, B.; Balázsik, K.; Szöllösi, G.; Felföldi, K.; Kun, I.; Bartók, M. Ultrasonics in Heterogeneous Metal Catalysis. Sonochemical Chemo- and Enantioselective Hydrogenations over Supported Platinum Catalysts. *Ultrason. Sonochem.* **1999**, *6*, 97–103.
35. Dehmlow, E. V.; Dehmlow, S. S. *Phase Transfer Catalysis*, 3rd ed.; VCH: Weinheim, 1993.
36. Starks, C. M.; Liotta, C. L.; Halpern, M. *Phase-Transfer Catalysis*; Chapman & Hall: New York, NY, 1994.
37. Sasson, Y.; Neumann, R. *Handbook of Phase-Transfer Catalysis*; Blackie Academic & Professional-Springer: London, 1997.
38. Nelson, A. Asymmetric Phase Transfer Catalysis. *Angew. Chem. Int. Ed.* **1999**, *38*, 1583–1585.

39. Shioiri, T. In *Handbook of Phase-Transfer Catalysis*; Sasson, Y., Neumann, R., Eds.; Blackie Academic & Professional-Springer: London, 1997. Chapter 14.
40. O'Donnell, M. J. In *Catalytic Asymmetric Synthesis*; Ojima, I., Ed.; 2nd ed.; Wiley-VCH: New York, NY-Weinheim, 2000. Chapter 10.
41. Shioiri, T.; Arai, S. In *Stimulating Concepts in Chemistry*; Vogtle, F., Stoddart, J. F., Shibasaki, M., Eds.; Wiley-VCH: Weinheim, 2000; pp. 123–143.
42. Mirza-Aghayan, M.; Etemad-Moghadam, G.; Zaparucha, A.; Berlan, J.; Loupy, A.; Koenig, M. Ultrasound Irradiation Effects on the Asymmetric Michael Reaction. *Tetrahedron: Asymmetry* **1995**, *6*, 2643–2646.
43. Manabe, K. Asymmetric Phase-Transfer Alkylation Catalyzed by a Chiral Quaternary Phosphonium Salt with a Multiple Hydrogen-Bonding Site. *Tetrahedron Lett.* **1998**, *39*, 5807–5810.
44. Jayaraman, S.; Kumaraguru, D.; Arockiam, J. B.; Paulpandian, S.; Rajendiran, B.; Siva, A. Highly Enantioselective Asymmetric Michael Addition Reactions with New Chiral Multisite Phase-Transfer Catalysts. *Synlett* **2014**, *25*, 1685–1691.
45. Beneto, A. J.; Sivamani, J.; Ashokkumar, V.; Balasaravanan, R.; Duraimurugan, K.; Siva, A. Highly Enantioselective Michael Addition Reactions with New Trimeric Chiral Phase Transfer Catalysts. *New J. Chem.* **2015**, *39*, 3098–3104.
46. Baldoli, C.; Del Buttero, P.; Licandro, E.; Papagni, A.; Pilati, T. Tricarbonyl(η6-Arene)Chromium(0) Complexes as Chiral Auxiliaries: Asymmetric Synthesis of β-Aminoesters and β-Lactams by Reformatskii Condensation. *Tetrahedron* **1996**, *52*, 4849–4856.
47. Woltersdorf, M.; Kranich, R.; Schmalz, H.-G. Enantioselective Synthesis of New C2-Symmetric Ferrocenylalkylamines Via Sonochemical Amination of 1-Ferrocenylalkyl Acetates. *Tetrahedron* **1997**, *53*, 7219–7230.
48. Hunt, J. C. A.; Lloyd, C.; Moody, C. J.; Slawin, A. M. Z.; Takle, A. K. Chiral Oxime Ethers in Asymmetric Synthesis. Part 4. Asymmetric Synthesis of N-Protected Amines and β-Amino Acids by the Addition of Organometallic Reagents to ROPHy/SOPHy-Derived Aldoximes. *J. Chem. Soc., Perkin Trans.* **1999**, *1* (23), 3443–3454.
49. Sendil, K.; Tekin, T.; Göksu, H.; Oguz, M.; Anil, B.; Gültekina, M. S. A Novel Method for the Synthesis of Newfangled Asymmetric Schiff Bases From α-Amino Acids Under Ultrasonic Conditions and in Aqueous Medium. *J. Chin. Chem. Soc.* **2016**, *63*, 808–817.
50. Xu, H.-C.; Chowdhury, S.; Ellman, J. A. Asymmetric Synthesis of Amines Using Tert-Butanesulfinamide. *Nat. Protoc.* **2013**, *8*, 2271–2280.
51. Zhou, W.; Nie, X.-D.; Zhang, Y.; Si, C.-M.; Zhou, Z.; Sun, X.; Wei, B.-G. A Practical Approach to Asymmetric Synthesis of Dolastatin 10. *Org. Biomol.Chem.* **2017**, *15*, 6119–6131.
52. Liu, X.; Hu, Y.-J.; Fan, J.-H.; Zhao, J.; Li, S.; Li, C.-C. Recent Synthetic Studies towards Natural Products Via [5 + 2] Cycloaddition Reactions. *Org. Chem. Front.* **2018**, *5*, 1217–1228.
53. Nugent, T. C.; El-Shazly, M. Chiral Amine Synthesis. Recent Developments and Trends for Enamide Reduction, Reductive Amination, and Imine Reduction. *Adv. Synth. Catal.* **2010**, *352*, 753–819.
54. Appa, R. M.; Lakshmidevi, J.; Prasad, S. S.; Muralidhar, B.; Naidu, B. R.; Narasimhulu, M.; Venkateswarlu, K. First Sonochemical, Simple and Solvent-Free Synthesis of Chiral Tert-Butanesulfinimines Using Silica Supported *p*-Toluenesulfonic Acid. *Synth. Commun.* **2019**, *49*, 56–64.
55. Kokel, A.; Schäfer, C.; Török, B. Microwave-assisted reactions in green chemistry. In *Encyclopedia of Sustainable Science and Technology*; Meyers, R. A., Ed.; Springer-Nature, 2018. https://doi.org/10.1007/978-1-4939-2493-6_1008-1.

56. Kokel, A.; Schäfer, C.; Török, B. Application of Microwave-Assisted Heterogeneous Catalysis in Sustainable Synthesis Design. *Green Chem.* **2017**, *19*, 3729–3751.
57. Varma, R. S. In *Microwaves: Theory and Application in Material Processing IV*; Clark, D. E., Sutton, W. H., Lewis, D. A., Eds.; American Ceramic Society: Westerville, Ohio, 1997; pp. 357–365.
58. Mingos, D. M. P. In *Microwave-Assisted Organic Synthesis*; Lidstrom, P., Tierney, J. P., Eds.; Blackwell: Oxford, 2004. Chapter 1.
59. Loupy, A. *Microwave in Organic Synthesis*, 2nd ed, Vol. 1; Wiley-VCH Verlag GmbH and Co. KGaA, 2006; pp. 362–451. Chapter 8.
60. Cho, H.; Török, F.; Török, B. Energy Efficiency of Heterogeneous Catalytic Microwave-Assisted Organic Reactions. *Green Chem.* **2014**, *16*, 3623–3634.
61. Trost, B. M.; VanVranken, D. L. Asymmetric Transition Metal–Catalyzed Allylic Alkylations. *Chem. Rev.* **1996**, *96*, 395–422.
62. Lu, Z.; Ma, S. M. Metal-Catalyzed Enantioselective Allylation in Asymmetric Synthesis. *Angew. Chem. Int. Ed.* **2008**, *47*, 258–297.
63. Kimura, M.; Horino, Y.; Mukai, R.; Tanaka, S.; Tamaru, Y. Strikingly Simple Direct α-Allylation of Aldehydes with Allyl Alcohols: Remarkable Advance in the Tsuji-Trost Reaction. *J. Am. Chem. Soc.* **2001**, *123*, 10401–10402.
64. Dos Santos, S.; Quignard, F.; Sinou, D.; Choplin, A. Allylic Substitution Catalysed by Silica-Supported Aqueous Phase Palladium (0) Catalysts. *Top. Catal.* **2000**, *13*, 311–318.
65. Kaiser, N. F. K.; Bremberg, U.; Larhed, M.; Moberg, C.; Hallberg, A. Microwave—Mediated Palladium—Catalyzed Asymmetric Allylic Alkylation; An Example of Highly Selective Fast Chemistry. *J. Organomet. Chem.* **2000**, *603*, 2–5.
66. Braga, A. L.; Silveira, C. C.; de Bolster, M. W. G.; Schrekker, H. S.; Wessjohann, L. A.; Schneider, P. H. Microwave-Accelerated Asymmetric Allylations Using Cysteine Derived Oxazolidine and Thiazolidine Palladium Complexes. *J. Mol. Catal. A Chem.* **2005**, *239*, 235–238.
67. Braga, A. L.; Vargas, F.; Sehnem, J. A.; Wessjohann, L. A. Microwave—Mediated Palladium—Catalyzed Asymmetric Allylic Alkylation Using Chiral Beta—Seleno Amides. *Eur. J. Org. Chem.* **2006**, *38*, 4993–4997.
68. Popa, D.; Puigjaner, C.; Gomez, M.; Benet-Buchholz, J.; Vidal-Ferran, A.; Pericas, M. A. Phosphinooxazolines Derived From 3-Amino-1,2-Diols: Highly Efficient Modular P-N Ligands. *Adv. Synt. Catal.* **2007**, *349*, 2265–2278.
69. Popa, D.; Marcos, R.; Sayalero, S.; Vidal-Ferran, A.; Pericos, M. A. Towards Continuous Flow, Highly Enantioselective Allylic Amination: Ligand Design, Optimization and Supporting. *Adv. Synth. Catal.* **2009**, *351*, 1539–1556.
70. Belda, O.; Moberg, C. Molybdenum-Catalyzed Asymmetric Allylic Alkylations. *Acc. Chem. Res.* **2004**, *37*, 159–167.
71. Belda, O.; Moberg, C. Bispyridylamides-Coordination Chemistry and Applications in Catalytic Reactions. *Coord. Chem. Rev.* **2005**, *249*, 727–740.
72. Litto, R. D.; Benessere, V.; Ruffo, F.; Moberg, C. Carbohydrate-Based Pyridine-2-Carboxamides for Mo-Catalyzed Asymmetric Allylic Alkylations. *Eur. J. Org. Chem.* **2009**, 1352–1356.
73. Wu, W.-Q.; Peng, Q.; Dong, D.-X.; Hou, X.-L.; Wu, Y.-D. A Dramatic Switch of Enantioselectivity in Asymmetric Heck Reaction by Benzylic Substituents of Ligands. *J. Am. Chem. Soc.* **2008**, *130*, 9717–9725.
74. Gilbertson, S. R.; Fu, Z.; Chiral, P. N-Ligands Based on Ketopinic Acid in the Asymmetric Heck Reaction. *Org. Lett.* **2001**, *3*, 161–164.
75. Loiseleur, O.; Meier, P.; Pfaltz, A. Chiral Phosphanyldihydrooxazoles in Asymmetric Catalysis: Enantioselective Heck Reactions. *Angew. Chem. Int. Ed. Engl.* **1996**, *35*, 200–202.

76. Loiseleur, O.; Hayashi, M.; Schmees, N.; Pfaltz, A. Enantioselective Heck Reactions Catalyzed by Chiral Phosphinooxazoline-Palladium Complexes. *Synthesis* **1997**, *1997*, 1338–1345.
77. Mazuela, J.; Pamies, O.; Dieguez, M. Biaryl Phosphite–Oxazoline Ligands From the Chiral Pool: Highly Efficient Modular Ligands for the Asymmetric Pd-Catalyzed Heck Reaction. *Chem. A Eur. J.* **2010**, *16*, 3434–3440.
78. Nilsson, P.; Gold, H.; Larhed, M.; Hallberg, A. The First Report of Catalytic Asymmetric Heck Reaction Using Pfaltz's PHOX and BINAP Ligands, Microwave-Assisted Enantioselective Heck Reactions: Expediting High Reaction Speed and Preparative Convenience. *Synthesis* **2002**, 1611–1614.
79. Kaukoranta, P.; Kallstorm, K.; Andersson, P. G. Microwave-Assisted Asymmetric Intermolecular Heck Reaction Using Phosphine-Thiazole Ligands. *Adv. Synth. Catal.* **2007**, *349*, 2595–2602.
80. Dounay, A. B.; Humphreys, P. G.; Overman, L. E.; Wrobleski, A. D. Total Synthesis of the Strychnos Alkaloid (+)-Minfiensine: Tandem Enantioselective Intramolecular Heck-Iminium Ion Cyclization. *J. Am. Chem. Soc.* **2008**, *130*, 5368–5377.
81. List, B.; Pojarliev, P.; Biller, W. T.; Martin, H. The Proline-Catalyzed Direct Asymmetric Three-Component Mannich Reaction: Scope, Optimization, and Application to the Highly Enantioselective Synthesis of 1,2-Amino Alcohols. *J. Am. Chem. Soc.* **2002**, *124*, 827–833.
82. Mitsumori, S.; Zhang, H.; Cheong, P. H.-Y.; Houk, K. N.; Tanaka, E.; Barbas, C. F., III. Direct Asymmetric Anti-Mannich-Type Reactions Catalyzed by a Designed Amino Acid. *J. Am. Chem. Soc.* **2006**, *128*, 1040–1041.
83. Rodriguez, B.; Bolm, C. Thermal Effects in the Organocatalytic Asymmetric Mannich Reaction. *J. Org. Chem.* **2006**, *71*, 2888–2891.
84. Pan, L.; Ding, X.; Ding, J.; Li, D.; Chen, J.; Zuo, X.; An, R. Design and Synthesis of L-Proline Derivatives as Enantioselective Organocatalysts for Synthesis of the (*S*)-Wieland-Miescher Ketone. *ChemistrySelect.* **2017**, *2*, 11999–12005.
85. Mosse, S.; Alexakis, A. Organocatalyzed Asymmetric Reactions Via Microwave Activation. *Org. Lett.* **2006**, *8*, 3577–3580.
86. Tanimoto, H.; Saito, R.; Chida, N. Formal Synthesis of (−)-Morphine from D-Glucal Based on the Cascade Claisen Rearrangement. *Tetrahedron Lett.* **2008**, *49*, 358–362 (and references cited therein).
87. Diaba, F.; Bonjoch, J. Asymmetric Synthesis of 2-Azabicyclo[3.3.1]Nonanes by a Microwave-Assisted Organocatalysed Tandem Desymmetrisation and Intramolecular Aldolisation. *Org. Biomol. Chem.* **2009**, 7, 2517–2519.
88. Landge, S. M.; Török, B. Highly Enantioselective Organocatalytic Addition of Ethyl Trifluoropyruvate to Ketones with Subzero Temperature Microwave Activation. *Catal. Lett.* **2009**, *131*, 432–439.
89. Wills, M. In *Modern Reduction Methods*; Andersson, P. G., Munslow, I. J., Eds.; Wiley-VCH: Weinheim, Germany, 2008; pp. 271–296.
90. Noyori, R.; Hashiguchi, S. Asymmetric Transfer Hydrogenation Catalyzed by Chiral Ruthenium Complexes. *Acc. Chem. Res.* **1997**, *30*, 97–102.
91. Lutsenko, S.; Moberg, C. Microwave-Mediated Ruthenium-Catalyzed Asymmetric Hydrogen Transfer. *Tetrahedron Asymmetry* **2001**, *12*, 2529–2532.
92. Jin, M.-J.; Sarkar, M. S.; Jung, J.-Y. Highly Effi Cient Microwave-Assisted Asymmetric Transfer Hydrogenation with SBA-15-Supported TsCHDA Chiral Ligands. *Stud. Surf. Sci. Catal.* **2007**, *165*, 455–458.
93. Balu, A. M.; Campelo, J. M.; Luque, R.; Romero, A. A. One-Step Microwave-Assisted Asymmetric Cyclisation/Hydrogenation of Citronellal to Menthols Using Supported Nanoparticles on Mesoporous Materials. *Org. Biomol. Chem.* **2010**, *8*, 2845–2849.

94. Pablo, Ó.; Guijarro, D.; Yus, M. Microwave-Enhanced Asymmetric Transfer Hydrogenation of N-(Tert-Butylsulfinyl)Imines. *Eur. J. Org. Chem.* **2014**, 7034–7038.
95. Gustafsson, T.; Hedenström, M.; Kihlberg, J. Synthesis of a C-Glycoside Analogue of β-D-Galactosyl Hydroxylysine and Incorporation in a Glycopeptide from Type II Collagen. *J. Org. Chem.* **2006**, *71*, 1911–1919.
96. Arena, G.; Barreca, G.; Carcone, L.; Cini, E.; Marras, G.; Nedden, H. G.; Rasparini, M.; Roseblade, S.; Russo, A.; Taddei, M.; Zanotti-Gerosac, A. Rhodium-Catalyzed Enantioselective Hydrogenation of (*E*)-Enol Acetate Acids. *Adv. Synth. Catal.* **2013**, *355*, 1449–1454.
97. Wan, S.; Gunaydin, H.; Houk, K. N.; Floreancig, P. E. An Experimental and Computational Approach to Defining Structure/Reactivity Relationships for Intramolecular Addition Reactions to Bicyclic Epoxonium Ions. *J. Am. Chem. Soc.* **2007**, *129*, 7915–7923.
98. Singh, S.; Guiry, P. J. Microwave-Assisted Synthesis of Substituted Tetrahydropyrans Catalyzed by $ZrCl_4$ and Its Application in the Asymmetric Synthesis of Exo- and Endo-Brevicomin. *J. Org. Chem.* **2009**, *74*, 5758–5761.
99. Fournier, J.; Lozano, O.; Menozzi, C.; Arseniyadis, S.; Cossy, J. Palladium-Catalyzed Asymmetric Allylic Alkylation of Cyclic Dienol Carbonates: Efficient Route to Enantioenriched γ-Butenolides Bearing an All-Carbon α-Quaternary Stereogenic Center. *Angew. Chem. Int. Ed.* **2013**, *52*, 1257–1261.
100. Dioos, B. M. L.; Jacobs, P. A. Microwave-Assisted Cr(Salen)-Catalysed Asymmetric Ring Opening of Epoxides. *J. Catal.* **2005**, *235*, 428–430.
101. Kureshy, R. I.; Singh, S.; Khan, N.-U. H.; Abdi, S.; Agrawal, H. R. S.; Mayani, V. J.; Jasra, R. V. Microwave-Assisted Asymmetric Ring Opening of Meso-Epoxides With Aromatic Amines Catalyzed by a Ti-S-(−)-BINOL Complex. *Tetrahedron Lett.* **2006**, *47*, 5277–5279.
102. Guryev, A. A.; Anokhin, M. V.; Averin, A. D.; Beletskaya, I. P. Polymer-Immobilized α,α-Bis[Bis-3,5-(Trifluoromethyl)Phenyl]Prolinol Silylether: Synthesis and Application in the Asymmetric a-Amination of Aldehydes. *Mendeleev Commun.* **2015**, *25*, 410–411.
103. Ghomia, J. S.; Zahedi, S. Power Ultrasound, Microwaves, and Nanomagnetite Organocatalyst: a Comparison Protocol in Anti-Selective Aldol and Mannich Reaction. *Polycyclic Aromat. Comp.* **2016**, *38*, 338–345.
104. O'Brien, T. F.; Bommoraju, T. V.; Hine, F. *Handbook of Chlor-Alkali Technology*; Springer: Dordecht, Netherlands, 2005.
105. American Chemical Society National Historic Chemical Landmarks. *Hall Process: Production and Commercialization of Aluminum*; 2020. http://www.acs.org/content/acs/en/education/whatischemistry/landmarks/aluminumprocess.html(Accessed 14 July 2020).
106. Seo, E. T.; Nelson, R. F.; Fritsch, J. M.; Marcoux, L. S.; Leedy, D. W.; Adams, R. N. Anodic Oxidation Pathways of Aromatic Amines. Electrochemical and Electron Paramagnetic Resonance Studies. *J. Am. Chem. Soc.* **1966**, *88*, 3498–3503.
107. Semmelhack, M. F.; Chou, C. S.; Cortes, D. A. Nitroxyl-Mediated Electrooxidation of Alcohols to Aldehydes and Ketones. *J. Am. Chem. Soc.* **1983**, *105*, 4492–4494.
108. Iversen, P. E.; Lund, H. Electrolytic Generation of Strong Bases I. Wittig Reaction. *Tetrahedron Lett.* **1969**, *10*, 3523–3524.
109. Uneyama, K.; Isimura, A.; Fujii, K.; Torii, S. Electrogenerated Acid as a Powerful Catalyst for Transformation of Epoxides to Ketones and Acetonides. *Tetrahedron Lett.* **1983**, *24*, 2857–2860.
110. Watkins, B. F.; Behling, J. R.; Kariv, E.; Miller, L. L. Chiral Electrode. *J. Am. Chem. Soc.* **1975**, *97*, 3549–3550.
111. Zhu, L.; Xiong, P.; Mao, Z.-Y.; Wang, Y.-H.; Yan, X.; Lu, X.; Xu, H.-C. Electrocatalytic Generation of Amidyl Radicals for Olefin Hydroamidation: Use of

Solvent Effects to Enable Anilide Oxidation. *Angew. Chem. Int. Ed.* **2016**, *55*, 2226–2229.
112. Fu, N.; Li, L.; Yang, Q.; Luo, S. Catalytic Asymmetric Electrochemical Oxidative Coupling of Tertiary Amines with Simple Ketones. *Org. Lett.* **2017**, *19*, 2122–2125.
113. Tang, F.; Chen, C.; Moeller, K. Electrochemistry and Umpolung Reactions: New Tools for Solving Synthetic Challenges of Structure and Location. *Synthesis* **2007**, 3411–3420.
114. Okada, Y.; Akaba, R.; Chiba, K. Electrocatalytic Formal [2 + 2] Cycloaddition Reactions between Anodically Activated Aliphatic Enol Ethers and Unactivated Olefins Possessing an Alkoxyphenyl Group. *Org. Lett.* **2009**, *11*, 1033–1035.
115. Kronenwetter, H.; Husek, J.; Etz, B.; Jones, A.; Manchanayakage, R. Electrochemical Pinacol Coupling of Aromatic Carbonyl Compounds in a [BMIM][BF_4]-H_2O Mixture. *Green Chem.* **2014**, *16*, 1489–1495.
116. Reckenthäler, M.; Griesbeck, A. G. Photoredox Catalysis for Organic Syntheses. *Adv. Synth. Catal.* **2013**, *355*, 2727–2744.
117. Chen, J.-R.; Hu, X.-Q.; Lu, L.-Q.; Xiao, W.-J. Visible Light Photoredox-Controlled Reactions of N-Radicals and Radical Ions. *Chem. Soc. Rev.* **2016**, *45*, 2044–2056.
118. Prier, C. K.; Rankic, D. A.; MacMillan, D. W. C. Visible Light Photoredox Catalysis With Transition Metal Complexes: Applications in Organic Synthesis. *Chem. Rev.* **2013**, *113*, 5322–5363.
119. Narayanam, J. M. R.; Stephenson, C. R. J. Visible Light Photoredox Catalysis: Applications in Organic Synthesis. *Chem. Soc. Rev.* **2011**, *40*, 102–113.
120. Huo, H.; Wang, C.; Harms, K.; Meggers, E. Enantioselective, Catalytic Trichloromethylation Through Visible-Light-Activated Photoredox Catalysis With a Chiral Iridium Complex. *J. Am. Chem. Soc.* **2015**, *137*, 9551–9554.
121. Ma, J.; Rosales, A. R.; Huang, X.; Harms, K.; Riedel, R.; Wiest, O.; Meggers, E. Visible-Light-Activated Asymmetric β-C-H Functionalization of Acceptor-Substituted Ketones with 1,2-Dicarbonyl Compounds. *J. Am. Chem. Soc.* **2017**, *139*, 17245–17248.
122. Huang, X.; Quinn, T. R.; Harms, K.; Webster, R. D.; Zhang, L.; Wiest, O.; Meggers, E. Direct Visible-Light-Excited Asymmetric Lewis Acid Catalysis of Intermolecular [2 +2] Photocycloadditions. *J. Am. Chem. Soc.* **2017**, *139*, 9120–9123.
123. Skubi, K. L.; Kidd, J. B.; Jung, H.; Guzei, I. A.; Baik, M.-H.; Yoon, T. P. Enantioselective Excited-State Photoreactions Controlled by a Chiral Hydrogen-Bonding Iridium Sensitizer. *J. Am. Chem. Soc.* **2017**, *139*, 17186–17192.
124. Nicewicz, D. A.; MacMillan, D. W. C. Merging Photoredox Catalysis With Organocatalysis: The Direct Asymmetric Alkylation of Aldehydes. *Science* **2008**, *322*, 77–80.
125. Welin, E. R.; Warkentin, A. A.; Conrad, J. C.; Macmillan, D. W. C. Enantioselective α-Alkylation of Aldehydes by Photoredox Organocatalysis: Rapid Access to Pharmacophore Fragments from β-Cyanoaldehydes. *Angew. Chem. Int. Ed.* **2015**, *54*, 9668–9672.
126. Capacci, A. G.; Malinowski, J. T.; McAlpine, N. J.; Kuhneand, J.; MacMillan, D. W. C. Direct, Enantioselective α-Alkylation of Aldehydes Using Simple Olefins. *Nat. Chem.* **2017**, *9*, 1073–1077.
127. Huang, X.; Webster, R. D.; Harms, K.; Meggers, E. Asymmetric Catalysis With Organic Azides and Diazo Compounds Initiated by Photoinduced Electron Transfer. *J. Am. Chem. Soc.* **2016**, *138*, 12636–12642.
128. Wei, G.; Zhang, C.; Bures, F.; Ye, X.; Tan, C.-H.; Jiang, Z. Enantioselective Aerobic Oxidative C(sp3)-H Olefination of Amines Via Cooperative Photoredox and Asymmetric Catalysis. *ACS Catal.* **2016**, *6*, 3708–3712.
129. Hou, M.; Lin, L.; Chai, X.; Zhao, X.; Qiao, B.; Jiang, Z. Enantioselective Photoredox Dehalogenative Protonation. *Chem. Sci.* **2019**, *10*, 6629–6634.

130. Liang, J.; Mundorff, E.; Voladri, R.; Jenne, S.; Gilson, L.; Conway, A.; Wong, J.; Huisman, G.; Truesdell, S.; Lalonde, J. Highly Enantioselective Reduction of a Small Heterocyclic Ketone: Biocatalytic Reduction of Tetrahydrothiophene-3-One to the Corresponding (*R*)-Alcohol. *Org. Process Res. Dev.* **2010**, *14*, 188–192.
131. Bornscheuer, U. Y.; Huisman, G. W.; Kazlauskas, R. J.; Lutz, S.; Moore, J. C.; Robins, K. Engineering the Third Wave of Biocatalysis. *Nature* **2012**, *485*, 185–194.
132. Biegasiewicz, K. F.; Cooper, S. J.; Emmanuel, M. A.; Miller, D. C.; Hyster, T. K. Catalytic Promiscuity Enabled by Photoredox Catalysis in Nicotinamide-Dependent Oxidoreductases. *Nat. Chem.* **2018**, *10*, 770–775.
133. Takacs, L. The Historical Development of Mechanochemistry. *Chem. Soc. Rev.* **2013**, *42*, 7649–7659.
134. Stuart, J. L.; Adams, C. J.; Bolm, C.; Braga, D.; Collier, P.; Friscic, T.; Grepioni, F.; Harris, K. D. M.; Hyett, G.; Jones, W.; Krebs, A.; Mack, J.; Maini, L.; Orpen, A. G.; Parkin, I. P.; Shearouse, W. C.; Steed, J. W.; Waddell, D. C. Mechanochemistry: Opportunities for New and Cleaner Synthesis. *Chem. Soc. Rev.* **2012**, *41*, 413–447.
135. McNaught, A. D.; Wilkinson, A. *Compendium of Chemical Terminology*; IUPAC Recommendations; IUPAC: Zürich, Switzerland, 1997.
136. Stolle, A.; Szuppa, T.; Leonhardt, S. E. S.; Ondruschka, B. Ball Milling in Organic Synthesis. Solutions and Challenges. *Chem. Soc. Rev.* **2011**, *40*, 2317–2329.
137. Rodríguez, B.; Bruckmann, A.; Rantanen, T.; Bolm, C. Solvent-Free Carbon-Carbon Bond Formations in Ball Mills. *Adv. Synth. Catal* **2007**, *349*, 2213–2233.
138. Thorwirth, R.; Stolle, A.; Ondruschka, B.; Wild, A.; Schubert, U. S. Fast, Ligand- and Solvent-Free Copper-Catalyzed Click Reactions in a Ball Mill. *Chem. Commun.* **2011**, *47*, 4370–4372.
139. Thorwirth, R.; Stolle, A.; Ondruschka, B. Fast Copper-, Ligand- and Solvent-Free Sonogashira Coupling in a Ball Mill. *Green Chem.* **2010**, *12*, 985–991.
140. Schneider, F.; Ondruschka, B. Mechanochemical Solid-State Suzuki Reactions Using an In Situ Generated Base. *ChemSusChem* **2008**, *1*, 622–625.
141. Toma, S.; Sebesta, R.; Meciarova, M. Organocatalytic Reactions Under Unusual Conditions. *Curr. Org. Chem.* **2011**, *15*, 2257–2281.
142. Bruckmann, A.; Krebs, A.; Bolm, C. Organocatalytic Reactions. Effects of Ball Milling, Microwave and Ultrasound Irradiation. *Green Chem.* **2008**, *10*, 1131–1141.
143. Burk, L.; Gliem, M.; Mülhaupt, R. Mechanochemical Routes to Functionalized Graphene Nanofillers Tuned for Lightweight Carbon/Hydrocarbon Composites. *Macromol. Mater. Eng.* **2019**, *304*, 1800496–1800582.
144. Hernandez, J. G.; Juaristi, E. Asymmetric Aldol Reaction Organocatalyzed by (*S*)-Proline-Containing Dipeptides: Improved Stereo Induction under Solvent-Free Conditions. *J. Org. Chem.* **2011**, *76*, 1464–1467.
145. Tanaka, K.; Asakura, A.; Muraoka, T.; Kalickib, P.; Urbanczyk-Lipkowska, Z. Asymmetric Direct Aldol Reactions Catalyzed by Chiral Amine Macrocycle–Metal(II) Complexes under Solvent-Free Conditions. *New J. Chem.* **2013**, *37*, 2851–2855.
146. Wang, Y.-F.; Chen, R.-X.; Wang, K.; Zhang, B.-B.; Li, Z.-B.; Xu, D.-Q. Fast, Solvent-Free and Hydrogen-Bonding-Mediated Asymmetric Michael Addition in a Ball Mill. *Green Chem.* **2012**, *14*, 893–895.
147. Jörres, M.; Mersmann, S.; Raabe, G.; Bolm, C. Organocatalytic Solvent-Free Hydrogen Bonding-Mediated Asymmetric Michael Additions under Ball Milling Conditions. *Green Chem.* **2013**, *15*, 612–616.
148. Staleva, P.; Hernández, J. G.; Bolm, C. Mechanochemical Copper-Catalyzed Asymmetric Michael-Type Friedel-Crafts Alkylation of Indoles With Arylidene Malonates. *Chem. A Eur. J.* **2019**, *25*, 9202–9205.

149. Yu, J.; Jiang, Z.; Su, W. Cross Dehydrogenative Coupling Reactions by Ball Milling. In *Ball Milling Towards Green Synthesis*; Ranu, B. C., Stolle, A., Eds.; RSC Green Chemistry Series, Vol. 31; 2015; pp. 96–113.
150. Li, Z.; Jiang, Z.; Su, W. Fast, Solvent-Free, Highly Enantioselective Three-Component Coupling of Aldehydes, Alkynes, and Amines Catalyzed by the Copper (II)Pybox Complex Under High-Vibration Ball-Milling. *Green Chem.* **2015**, *17*, 2330–2334.

CHAPTER SIXTEEN

High hydrostatic pressure in food industry applications

Kazutaka Yamamoto
Head of Laboratory, Food Process Engineering, Division of Food Processing and Distribution, Food Research Institute, National Agriculture and Food Research Organization, Tsukuba, Ibaraki, Japan

1 Introduction

Mankind had been processing foods by using heat for millennia. Food processing and cooking were almost equal to the active use of heat or temperature as a thermal parameter. The thermal process gives foods unique flavors via roasting and grilling and/or extended shelflives via pasteurization. In modern society, people started to use microwave and electric ovens and induction heating, which are thermal processors replacing fire at home.

In 1987 Dr. Rikimaru Hayashi suggested in a paper in Japanese[1] to apply high hydrostatic pressures (HHPs) of several hundred MPa ($\sim$several thousand atm) for food processing, and soon after, pressure joined food processing as a nonthermal parameter. As a comparison, an underwater depth of 10,000 m, which is found at the deepest bottom of the Challenger Deep of the Mariana Trench in the Pacific Ocean, is comparable with 100 MPa. In HHP food processing, pressures of 100–600 MPa, especially 600 MPa, have been adopted by the food industry. Thereafter, research and development on the application of HHP in food processing has become particularly active in Japan. Following this stream the research and development on HHP food processing has also spread out to the world.

Much earlier than the dawn of HHP food processing, application of low pressure in treating food could be seen during the era of Napoleon Bonaparte. The idea was originated from the bottled foods invented by Nicolas Appert at that time. Pressure seemed to act as a processing parameter after the glass-bottled foods became canned foods. Foods were hermetically sealed in a can and then heated. Vapor in the can raised the inner pressure to several atm, and the foods were pasteurized/sterilized. Based on the canned food technology, sterilized foods have been marketed as canned and retort foods.

Nontraditional Activation Methods in Green and Sustainable Applications
https://doi.org/10.1016/B978-0-12-819009-8.00010-4
© 2021 Elsevier Inc.
All rights reserved.

Similarly, cooking time can be reduced by home pressure cooker. These technologies are characterized as the passive use of pressure where a pressure of several atm is generated by heating in a closed system and the pressure plays a peripheral role next to heat.

The effect of high static/hydrostatic pressure on living organisms and biological materials has been studied as a basic science since late 19th century.[2] Thereafter, phase diagram of water (1912)[3] and denaturation of egg white albumin (1914)[4] were presented by Prof. P. W. Bridgman, the father of high-pressure science and technology. Relationship between high pressure treatment and microbial preservability of fruits and vegetables was reported in 1914.[5] More than a half century later, the aforementioned suggestion[1] ignited the research and development of HHP toward its commercial applications in the food industry, and, thereafter, high-pressure science and technology with regard to food and biology have developed with dramatic pace.

Aside from HHP food processing, application of high dynamic pressure (HDP) including shock waves has been studied toward commercialization.[6] Therefore, attention should be paid to distinguish HHP or high static pressure (HSP) and HDP in food processing. An acronym of HPP, which stands for high pressure processing, is often used in the food industry. However, it includes both HHP/HSP and HDP, and it may lead food technologists to confusions in the future.

2 Difference between HHP and heat

In food processing, heat accelerates chemical reactions while producing new components related to color and flavor or damaging beneficial components.

In contrast, HHP does not accelerate chemical reactions in general, and it instead induces physical changes in molecules microscopically and/or macroscopically. For instance, from a microscopic aspect, HHP restricts molecular motions while breaking hydrogen bonds toward filling gaps among packed molecules. From a macroscopic aspect, HHP dissolves air bubbles in gas-liquid dispersion system into the liquid phase. HHP food processing utilizes those sorts of physical changes.

HHP process does not necessarily require active heating, although passive heating by adiabatic compression and passive cooling by adiabatic decompression are inevitable.[7] Therefore, HHP process is neither a nonheating process nor, strictly speaking, a nonthermal process. However, in a broad sense, it is, as well as irradiation and low temperature plasma process, categorized into a group of nonthermal processes due to its passive aspect of heating.

Change in temperature during compression or decompression in HHP food processing depends on the rate of compression or decompression, dimensions of the HHP vessel, compressibility of pressure medium, and others. Water, mostly used as pressure medium for food processing, is compressed by a few percent, which is significantly lower than that in the case of a gas. An ideal gas can be compressed by 1/1000 when compressed under 1000 atm (c. 100 MPa). Accordingly the heat generated by adiabatic compression of water will be much smaller than that of gases. However, in some specific HHP treatments, a relatively large temperature increase by c. 10°C/ 100 MPa can be observed. Some may concern about the effect of coexisting air bubbles in pouched foods during the adiabatic heating. However, most of the raw food materials contain sufficient amount of water, and it dissolves air. Therefore, the heating due to gas compression is reduced significantly. Other than the adiabatic heating, air bubbles in a pouched food may be problematic due to the prolonged compression time and thus decreased number of process cycles. The food industry, in general, has a mission to process foods in a large amount at low price, and thus it is wise to minimize the amount of air in the pouched food.

Thermal food preservation requires relatively high energy consumption, since heating and cooling processes need continuous energy input and removal of heat, respectively. On the other hand, HHP processes are environmentally friendly as compared with thermal process: once pressure is raised, no energy consumption is required to hold the pressure without pressure leakage.

3 Research and development for HHP food processing

The effect of heat in food processing on microbial inactivation and quality has been conventionally and intensively studied, and an extensive amount of useful information is available. However, the effect of pressure, especially HHP, on the characteristics of foods is significantly less studied than that of heat.

Industrial high pressure installations such as hot isostatic press (HIP) and cold isostatic press (CIP) have been modified and utilized as HHP food processors after the suggestion by Dr. Hayashi in 1987.[1] Implementation of low temperature control and hygienic processing into HHP food processors were studied thereafter.

The first commercial HHP-processed foods in the world were marketed as fruit products such as fruit jams in 1990 in Japan[8, 9] (Fig. 1). The HHP process was not aiming at inactivation of endogenous enzymes but microbial control, and thus the products were distributed and stored under refrigerated conditions so that the enzymes did not act on the pectic substances, which contribute to the texture of the fruit pastes. The HHP-processed fruit products had long shelf lives while retaining the fresh attributes, and thus they attracted much attention as novel foods.

Since then, HHP food processing has been studied mostly for microbial control of foods with retained freshness. One of the early applications was rice cake mixed with a green herb (Japanese mugwort: *Artemisia indica* var. *maximowiczii*). Fruits and vegetables including herbs are often contaminated with microbes, and thus they are HHP-treated for enhanced food safety instead of heat treatment, which damages fresh color, flavor, and nutrients of the herbs. HHP-treated rice products[11] are also commercialized by a Japanese company (Fig. 2). They treated rice with HHP for improved water impregnation into the grains so to improve the texture. It is

Fig. 1 Strawberry, apple, and blueberry jams as the first commercial HHP-processed foods in the world[10]: kiwifruit jam was in the line at the first commercialization, and the blueberry jam replaced it later.

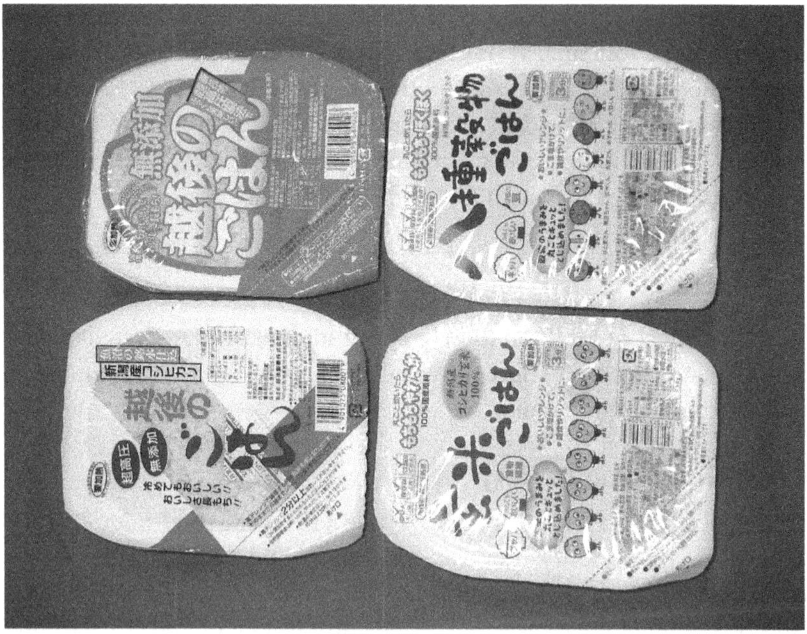

Fig. 2 Retort rice products cooked after HHP-enhanced water impregnation.[10]

noteworthy that the HHP treatment is not aiming at pressure-induced gelatinization of rice starch but the water impregnation into the rice grains.

HHP food processing in Japan was mostly aiming at replacing conventional pasteurization as a microbial control in its early applications, and an extensive effort was devoted to the research and development on microbial inactivation. However, in the Japanese food industry, it was gradually recognized that spores of bacteria such as *Bacilli* and *Clostridia* are quite resistant to HHP pasteurization and that the initial investment for HHP food processors was too high for small and medium-sized food companies, which dominate Japanese food industry in terms of the number. Accordingly, many of the interested parties have given up the research and development and/or withdrawn from the market. Thereafter, HHP food processing seemed to lose the early enthusiasms in Japan, while some food companies in other countries continued to work on the development intensively.

On the other hand, the effect of HHP on various organisms and biomolecules has been studied steadily in the world. In fact, international conferences specialized on the effect of high pressure on bioscience and biotechnology have been organized. Other than the International Conference on High Pressure Bioscience and Biotechnology (HPBB), HHP food processing has been discussed as an emerging technology in other international academic fora such as Institute of Food Technologists (IFT), International Association for the Advancement of High Pressure Science and Technology/Association Internationale pour l'Avancement de la Recherche et de la Technologie aux Hautes Pressions (AIRAPT), and the European High Pressure Research Group (EHPRG).

4 Effect of HHP on food

The effect of HHP on food has been studied from several viewpoints such as microbial food safety, physical changes in food components, and food nutrition and functionality.

4.1 Microbial food safety

Food safety is the most important in food processing, and microbial food safety is the first priority. Ensuring microbial food safety is often referred to as pasteurization and sterilization. However, it is recommended to describe the phenomena as lethal inactivation (leading to death) and sublethal inactivation (leading to injury), since the events of microbial injury and recovery should be considered, especially in pasteurization. When the microbes are lethally inactivated, they are judged as dead, and thus pasteurization (partial population death) and sterilization (total population death) describe the phenomenon properly. On the other hand, when the microbes are sublethally inactivated, they are considered to be injured.

Studies on HHP inactivation of microbes, especially bacteria, have a long history in HHP food science since 1899,[2] and the knowledge has drastically increased since the first commercialization of HHP food processing in 1990.[8, 9] In the past decades, HHP-injured bacteria that are not detected immediately after HHP treatment have been studied intensively to reveal their recovery behaviors during the storage of HHP-processed foods.[12]

HHP food processing inactivates microbes lethally and/or sublethally while minimizing the degradation of flavor, color, and nutritional and functional components, and it enables the production of minimally processed high quality, safe foods. As commercially available HHP-processed products aiming at microbial inactivation, HHP-processed meat products followed the HHP-processed fruit products in 1998.[13] In conventional large-scale meat processing, thermal process has been indispensable for primary pasteurization objectives where pathogenic microbes are inactivated. Following the primary pasteurization, block hams are sliced, chain sausages are cut into pieces, and the resulting slices and sausage pieces are pouched. In these processes, spoilage microbes such as lactic acid bacteria may contaminate the products, and thus food additives are used to suppress their growth and to extend the shelf life. Lactic acid bacteria may produce unfavorable acids and polysaccharides, which reduce pH and produce viscous materials, respectively. As a significant green advantage of the HHP treatment, the food additives can be replaced with HHP process as a secondary pasteurization method, which may inactivate the spoilage microbes and extend the shelf life without food additives. Recently the demand for HHP-processed beverages such as juice, smoothie, and coconut water are increasing rapidly and large-scale HHP food processors (e.g., 420 L, 525 L, and 2 × 525 L) have been installed for the commercial production. The market of HHP-processed beverages has been continuously growing over these years in many countries such as the United States, European countries, Korea, and Taiwan. As of 2020, there are no international standards for trading HHP-processed foods, and, therefore, it should be requested that the Codex Alimentarius Commission establish relevant standards and recommendations, which would contribute to avoid international controversy on non-tariff barrier issues raised by the World Trade Organization.

Bacterial spores are resistant to a commercial HHP (600 MPa) at low temperatures (e.g., 10°C), but combinations of HHP and high temperatures (e.g., 80°C) may inactivate the spores.[14] Furthermore, shock wave, which can be generated when an HHP is released instantly, is reported to be effective to inactivate bacterial spores.[15] However, no commercial application of shock wave has been found in the food industry. On the other hand, it is reported that germination of bacterial spores can be induced by medium HHPs (MHHPs) of 100–200 MPa or HHPs of 500–600 MPa[16]. Once the spores germinate to vegetative cells, they can be inactivated by conventional thermal processes. The same is applicable to non-spore-formers such as *Escherichia coli* and *Listeria monocytogenes*. Therefore, the resulting vegetative cells can be killed after spore germination is induced by MHHP. Another option is "suicidal germination" where MHHP induces spore germination into vegetative cells at medium or high temperature at around 55–75°C, which kills the germinated cells upon germination.[17]

The effect of HHP on microbes has been studied intensively to understand the mechanisms of HHP-induced microbial inactivation. Recent developments in genomics, transcriptomics, proteomics, and metabolomics may contribute to identify specific biomolecules that are upregulated or downregulated by HHP treatment. Those methodologies may be applied to elucidate the pressure tolerance of deep-sea microorganisms and others.

4.2 Physical changes in food components

HHP treatment may induce microscopic changes of food biopolymers such as starch gelatinization, protein denaturation, and lipid phase transition.

Starch is a major food component, contributing as energy source and gel-forming texture modifier. Heat gelatinization of starch can be induced when heated sufficiently in the presence of water. Likewise, pressure gelatinization can be induced when it is HHP-treated. The effect of pressure and temperature on various starches of different botanical origins has been studied since 1929.[18] Susceptibility to enzymatic degradation of HHP-treated starch is enhanced,[19] and HHP-treated starch shows unique viscoelastic behaviors.[20] Retrogradation, where gelatinized starch recrystallizes and stiffens, is a slow event for heat-gelatinized starch. However, rapid retrogradation[21] of HHP-gelatinized starch can be observed immediately after an HHP process. Systematic understanding of pressure gelatinization and rapid retrogradation was enhanced by depicting state diagrams of HHP-treated starch-water mixtures.[22, 23] Especially, state diagrams of HHP-treated mixtures of potato starch and water at various water contents[24–26] will help food technologists establish desired operating conditions to process HHP-treated starch.

Various food-based proteins have been studied in terms of HHP-induced physical changes: gelation of milk proteins such as α-lactoglobulin,[27] β-lactoglobulin,[27] and whey protein isolate[28]; gelation of egg protein[27]; and gel-sol transition of actomyosin.[29, 30]

Lipids are also important food components in high fat foods such as chocolates and are indispensable substances in the cell membrane. HHP can induce various phase transitions of lipids, and the phase transitions can be controlled by temperature and pressure: solid-liquid phase transition of oils such as cacao butter, soybean oil, and palm oil[31] and gel-liquid crystal transition of double-layered membrane of phospholipids.[32]

4.3 Physical changes in foods

Macroscopic physical changes can be observed mostly in MHHP treatments: freezing and thawing at subzero temperatures, enhanced water impregnation, extraction, and shucking of bivalves and crustaceans.

Ice I is the most popular ice crystal among others, since pure water freezes into Ice I at 0°C under the atmospheric pressure of 0.1 MPa. According to the phase diagram of water,[3] ice I at −10°C can be thawed when pressurized over 100 MPa. In contrast, pure water at higher than 0°C at 100 MPa can be frozen when it is cooled to lower than −10°C at 100 MPa. As practical applications of the phase transition of water, pressure-shifted thawing of tofu[33] and pressure-induced freeze denaturation of agar and carrageenan gels[34] were reported. It should be noted that water in food is usually not chemically pure, but it dissolves many ions and other components.

The traditional process to impregnate dry grains with water is time consuming and space demanding. Dry soybean has to be soaked in water for a long time of ca. 16–18 h before other successive processes such as steaming. HHP-accelerated water impregnation can shorten the time. HHP can reduce soaking time of brown rice before its cooking to 30 min from 12 h[35]. Another advantage of the HHP impregnation is that it minimizes microbial contamination, while traditional soaking might allow microbial growth.

HHP treatment may reduce allergens via extraction. Hypoallergenic cooked rice and wheat bread were developed by HHP-enhanced extraction of rice and wheat allergens, respectively.[36] It is also expected to promote digestions of allergens with proteolytic enzymes while unfolding the proteins under moderate HHP. Partially digested allergens might be used for oral desensitization.[37]

HHP can open bivalves and detach their flesh,[38] although the mechanism is not yet confirmed. However, it might be ascribed to the difference between high compressibility of flesh and low compressibility of shell that may separate the soft flesh including adductor from the hard shell. HHP-shucked oysters, clams, and lobsters have been commercialized for raw servings and further processes. The shucking process has some advantages over conventional one. Conventional shucking causes contamination of shell pieces, which are not perfectly avoidable even by introducing x-ray monitoring. However, the risk of shell piece contamination is significantly reduced by adopting HHP shucking. Conventional shucking requires time and labor of shuckers, but HHP shucking can minimize both. Recently, it is getting difficult to recruit shuckers in Japan since they are getting old, and younger workers do not wish to work in this profession. Accordingly, HHP shucking may draw increasing attention in the near future.

Some obligatory piezophiles, which can optimally grow under 100 MPa, are found in the deep sea.[39] However, growth of most miscellaneous microbes are suppressed at an MHHP of 60.8 MPa (600 atm),[40] although some lactic acid bacteria and heat-resistant bacteria can grow under those conditions.[41] Meanwhile, high concentration of salt may inhibit enzyme activities in food, and thus fermentation of salty products such as soy sauce and miso (Japanese fermented soybean paste) requires months of fermentation and maturation periods. However, the enzyme activities are drastically improved under MHHP where no salt is required to inhibit unfavorable microbial growth during enzymatic degradation of perishable food materials. Taking advantage of the principle, novel methods were suggested to produce fish sauce or soy sauce without adding salt, which is indispensable in conventional process of fish sauce and soy sauce fermentations.[42, 43] Since yeasts are not able to grow and no yeast fermentation occurs under MHHP, characteristic flavors of fish and soy sauces are not produced. However, the products with no added salt may contribute to improve the quality of life of the patients suffering from kidney disease. The MHHP-accelerated enzymatic process has been applied to produce various broths and hydrolysates for food and cosmetic uses.

4.4 Biochemical changes in foods

The first commercial HHP-processed foods were fruit products with fresh attributes and long shelf life. For instance, HHP-processed jam[9] retained significantly high level of flavor compounds such as linalool and 2-methylbutyric acid while suppressing the increase in unfavorable furfural during refrigeration.

Enzymes, playing important roles in biochemical changes in HHP-processed foods, may be inactivated by HHP and heat, depending on the levels of pressure and temperature, respectively.[44] Combination of HHP and heat may enhance the inactivation of enzymes such as alkaline phosphatase, lipoxygenase, myrosinase, hydroperoxide lyase, β-glucosidase, pectin methylesterase, peroxidase, trimethylamine-N-oxide demethylase, polygalacturonase, polyphenol oxidase, and xylanase. The inactivation of enzymes is mostly irreversible, and the inactivation efficacy is largely dependent on the enzyme types and sources. Denaturation of enzymes may occur reversibly at an MHHP of less than 200 MPa, being followed by slow refolding process after pressure release. At pressures beyond 300 MPa, irreversible protein denaturation may take place.

5 HHP food processing equipment

In MHHP and HHP food processing, MHHPs of 100–200 MPa and HHPs of 400–600 MPa are applied in general. Industrial installations were modified for food use since the 1990s. Two options are available in terms of the compression type. Direct compression type intrudes a piston into the vessel so as to reduce the volume and thus raise the pressure. Indirect compression type instruments adopt exterior compression system that pressurizes the pressure medium using one or more pressure intensifiers and intrudes the pressurized medium into the vessel for compression. The former is sometimes used for laboratory studies, and the latter is for industrial applications. As industrial applications, vertical and horizontal types are commercialized (Fig. 3). In the early times of HHP food processing, the vertical type was dominant. However, the horizontal type dominates recent installments with large vessels of more than 100 L. Food processing has a mission to produce a large amount product at low cost, and thus HHP food processing requires

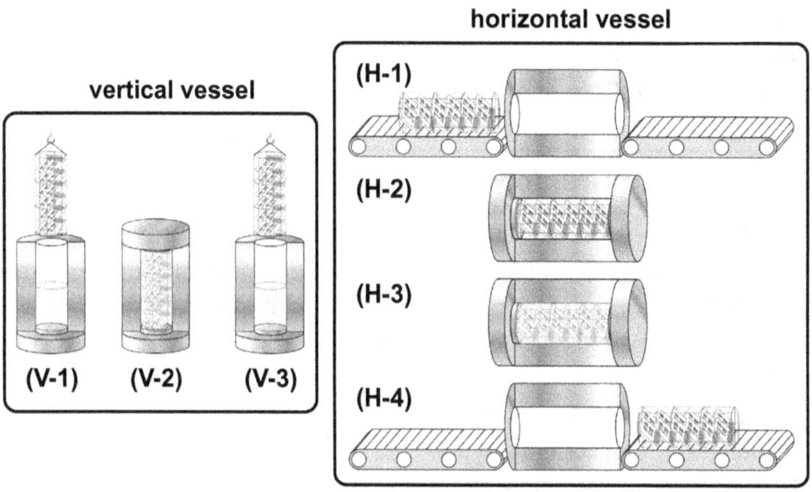

Fig. 3 HHP food processing equipment of vertical and horizontal installations.[45] V-1, soaking the basket containing bagged foods into water (pressure medium) in the vessel; V-2, HHP treatment; V-3, taking out the basket; H-1, conveying the basket containing bagged foods to the vessel; H-2, filling the vessel with water using an ordinary water pump; H-3, HHP treatment using an HHP pump; H-4, recovering water into a reservoir for reuse after pressure release and conveying out the basket.

large vessels, which are expensive as compared with conventional processors. For increasing the vessel volume, it is required to increase either the diameter or the length. Larger diameter requests exponentially thick vessel wall, raising the investment. Longer vessel may ensure larger volume and higher height, requiring a ceiling by at least two times higher than the vessel to lift the food cage over the vessel. Therefore, a large vessel of horizontal type is more realistic, and accordingly, its industrial application is growing in recent years. In the horizontal equipment, water as pressure medium must be drained from the vessel after each treatment. However, a high-speed pump recycles most of the drained water, and the loss of water is minimized. As of September 2020, the largest volume of one HHP food processor is 525 L.

Large HHP food processing equipment is ideal for food processing. However, small and medium-sized companies dominate the food industry in terms of the number of the companies. Therefore, it is very difficult or often impossible for most of these companies to install such an expensive equipment into their factories. Solution may be toll processing or tolling system where they can use HHP food processing equipment without capital investments for the equipment. All they have to do is to pay for the processing per volume, batch, or time. HHP toll processing of food is recently growing in the world. Toll processing will be a key for further dissemination of HHP food processing equipment in the world.

6 Cost of HHP food processing[46]

Initial investment for HHP equipment may be estimated from 0.6 to 4 million US dollars. The capital cost of the equipment dominates the investment (75%–80%), and operating costs must be considered: labor (5%–10%), utilities (2%–4%), and space (1%–2%). Volume efficiency of an HHP vessel can be estimated to be 60 % approximately, and thus a vessel of 100 L size may be filled with packages totaling 60 L. However, the efficiency may vary depending on the packaging types. The least efficient types are MAP trays (volume efficacy: 25%–40%), and the most efficient ones are vacuum-sealed bag-in-box big pouches (volume efficacy: 75%–85%). Number of process cycles can be estimated from 8 to 10 cycles/h in an HHP process at 600 MPa for a holding time of 3 min. Processing costs are estimated to range from 0.14 euro/kg (55 L vessel) to 0.071 euro/kg (525 L vessel) with 5-year amortization, 300 working days/year, 16 h/day, 60% volume efficiency, and costs for wear parts and utilities. Thermal processes require continuous heating or freezing, resulting in high energy input. In contrast, HHP processes do not require intensive heating nor freezing but relatively low energy for compression and refrigeration, although the initial investment to the equipment is relatively high. Accordingly, HHP food processing is considered to be one of green technologies and environmentally friendly.

7 Concluding remarks

HHP treatment of food is a contemporary green way of food processing. It has several advantages as compared with traditional protocols: most of all, it requires less energy than thermal processing, while it provides us with high quality food of fresh attributes. Therefore, further dissemination of the technology may contribute to the society with saved energy and improved quality of life by HHP-processed high quality foods. Further study will be required for solving conventional problems in food processing and, in addition, for creating novel properties toward high quality safe foods, which will meet consumers' demands and will open doors to new markets.

Acknowledgments

The author is grateful to the current and past members of his research group for their contributions to extend knowledge in HHP food science and technology and their supports for the research and development.

References

1. Hayashi, R. Possibility to Apply High Pressure to Cooking, Processing, Pasteurization, and Preservation. *Shokuhin to Kaihatsu (Food Process. Ingredients)* **1987**, *22*, 55–62 (in Japanese).
2. Hite, B. H. The Effect of Pressure in the Preservation of Milk. *Bull. W. Va. Univ. Agric. Exp. Stn.* **1899**, *58*, 15–35.
3. Bridgman, P. W. Water, In the Liquid and Five solid Forms, Under Pressure. *Proc. Am. Acad. Arts Sci.* **1912**, 7, 441–558.
4. Bridgman, P. W. The Coagulation of Albumen by Pressure. *J. Biol. Chem.* **1914**, *19*, 511–512.
5. Hite, B. H.; Giddings, N. J.; Weakley, C. E. The Effects of Pressure on Certain Micro-organisms Encountered in the Preservation of Fruits and Vegetables. *Bull. W. Va. Univ. Agric. Exp. Stn.* **1914**, *146*, 3–67.
6. Iwasa, S.; Maehara, H.; Nishi, M.; Tanaka, S.; Watanabe, T.; Itoh, S.; Hokamoto, K. Food Processing Trials Using Underwater Shock Wave by Electrical Discharge. *Mater. Sci. Forum* **2013**, *767*, 223–228.
7. Grauwet, T.; Van der Plancken, I.; Vervoort, L.; Hendrickx, M.; Van Loey, A. High-Pressure Processing Uniformity. In *High Pressure Processing of Foods*; Balasubramaniam, V. M., Barbosa-Cánovas, G. V., Lelieved, H. L. M., Eds.; Springer: New York, 2016; pp. 253–268.
8. Horie, Y.; Kimura, K.; Ida, M.; Yoshida, Y.; Ohki, K. Jam Preparation by Pressurization. *Nippon Nogeikagaku Kaishi (J. Agric. Chem. Soc. Jpn.)* **1991**, *65*, 975–980 (in Japanese).

9. Kimura, K.; Ida, M.; Yoshida, Y.; Ohki, K.; Fukumoto, T.; Sakui, N. Comparison of Keeping Quality Between Pressure-processed Jam and Heat-processed Jam: Changes in Flavor Components, Hue, and Nutrients During Storage. *Biosci. Biotechnol. Biochem.* **1994**, *58*, 1386–1391.
10. Yamamoto, K.; Koseki, S. Food Processing by High Hydrostatic Pressure. *Nougyou Gijutsu (J. Agric. Sci.)* **2008**, *63*, 69–77 (in Japanese).
11. Yamazaki, A.; Sasagawa, A. Development of Rice Food Products Processed by High Pressure Treatment. *Nippon Nogeikagaku Kaishi (J. Agric. Chem. Soc. Jpn.)* **2000**, *74*, 619–623 (in Japanese).
12. Koseki, S.; Yamamoto, K. Recovery of *Escherichia coli* ATCC 25922 in Phosphate Buffered Saline After Treatment With High Hydrostatic Pressure. *Int. J. Food Microbiol.* **2006**, *110*, 108–111.
13. Jofré, A.; Serra, X. Processing of Meat Products Utilizing High Pressure. In *High Pressure Processing of Foods*; Balasubramaniam, V. M., Barbosa-Cánovas, G. V., Lelieved, H. L. M., Eds.; Springer: New York, 2016; pp. 591–623.
14. Daryaei, H.; Yousef, A. E.; Balasubramaniam, V. M. Microbiological Aspects of High-pressure Processing of Food: Inactivation of Microbial Vegetative Cells and Spores. In *High Pressure Processing of Foods*; Balasubramaniam, V. M., Barbosa-Cánovas, G. V., Lelieved, H. L. M., Eds.; Springer: New York, 2016; pp. 271–294.
15. Hayakawa, I.; Furukawa, S.; Midzunaga, A.; Horiuchi, H.; Nakashima, T.; Fujio, Y.; Yano, Y.; Ishikura, T.; Sasaki, K. Mechanism of Inactivation of Heat-tolerant Spores of *Bacillus stearothermophilus* IFO 122550 by Rapid Decompression. *J. Food Sci.* **1998**, *63*, 371–374.
16. Black, E. P.; Wei, J.; Atluri, S.; Cortezzo, D. E.; Koziol-Dube, K.; Hoover, D. G.; Setlow, P. Analysis of Factors Influencing the Rate of Germination of Spores of *Bacillus subtilis* by Very High Pressure. *J. Appl. Microbiol.* **2007**, *102*, 65–76.
17. Morimatsu, K.; Nakaura, Y.; Inaoka, T.; Kimura, K.; Yamamoto, K. Effects of Solution pH and Ions on Suicidal Germination of *Bacillus subtilis* Spores Induced by Medium High Temperature-medium High Hydrostatic Pressure Treatment. *Biocontrol Sci.* **2019**, *24* (3), 167–172.
18. Yamamoto, K.; Kawai, K.; Fukami, K.; Koseki, S. Pressure Gelatinization of Potato Starch. *Food* **2009**, *3* (SI 1), 57–66.
19. Hayashi, R.; Hayashida, A. Increased Amylase Digestibility of Pressure-Treated Starch. *Agric. Biol. Chem.* **1989**, *53*, 2543–2544.
20. Katopo, H.; Song, Y.; Jane, J. L. Effect and Mechanism of Ultrahigh Hydrostatic Pressure on the Structure and Properties of Starches. *Carbohydr. Polym.* **2002**, *47*, 233–244.
21. Stute, R.; Klingler, R. W.; Boguslawski, S.; Eshtiaghi, M. N.; Knorr, D. Effects of High Pressure Treatment on Starches. *Starch/Stärke* **1996**, *48*, 399–408.
22. Douzals, J. P.; Perrier-Cornet, J. M.; Coquille, J. C.; Gervais, P. Pressure-temperature Phase Transition Diagram for Wheat Starch. *J. Agric. Food Chem.* **2001**, *49*, 873–876.
23. Bauer, B. A.; Knorr, D. The impact of Pressure, Temperature and Treatment Time on Starches: Pressure-induced Starch Gelatinisation as Pressure Time Temperature Indicator for High Hydrostatic Pressure Processing. *J. Food Eng.* **2005**, *68*, 329–334.
24. Kawai, K.; Fukami, K.; Yamamoto, K. State Diagram of Potato Starch–water Mixtures Treated With High Hydrostatic Pressure. *Carbohydr. Polym.* **2007**, *67*, 530–535.
25. Kawai, K.; Fukami, K.; Yamamoto, K. Effect of Temperature on Gelatinization and Retrogradation in High Hydrostatic Pressure Treatment of Potato Starch-Water Mixtures. *Carbohydr. Polym.* **2012**, *87*, 314–321.
26. Buckow, R.; Yamamoto, K. Pressure Gelatinization of Starch. In *High Pressure Processing of Foods*; Balasubramaniam, V. M., Barbosa-Cánovas, G. V., Lelieved, H. L. M., Eds.; Springer: New York, 2016; pp. 433–460.

27. Dumoulin, M.; Hayashi, R. High Pressure, a Unique Tool for Food Texturization. *Food Sci. Technol. Int. Tokyo* **1998**, *4*, 99–113.
28. Kanno, C.; Mu, T. H.; Hagiwara, T.; Ametani, M.; Azuma, N. Gel formation from Industrial Milk Whey Proteins Under Hydrostatic Pressure: Effect of Hydrostatic Pressure and Protein Concentration. *J. Agric. Food Chem.* **1998**, *46*, 417–424.
29. Ikeuchi, Y.; Tanji, H.; Kim, K.; Suzuki, A. Dynamic Rheological Measurements on Heat-Induced Pressurized Actomyosin Gels. *J. Agric. Food Chem.* **1992**, *40*, 1751–1755.
30. Ikeuchi, Y.; Tanji, H.; Kim, K.; Suzuki, A. Mechanism of Heat-Induced Gelation of Pressurized Actomyosin: Pressure-Induced Changes in Actin and Myosin in Actomyosin. *J. Agric. Food Chem.* **1992**, *40*, 1756–1761.
31. Yasuda, A.; Mochizuki, K.; Shimizu, T.; Suzuki, A. The Behaviors of Triglycerides Under High Pressure. In *Kouatsu Kagaku To Kaatsu Shokuhin (High Pressure Science for Food)*; Hayashi, R., Ed.; San-Ei Shuppan: Kyoto, Japan, 1991; pp. 176–180 (in Japanese with English abstract).
32. Goto, M.; Endo, T.; Yano, T.; Tamai, N.; Kohlbrecher, J.; Matsuki, H. Comprehensive Characterization of Temperature- and Pressure-Induced Phase Transitions for Saturated Phosphatidylcholines Containing Longer Chain Homologs. *Colloids Surf. B: Biointerfaces* **2015**, *128*, 389–397.
33. Fuchigami, M. Texture and Structure of High-Pressure-Frozen Foods. *Kouatsuryoku No Kagaku To Gijutsu (Rev. High Pressure Sci. Technol.)* **1999**, *9*, 191–198 (in Japanese with English abstract).
34. Fuchigami, M.; Teramoto, A. Texture and Structure of High-Pressure-Frozen Food Gels. In *Advances in High Pressure Bioscience and Biotechnology*; Ludwig, H., Ed.; Springer-Verlag: Berlin/Heidelberg, Germany, 1999; pp. 493–496.
35. Yamazaki, A. Story of Development of High-Pressure Processed Foods, Part 2—A Method of Producing Rapidly Cookable Processed Unpolished Rice. *Kouatsuryoku No Kagaku To Gijutsu (Rev. High Pressure Sci. Technol.)* **1997**, *6*, 182–186 (in Japanese with English abstract).
36. Yamazaki, A.; Sasagawa, A. Story of Development of High-Pressure Processed Foods, Part 3—A Method of Development of Low Allergenic Cooked Rice and Bread. *Kouatsuryoku No Kagaku To Gijutsu (Rev. High Pressure Sci. Technol.)* **1997**, *6*, 236–241 (in Japanese with English abstract).
37. Odani, S.; Kanda, Y.; Hara, T.; Matsuno, M.; Suzuki, A. Effects of High Hydrostatic Pressure Treatment on the Allergenicity and Structure of Chicken Egg White Ovomucoid. In *High Pressure Bioscience and Biotechnology, Proceedings of the 4th International Conference on High Pressure Bioscience and Biotechnology*; Japanese Research Group of High Pressure Bioscience and Biotechnology: Japan, 2007; pp. 252–258.
38. Martin, D. E.; Hall, S. G. Oyster Shucking Technologies: Past and Present. *Int. J. Food Sci. Technol.* **2006**, *41*, 223–232.
39. Kusube, M.; Kyaw, T. S.; Tanikawa, K.; Chastain, R. A.; Hardy, K.; Cameron, J.; Bartlett, D. H. *Colwellia marinimaniae* sp. nov., a Hyperpiezophilic Species Isolated From an Amphipod Within the Challenger Deep, Mariana Trench. *Int. J. Syst. Evol. Microbiol.* **2017**, *67*, 824–831.
40. ZoBell, C. E.; Johnson, F. H. The Influence of Hydrostatic Pressure on the Growth and Viability of Terrestrial and Marine Bacteria. *J. Bacteriol.* **1949**, *57*, 179–189.
41. Aoyama, Y.; Shigeta, Y.; Okazaki, T.; Hagura, Y.; Suzuki, K. Growth Inhibition of Microorganisms by Hydrostatic Pressure. *Food Sci. Technol. Res.* **2004**, *10*, 268–272.
42. Okazaki, T.; Shigeta, Y.; Aoyama, Y.; Namba, K. Autolysis of Unsalted Fish Protein Under Pressurization. *Fish. Sci.* **2003**, *69*, 1257–1262.
43. Aoyama, Y.; Shigeta, Y.; Okazaki, T.; Hagura, Y.; Suzuki, K. Non-thermal Inactivation of *Bacillus* Spores by Pressure-holding. *Food Sci. Technol. Res.* **2005**, *11*, 324–327.

44. Oey, I. Effects of High Pressure on Enzymes. In *High Pressure Processing of Foods*; Balasubramaniam, V. M., Barbosa-Cánovas, G. V., Lelieved, H. L. M., Eds.; Springer: New York, 2016; pp. 391–431.
45. Yamamoto, K.; Koseki, S. Shokuhin To Youki [Food & Packaging]. In Vol. 48; *4th International Conference on High Pressure Bioscience and Biotechnology: Trends In Food Processing Technologies Utilizing High Pressure*; 2007; pp. 150–154 (in Japanese).
46. Balasubramaniam, V. M.; Barbosa-Cánovas, G. V.; Lelieveld, H. L. M. High-pressure Processing Equipment for the Food Industry. In *High Pressure Processing of Foods*; Balasubramaniam, V. M., Barbosa-Cánovas, G. V., Lelieved, H. L. M., Eds.; Springer: New York, 2016; pp. 39–65.

Index

Note: Page numbers followed by *f* indicate figures, *t* indicate tables, and *s* indicate schemes.

D

N

O

T

U

V

www.ingramcontent.com/pod-product-compliance
Lightning Source LLC
Chambersburg PA
CBHW071653130426
42875CB00127B/1841